The Fungal Spore and Disease Initiation in Plants and Animals

Edited by
Garry T. Cole
The University of Texas at Austin
Austin, Texas

and
Harvey C. Hoch
Cornell University
New York State Agricultural Experiment Station
Geneva, New York

Plenum Press ● New York and London

Library of Congress Cataloging-in-Publication Data

The Fungal spore and disease initiation in plants and animals / edited
  by Garry T. Cole and Harvey C. Hoch.
      p.   cm.
    Includes bibliographical references and index.
    ISBN 0-306-43454-7
    1. Mycoses--Pathogenesis. 2. Fungal diseases of plants-
  -Pathogenesis. 3. Fungi--Spores.   I. Cole, Garry T., 1941-   .
  II. Hoch, Harvey C.
    [DNLM: 1. Spores, Fungal--pathogenicity.   QW 180 F9818]
  QR245.F854   1991
  632'.4--dc20
  DNLM/DLC
  for Library of Congress   90-14316
                                                            CIP

ISBN 0-306-43454-7

© 1991 Plenum Press, New York
A Division of Plenum Publishing Corporation
233 Spring Street, New York, N.Y. 10013

All rights reserved

No part of this book may be reproduced, stored in a retrieval system, or transmitted
in any form or by any means, electronic, mechanical, photocopying, microfilming,
recording, or otherwise, without written permission from the Publisher

Printed in the United States of America

Dedicated
to our students
and colleagues;
their diligence and inspiration
are deeply appreciated.

Contributors

James R. Aist • Department of Plant Pathology, Cornell University, Ithaca, New York 14853

Drion G. Boucias • Department of Entomology and Nematology, University of Florida, Gainesville, Florida 32611-0711

William R. Bushnell • USDA–ARS Cereal Rust Laboratory, University of Minnesota, St. Paul, Minnesota 55108

A. K. Charnley • School of Biological Sciences, University of Bath, Bath, Avon BA2 7AY, England

Garry T. Cole • Department of Botany, University of Texas, Austin, Texas 78713

Pierre J. G. M. De Wit • Department of Phytopathology, Wageningen Agricultural University, 6700 EE Wageningen, The Netherlands

R. J. Dillon • School of Biological Sciences, University of Bath, Bath, Avon BA2 7AY, England

Dennis M. Dixon • Laboratories for Mycology, Wadsworth Center for Laboratories and Research, New York State Department of Health, Albany, New York 12201-0509

Lynn Epstein • Department of Plant Pathology, University of California, Berkeley, California 94720

Iwao Furusawa • Laboratory of Plant Pathology, Faculty of Agriculture, Kyoto University, Kyoto 606, Japan

Robert C. Garber • Biotechnology Center and Department of Plant Pathology, Ohio State University, Columbus, Ohio 43210. *Present address*: Springer-Verlag, New York, New York 10010

Randall E. Gold • BASF AG, Agricultural Research Station, D-6703 Limburgerhof, Federal Republic of Germany

Tadayo Hashimoto • Department of Microbiology, Loyola University Stritch School of Medicine, Maywood, Illinois 60153

H. C. Hoch • Department of Plant Pathology, Cornell University, New York State Agricultural Experiment Station, Geneva, New York 14456

Matthieu H. A. J. Joosten • Department of Phytopathology, Wageningen Agricultural University, 6700 EE Wageningen, The Netherlands

Michael J. Kennedy • The Upjohn Company, Kalamazoo, Michigan 49001

Theo N. Kirkland • Veterans Administration Medical Center, San Diego, California 92161

Wolfram Köller • Department of Plant Pathology, Cornell University, New York State Agricultural Experiment Station, Geneva, New York 14456

Yasuyuki Kubo • Laboratory of Plant Pathology, Faculty of Agriculture, Kyoto University, Kyoto 606, Japan

Joseph Kuć • Department of Plant Pathology, University of Kentucky, Lexington, Kentucky 40546-0091

J. P. Latgé • Mycology Unit, Pasteur Institute, 75015 Paris, France

Nageswara Rao Madamanchi • Department of Plant Pathology, Physiology and Weed Science, Virginia Polytechnic Institute and State University, Blacksburg, Virginia 24061-0331

Kurt Mendgen • Department of Plant Pathology, University of Constance, D-7750 Constance, Federal Republic of Germany

Ralph L. Nicholson • Department of Botany and Plant Pathology, Purdue University, West Lafayette, Indiana 47906

Anton Novacky • Department of Plant Pathology, University of Missouri, Columbia, Missouri 65211

Frank C. Odds • Department of Bacteriology and Mycology, Janssen Research Foundation, B-2340 Beerse, Belgium

S. Paris • Mycology Unit, Pasteur Institute, 75015 Paris, France

Jacquelyn C. Pendland • Department of Entomology and Nematology, University of Florida, Gainesville, Florida 32611-0711

Annemarie Polak • F. Hoffmann–La Roche Ltd., CH 4002 Basel, Switzerland

Donald W. Roberts • Insect Pathology Resource Center, Boyce Thompson Institute, Cornell University, Ithaca, New York 14853

Maura D. Robertson • Institute of Occupational Medicine, Edinburgh EH8 9SU, Scotland

Ingrid M. J. Scholtens-Toma • Department of Phytopathology, Wageningen Agricultural University, 6700 EE Wageningen, The Netherlands

M. G. Smart • Northern Regional Research Center, Agricultural Research Service, U.S. Department of Agriculture, Peoria, Illinois 61604. *Present address*: Kraft General Foods, Inc., Glenview, Illinois 60025

David R. Soll • Department of Biology, University of Iowa, Iowa City, Iowa 52242

R. J. St. Leger • Boyce Thompson Institute, Cornell University, Ithaca, New York 14853

R. C. Staples • Boyce Thompson Institute, Cornell University, Ithaca, New York 14853

Paul J. Szaniszlo • Department of Microbiology, University of Texas, Austin, Texas 78713

Alayn R. Waldorf • Department of Biological Sciences, California State University, Hayward, California 94542

Preface

This treatise is focused on early aspects of fungal pathogenesis in plant and animal hosts. Our aim in choosing the topics and contributors was to demonstrate common approaches to studies of fungal–plant and fungal–animal interactions, particularly at the biochemical and molecular levels. For example, the initial events of adhesion of fungal spores to the exposed surface tissues of the host are essential for subsequent invasion of the plant or animal and establishment of pathogenesis. A point of consensus among investigators who have directed their attention to such events in plants, insects, and vertebrates is that spore adhesion to the host cuticle or epithelium is more than a simple binding event. It is a complex and potentially pivotal process in fungal–plant interactions which "may involve the secretion of fluids that prepare the infection court for the development of morphological stages of the germling" and subsequent invasion of the host (Nicholson and Epstein, Chapter 1). The attachment of the fungal propagule to the arthropod cuticle is also "mediated by the chemical components present on the outer layer of the spore wall and the epicuticle. . . . Initial attachment may be reinforced further by either the active secretion of adhesive materials or the modification of spore wall material located at the [fungal spore–arthropod] cuticle interface (Boucias and Pendland, Chapter 5). The parallel nature of the research approaches and comparability of experimental methods used for examinations of *spore attachment* to and *invasion* of both plant and animal hosts is illustrated in the first eight chapters of this volume.

The plant and insect cuticle, and epithelial surfaces of vertebrates, represent the principal passive barriers to fungal invasion of respective hosts. It is intuitive that many fungi actively breach these passive barriers. Evidence has been presented in Chapters 10, 12, and 13 that specific hydrolytic enzymes produced by certain fungal pathogens of plants and animals are associated with cuticular/epithelial penetration. Common goals are underscored by the authors of these chapters, namely the isolation and characterization of these key enzymes (virulence factors) and identification of compounds that selectively inhibit the enzyme activity *in vivo*. It is also evident from these and other chapters in this book that address the topic of *fungal spore products and pathogenesis* that future research must focus on the molecular basis of synthesis and release of substances derived from the pathogen and host which are involved in fungal penetration. Results of these exciting new approaches to old problems will contribute significantly to our understanding of the early mechanisms of host invasion (Odds, Chapter 13).

Although there are clear differences in the nature of *host response to early fungal invasion* of plants and animals, an understanding of host "defenses and how the fungus avoids or overcomes them is rudimentary" (Aist and Bushnell, Chapter 15). Parallels of oxygen free radical production by both plant and animal defense systems are reported in Chapters 17 and 20, respectively. Discussions of active and passive mechanisms by which fungal spores or highly specialized infection structures modify, suppress, or resist the battery of plant and animal host defenses are presented in Chapters 15, 19, and 21.

The development of recombinant DNA technology has introduced new and potentially rewarding possibilities for investigators of fungal pathogenesis of plants and animals. Under the topic *"Molecular Aspects of Disease Initiation"* discusssions of fungal genes responsible for putative pathogenic factors are presented (Chapters 22 and 23). The application of molecular probes to assess strain relatedness and explore basic questions of fungal epidemiology and pathogenesis is also examined in Chapter 23. These chapters again emphasize the similarities of experimental design which plant pathologists and medical mycologists have devised for investigations of molecular aspects of fungal–host interactions. At least three, not mutually exclusive, approaches for future research on fungal pathogenicity genes emerge from these discussions. "The first, and most developed to date, is to study the regulation of a gene in the organism from which it was isolated. . . . A second approach is to transfer genes into nonpathogens or into pathogens with different hosts. . . . A third and extremely underexploited route to the problem of isolating fungal pathogenicity genes is through mutant analysis" (Garber, Chapter 22).

We are most grateful for the contributions of the authors as well as the support of many other individuals who provided assistance during the preparation of this treatise.

Garry T. Cole
Harvey C. Hoch

Austin, Texas
Geneva, New York

INTRODUCTION

Fungal–Host Interactions Opportunities for Interdisciplinary Research

Donald W. Roberts

This book provides, for the first time, the opportunity for researchers and educators interested in fungal diseases of animals (vertebrates and invertebrates) and plants to examine in one volume the salient features of disease initiation in both types of hosts. Are there sufficient commonalities between fungal infections of the two host groups to make the information from fungal–plant interactions enlightening to specialists of animal mycoses, and vice versa? The answer, in my opinion, is an unqualified "yes." It is probably safe to suppose that the great majority of readers of the following pages will be specialists actively involved with either plants or animals, but not both. As such, the natural tendency will be to view the book as a comprehensive, current review of their specialty which happens to be cluttered with intervening, peripheral treatises on other host systems. This view is diametrically opposed to that of the Editors (see Preface), and to me (see Roberts and Aist, 1984). With few exceptions, there is little contact between students of animal and plant mycoses. This book was conceived as one tool to facilitate comparing basic knowledge, current problems, and methods of research on fungal infective units and early disease events between the two host groups.

I recommend that readers *first* read the chapters on the host group foreign to them. It is very likely that this will provide new insights and methods for research in the area of primary interest. Of equal importance, each treatise specifies a number of unresolved but important questions pertaining to that chapter's topic—and readers working with other systems may have approaches based on their own experience to resolve some of these questions. If the book succeeds in its goals, it will stimulate cross-disciplinary collaboration and incite significant research progress in mycoses of both host groups.

The study of fungi is justified on many grounds. For example, fungi are eukaryotic microorganisms which usually are haploid most of their life cycle, making them powerful tools for basic genetic and biochemical studies. They tend to secrete, rather than sequester, metabo-

Donald W. Roberts • Insect Pathology Resource Center, Boyce Thompson Institute, Cornell University, Ithaca, New York 14853.

lites, and this enhances their value as industrial fermentation agents. Fungi produce pharmacologically important and edible metabolites and, in some cases, the fungal biomass is itself a
foodstuff. Although bacteria were the first tools of the modern molecular geneticists, methods to
utilize yeasts in molecular biology were soon developed and pioneering research has been done
with these microorganisms. More recently, filamentous fungi have yielded to transformation
attempts, and, as described in the next-to-last chapter of this book, they have been added to the list
of organisms available as tools for molecular genetics. From the viewpoint of man, the above are
all beneficial characteristics which can and are being exploited. Fungi, however, can be detrimental to man. Some infest improperly stored grain and nuts and produce mycotoxins, and there has
been considerable mycological research directed at first understanding, and then reducing, the
hazard of these compounds to man and domesticated animals.

Despite the importance of the above (and other) activities of fungi which have stimulated
research on them, the single activity which probably is used more than any other to justify
mycological and mycologically related research is that of pathogenicity of some fungi to other
organisms. Their hosts include other fungi, vascular plants, invertebrates, and both warm- and
cold-blooded vertebrates. Most plant diseases are caused by fungi (Aist, 1984). All classes of the
Eumycotina include at least a few plant pathogens (Alexopoulos and Mims, 1979; Agrios, 1988;
Strobel and Mathre, 1970). At least 90 genera and more than 700 species of fungi have been
identified as closely associated with invertebrates. Virtually every major fungal taxonomic group
except the higher basidiomycetes and dematiaceous Hyphomycetes has members pathogenic to
invertebrates, primarily insects (Roberts and Humber, 1981). The numbers of fungal species
which are primarily pathogenic, as opposed to adventitiously pathogenic, to warm-blooded
animals are rather small. There are only about 20 fungi which routinely cause systemic mycoses
in these animals, and about a dozen associated with subcutaneous disease. In addition, there are
approximately 20 dermatophytes which are restricted to the outer (keratin-containing) layers of
the skin (Rippon, 1988). The latter are not generally considered a threat to life, whereas the
former are very much so. The recent dramatic increase in numbers of immunosuppressed
medical patients, either from leukemia, from intentional immunosuppression with chemicals for
such medical events as organ transplants, or from the immunosuppression associated with AIDS
cases, has been followed by a sharply escalated number of fungal infections (Campbell and
White, 1989). In fact, esophageal candidosis and cryptococcosis of the central nervous system
are valuable indicators of AIDS. Additional fungal diseases frequently found in HIV-positive
patients include histoplasmosis and candidosis of the bronchi or lungs. The range of mycoses in
AIDS patients, however, is much wider than the four listed here. Almost all of the well-
recognized fungal pathogens of man have been described from these patients, in addition to a
number of very rare types. The sharp increase in human mycoses has engendered a new sense of
urgency to understanding and controlling these diseases.

In addition to protecting man and his domestic animals, the pathogenic processes of fungi
have been studied to devise methods of protection in other hosts upon which man depends, such
as food plants, and beneficial invertebrates (primarily honeybee, silkworm, and shrimp). The
scientific specialties of plant pathology and medical mycology basically originated around the
protection concept. In addition, the potentially important role of fungi in control of pest
organisms has stimulated concerted efforts to discover and develop virulent strains of fungi
pathogenic to pest insects and pest plants (weeds). This pest-control concept has been an
important force in the development of insect pathology, and recently it has played a significant
role in plant pathology where fungi are being selected and/or engineered for weed and plant
pathogen (primarily fungal) control.

Regardless of the motivation for study, the type of host, or the fungal pathogen, there is one
common event underlying the understanding of all fungal diseases, viz. the initiation of the
infection event. As pointed out by the Editors in the Preface, the early events of infection can be

divided into three major areas: (1) spore attachment and invasion, (2) fungal spore products and pathogenesis, and (3) host response to early fungal invasion. They delineate one more area to emphasize its emerging power and utility in understanding disease initiation: (4) molecular aspects of disease initiation. The reader is referred to the Preface for a brief discussion of these topics. It is understandable that all topics are not examined at the same depth for each host group. For example, the lack of a good animal model for dermatophytes has prevented clear-cut, ultrastructure-level elucidation of the events associated with host invasion. In fact, according to Hashimoto (Chapter 8), the first convincing evidence that the presumed infective units of dermatophytes (i.e., arthroconidia) are indeed infectious was not published until 1987.

It is not my intent in this introductory chapter to provide reviews or summaries of the individual treatises that follow, but rather to provide some brief comments on a few of many possible examples of similarities and differences in findings with the various host groups.

It is apparent that almost all of the contributors have little firsthand knowledge of fungal diseases of organisms outside their primary area of study. The literature cited seldom includes references to infections of other host groups. This is almost completely the case with plant pathologists and medical mycologists. Invertebrate pathologists, on the other hand, do cite some plant pathology studies—probably because of the basic similarities of the fungi involved and common characteristics of the hosts' outer surfaces (hard, wax-covered), which is the normal invasion site. Also, mycological plant pathologists probably outnumber mycological invertebrate pathologists by more than ten to one, and this is reflected in the numbers of publications on invasion processes in these two host types. Invertebrate pathologists have undoubtedly gained from their borrowing of ideas and methods from other areas. The clear message is that the same potential for cross-fertilization of ideas exists for those working with plants and vertebrates. An example is the discussion by Kennedy (Chapter 7) on the adhesion of *Candida* blastospores to the gastrointestinal tract of vertebrates. For plant or invertebrate pathologists, comments on this yeastlike fungus in a vertebrate-specific organ would be assumed of little relevance to their host systems—although the plant-pathologist authors, Nicholson and Epstein (Chapter 1), do cite studies describing lectin-mediated binding of *Candida* to human epithelium. Nevertheless, in trying to explain the events leading to human gut infections, Kennedy presents concepts of importance to understanding adhesion of any small, charged, chemically active body to another structure. He invokes the classic lyophobic colloid theory (Derjagiun and Landau, 1941; Verwey and Overbeek, 1948) (DLVO theory) and its derivations (e.g., Jones and Isaacson, 1983) to explain adhesive interactions between the surfaces of blastospores and the gut mucosa, which theoretically should repel one another since they are both negatively charged. He suggests that the DLVO theory describes long-range (> 10 nm) adhesive interactions, whereas close-range interactions (< 1 nm) are regulated by adhesion-receptor binding.

A large number of research opportunities are mentioned in this book which probably are amenable to successful resolution by experts of other host/pathogen systems. As mentioned previously, virtually all the authors point out lacunae in knowledge concerning their research areas. Students, professors, and research scientists alike will be well advised to carefully search these pages for research opportunities with mycoses of both plants and animals. This may lead to collaborations with specialists working with hosts new to them. I can attest from personal experience that such collaborative research can have synergistic, favorable effects on both research teams (for recent results of work on an insect pathogen in collaboration with a plant pathologist, Richard C. Staples, see Goettel *et al.*, 1989, 1990; St. Leger *et al.*, 1989a–1990).

Characterizing and modifying nucleic acids of pathogenic fungi, without doubt, will provide new insights into processes of disease initiation and development. Molecular approaches currently available to students of pathogenic fungi are described by Garber for phytopathogens (Chapter 22) and by Soll for *Candida* infecting man (Chapter 23). Cloning by transformation is a particularly powerful tool for working with components of disease, e.g., specificity and virulence, where the biochemical bases of the traits are not known. The ability afforded by this approach to isolate genes in the absence of knowledge about gene products is, as stated by Garber, "invaluable."

A clearly expressed plea to students of pathogenesis to utilize the powerful modern molecular biology tools in their investigations, but to be certain that these studies are based on sound biological information, constitutes the final paragraph of Soll's contribution (Chapter 23). This, in my opinion, should be read, and reread periodically, to encourage relevance to the "real" world of virulence in molecular genetics experiments.

Dimorphism (the production of yeast and filamentous growth forms by a fungus) has been noted for several entomopathogenic fungi. The yeast phase is almost exclusively produced in the hemocoel of the arthropod host; but one species, *Nomuraea rileyi*, readily produces the yeast phase on agar media—switching after a few days to filamentous growth. Neither the importance of this phenomenon to virulence nor the underlying mechanisms involved in determining which type of growth will occur have been subjected to close scrutiny with entomopathogenic fungi. Dimorphism and phenotypic switching of *Candida* are important, however, to *Candida* strains pathogenic to humans; and these topics have been, and continue to be, very active research areas with this organism (Soll, Chapter 23). There are obvious, but to date underexploited, opportunities for research on "blastospore"-producing entomopathogenic fungi based on *Candida* findings and methods.

Metabolites produced by pathogenic fungi during their growth are suspected to have important roles in disease development. Even though these metabolites are numerous and some are fully described chemically, their biological effects on the host usually are still unknown. As mentioned by Garber (Chapter 22), some toxic peptides produced by plant pathogenic fungi are assembled by specific synthetase enzymes rather than by the normal ribosomal system. The very existence of these specific synthetase enzymes suggests strongly that the fungi find the resulting peptides of significant survival value. Their role in pathogenesis is best documented for certain plant pathogens (Deol *et al.*, 1978; Macko, 1983; Macko, *et al.*, 1985; Wolpert *et al.*, 1985; Springer, *et al.*, 1984). Beauvericin, bassianolide, and destruxins, insect-toxic depsipeptides produced by the entomopathogenic fungi *Verticillium lecanii*, *Beauveria bassiana*, and *Metarhizium anisopliae*, respectively, also probably are synthesized by the enzymatic method (Grove and Pople, 1980; Roberts, 1981). The destruxins are particularly relevant to this book in that they are toxic to both insects and plants (Bains and Tewari, 1987; Gupta *et al.*, 1990). Although the methods are technically demanding, it is probable that the examination of the synthesis of other active peptides produced by pathogenic fungi will reveal synthetase enzymes. The next hurdle will be to modify these enzymes by molecular genetics methodologies. This will be of great utility in defining the role of the toxic peptides in disease and, perhaps, in producing fungal strains with increased or reduced virulence—according to their intended use.

Many, and probably most, toxins produced by pathogenic fungi are not peptides. For example, cytochalasins C and D, two members of a biologically active class of compounds, were first isolated from cultures of an entomopathogenic fungus, *M. anisopliae*. Their function in insect disease has not been elucidated. Accordingly, the comment by Robertson (Chapter 21) that *Aspergillus fumigatus* conidia release a diffusible substance into their hosts' lungs which, although not chemically defined, is similar to cytochalasins in biological activity—viz., antiphagocytic—immediately suggests an appropriate approach to defining the role of cytochalasins produced by *M. anisopliae* in insects. It should be mentioned that depsipeptides (destruxins) produced by *M. anisopliae* have been reported to inhibit insect phagocytes (Huxham *et al.*, 1989).

The theme of this book is the similarities of fungal infections, regardless of host. It is patently clear in the following chapters that such similarities abound in all four subdivisions of the infection process around which the book is organized. Also, there is extensive commonality of concepts and methods, and therefore the potential for truly meaningful dialogue between pathologists working with very different hosts certainly exists. As mentioned previously, such recognition of common ground will facilitate development of all areas of the understanding of mycoses. In addition, there are, without doubt, some important differences between the hosts, pathogenic fungi, and fungus/host interactions. For example, although many of the fungi which infect plants and invertebrates fall within the same taxonomic groups, there is very little overlap at the fungal genus level, and virtually none at the species level. It follows that the characteristics needed by fungi to successfully establish disease in plants must be fundamentally different in some ways from those needed to infect animals. These differences, when clearly delineated, will indicate probable key virulence characters for pathogens of the two host groups. One approach to identifying specialized pathogenicity traits will be to compare available knowledge on all pathogenic fungi. Another function of the information in this book, therefore, could be its use by students engaged in the study of evolution of pathogenicity in fungi.

REFERENCES

Agrios, G. N., 1988, *Plant Pathology,* 3rd ed., Academic Press, New York.

Aist, J. R., 1984, A survey of plant diseases caused by fungi that attack cuticularized plant surfaces, in: *Infection Processes of Fungi* (D. W. Roberts and J. R. Aist, eds.), The Rockefeller Foundation, New York, pp. 13–15.

Alexopoulos, C. J., and Mims, C. W., 1979, *Introductory Mycology,* 3rd ed., Wiley, New York.

Bains, P. S., and Tewari, J. P., 1987, Purification, chemical characterization and host-specificity of the toxin produced by *Alternaria brassicae, Physiol. Mol. Plant Pathol.* 30:259–271.

Campbell, C. K., and White, G. C., 1989, Fungal infection in 'AIDS' patients, *Mycologist* 3:7–9.

Deol, B. S., Ridley, D. D., and Singh, P., 1978, Isolation of cyclodepsipeptides from plant pathogenic fungi, *Aust. J. Chem.* 31:1397–1399.

Derjagiun, B. V., and Landau, L., 1941, Theory of the stability of strongly charged lyophobic sols and of the adhesion of strongly charged particles in solutions of electrolytes, *Acta Physicochim. URSS* 14:633–662.

Goettel, M. S., St. Leger, R. J., Rizzo, N. W., Staples, R. C., and Roberts, D. W., 1989, Ultrastructural localization of a cuticle-degrading protease produced by the entomopathogenic fungus *Metarhizium anisopliae* during penetration of host cuticle, *J. Gen. Microbiol.* 135:2233–2239.

Goettel, M. S., St. Leger, R. J., Bhairi, S., Jung, M. K., Oakley, B. R., Roberts, D. W., and Staples, R. C., 1990, Pathogenicity and growth of *Metarhizium anisopliae* stably transformed to benomyl resistance, *Curr. Genet.* 17:129–138.

Grove, J. F., and Pople, M., 1980, The insecticidal activity of beauvericin and the enniatin complex, *Mycopathologia* 70:103–105.

Gupta, S., Roberts, D. W., and Renwick, J. A. A., 1989, Insecticidal cyclodepsipeptides from *Metarhizium anisopliae, J. Chem. Soc. Perkin Trans.* I:2347–2357.

Huxham, I. M., Lackie, A. M., and McCorkindale, N. J., 1989, Inhibitory effects of cyclodepsipeptides, destruxins, from the fungus *Metarhizium anisopliae,* on cellular immunity in insects, *J. Insect Physiol.* 35:97–105.

Jones, G. W., and Isaacson, R. E., 1983, Proteinaceous bacterial adhesins and their receptors, *CRC Crit. Rev. Microbiol.* 10:229–260.

Macko, V., 1983, Structural aspects of toxins, in: *Toxins and Plant Pathogenesis* (J. M. Daly and B. J. Deverall, eds.), Academic Press, New York, pp. 41–80.

Macko, V., Wolpert, T. J., Acklin, W., Jaun, B., Seibl, J., Meili, J., and Arigoni, D., 1985, Characterization of victorin C, the major host-selective toxin from *Cochliobolus victoriae*: Structure of degradation products, *Experientia* 41:1366–1370.

Peeters, H., Zocher, R. and Kleinkauf, H., 1988, Synthesis of beauvericin by a multifunctional enzyme, *J. Antibiot.* 41:352–259.

Rippon, J. W., 1988, *Medical Mycology: The Pathogenic Fungi and the Pathogenic Actinomycetes,* 3rd ed., Saunders, Philadelphia.

Roberts, D. W., 1981, Toxins of entomopathogenic fungi, in: *Microbial Control of Pests and Plant Diseases: 1970–1980* (H. D. Burges, ed.), Academic Press, New York, pp. 441–464.

Roberts, D. W., and Aist, J. R., eds., 1984, *Infection Processes of Fungi,* The Rockefeller Foundation, New York.

Roberts, D. W., and Humber, R. A., 1981, Entomopathogenic fungi, in: *The Biology of Conidial Fungi*, Volume 2 (G. T. Cole and B. Kendrick, eds.), Academic Press, New York, pp. 201–236.

Springer, J. P., Cole, R. J., Dorner, J. W., Cox, R. H., Richard, J. L., Barnes, C. L., and van der Helm, D., 1984, Structure and conformation of roseotoxin B, *J. Am. Chem. Soc.* **106**:2388–2392.

St. Leger, R. J., Butt, T. M., Goettel, M. S., Staples, R. C., and Roberts, D. W., 1989a, Production *in vitro* of appressoria by the entomopathogenic fungus *Metarhizium anisopliae*, *Exp. Mycol.* **13**:274–288.

St. Leger, R. J. Butt, T. M., Staples, R. C., and Roberts, D. W. 1989b, Synthesis of proteins including a cuticle-degrading protease during differentiation of the entomopathogenic fungus *Metarhizium anisopliae*, *Exp. Mycol.* **13**:253–262.

St. Leger, R. J., Roberts, D. W., and Staples, R. C., 1989c, Calcium and calmodulin mediated protein synthesis and protein phosphorylation during germination, growth and protease production by *Metarhizium anisopliae*, *J. Gen. Microbiol.* **135**:2141–2154.

St. Leger, R. J., Roberts, D. W., and Staples, R. C., 1989d, Novel GTP-binding proteins in plasma membranes of the fungus *Metarhizium anisopliae*, *Biochem. Biophys. Res. Commun.* **164**:562–566.

St. Leger, R. J., Laccetti, L. B., Staples, R. C., and Roberts, D. W., 1990, Protein kinases in the entomopathogenic fungus *Metarhizium anisopliae*, *J. Gen. Microbiol.* **136**:1401–1411.

Strobel, G. A., and Mathre, D. E., 1970, *Outlines of Plant Pathology*, Van Nostrand–Reinhold, Princeton, N.J.

Verwey, E. J. W., and Overbeek, J. T. G., 1948, *Theory of the Stability of Lyophobic Colloids*, Elsevier, Amsterdam.

Wolpert, T. J., Macko, M., Acklin, W., Jaun, B., Seibl, J., Meili, J., and Arigoni, D., 1985, Structure of victorin C, the major host-selective toxin from *Cochliobolus victoriae*, *Experientia* **41**:1524–1529.

Contents

III. Host Response to Early Fungal Invasion

15. *Invasion of Plants by Powdery Mildew Fungi, and Cellular Mechanisms of Resistance* 321
    James R. Aist and William R. Bushnell

    1. Introduction 321
    2. Primary Germ Tubes 324
    3. Cytoplasmic Aggregation 325
    4. Halo Formation 326
    5. Papilla Formation 327
    6. Haustoria 330
    7. Hypersensitive Response 333
    8. Levels of Specificity 335
    9. Concluding Statement 339
    10. References 339

16. *Induced Systemic Resistance in Plants* 347
    Nageswara Rao Madamanchi and Joseph Kuć
    1. Introduction 347
    2. The Phenomenon 348
    3. Mechanisms Activated 350
    4. Gene Expression as it Relates to Induced Systemic Resistance—Applications of Molecular Biology 355
    5. Concepts and Applications for the Future 356
    6. References 357

17. *The Plant Membrane and Its Response to Disease* 363
    Anton Novacky

    1. Introduction 363
    2. Membrane Structure and Function 364
    3. Interaction with the Plasma Membrane H^+-ATPase 364
    4. Interaction with a Single Ion Channel 367
    5. The Plasma Membrane as a Possible Toxin Target 368
    6. Recognition of an Incompatible Pathogen 372
    7. Conclusion 374
    8. References 375

18. *The Fungal Spore: Reservoir of Allergens* 379
    J. P. Latgé and S. Paris

    1. Introduction 379
    2. Methodology Available to Study Fungal Spore Allergy 380
    3. Characterization of Spore Allergens 388
    4. Other Fungi 396
    5. Perspectives for Future Research 397
    6. References 398

IV. Molecular Aspects of Disease Initiation

*The Fungal Spore and
Disease Initiation in
Plants and Animals*

I

Spore Attachment and Invasion

1

Adhesion of Fungi to the Plant Surface

Prerequisite for Pathogenesis

Ralph L. Nicholson and Lynn Epstein

1. INTRODUCTION

The initial process of attachment of fungal propagules to a host plant is essential to the successful establishment of pathogenesis. Attachment may be involved in recognition of the host surface, serve as a base around which the infection court can be altered, and may include adhesion of the propagule. It was considered initially that attachment was purely a chance event resulting from physical entrapment of the propagule or germling. We know now that attachment involves an active process of secretion of adhesive materials by the fungus that in some cases are highly specific for the recognition of, and binding to, a particular host species. Adhesive production may occur at a specific stage of conidium or germling development, but may best be considered a general phenomenon for the establishment of the fungus prior to penetration.

Adhesion implies a simple binding event. However, we will point out that this process is more complex and may involve the secretion of fluids which prepare the infection court for the development of morphological stages of the germling (eg., appressoria or attachment organs) that are a prerequisite to penetration of the host. Frequently, research has shown that appressorial related adhesives are commonly observed in fungi (Edwards and Allen, 1970; Emmett and Parbery, 1975). Thus, it was assumed that adhesion was specific to the formation of the appressorium. In fact, adhesion and surface recognition may occur much earlier depending on the fungal species. Evidence shows that extracellular mucilages or adhesives are common on fungal germlings (Nicholson, 1984) and in some cases may even arise from the nongerminated conidium (Paus and Raa, 1973; Hamer et al., 1988).

Extracellular fluids or matrices were first reported by Ward (1888) who described them as films that surrounded the organs of attachment of a *Botrytis* sp. Subsequent reports demonstrated that numerous fungi have an extracellular matrix or mucilage associated with their germ tubes (Jacobi et al., 1982; Hau and Rush, 1982; Locci, 1969). Evans and colleagues (1981, 1982)

Ralph L. Nicholson • Department of Botany and Plant Pathology, Purdue University, West Lafayette, Indiana 47906.   *Lynn Epstein* • Department of Plant Pathology, University of California, Berkeley, California 94720.

demonstrated that in *Bipolaris maydis* a multilayered extracellular sheath of mucilage appeared prior to the germination of the conidium and at the site of germ tube emergence. More recently, Kunoh and colleagues (1988) demonstrated that *Erysiphe graminis* conidia release a liquid film within minutes of contact with a barley leaf. In the rust fungus *Uromyces appendiculatus* the entire length of the basidiospore germ tube exudes a film that appears attached to the surface of the host (Gold and Mendgen, 1984). Similarly, *Gymnosporangium juniperi-virginianae* basidiospore germlings produce a diffuse, filamentous network of material (Fig. 1; see Fig. 16 of Chapter 4) that subsequently changes to an electron-dense layer binding the entire germling to the host surface (Mims and Richardson, 1989). During zoospore encystment in *Pythium ultimum*, an adhesive is released that binds the cyst to the host surface (Grove and Bracker, 1978); a similar mechanism is known for *Plasmodiophora brassicae* (Aist and Williams, 1971). In some instances, notably with *Phytophthora* and *Pythium* spp., adhesion of the zoospore is an event that is not only host specific, but also specific to certain zones of the root, indicating a level of host recognition at an extremely early stage of the pending host–parasite interaction (Goode, 1956; Hinch and Clarke, 1980; Longman and Callow, 1987; Mitchell and Deacon, 1987). This early form of attraction and attachment has also been described for zoospores of the parasitic chytrid *Rozella* (Held, 1974). In nematophagous fungi the release of adhesive materials is known to be stimulated by the presence of the nematode host (Nordbring-Hertz, 1988). Thus, the release of extracellular mucilages or matrices appears to span a broad spectrum of fungi.

2.   INITIATION OF THE INFECTION PROCESS, THE PRODUCTION OF EXTRACELLULAR MATRICES AND ADHESION

2.1.   Preparation of the Infection Court

Investigations by Staub and co-workers (1974) showed that the barley cuticle was eroded by germlings of *Erysiphe graminis*, suggesting that cutin and interspersed wax polymers were enzymatically degraded as the pathogen grew along the leaf surface. *E. graminis* was capable of eroding the surface cuticle of both barley and cucumber whereas *E. cichoracearum* only eroded the cucumber cuticle. Thus, specificity for the enzymatic degradation of cuticle components of a particular host may contribute to the ultimate specificity of the interaction. Whether this degradation is required for surface recognition is unknown, but investigations by Kunoh and co-workers (1988) and Nicholson and co-workers (1988) have shown that *E. graminis* conidia alter the host surface up to 2 hr prior to conidium germination. Within 2 min after contact with the host surface, the conidium released a liquid film which was dispersed in a 15- to 25-μm zone outward from the body of the conidium. This film contained three highly active esterase enzymes. Scanning electron microscopy demonstrated the erosion of cuticular waxes adjacent to, and underlying the conidium within 20 min of liquid release. Subsequent studies involving treatment of the barley leaf surface with the partially purified esterases have shown that the enzyme preparation partially degrades the surface waxes of the leaf (Kunoh and Nicholson, unpublished). Importantly, the growth of the appressorial germ tube appears to be limited to the zone of deposition of the liquid film. These events imply a need for the preparation of the infection court to provide a surface that is recognized by the pathogen as amenable to penetration.

As indicated above, several studies suggest that preparation of the infection court involves active dissolution of the host cuticle by foliar pathogens. An unusual case of adhesion involves *Zygophiala jamaicensis*, the causal agent of flyspec of several fruits (Nasu and Kunoh, 1986, 1987; Nasu *et al.*, 1986). This organism does not penetrate the host tissue and survives solely on the wax and cutin components of the fruit cuticle. The fungus first erodes away surface waxes and then becomes firmly embedded within the cuticle with an extensive thallus tightly appressed to the plant surface. An exceptional amount of mucilage is secreted by the mycelium and effects a

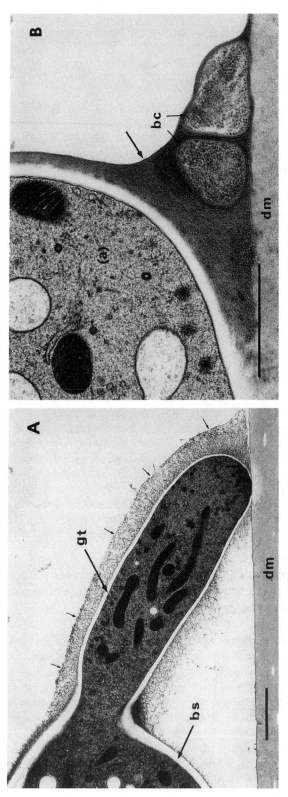

Figure 1.   Transmission electron micrograph of a basidiospore germling of the rust *Gymnosporangium juniperi-virginianae* prepared by freeze substitution. (A) A germ tube (gt) arising from a basidiospore (bs) is touching against the surface of a dialysis membrane (dm). The germ tube appears to be attached to the membrane at the point of contact by an extracellular matrix that extends up the germ tube to the basidiospore (arrows). Bar = 3 μm. (B) Higher magnification of the extracellular matrix (arrow) associated with binding of a germling appressorium (a) to the contact surface. The matrix, which appears fibrillar, has partially covered two bacterial cells (bc) on the surface of the dialysis membrane (dm). Bar = 0.5 μm. Micrographs courtesy of Charles Mims. B reprinted from *Protoplasma* **148**:111–119.

zone of cuticular erosion in advance of, and surrounding the hyphae (Fig. 2). As in the above cases, the mucilage contained high levels of esterase activity (Nasu, unpublished), probably of the cutinase class.

2.2. Characteristics of the Host Surface

Characteristics of the host surface that mediate successful adhesion are both chemical and morphological. These have been reviewed recently and will be addressed only briefly here. The most pertinent reviews are those of Hoch and Staples (1987), Wynn (1981), and Wynn and Staples (1981) who detail factors that condition thigmodifferentiation with emphasis on germlings of the rust fungi. The most notable morphological requirement described to date is that of the height of the guard cell lip of bean that triggers the initiation of appressorium formation in *U. appendiculatus* (Hoch and Staples, 1987; Chapter 2, this volume).

Aside from morphological determinants that trigger fungal differentiation is the larger question of the chemistry of the host surface that may either evoke the release of an adhesive or, more typically, be detected accurately for the process of binding. Because surface recognition is often highly specific, it might be assumed that it represents an early event in the specificity of the host–parasite interaction. In the nematophagous fungi, chemical recognition by the fungus of the prey appears to be well developed and to control even the formation of the fungal morphological structures specific for trapping the nematode. Such induction was first demonstrated by Nordbring-Hertz (1973) with peptides that induce trap structure formation in *Arthrobotrys oligospora*. In turn, it is only on the hyphal traps that the adhesive for nematode binding is found (Veenhuis *et al.*, 1985).

In many instances the associated events that mediate adhesion are known to involve the highly specific binding of lectins and haptens, either of which occur on the pathogen or host. Encystment of *Phytophthora cinnamomi* zoospores was found to be specifically induced by the lectin concanavalin A (Hardham and Suzaki, 1986). As a chemotactic response is involved in active movement of zoospores to the host root surface, it remains a question whether the lectin is released by the root into the surrounding rhizosphere or whether the lectin–hapten binding occurs only on the root surface. Regardless of the events that mediate encystment, the outcome is the same with the rapid loss of zoospore motility, exocytosis of the adhesive contents of surface vesicles, and finally adhesion to the root itself (Carlile, 1983; Sing and Bartnicki-Garcia, 1975a,b). In other cases it has been shown that binding to the root surface occurs in specific root zones, again suggesting the presence of haptens only on specific sites of the root (Longman and Callow, 1987).

3. COMPOSITION AND SIGNIFICANCE OF NONADHESIVE COMPOUNDS IN THE EXTRACELLULAR MATRIX OF COLLETOTRICHUM GRAMINICOLA

Aside from the obvious function of attachment mediated by adhesives, little attention has been given to the properties and function of extracellular fungal matrices. A notable exception is a mucilage with remarkable protective properties that is produced during the sporulation of *Colletotrichum graminicola*. Conidia are produced, in culture and acervuli, in masses embedded in a water-soluble mucilage (Fig. 3) composed primarily of high-molecular-weight glycoproteins (Nicholson and Moraes, 1980; Ramadoss *et al.*, 1985). The mucilage is produced only at the time of sporulation and its synthesis, as well as sporulation, is light dependent. The principle component is a high-molecular-weight complex of glycoproteins composed of mannose, rhamnose, galactose, and glucose (66, 22, 10, and 2 mole%, respectively). As expected, the mucilage exhibits a high degree of affinity for Con A. The peptide components of the glycoproteins are

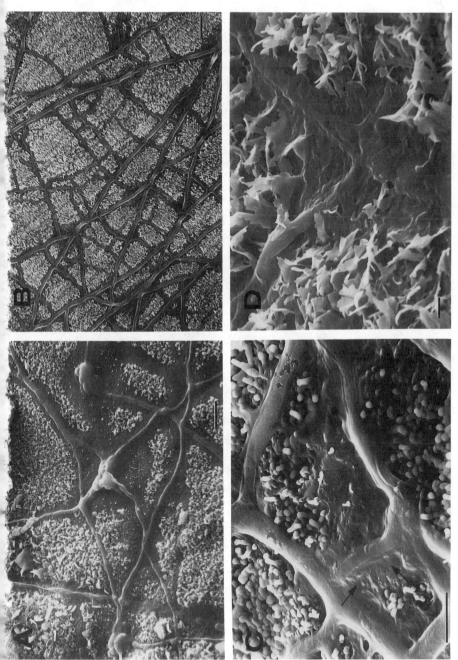

Figure 2. Growth of *Zygophiala jamaicensis* on the surface of various fruits. (A, B) Erosion of host cuticular components shown as areas free of surface waxes surrounding hyphae growing on plum (A) and Japanese persimmon fruits (B). (C) Production of extensive mucilage by the fungus which appears as a film (arrows) surrounding hyphae growing on Chinese quince. (D) Dissolution or erosion of cuticular waxes of grapefruit in areas adjacent to the fungal mycelium. Scale bars: A = 10 μm, B = 20 μm, C = 5 μm, D = 1 μm. Micrographs courtesy of Hitoshi Kunoh; A, B, and C reprinted from *Plant Dis.* **71**:361–364; D reprinted from *Ann. Phytopathol. Soc. Jpn.* **52**:466–474.

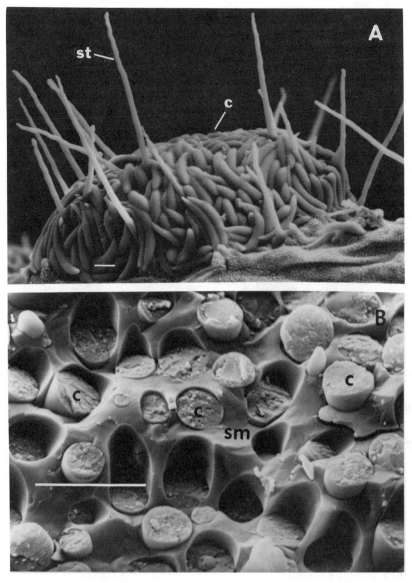

Figure 3. Sporulation of *Colletotrichum graminicola* on the surface of an infected corn leaf. (A) A fully developed acervulus of the fungus 24 hr after exposure of the leaf tissue to 100% relative humidity. The mass of conidia (c) is interspersed with sterile setae (st). Bar = 10 μm. (B) The cut surface of a conidial mass from an acervulus photographed 24 hr after exposure of the leaf to 100% relative humidity. The cut surfaces of numerous conidia (c) are visible. The conidia are embedded within, and surrounded by the spore mucilage (sm.) Bar = 10 μm. Micrographs reproduced from *Physiol. Mol. Plant Pathol.* **35:**243–252.

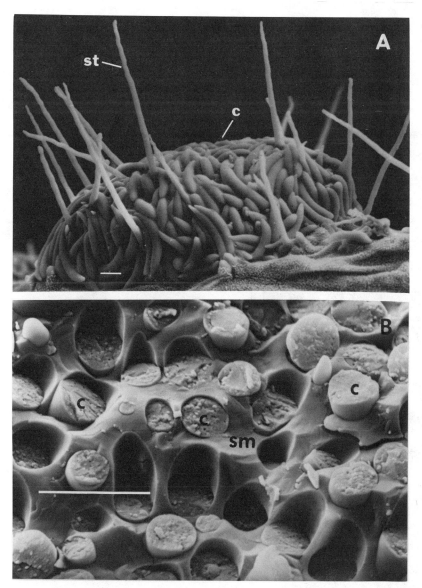

Figure 3. Sporulation of *Colletotrichum graminicola* on the surface of an infected corn leaf. (A) A fully developed acervulus of the fungus 24 hr after exposure of the leaf tissue to 100% relative humidity. The mass of conidia (c) is interspersed with sterile setae (st). Bar = 10 μm. (B) The cut surface of a conidial mass from an acervulus photographed 24 hr after exposure of the leaf to 100% relative humidity. The cut surfaces of numerous conidia (c) are visible. The conidia are embedded within, and surrounded by the spore mucilage (sm.) Bar = 10 μm. Micrographs reproduced from *Physiol. Mol. Plant Pathol.* **35**:243–252.

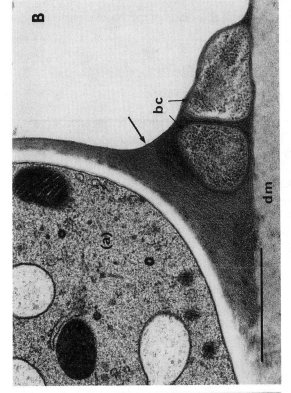

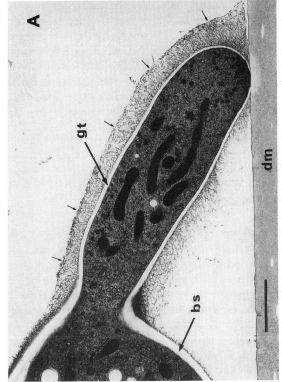

Figure 1. Transmission electron micrograph of a basidiospore germling of the rust *Gymnosporangium juniperi-virginianae* prepared by freeze substitution. (A) A germ tube (gt) arising from a basidiospore (bs) is touching against the surface of a dialysis membrane (dm). The germ tube appears to be attached to the membrane at the point of contact by an extracellular matrix that extends up the germ tube to the basidiospore (arrows). Bar = 3 μm. (B) Higher magnification of the extracellular matrix (arrow) associated with binding of a germling appressorium (a) to the contact surface. The matrix, which appears fibrillar, has partially covered two bacterial cells (bc) on the surface of the dialysis membrane (dm). Bar = 0.5 μm. Micrographs courtesy of Charles Mims. B reprinted from *Protoplasma* **148**:111–119.

zone of cuticular erosion in advance of, and surrounding the hyphae (Fig. 2). As in the above cases, the mucilage contained high levels of esterase activity (Nasu, unpublished), probably of the cutinase class.

2.2. Characteristics of the Host Surface

Characteristics of the host surface that mediate successful adhesion are both chemical and morphological. These have been reviewed recently and will be addressed only briefly here. The most pertinent reviews are those of Hoch and Staples (1987), Wynn (1981), and Wynn and Staples (1981) who detail factors that condition thigmodifferentiation with emphasis on germlings of the rust fungi. The most notable morphological requirement described to date is that of the height of the guard cell lip of bean that triggers the initiation of appressorium formation in *U. appendiculatus* (Hoch and Staples, 1987; Chapter 2, this volume).

Aside from morphological determinants that trigger fungal differentiation is the larger question of the chemistry of the host surface that may either evoke the release of an adhesive or, more typically, be detected accurately for the process of binding. Because surface recognition is often highly specific, it might be assumed that it represents an early event in the specificity of the host–parasite interaction. In the nematophagous fungi, chemical recognition by the fungus of the prey appears to be well developed and to control even the formation of the fungal morphological structures specific for trapping the nematode. Such induction was first demonstrated by Nordbring-Hertz (1973) with peptides that induce trap structure formation in *Arthrobotrys oligospora*. In turn, it is only on the hyphal traps that the adhesive for nematode binding is found (Veenhuis *et al.*, 1985).

In many instances the associated events that mediate adhesion are known to involve the highly specific binding of lectins and haptens, either of which occur on the pathogen or host. Encystment of *Phytophthora cinnamomi* zoospores was found to be specifically induced by the lectin concanavalin A (Hardham and Suzaki, 1986). As a chemotactic response is involved in active movement of zoospores to the host root surface, it remains a question whether the lectin is released by the root into the surrounding rhizosphere or whether the lectin–hapten binding occurs only on the root surface. Regardless of the events that mediate encystment, the outcome is the same with the rapid loss of zoospore motility, exocytosis of the adhesive contents of surface vesicles, and finally adhesion to the root itself (Carlile, 1983; Sing and Bartnicki-Garcia, 1975a,b). In other cases it has been shown that binding to the root surface occurs in specific root zones, again suggesting the presence of haptens only on specific sites of the root (Longman and Callow, 1987).

3. COMPOSITION AND SIGNIFICANCE OF NONADHESIVE COMPOUNDS IN THE EXTRACELLULAR MATRIX OF COLLETOTRICHUM GRAMINICOLA

Aside from the obvious function of attachment mediated by adhesives, little attention has been given to the properties and function of extracellular fungal matrices. A notable exception is a mucilage with remarkable protective properties that is produced during the sporulation of *Colletotrichum graminicola*. Conidia are produced, in culture and acervuli, in masses embedded in a water-soluble mucilage (Fig. 3) composed primarily of high-molecular-weight glycoproteins (Nicholson and Moraes, 1980; Ramadoss *et al.*, 1985). The mucilage is produced only at the time of sporulation and its synthesis, as well as sporulation, is light dependent. The principle component is a high-molecular-weight complex of glycoproteins composed of mannose, rhamnose, galactose, and glucose (66, 22, 10, and 2 mole%, respectively). As expected, the mucilage exhibits a high degree of affinity for Con A. The peptide components of the glycoproteins are

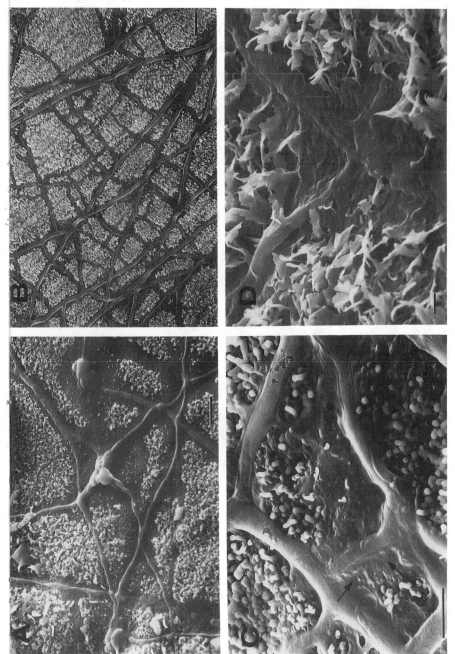

Figure 2. Growth of *Zygophiala jamaicensis* on the surface of various fruits. (A, B) Erosion of host cuticular components shown as areas free of surface waxes surrounding hyphae growing on plum (A) and Japanese persimmon fruits (B). (C) Production of extensive mucilage by the fungus which appears as a film (arrows) surrounding hyphae growing on Chinese quince. (D) Dissolution or erosion of cuticular waxes of grapefruit in areas adjacent to the fungal mycelium. Scale bars: A = 10 μm, B = 20 μm, C = 5 μm, D = 1 μm. Micrographs courtesy of Hitoshi Kunoh; A, B, and C reprinted from *Plant Dis.* **71**:361–364; D reprinted from *Ann. Phytopathol. Soc. Jpn.* **52**:466–474.

approximately 50% hydrophobic and 23% hydroxylic amino acids, with very low levels of aromatic amino acids. A substantial level of free uracil ($0.3\ \mu M$) is also present. Trace amounts of UV-absorbing material(s) (309 nm) are present and it has been shown to function in some instances as a stimulant of germination whereas in others it represses germination. Although this effect may be concentration dependent, it seems more likely that its expression depends on the degree of spore maturity (Nicholson, unpublished).

Other components of the mucilage that appear essential to fungal survival include enzymes, some of which have been identified as nonspecific esterase, invertase, β-glucosidase, peroxidase, and a DNase, possibly of the endonuclease class (Ramadoss *et al.*, 1985; Bergstrom and Nicholson, 1981; Snyder and Nicholson, 1988). The fungus also has the capacity to produce a cutinase (Kolattukudy, 1985a). The presence of similar enzymes in extracellular matrices of other fungi has been reported subsequently (Lewis and Cooke, 1985).

As the mucilage is carried with conidia to the leaf surface, one can easily envision a role for these enzymes, especially the esterase and cutinase, as a prerequisite to establishing the infection court and to adhesion and penetration. The specific function of factors that stimulate germination is unclear but may be related to the dispersal of the mature spore from the acervulus. It is known that spore maturation to the level of germinability requires 3 days. No function is known for uracil as it does not alter germination or infectivity.

Of perhaps greater significance is the presence of three proline-rich proteins in the *C. graminicola* mucilage that bind the phenylpropanoids *p*-coumaric and ferulic acid which are toxic to the fungus (Nicholson *et al.*, 1986). The proteins are envisioned to function in protection of conidia from toxic phenols produced in necrotic leaf tissue in response to infection (Lyons *et al.*, 1987). When the fungus sporulates on the leaf it must be carried in a film of liquid water to healthy leaf tissue for the process of secondary spread of conidia. The phenols, which leach readily into water from the necrotic tissue, are assumed to be bound by the proline-rich proteins and the fungus protected from their inhibitory effects (Nicholson *et al.*, 1986, 1989). Such a mechanism would resemble that described for a proline-rich protein from the mouse submandibular gland which protects the animal from the deleterious effects of dietary tannins (Mehansho *et al.*, 1985). Comparison of the binding affinities of the purified mouse protein with the partially purified proline-rich proteins from *C. graminicola* demonstrated that the fungal proteins had a fivefold greater affinity for tannin than did the mouse protein (Nicholson *et al.*, 1986).

The function of extracellular matrices that extend beyond adhesion remain somewhat of a mystery except for those studies outlined above that demonstrate a role in the preparation of the infection court, a phenomenon that seems to be mediated by enzymes (Kunoh *et al.*, 1988; Nicholson *et al.*, 1988). A striking feature of the extracellular matrix of *C. graminicola* is its antidesiccant property (Nicholson and Moraes, 1980). In the presence of mucilage, conidia survive in a dried state for several months whereas if mucilage is removed, viability is lost within hours. Upon drying, the mucilage forms a thin film ($<0.1\ \mu m$) around conidia (Fig. 4) but the mechanism through which the antidesiccant property of the film is effected is unknown.

4. FUNCTIONAL SIGNIFICANCE OF ADHESION AND EXTRACELLULAR MATRICES

4.1. Composition of Adhesive Compounds

There is no evidence that fungi share a common adhesive compound or a common mechanism for adhesion. The surface of fungal germ tubes and hyphae are ensheathed by a protein, or more typically, glycoprotein/carbohydrate matrix (Bonfante-Fasolo and Perotto, 1986; Onyile *et al.*, 1982; Evans *et al.*, 1982; Mendgen *et al.*, 1985). There appear to be two classes of adhesives: polysaccharides and proteins or glycoproteins. On *Neurospora crassa* (Reissig *et al.*, 1975) and

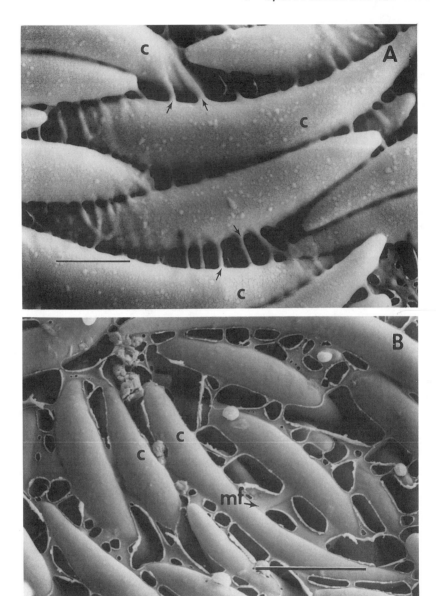

Figure 4. Surface view of conidia of *Colletotrichum graminicola* produced on an infected corn leaf. (A) Strands of mucilage (arrows) extending between conidia (c) that are in a hydrated state. Bar = 10 μm. (B) Surface view of conidia (c) where the conidial mass was allowed to dry under ambient conditions for 15 min in the laboratory. The mucilage that surrounds conidia has dried to a thin film (mf). Bar = 10 μm. A reprinted from *Phytopathology* **76:**1315–1318; B reprinted from *Physiol. Mol. Plant Pathol.* **35:**243–252.

Bipolaris sorokiniana (Pringle, 1981), a galactosaminoglycan appears to serve as a surface-binding compound while on *C. graminicola* appressoria, a "hemicellulose" is apparently involved in adhesion (Lapp and Skoropad, 1978). A protein or glycoprotein appears to adhere bean rust germlings (Epstein *et al.*, 1985), *Buergenerula spartinae* hyphopodia (Onyile *et al.*, 1982), *Phytophthora cinnamomi* and *P. palmivora* zoospores (Hinch and Clarke, 1980; Sing and Bartnicki-Garcia, 1975b), and *Candida albicans* cells (McCourtie and Douglas, 1985).

In some cases where fungi adhere to roots, nematodes, other fungi, or animal cells, adhesion is mediated by lectin binding to specific haptens. Firm fungal/nematode attachment is required for parasitism by nematophagous fungi (Nordbring-Hertz, 1988). The surface of nematodes has been shown to be interspersed with specific saccharides, often specific to stages of nematode development (Forrest and Robertson, 1986). Borrebaeck *et al.* (1984) demonstrated and partially characterized a carbohydrate-binding protein (approximately 20kDa) with specificity for *N*-acetyl-D-galactosamine (present on the nematode surface) from trap structures of *Arthrobotrys oligospora*. Importantly, synthesis of the lectin was shown to be developmentally regulated, occurring only on the surface of trap-bearing hyphae formed in response to the presence of the nematode. The firm adhesion of the fungus to the nematode was further demonstrated by transmission electron microscopy that showed the presence of an adhesive only on the fungal trap, and that penetration of the nematode occurred specifically at sites of adhesive–nematode contact (Veenhuis *et al.*, 1985).

Another nematophagous species, *Meria coniospora*, also adheres to its nematode host through lectin-mediated binding. In this case, however, it is the fungal conidium rather than a specialized trap structure that binds to the nematode. The conidium possesses a knoblike structure at its apices that is covered by a layer of mucilage that attaches the conidium to the nematode host (Saikawa, 1982). Adhesion of the fungus to specific nematode species and either to the cephalic region of both males and females or to the tail of males was mediated by a fungal lectin specific for *N*-acetylneuraminic acid (sialic acid) (Jansson and Nordbring-Hertz, 1983, 1984). Subsequent studies suggested that differences in steric configuration of sialic acid residues on the nematode surface may serve as different recognition stimuli for adhesion and infection (Jansson *et al.*, 1985).

With species of the mycoparasite *Trichoderma*, Elad and co-workers (1983a,b) have demonstrated that lectin binding probably accounts for the specificity of the identification of host fungi by the parasite. Unlike the case with the nematophagous fungal interactions, it is the host fungi (*Rhizoctonia solani* and *Sclerotium rolfsii*) that possess surface lectins and the parasites that possess surfaces rich in galactose residues that serve as haptens. Electron micrographs of these fungal interactions also showed extensive zones of erosion and sites of penetration of the host surface. Mucilage-like deposits extending apparently from the parasite to the host were also prevalent (Elad *et al.*, 1983b). Whether these deposits are important to the adhesion of the organisms is unknown but they do resemble deposits from appressoria and germ tubes that have often been associated with sites of appressorial adhesion (Nicholson, 1984).

Lectin binding sites, specifically for glucose and mannose, were similarly detected in the surface of the fungal wall and in an extracellular material surrounding the wall of ericoid mycorrhizal fungi (Bonfante-Fasolo and Perotto, 1986; Perotto and Bonfante-Fasolo, 1985). In *Hymenoscyphus ericae*, specific Con A binding sites were localized in an extracellular material that radiated from the fungal wall and the abundance of binding sites was markedly increased by growth of the fungus in the presence of host roots. A noninfective strain of the fungus essentially lacked the hapten regardless of the presence of the host (Bonfante-Fasolo *et al.*, 1987). The extracellular fungal matrix, rich in mannose binding sites, was a prerequisite for mycorrhizal establishment, and the binding sites and extracellular matrix disappeared from the fungus once penetration of the host occurred (Bonfante-Fasolo, 1988). This apparent induction of hapten binding sites at the time of recognition and establishment of adhesion closely resembles the events that occur with the induction of adhesion in *Candida albicans* to human epithelium, a process that is also induced by contact, and mediated by the lectin binding of glucose and mannose haptens on the fungal surface (Tronchin et al., 1984; Brawner and Cutler, 1986).

4.2. Significance of Adhesion

4.2.1. Dissemination

Fungal adhesion to a plant surface probably serves several functions. Although not commonly reported in the literature, adhesive compounds may be involved in dissemination of "sticky" propagules in some vector-transmitted diseases (Ingold, 1978). An unusual case of dispersal has been described for *Sarcinulella banksiae* in which conidia are produced in association with a mucilage (Sutton and Alcorn, 1983). Passage of the conidia through the fungal periphyses results in conidial aggregation and the mucilage forms a rigid cylindrical matrix in which conidia are embedded. The binding together of clusters of conidia of *C. graminicola* by the spore mucilage allows this fungus to be dispersed by wind as dry, particulate matter (Nicholson and Moraes, 1980).

4.2.2. Prevention of Fungal Displacement

Probably the most common function of adhesion is to prevent displacement by wind or water. While "underwater" glues require special chemical features (Waite, 1983), aquatic environments often necessitate firm attachment of a fungus to its potential substrate (Ingold, 1978; Rees and Jones, 1984). It is not surprising that the best studied examples of adhesion of plant pathogens are Oomycetes, or "water molds" in the genera *Phytophthora* and *Pythium*. These fungi include many soilborne plant pathogens that flourish in wet soil. Infection occurs after the flagellated zoospores swim in a water film to the root surface, encyst and adhere to the root, and then penetrate into the plant tissue (Gubler and Hardham, 1988; Hinch and Clarke, 1980; Longman and Callow, 1987; Mitchell and Deacon, 1987; Sing and Bartnicki-Garcia, 1975a,b). There are also examples of adhesion of phytopathogens to aquatic plants (Onyile et al., 1982); the best studied example is of attachment of *Magnaporthe grisea* conidia to rice plants (Hamer et al., 1988). Here, an adhesive compound is released upon conidial hydration.

The capability to adhere in water may actually be important for a wide variety of foliar pathogens. Many fungal spores, such as *Colletotrichum* spp., are dispersed in water. Fungi that can attach to the leaf surface may have a selective advantage. Nonadherent fungi would probably remain suspended in the water droplet and then roll onto the soil when the water drops off the leaf. Nonadherent fungi could also be washed off the leaf surface with the next rain droplet. Water is also a necessary component of the environment of most germinating spores and germination generally requires "free water." Hence, it is not surprising that fungi such as the foliar bean rust fungus, *Uromyces appendiculatus*, can grow and attach while completely submerged in water (Epstein et al., 1987).

4.2.3. Morphogenesis

Contact stimuli are important in the movement (thigmotropism) and development (thigmo-differentiation) of fungi such as the bean rust fungus, *U. appendiculatus* (Wynn and Staples, 1981). Urediospore germlings grow across the leaf surface in an oriented manner (S. Dickinson, 1969); apparently this provides an increased probability that the fungus will cross a leaf stomate, the site of penetration (Wynn, 1981). When a stomate is recognized, the fungus forms several specialized structures (including an appressorium), which allow infection to occur. Both oriented growth and stomatal recognition are a response to the physical topography of the plant surface (Wynn, 1976; Hoch et al., 1987). If the fungus is not well attached to the surface, there is neither oriented growth nor appressorial induction (Epstein *et al.*, 1987), presumably because the fungus does not receive appropriate signals from the surface. Hence, adhesion seems essential for pathogenicity of this obligate parasite. In addition to many other rusts, adhesion may be essential for the induction of appressoria (thigmodifferentiation) in other fungi, notably species of *Colletotrichum*. Generally, the germ tubes of the anthracnose fungi grow attached to the leaf surface (Landes and Hoffmann, 1979). After contact with the junction between cells (Preece *et al.*, 1967), the fungus forms an appressorium.

4.2.4. Host–Pathogen Interactions

In addition to being involved in the induction of infection structures, adhesion seems to be required for penetration by the mature infection structure. Indeed, appressoria are often defined as adhesive structures (Emmett and Parbery, 1975) and a "mucilaginous substance" between the fungus and the substrate is often seen in electron micrographs (Uchiyama *et al.*, 1979). Other attachment structures, such as hyphopodia (Onyile *et al.*, 1982) and infection cushions (Armentrout and Downer, 1987), are also tightly adhered to the plant surface.

The adhesive material may also be important for other host–pathogen interactions. Adhesives, or other extracellular compounds, may enhance the physical contact between fungal toxins or enzymes which result in chemical alteration or debilitation of the host. In addition, these extracellular materials may also prevent the movement of enzymes or toxins outside of a specified area occupied by the fungus and thereby restrict physiological activity of the fungus to a region in which it can best compete and most efficiently utilize or recognize the host's soluble degradation products.

4.2.5. Adhesion in Other Systems

The need for adhesion extends beyond that of the host–parasite relationship. An example of particular importance is the sexual agglutination of yeast cells. Mating types may produce complementary oligopeptide sex pheromones (as in *Saccharomyces*) or glycoproteins (as in *Hansenula*) that increase the adhesion of cells of the opposite mating type (Terrance *et al.*, 1987; Crandall and Brock, 1968). Similarly, cell recognition in pollen–stigma relationships involves lectins, as blocking of stigma receptor sites by Con A prevents pollen tube penetration (Knox *et al.*, 1976).

Ahmadjian and Jacobs (1981) have suggested that in lichens the relationship of symbionts more closely resembles a controlled parasitism than a true symbiosis. In a system designed to induce lichen formation artificially, it was found that the mycobiont *Cladonia cristatella* successfully established relationships with several species of the alga *Trebouxia*. It was from the alga, however, that contact adhesion was established as the fungus even encircled glass beads, and appeared to exhibit little specificity. Only after initiation of the presymbiotic state of lichen formation was the fungus found to exude an extracellular gelatinous matrix that presumably was involved in interorganismal attachment (Ahmadjian and Jacobs, 1983).

5. MECHANISMS OF FUNGAL ADHESION TO PLANT SURFACES

5.1. Production of Fungal Adhesive Compounds

In *Phytophthora* spp. zoospores (Sing and Bartnicki-Garcia, 1975a,b), *Magnaporthe grisea* conidia (Hamer *et al.*, 1988), and *Dilophospora alopecuri* (Bird and McKay, 1987), the adhesive material appears to be "prepackaged" in, and exuded from the spore. Using monoclonal antibodies, Gubler and Hardham (1988) established that the adhesive material in *Phytophthora cinnamomi* zoospores is localized in small vesicles in the cell periphery. Previous, less definitive studies had postulated that the adhesive material was localized in large peripheral vesicles (Sing and Bartnicki-Garcia, 1975b), but this is not the case. Early in the encystment process, the high-molecular-weight adhesive glycoproteins are released. In contrast to *P. cinnamomi* where the adhesive material covers the nascent cyst, in *M. grisea*, the adhesive is localized within the spore tip. After hydration, the mucilage is discharged from the spore apex and remains localized at the tip (Fig. 5); no metabolism is required (Hamer *et al.*, 1988). Unlike *M. grisea*, with *Colletotrichum lindemuthianum* conidia (Young and Kauss, 1984) and *Nectria haematococca* mating population (MP) I macroconidia (Jones and Epstein, 1989), spore metabolism is required for adhesion, and in *N. haematococca* protein synthesis is required prior to spore attachment (Jones and Epstein, 1989).

Little is known about the process of synthesis and deposition of adhesives in either germlings or appressoria. For some fungi, microscopic evidence suggests that adhesion occurs specifically either at the point of germ tube emergence (Jacobi *et al.*, 1982), in specific regions along the hyphae (Akai and Ishida, 1968; Hawker and Hendy, 1963), or at the germ tube tip just prior to penetration (McKeen, 1974). In other fungi, adhesion appears to occur along with hyphal elongation (Epstein *et al.*, 1987). Nicholson (1984) suggested that adhesion may occur specifically either in response to contact stimuli, or at bends and other points of stress in the fungal wall. Specific antibodies for adhesive compounds (Gubler and Hardham, 1988) could be used to determine the surface distribution of adhesive material.

5.2. Attachment of Fungi to Plant Surfaces

There is no evidence that all fungi bind to plant surfaces by the same mechanism(s) or that several binding mechanisms are not involved concurrently. Daniels (1980), Waite (1983), and Lippincott and Lippincott (1984) have summarized the theoretical organism–substratum binding mechanisms; Kolattukudy (1980) and Moody and co-workers (1988) have summarized the

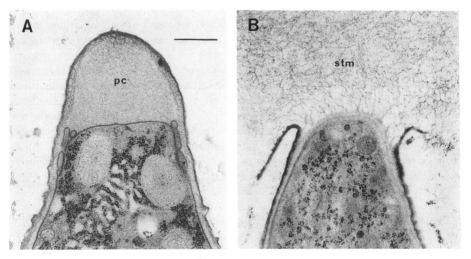

Figure 5. Apical region of two different freeze-substituted conidia of *Magnaporthe grisea* illustrating an intact (A) and ruptured (B) periplasmic compartment of spore tip mucilage. pc, apical periplasmic compartment; stm, spore tip mucilage released after rupture of the periplasmic compartment. Bar = 0.5 μm. Micrographs courtesy of Richard J. Howard. Reprinted from *Science* **239**:288–290. Copyright 1988 by the AAAS.

chemistry of roots and shoots, respectively. The most obvious difference between the surfaces of roots and shoots is that shoots are generally covered with waxy, hydrophobic materials such as cutin. Roots are covered with a hydrophilic slime composed primarily of carbohydrate. Some fungi, such as *N. haematococca* MP I macroconidia, adhere to both hydrophobic and hydrophilic surfaces (Hickman and Epstein, 1988).

5.2.1. Model Surfaces of Plants

There are obvious advantages of using a flat, chemically defined surface on which to study adhesion: microscopy, chemical characterization and quantification generally are simplified. Polystyrene (Jones and Epstein, 1989; Young and Kauss, 1984), polyethylene (Epstein *et al.*, 1985, 1987), and Teflon (Hamer *et al.*, 1988) have been used to represent the aerial portions of plants which are generally hydrophobic. Regardless of similarities in hydrophobicity, in comparison to plastics, natural plant surfaces obviously have more varied topography (Hallam and Juniper, 1971; Wynn, 1981) and chemical sites (Kolattukudy, 1980) for attachment. Indeed, microorganisms are often concentrated in depressions on the leaf surface (Campbell, 1985). In addition, microorganisms may alter the leaf surface by enzymatic activity. Cutinase activity (Dickman and Patil, 1986); Kolattukudy, 1985b; Woloshuk and Kolattukudy, 1986) could theoretically produce a surface with different properties for attachment. Using scanning electron microscopy, several researchers have observed that germlings can leave an imprint in the plant cuticle (Garcia-Arenal and Sagasta, 1980; Hau and Rush, 1982; Staub *et al.*, 1974). It is possible that these imprints are formed during specimen processing. If so, the imprints may indicate that the force of fungal–cuticle attachment is greater than the force of cuticular binding to itself (C. Dickinson, 1986). Alternatively, the imprints may be the result of enzymatic erosion of the cuticle (Nicholson, 1984; Kunoh *et al.*, 1988; Nicholson *et al.*, 1988). If so, the fungi would be essentially embedded in the plant surface and cuticular erosion could, theoretically, enhance adhesion either by serving as a physical cast or by providing new sites for chemical bonding.

Model surfaces are not generally used to study fungal attachment to roots; in at least some cases, fucosyl residues in the root mucilage are needed for fungal attachment (Hinch and Clarke, 1980), and such a synthetic surface is not readily available. However, ^{14}C-labeled mucilage has been isolated from roots and used *in vitro* to study attachment (Gould and Northcote, 1986).

5.2.2. Chemical Bonds Involved in Fungal Adhesion to Plants

5.2.2a. Hydrophobic Bonds. Several methods are available for measuring the hydrophobicity of plant and fungal cell surfaces (Mozes and Rouxhet, 1987). The major evidence that hydrophobic bonds are involved in attachment of fungi to plants is simply that aerial fungal pathogens, such as *M. grisea* conidia (Hamer *et al.*, 1988) and *U. appendiculatus* germlings (Epstein *et al.*, 1987), bind tenaciously to extremely hydrophobic, inert surfaces such as Teflon or polyethylene, respectively. In the case of *M. grisea*, the conidia bound more tightly to Teflon than to glass, a less hydrophobic surface (Hamer *et al.*, 1988). Young and Kauss (1984) suggested that hydrophobic binding was involved in attachment of *C. lindemuthianum* conidia to bean hypocotyls on the basis of: (1) attachment to polystyrene, (2) reduced attachment to waxless hypocotyls, (3) reduced attachment in the presence of the surfactant Tween 80, and (4) increased attachment in the presence of mono- and divalent cations.

5.2.2b. Lectinlike Bonds. Lectinlike interactions appear involved in several examples of fungal/root adhesion (Gubler and Hardham, 1988; Hinch and Clarke, 1980; Sing and Bartnicki-Garcia, 1975b) and fungal hyperparasitism (Barak *et al.*, 1985; Borrebaeck *et al.*, 1984; Elad *et al.*,1983a; Jansson and Nordbring-Hertz, 1984; Manocha, 1984). With *Phytophthora* spp. and

Pythium cysts, fungal glycoproteins (Gubler and Hardham, 1988; Sing and Bartnicki-Garcia, 1975b) bind to fucosyl residues in root mucilage (Hinch and Clarke, 1980; Longman and Callow, 1987).

5.2.2c. Other Bonds. Presumably, the adhesive compound is released into the wall in a soluble form and, in some cases, becomes less water-soluble extracellularly (Mims and Richardson, 1989). The mechanism(s) of the insolubilization of the extracellular matrix of plant pathogenic fungi is unknown; however, oxidative or enzymatic cross-linking are possible. For example, plant cell wall material appears to be extracellularly cross-linked by peroxidase (Cooper and Varner, 1983), and an adhesive protein from mussels is apparently extracellularly cross-linked by a catechol oxidase (Waite, 1985). The insolubilization of the adhesive material could be due either to cross-linking to itself, or to other extracellular components. Hence, some fungi could conceivably bind covalently to reactive sites on the plant surface. The presence of peroxidase and polyphenoloxidase in the spore mucilage of *C. graminicola* may be an indication that such mechanisms for insolubilization exist (Nicholson, unpublished).

6. METHODOLOGY FOR THE STUDY OF EXTRACELLULAR MATRICES AND ADHESIVES

6.1. Visualization

Fungal extracellular matrices are often visible with light microscopy. In cases where the extracellular material is abundant and viscous, as with conidia produced in slime, unstained matrices can be seen readily using phase contrast or differential interference contract optics. Where the matrix is less abundant, visualization may require the use of stains. Methods used include negative staining (Evans *et al.*, 1982) and histochemical stains (Rohringer *et al.*, 1982). More recently, fluorescently labeled lectins (Hamer *et al.*, 1988) and antibodies (Hardham *et al.*, 1986) have been used to visualize the fungal cell surface.

Fungal extracellular matrices are not readily apparent with light or fluorescence microscopy with some fungi with well-attached propagules, germlings, or infection structures. In rust fungi, a relatively thin layer of extracellular material has been seen with electron microscopy (Beckett and Porter, 1988; Hoch and Staples, 1983). In such cases where the enhanced resolution of electron microscopy is desired, tissue fixed by freeze substitution appears to retain more extracellular material than does tissue fixed by the traditional chemical means (Hamer *et al.*, 1988; Howard and Aist, 1979; Hoch and Howard, 1980; Hoch and Staples, 1983; Epstein *et al.*, 1987). Freeze substitution or other cryotechniques may be particularly advantageous when the extracellular compounds are water soluble; water-soluble adhesives and matrices have been the most difficult to localize. Recent work in the laboratory of Dr. W. Wynn (Chubal, 1987) has shown that mucilage may be stabilized with the cationic detergent cetylpyridinium chloride or the polycationic stain ruthenium red (Fig. 6).

6.2. Quantification of Adhesion

Most commonly, adhesion has been quantified by microscopically counting the number of fungal units (spores, germlings, or appressoria) remaining on a microscope slide or membrane after washing (Hamer *et al.*, 1988; Epstein *et al.*, 1985, 1987). While assessing units per field of view is straightforward, the process is relatively time-consuming and does not lend itself to quantification of fungal material on all plant surfaces.

There are a wider variety of methods for estimating populations of spores than hyphae. For spore attachment assays, spores can be incubated on either a plant or an inanimate model

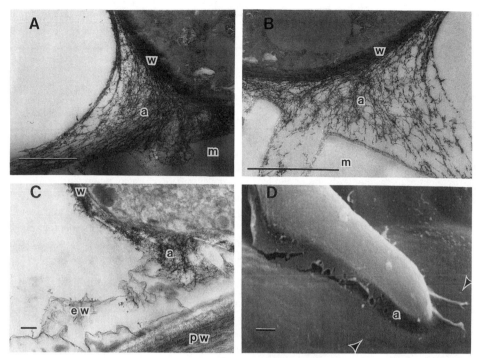

Figure 6. Transmission (TEM) and scanning (SEM) electron micrographs of germ tubes of *Puccinia sorghi* growing on artificial membranes and host plant cuticles, illustrating the use of polycations as fixatives to visualize the adhesive material at the interface between the fungus and the surfaces. a, adhesive material; ew, epicuticular wax layer; m, polycarbonate membrane; pw, plant epidermal cell wall; w, fungal germ tube wall. (A) TEM of transverse section of germ tube (upper right) on Nucleopore polycarbonate membrane (0.4-μm pore size). Fixation was in glutaraldehyde and postfixation in osmium tetroxide, both containing 0.01% cuprolinic blue to visualize the adhesive material. Note that the adhseive material has flowed into the pores in the membrane. Bar = 1.0 μm. (B) Similar to panel A except that the glutaraldehyde contained 0.05% cetylpyridinium chloride and the osmium tetroxide contained 0.01% ruthenium red for fixation of adhesive material. Bar = 1.0 μm. (C) TEM of transverse section of germ tube (upper right) on young corn leaf covered with epicuticular wax. Fixation of the adhesive material was with ruthenium red in both the glutaraldehyde and osmium tetroxide solutions. Note the movement of the adhesive material into the wax layer. Bar = 0.1 μm. (D) SEM of germ tube tip on mature corn leaf without epicuticular wax. Cetylpyridinium chloride was added to the glutaraldehyde only. The adhesive material is partly broken as a result of the stress applied to the specimen during drying and placing under vacuum. The distance that the adhesive material flowed away from the germ tube in a very thin film can be seen by the faint line (arrowheads) marking its outer edge. Bar = 1.0 μm. Micrographs courtesy of Willard K. Wynn.

substrate. After rinsing the surface, the spores in the wash can be counted with a hemacytometer (Smith and Cruickshank, 1987), Coulter counter (Young and Kauss, 1984) or in a scintillation counter, if the spores are radioactively labeled beforehand (Hickman and Epstein, 1988). By comparing the quantity of spores added originally with the quantity of spores recovered in the wash, the number of adherent spores can be calculated.

We know very little about how tightly a given fungal unit is attached to its plant surface. Using Teflon as a model surface, Hamer *et al.* (1988) incubated *M. grisea* conidia in a flow cell and determined the shear force required to dislodge conidia. It appears that different fungi attach with varying amounts of force, and are probably attached via different compounds. For example *N. haematococca* MP I macroconidia can be removed from polystyrene with far less force than can *M. grisea* conidia (Hamer *et al.*, 1988). It is also apparent that adhesion is a component of fungal development and that each stage of development may have different adhesive properties.

6.3. Identification of Function of Extracellular Compounds

While progress has been made in the biochemical characterization of the fungal extracellular matrix, we know relatively little about the functional significance of each of these components. For example, which of the extracellular compounds are specifically associated with adhesion? Isolation of mutants would facilitate the characterization of compound(s) of interest and an assessment of the role of each compound in fungal development and plant pathogenesis. Hamer and colleagues (1989) isolated *M. grisea* mutants with nonadhesive conidia; these "Smo" mutants also have abnormal conidial morphology and aberrations in appressorium formation and number of conidia. Other fungal mutants with altered wall composition and morphology have been isolated; these include the plant pathogenic *Rhizoctonia solani* (Shibata *et al.*, 1986) and the saprophytic *Aspergillus nidulans* (Claverie-Martin *et al.*, 1988) and *Neurospora crassa* (Selitren-nikoff *et al.*, 1979).

Although adhesionless mutants would provide a powerful analytical tool, such mutants may be difficult to obtain. The rusts, for example, are basically obligate parasites; this confounds standard microbiological procedures. Since development, and hence survival of the bean rust fungus on its plant host appears to require adhesion (Epstein *et al.*, 1985, 1987), conditional adhesionless mutations would have to be selected. Rather than use mutants, Epstein and colleagues (1985, 1987) used enzymes to identify specific extracellular components of the fungal extracellular matrix which affected adhesion. Several proteases including trypsin, but not glycosidases or lipases, removed adherent germlings and inhibited appressorial formation on a normally inductive surface. It was concluded that an extracellular protein(s) was involved in adhesion and the surface-induced differentiation of bean rust germlings. Appropriate controls included utilization of trypsin inhibitor (which annuls trypsin activity, but not activity of contaminating enzymes) and demonstration that enzymatic activity was limited to the extracellular matrix (by measuring physiological processes such as germination and germ tube elongation). Enzymatic analyses have also been used to identify adhesive materials of *C. graminicola* appressoria (Lapp and Skoropad, 1978) and *Buergenerula spartinae* hyphopodia (Onyile *et al.*, 1982). In these reports, however, the enzymes used were probably too impure for adequate identification of the adhesive compounds. Antibodies might also be used to identify the function of specific extracellular compounds, if specific antibodies can be identified that annul the function of the particular compound.

7. CONCLUSIONS

As in other eukaryotic organisms, we are beginning to appreciate the role of the extracellular matrix in organismal development and host–pathogen interactions. In the case of phytopathogenic fungi, we know that early development of the fungus on the plant surface is often associated with the development of adhesiveness of spores, germlings, and infection structures. The magnitude of importance of adhesion is unknown. In addition to providing adhesive material, the fungal extracellular matrix has other roles in events that prepare the infection court as well as in survival of the organism (Nicholson and Moraes, 1980; Nicholson *et al.*, 1986, 1989). Future research should establish the structure, function, and importance of each extracellular compound in fungal morphogenesis and in the establishment of host–pathogen relationships.

ACKNOWLEDGMENTS. This review is article number 11854 of the Purdue University Agricultural Experiment Station. The work was supported in part by grants to R.L.N. from the National Science Foundation (PCM-8404910) and Indiana Corporation for Science and Technology and to L.E. from the Herman Frasch Foundation (Grant 0162 HF) and the Biomedical Research Support Program, Division of Research Resources, National Institutes of Health BRSG Grant 2-S07-RR07006.

8. REFERENCES

Ahmadjian, V., and Jacobs, J. B., 1981, Relationship between fungus and alga in the lichen *Cladonia cristatella* Tuck, *Nature* **289**:169–172.

Ahmadjian, V., and Jacobs, J. B., 1983, Algal–fungal relationships in lichens: Recognition, synthesis, and development, in: *Algal Symbiosis* (L. J. Goff, ed.), Cambridge University Press, London, pp. 147–172.

Aist, J. R., and Williams, P. H., 1971, The cytology and kinetics of cabbage root hair penetration by *Plasmodiophora brassicae, Can. J. Bot.* **49**:2023–2034.

Akai, S., and Ishida, N., 1968, An electron microscopic observation on the germination of conidia of *Colletotrichum lagenarium, Mycopathol. Mycol. Appl.* **34**:337–345.

Armentrout, V. N., and Downer, A. J., 1987, Infection cushion development by *Rhizoctonia solani* on cotton, *Phytopathology* **77**:619–623.

Barak, R., Elad, Y., Mirelman, D., and Chet, I., 1985, Lectins: A possible basis for specific recognition in the interaction of *Trichoderma* and *Sclerotium rolfsii, Phytopathology* **75**:458–462.

Beckett, A., and Porter, R., 1988, The use of complementary fractures and low-temperature scanning electron microscopy to study hyphal–host cell surface adhesion between *Uromyces vicae-fabae* and *Vicia faba, Can. J. Bot.* **66**:645–652.

Bergstrom, G. C., and Nicholson, R. L., 1981, Invertase in the spore matrix of *Colletotrichum graminicola, Phytopathol. Z.* **102**:139–147.

Bird, A. F., and McKay, A. C., 1987, Adhesion of conidia of the fungus *Dilophospora alopecuri* to the cuticle of the nematode *Anguina agrostis*, the vector in annual ryegrass toxicity, *Int. J. Parasitol.* **17**:1239–1247.

Bonfante-Fasolo, P., 1988, The role of the cell wall as a signal in mycorrhizal associations, in: *Cell to Cell Signals in Plant, Animal and Microbial Symbiosis* (S. Scannerini, D. Smith, P. Bonfante-Fasolo, and V. Gianinazzi-Pearson, eds.), Springer-Verlag, Berlin, pp. 219–235.

Bonfante-Fasolo, P., and Perotto, S., 1986, Visualization of surface sugar residues in mycorrhizal ericoid fungi by fluorescein conjugated lectins, *Symbiosis* **1**:269–288.

Bonfante-Fasolo, P., Perotto, S., Testa, B., and Faccio, A., 1987, Ultrastructural localization of cell surface sugar residues in ericoid mycorrhizal fungi by gold-labeled lectins, *Protoplasma* **139**:25–35.

Borrebaeck, C. A. K., Mattiasson, B., and Nordbring-Hertz, B., 1984, Isolation and partial characterization of a carbohydrate-binding protein from a nematode-trapping fungus, *J. Bacteriol.* **159**:53–56.

Brawner, D. L., and Cutler, J. E., 1986, Ultrastructural and biochemical studies of two dynamically expressed cell surface determinants on *Candida albicans, Infect. Immun.* **51**:327–336.

Campbell, R., 1985, *Plant Microbiology*, Arnold, London.

Carlile, M. J., 1983, Motility, taxis, and tropism in *Phytophthora*, in: *Phytophthora: Its Biology, Taxonomy, Ecology, and Pathology* (D. C. Erwin, S. Bartnicki-Garcia, and P. H. Tsao, eds.), American Phytopathological Society, St. Paul, pp. 95–107.

Chubal, R., 1987, Adhesion and growth away from surfaces by urediniospore germ tubes of *Puccinia sorghi*, Ph.D. dissertation, University of Georgia, Athens.

Claverie-Martin, F., Diaz-Torres, M. R., and Geoghegan, M. J., 1988, Chemical composition and ultrastructure of wild-type and white mutant *Aspergillus nidulans* conidial walls, *Curr. Microbiol.* **16**:281–287.

Cooper, J. B., and Varner, J. E., 1983, Insolubilization of hydroxyproline-rich cell wall glycoprotein in aerated carrot root slices, *Biochem. Biophys. Res. Commun.* **112**:161–167.

Crandall, M. A., and Brock, T. D., 1968, Molecular basis of mating in the yeast *Hansenula wingei, Bacteriol. Rev.* **32**:139–163.

Daniels, S. L., 1980, Mechanisms involved in sorption of microorganisms to solid surfaces, in: *Adsorption of Microorganisms to Surfaces* (G. Bitton and K. C. Marshall, eds.), Wiley, New York, pp. 7–58.

Dickinson, C. H., 1986, Adaptations of micro-organisms to climatic conditions affecting aerial plant surfaces, in:

Microbiology of the Phyllosphere (N. J. Fokkema and J. Van den Heuvel, eds.), Cambridge University Press, London, pp. 77–100.

Dickinson, S., 1969, Studies in the physiology of obligate parasitism. VI. Directed growth, *Phytopathol. Z.* **66**: 38–49.

Dickman, M. B., and Patil, S. S., 1986, Cutinase deficient mutants of *Colletotrichum gloeosporioides* are nonpathogenic to papaya fruit, *Physiol. Mol. Plant Pathol.* **28**:235–242.

Edwards, H. H., and Allen, P. J., 1970, A fine-structure study of the primary infection process during infection of barley by *Erysiphe graminis* f. sp. *hordei*, *Phytopathology* **60**:1504–1509.

Elad, Y., Barak, R., and Chet, I., 1983a, Possible role of lectins in mycoparasitism, *J. Bacteriol.* **154**:1431–1435.

Elad, Y., Chet, I., Boyle, P., and Henis, Y., 1983b, Parasitism of *Trichoderma* spp. on *Rhizoctonia solani* and *Sclerotium rolfsii*—Scanning electron microscopy and fluorescence microscopy, *Phytopathology* **73**:85–88.

Emmett, R. W., and Parbery, D. G., 1975, Appressoria, *Annu. Rev. Phytopathol.* **13**:147–167.

Epstein, L., Laccetti, L., Staples, R. C., Hoch, H. C., and Hoose, W. A., 1985, Extracellular proteins associated with induction of differentiation in bean rust uredospore germlings, *Phytopathology* **75**:1073–1076.

Epstein, L., Laccetti, L., Staples, R. C., and Hoch, H. C., 1987, Cell–substratum adhesive protein involved in surface contact responses of the bean rust fungus, *Physiol. Mol. Plant Pathol.* **30**:373–388.

Evans, R. C., Stempen, H., and Stewart, S. J., 1981, Development of hyphal sheaths in *Bipolaris maydis* race T, *Can. J. Bot.* **59**:453–459.

Evans, R. C., Stempen, H., and Frasca, P., 1982, Evidence for a two-layered sheath on germ tubes of three species of *Bipolaris*, *Phytopathology* **72**:804–807.

Forrest, J. M. S., and Robertson, W. M., 1986, Characterization and localization of saccharides on the head region of four populations of the potato cyst nematode *Globodera rostochiensis* and *G. pallida*, *J. Nematol.* **18**: 23–26.

Garcia-Arenal, F., and Sagasta, E. M., 1980, Scanning electron microscopy of *Botrytis cinerea* penetration of bean (*Phaseolus vulgaris*) hypocotyls, *Phytopathol. Z.* **99**:37–42.

Gold, R. E., and Mendgen, K., 1984, Cytology of basidiospore germination, penetration, and early colonization of *Phaseolus vulgaris* by *Uromyces appendiculatus* var. *appendiculatus*, *Can. J. Bot.* **62**:1989–2002.

Goode, P. M., 1956. Infection of strawberry roots by zoospores of *Phytophthora fragariae*, *Trans. Br. Mycol. Soc.* **39**:367–377.

Gould, J., and Northcote, D. H., 1986, Cell–cell recognition of host surfaces by pathogens: The adsorption of maize (*Zea mays*) root mucilage by surfaces of pathogenic fungi, *Biochem. J.* **233**:395–405.

Grove, S. N., and Bracker, C. E., 1978, Protoplasmic changes during zoospore encystment and cyst germination in *Pythium aphanidermatum*, *Exp. Mycol.* **2**:51–98.

Gubler, F., and Hardham, A. R., 1988, Secretion of adhesive material during encystment of *Phytophthora cinnamomi* zoospores, characterized by immunogold labeling with monoclonal antibodies to components of peripheral vesicles, *J. Cell Sci.* **90**:225–235.

Hallam, N. D., and Juniper, B. E., 1971, The anatomy of the leaf surface, in: *Ecology of Leaf Surface Microorganisms* (T. F. Preece and C. H. Dickinson, eds.), Academic Press, New York, pp. 3–37.

Hamer, J. E., Howard, R. J., Chumley, F. G., and Valent, B., 1988, A mechanism for surface attachment in spores of a plant pathogenic fungus, *Science* **239**:288–290.

Hamer, J. E., Valent, B., and Chumley, F. G., 1989, Mutations at the *SMO* genetic locus affect the shape of diverse cell types in the rice blast fungus, *Genetics* **122**:351–361.

Hardham, A. R., and Suzaki, E., 1986, Encystment of zoospores of the fungus, *Phytophthora cinnamomi*, is induced by specific lectin and monoclonal antibody binding to the cell surface, *Protoplasma* **133**:165–173.

Hardham, A. R., Suzaki, E., and Perkin, J. L., 1986, Monoclonal antibodies to isolate-, species-, and genus-specific components on the surface of zoospores and cysts of the fungus *Phytophthora cinnamomi*, *Can. J. Bot.* **64**:311–321.

Hau, F. C., and Rush, M. C., 1982, Preinfectional interactions between *Helminthosporium oryzae* and resistant and susceptible rice plants, *Phytopathology* **72**:285–292.

Hawker, L. E., and Hendy, R. J., 1963, An electron-microscope study of germination of conidia of *Botrytis cinerea*, *J. Gen. Microbiol.* **33**:43–46.

Held, A. A., 1974, Attraction and attachment of zoospores of the parasitic chytrid *Rozella allomycis* in response to host-dependent factors, *Arch. Microbiol.* **95**:97–114.

Hickman, M. J., and Epstein, L., 1988, *Nectria haematococca* macroconidia attach to plant surfaces, *Phytopathology* **78**:1523 (Abstr.).

Hinch, J. M., and Clarke, A. E., 1980, Adhesion of fungal zoospores to root surfaces is mediated by carbohydrate determinants of the root slime, *Physiol. Plant Pathol.* **16**:303–307.

Hoch, H. C., and Howard, R. J., 1980, Ultrastructure of freeze-substituted hyphae of the basidiomycete *Laetisaria arvalis*, *Protoplasma* **103**:281–297.

Hoch, H. C., and Staples, R. C., 1983, Ultrastrucural organization of the nondifferentiated uredospore germling of *Uromyces phaseoli*, *Mycologia* **75**:795–824.

Hoch, H. C., and Staples, R. C., 1987, Structural and chemical changes among the rust fungi during appressorium development, *Annu. Rev. Phytopathol.* **25**:231–247.

Hoch, H. C., Staples, R. C., Whitehead, B., Comeau, J., and Wolf, E. D., 1987, Signaling for growth orientation and cell differentiation by surface topography in *Uromyces*, *Science* **235**:1659–1662.

Howard, R. J., and Aist, J. R., 1979, Hyphal tip cell ultrastructure of the fungus *Fusarium*: Improved preservation by freeze-substitution, *J. Ultrastruct. Res.* **66**:224–234.

Ingold, C. T., 1978, Role of mucilage in dispersal of certain fungi, *Trans. Br. Mycol. Soc.* **70**:137–140.

Jacobi, W. R., Amerson, H. V., and Mott, R. L., 1982, Microscopy of cultured loblolly pine seedlings and callus inoculated with *Cronartium fusiforme*, *Phytopathology* **72**:138–143.

Jansson, H.-B., and Nordbring-Hertz, B., 1983, The endoparasitic nematophagous fungus *Meria conispora* infects nematodes specifically at the chemosensory organs, *J. Gen. Microbiol.* **129**:1121–1126.

Jansson, H.-B., and Nordbring-Hertz, B., 1984, Involvement of sialic acid in nematode chemotaxis and infection by an endoparasitic nematophagous fungus, *J. Gen. Microbiol.* **130**:39–43.

Jansson, H.-B., Jeyaprakash, A., and Zuckerman, B. M., 1985, Differential adhesion and infection of nematodes by the endoparasitic fungus *Meria coniospora* (*Deuteromycetes*), *Appl. Environ. Microbiol.* **49**:552–555.

Jones, M. J., and Epstein, L., 1989, Adhesion of *Nectria haematococca* macroconidia. *Physiol. Mol. Plant Pathol.* **35**:453–461.

Knox, R. B., Clarke, A., Harrison, S., Smith, P., and Marchalonis, J. J., 1976, Cell recognition in plants: Determinants of the stigma surface and their pollen interactions, *Proc. Natl. Acad. Sci. USA* **73**:2788–2792.

Kolattukudy, P. E., 1980, Biopolyester membranes of plants: Cutin and suberin, *Science* **208**:990–1000.

Kolattukudy, P. E., 1985a, Enzymatic penetration of the plant cuticle by fungal pathogens, *Annu. Rev. Phytopathol.* **23**:223–250.

Kolattukudy, P. E., 1985b, Cutinases from fungi and pollen, in: *Lipases* (B. Borgstrom and H. L. Brockman, eds.), Elsevier, Amsterdam, pp. 471–504.

Kunoh, H., Yamaoka, N., Yoshioka, H., and Nicholson, R. L., 1988, Preparation of the infection court by *Erysiphe graminis*. I. Contact-mediated changes in morphology of the conidium surface, *Exp. Mycol.* **12**:325–335.

Landes, M., and Hoffmann, G. M., 1979, Zum Keimungs-und Infektionsverlauf bei *Colletotrichum lindemuthianum* auf *Phaseolus vulgaris*, *Phytopathol. Z.* **95**:259–273.

Lapp, M. S., and Skoropad, W. P., 1978, Nature of adhesive material of *Colletotrichum graminicola* appressoria, *Trans. Br. Mycol Soc.* **70**:221–223.

Lewis, I., and Cooke, R. C., 1985, Conidial matrix and spore germination in some plant pathogens, *Trans. Br. Mycol. Soc.* **84**:661–667.

Lippincott, J. A., and Lippincott, B. B., 1984, Concepts and experimental approaches in host–microbe recognition, in: *Plant–Microbe Interactions: Molecular and Genetic Perspectives, Volume 1*, (T. Kosuge and E. W. Nester, eds.), Macmillan Co., New York, pp. 195–214.

Locci, R., 1969, Scanning electron microscopy of *Helminthosporium oryzae* on *Oryza sativa*, *Riv. Patol. Veg. Secc. IV* **5**:179–195.

Longman, D., and Callow, J. A., 1987, Specific saccharide residues are involved in the recognition of plant root surfaces by zoospores of *Pythium aphanidermatum*, *Physiol. Mol. Plant Pathol.* **30**:139–150.

Lyons, P. C., Hipskind, J. D., and Nicholson, R. L., 1987, Changes in methanol-soluble and wall-bound phenylpropanoids in maize in response to infection, *Phytopathology* **77**:1751 (Abstr.).

McCourtie, J., and Douglas, L. J., 1985, Extracellular polymer of *Candida albicans*: Isolation, analysis and role in adhesion, *J. Gen. Microbiol.* **131**:495–503.

McKeen, W. E., 1974, Mode of penetration of epidermal cell walls of *Vicia faba* by *Botrytis cinerea*, *Phytopathology* **64**:461–467.

Manocha, M. S., 1984, Cell surface characteristics of *Mortierella* species and their interaction with a mycoparasite, *Can. J. Microbiol.* **30**:290–298.

Mehansho, H., Clements, S., Sheares, B.T., Smith, S., and Carlson, D. M., 1985, Induction of proline-rich glycoprotein synthesis in mouse salivary glands by isoproterenol and by tannins, *J. Biol. Chem.* **260**:4418–4423.

Mendgen, K., Lange, M., and Bretschneider, K., 1985, Quantitative estimation of the surface carbohydrates on the infection structures of rust fungi with enzymes and lectins, *Arch. Microbiol.* **140**:307–311.

Mims, C. W., and Richardson, E. A., 1989, Ultrastructure of appressorium development by basidiospore germlings of the rust fungus *Gymnosporangium juniperi-virginianae*, *Protoplasma* **148**:111–119.

Mitchell, R. T., and Deacon, J. W., 1987, Differential adhesion of zoospore cysts of *Pythium* on roots of gaminaceous and nongraminaceous plants, *Trans. Br. Mycol. Soc.* **88**:401–403.

Moody, S. F., Clarke, A. E., and Bacic, A., 1988, Structural analysis of secreted slime from wheat and cowpea roots, *Phytochemistry* **27**:2857–2861.

Mozes, N., and Rouxhet, P. G., 1987, Methods for measuring hydrophobicity of microorganisms, *J. Microbiol. Methods* **6**:99–112.

Nasu, H., and Kunoh, H., 1986, A unique sclerotium-like structure produced by *Zygophiala jamaicensis* on grape berries, *Trans. Mycol. Soc. Jpn.* **27**:225–233.

Nasu, H., and Kunoh, H., 1987, Scanning electron microscopy of flyspec of apple, pear, Japanese persimmon, plum, Chinese quince, and pawpaw, *Plant Dis.* **71**:361–364.

Nasu, H., Hatamoto, M., and Kunoh, H., 1986, Behavior of causal fungus and process of lesion on flyspek-affected berries of grape, *Ann. Phytopathol. Soc. Jpn.* **52**:445–452.

Nicholson, R. L., 1984, Adhesion of fungi to the plant cuticle, in: *Infection Processes in Fungi* (D. W. Roberts and J. R. Aist, eds.), Rockefeller Foundation, New York, pp. 74–89.

Nicholson, R. L., and Moraes, W. B. C., 1980, Survival of *Colletotrichum graminicola*: Importance of the spore matrix, *Phytopathology* **70**:255–261.

Nicholson, R. L., Butler, L. G., and Asquith, T. N., 1986, Glycoproteins from *Colletotrichum graminicola* that bind phenols: Implications for survival and virulence of phytopathogenic fungi, *Phytopathology* **76**:1315–1318.

Nicholson, R. L., Yoshioka, H., Yamaoka, N., and Kunoh, H., 1988, Preparation of the infection court by *Erysiphe graminis*. II. Release of esterase enzyme from conidia in response to a contact stimulus, *Exp. Mycol.* **12**:336–349.

Nicholson, R. L., Hipskind, J., and Hanau, R. M., 1989, Protection against phenol toxicity by the spore mucilage of *Colletotrichum graminicola*, an aid to secondary spread, *Physiol. Mol. Plant Pathol.* **35**:243–252.

Nordbring-Hertz, B., 1973, Peptide-induced morphogenesis in the nematode-trapping fungus *Arthrobotrys oligospora*, *Physiol. Plant.* **29**:223–233.

Nordbring-Hertz, B., 1988, Ecology and recognition in the nematode–nematophagous fungus system, in: *Advances in Microbial Ecology*, Volume 10 (K. C. Marshall, ed.), Plenum Press, New York, pp. 81–114.

Onyile, A. B., Edwards, H. H., and Gessner, R. V., 1982, Adhesive material of the hyphopodia of *Buergenerula spartinae*, *Mycologia* **74**:777–784.

Paus, F., and Raa, J., 1973, An electron microscope study of infection and disease development in cucumber hypocotyls inoculated with *Cladosporium cucumerinum*, *Physiol. Plant Pathol.* **3**:461–464.

Perotto, S., and Bonfante-Fasolo, P., 1985, Lectin binding on surfaces of ericoid mycorrhizal fungi, in: *Physiological and Genetic Aspects of Mycorrhizae* (V. Gianinazzi Pearson and S. Gianinazzi, eds.), INRA Press, Paris, pp. 581–585.

Preece, T. F., Barnes, G., and Bayley, J. M., 1967, Junctions between epidermal cells as sites of appressorium formation by plant pathogenic fungi, *Plant Pathol.* **16**:117–118.

Pringle, R. B., 1981, Nonspecific adhesion of *Bipolaris sorokiniana* sporelings, *Can. J. Plant Pathol.* **3**:9–11.

Ramadoss, C. S., Uhlig, J., Carlson, D. M., Butler, L. G., and Nicholson, R. L., 1985, Composition of the mucilaginous spore matrix of *Colletotrichum graminicola*, a pathogen of corn, sorghum, and other grasses, *J. Agric. Food Chem.* **33**:728–732.

Rees, G., and Jones, E. B. G., 1984, Observations on the attachment of spores of marine fungi, *Bot. Mar.* **27**:145–160.

Reissig, J. L., Lai, W. H., and Glasgow, J. E., 1975, An endogalactosaminidase from *Streptomyce griseus*, *Can. J. Biochem.* **53**:1237–1249.

Rohringer, R., Ebrahim-Nesbat, F., and Heitefuss, R., 1982, Ultrastructural and histochemical studies on mildew of barley (*Erysiphe graminis* DC. f. sp. *hordie* Marchal). I. The cell envelope of conidia, germ tubes, and appressoria, *Phytopathol. Z.* **104**:325–336.

Saikawa, M., 1982, An electron microscope study of *Meria coniospora*, an endozoic nematophagous hyphomycete, *Can. J. Bot.* **60**:2019–2023.

Selitrennikoff, C. P., Slemmer, S. F., and Nelson, R. E., 1979, Cell surface changes associated with the loss of cell–cell recognition in *Neurospora*, *Exp. Mycol.* **3**:363–373.

Shibata, M., Uyeda, M., Kido, Y., Akamatsu, T., Nagayoshi, K., and Nagasato, T., 1986, Morphological mutant of *Rhizoctonia solani* and its cell wall component, *Agric. Biol. Chem.* **50**:915–921.

Sing, V. O., and Bartnicki-Garcia, S., 1975a, Adhesion of *Phytophthora palmivora* zoospores: Electron microscopy of cell attachment and cyst wall fibril formation, *J. Cell Sci.* **18**:123–132.

Sing, V. O., and Bartnicki-Garcia, S., 1975b, Adhesion of *Phytophthora palmivora* zoospores: Detection and ultrastructural visualization of concanavalin A-receptor sites appearing during encystment, *J. Cell Sci.* **19**:11–20.

Smith, M. M., and Cruickshank, I. A. M., 1987, Dynamics of conidial adhesion and elicitor uptake in relation to pisatin accumulation in aqueous droplets on the endocarp of pea pods, *Physiol. Mol. Plant Pathol.* **31**:315–324.

Snyder, B. A., and Nicholson, R. L., 1988, Nuclease activity in the extracellular spore matrix of *Colletotrichum* spp., *Phytopathology* **78**:1588 (Abstr.).

Staub, T., Dahmen, H., and Schwinn, F. J., 1974, Light and scanning electron microscopy of cucumber and barley powdery mildew on host and nonhost plants, *Phytopathology* **64**:364–372.

Sutton, B. C., and Alcorn, J. L., 1983, *Sarcinulella banksiae gen. et sp. nov.*, a coelomycete with a unique method of conidial dispersal, *Mycotaxon* **16**:557–564.

Terrance, K., Heller, P., Wu, Y.-S., and Lipke, P. N., 1987, Identification of glycoprotein components of alpha-agglutinin, a cell adhesion protein from *Saccharomyces cerevisiae*, *J. Bacteriol.* **169**:475–482.

Tronchin, G., Poulain, D., and Vernes, A., 1984, Cytochemical and ultrastructural studies of *Candida albicans*. III. Evidence for modifications of the cell wall coat during adherence to human buccal epithelial cells, *Arch. Microbiol.* **139**:221–224.

Uchiyama, T., Ogasawara, N., Nanba, Y., and Ito, H., 1979, Conidial germination and appressorial formation of the plant pathogenic fungi on the coverglass or cellophane coated with various lipid components of plant waxes, *Agric. Biol. Chem.* **43**:383–384.

Veenhuis, M., Nordbring-Hertz, B., and Harder, W., 1985, An electron-microscopical analysis of capture and initial stages of penetraion of nematodes by *Arthrobotrys oligospora*, *Antonie van Leeuwenhoek J. Microbiol. Serol.* **51**:385–398.

Waite, J. H., 1983, Adhesion in byssally attached bivalves, *Biol. Rev.* **58**:209–231.

Waite, J. H., 1985, Catechol oxidase in the byssus of the common mussel, *Mytilus edulis* L., *J. Mar. Biol. Assoc. U.K.* **65**:359–371.

Ward, H. M., 1888, A lily-disease, *Ann. Bot.* **2**:319–382.

Woloshuk, C. P., and Kolattukudy, P. E., 1986, Mechanism by which contact with plant cuticle triggers cutinase gene expression in the spores of *Fusarium solani* f. sp. *pisi*, *Proc. Natl. Acad. Sci. USA* **83**:1704–1708.

Wynn, W. K., 1976, Appressorium formation over stomates by the bean rust fungus: Response to a surface contact stimulus, *Phytopathology* **66**:136–146.

Wynn, W. K., 1981, Tropic and taxic responses of pathogens to plants, *Annu. Rev. Phytopathol.* **19**:237–255.

Wynn, W. K., and Staples, R. C., 1981, Tropisms of fungi in host recognition, in: *Plant Disease Control: Resistance and Susceptibility* (R. C. Staples and G. H. Toenniessen, eds.), Wiley, New York, pp. 45–69.

Young, D. H., and Kauss, H., 1984, Adhesion of *Colletotrichum lindemuthianum* spores to *Phaseolus vulgaris* hypocotyls and to polystyrene, *Appl. Environ. Microbiol.* **47**:616–619.

2

Signaling for Infection Structure Formation in Fungi

H. C. Hoch and R. C. Staples

1. INTRODUCTION

Fungal plant pathogens establish parasitic relationships with their respective hosts primarily for the perpetuation of the pathogen. Before such relationships can develop, several recognition steps must occur during the early stages of the association. Most important are perception of the correct direction of hyphal growth or movement (as with zoospore taxis) of the pathogen to an appropriate infection site, recognition of that site, and recognition that a successful interaction has occurred between the fungus and host. A successful host–parasite relationship will be aborted should any of these steps fail. Other important tropic recognition steps have been discussed in detail by Wynn and Staples (1981). Of particular relevance to the topic of this chapter are the first two steps, recognition by the fungus of the right place and time for infection structure formation. Some fungal propagules, such as uredospore germlings of many of the rust fungi, e.g., *Uromyces appendiculatus*, *Puccinia graminis tritici*, and *P. sorghi*, must seek and recognize leaf stomata, the only site that triggers appressorium formation (the first in a series of specialized infection structures produced by these organisms). Furthermore, these pathogens only enter the host through the stomatal aperture. Similarly, the causal agent of downy mildew of grape, *Plasmopara viticola*, produces zoospores that generally swim toward stomata through which ingress into the host is achieved. In this case the pathogen exhibits a taxic movement, a relatively uncommon feature for most filamentous fungal pathogens. Other nonstomatal penetrating fungal pathogens may or may not exhibit preferences for specific sites. Many of the foliar pathogens are capable of forming appressoria at almost any place on the host (and nonhost). Spores germinate and grow without any apparent preference for direction and develop appressoria at times and places determined by environmental, physiological, genetic, and/or temporal factors. Fungi, such as *Erysiphe polygoni* and *Colletotrichum* spp., develop appressoria more frequently over or near the anticlinal wall of the epidermis. Preference for these latter sites is likely mediated by surface nutrients, leaked from within the leaf tissue. In sum, then, most fungal plant pathogens exhibit

H. C. Hoch • Department of Plant Pathology, Cornell University, New York State Agricultural Experiment Station, Geneva, New York 14456.   R. C. Staples • Boyce Thompson Institute, Cornell University, Ithaca, New York 14853.

some type of tropic response. Several recent reviews (e.g., Emmett and Parbery, 1975; Hoch and Staples, 1987; Staples and Macko, 1984; Wynn and Staples, 1981; Wynn, 1981; Staples *et al.*, 1985b) have dealt, in part, with tropic responses in fungi, and the reader is directed to those works for additional detail.

Answering the questions of how fungal cells sense growth direction and the correct site for the infection event provides important knowledge necessary to develop unique and effective disease control strategies. Once the mechanisms involved in signal perception are known in greater detail, it may be possible to design specific inhibitors that interrupt the recognition process, or to engineer host surfaces that lack proper signals. Thus, the purpose of this work is to provide a summation of what we know about how fungi interact with the plant host during the very early stages of host–parasite establishment. As will become evident, there is a reasonable amount of information about the signals involved in growth orientation and in site recognition for some fungi, but we are only beginning to delve into the mechanisms of signal reception and response.

1.1. Diversity of Infection Structures

Fungi are armed with a wide array of specialized structures used in preparation for invasion of the host. The pathogens have evolved to such an extent that, under normal circumstances, successful penetration of the host cannot be achieved without these structures, and evolutionary development has probably occurred as a mechanism by which the pathogen overcame barriers produced by the host to invasion.

The simplest means of host invasion is through the production of toxic compounds and/or enzymes by the fungus, causing the host cell to die. The pathogen then grows into the necrotic cell by enzymatic digestion and/or through the application of mechanical pressure of the growing hyphal apex to the outer cell barrier of the host. Ingress into the host cell can also begin before the host cell has become necrotic (see Aist, 1981). Prominent specialized infection structures are not produced by these fungi. Most fungi, however, develop specialized structures termed *appressoria*, first described by Frank (1883), and recently reviewed extensively by Emmett and Parbery (1975). Appressoria usually are differentiated hyphal apices that have become somewhat enlarged, and provide the pathogen with the capacity to adhere to the host surface in preparation for subsequent invasion. Appressoria, however, may not be morphologically distinct from the hypha (Emmett and Parbery, 1975). Alternatively, they may be quite complex with multilobed features, with or without septa. In either case, a penetration peg(s) arises from the appressorium on the side adjacent to the host surface after a period of maturation. It is these latter infection structures that actually penetrate the host. Still another group of fungi (e.g., *Botrytis* spp., *Rhizoctonia* spp., *Sclerotinia* spp., *Sclerotium* spp.) frequently develop "infection cushions" that are compact aggregations of highly branched hyphae from which the fungus sends infection pegs into the host. As has been observed with *R. solani*, some fungi have the ability to develop different kinds of initial infection structures. Infection structures of this fungus vary from none in which hyphae penetrate the host plant directly via stomates, although rarely (Dodman *et al.*, 1968; Murray, 1982), to the development of lobate appressoria (Dodman *et al.*, 1968; Flentje, 1957; Marshall and Rush, 1980; Murray, 1982), to complex dome-shaped infection cushions (Armentrout and Downer, 1987; Armentrout *et al.*, 1987) (see Fig. 6). Single infection pegs arise from appressoria, while multiple infection pegs develop from infection cushions. Entrance into the stomate probably represents errant hyphae, while development of appressoria and infection cushions is dependent on the host and the anastomosis group of *R. solani*. Many obligate pathogens, including the rust fungi (Hoch and Staples, 1987; Littlefield and Heath, 1979) and the powdery mildew fungi (see Chapter 15), progress through a specific sequence of events before host invasion is completed (Figs. 1–3). The diversity that exists in the types of infection structures fungi use to breach their host surfaces is obvious.

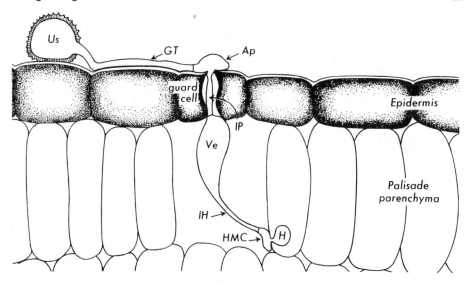

Figure 1. Schematic depiction of the infection structures of *Uromyces appendiculatus* formed during ingress of a host leaf surface. Appressorium, Ap; germ tube, GT; haustorium, H; haustorial mother cell, HMC; infection hypha, IH; infection peg, IP; uredospore, Us; vesicle, Ve. (Staples *et al.*, 1985b).

1.2. Cell Growth and Differentiation into Infection Structures

Hyphal growth is well documented to result from cell wall and membrane material being transported to and deposited at the cell apex (Howard, 1981; Grove and Bracker, 1970; Grove *et al.*, 1970, and references therein; Hill and Mullins, 1980). It is generally not viewed as occurring through extension or stretching of the cell, as is the situation in the region of elongation for roots. Apical vesicles fusing with the plasma membrane contribute membrane material, wall precursors, and various other components necessary for polarized growth. Yet, somewhat of a mystery is exactly how these materials are so directed and maintained in a cellular position as to be deposited only at the tip of the cylindrically shaped hyphal cell. We know that the cluster of heterogeneous apical vesicles (= Spitzenkörper, in Ascomycetes and Basidiomycetes) (Brunswik, 1924; Girbardt, 1955, 1957; Grove and Bracker, 1970) positioned in the apex is involved in growth direction. The position of this apical body is maintained, in part, by cytoskeletal elements, e.g., actin, since pharmacological treatments known to depolymerize actin result in a ballooning of the cell apex (Grove and Sweigard, 1980; Tucker *et al.*, 1986), leading to loss of polarized tip growth.

Appressorium formation begins when the germ tube or hyphal apex ceases forward growth and enlarges. One of the first detectable events in appressorium formation, beyond cessation of forward growth, is the dispersal of the cluster of apical vesicles from that of a highly ordered and compact apical position to one in which they become distributed peripherally in the general region of the cell tip (Staples and Macko, 1984). The net result is a ballooning of the hyphal apex essentially through lateral growth as the vesicles fuse with the plasma membrane.

Appressorium formation varies with the pathogen. In the rust uredospore germling, for example, appressoria develop as a single lobate infection apparatus into which all of the germ tube cytoplasm migrates. A septum is formed delimiting the appressorium from the emptied germ tube. During this time a single round of mitotic nuclear division ensues simultaneously in both haploid nuclei, and is completed before septum formation is finished. Early in the development of

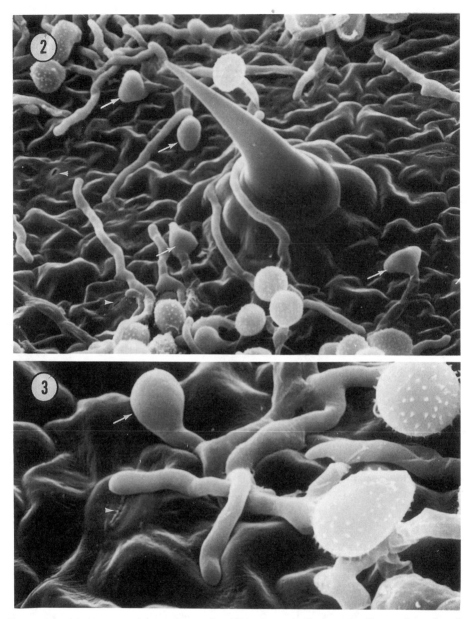

Figures 2 and 3. Scanning electron micrographs of *Uromyces appendiculatus* germlings on the surface of a bean leaf. Germ tubes of the fungus grow toward leaf stomata (arrowheads), the sites over which appressoria (arrows) are formed and through which the host is invaded. ×530 and ×1300, respectively.

the appressorium, the microtubule and microfilament (F-actin) cytoskeleton is rearranged from being oriented mostly parallel to the longitudinal axis of the hypha, to being oriented parallel to the inducing signal (perpendicular to the cell's longitudinal axis; Hoch *et al.*, 1986). The entire process from cessation of forward growth to completion of the appressorium (septum completion and nuclear division) occurs within 50–60 min, depending on environmental conditions. The appressorium "matures" for another 30–60 min before the penetration peg is initiated. During this time the cell is progressing through important physiological changes, including gene expression and the production of differentiation-related proteins (Bhairi *et al.*, 1989; Staples and Hoch, 1988; Staples and Macko, 1984; Staples *et al.*, 1988; see also Section 4). Appressorium development in powdery mildews is entirely different and is reviewed in Chapter 15. In some fungi, such as *Venturia inequalis* (causal agent of apple scab) and *Microcyclus ulei* (South American leaf spot of rubber), appressorium formation occurs almost immediately upon germination of conidia and ascospores. The germ tube can be so short that it is not easily discernible from the spore or appressorium (Liyanage and Hoch, unpublished; Maeda, 1970; Nushbaum and Keitt, 1938) (Figs. 4, 5). In *Colletotrichum* spp., *Phyllosticta* spp., and others, the appressorium is frequently multilobed with several septa (Sutton, 1962). Development of appressoria in most fungi is similar with regard to the initial changes involved in growth cessation and apical vesicle redistribution but deviates in subsequent morphological, pigmentation (see Chapter 9), and functional specializations, according to the species.

2. SIGNALING FOR GROWTH ORIENTATION

Most fungal pathogens are indiscriminate as to a specific infection site, and as such growth orientation is not relevant even though they may exhibit directional growth under specific circumstances, particularly *in vitro*. For many fungi, some degree of infection site selectivity is exhibited, and growth is necessarily oriented toward these preferred sites by either chemical or

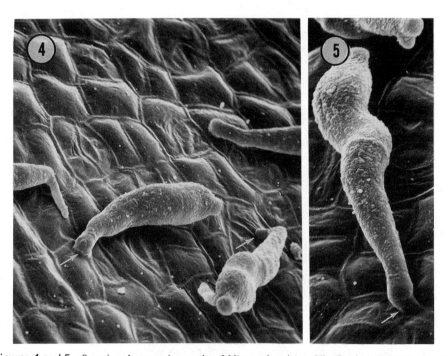

Figures 4 and 5. Scanning electron micrographs of *Microcyclus ulei* conidia (South American leaf spot of rubber). Appressorium (arrows) formation occurs almost immediately upon germination of the conidia and contact with the host, resulting in very short germ tubes (Liyanage and Hoch, unpublished). ×2000 and ×2600, respectively.

physical signals. Discerning whether or not fungal propagules respond to chemical or physical stimuli has often not been an easy undertaking. In addition to chemical and physical signals, light has been shown to influence not only spore germination in the rust (Knights and Lucas, 1980; Chang and Calpouzos, 1973) and other fungi, but also the direction of growth, sporulation, and pigment production. Depending on the fungus, light causes either negative or positive phototropic responses (Tan, 1978). Light does not, however, play a primary role in directing growth for subsequent infection, and thus is not considered relevant to the topics discussed here.

2.1. Role of Chemical Signals

Chemotropic responses have been observed in many fungi (Staples *et al.*, 1984; Wynn and Staples, 1981). Perhaps the more clearly proven examples where responses to chemical signals have been exhibited are with germlings of encysted zoospores. Encysted zoospores of *Phytophthora* and *Pythium* spp. almost always germinated from the side of the spore nearest the incipient infection site (Hickman, 1970; Royle and Hickman, 1964). Root exudates, whether from intact roots or from extract-filled glass capillaries, served as chemical attractants for spore germination and subsequent growth. Similarly, specific ions (e.g., H^+) can serve as a growth orienting factor (Ho and Hickman, 1967). Motile zoospores, of course, are well known to be actively attracted toward plant extracts (Hickman and Ho, 1966) and particularly toward stomata and root tips. Zoospores of the downy mildews, e.g., *Plasmopara viticola* (see Hickman and Ho, 1966), *Pseudoperonospora humuli* (Royle and Thomas, 1971), and *Pseudoperonospora cubensis* (Cohen and Eyal, 1980), invade their hosts almost exclusively through the stomata. Interestingly, germlings of conidial peronosporaceous pathogens can penetrate either through the cuticle directly or via stomata (Cohen, 1981; Yeh and Frederiksen, 1980).

Directed growth of *Uromyces viciae-fabae* germ tubes has been correlated with the pH gradients that surround leaf stomata (Edwards and Bowling, 1986). A correlation was made between the size of the stomatal aperture and the ability of uredospore germ tubes to find stomata on the host, *Vicia fava*, as well as on some nonhost plants. It was determined that pH gradients were present around closed, but not around open stomata, and was related to cessation of proton pumping. Fewer germ tubes were located over stomata when they were open. It would have been informative to determine whether or not a simple pH gradient would cause growth orientation of the germlings. While pH may have some role in directing the growth of uredospore germlings, clearly it does not represent the entire mechanism of signaling since it has been well established that surface features are at least equally effective (see Section 2.2).

Gaseous emissions from stomata are also likely involved in directing fungal cell growth. Gradients of oxygen and water have been implicated in guiding both zoospore taxis and conidial germling growth to these sites. For example, penetration of sugar beet by *Cercospora beticola* was related to water gradients around the stomata, a form of hydrotropism (Rathaiah, 1977). Conidia of *C. beticola* not only exhibited an initiation of penetrant germ tubes over stomata, but also produced branches from the conidium on the side toward the stomata. These effects were enhanced when the leaf's environment was interrupted with a dry period. Peterson and Walla (1978) noted a similar hydrotropism of a needle blight pathogen (*Dothistroma pini*) toward pine stomata. They surmised that another pine pathogen, *Scirrhia acicola*, also behaved hydrotropically.

Of the multitudes of fungal plant pathogens, only a minority of them are directed to grow toward stomata. These fungi, however, have received considerable attention because of the very specific and easily discernible site that they seek, let alone the immense economic importance that many of them cause through disease (particularly rusts and downy mildews). Much more common are the pathogens that exhibit little preference for specific sites or those that preferentially grow toward anticlinal walls. The attraction toward anticlinal walls has been thought to be due primarily to the "leakiness" of the area to exudates from within the leaf both because the cuticular waxes are thinner in this region (Schönherr and Bukovac, 1970) and because the anticlinal walls serve as a wick for the exudates. Hyphae of *R. solani* show preferential growth habits along the anticlinal walls of cotton seedling hypocotyls (Armentrout and Downer, 1987) that may or may not be attributed to chemical exudates (Figs. 6, 7). *Botrytis cinerea* conidia germinate and grow with a preference toward anticlinal wall junctures stimulated by onion leaf topography but greatly enhanced by the chemicals present (Clark and Lorbeer, 1976a,b).

2.2. Role of Physical Signals

Thigmotropism or contact tropism (Staples *et al.*, 1984; Wynn and Staples, 1981) of germ tubes and hyphae are much more readily recognized that are chemical tropisms, and thus have been documented more frequently. One criterion that takes precedence in order for contact tropism to occur is that of adhesion (see Chapter 1); without proper cell-to-substrate contact, such tropisms fail (Epstein *et al.*, 1985). Growth of fungal cells in response to substrate features occurs in a very wide range of fungi. The rust fungi, in particular, have been studied in detail for their thigmotropic responses. Johnson (1934) was perhaps the first to document (although he referenced M. Newton as having made a similar observation) directional growth of the rust fungi, *Puccinia graminis tritici* and *P. helianthi*, on leaves of wheat and sunflower, respectively. Germ tubes of these fungi grew at right angles to the depressions on the epidermis created at the anticlinal junctures. It was Johnson who surmised that such growth habits of uredospore germlings on wheat leaves favored the fungus in locating stomata, since stomates on graminaceous host are arranged in longitudinal rows, and that the chance of the germling encountering a stomate is thus increased. Later, Dickinson (1949a,b, 1969, 1970, 1971, 1972, 1977, 1979), in an impressive series of papers, studied in greater detail the "signals" and responses of not only rust (e.g., *P. recondita*, *P. cornata*, *P. striiformis*, *P. graminis*, *P. sorghi*), but also powdery mildew fungi. Working mostly with artificial substrates, he demonstrated that growth occurred at right angles to ridges or grooves, similar to what occurs at the anticlinal wall junctures on plants. Wynn (1976) illustrated that germlings of rust fungi grew perpendicular to the epidermal wall junctures of leaf cells replicated by plastic, thus alleviating any chemical influences that might have been derived from the host. Numerous other workers have subsequently employed plastic replicas of leaf surfaces to demonstrate fungal growth orientation. Recently, Staples and co-workers (1983b) have shown that *P. graminis* would grow oriented perpendicular to topographically undefined grooves and ridges in polyethylene, polystyrene, and aluminum prior to forming appressoria. *Uromyces appendiculatus*, likewise, grows oriented on leaves (Hoch *et al.*, 1987b; Pring, 1980; Wynn, 1976) and a variety of substrates having patterned topographies (Staples *et al.*, 1985c; Epstein *et al.*, 1985). Approximate dimensions of the topographical features have been determined for growth orientation (Dickinson, 1949a); however, Hoch and colleagues (1987b) have recently defined more exactly the size of the signals that orient germling growth in *U. appendiculatus*. Using electron beam lithography, grooves of precise dimensions were microfabricated into silicon wafers from which polystyrene replicas were produced. The best signal for growth orientation was found to be ridges or grooves in the substrate spaced 0.5 μm to 15.0 μm apart. The regularity of germ tube orientation diminished as the spacing increased, and spacing beyond 15.0 μm was not effective in maintaining oriented growth. Ridges spaced 0.5 μm apart resulted in very straight germ tubes, unlike that found in nature. In addition to the epidermal cell junctures, the pattern of epicuticular wax depositions has been proposed as the signal orientating germ tube growth of *P. graminis tritici* (Lewis and Day, 1972). Also, cuticular ridges frequently seen in scanning electron micrographs have been thought to represent one of the signals that guides germ tube growth, although caution needs to be stressed in this interpretation, since many of these ridges are most likely artifacts derived during specimen preparation.

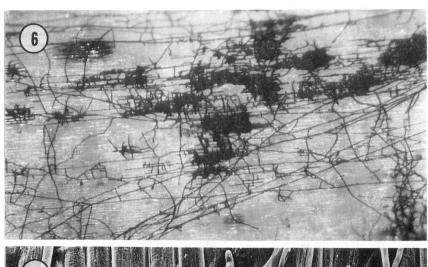

Figures 6 and 7.   (6) Infection cushions of *Rhizoctonia solani* on the surface of a cotton seedling hypocotyl. It is from these compact aggregates of specialized hyphae that multiple penetrations of the host occur. (7) Early development of *R. solani*. prior to infection cushion development, on the surface of a cotton hypocotyl. Many hyphae grow parallel to the longitudinal axis of the epidermal cells, especially at the anticlinal wall junctures. ×35 and ×750, respectively. Reprinted with permission from Armentrout *et al.* (1987) and Armentrout and Downer (1987).

2.3. Mechanisms Involved in Growth Orientation

For most fungi, the mechanism(s) involved in signal reception for growth orientation are not much better understood today than they were when such phenomena were first observed. It is well known that many fungi recognize and respond to hormones (see Crandall *et al.*, 1977; Gooday, 1974; Turian, 1978) for the signaling and development of reproductive structures and for growth toward nutrient sources (Musgrave *et al.*, 1977), and thus it is quite conceivable that similar chemical receptors operate in growth orientation. The receptor mechanisms for chemical responses remain poorly understood.

Hyphal growth and direction of growth have been correlated with the presence and position of the Spitzenkörper in the cell apex since it was first observed (Brunswik, 1924). This cluster of specialized vesicles has been the subject of numerous studies. Girbardt (1955, 1957) observed that in cells no longer elongating, the Spitzenkörper was absent, but reappeared prior to resumption of growth. He, and others (e.g., Grove *et al.*, 1970; Grove and Bracker, 1970; Howard, 1981), further noted that the position of the Spitzenkörper, or its equivalent in Phycomycetes, could be correlated to the direction that the cell would curve. Since cell growth occurs through the fusion of apical vesicles with the plasma membrane at the cell apex coincidental with deposition of their contents, it is relatively easy to understand how the cell extends. More difficult, however, is the understanding of the mechanism(s) involved in determining the position of the cluster of apical vesicles that guide cell direction. To date, the evidence that the cluster of apical vesicles respond to thigmosensitive receptors is only correlative. It has been documented, however, that the apical region is very sensitive to touch and that the cell will cease growth or change direction of growth when slightly perturbed (Castle, 1942; Middlebrook and Preston, 1952). Clearly, many fungi, as exemplified by *U. appendiculatus* and *P. graminis*, grow in a direction dictated by the topography of the substrate (Hoch *et al.*, 1987b; Wynn, 1976). Dickinson (1969, 1977) in his studies with *P. coronata* thought that growth direction was related to the thigmotropic response that arose from the interaction of stress placed on microfibrils of the germ tube wall that resulted from its adherence to the substrate. Stress on the cell's microfibrils was hypothesized to be moderated by the orientation of the substrate macromolecular topography.

3. SIGNALS FOR INFECTION STRUCTURE FORMATION

The site of infection structure formation varies with the particular pathogen–host relationship, and as discussed above at least two prominent sites have received considerable attention—location near or at anticlinal walls and over stomates. Both chemical and physical signals have roles (either alone, together, or in combination with other factors, e.g., light, age, humidity) in triggering infection structures at these sites. In several situations, evidence for a single type of signal as an effective inducer has been obtained, e.g., plastic replicas of leaf stomata or artificial substrates with specific topographies. However, this only indicates that the replica or substrate is an effective inducer *in vitro*. It does not indicate that surface topography is the sole operative signal in nature where chemical signals may also be functional, and perhaps synergistic. Armentrout and co-workers (1987), as an example, discussed the role of contact signals in inducing infection cushions in *R. solani* and that chemical stimuli enhanced the effect. Thus, while a single and "pure" stimulus can trigger infection structure formation, we should bear in mind that such triggering in nature may be complex—it certainly is not easily discernible.

3.1. Chemical Signals

Chemical signals external to the fungal cell frequently have been noted to trigger, or at least enhance, infection structure formation. Plant extracts are the most frequently cited chemical factor inducing infection cushion formation in *R. solani* (Flentje *et al.*, 1963; Marshall and Rush, 1980; Stockwell and Hanchey, 1983) as well as in *Sclerotium cepiyorum* (Steward *et al.*, 1989). Likewise, leaf diffusates were demonstrated to not only stimulate lateral germination of *Botrytis* conidia, but also to enhance appressorium formation (Clark and Lorbeer, 1976a,b). Guttation fluids appear to promote appressorium formation in *Magnaporthe grisea* (Frossard, 1981) as well as in *Cochliobolus* spp. (Endo and Amacher, 1964). One subgroup of anthracnose fungi develop appressoria in response to chemicals secreted by ripening fruit, e.g., *Colletotrichum piperatum* (Grover, 1971).

Chemostimulation for appressorium formation in the rust fungi has been documented for a number of species *in vitro*. Allen (1957), French and Weintraub (1957), and later others (Maheshwari *et al.*, 1967; Allen and Dunkle, 1970) were the first to show that volatile fractions from uredospores of *P. graminis* would induce appressoria and vesicles in germlings of the same fungus. Macko and co-workers (1978) subsequently determined that acrolein (2-propenal) was the active component extracted from uredospores. French and Weintraub (1957) also showed that nonanol (pelargonaldehyde) would stimulate a low level of appressorium production. Grambow and colleagues (Grambow, 1977, 1979; Grambow and Grambow, 1978; Grambow and Riedel, 1977) believe that certain volatiles and phenolic compounds emitted from the stomatal region are responsible for appressorium induction, and in fact have demonstrated that some of these compounds (e.g., *trans*-2-hexen-1-ol, *cis*-3-hexen-1-ol) in proper combinations can be effective *in vitro*.

Other chemicals and ions, such as simple sugars and K^+, will effectively induce differentiation of *U. appendiculatus* uredospore germlings *in vitro* (Kaminskyj and Day, 1984a,b; Staples *et al.*, 1983a, 1985a). *U. appendiculatus* germlings that are chemically induced to form appressoria do so only when the germlings grow aerially, away from the solutions or agar media containing the triggering chemicals, e.g., K^+, Ca^{2+}, or sucrose (Hoch *et al.*, 1987a). The infection structures develop on the aerial germ tubes. Germ tubes on the surface or submerged in the solutions develop appressoria, but only in response to thigmotropic signals inherent in submerged membranes inductive for differentiation. In addition to these agents, cyclic nucleotides (e.g., cAMP, cGMP), inhibitors of phosphodiesterase [e.g., 3-isobutyl-1-methyl xanthine, 4-(3-butoxy)-4-methylbenzyl-2-imidazolidinone], and putative stimulators of adenylate cyclase (e.g., NaF, adenosine, 2'-deoxyadenosine, forskolin) induce appressorium development in *U. appendiculatus* (Hoch and Staples, 1984).

3.2. Topographical Signals

Considerable evidence exists for the role of surface features in signaling infection structure formation. Since the early 1900s it has been recognized that many plant pathogens, e.g., *Colletotrichum* spp., *Sclerotinia* spp., *Gloeosporium* spp., and *Puccinia* spp., develop appressoria in response to "hard" surfaces (Boyle, 1921; Brown and Harvey, 1927; Dickinson, 1949b; Hasselbring, 1906; Jenkins and Winstead, 1964). A range of materials, such as agar, gelatin, various plastics, and waxes, have been used to create artificial surfaces with varying degrees of hardness. Because of the diversity of these materials, it often has been difficult to conclude beyond a doubt that hardness, and not some other associated property, was the responsible factor. We would venture that in most instances, hardness *per se* is not the crucial factor, primarily because fungi in general do not possess mechanisms for determining hardness. Rather, other surface properties are likely involved such as rigidity (Emmett and Parbery, 1975), porosity of the substrate to ions and metabolites, and surface energy, i.e., hydrophobicity of the surface. Porosity of the surface may be quite important. For example, if fungus-leaked compounds are not allowed to diffuse away, appressorium formation may be signaled. Such would be the instance for most waxes and plastics where infection structures are readily induced upon initial contact. Hard agars (>6% agar) similarly might impede diffusion enough to trigger cell differentiation. Surfaces with low surface energy (relatively hydrophobic) have been shown to be optimal for *M. grisea* spores (Hamer *et al.*, 1988) and *U. appendiculatus* germling adhesion (see Chapter 1; Hoch, unpublished). Other physical properties, for which hardness is coincidental, may also be influential on induction of appressoria.

Fungi, such as *Colletotrichum* spp. (Akai *et al.*, 1967; Anderson and Walker, 1962; Lapp and Skoropad, 1978), *Erysiphe polygoni*, *Peronospora parasitica* (Preece *et al.*, 1967), *Cochliobolus* spp. (Hau and Rush, 1979; Wynn and Staples, 1981), *Endocronartium harknessii* (Hopkin *et al.*, 1988), and *Botrytis* spp. (Clark and Lorbeer, 1976a,b), form appressoria preferentially at the anticlinal groove formed by the juncture of epidermal cells. Evidence that appressoria are initiated in these fungi by topographical signals has been obtained both directly from observations of plant material and from plastic replicas of leaf and hypocotyl surfaces. Lapp and Skoropad (1978) clearly showed that appressoria of *C. graminicola* formed preferably over the anticlinal walls from both intact leaf material as well as from various plastic replicas of leaf surfaces. Well over 50% of the appressoria were formed within 3 μm or less of the groove. Since there was no statistical difference between the number of appressoria formed on leaf material versus plastic replicas, they concluded that the signal was controlled by the topography of the leaf surface and not by localized host exudates. Similar observations have been made for many other fungi (see Wynn and Staples, 1981).

In addition to the chemical stimulus discussed above for promoting infection cushion formation in *R. solani*, contact stimuli have also been found (Da Silva and Wood, 1964; Gladders and Coley-Smith, 1977; Murray, 1982). Recently, Armentrout and co-workers (1987) examined this aspect further using cotton hypocotyls and polystyrene replicas of similar hypocotyls, and concluded that while chemodifferentiation may be important in some aspects of infection cushion development, e.g., providing nutrients for mucilage production so the fungus can adhere to the substrate to receive effective stimulation, thigmodifferentiation was more important in the early phases of development.

Stomatal penetrating fungi represent a group of pathogens that have evolved extremely sensitive and precise mechanisms for perceiving the correct site to develop infection structures. By far the largest group of stomatal penetrating fungi studied are the rusts, although other pathogens also exhibit specific preferences for stomata, e.g., some of the downy mildews and *Cercospora* leaf spot of beet. It is primarily the uredospore stage of the rust that develops appressoria over stomata, and then not all rust species show this preference. For example, soybean rust, *Phakospora pachyrhizi* (Bonde *et al.*, 1976; Keogh *et al.*, 1980; Koch *et al.*, 1983), the rust pathogen of *Caesalpinia pulcherrima*, *Ravenelia humphreyana* (Hunt, 1968), and rose apple rust, *Puccinia psidii* (Hunt, 1968), all penetrate the host directly via the cuticle. Appressoria of *P. pachyrhizi* develop over anticlinal walls or over the center of the epidermal cells, but rarely over stomata. When development is over stomata, penetration is through one of the guard cells rather than through the stomatal aperture (Koch *et al.*, 1983).

The precision with which most rust uredospore germlings target stomata has been demonstrated by a number of researchers as discussed in previous sections. Wynn (1976) was the first to use plastic replicas of leaves to demonstrate that *U. appendiculatus* would form appressoria equally well over plastic stomatal impressions as on native leaf stomata. The fact that such replicas were inert and free of any leaf "chemicals" provides quite sound evidence that appressoria are triggered by physical features that directly mimic corresponding leaf signals. Of course, Dickinson (1949a,b) earlier demonstrated that a variety of plastic substrates possessing certain topographical features (e.g., scratches, crystal imprints, craters) would initiate appressorium development. Whatever the surface, it has been well documented that these fungi differentiate because of specific topographical signals inherent on the surface (Bourett *et al.*, 1987; Staples *et al.*, 1983b; Staples and Hoch, 1982; Wynn, 1976) and not because of chemical factors present on the substrate. Using electron beam-lithographed, microfabricated silicon wafers, the topographical signal was recently further defined to be a simple, but sharp change in substrate elevation (e.g., a ridge or groove) on the order of 0.5 μm (Hoch *et al.*, 1987b) (Fig. 8). Ridge elevations greater than 1.0 μm or less than 0.25 μm were not particularly inductive. Detailed examination of the stomatal region revealed guard cells with "lips" of nearly the same height (average = 0.48 μm). Such site recognition is based solely on thigmotropic sensing. Also, it was apparent that the site of initial signal reception for the germling is in the apical region of the hypha, near the hypha–substrate interface. It is obvious that "signal" recognition for appressorium initiation is contingent upon the rust germling having intimate contact with the surface topography. Such contact for thigmotropic sensing was proposed by Wynn (1976) following observations of germling growth on "waxless" leaf surfaces; and more recently further elucidated by Epstein and co-workers (1985, 1987; see also Chapter 1). In the absence of suitable topographical signals, uredospore germ tube growth continues (16–24 hr) until endogenous nutrient reserves are depleted.

In addition to *U. appendiculatus*, many of the other stomatal penetrating rust pathogens have been shown to respond to a variety of artificial substrates and form appressoria. A summary of some of these fungi and the inductive surfaces has been published (Wynn and Staples, 1981). The topographical features that signal appressorium formation in these rust thus far have not been elucidated with precision as they have been for *U. appendiculatus*; it would be informative to test them on the microfabricated substrates.

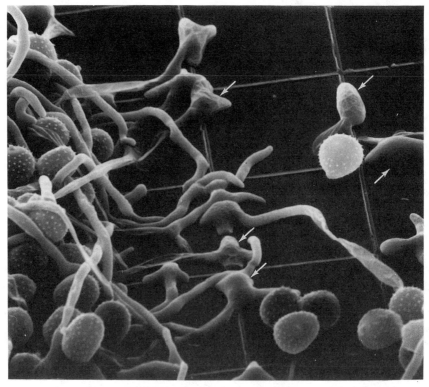

Figure 8.   Uredospore germlings of *Uromyces appendiculatus* growing on the surface of a polystyrene replica having ridges 0.5 μm in height. Such topographical features signal for appressorium (arrows) development. ×700.

Conidia of *Peronosclerospora sorghi*, the causal agent of downy mildew of sorghum, germinate on leaf surfaces and grow toward stomates where they develop appressoria (Yeh and Frederiksen, 1980). Little is known about the leaf surface feature that signals for infection structure initiation in this and related pathogens. It would be quite interesting to determine whether or not abrupt changes in surface topography serve as the signal as they do for many of the rust uredospore germlings. During the course of *P. sorghi* germling growth toward stomata, the hyphae exhibit swellings at or near junctions of the epidermal cells. Such swellings are similar to those observed for the rust fungi when a "partial signal" was received, presumably for appressorium initiation (Dickinson, 1949a,b; Johnson, 1934; Lewis and Day, 1972; Tucker *et al.*, 1986; Wynn, 1981). Wynn (1976) surmised that such incomplete development of infection structures by the signal "suggest[s] that the contact stimulus may be quantitative," meaning that the cell receptor product is possibly accumulative and, at the least, also quantitative.

3.3. Temperature Signals

An abrupt change in temperature will trigger appressorium formation in several of the rust fungi. Such inducers, however, do not likely have a role in nature, but serve as model systems for investigating the processes of cell differentiation (appressorium development) *in vitro*. Uredospore germlings of *Puccinia coronata*, and *P. graminis tritici* develop appressoria, penetration pegs, and vesicles following a brief rise in temperature of about 10°C (Dunkle *et al.*, 1969; Kim *et al.*, 1982; Maheshwari *et al.*, 1967; Mendgen and Dressler, 1983; Wanner *et al.*, 1985). Most rust species, however, have not been induced to develop infection structures with a heat shock (see Wynn and Staples, 1981). Only recently have appressoria been induced to develop in *U. appendiculatus* (Hoch *et al.*, 1986; Staples and Hoch, 1982; Staples *et al.*, 1989). The stressed germlings undergo one or two rounds of nuclear division, and the infection structures that develop have a normal morphology.

3.4. Mechanisms Involved in Signal Perception

Very little is known about how pathogenic fungi perceive and/or transmit signals for infection structure formation. It is an area that is awaiting innovative research approaches. To date, most research has centered on the gathering of observational and correlative data. With the introduction of new methodologies used in contemporary cell biology and generic manipulative techniques fast becoming commonplace, we will quite likely see many answers to the question of how signals are perceived. Already, in mammalian cell systems, and to some extent in higher plant systems, signaling processes are being elucidated and may serve as guides for mycological research. Several recent and comprehensive reviews on transmembrane signaling and cellular communication have been published (e.g., Berridge, 1985, 1987; Nishizuka, 1984; Gilman, 1987; Stryer and Bourne, 1986; Kikkawa and Nishizuka, 1986), mostly concerning mammalian systems. Two major signal pathways are known in mammalian cells: one employs cyclic AMP, and the other is a combination of second messengers including Ca^{2+}, inositol triphosphate, and diacylglycerol. The pathways have much in common, and growing evidence suggests that there is cross talk between them. A receptor at the surface of the cell transmits information through the plasmalemma and into the cell by a family of G-proteins (proteins that bind GTP) which activate an amplifier, either adenylate cyclase or phospholipase C. These in turn start phosphorylation events to activate target proteins, e.g., trehalase (Thevelein, 1984), which alters cellular activities in response to the external signal. The signaling systems used by cells often display extensive heterogeneity, and many variations exist from tissue to tissue.

Higher plants have been studied for the presence of transmembrane signaling systems using the studies from mammalian systems as a model. It has been possible to show that metabolites of the phosphatidylinositol cycle are present in extracts of pulvinus cells from *Samanea* (Morse *et al.*, 1987), and suspension cultures of carrot (Boss *et al.*, 1985). GTP-binding proteins have been detected in extracts of *Lemna* (Hasunuma and Funadera, 1987). The operation of a Ca^{2+}-based secondary messenger system in higher plants seems likely; however, a cyclic nucleotide cascade may not be as functional as in mammalian cells.

As already discussed, fungi respond morphologically to a very wide array of physical and chemical signals, including touch, light, pheromones, and nutrients. The mode of signal reception is poorly understood, however, except in *Dictyostelium* where a complex transmembrane signaling system for cyclic AMP has been elucidated (Gerisch, 1987). The bulk of other studies concerning transmembrane signaling pathways in fungi appear to have been done using the yeast, *Saccharomyces cerevisiae*. For example, it has been shown that yeast cells possess the entire cyclic AMP-based regulatory cascade which apparently is required for cell-cycle progression (Thorner, 1982). In addition, polyphosphoinositides have been known for years in fungi (Mitchell, 1975), although appreciation of their significance has only been recent (Hasunuma *et al.*, 1987; Favre and Turian, 1987; Europe-Finner and Newell, 1985; Dahl and Dahl, 1985). The involvement of the cyclic AMP cascade in the control of dormancy and induction of germination in fungal spores has been reviewed by Thevelein (1984), and the list of fungi apparently having this cascade has been widened considerably since then. Evidence for membrane-sited receptors, e.g., the G-proteins, and phosphatidylinositol bisphosphate and protein kinase C, is less extensive. In yeast, the *ras* genes have been cloned and thoroughly characterized chemically and genetically, and it has been shown that these genes code for the alpha-subunit of kinase-regulatory G-proteins which govern adenylate cyclase activity (Frankel, 1985; Toda *et al.*, 1985; Reymond *et al.*, 1985). G-proteins have been partially characterized from the filamentous fungus, *Neurospora crassa*; however, the details of these systems, as well as knowledge of their receptors, are still elusive (Hasunuma *et al.*, 1987).

3.4.1. Chemical-Mediated Receptors

Although many chemical signals start mitosis, frequently a precursor or coincidental event with infection structure formation in fungi, the mechanisms involved in initiating this process are not clear. Second messengers are possibly involved in transmitting signals for appressorium formation in rust uredospore germlings. One possible messenger complex is calcium and its calcium-regulating protein, calmodulin. Fluctuating Ca^{2+} levels in the cytoplasm might trigger other cytoplasmic changes such as microtubule depolymerization that, in turn, may influence appressorium initiation. Other possible second messengers are cyclic nucleotides. Interestingly, exogenously supplied cyclic AMP or cyclic GMP induced *U. appendiculatus* germlings to undergo one round of mitosis and to form septa, processes normally associated with appressorium formation (Hoch and Staples, 1984). Appressoria are not formed, unless very low levels of cyclic AMP are employed, and the septa are abnormally positioned (Hoch, unpublished). Phosphodiesterase inhibitors, which cause a rise in cellular cyclic AMP, induced moderate levels of appressoria. Little is known about the involvement of cyclic AMP and cyclic GMP in appressorium development, but the uredospore germling contains three cyclic nucleotide binding peptides, proteins that bind either cyclic AMP or cyclic GMP (Epstein *et al.*, 1989). Furthermore, uredospores contain a peptide that is phosphorylated in response to either cyclic AMP or cyclic GMP.

The involvement of K^+ in *Uromyces* appressorium formation possibly indicates a transmembrane signaling response, involving membrane potentials. Potassium appears to accelerate protein synthesis and the functioning of a K^+/H^+ antiport (Staples *et al.*, 1985a). Vanadate (K^+/H^+ ATPase pump inhibitor) prevented appressorium formation, while ouabain, known not to be effective in most plant systems, was ineffective (Hoch, unpublished).

3.4.2. Thigmotropic Receptors

There should be no doubt that many fungi can perceive and respond to topographical signals. However, as Wynn and Staples (1981) carefully pointed out in their review, what we consider to be contact sensing is really chemical sensing since at some place between the cell and the substrate a chemical interaction of some type must occur—we just do not known how to recognize it with current methods. Topographic features activate a chemical transducer somewhere in the cell. As described in the preceding sections, such perception has most clearly been documented in *U. appendiculatus*. Hoch and colleagues (Bourett *et al.*, 1987; Hoch *et al.*, 1986, 1987c; Staples and Hoch, 1982) have pursued the idea that the microtubule (and perhaps microfilament) cytoskeleton may be involved in topographic sensing. Clearly the most inductive signal is provided by a sharp change in substrate elevation on the order of 0.5 μm (Hoch *et al.*, 1987b; Bourett *et al.*, 1987). When a germling grows over such a feature, a permanent indentation is formed in the cell. Microtubules could function as the primary signal receptor by not being able to continue polymerizing beyond the indentation, even though the cell continues to grow for a short distance beyond. It was previously shown that the majority of cytoskeletal microtubules are intimately associated with, and most likely "glued" via other cytoskeletal proteins to, the plasma membrane (Hoch and Staples, 1983). Furthermore, it is well known that microtubules cannot bend sharply (e.g., Crossin and Carney, 1981). Thus, and indentation of 0.5 μm could prevent continued microtubule polymerization, and directly signal for cell differentiation because a population of microtubules would not be able to continue to elongate (Otto *et al.*, 1981; Hoch *et al.*, 1987c). It is quite conceivable that microtubules may have only a secondary role in signal reception (viz., only be involved in mediation of the signal to the nucleus and/or in architectural changes of the germ tube tip resulting in appressorium formation). Certainly, an intact microtubule cytoskeleton is necessary for the development of the appressorium (Hoch *et al.*, 1987c). When uredospore germlings were treated with microtubule-depolymerizing drugs at rates that permitted normal growth, no appressoria were formed on germlings under normally inductive conditions; removal of the agents allowed appressorium development. Alternatively, if the microtubule cytoskeleton is so stabilized, as with D_2O or taxol, that it cannot respond to signals, then appressorium formation is also prevented (Hoch *et al.*, 1986).

4. DIFFERENTIATION PROTEINS—A CONSEQUENCE OF SIGNAL RECEPTION

The obvious result of signal reception is cell response, expressed through such morphological changes as growth or infection structure formation. More subtle and equally profound are those changes involving alterations of the cell at the molecular level—synthesis of differentiation-related proteins.

In the rust fungi, and in particular *U. appendiculatus*, development of the infection structures is accompanied by the synthesis of at least 15 differentiation-related (dr) proteins, i.e., proteins not present in the germling until differentiation is induced (Staples *et al.*, 1988; Huang and Staples, 1982). A downshift in the synthesis of some proteins also occurs during infection structure development, and this reduction perhaps has a role in the development process. With current levels of detection and methodologies, the dr proteins have been observed, for the most part, toward the end of appressorium development; all 15 can be detected at the end of the morphogenetic pause when the infection peg arises, a prelude to a second round of nuclear division. Synthesis of the dr proteins is accompanied by the accumulation of mRNA specific for a small group of INF genes (Staples and Hoch, 1988), apparently as a result of an upshift in gene expression. Expression of all six genes cloned so far coincides with appearance of the dr proteins;

however, no evidence has been presented that these genes are identical with the proteins. The INF24 gene has been sequenced and the gene product identified as an approximately 20-kDa polypeptide (Bhairi *et al.*, 1989). Gene expression begins at about 4½ hr after the start of germination on inductive substrates, i.e., at the end of appressorium development. The enhanced level of mRNA was maintained throughout development of the infection structures.

Despite the cytological and functional similarities of the thigmo- and temperature-induced infection structures in the rust fungi (Hoch *et al.*, 1986; Staples *et al.*, 1989; Maheshwari *et al.*, 1967; Wanner *et al.*, 1985; Shaw *et al.*, 1985; Bourett *et al.*, 1987), the pattern of proteins synthesized during development induced by heat shock is quite different from that induced thigmotropically. The 15 thigmotropically induced peptides in *U. appendiculatus*, which with one exception are in the weight range of 14 to 26 kDa (Epstein *et al.*, 1989; Staples *et al.*, 1988), have not been detected in differentiated germlings induced by heat shock (Staples *et al.*, 1989). Heat shock induced the synthesis of at least six distinctly different heat-stress peptides. In a study on the effect of heat shock on protein synthesis in *M. lini* germlings, Shaw and colleagues (1985) reported the presence of seven new proteins which had approximate molecular masses of 17, 18, 20, 30, 44, 71, and 84 kDa. Except for the 84-kDa peptide, the *U. appendiculatus* peptides had molecular masses which resembled them. In contrast, only two polypeptides, those of about 20 and 30 kDa, respectively, were synthesized by *P. graminis tritici* germlings in response to heat shock (Kim *et al.*, 1982; Wanner *et al.*, 1985). The functions of heat-shock proteins in any tissue are not known (Schlesinger, 1986), although a 70-kDa protein is thought to have a role in protecting the cell against the stress effects of heat perhaps by disrupting inappropriate protein–protein interactions (Pelham, 1988).

In *U. appendiculatus*, synthesis of the 15 dr proteins occurs detectably only when differentiation is induced thigmotropically, while the heat-stress proteins are synthesized regardless of whether or not the infection structures develop (Staples *et al.*, 1989). This suggests that contact with a thigmotropically inductive signal causes the synthesis of a unique set of proteins which may have a role in germling response to a thigmotropic signal; alternatively, the heat shock may have caused a rapid degradation of the thigmo-induced proteins. Further research should elucidate the function of the dr proteins; however, it seems possible that uredospore germlings may have several inducer-responsive pathways to such triggers as K^+ (Staples *et al.*, 1983a), heat shock, and thigmo-sensing. Either could start appressorium development independently, or start a common differentiation pathway. In view of the apparent absence of the dr proteins from heat-shocked germlings, it seems unlikely that heat shock starts a pathway that is entirely in common with the thigmotropic pathway, a conclusion supported by studies on the expression of the INF genes induced by heat shock (Bhairi *et al.*, 1989). mRNA specific for INF24 and INF56 accumulates as the appressorium is completed; however, mRNA specific for INF88 did not increase as the infection structures developed. Thus, one of the genes appears to be expressed only in response to a thigmotropic signal, and mRNA specific for the dr proteins may fail to accumulate significantly in heat-shocked germlings.

Efforts to study the process of appressorium induction and development in the anthracnose fungi have led to conflicting conclusions by groups using different species of anthracnose fungi. For example, Staples and co-workers concluded from studies using *Colletotrichum truncatum* (Staples *et al.*, 1976) that germinating conidia prepare to produce appressoria by altering the messenger program of their germ tube nuclei. Using *C. lagenarium*, Suzuki and co-workers (1981, 1982) reported that when protein synthesis was completely inhibited by cycloheximide after 40 min of germination, appressoria matured in structure but not in function, i.e., were unable to penetrate membranes. The latter authors concluded that proteins required for appressorial morphogenesis were already present in dormant conidia, and that a *de novo* synthesis of proteins during germination was unnecessary to appressorium development (Suzuki *et al.*, 1981). In recent studies, Staples prepared one-dimensional PAGE of conidia germinated on noninductive agar surfaces (Bhairi *et al.*, 1990) in order to compare their results with those reported by Suzuki and co-workers (1981), and found as they did, that there were no apparent changes among the proteins in the critical 3- to 4-hr period when the protruded germ tube was beginning to swell to form the appressorium. Reexamined by two-dimensional PAGE, however, Staples observed at least ten peptides that were not present in nondifferentiating germlings. Some of the proteins have been shown to be related to accompanying changes in the mRNA pool, i.e., gene expression (Bhairi *et al.*, 1990). Thus, significant changes in protein complexity occur during the time when the appressorium develops that are not related to spore germination and that occur well before pigmentation of the appressorium at 17 hr after the start of spore germination. Further studies to understand the significance of these changes and the roles of the proteins in differentiation will require new approaches including an analysis of the differentiation-specific DNA clones.

5. CONCLUSION

From the time a fungal propagule germinates on an appropriate host, sensing for the correct site to mount an invasion is paramount to the success of the perpetuation of that pathogen. For some pathogens, ingress into the host can be achieved only through exceedingly specific avenues, e.g., stomata, and to reach these sites the pathogen uses chemo- and thigmo-sensing. Sensing in many instances is so precise that relatively little energy is expended in arriving at the infection court. For other pathogens, sensing may be just as precise, but the specificity of the incipient site of invasion may be less demanding, e.g., juncture of the anticlinal walls. And yet, for other pathogens, a parasitic relationship with an appropriate host can be established at almost any place. In the last few years, considerable insight into the biology of infection structure initiation of a few pathogens, *Colletotrichum* spp., some powdery mildew fungi, *U. appendiculatus*, and *Puccinia* spp., in particular, has been made. Hopefully, through the use of contemporary research tools, methods, and approaches we are on the threshold of arriving at some "final" answers that can be put to use toward lasting disease control strategies.

Since fungi were first closely examined by deBary, Buller, Reinhardt, and others, many persistent questions remain regarding the biology of hyphal growth and infection structure initiation. Fundamental to understanding how pathogens find and recognize the infection court are several basic, yet poorly comprehended cell functions ripe for diligent and innovative research. Very little is known, for example, about how hyphal cells and germ tubes sense for growth direction, much less what within the cell controls direction. Is it the cluster of apical vesicles as has been suggested by some, or do these organelles simply provide material for growth? If the apical cluster of vesicles (Spitzenkörper) is involved in directing growth, then what "reins" this body in a purposeful direction? Is it the wall microfibrils as suggested by Dickinson? Considerable progress was made in the 1960s and 1970s regarding hyphal cell organization and physiology; perhaps now is the time to elucidate mechanisms of signal reception. Based on our knowledge of mammalian, as well as a (very) few plant and fungal, cell receptor systems, we know that many receptors are associated with the plasma membrane; they have been reasonably well characterized chemically and structurally. Chemical receptors for fungal growth orientation could be similarly characterized. Perhaps even more intriguing is how some fungi, such as *U. appendiculatus*, sense topography. How does this rust pathogen discern a change in substrate elevation of 0.5 μm from one of 0.2 μm or less, and respond only to the higher topography to initiate appressorium formation?

Considerable research needs to be undertaken if we are to understand the fungi that we live with, and for many of these questions, the tools and know-how exists. We simply need to go to it!

ACKNOWLEDGMENTS.   The research reviewed herein by the authors was supported in part by grants from the U.S. National Science Foundation, the USDA–CRGO, and the Whitehall Foundation.

6. REFERENCES

Aist, J. R., 1981, Development of parasitic conidial fungi in plants, in: *Biology of Conidial Fungi*, Volume 2 (G. T. Cole and B. Kendrick, eds.), Academic Press, New York, pp. 75–110.

Akai, S., Fukutomi, M., Ishida, N., and Kunoh, H., 1967, An anatomical approach to the mechanism of fungal infections in plants, in: *The Dynamic Role of Molecular Constituents in Plant–Parasite Interaction* (C. J. Mirocha and I. Uritani, eds.), American Phytopathological Society, St. Paul, pp. 1–20.

Allen, P. J., 1957, Properties of a volatile fraction from urediospores of *Puccinia graminis* var *tritici* affecting their germination and development. I. Biological activity. *Plant Physiol.* **32:**385–389.

Allen, P. J., and Dunkle, L. D., 1970, Natural activators and inhibitors of spore germination. In: Morphological and Biochemical Events in Plant-Parasite Interactions (S. Akai, and S. Ouchi, eds.), pp. 23–58. Phytopathological Society of Japan, Tokyo.

Anderson, J. L., and Walker, J. C., 1962, Histology of watermelon anthracnose, *Phytopathology* **52:**650–653.

Armentrout, V. N., and Downer, A. J., 1987, Infection cushion development by *Rhizoctonia solani* on cotton, *Phytopathology* **77:**619–623.

Armentrout, V. N., Downer, A. J., Grasmick, D. L., and Weinhold, A. R., 1987, Factors affecting infection cushion development by *Rhizoctonia solani* on cotton, *Phytopathology* **77:**623–630.

Berridge, M. J., 1985, The molecular basis of communication within the cell, *Sci Am.* **235:**142–152.

Berridge, M. J., 1987, Inositol triphosphate and diacylglycerol: Two interacting second messengers, *Annu. Rev. Biochem.* **56:**159–193.

Bhairi, S. M., Staples, R. C., Freve, P., and Yoder, O. C., 1989, Characterization of an infection structure specific gene from the rust fungus, *Uromyces appendiculatus*, *Gene* **81:**237–243.

Bhairi, S., Buckley, E. H., and Staples, R. C., 1990, Protein synthesis and gene expression during appressorium formation in *Gomerella magna*, *Exp. Mycol.* (in press).

Bonde, M. R., Melching, J. S., and Bromfield, K. R., 1976, Histology of the suscept–pathogen relationship between *Clycine max* and *Phakopsora pachyrhizi*, the cause of soybean rust, *Phytopathology* **66:**1290–1294.

Boss, M. J., Crain, R. C., and Satter, R. L., 1985, Polyphosphoinositides are present in plant tissue culture cells, *Biochem. Biophys. Res. Commun.* **132:**1018–1023.

Bourett, T. M., Hoch, H. C., and Staples, R. C., 1987, Association of the microtubule cytoskeleton with the thigmotropic signal for appressorium formation in *Uromyces*, *Mycologia* **79:**540–545.

Boyle, C., 1921, Studies on the physiology of parasitism. IV. Infection by *Sclerotinia libertiana*, *Ann. Bot.* **35:**337–347.

Brown, W., and Harvey, C. C., 1927, Studies on the physiology of parasitism. X. On the entrance of parasitic fungi into the host plant, *Ann. Bot.* **41:**643–662.

Brunswik, H., 1924, Untersuchungen über die Geschlechts und Kernverhältnisse bei den Hymenomyceten Gattung *Coprinus*, *Bot. Ann. K. Goebel.* **5.**

Castle, E. S., 1942, Spiral growth and reversal of spiraling in phycomyces, and their gearing on primary wall structure, *Am. J. Bot.* **29:**664–672.

Chang, H. S., and Calpouzos, L., 1973, Phototropism of the uredospore germ tubes of *Puccinia graminis tritici*, *Bot. Bull. Acad. Sin.* **14:**35–40.

Clark, C. A., and Lorbeer, J. W., 1976a, Comparative histopathology of *Botrytis squamose* and *B. cinerea* on onion leaves, *Phytopathology* **66:**1279–1289.

Clark, C. A., and Lorbeer, J. W., 1976b, The development of *Botrytis squamose* and *B. cinerea* on surface of onion leaves as affected by exogenous nutrients and epiphytic bacteria, in: *Advances in the Microbiology of the Aerial Surfaces of Plants* (C. H. Dickinson, and T. F. Preece, eds.), Academic Press, New York, pp. 607–625.

Cohen, Y., 1981, The processes of infection of downy mildews on leaf surfaces, in: *Microbial Ecology of the Phylloplane* (J. P. Blakeman, ed.) Academic Press, New York, pp. 113–133.

Cohen, Y., and Eyal, H., 1980, Effects of light during infection on the incidence of downy mildew (*Pseudoperonospora cubensis*) on cucumbers, *Physiol Plant Pathol.* **17:**53–62.

Crandall, M., Egel, R., and Mackay, V. L., 1977, Physiology of mating in three yeasts, *Adv. Microb. Physiol.* **15:**307–398.

Crossin, K. L., and Carney, D. H., 1981, Evidence that microtubule depolymerization early in the cell cycle is sufficient to initiate DNA synthesis, *Cell* **23:**61–71.

Dahl, J. S., and Dahl, C. E., 1985, Stimulation of cell proliferation and polyphophoinositide metabolism in *Saccharomyces cerevisiae* GL7 by ergosterol, *Biochem. Biophys. Res. Commun.* **133:**844–850.

Da Silva, R. L., and Wood, R. K. S., 1964, infection of plants by *Corticium solani* and *Corticium praticola*—Effect of plant exudates, *Trans. Br. Mycol. Soc.* **47:**15–24.

Dickinson, S., 1949a, Studies in the physiology of obligate parasitism. I. The stimuli determining the direction of growth of the germ-tubes of rust and mildew spores, *Ann. Bot.* **13**:89–104.

Dickinson, S., 1949b, Studies in the physiology of obligate parasitism. II. The behaviour of the germ-tubes of certain rusts in contact with various membranes. *Ann. Bot.* **13**:219–236.

Dickinson, S., 1979, Growth of *Erysiphe graminis* on artificial membranes. *Physiol. Plant Pathol.* **15**:219–221.

Dickinson, S., 1969, Studies in the physiology of obligate parasitism. VI. Directed growth, *Phytopathol. Z.* **66**:38–49.

Dickinson, S., 1970, Studies in the physiology of obligate parasitism. VII. The effect of a curved thigmotropic stimulus, *Phytopathol. Z.* **69**:115–124.

Dickinson, S., 1971, Studies in the physiology of obligate parasitism. VIII. An analysis of fungal responses to thigmotropic stimuli, *Phytopathol. Z.* **70**:62–70.

Dickinson, S., 1972, Studies in the physiology of obligate parasitism. IX. The measurement of a thigmotropic stimulus, *Phytopathol. Z.* **73**:347–358.

Dickinson, S., 1977, Studies in the physiology of obligate parasitism. X. Induction of response to a thigmotropic stimulus, *Phytopathol. Z.* **89**:97–115.

Dodman, R. L., Barker, K. R., and Walker, J. C., 1968, A detailed study of the different modes of penetration by *Rhizoctonia solani*, *Phytopathology* **58**:1271–1276.

Dunkle, L. D., Maheshwari, R., and Allen, P. J., 1969, Infection structures from rust urediospores: Effect of RNA and protein synthesis inhibitors, *Science* **163**:481–482.

Edwards, M. C., and Bowling, D. J. F., 1986, The growth of rust germ tubes towards stomata in relation to pH gradients, *Physiol. Mol. Plant Pathol.* **29**:185-196.

Emmett, R. W., and Parbery, D. G., 1975, Appressoria, *Annu. Rev. Phytopathol.* **13**:147–167.

Endo, R. M., and Amacher, R. H., 1964, Influence of guttation fluid on infection structures of *Helminthosporium sorokinianum*, *Phytopathology* **54**:1327–1334.

Epstein, L., Laccetti, L., Staples, R. C., Hoch, H. C., and Hoose, W. A., 1985, Extracellular proteins associated with induction of differentiation in bean rust uredospore germlings, *Phytopathology* **75**:1073–1076.

Epstein, L., Laccetti, L. B., Staples, R. C., and Hoch, H. C., 1987, Cell–substratum adhesive protein involved in surface contact responses of the bean rust fungus, *Physiol. Mol. Plant Pathol.* **30**:373–388.

Epstein, L., Staples, R. C., and Hoch, H. C., 1989, Cyclic AMP, cyclic GMP, and bean rust uredospore germlings, *Exp. Mycol.* **13**:100–104.

Europe-Finner, G. N., and Newell, P. C., 1985, Inositol 1,4,5-trisphosphate induces cyclic GMP formation in *Dictyostelium discoideum*, *Biochem. Biophys. Res. Commun.* **130**:1115–1122.

Favre, B., and Turian, G., 1987, Identification of a calcium- and phospholipid-dependent protein kinase (protein kinase C) in *Neurospora crassa*, *Plant Sci.* **49**:15–21.

Flentje, N. T., 1957, Studies on *Pellicularia filamentosa* (Pat.) Rogers. III. Host penetration and resistance, and strain specialization, *Trans. Br. Mycol. Soc.* **40**:322–336.

Flentje, N. T., Dodman, R. L., and Kerr, A., 1963, The mechanism of host penetration by *Thanatephorus cucumeris*, *Aust. J. Biol. Sci.* **16**:784–799.

Frank, A. B., 1883, Uber einige neue und weniger bekannte Pflanzenkrankheiten, *Ber. Dtsch. Bot. Ges.* **1**:I 29–34, II 58–63.

Frankel, D. G., 1985, On *ras* gene function in yeast, *Proc. Natl. Acad. Sci. USA* **82**:4740–4744.

French, R. C., and Weintraub, R. L., 1957, Pelargonaldehyde as an endogenous germination stimulator of wheat rust spores. *Arch. Biochem. Biophys.* **72**:235–237.

Frossard, R., 1981, Effect of guttation fluids on growth of micro-organisms on leaves, in: *Microbial Ecology of the Phylloplane* (J. P. Blakeman, ed.), Academic Press, New York, pp. 213–226.

Gerisch, G., 1987, Cyclic AMP and other signals controlling cell development and differentiation in *Dictyostelium*, *Annu. Rev. Biochem.* **56**: 853–879.

Gilman, A. G., 1987, G proteins: Transducers of receptor-generated signals, *Ann. M. Rev. Biochem.* **56**:615–649.

Girbardt, M., 1955, Lebendbeobachtungen an *Polystictus versicolor* (L), *Flora (Jena)* **142**:540–563.

Girbardt, M., 1957, Der Spitzenkörper von *Polystictus versicolor* (L), *Planta* **50**:47–59.

Gladders, P., and Coley-Smith, J. R., 1977, Infection cushion formation in *Rhizoctonia tuliparum*, *Trans. Br. Mycol. Soc.* **68**:155–118.

Gooday, G. W., 1974, Fungal sex hormones, *Annu. Rev. Biochem.* **43**:35–49.

Grambow, H.-J., 1977, The influence of volatile leaf constituents on the in vitro differentiation and growth of *Puccinia graminis* f. sp. *tritici*, *Z. Pflanzenphysiol* **85**:361–372.

Grambow, H.-J., 1979, *A New Perspective of the Obligate Parasitism of the Wheat Rust Fungus Puccinia graminis f. sp. tritici*, Habilitation thesis to RWTH, Aachen.

Grambow, H.-J., and Grambow, G. E., 1978, The involvement of epicuticular and cell wall phenols of the host plant in the in vitro development of *Puccinia graminis* f. sp. *tritici*, *Z. Pflanzenphysiol.* **90**:1–9.

Grambow, H.-J., and Riedel, S., 1977, The effect of morphogenically active factors from host and nonhost plants on the in vitro differentiation of infection structures of *Puccinia graminis* f. sp. *tritici*, *Physiol. Plant Pathol.* **11**:213–224.

Grove, S. N., and Bracker, C. E., 1970, Protoplasmic organization of hyphal tips among fungi: Vesicles and Spitzenkörper, *J. Bacteriol.* **104**:989–1009.

Grove, S. N., and Sweigard, J. A., 1980, Cytochalasin A inhibits spore germination and hyphal tip growth in *Gilbertella persicari*, *Exp. Mycol.* **4**:239–250.

Grove, S. N., Bracker, C. E., and Morre, D. J., 1970, An ultrastructural basis for hyphal tip growth in *Phythium ultimum*, *Am. J. Bot.* **57**:245–266.

Grover, R. K., 1971, Participation of host exudate chemicals in appressorium formation by *Colletotrichum piperatum*, in: *Ecology of Leaf Surface Microorganisms* (T. F. Preece and C. H. Dickinson, eds.), Academic Press, New York, pp. 509–518.

Hamer, N. D., Howard, R. J., Chumley, F. G., and Valent, B., 1988, A mechanism for surface attachment in spores of a plant pathogenic fungus, *Science* **239**:288–290.

Hasselbring, H., 1906, The appressoria of the anthracnoses, *Bot. Gaz.* **42**:135–142.

Hasunuma, K., and Fundera, K., 1987, GTP-binding protein(s) in green plant, *Lemna paucicostata*, *Biochem. Biophys. Res. Commun.* **143**:908–912.

Hasunuma, K., Miyamoto-Shinohara, Y., and Furukawa, K., 1987, Partial characterization of GTP-binding proteins in *Neurospora*, *Biochem. Biophys. Res Commun.* **146**:1178–1183.

Hau, F. C., and Rush, M. C., 1979, Leaf surface interactions between *Cochliobolus miyabeanus* and susceptible and resistance rice cultivars, *Phytopathology* **69**:527.

Hickman, C. J., 1970, Biology of *Phytophthora* zoospores, *Phytopathology* **60**:1128–1135.

Hickman, C. J., and Ho, H. H., 1966, Behaviour of zoospores in plant pathogenic Phycomycetes, *Annu. Rev. Phytopathol.* **4**:195–220.

Hill, T. W., and Mullins, J. T., 1980, Hyphal tip growth in *Achlya*. I. Cytoplasmic organization, *Can. J. Bot.* **26**:1132–1140.

Ho, H. H., and Hickman, C. J., 1967, Factors governing zoospore responses of *Phytophthora megasperma* var. *Sojae* to plant roots, *Can. J. Bot.* **45**:1983–1994.

Hoch, H. C., and Staples, R. C., 1983, Ultrastructural organization of the non-differentiated uredospore germling of *Uromyces phaseoli*, *Mycologia* **75**:795–824.

Hoch, H. C., and Staples, R. C., 1984, Evidence that cyclic AMP initiates nuclear division and infection structure formation in the bean rust fungus, *Uromyces phaseoli*, *Exp. Mycol.* **8**:37–46.

Hoch, H. C., and Staples, R. C., 1987, Structural and chemical changes among the rust fungi during appressorium formation, *Annu. Rev. Phytopathol.* **25**:231–247.

Hoch, H. C., Bourett, T., and Staples, R. C., 1986, Inhibition of cell differentiation in *Uromyces* with D_2O and taxol, *Eur. J. Cell Biol.* **41**:290–297.

Hoch, H. C., Staples, R. C., and Bourett, T. M., 1987a, Chemically induced appressoria in *Uromyces appendiculatus* are formed aerially, apart from the substrate, *Mycologia* **79**:418–424.

Hoch, H. C., Staples, R. C., Whitehead, B., Comeau, J., and Wolf, E. D., 1987b, Signaling for growth orientation and cell differentiation by surface topography in *Uromyces*, *Science* **235**:1659–1662.

Hoch, H. C., Tucker, B. E., and Staples, R. C., 1987c, An intact microtubule cytoskeleton is necessary for mediation of the signal for cell differentiation in *Uromyces*, *Eur. J. Cell Biol.* **45**:209–218.

Hopkin, A. A., Reid, J., Hiratsuka, Y., and Allen, E., 1988, Initial infection and early colonization of *Pinus contorta* by *Endocronartium harnessii* (western gall rust), *Can. J. Plant Pathol.* **10**:221–227.

Howard, R. J., 1981, Ultrastructural analysis of hyphal tip cell growth in fungi: Spitzenkörper, cytoskeleton, and endomembranes after freeze-substitution, *J. Cell Sci.* **48**:89–103.

Huang, B. F., and Staples, R. C., 1982, Synthesis of proteins during differentiation of the bean rust fungus, *Exp. Mycol.* **6**:7–14.

Hunt, P., 1968, Cuticular penetration by germinating uredospores, *Trans. Br. Mycol. Soc.* **51**:103–112.

Jenkins, S. F., and Winstead, N. N., 1964, *Glomerella magna*, cause of a new anthracnose of cucurbits, *Phytopathology* **54**:452–454.

Johnson, T., 1934, A tropic response in germ tubes of urediospores of *Puccinia graminis tritici*, *Phytopathology* **24**:80–82.

Kaminskyj, S. G. W., and Day, A. W., 1984a, Chemical induction of infection structures in rust fungi. I. Sugars and complex media, *Exp. Mycol.* **8**:63–72.

Kaminskyj, S. G. W., and Day, A. W., 1984b, Chemical induction of infection structures in rust fungi. II. Inorganic ions, *Exp. Mycol.* **8**:193–201.

Keogh, R. C., Deverall, B. J., and McLeod, S., 1980, Comparison of histological and physiological responses to *Phakopsora pachyrhizi* in resistant and susceptible soybean, *Trans. Br. Mycol. Soc.* **74**:329–333.

Kikkawa, U., and Nishizuka, Y., 1986, The role of protein kinase C in transmembrane signaling, *Annu. Rev. Cell Biol.* **2:**149–178.

Kim, W. K., Howes, N. K., and Rohringer, R., 1982, Detergent-soluble polypeptides in germinated uredospores and differentiated uredosporelings of wheat stem rust, *Can. J. Plant Pathol.* **4:**328–333.

Knights, I. K., and Lucas, J. A., 1980, Photosensitivity of *Puccinia graminis* f. sp. *tritici* urediniospores in vitro and on the leaf surface, *Trans. Br. Mycol. Soc.* **74:**543–549.

Koch, E., Ebrahim-Nesbat, F., and Hoppe, H. H., 1983, Light and electron microscopic studies on the development of soybean rust *Phakopsora-pachyrhizi* in susceptible soybean leaves, *Phytopathol. Z.* **106:**302–320.

Lapp, M. S., and Skoropad, W. P., 1978, Location of appressoria of *Colletotrichum graminicola* on natural and artificial barley leaf surfaces, *Trans. Br. Mycol. Soc.* **70:**225–228.

Lewis, B. G., and Day, J. R., 1972, Behaviour of urediospore germ tubes of *Puccinia graminis tritici* in relation to the fine structure of wheat leaf surfaces, *Trans. Br. Mycol. Soc.* **58:**139–145.

Littlefield, L. J., and Heath, M. C., 1979, *Ultrastructure of Rust Fungi,* Academic Press, New York.

Maeda, K. M., 1970, An ultrastructural study of *Venturia inaequalis* (Cke.) Wint. infection of *Malus* hosts, M.S. thesis, Purdue University, Lafayette.

Macko, V., Renwick, J. A. A., and Rissler, J. F., 1978, Acrolein induces differentiation of infection structures in the wheat stem rust fungus. *Science* **199:**442–443.

Maheshwari, R., Hildebrandt, A. C., and Allen, P. J., 1967, The cytology of infection structure development in uredospore germ tubes of *Uromyces phaseoli* var. *typica* (Pers.) Wint, *Can. J. Bot.* **45:**447–450.

Marshall, D. S., and Rush, M. C., 1980, Infection cushion formation on rice sheaths by *Rhizoctonia solani,* *Phytopathology* **70:**947–950.

Mendgen, K., and Dressler, E., 1983, Culturing *Puccinia coronata* on a cell monolayer of the *Avena sativa* coleoptile, *Phytopathol. Z.* **108:**226–234.

Middlebrook, M. J., and Preston, R. D., 1952, Spiral growth and spiral structure. III. Wall structure in the growth zone of *Phycomyces, Biochim. Biophys. Acta* **9:**32–48.

Mitchell, R. H., 1975, Inositol phospholipids and cell surface receptor function, *Biochim. Biophys. Acta* **415:**81–92.

Morse, M. J., Crain, R. C., and Satter, R. L., 1987, Phosphatidylinositol cycle metabolites in *Samanea saman* pulvini, *Plant Physiol.* **83:**640–644.

Murray, D. I. L., 1982, Penetration of barley root and coleoptile surfaces by *Rhizoctonia solani, Trans. Br. Mycol. Soc.* **79:**354–360.

Musgrave, A., Loes, E., Scheffer, R., and Oehlers, E., 1977, Chemotropism of *Achyla bisexualis* germ hyphae to casein hydrolysate and amino acids, *J. Gen. Microbiol.* **101:**65–70.

Nishizuka, Y., 1984, Protein kinases in signal transduction, *Trends Biol. Sci.* **9:**163–167.

Nushbaum, C. J., and Keitt, G. W., 1938, A cytological study of host–parasite relations of *Venturia inaequalis* on apple leaves, *J. Agric. Res.* **56:**595–618.

Otto, A. M., Ulrich, M., Zumbe, A., and DeAsua, L. J., 1981, Microtubule-disrupting agents affect two different events regulating the initiation of DNA synthesis in Swiss 3T3 cells, *Proc. Natl. Acad. Sci. USA* **73:**3063–3067.

Pelham, H., 1988, Coming in from the cold, *Nature* **332:**776–777.

Peterson, G. W., and Walla, J. A., 1978, Development of *Dothistroma pini* upon and within needles of Australian and Ponderosa pines in eastern Nebraska, *Phytopathology* **68:**1422–1430.

Preece, T. F., Barnes, G., and Bayley, J. M., 1967, Junction between epidermal cells as sites of appressorium formation by plant pathogenic fungi, *Plant Pathol.* **16:**117–118.

Pring, R. J., 1980, A fine-structural study of the infection of leaves of *Phaseolus vulgaris* by uredospores of *Uromyces phaseoli, Physiol. Plant Pathol.* **17:**269–276.

Rathaiah, Y., 1977, Stomatal tropism of *Cercospora beticola* in sugarbeet, *Phytopathology* **67:**358–362.

Reymond, C. D., Nellen, W., and Firtel, R. A., 1985, Regulated expression of *ras* gene constructs in *Dictyostelium* transformants, *Proc. Natl. Acad. Sci. USA* **82:**7005–7009.

Royle, D. J., and Hickman, C. J., 1964, Analysis of factors governing in vitro accumulation of zoospores of *Phythium aphanidermatium* on roots. I. Behavior of zoospores, *Can. J. Bot.* **10:**151–162.

Royle, D. J., and Thomas, G. G., 1971, The influence of stomatal opening on the infection of hop leaves by *Pseudoperonospora humuli, Physiol. Plant Pathol.* **1:**329–343.

Schlesinger, M. J., 1986, Heat-shock proteins: The search for functions, *J. Cell Biol.* **103:**321–325.

Schönherr, J., and Bukovac, M. J., 1970, Preferential polar pathways in the cuticle and their relationship to ectodesmata, *Planta* **92:**189–201.

Shaw, M., Boasson, R., and Scrubb, L., 1985, Effect of heat shock on protein synthesis in flax rust uredosporelings, *Can. J. Bot.* **63:**2069–2076.

Staples, R. C., and Hoch, H. C., 1982, A possible role for microtubules and microfilaments in the induction of nuclear division in bean rust uredospore germlings, *Exp. Mycol.* **6:**293–302.

Staples, R. C., and Hoch, H. C., 1988, Preinfection changes in germlings of a rust fungus induced by host contact, in: *Biotechnology for Crop Protection* (P. A. Hedin, J. J. Menn, and R. M. Hollingworth, eds.), American Chemical Society, Washington, DC. pp. 82–93.

Staples, R. C., and Macko, V., 1984, Germination of urediospores and differentiation of infection structures in: *The Cereal Rusts*, Volume 1 (W. R. Bushnell and A. J. Roelfs, eds.) Academic Press, New York, pp. 255–289.

Staples, R. C., Laccetti, L., and Yaniv, Z., 1976, Appressorium formation and nuclear division in *Colletotrichum truncatum*, *Arch. Microbiol.* **109:**75–84.

Staples, R. C., Grambow, H. J., and Hoch, H. C., 1983a, Potassium ion induces rust fungi to develop infection structures, *Exp. Mycol.* **7:**40–46.

Staples, R. C., Grambow, H. J., Hoch, H. C., and Wynn, W. K., 1983b, Contact with membrane grooves induces wheat stem rust uredospore germlings to differentiate appressoria but not vesicles, *Phytopathology* **73:**1436–1439.

Staples, R. C., Macko, V., Wynn, W. K., and Hoch, H. C., 1984, Terminology to describe the differentiation response by germlings of fungal spores, *Phytopathology* **74:**380.

Staples, R. C., Hassouna, S., and Hoch, H. C., 1985a, Effect of potassium on sugar uptake and assimilation by bean rust germlings, *Mycologia* **77:**248–252.

Staples, R. C., Hoch, H. C., and Epstein, L., 1985b, The development of infection structures by the rust and other fungi, *Microbiol. Sci.* **2:**193–198.

Staples, R. C., Hoch, H. C., Epstein, L., Laccetti, L., and Hassouna, S., 1985c, Recognition of host morphology by rust fungi: Responses and mechanisms, *Can. J. Plant Pathol.* **7:**314–322.

Staples, R. C., Yoder, O. C., Hoch, H. C., Epstein, L. and Bhairi, S., 1988, Gene expression during infection structure development by germlings of the rust fungus, in: *Biology and Molecular Biology of Plant–Pathogen Interactions* (J. A. Bailey, ed.), Springer-Verlag, Berlin, pp. 331–341.

Staples, R. C., Hoch, H. C., Freve, P., and Bourett, T. M., 1989, Heat shock induced development of infection structures by bean rust uredospore germlings, *Exp. Mycol.* **13:**149–157.

Steward, A., Backhouse, D., Sutherland, P. W., and Fullerton, R. A., 1989, The development of infection structures of *Sclerotium cepivorum* on onion, *J. Phytopathol.* **126:**22–32.

Stockwell, V., and Hanchey, P., 1983, The role of the cuticle in resistance of beans to *Rhizoctonia solani*, *Phytopathology* **73:**1640–1642.

Stryer, L., and Bourne, H. R., 1986, G proteins: A family of signal transducers, *Annu. Rev. Cell Biol.* **2:**391–419.

Sutton, B. C., 1962, *Colletrichum dematium,* and *C. trichellum*, *Trans. Br. Mycol. Soc.* **45:**222–232.

Suzuki, K., Furusawa, I., Ishida, N., and Yamamoto, M., 1981, Protein synthesis during germination and appressorium formation of *Colletotrichum lagenarium* spores, *J. Gen. Microbiol.* **124:**61–69.

Suzuki, K., Furusawa, I., Ishida, N., and Yamamoto, M., 1982, Chemical dissolution of cellulose membranes as a prerequisite for penetration from appressoria of *Colletotrichum lagenarium*, *J. Gen Microbiol.* **128:**1035–1039.

Tan, K. K., 1978, Light induced fungal development, in: *The Filamentous Fungi*, Volume 3 (J. E. Smith and D. R. Berry, eds.), Wiley, New York, pp. 334–357.

Thevelein, J. M., 1984, Activation of trehalase by membrane-depolarizing agents in yeast vegetative cells and ascospores, *J. Bacteriol.* **158:**337–339.

Thorner, J., 1982, An essential role for cyclic AMP in growth control: The case for yeast, *Cell* **30:**5–6.

Toda, T., Uno, I., Ishikawa, T., Powers, S., Kataoka, T., Broek, D., Cameron, S., Broach, J., Matsumoto, K., and Wigler, M., 1985, In yeast, RAS proteins are controlling elements of adenylate cyclase, *Cell* **40:**27–36.

Tucker, B. E., Hoch, H. C., and Staples, R. C., 1986, The involvement of F-actin in *Uromyces* cell differentiation: The effect of cytochalasin E and phalloidin, *Protoplasma* **135:**88–101.

Turian, G., 1978, Sexual morphogenesis in the Ascomycetes, in: *The Filamentous Fungi*, Volume 3 (J. E. Smith and D. R. Berry, eds.), Wiley, New York, pp. 315–333.

Wanner, R., Forster, H., Mendgen, K., and Staples, R. C., 1985, Synthesis of differentiation-specific proteins in germlings of the wheat stem rust fungus after heat shock, *Exp. Mycol.* **9:**279–283.

Wynn, W. K., 1976, Appressorium formation over stomates by the bean rust fungus: Response to a surface contact stimulus, *Phytopathology* **66:**136–146.

Wynn, W. K., 1981, Tropic and taxic responses of pathogens to plants, *Annu. Rev. Phytopathol.* **19:**237–255.

Wynn, W. K., and Staples, R. C., 1981, Tropisms of fungi in host recognition, In: *Plant Disease Control: Resistance and Susceptibility* (R. C. Staples and G. H. Toenniessen, eds.), Wiley–Interscience, New York, pp. 45–69.

Yeh, Y., and Frederiksen, R. A., 1980, Sorghum downy mildew: Biology of systemic infection by conidia and of a resistant response in sorghum, *Phytopathology* **70:**372–376.

3

The Plant Cell Wall as a Barrier to Fungal Invasion

M. G. Smart

1. INTRODUCTION

The subject of this chapter is the plant cell wall. Cell walls are fascinatingly complex formations, variable in structure and composition depending upon their location in the plant body, their age, their genus, and the environment. Their composition is so complex that it is still not completely understood (McNeil *et al.*, 1984). Cell walls serve many functions including a role as a barrier to fungal penetration, as implied in the title of this chapter. Cell wall composition is reflected in all these roles. For example, lignification provides the mechanical strength for aerial growth of plants and may incidentally render the cell wall resistant to attack by most potential pathogens. But in other situations, the normal requirement of cell wall function may be in conflict with defense: The cell walls of the "zone of elongation" in roots are expanding, presumably under hormonal influence, and, hence, must be relatively weak if growth occurs by bond cleavage and interpolation of new wall material. This is the precise region of the root most susceptible to invasion by *Phytophthora* spp. and other soil-borne pathogens. This potential for conflict between different functions of cell walls should always be kept in mind as it may profoundly influence the success or failure of defense responses.

This chapter will focus on the cell wall in its role as a barrier to fungal ingress. Walls may be constitutively resistant to penetration because of their native composition. Or walls may be modified more or less extensively, in both content and spatially, by the formation of halos, wall appositions or the wider responses of cicatrix formation and periderm induction. Before discussing these phenomena, it is instructive to briefly review the macromolecular constituents of cell walls, especially as they relate to pathogenesis.

Much of the experimental work investigating the role of cell wall (structural) responses has been performed with the powdery mildews. Because Aist and Bushnell review this work in Chapter 15 it will only be touched upon here. Similarly, Köller (Chapter 10) reviews the role of the

M. G. Smart • Northern Regional Research Center, Agricultural Research Service, U.S. Department of Agriculture, Peoria, Illinois 61604. Present address: Kraft General Foods, Inc., Glenview, Illinois 60025. The mention of firm names or trade products does not imply that they are endorsed or recommended by the U.S. Department of Agriculture over other firms or similar products not mentioned.

plant cuticle in defense. Finally, the subject of this chapter has been reviewed at length (Aist, 1976, 1983; Heath, 1980a; Ride, 1983; Sherwood and Vance, 1982) and the reader is encouraged to read these other perspectives.

2. STRUCTURE OF THE PLANT CELL WALL

Plant cell walls are exceedingly complex structures. They encompass all the macro-molecules outside the plasmalemma (Colvin, 1981) and also contain ions, water, lipids, and waxes to varying degrees (Esau, 1965; Wardrop, 1971). Usually, cell walls are considered to be involved in the control of cell expansion and therefore they influence cell shape and, hence, plant morphogenesis (Colvin, 1981; McNeil *et al.*, 1984). They are credited, too, with the prevention of pathogen ingress, the evidence for which I shall examine in Sections 3, 4, and 5.

It is self-evident in a hand section of any plant organ stained with toluidine blue 0 that cell walls can differ radically from one cell to the next. By implication, even this relatively crude indicator tells us that wall composition can vary drastically over short distances (and through time). Therefore, the sophisticated techniques available to investigators of the biochemistry of cell walls have been used on simple model systems, such as callus tissues, suspension cultures, or even protoplasts regenerating walls (see references in McNeil *et al.*, 1984; Kato, 1981). Investigators then minimize the problem associated with biochemical studies: the averaging effects of *en masse* wall isolation.

Cell walls also differ individually according to the distance from the middle lamella and their age. During cytokinesis, the phragmoplast elaborates the first wall between the two new daughter cells, the cell plate (Bajer, 1968; Bajer and Mole-Bajer, 1972), which will become the middle lamella (Willison, 1981). After cell division is complete, but while the cell is still growing, the protoplast lays down the "primary" cell wall (Esau, 1965). Many cell types show no further development whereas in others, lignification and (or) suberization processes occur. Lignification begins (and may end) at the middle lamella and then extends into the primary wall, displacing water to make the cells more rigid and able to tolerate mechanical stress (Wardrop, 1971). Finally, a secondary wall may be deposited inside the primary wall after cell expansion ceases. This final wall may be composed of ultrastructurally discernible layers which may or may not subsequently be lignified (Colvin, 1981; Esau, 1965). The walls of epidermal cells are particularly relevant because they are the first line of defense in many pathogen-suscept interactions. Unfortunately, their structure is known only from histochemical tests which show that, in the absence of stress, they rarely possess a secondary wall. The primary wall is not lignified until the development of the secondary plant body, if ever. Walls of leaf epidermal cells, in particular, are rarely lignified (Esau, 1965).

The complexity of even the simpler cell walls of model systems has generated an enormous literature which is beyond the scope of this chapter—although it is hardly irrelevant! Readers are referred to reviews by Colvin (1981), Fincher and Stone (1981), Kato (1981), Labovitch (1981), Lamport and Epstein (1983), or Preston (1979) for an introduction to the literature.

In the paragraphs that follow, I will briefly recount the commonly isolated macromolecular constituents of the cell wall and indicate their putative or potential functions in the context of disease resistance.

2.1. Chemistry

2.1.1. Carbohydrates

Most published work on cell walls has used nonlignified, primary walls from suspension cultures (McNeil *et al.*, 1984). The structure of secondary walls is less well understood although it is known that they contain proportionally more cellulose which is of a higher degree of polymerization (Preston, 1979; McNeil *et al.*, 1984). Nevertheless, cellulose is the most common constituent of the primary cell wall, forming 20 to 30% of the dry weight. Cellulose is an almost pure homopolymer of (1→4)-β-linked glucosyl residues. The resulting chains are aggregated into microfibrils that are semicrystalline and visible ultrastructurally (Roelefson, 1965). Not all cell walls have these high proportions of cellulose. Monocotyledons, particularly grasses, have certain cell types with less than 5.0% w/w cellulose (Fincher and Stone, 1981). Most often these walls are from the endosperm. Cellulose provides a framework for the matrix polysaccharides which fill the spaces between neighboring microfibrils. Microfibrils have enormous tensile strength (Esau, 1965) but have not been proposed as a participant in the resistance of walls to pathogens.

The next most common component of dicotyledonous cell walls is a xyloglucan (McNeil *et al.*, 1984). The (1→4)-β-linked glucosyl backbone residues are heavily substituted with α-linked xylosyl residues. Although monocotyledons possess xyloglucans, they are only a minor component: Their place is taken by arabinoxylans. That is, the (1→4)-β-linked xylosyl backbone residues are substituted with arabinose side chains. Up to 50% of the xylosyl residues are acetylated (Bouveng, 1961), a factor of, perhaps, some significance in the resistance of grass cell walls to enzyme attack (Bacon *et al.*, 1975, 1976; Morris and Bacon, 1977). The last authors found a good correlation between the digestibility of grass cell walls and their acetyl content; the correlation between "lignification" and acetyl content was not as great.

Monocotyledons also have varying amounts of β-glucans, particularly among members of the Poaceae. β-Glucans are mixed-linkage (1→3)-β- and (1→4)-β-linked glucosyl residues (McNeil *et al.*, 1984). The enormous potential for variation in the linkage pattern of the backbone and the equally enormous potential variation in side chain substituent groups give β-glucans informational value in the regulation of plant development, recognition phenomena, and phytoalexin elicitation (McNeil *et al.*, 1984). Additionally, modification of the details of backbone linkages or side chain substitution, including their acetylation patterns in β-glucans (and xyloglucans), may be involved in the production of wall alterations occurring in response to attempted penetration by fungi. These modifications could sterically hinder pathogen-derived enzymes, or reduce the permeability of the wall to nutrients needed by the invading fungus. Mature cell walls do not possess callose [(1→3)-β-D-glucans] in the sense that this term is usually used. The first products of the phragmoplast are callosic, but this material is quickly replaced as the new walls develop after cytokinesis (Fulcher *et al.*, 1976). Callose is one of the components of wall modifications most often cited as being involved in resistance.

The final major carbohydrate class I wish to mention here are the acidic polysaccharides, usually identified with the pectic fraction. The rhamnogalacturonans (I and II) are very different polymers identified from both monocotyledon and dicotyledon cell walls. Rhamnogalacturonan I is a linear backbone of alternating 2-linked rhamnosyl and 4-linked galactosyluronic acid residues (McNeil *et al.*, 1984). Rhamnoagalacturonan II is more complex and appears to be composed largely of heptasaccharide units. These pectic fractions—and other matrix polysaccharides of the cell wall—have been implicated in the induction of changes in the cytoplasm, including phytoalexin elicitation (see review by Bell, 1981). Apart from interactions with microorganisms, the pectic fraction is the major component of the middle lamella and, by the cross-linking effect of calcium ions, becomes the cell-to-cell "glue" (Colvin, 1981).

2.1.2. Proteins

The protein content of cell walls varies with cell type and environmental conditions but is generally less than 10% (McNeil *et al.*, 1984). The proteins can be divided into at least three classes based upon their function. The extensins are highly unusual glycoproteins which, in addition to hydroxyproline, contain dityrosine residues (Lamport and Epstein, 1983; Tierney and

Varner, 1987; McNeil *et al.*, 1984). The extreme insolubility of extensin has led to the proposal that it has a structural function (Lamport and Epstein, 1983). It has been implicated in the resistance of cell walls to fungal ingress since concentrations of extensin (as hydroxyproline-rich glycoproteins) increase during challenge by pathogens (Esquerre-Tugaye *et al.*, 1979). This role will be examined in detail below (Section 4). The second major group of proteins are the arabinogalactan proteins (AGPs). As the name suggests, these proteins are also glycoproteins. Unlike extensin, which is 50% protein, the protein content of AGPs is 10% (Clarke *et al.*, 1979). AGPs contain hydroxyproline but not dityrosine and are freely soluble (Clarke *et al.*, 1979). There has been some speculation as to their function, particularly in flower styles during pollination, but in effect their function is unknown. The final group of proteins routinely found in the cell wall are enzymes. Peroxidase appears to have both many isozymes and multiple functions in the wall. It is involved in lignification (Higuchi, 1981), in the formation of the dityrosine residues thought to be responsible for extensin's insolubility (McNeil *et al.*, 1984), in the formation of diferuloyl cross-links in grass arabinoxylans (Markwalder and Neukom, 1976), and even in auxin metabolism (Lamport and Catt, 1981). Other polysaccharidases such as cellulase (Fan and MacLachlan, 1967), $(1\rightarrow3)$-β-glucanase (Heyn, 1969), and polygalacturonase (Wallner and Walker, 1975) have been isolated from cell walls. These and other wall-degrading enzymes are involved in various aspects of cell wall metabolism. For example, polygalacturonase is involved in fruit softening during ripening. Other polysaccharidases are involved in the expansion growth of cells, in abscission, and in xylem vessel differentiation (see Fincher and Stone, 1981; Lamport and Catt, 1981).

2.1.3. *Phenolic Derivatives*

Derivatives of cinnamic acid are common in cell walls, the most common form being lignin. Higuchi (1981) has reviewed the process of lignification and it is demonstrable that lignified tissues are refractory to enzyme degradation. The reason for this lies in the nonenzymatic, free-radical-mediated condensation which is the final step in lignin formation (Higuchi, 1981). Perhaps of equal importance in the structure of monocotyledonous cell walls is the cross-linking of arabinoxylans by diferulate bridges (Markwalder and Neukom, 1976). Such a process has been held responsible for the oxidative gelation of wheat flour (Markwalder and Neukom, 1976). It has been suggested that diferuloyl cross-links render cell walls difficult to digest enzymatically (Fulcher *et al.*, 1972; Shimony and Friend, 1976). This attractive notion has been suggested, along with lignification itself, as an induced disease-resistance mechanism although there is little evidence for it (Section 3).

There is some circumstantial evidence against ferulic acid conferring resistance to enzymes in grass cell walls. During germination, cereal embryos initiate a complex sequence of *de novo* enzyme synthesis culminating in their subsequent secretion from the aleurone layer (Taiz and Jones, 1970; Ashford and Jacobsen, 1977). The substrates for the secreted enzymes are the reserves found in the endosperm: lipid, starch, protein, and cell walls. The products, monosaccharides and mono- or dipeptides, are assimilated by the scutellum for the development of the embryo. The cell walls of both the aleurone layer and the scutellum are autofluorescent in ungerminated wheat and barley and they contain ferulic acid (Taiz and Jones, 1970; Fulcher *et al.*, 1972; Smart and O'Brien, 1979a,b). In the wheat aleurone layer, one in every 250 xylose residues of the wall arabinoxylan is feruloylated (Bacic and Stone, 1981), a degree of substitution thought to be too low to account for the observed resistance of aleurone cell walls to enzymatic attack (Fincher and Stone, 1981). Indeed, the aleurone cell wall is recognizably two-layered ultrastructurally and the outer layer—despite autofluorescence and ferulic acid content—is degraded during germination (Taiz and Jones, 1970; Fulcher *et al.*, 1972). Observations on wheat, barley, oat, and *Lolium* scutella during germination showed that the autofluorescence of the epithelial cell walls is lost, temporarily, during the time of cell expansion. Yet these walls remain intact in a "soup" of hydrolytic enzymes (Smart and O'Brian, 1979b).

How do these inner walls of the aleurone layer and epithelial walls of the scutellum survive? There is little difference in the polysaccharide composition of aleurone and endosperm cell walls (Fincher and Stone, 1981) and we have seen that ferulic acid is only doubtfully responsible. Perhaps another macromolecule (extensin?) or modified acetylation patterns may be responsible (see Smart and O'Brien, 1979a). Whatever the ultimate reason, the behavior of the aleurone layer and scutellum cell walls in the germination of cereals should be kept in mind when one reads about the putative resistance of other cell walls to fungal ingress due to phenolic (nonlignin) infusion. Quantitative data on phenolic and acetyl concentrations would be useful in making judgements about their role in resistance. Such data might be difficult to obtain from highly localized areas around penetration sites but are feasible with techniques such as microspectro-fluorimetry.

2.2. Physical Structure

From a pathogen–plant viewpoint, there are two important physical attributes of cell walls: mechanical strength and permeability to water, ions, and larger molecules (e.g., enzymes, elicitors). Largely due to their cellulose content, cell walls are enormously strong under tensile stress (Esau, 1965) and, once lignified, are resistant to compressive forces (Wardrop, 1971). These mechanical demands which are imposed by the physical environment and plant morphology can only make mechanical penetration of walls by fungi more difficult. But recall that most direct penetrations occur in primary cell walls, usually the outer periclinal cell wall of the epidermis. These walls are rarely lignified at the site and time of successful penetration. The autofluorescent cell walls of grass leaves involve ferulic acid, not lignin (see Harris and Hartley, 1976, for the distribution of ferulic acid in grasses). Many authors have observed host cell wall microfibril disorientation around penetration pegs, claiming this as evidence for the use of mechanical force in penetration (e.g., Chou, 1970; Nemec, 1971). The reader is referred to the chapter by Kubo and Furusawa (Chapter 9) for a discussion of this topic.

Nonlignified, primary cell walls are variously reported to have pores which allow diffusion of molecules of up to 50,000 daltons, (Gaff *et al.*, 1964; Carpita *et al.*, 1979; Crowley, 1977; Tepfer and Taylor, 1981). Of course, different cell walls must have different permeabilities depending on location, age, and other factors. The porosity of the gel matrix is a function of "junction zones" in the matrix polysaccharides (Rees and Welch, 1977) and must be influenced by local pH values, a consideration in pathogenesis.

The porosity of the wall is important because of its potential role in recognition phenomena, phytoalexin elicitation (Ralton *et al.*, 1987), enzymatic penetration of walls, and fungal nutrition (Aist, 1983; Ride, 1983). Unfortunately, permeability studies are fraught with technical and theoretical difficulties, not least because the permeability found will be a function of pH, thus making it mandatory to disturb the cells as little as possible.

3. RESISTANCE OF NATIVE CELL WALLS TO FUNGI

When pathogen propagules come into contact with the epidermis, some, such as the uredospore germlings of rust fungi, avoid it by penetrating stomata, others require wounds before they can attack the plant. But many potential pathogens penetrate the epidermis directly, either entering the epidermal cells (e.g., the powdery mildew fungi) or penetrating between epidermal cells to enter cells in the tissues below (e.g., the phycomycetes on roots). Is there any evidence that such walls can constitutively resist penetration? We shall examine this question in the following paragraphs.

Sherwood and Vance (1980) inoculated 12 species in 11 tribes of the Poaceae with an incompatible isolate of *Curvularia lunata*. In nontreated tissues, the number of penetration attempts per 100 conidia varied from 153 on oats to 314 on reed canary grass leaves. When the tissues were treated with cycloheximide, a potent inhibitor of protein synthesis, the incidence of penetration attempts did not change appreciably (although their success rate did improve). The authors concluded that an unknown preformed factor was involved in some of the penetration failures. In elaborate binary analysis experiments, Johnson *et al.* (1979, 1982) found similar penetration failures for powdery mildew fungi.

Such constitutive resistance of cell walls to penetration is usually ascribed to lignin and (or) suberin. In grasses, "lignin" most probably means ferulic acid esterified into the wall matrix. Earlier workers have noticed the restriction of lesions in leaves and hypocotyls by fibers and lignified bundle sheaths (Griffey and Leach, 1965; Hursh, 1924). This author and co-workers (M. G. Smart, D. T. Wicklow and R. W. Caldwell, unpublished) have noticed the exclusion of *Aspergillus flavus* hyphae from the vascular bundles of the maize rachis at the (lignified) bundle sheath. All such observations suggest, but do not prove, that lignified walls, including feruloylated walls in grasses (or suberized in the vascular bundle sheath; see O'Brien and Kuo, 1975), may be involved in constitutive resistance. Experimental data are lacking entirely.

Similar associations of chemical content of walls and restricted penetration or lesion size exist for silicon and calcium. Bateman and Lumsden (1965) correlated the increased resistance of older bean hypocotyls attacked by *Rhizoctonia solani* with increased calcium content in the pectic fraction. They proposed that increasingly calcium cross-linked galacturonic acid residues reduced cell wall susceptibility to enzyme attack, thus conferring resistance. Direct evidence for the inhibitory effect of increased calcium content on polygalacturonase activity, *in vitro*, has been reported (Conway *et al.*, 1988). The authors found that the enzyme activity of a polygalacturonase from *Penicillium expansum* was inhibited by about 60% when cell walls from calcium-infiltrated apples were used as substrate.

Evidence for a constitutive role for silicon in resistance is less direct. Volk and co-workers (1958) and Akai and Fukutomi (1980) have correlated silicon supplementation with increased resistance of rice to *Pyricularia oryzae*. Apparently, the silicon is deposited in a layer beneath the cuticle in rice (Yoshida *et al.*, 1962), a condition which may not occur in other grasses.

The actual reason for such penetration failures in the absence of a macroscopically recognizable host response is not clear. In addition to the reason that the walls may constitutively resist penetration, other possibilities have been suggested. Either the hypha of a particular failed propagule does not possess the necessary enzymes for penetration (Ride, 1983) or the hypha was inherently unable to attempt penetration for nutritional or other reasons (Aist, 1983). There may also be adhesion problems resulting in failure of recognition processes on the part of the fungus. Sorting out these factors experimentally would be difficult but localization of pathogen-derived wall-degrading enzymes at the hyphal tip would be a useful start.

Evidence against constitutive factors conferring resistance to pathogens has been published. As tissues age, they often cease to be susceptible to attack by certain pathogens. Hayes and Jones (1966) reported that oat plants fully susceptible to penetration by particular races of *Erysiphe graminis* f. sp. *avenae* while still seedlings became increasingly resistant with age. This has been confirmed elsewhere (Hite *et al.*, 1977). The possible reasons for this effect—including papilla formation, negative germination effects, and constitutive resistance of older cell walls—have recently been investigated (Douglas *et al.*, 1984). The authors used three oat cultivars, one without adult plant resistance and two with it, and found that there were no significant differences between the incidence of oat powdery mildew penetration pegs (without further pathogen development) on young or old leaves of any cultivar. They concluded that constitutive defense (due to the cell wall) was not involved in adult plant resistance. Note that this work involved the epidermal cells only of the oat plant. These walls are neither lignified nor, apparently, infused with silicon although they probably contain diferulate bridges (see Harris and Hartley, 1976). We are left with the conclusion that, where natural development of cell walls includes lignification or silicification, constitutive resistance may well be a factor in resistance. In walls which remain unmodified, chiefly in the primary plant body, the evidence for a constitutive resistance of cell walls to penetration is weak and no one component seems to have been implicated clearly.

4. MODIFICATION OF CELL WALLS IN RESISTANCE TO FUNGI

Cells respond to the presence of fungi even before penetration activities begin (Zeyen and Bushnell, 1979; Heath, 1981a). This implies a chemical (Aist, 1983) or, perhaps, electrochemical stimulus. The host cell usually responds first with the formation of a cytoplasmic aggregate (Aist, 1976; Aist and Israel, 1977a,b; Bushnell and Berquist, 1975). Aggregates are small volumes of cytoplasm which form directly beneath the point of incipient penetration. Ultrastructurally, they contain dictyosomes, rough endoplasmic reticulum, vesicles, and, presumably, the enzymes and precursors necessary for wall modification. I am unaware of any enzyme (or wall component) localization studies in cytoplasmic aggregates, however. The reader is referred to Chapter 15 for a full discussion of this topic in powdery mildews, which are taken as representative of all diseases in which aggregates occur.

Whatever the precise role of the aggregate, coincident to their formation two types of wall modification often occur: halo formation and papilla deposition (see Fig. 1). Halos are roughly circular areas (when viewed from above) of the cell wall which form around the point of incipient fungal penetration (McKeen *et al.*, 1968; Sargant and Gay, 1977). Papillae are generally smaller in diameter than halos and, unlike them, are hemispherical deposits of new wall material apposed between the old wall and the plasmalemma (Aist, 1976, 1983; de Bary, 1863). The two may or may not both occur in an interaction. In this section, these localized cell wall modifications are discussed in relation to their composition and function. For some host–parasite interactions, the deposition of certain materials is more extensive than a "halo" even affecting neighboring cells. For the purposes of this discussion, no distinction will be made between the two since they are often not clearly distinguished in the literature.

4.1. Halos as Wall Modifications

Many substances have been localized in halos; among them are lignin, "callose," lipids, silicon, and reducing sugars (Bird and Ride, 1981; Defosse, 1971, 1976; Hammerschmidt and Kuć, 1982; Kunoh and Ishizaki, 1975a,b; McKeen *et al.*, 1968; Ride and Pearce, 1979; Sargant and Gay, 1977; Sherwood and Vance, 1976). Tests for some common cell wall components have been negative in certain host–parasite interactions, cutin, cellulose, and pectin being examples (Kunoh and Akai, 1969; McKeen *et al.*, 1968). These latter results have been taken as indicating halos are regions of localized wall dissolution, although Sargant and Gay (1977) point out that infusion of lipids and silicon into the wall may mask histochemically reactive wall sites, leading to false-negative results. In some cases there is direct evidence for halos as sites of wall dissolution. Pappelis and collaborators (Mayama and Pappelis, 1967; Russo and Pappelis, 1981) used quantitative interference microscopy on onion leaves attacked either by *Botrytis* sp. or by *Colletotrichum circinans*. They found that, if the interaction was susceptible, mass was lost from the halo area; if resistant, mass was gained. Their results remind us that similar structures, in this case halos, may not be the same chemically, despite similar appearances, and therefore may not have the same function in resistant or susceptible interactions.

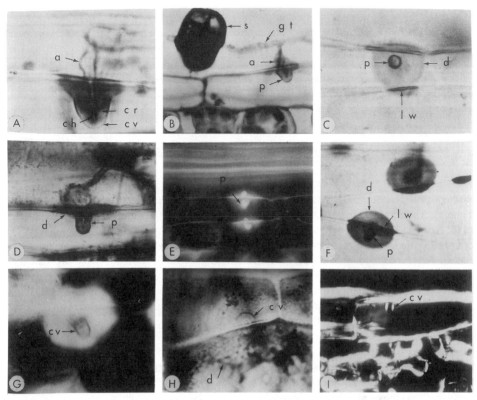

Figure 1.   Halos (d) and papillae (p) in epidermal strips of reed canary grass leaves attacked by filamentous fungi. The halo is most obvious in panel F, which is a view from above. Variously treated tissues. A, stained with acetophenol-cotton blue; B, stained with safranin and fast green; C, stained with phloroglucinol-HC1; D, stained with lacmoid; E, stained with aniline blue and viewed by fluorescence microscopy; F, stained with glycerine-ferricyanide; G, H, stained with IKI-H_2SO_4; I, viewed through crossed polarizers. (From Sherwood and Vance, 1976, with permission.)

Various reports have shown an infusion of electron-dense material in cell walls coincident with attempted penetration (Heath, 1979; Kunoh and Ishizaki, 1975a,b; Sargant and Gay, 1977). This material is, in part at least, due to the accumulation of insoluble silicon in the halo (Carver *et al.*, 1987; Heath 1979, 1981a; Kunoh and Ishizaki, 1975a,b). Plants actively accumulate silicon as monosilicic acid (Lewin and Remann, 1969), but its passage through the plant is said to be passive in the transpiration stream, that is, apoplastic. It is not clear where silicon enters the symplast. Nevertheless, whatever its immediate origin at penetration sites, silicon accumulates there in response to the fungus (Carver *et al.*, 1987; Heath, 1980b, 1981a,b; Kunoh and Ishizaki, 1975a,b; Zeyen *et al.*, 1983). How insoluble silicon may affect fungal penetration is still uncertain and is examined below.

The supplementation of soil with silicon and its effect on increasing the resistance of rice to the blast fungus, *Pyricularia oryzae*, has been noted above (Volk *et al.*, 1958; Akai and Fukutomi, 1980). Heath and co-workers have obtained correlative evidence for an active role for silicon in protection of bean plants from attack by *Uromyces vignae* (the cowpea rust fungus) for which beans are nonhosts (Heath, 1981a,b; Stumpf and Heath, 1985). Bean plants preinfused with exudates from germlings of the bean rust fungus or with extracts from bean rust fungus-infected plants suppress silicon deposition in plants subsequently inoculated with the nonpathogen, *U. vignae* (Heath, 1981b), thus implicating the element in resistance—although papillae were suppressed also. Stumpf and Heath grew silicon-depleted bean plants and normal silicon-containing plants (Stumpf and Heath, 1985). By means of carefully correlated light and fluorescence microscopy combined with energy-dispersive analysis of X rays (see Zeyen *et al.*, 1983), they determined that similar, but low, numbers of infection hyphae formed haustoria in mesophyll cells (primary, normally unlignified cell walls) in both normal silicon and silicon-depleted plants. When silicon-minus plants were treated with heat shock or infused with extracts from bean rust fungus-infected plants, greater numbers of haustoria formed, indicating some factor other than silicon is involved in resistance. The authors concluded that silicon was probably not the primary determinant of resistance but did not rule out a role for it in defense. Carver *et al.* (1987) came to a similar conclusion in barley attacked by *Erysiphe graminis* f. sp. *hordei* and further showed that, in compatible inoculations, the timing of silicon deposition is involved in penetration failure. Finally, although the evidence for some role for silicon in resistance is good, Sargant and Gay (1977) believe that its accumulation around penetration pegs (along with lipid infusions) is a wound response related to the control of water loss. The separation of silicon as a defense response and as a wound response has not been clearly resolved.

Lignin has been identified often in halos and across broader areas of the cell wall after wounding or inoculation with pathogens and nonpathogens (for examples see: Bird and Ride, 1981; Hammerschmidt, 1984; Hammerschmidt and Kuć, 1982; Ride, 1975; Ride and Barber, 1987; Ride and Pearce, 1979; Sherwood and Vance, 1976, 1980). In barley leaves and coleoptiles attacked by *E. graminis* f. sp. *hordei*, however, halos are not lignified (where they occur) (Mayama and Shishiyama, 1978; Smart *et al.*, 1986a), although they are autofluorescent.

The deposition of lignin in wheat leaves in response to wounding and subsequent inoculation by various fungi, and the role lignin may play in resistance in grasses have been investigated extensively by Ride and co-workers (Bird and Ride, 1981; Pearce and Ride, 1980, 1982; Ride, 1975, 1983; Ride and Barber, 1987; Ride and Pearce, 1979). Filamentous fungi (including nonpathogens) and various abiotic elicitors, especially chitin, rapidly induce lignin formation in a ring covering up to several cells when the inoculum is placed on compression wounds. Whereas nonpathogens are restricted to the reaction zone, pathogens such as *Septoria* slowly spread from the wound (Bird and Ride, 1981; Pearce and Ride, 1980; Ride, 1975). Recently, Ride and Barber (1987) induced lignification in wheat leaves with either chitosan or *Botrytis cinerea* and subsequently challenge inoculated with *Fusarium graminearum* or *Penicillium oxalicum*. Growth of the two pathogens was inversely correlated with lignification (the latter measured quantitatively; Barber and Ride, 1988). Suppression of the lignification response with the metabolic inhibitor cycloheximide resulted in susceptibility to nonpathogens but, interestingly, attempts to duplicate this with α-aminooxy acetate or α-aminooxy-β-propionate were unsuccessful. The authors liken lignification in wheat to the accumulation of phytoalexins in dicotyledons in response to pathogen attack, although they state that factors other than lignification "cannot be ruled out" in resistance (Ride and Barber, 1987).

Lignin is also involved in other host–parasite situations. When the lower leaves of cucumbers are inoculated with *Colletotrichum lagenarium*, systemic resistance is induced in younger leaves (Caruso and Kuć, 1979; Kuć *et al.*, 1975). Challenge inoculation of the upper plant with another dose of *C. lagenarium* or, indeed, other pathogens, dramatically reduces the disease incidence by reducing epidermal penetrations (Richmond *et al.*, 1979). Associated with these reductions is an appearance of lignin (detected histochemically) (Hammerschmidt and Kuć, 1982). Dean and Kuć (1987) vacuum infiltrated radiolabeled lignin precursors into cucumber leaf disks from protected and unprotected leaves. Incorporation into lignin was inferred from the recovery of [14C]vanillin and [14C]-*p*-hydroxy benzaldehyde following alcoholic nitrobenzene oxidation. Protected leaf disks (those with prior inoculation) incorporated precursors faster and accumulated more label than did unprotected disks. The workers concluded that rapid lignification is a component of the systemic resistance response (Dean and Kuć, 1987).

The evidence for some role of lignin in resistance in cucumbers and wheat is thus strong, there being good correlative and some experimental evidence in both cases. We must remember, however, that these experiments relate to *challenge inocula*, and not primary inocula. That is, the plants are "sensitized" by some pretreatment before the role of lignin is investigated. In fact, Ride and Barber (1987) themselves note that, without wounding, challenge inoculum is unimpeded by halos and papillae induced by *Botrytis*.

Another cell response which is not particularly confined to halos, although it includes them, is the accumulation of protein. In particular, the hydroxyproline-rich glycoproteins (HPRGs) have been shown to increase in response to a wide variety of pathogens (Mazau and Esquerre-Tugaye, 1986; see Chapter 11) in both susceptible and resistant situations. HPRGs are normal constituents of cell walls (Tierney and Varner, 1987) and have been proposed to be a major structural element of the cell wall (Lamport and Epstein, 1983). HPRGs increase faster in melons in which resistance has been induced (Esquerré-Tugayé et al., 1979), paralleling the lignin story of Kuć and others (above). Interestingly, there were no significant increases in HPRG levels of monocotyledons upon pathogen inoculation (Mazau and Esquerre-Tugaye, 1986). As far as I am aware, experimental manipulation of this response is lacking; its discovery is so new that workers are still exploring the limits of the phenomenon. In fact, there may be a connection between lignification and HPRG enhancement in certain situations (see Hammerschmidt et al., 1984).

4.2. Apposition of Wall-like Materials: Composition and Function

Papillae are wall appositions deposited in localized areas between the plasmalemma and the cell wall at the site of attempted penetration (Aist, 1976). They are heterogeneous in texture and staining properties at the ultrastructural level (Aist, 1976; Bushnell and Berquist, 1975) but are large enough to be visible easily at the light microscope level of resolution (de Bary, 1863). It was de Bary who first suggested that papillae may function in disease resistance after he noticed the co-incidence of penetration failure and papilla formation (de Bary, 1863). Many others have since correlated resistance (the cessation of fungal growth) with papilla formation (for examples in a wide range of host–parasite interactions, see Aist and Israel, 1977a,b; Aist et al., 1979; Akai et al., 1968; Allen and Friend, 1983; Coffey and Wilson, 1983; Defosse, 1971, 1976; Heath, 1979; Mayama and Shishiyama, 1978; Ride, 1983; Ride and Pearce, 1979; Sherwood and Vance, 1980). However, we need to establish the components and properties of papillae in order to identify candidate molecules involved in resistance.

One is struck by the compositional heterogeneity of papillae in the literature (see also Fig. 1). In particular, papillae seem to result from the compression of cell wall ontogeny into a relatively short time span. Thus, papillae often simultaneously contain wall materials characteristic of young, primary walls and of older, even secondary, walls. In the first category, many papillae contain "callose" (but the reader must bear in mind that few investigators do more than stain with aniline blue) (Aist and Williams, 1971; Sherwood and Vance, 1976; Skou, 1982). More definitive identifications of callose have been made (Beckman et al., 1982; Hackler and Hohl, 1982; Hinch and Clarke, 1982; Smart et al., 1986a). Pectic substances, too, have been identified histochemically (Sherwood and Vance, 1976; Smart et al., 1986a). Components characteristic of older cell walls identified in papillae by histochemical or biochemical means are: lignin (Hammerschmidt, 1984; Mayama and Shishiyama, 1978; Ride, 1975; Ride and Pearce, 1979; Ride and Barber, 1987; Sherwood and Vance, 1976), suberin (Fellows, 1928), and protein (Esquerre-Tugaye et al., 1979; Mazau and Esquerre-Tugaye, 1986).

It has long been though that callose is a ubiquitous component of papillae (Aist, 1976; Mangin, 1895). As has been alluded to above, there are some problems associated with its identification, problems which are seldom addressed. It is useful to review these problems briefly here since their resolution provide instructive lessons in the interpretation of histochemical results. The most common method of identifying callose is by staining tissues with aniline blue and examining them at high pH with near-UV radiation and appropriate filters (Allen and Friend, 1983; Aspinall and Kessler, 1957; Beckman et al., 1982; Eschrich and Currier, 1964; Hachler and Hohl, 1982; Knox and Heslop-Harrison, 1970). Under these conditions, callose fluoresces but there is also some nonspecific background staining present (Faulkner et al., 1973; Smith and McCully, 1978). Tissues should be checked for autofluorescence, particularly if lignin (or other phenolic material) is present. Prior treatment of tissues with the periodic acid–Schiff reagent eliminates autofluorescence and nonspecific binding, thus increasing the power of aniline blue as an identifier (Smith and McCully, 1978). The use of enzymes which can degrade callose confirms its presence (Hinch and Clarke, 1982; Smart et al., 1986b). But what is "callose"? If the treatments just described are performed, one can conclude only that $(1\rightarrow3)$-β-D-glucosyl linkages are present (Smart et al., 1986b). For many years the only biochemical characterization of aniline blue deposits was that of Aspinall and Kessler (1957) who isolated the plugs formed in sieve tubes of grape phloem at winter dormancy. They found that the plugs were almost pure $(1\rightarrow3)$-β-D-glucans (Aspinall and Kessler, 1957). Since then, Vithanage and co-workers (1980) have isolated similarly aniline blue-positive plugs formed in *Secale* pollen after self-pollination of the flower. These deposits are less than 10% w/v $(1\rightarrow3)$-β-D-glucan. Thus, similarity of aniline blue staining is no indicator of $(1\rightarrow3)$-β-D-glucan content. The "family" of aniline blue-positive compounds (which we may call "callose" as a collective noun) will not be chemically identical. Such a lack of chemical identity has obvious implications in the resulting physical properties claimed for callose.

The chief physical attribute of callose is its supposed impermeability and this has been suggested to inhibit the passage of small molecules and ions in papillae (Allen and Friend, 1983; Beckman, 1980; Skou, 1982). Unfortunately, the only basis for this claim is work by Knox and Heslop-Harrison (1970) on developing pollen grains. These workers found that fluorescein diacetate was excluded from the protoplast only when the surrounding walls were "callosic" (i.e., "aniline blue-positive"). The chemical composition of this wall is not known and the correlation may be fortuitous. We are reminded of the warning of O'Brien (1972) about the apparent ease with which inconclusive, but well-cited, results become truisms in science. What proportion of $(1\rightarrow3)$-β-D-glucans is necessary to reduce porosity in papillae? We don't know. The only experimental investigation of permeability in papillae, which had been shown to contain callose, failed to show impermeability to small ions and molecules of less than 390 daltons (Smart et al., 1986b, 1987).

The demonstration of callose, $(1\rightarrow3)$-β-D-glucans, is most specific using combined periodic acid–Schiff with aniline blue and specific $(1\rightarrow3)$-β-glucanase treatment (Smart et al., 1986b) but other methods have been employed, usually using lacmoid or resorcinol blue (Skou, 1982; Eschrich and Currier, 1964), but these stains must be differentiated with acid for any reasonable specificity (O'Brien and McCully, 1981; see Sherwood and Vance, 1976; Smart et al., 1986a).

Another usual suggestion is that callose sequesters toxins, including phenolics (Aist, 1976; Allen and Friend, 1983; Ride, 1983), resulting in the resistance of papillae, and hence the host, to penetration. It is true that treatment of barley coleoptile papillae with laminarinase [a $(1\rightarrow3)$-β-glucanase] will remove papilla, but not cell wall autofluorescence, thereby indicating some chemical interaction between callose and the autofluorescent moiety. Nevertheless, removal of callose and autofluorescence from preformed, impenetrable papillae (Aist et al., 1979) leaves them unassailable (Smart et al., 1986b). Callose is therefore not necessary for the expressed resistance. Perhaps other experimental evidence for or against a role for callose in resistance will be forthcoming.

As is true for halos, lignin is thought to play a direct role in the observed correlation of penetration failure and papilla incidence. The same relatively imprecise histochemical techniques that are used for halos are usually employed to identify lignin (or its absence) in papillae (Bird and Ride, 1981; Hachler and Hohl, 1982; Hammerschmidt, 1984; Ride, 1975; Ride and Pearce, 1979; Sherwood and Vance, 1976; Smart *et al.*, 1986a; Vance *et al.*, 1980). Barber and Ride (1988) recently refined techniques for the biochemical identification of lignin. The argument relevant to halos and wider responses (above) applies here and, in essence, these workers believe that lignin in papillae has a substantial role in host resistance to penetration. This subject has been reviewed in depth (Vance *et al.*, 1980). With reference to lignified papillae in particular, Ride and Barber (1987) do not regard them as sites of successful *induced* resistance after initial inoculation with a nonpathogen. This is presumably because the area occupied by the papillae is small in comparison with that of the upper epidermal periclinal wall as a whole.

Other substances have been reported in papillae, as for halos. The data for silicon and suberin have been detailed above and will not be repeated here. Protein (Esquerre-Tugaye *et al.*, 1979; Mazau and Esquerre-Tugaye, 1986), lipid (Fellows, 1928), and other carbohydrates (Defosse, 1971, 1976; Hachler and Hohl, 1982; Mayama and Shishiyama, 1978; Ride and Pearce, 1979; Sherwood and Vance, 1976) have all been reported. The HPRG of Esquerre-Tugaye and co-workers has been discussed above while lipids are seldom seen in papillae, except as degenerative membrane fragments (Aist, 1983). As far as I know, general carbohydrates have never been proposed as the cause of resistance of papillae to penetration. The infinite possibilities of bond types and acetylation patterns, mentioned in Section 2.1, should be remembered. Experimental studies on the role of papillae, as a whole, are almost exclusively confined to powdery mildew diseases and rusts and the reader is referred to Chapter 15 by Aist and Bushnell for discussion.

5. COOPERATIVE RESISTANCE

Contiguous cells often respond coordinately to wounding or attempted penetration by filamentous fungi. The work of Ride and co-workers has already been discussed at some length (Section 4.1). The cooperative nature of cell responses is also evident in vascular wilt diseases, pathogenesis in the secondary plant body (secondary roots and stems) and in the shot-hole diseases of leaves. Much of this work is necessarily descriptive and correlative; experimental evidence for or against a particular process is largely lacking. The discussion which follows is therefore brief and parts have been reviewed elsewhere (Beckman and Talboys, 1981).

5.1. Cooperative Barrier Zones

The erection of cooperative barriers to wall-off wounds and pathogen-caused lesions is a common occurrence in a wide range of plants (see Fig. 2; Pearce, 1987). Generally, healthy cells at the edge of wounds begin to thicken their walls and, in angiosperms at least, lignify and (or) suberize them (Biggs, 1984, 1987; Cunningham, 1928; Pearce, 1987; Pearce and Rutherford, 1981; Woodward and Pearce, 1988a). This cicatrix, as it is properly called (Cunningham, 1928; Esau, 1977), seals off the underlying healthy tissue. This is demonstrable with the "F-F" test which uses ferric chloride and ferricyanide ions in a sequence at acid pH. The test purportedly demonstrates wall impermeability (Mullick, 1975) at sites of no color. The acidic conditions of the test, however, would be expected to dramatically alter the physical characteristics of all walls from those in their native (i.e., wounded) state: Walls with ferric ferricyanide coloration are not necessarily permeable *in vivo*. In gymnosperms, "impermeable" cicatrix walls are not lignified or suberized (Mullick, 1975). This layer in gymnosperms is referred to as the "non-suberized impermeable layer (NIT)" by Mullick (1975, 1977). However, Woodward and Pearce (1988a) have evidence that phenolics may mask suberin deposits in these walls, at least in Sitka spruce.

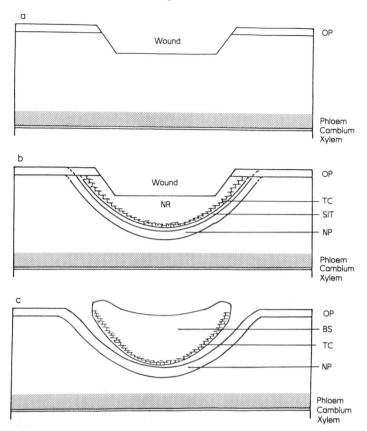

Figure 2. Sequence of events leading to wound healing in Sitka spruce. Intact cells (TC) at the wound's edge thicken, and often suberize, their cell walls about 24–36 days after wounding (panel b). A new periderm forms also (NP), one of the products being the suberized impermeable tissue (SIT). The NP reestablishes continuity with the old periderm (OP). Eventually (panel c), the necrotic area of the wound (NP) sloughs off as a scale (BS). (From Woodward and Pearce, 1988, with permission.)

After the formation of the cicatrix, a new cambium appears beneath (away from) the wound/ pathogen (Biggs, 1985; Cunningham, 1928; Mullick, 1975; Pearce and Rutherford, 1981; Woodward and Pearce, 1988a). The position of this cambium is presumably under hormonal control and ultimately it restores the continuity of phellem and phelloderm, thus usually (but not always) walling off the wound or pathogen (Biggs, 1987; Bostock and Middleton, 1987; Cunningham, 1928; Jewell, 1988; Pearce and Rutherford, 1981; Woodward and Pearce, 1988a). Mullick termed this cambium the "necrophylactic periderm" (Mullick, 1975, 1977).

The coincidence of cessation of fungal growth and suberization (or lignification since in some cases the two processes may be synonymous; see Aist, 1983) implies a role in resistance. Indeed the correlation is extremely good (see references above) but experimental evidence is lacking. Pearce and Rutherford (1981) have shown that *Stereum gausapatum*, a pathogen of oak, induces suberization in cells which it cannot then penetrate and that, *in vitro*, *S. gausapatum* is less able to utilize suberized cell walls than suberin-extracted walls. Of course, other responses such as phenolic accumulation (apart from lignification) or other secondary metabolites may be involved in resistance (e.g., stilbenes in spruce; Woodward and Pearce, 1988b). Necrophylactic periderms and, perhaps, cicatrices would then be relegated to roles in wound healing (Mullick, 1977). In this context, more aggressive pathogens are known which appear to repeatedly penetrate induced cicatrices and NITs (Biggs, 1987; Jewell, 1988).

5.2. Vascular Wilt Diseases

Symptoms of vascular wilt diseases, yellowing of leaves and wilting, are caused in part by the disruption of the transpiration stream. Pathogen propagules in the xylem vessels cause gum formation and tylose development whereas wall appositions form in the xylem parenchyma on a structural level and phytoalexin accumulation accompanied by hormonal changes occur on a biochemical level (Beckman and Talboys, 1981; Harrison and Beckman, 1987). The diseases are obviously quite complex and I intend to focus only on their structural aspects here. Vascular wilt diseases are widespread in plants, including economically important crops (such as banana, tomato, and beans). They are caused by members of the species *Verticillum albo-atrum*, *Fusarium oxysporum*, and *Ceratocystis* spp.

These pathogens generally invade plants at the roots. Bishop and Cooper (1983a) showed that in resistant and susceptible tomato cultivars, there are few differences in host responses to invasion by *Fusarium* spp. or *Verticillum albo-atrum*, at least until the stele is reached. The pathogens penetrate between the anticlinal walls of the epidermal cells and grow, largely intercellularly, through the cortex to the stele. Although mature endodermal cells are the site of many penetration failures (presumably because of the suberized, preformed, Casparian band), the pathogens can penetrate sometimes and, in any case, can grow up the stele from penetrations closer to the root apex where the Casparian band is not mature (Bishop and Cooper, 1983a). It is not until tyloses appear that vertical resistance is expressed, generally in the aerial organs (Bishop and Cooper, 1983b, 1984).

Tyloses are outgrowths of xylem parenchyma cells pushing through the pit fields into the lumina of xylem vessels (VanderMolen *et al.*, 1987). These cells have thin walls initially but usually thicken, lignify and (or) suberize them (Bishop and Cooper, 1984; VanderMolen *et al.*, 1987). Once the vessel lumen is occluded in this fashion, the transpiration stream ceases to flow in that file. If the occlusion is timely (Bishop and Cooper, 1984), the pathogen is confined, and the plant resistant. Usually the hyphae do not penetrate or colonize xylem parenchyma cells or neighboring vessels until late in disease development (Bishop and Cooper, 1983b, 1984). Why this should be so is uncertain but a "coating layer" of electron-dense material is often found deposited on lateral (periclinal) vessel walls and pit fields (Bishop and Cooper, 1983a; Newcombe and Robb, 1988; Robb *et al.*, 1987). This material is said to contain phenolics or even suberin and in alfalfa and tomatoes is thought to be responsible for the lateral resistant response (Robb *et al.*, 1987). The authors claim from time course evidence that resistance precedes gelation and tylose formation by some hours. The complexities of these arguments are beyond the scope of this chapter except that the evidence for the involvement of structural responses in resistance is quite strong, albeit circumstantial.

6. CONCLUDING REMARKS

The evidence for the involvement of cell walls and their various modifications in resistance is quite strong at the levels of correlative data and event timings. As Aist (1983) explained clearly, in order to prove a role in resistance for any of the factors we have discussed (halos, papillae, cicatrices, NITs)—or their components—one must somehow remove them, singly, by mutation, DNA manipulation, inhibition, or chemical removal and then retest the resistance correlations and (or) timing. Before doing this, the original resistance correlations and timing need to be firmly established. These initial studies are crucial and, as always, the more quantitative they are, the better our confidence in them since such data are amenable to statistical analysis. It is no accident that most quantitative data have been gathered from *in vivo* observations of the epidermal parasites, the powdery mildews, and from cleared leaves of rust fungus-infected plants. The superb optics obtainable in living coleoptiles or cleared leaves have facilitated great advances in our understanding of these diseases (see Chapter 15). But especially in trees we have to infer timing of events and their correlation with resistance from carefully planned and executed microscopy of fixed, embedded material since *in vivo* microscopy is obviously impossible.

It is particularly exciting that the recent advances in immunobiology, lectins, and molecular biology have created the possibility of a "new" histochemistry, more precise and more specific than was generally true of classical histochemistry. The carbohydrate structure of the wall is amenable to dissection with lectins which becomes a powerful technique if used in combination with enzyme hydrolysis. Monoclonal antibodies can be produced, in theory, from virtually any macromolecule, provided it can be purified sufficiently. With these two methods it should be possible to discover subtle differences in the composition of walls and their modifications at encounter sites in near-isogenic resistant and susceptible interactions. But perhaps the most exciting possibilities are afforded by the molecular technique of *in situ* hybridization. The cDNA-mediated localization of mRNA should allow more accurate inferences to be made about the timing of specific defense responses, lignification for example, by revealing the timing and location of translation. Of course, the necessary clones require an enormous amount of work but their generation should help unite biochemistry with microscopy, allowing a new perspective in disease resistance studies.

ACKNOWLEDGMENTS. I thank June Stewart for her critical reading of the manuscript and Marsha Ebener for her fast, accurate typing. As always, the kidney donation has made this possible.

7. REFERENCES

Aist, J. R., 1976, Papillae and related wound plugs of plant cells, *Annu. Rev. Phytopathol.* **14**:145–163.

Aist, J. R., 1983, Structural responses as resistance mechanisms, in: *The Dynamics of Host Defense* (J. A. Bailey and B. J. Deverall, eds.), Academic Press, New York, pp. 33–70.

Aist, J. R., and Israel, H. W., 1977a, Timing and significance of papilla formation during host penetration by *Olpidium brassicae*, *Phytopathology* **67**:187–194.

Aist, J. R., and Israel, H. W., 1977b, Papilla formation: Timing and significance during penetration of barley coleoptiles by *Erysiphe graminis hordei*, *Phytopathology* **67**:455–461.

Aist, J. R., and Williams, P. H., 1971, The cytology and kinetics of cabbage root hair penetration by *Plasmodiophora brassicae*, *Can. J. Bot.* **49**:2023–2034.

Aist, J. R., Kunoh, H. and Israel, H. W., 1979, Challenge appressoria of *Erysiphe graminis* fail to breach preformed papillae of a compatible barley cultivar, *Phytopathology* **69**:1245–1250.

Akai, S., and Fukutomi, M., 1980, Preformed internal physical defenses, in: *Plant Disease: An Advanced Treatise* (J. G. Horsfall and E. B. Cowling, eds.), Academic Press, New York, pp. 139–159.

Akai, A., Kunoh, H., and Fukutomi, M., 1968, Histochemical changes in the epidermal cell wall of barley leaves infected with *Erysiphe graminis hordei*, *Mycopathol. Mycol. Appl.* **35**:175–180.

Allen, F. H. E., and Friend, J., 1983, Resistance of potato tubers to infection by *Phytophthora infestans*: A structural study of haustorial encasement, *Physiol. Plant Pathol.* **22**:282–292.

Ashford, A. E., and Jacobsen, J. V., 1974, Cytochemical localization of phosphate in barley aleurone cells: The pathway of gibberellic acid-induced enzyme release, *Planta* **120**:81–105.

Aspinall, G., and Kessler, G., 1957, The structure of callose from the grape vine, *Chem. Ind. (London)* **1957**:1296.

Bacic, A., and Stone, B. A., 1981, Chemistry and organization of aleurone cell wall component from wheat and barley, *Aust. J. Plant Physiol.* **8**:475–495.

Bacon, J. S., Gordon, A. H., Morris, E. J., and Farmer, V. C., 1975, Acetyl groups in cell-wall preparations from higher plants, *Biochem. J.* **149**:485–487.

Bacon, J. S., Gordon, A. H., and Hay, A. J., 1976, Acetyl groups in dietary polysaccharides, *Proc. Nutr. Soc.* **35**:93A–94A.

Bajer, A. S., 1968, Fine structure studies on phragmoplast and cell plate formation, *Chromosoma* **24**:383–417.

Bajer, A. S., and Mole-Bajer, J., 1972, Spindle dynamics and chromosome movement, *Int. Rev. Cytol. Suppl.* **3**.

Barber, M. S., and Ride, J. P., 1988, A quantitative assay for induced lignification in wounded wheat leaves and its use to survey potential elicitors of the response, *Physiol. Mol. Plant Pathol.* **32**:185–197.

Bateman, D. F., and Lumsden, R. D., 1965, Relation of calcium content and nature of pectic substances in bean hypocotyls of different ages to susceptibility to an isolate of *Rhizoctonia solani*, *Phytopathology* **55**:734–738.

Beckmann, C. H., 1980, Defenses triggered by the invader: physical defenses, in: *Plant Disease: An Advanced Treatise* (J. G. Horsfall and E. B. Cowling, eds.), Academic Press, New York, pp. 224–245.

Beckman, C. H., and Talboys, P. W., 1981, Anatomy of resistance, in: *Fungal Wilt Diseases of Plants* (M. E. Mace, A. A. Bell, and C. H. Beckman, eds.), Academic Press, New York, pp. 487–521.

Beckman, C. H., Mueller, W. C., Tessier, B. J., and Harrison, N. A., 1982, Recognition and callose deposition in response to vascular infection in fusarium-wilt resistant or susceptible tomato plants, *Physiol. Plant Pathol.* **20**:1–10.

Bell, A. A., 1981, Biochemical mechanisms of disease resistance, *Annu. Rev. Plant Physiol.* **32**:21–81.

Biggs, A. R., 1984, Boundary-zone formation in peach bark in response to wounds and *Cytospora leucostoma* infection, *Can. J. Bot.* **62**:2814–2821.

Biggs, A. R., 1985, Suberized boundary zones and the chronology of wound response in tree bark, *Phytopathology* **75**:1191–1195.

Biggs, A. R., 1987, Occurrence and location of suberin in wound reaction zones of xylem in 17 tree species, *Phytopathology* **77**:718–725.

Bird, P. M., and Ride, J. P., 1981, The resistance of wheat to *Septoria nodorum*: Fungal development in relation to host lignification, *Physiol. Plant Pathol.* **19**:289–299.

Bishop, C. D., and Cooper, R. M., 1983a, An ultrastructural study of root invasion in three vascular wilt diseases, *Physiol. Plant Pathol.* **22**:15–27.

Bishop, C. D., and Cooper, R. M., 1983b, An ultrastructural study of vascular colonization in three vascular wilt diseases. I. Colonization of susceptible cultivars, *Physiol. Plant Pathol.* **23**:323–343.

Bishop, C. D., and Cooper, R. M., 1984, An ultrastructural study of vascular wilt diseases. II. Colonization of resistant cultivars, *Physiol. Plant Pathol.* **24**:277–289.

Bostock, R. M., and Middleton, G. E., 1987, Relationship of wound periderm formation to resistance to *Ceratocystis fimbriata* in almond bark, *Phytopathology* **77**:1174–1180.

Bouveng, H. O., 1961, Phenyl iso-cyanate derivatives of carbohydrates. II. O-ethyl groups in birch xylan, *Acta Chem. Scand.* **15**:96–100.

Bushnell, W. R., and Berquist, S. E., 1975, Aggregation of host cytoplasm and the formation of papillae and haustoria in powdery mildew of barley, *Phytopathology* **65**:310–318.

Carpita, N., Sabularse, D., Montezinos, D., and Delmer, D. P., 1979, Determination of the pore size of the cell walls of living plant cells, *Science* **205**:1144–1147.

Caruso, F. L., and Kuc, J., 1979, Induced resistance of cucumber to anthracnose and angular leaf spot by *Pseudomonas lachrymans* and *Colletotrichum lagenarium*, *Physiol. Plant Pathol.* **14**:191–201.

Carver, T. L. W., Zeyen, R. J., and Ahlstrand, G. G., 1987, The relationship between insoluble silicon and success or failure of attempted primary penetration by powdery mildew (*Erysiphe graminis*) germlings on barley, *Physiol. Mol. Plant Pathol.* **31**:133–148.

Chou, C. K., 1970, An electron-microscope study of host penetration and early stages of haustorium formation of *Peronospora parasitica* (fr.) Tul. on cabbage cotyledons, *Ann. Bot.* **34**:189–204.

Clarke, A. E., Anderson, R. L., and Stone, B. A., 1979, Form and function of arabinogalactans and arabinogalactan-proteins, *Phytochemistry* **1**:175–178.

Coffey, M. D., and Wilson, U. E., 1983, An ultrastructural study of the late blight fungus *Phytophthora infestans* and its interaction with the foliage of two potato cultivars possessing different levels of general (field) resistance, *Can. J. Bot.* **61**:2669–2685.

Colvin, J. R., 1981, Ultrastructure of the plant cell wall: Biophysical viewpoint, in: *Encyclopedia of Plant Physiology, Plant Carbohydrates II, Extracellular Carbohydrates* (W. Tanner and F. A. Loweus, eds.), New Series, Volume 13B, Springer-Verlag, Berlin.

Conway, W. S., Gross. K. C., Boyer, C. D., and Sams, C. E., 1988, Inhibition of *Penicillium expansum* polygalacturonase activity by increased apple cell wall calcium, *Phytopathology* **78**:1052–1055.

Crowley, S. H., 1977, Translocation, in: *Systematic Fungicides*, 2nd ed. (R. W. Marsh, ed.), Longman, London.

Cunningham, H. S., 1928, A study of the histologic changes in leaves caused by certain leaf-spotting fungi, *Phytopathology* **18**:717–752.

Dean, R. A., and Kuc, J., 1987, Rapid lignification in response to wounding and infection as a mechanism for induced systemic protection in cucumber, *Physiol. Mol. Plant Pathol.* **31**:69–81.

de Bary, A., 1863, Recerches sur le developpement de quelques champignons parasites, *Ann. Sci. Nat. Bot. Biol. Veg.* **20**:5–148.

Defosse, L., 1971, Recherches histochemiques sur la penetration du *Cercosporella herpotrichoides* fron. dans les gaines foliaires des cereales, *Phytopathol. Z.* **70**:1–10.

Defosse, L., 1976, Recherches histochemiques sur la nature des reactions parietales formees en reponse a la penetration de *Cercosporella herpotrichoides* fron. dans le coleoptile du ble, *Parasitica* **32**:147–157.

Douglas, S. M., Sherwood, R. T., and Lukezic, F. L., 1984, Effect of adult plant resistance on primary penetration of oats by *Erysiphe graminis* f. sp. *avenae*, *Physiol. Plant Pathol.* **25**:219–228.

Esau, K., 1965, *Plant Anatomy*, 2nd ed., Wiley, New York.

Esau, K., 1977, *Anatomy of Seed Plants*, 2nd ed., Wiley, New York.

Eschrich, W., and Currier, H. B., 1964, Identification of callose by its diachrome and fluorochrome reactions, *Stain Technol.* **39**:303–307.

Esquerré-Tugayé, M. T., Lafitte, C., Mazau, D., Toppan, A., and Touze, A., 1979, Cell surfaces in plant–microorganism interactions. II. Evidence for the accumulation of hydroxyproline-rich glycoproteins in the cell wall of diseased plants as a defense mechanism, *Plant Physiol.* **64**:320–326.

Fan, D. F., and MacLachlan, G. A., 1967, Studies on the regulation of cellulase activity and growth in excised pea epicotyl sections, *Can. J. Bot.* **45**:1837–1844.

Faulkner, G., Kimmins, W. C., and Brown, R. G., 1973, The use of fluorochromes for the identification of $\beta(1\rightarrow3)$ glucans, *Can. J. Bot.* **51**:1503–1504.

Fellows, H., 1928, Some chemical and morphological phenomena attending infection of the wheat plant by *Ophiobolus graminis*, *J. Agric. Res.* **37**:647–661.

Fincher, G. B., and Stone, B. A., 1981, Metabolism of noncellulosic polysaccharides, in: *Encolpedia of Plant Physiology, Plant Carbohydrates II, Extracellular Carbohydrates* (W. Tanner and F. A. Loweus, eds.), New Series, Volume 13B, Springer-Verlag, Berlin.

Fulcher, R. G., O'Brien, T. P., and Lee, J. W., 1972, Observations on the aleurone layer. I. Conventional and fluorescence microscopy of the cell wall with emphasis on phenol–carbohydrate complexes in wheat, *Aust. J. Biol. Sci.* **25**:23–34.

Fulcher, R. G., McCully, M. E., Setterfield, G., and Sutherland, J., 1976, β-1,3-glucans may be associated with cell plate formation during cytokinesis, *Can. J. Bot.* **54**:539–542.

Gaff, D. F., Chambers, T., and Markus, K., 1964, Studies on the extrafascicular movement of water in the leaf, *Aust. J. Biol. Sci.* **17**:581.

Griffey, R. T., and Leach, J. G., 1965, The influence of age on the development of bean anthracnose lesions, *Phytopathology* **55**:915–918.

Hachler, H., and Hohl, H. R., 1982, Histochemistry of papillae in potato tuber tissue infected with *Phytophthora infestans*, *Bot. Helv.* **92**:23–31.

Hammerschmidt, R., 1984, Rapid deposition of lignin in potato tuber tissue as a response to fungi non-pathogenic on potato, *Physiol. Plant Pathol.* **24**:33–42.

Hammerschmidt, R., and Kuć, J., 1982, Lignification as a mechanism for induced systemic resistance in cucumber, *Physiol. Plant Pathol.* **20**:61–71.

Hammerschmidt, R., Lamport, D. T. A., and Muldoon, E. P., 1984, Cell wall hydroxyproline enhancement and lignin deposition as an early event in the resistance of cucumber to *Cladosporium cucumerincum*, *Physiol. Plant Pathol.* **24**:43–47.

Harris, P. J., and Hartley, R. D., 1976, Detection of bound ferulic acid in cell walls of Gramineae by ultraviolet fluorescence microscopy, *Nature* **259**:508–510.

Harrison, N. A., and Beckman, C. H., 1987, Growth inhibitors associated with fusarium wilt of tomato, *Physiol. Mol. Plant Pathol.* **30**:401–420.

Hayes, J. D., and Jones, I. T., 1966, Variation in the pathogenicity of *Erysiphe graminis* D.C. f. sp. *avenae* and its relation to the development of mildew resistant oat cultivars, *Euphytica* **15**:80–86.

Heath, M., 1979, Partial characterization of the electron-opaque deposits formed in the non-host plant, French bean, after cowpea rust infection, *Physiol. Plant Pathol.* **15**:141–148.

Heath, M., 1980a, Reactions of non-suscepts to fungal pathogens, *Annu. Rev. Phytopathol.* **18**:211–236.

Heath, M., 1980b, Effects of infection by compatible species or injection of tissue extracts on the susceptibility of nonhost interactions with rust fungi, *Phytopathology* **70**:356–360.

Heath, M., 1981a, The suppression of the development of silicon-containing deposits in French bean leaves by exudates of the bean rust fungus and extracts from bean-rust infected tissue, *Physiol. Plant Pathol.* **18**:149–155.

Heath, M., 1981b, Insoluble silicon in necrotic cowpea cells following infection with an incompatible isolate of the cowpea rust fungus, *Physiol. Plant Pathol.* **19**:273–276.

Heyn, A. N. J., 1969, Glucanase activity in coleoptiles of Avena, *Arch. Biochem. Biophys.* **132**:442–449.

Higuchi, T., 1981, Biosynthesis of lignin, in: *Encyclopedia of Plant Physiology, Plant Carbohydrates II, Extracellular Carbohydrates* (W. Tanner and F. A. Loweus, eds.), New Series, Volume 13B, Springer-Verlag, Berlin.

Hinch, J. M., and Clarke, A. E., 1982, Callose formation in *Zea mays* as a response to infection with *Phytophthora cinnamomi*, *Physiol. Plant Pathol.* **21**:113–124.

Hite, R. E., Sherwood, R. T., and Marshall, H. G., 1977, Adult plant resistance to powdery mildew in "Dal" oats, *Plant Dis. Rep.* **61**:273–277.

Hursh, C. R., 1924, Morphological and physiological studies on the resistance of wheat to *Puccinia graminis tritici*, Erikss. and Henn., *J. Agric. Res.* **27**:381–411.

Jewell, F. F., 1988, Histopathology of fusiform-rust inoculated progeny from (Shortleaf × Slash) × Shortleaf pine crosses, *Phytopathology* **78**:396–402.

Johnson, L. E. B., Bushnell, W. R., and Zeyen, R. J., 1979, Binary pathways for analysis of powdery mildew infection and host response in populations of powdery mildew fungi, *Can. J. Bot.* **57**:497–511.

Johnson, L. E. B., Bushnell, W. R., and Zeyen, R. J., 1982, Defense patterns in nonhost higher plant species against two powdery mildew fungi. I. Monocotyledonous species, *Can. J. Bot.* **60**:1068–1083.

Kato, K., 1981, Ultrastructure of the plant cell wall: Biochemical viewpoint, in: *Encyclopedia of Plant Physiology Plant Carbohydrates II, Extracellular Carbohydrates* (W. Tanner and F. A. Loweus, eds.), New Series, Volume 13B, Springer-Verlag, Berlin.

Knox, R. B., and Heslop-Harrison, J., 1970, Direct demonstration of the low permeability of the angiosperm meiotic tetrad using a fluorogenic ester, *Z. Pflanzenphysiol.* **62**:451–459.

Kuć, J., Shockley, G., and Kearney, K., 1975, Protection of cucumber against *Colletotrichum lagenarium* by *Colletotrichum lagenarium*, *Physiol. Plant Pathol.* **7**:195–199.

Kunoh, H., and Akai, S., 1969, Histochemical observation of the halo on the epidermal wall of barley leaves attacked by *Erysiphe graminishordei*, *Mycopath. Mycol. Appl.* **37**:113–118.

Kunoh, H., and Ishizaki, H., 1975a, Silicon levels near penetration sites of fungi on wheat, barley, cucumber and morning glory leaves, *Physiol. Plant Pathol.* **5**:283–287.

Kunoh, H., and Ishizaki, H., 1975b, Composition analysis of 'halo' area of barley leaf epidermis incited by powdery mildew infection, *Ann. Phytopathol. Soc. Jpn.* **41**:33–39.

Labovitch, J. M., 1981, Cell wall turnover in plant development, *Annu. Rev. Plant Physiol.* **32**:385–406.

Lamport, D. T. A., and Catt, J. W., 1981, Glycoproteins and enzymes of the cell wall, in: *Encyclopedia of Plant Physiology, Plant Carbohydrates II, Extracellular Carbohydrates* (W. Tanner and F.A. Loweus, eds.), New Series, Volume 13B, Springer-Verlag, Berlin.

Lamport, D. T. A., and Epstein, L., 1983, A new model for the primary cell wall: A concatenated extensin–cellulose network, in: *First International Congress of Plant Molecular Biology*, University of Georgia Center for Continued Education, Athens, p. 91.

Lewin, J., and Remann, B. E. F., 1969, Silicon and plant growth, *Annu. Rev. Plant Physiol.* **20**:289–304.

McKeen, W. E., Smith, R., and Bhattacharya, P. H., 1968, Alterations of the host cell wall surrounding the infection peg of powdery mildew fungi, *Can. J. Bot.* **47**:701–706.

McNeil, M., Darvill, A. G., Fry, S. C., and Albersheim, P., 1984, Structure and function of the primary cell walls of plants, *Annu. Rev. Biochem.* **53**:625–663.

Mangin, L., 1895, Recherches sur les Peronosporees, *Bull. Soc. Hist. Nat. Autun.* **8**:55–108.

Markwalder, H. U., and Neukom, H., 1976, Diferulic acid as a possible crosslink in hemicelluloses from wheat germ, *Phytochemistry* **15**:836–837.

Mayama, S., and Pappelis, A. J., 1967, Application of interference microscopy to the study of fungal penetration of epidermal cells, *Phytopathology* **57**:1300–1302.

Mayama, S., and Shishiyama, J., 1978, Localized accumulation of fluorescent and UV-absorbing compounds at penetration sites in barley leaves infected with *Erysiphe graminis hordei*, *Physiol. Plant Pathol.* **13**:347–354.

Mazau, D., and Esquerre-Tugaye, M. T., 1986, Hydroxyproline-rich glycoprotein accumulation in the cell walls of plants infected by various pathogens, *Physiol. Plant Pathol.* **29**:147–157.

Morris, E. J., and Bacon, J. S., 1977, The fate of acetyl groups and sugar components during the digestion of grass cell walls in sheep, *J. Agric. Sci.* **89**:327–340.

Mullick, D. B., 1975, A new tissue essential to necrophylactic periderm formation in the bark of four conifers, *Can. J. Bot.* **53**:2443–2457.

Mullick, D. B., 1977, The non-specific nature of defense in bark and wood during wounding, insect and pathogen attack, *Recent Adv. Phytochem.* **11**:395–442.

Nemec, S., 1971, Mode of entry of *Pythium perniciosum* into strawberry roots, *Phytopathology* **61**:711–714.

Newcombe, A. G., and Robb, J., 1988, Vascular coating secretion: A resistance response in verticillium wilt of alfalfa, *Physiol. Mol. Plant Pathol.* **33**:47–58.

O'Brien, T. P., 1972, The cytology of cell-wall formation in some eukaryotic cells, *Bot. Rev.* **38**:87–117.

O'Brien, T. P., and Kuo, J., 1975, Development of the suberized lamella in the mestome sheath of wheat leaves, *Aust. J. Bot.* **23**:783–794.

O'Brien, T. P., and McCully, M. E., 1981, The study of plant structure: Principles and selected methods, *Termacarphi*, Melbourne.

Pearce, R. B., 1987, Antimicrobial defences in secondary tissues of woody plants, in: *Fungal Infection of Plants* (G. F. Pegg and P. G. Ayres, eds.), Cambridge University Press, London.

Pearce, R. B., and Ride, J. P., 1980, Specificity of induction of the lignification response in wounded wheat leaves, *Physiol. Plant Pathol.* **16**:197–204.

Pearce, R. B., and Ride, J. P., 1982, Chitin and related compounds as elicitors of the lignification response in wounded wheat leaves, *Physiol. Plant Pathol.* **20**:119–123.

Pearce, R. B., and Rutherford, J., 1981, A wound associated suberized barrier to the spread of decay in the sapwood of oak (*Quercus robur* L.), *Physiol. Plant Pathol.* **19**:359–370.

Preston, R. D., 1979, Polysaccharide conformation and cell wall function, *Annu. Rev. Plant Physiol.* **30**:55–78.

Ralton, J. E., Smart, M. G., and Clarke, A. E., 1987, Recognition and infection processes in plant–pathogen interactions, in: *Plant–Microbe Interactions: Molecular and Genetic Perspectives*, Volume 2 (T. Kosuge and E. W. Nester, eds.), Macmillan Co., New York.

Rees, D. A., and Welsh, E. J., 1977, Secondary and tertiary structure of polysaccharides in solutions and gels, *Angew. Chem.* **16**:214–224.

Richmond, S., Kuć, J., and Elliston, J. E., 1979, Penetration of cucumber leaves by *Colletotrichum lagenarium* is reduced in plants systemically protected by previous infection with the pathogen, *Physiol. Plant Pathol.* **14**:329–338.

Ride, J. P., 1975, Lignification in wounded wheat leaves in response to fungi and its possible role in resistance, *Physiol. Plant Pathol.* **5**:125–134.

Ride, J. P., 1983, Cell walls and other structural barriers in defence, in: *Biochemical Plant Pathology* (J. A. Callow, ed.), Wiley, New York, pp. 215–236.

Ride, J. P., and Barber, M. S., 1987, The effect of various treatments on induced lignification and the resistance of wheat to fungi, *Physiol. Plant Pathol.* **31**:349–360.

Ride, J. P., and Pearce, R. B., 1979, Lignification and papilla formation at sites of attempted penetration of wheat leaves by non-pathogenic fungi, *Physiol. Plant Pathol.* **15**:79–92.

Robb, J., Powell, D. A., and Street, P. F. S., 1987, Time course of wall-coating secretion in *Verticillium*-infected tomatoes, *Physiol. Mol. Plant Pathol.* **31**:217–226.

Roelefson, P. A., 1965, Ultrastructure of the wall in growing cells and its relation to the direction of the growth, *Adv. Bot. Res.* **2**:69–149.

Russo, V. M., and Pappelis, A. J., 1981, Observations of *Colletotrichum circinans* f. *dematium* on *Allium cepa*: Halo formation and penetration of epidermal cells, *Physiol. Plant Pathol.* **19**:127–136.

Sargant, J. A., and Gay, J. L., 1977, Barley epidermal apoplast structure and modification by powdery mildew contact, *Physiol. Plant Pathol.* **11**:195–206.

Sherwood, R. T., and Vance, C. P., 1976, Histochemistry of papillae formed in reed canary grass leaves in response to noninfecting pathogenic fungi, *Phytopathology* **66**:503–510.

Sherwood, R. T., and Vance, C. P., 1980, Resistance to fungal penetration in Gramineae, *Phytopathology* **70**:273–279.

Sherwood, R. T., and Vance, C. P., 1982, Initial events in the epidermal layer during penetration, in: *Plant Infection: The Physiological and Biochemical Basis* (Y. Asada, W. R. Bushnell, S. Ouchi, and C. P. Vance, eds.), Japan Scientific Societies Press, Tokyo, and Springer-Verlag, Berlin.

Shimony, C., and Friend, J., 1976, Ultrastructure of the interaction between *Phytophthora infestans* and tuber slices of resistant and susceptible cultivars of potato (*Solanum tuberosum* L.) Orion and Majestic, *Isr. J. Bot.* **25**:174–183.

Skou, J. P., 1982, Callose formation responsible for the powdery mildew resistance in barley with genes in the ml-o-locus, *Phytopathol. Z.* **104**:90–95.

Smart, M. G., and O'Brien, T. P., 1979a, Observations on the scutellum. III. Ferulic acid as a component of the cell wall in wheat and barley, *Aust. J. Plant Physiol.* **6**:485–491.

Smart, M. G., and O'Brien, T. P., 1979b, Observations on the scutellum. II. Histochemistry and autofluorescence of the cell wall in mature grain and during germination of wheat, barley, oats and ryegrass, *Aust. J. Bot.* **27**:403–411.

Smart, M. G., Aist, J. R., and Israel, H. W., 1986a, Structure and function of wall appositions. 1. General histochemistry of papillae in barley coleoptiles attacked by *Erysiphe graminis* f. sp. *hordei*, *Can. J. Bot.* **64**:793–801.

Smart, M. G., Aist, J. R., and Israel, H. W., 1986b, Structure and function of wall appositions. 2. Callose and the resistance of oversize papillae to penetration by *Erysiphe graminis* f. sp. *hordei*, *Can. J. Bot.* **64**:802–804.

Smart, M. G., Aist, J. R., and Israel, H. W., 1987, Some exploratory experiments on the permeability of papillae induced in barley coleoptiles by *Erysiphe graminis* f. sp. *hordei*, *Can. J. Bot.* **65**:745–749.

Smith, M. M., and McCully, M. E., 1978, A critical evaluation of the specificity of aniline blue-induced fluorescence, *Protoplasma* **95**:229–254.

Stumpf, M. A., and Heath, M., 1985, Cytological studies of the interactions between the cowpea rust fungus and silicon-depleted French bean plants, *Physiol. Plant Pathol.* **27**:369–385.

Taiz, L., and Jones, R. L., 1970, Gibberellic acid, β-1,3-glucanase and the cell walls of barley aleurone layers, *Planta* **92**:73–84.

Tepfer, M., and Taylor, I. E. P., 1981, The permeability of plant cell walls as measured by gel filtration chromatography, *Science* **213**:761–763.

Tierney, M. L., and Varner, J. E., 1987, The extensins, *Plant Physiol.* **84**:1–2.

Vance, C. P., Kirk, T. K. and Sherwood, R. T., 1980, Lignification as a mechanism of disease resistance, *Annu. Rev. Phytopathol.* **18**:259–288.

VanderMolen, G. E., Beckman, C. H., and Rodehorst, E., 1987, The ultrastructure of tylose formation in resistant banana following inoculation with *Fusarium oxysporum* f. sp. *cubense*, *Physiol. Mol. Plant Pathol.* **31**:185–200.

Vithanage, H. I. M. U., Gleeson, P. A., and Clarke, A. E., 1980, The nature of callose produced during self-pollination in *Secale cereale*, *Planta* **148**:498–505.

Volk, R. J., Kahn, R. P., and Weintraub, R. L., 1958, Silicon content of the rice plant as a factor influencing its resistance to infection by the blast fungus, *Pyricularia oryzae*, *Phytopathology* **48**:179–184.

Wallner, S. J., and Walker, J. E., 1975, Glycosidases in cell wall-degrading extracts of ripening tomato fruits, *Plant Physiol.* **55**:94–98.

Wardrop, A. B., 1971, Lignins in the plant kingdom. Occurrence and formation in plants, in: *Lignins: Formation, Structure and Reactions* (K. V. Sarkanen and C. H. Ludwig, eds.), Wiley, New York.

Willison, J. H. M., 1981, Secretion of cell wall material in higher plants, in: *Encyclopedia of Plant Physiology, Plant Carbohydrates II, Extracellular Carbohydrates* (W. Tanner and F. A. Loewus, eds.), New Series, Volume 13B, Springer-Verlag, Berlin.

Woodward, S., and Pearce, R. B., 1988a, Wound-associated responses in Sitka spruce root bark challenged with *Phaseolus schweinitzii*, *Physiol. Mol. Plant Pathol.* **33**:151–162.

Woodward, S., and Pearce, R. B., 1988b, The role of stilbenes in resistance of Sitka spruce (*Picea sitchensis* (Bong.) Carr) to entry of fungal pathogens, *Physiol. Mol. Plant Pathol.* **33**:127–149.

Yoshida, S., Ohrishi, Y., and Kitagishi, K., 1962, Histochemistry of silicon in rice plant. III. The presence of a cuticle–silica double layer in epidermal tissue, *Soil Sci. Plant Nut.* **8**:1–5.

Zeyen, R. J., and Bushnell, W. R., 1979, Papilla response of barley epidermal cells caused by *Erysiphe graminis*: Rate and method of deposition determined by microcinematography and transmission electron microscopy, *Can. J. Bot.* **57**:898–913.

Zeyen, R. J., Carver, T. L., and Ahlstrand, G. G., 1983, Relating cytoplasmic detail of powdery mildew infection to presence of insoluble silicon by sequential use of light microscopy. SEM and X-ray microanalysis, *Physiol. Plant Pathol.* **22**:101–108.

4

Rust Basidiospore Germlings and Disease Initiation

Randall E. Gold and Kurt Mendgen

1. INTRODUCTION

1.1. General

The rust fungi are potentially dangerous plant disease organisms. Infections caused by dikaryotic urediniospores play a major role in limiting agricultural production. Consequently, research has focused on infections of this type and has been the central theme of numerous reviews (Littlefield and Heath, 1979; Bushnell and Roelfs, 1984; Mendgen *et al.*, 1988; Hoch and Staples, this volume). Research on basidiospores and basidiospore-derived infections has received significantly less attention. This may be partially due to their lesser economic importance, but is also a result of the lack of reliable methods to activate teliospore germination, and therefore basidiospore production, of agriculturally important rusts (Petersen, 1974; Mendgen, 1984).

This chapter builds upon existing brief reviews of basidiospore germination (Petersen, 1974; Mendgen, 1984), host penetration (Littlefield and Heath, 1979), and the morphology of intracellular structures (Harder and Chong, 1984). Our approach relies on a comprehensive illustration of the events and structures important in the initiation of basidiospore-derived infections.

1.2. Terminology

The terminology suggested by Littlefield and Heath (1979) will be followed with the exception that intracellular infection structures, often designated as monokaryotic haustoria, will be called intracellular hyphae (Rijkenberg and Truter, 1972; Gold and Mendgen, 1984a). The extracellular, slimelike substance commonly associated with infection structures will be termed the extracellular matrix (see Gold and Mendgen, 1984b; Mims and Richardson, 1989; Nicholson and Epstein, this volume).

Randall E. Gold • BASF AG, Agricultural Research Station, D-6703 Limburgerhof, Federal Republic of Germany.   *Kurt Mendgen* • Department of Plant Pathology, University of Constance, D-7750 Constance, Federal Republic of Germany.

2. BASIDIOSPORE MORPHOLOGY AND GERMINATION

Knowledge of the morphology and germination of the rust basidiospores dates back to Gasparrini (1848) and the early accounts by Tulasne (1854), de Bary (1863, 1866), Reess (1865, 1870), and Sappin-Trouffy (1896). Basidiospores, also called sporidia in the earlier literature, are the only spore type in the rust life cycle that does not form after a preliminary infection event. Instead, they generally arise after the germination of diploid teliospores or, less commonly, in endocyclic species, after germination of aecioid teliospores (Petersen, 1974). The metabasidium generally gives rise to four, sometimes two, basidiospores on sterigmata (Figs. 1–3).

2.1. Morphology

2.1,1. Macroscopic

Masses of fresh basidiospores range from yellow to red (Reed and Crabill, 1915; Snow and Froelich, 1968) due to the presence of pigmented lipid droplets that may protect against UV radiation. Hyaline or white basidiospores have also been reported (Diner and Mott, 1982a; Gold and Mendgen, 1984c).

2.1.2. Microscopic

Mature basidiospores vary considerably in morphology, even within a single species. However, reniform (e.g., Fig. 4), round (e.g., Fig. 10), or ovate-elliptical (e.g., Fig. 2) are

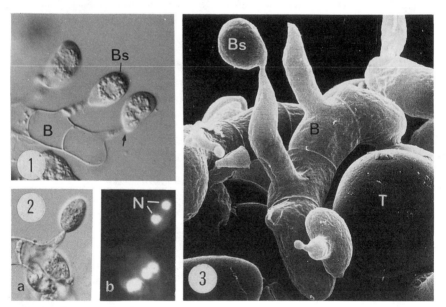

Figure 1. Metabasidium (B) of *Uromyces appendiculatus*. Basidiospores (Bs) are borne at the tip of the sterigmata (arrow). (750×.)

Figure 2. Light (a) and fluorescence (b) micrographs of fixed, DAPI-stained basidiospores of *Uromyces appendiculatus* containing one or two nuclei (N) (620×). (From Gold and Mendgen, 1984c.)

Figure 3. Germinated teliospore (T) of *Gymnosporangium fuscum* with a septate metabasidium (B) and basidiospores (Bs). The labeled basidiospore is still attached to the sterigma. (1400×.) (Courtesy of B. Metzler, unpublished.)

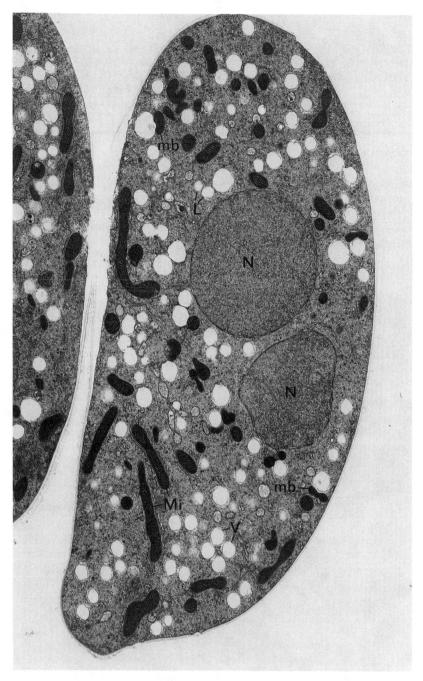

Figure 4. Longitudinal section through a basidiospore of *Gymnosporangium juniperi-virginianae* prepared for TEM by freeze substitution. Nuclei (N), lipid bodies (L), mitochondria (Mi), microbodies (mb), and vacuoles (V) are shown. (12,070×.) (From Mims *et al.*, 1988.)

common shapes. Sizes range from 3 × 4 μm (e.g., *Cronartium flaccidum*) to 14 × 25 μm (e.g., *Coleosporium tussilaginis*). All basidiospores that have been examined by conventional SEM are smooth or slightly wrinkled (e.g., Fig. 25) and bear a prominent apiculus (Fig. 26).

When examined by TEM, the spore wall appears thin (0.05–0.25 μm) in transverse section and consists of a single or a double layer. In the hilar region a thin, electron-dense outer layer is more pronounced in several species (Metzler, 1982; Bauer and Oberwinkler, 1988; Mims, 1977; Mims *et al.*, 1988). Preformed germ pores are not present. Basidiospores have a densely packed protoplasm and typically contain two nuclei, microtubules, and endoplasmic reticulum (ER) (Fig. 4). Filasomes and tubular–vesicular complexes (TVC) were detected in germinating basidiospores preserved by ultrarapid freezing and freeze substitution protocols (Mims and Richardson, 1989). TVC seem to include the Golgi apparatus of the cell (Welter *et al.*, 1988) and were tentatively distinguished as TVC 1 and TVC 2 (Figs. 5, 6). TVC 1 may be derived from the ER and has only been found in fungal structures growing within the leaf. TVC 1 seems to be involved in secretion (Knauf and Mendgen, 1988). TVC 2 resembles single cisternae of the dictyosomes. An obvious spatial relationship between vacuoles, lipid bodies, and microbodies interconnected by strands of ER has been observed (Fig. 4) (Gold and Mendgen, 1984a; Mims *et al.*, 1988). Such pronounced organization is presumably related to the metabolic activities of germination, particularly the utilization of lipid reserves (Mendgen, 1973).

2.2. Nuclear Condition

Basidiospores typically contain two haploid, homokaryotic nuclei at maturity. During teliospore germination the diploid nucleus undergoes meiosis, followed by a mitosis, usually after each daughter nucleus has migrated into a basidiospore initial (Fig. 2) (Gold and Mendgen, 1984c). Binucleate basidiospores were first observed by Poirault and Raciborski (1895) and Sappin-Trouffy (1896), a condition confirmed by subsequent reports (Hoffman, 1912; Ashworth,

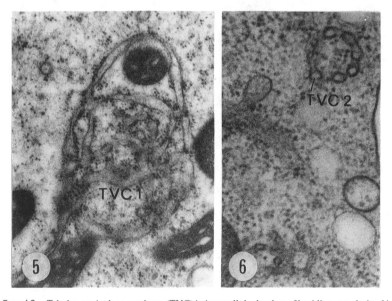

Figures 5 and 6. Tubular–vesicular complexes (TVC) in intercellular hyphae of basidiospore-derived infections of *Uromyces appendiculatus. Figure 5.* TVC 1 may be part of the endoplasmic reticulum. (35,000×.) *Figure 6.* TVC 2 appears similar to dictyosomes. (60,000×.) (Courtesy of L. Bruscaglioni, unpublished.)

1935; Savile, 1939; Thirumalachar, 1939; Kapooria, 1971; Gopinathan Nair, 1972; Bauer, 1986; Mims *et al.*, 1988; see Allen, 1933; Petersen, 1974; Anikster, 1983; Gold and Mendgen, 1984c, for additional references). Other studies have described species with simplified two- or three-celled metabasidia (Petersen, 1974) and have also revealed unique alternative development patterns, including binucleate, heterokaryotic (self-fertile) (Lindfors, 1924; Thirumalachar, 1945; Anikster and Wahl, 1985) and quadranucleate (Olive, 1949; Kapooria, 1968; Kohno *et al.*, 1975) basidiospores. Anikster and co-workers (1980) and Petersen (1974) have summarized the numerous variations and consequences of functionally heterokaryotic basidiospores. Recent quantitative studies (Kapooria and Zadoks, 1973; Hansen and Patton, 1975; Anikster, 1983; Gold and Mendgen, 1984c) and numerous descriptive or semiquantitative reports confirm that binucleate spores are clearly the predominant form.

2.3. Germination

After being deposited on their host, basidiospores can germinate within a few hours. Germination may follow two characteristic forms (Fig. 7): either directly via germ tubes (Fig. 8) or indirectly via formation of secondary spores (Fig. 9). The pattern followed is dependent on environmental conditions and the host surface (Ragazzi and Dellavalle Fedi, 1982; Desprez-Loustau, 1986).

2.3.1. Influence of Water

Of the basic environmental factors influencing spore germination, water, in the form of humidity and surface moisture, appears to be the most crucial. High humidity is generally needed (Cotter, 1932; Hansen and Patton, 1975); however, the conditions required are difficult to generalize. For example, basidiospores of *Cronartium ribicola* (Hansen and Patton, 1975), *Puccinia horiana* (Kapooria and Zadoks, 1973), and *Puccinia malvacearum* (Ashworth, 1931) all germinate well on both water agar and dry glass slides enclosed in a moist chamber. In contrast, basidiospores of *Melampsora pinitorqua* on 1.5% water agar germinate equally well at humidities

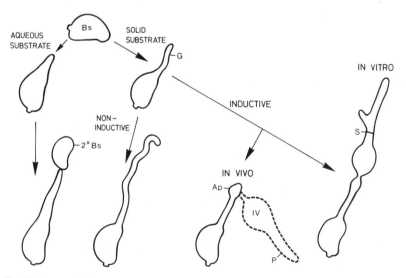

Figure 7. Diagrammatic illustration of basidiospore germination and development on various substrata. Ap, appressorium; Bs, basidiospore; G, germ tube; IV, intraepidermal vesicle; P, primary hypha; S, septum. (Modified after Bauer, 1986.)

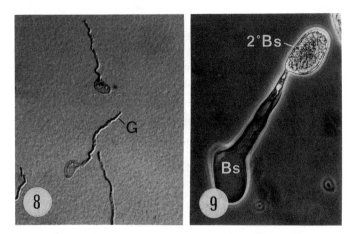

Figure 8. Basidiospore germlings of *Uromyces appendiculatus* on 2% water agar. Note long, thin, and wavy germ tubes (G). (290×.) (From Freytag *et al.*, 1988.)

Figure 9. Secondary basidiospore (2°Bs) of *Coleosporium tussilaginis* arises from the basidiospore (Bs) in a water droplet on 0.2% water agar. (540×.) (From Bauer, 1986.)

ranging from 35 to 100%, but fail to germinate under any humidity conditions when incubated on glass slides (Desprez-Loustau, 1986). Differing sensitivities to surface properties of glass and/or differences in spore water content may explain these results.

Basidiospores are extremely sensitive to desiccation. When exposed to unfavorable conditions, such as low humidity or direct sunlight, they quickly lose viability (Spaulding and Rathbun-Gravatt, 1926; MacLachlan, 1935; Blank and Leathers, 1963; Gould and Shaw, 1969; Snow, 1968a). Excessive free water is also generally inhibitory to germination (Desprez-Loustau, 1986) or leads to the formation of secondary basidiospores (Figs. 7, 9) (Hansen and Patton, 1975; Bauer, 1986; Mims and Richardson, 1989). A basidiospore can produce up to six successive secondary spores (Petersen, 1974; Bauer, 1986), all equally pathogenic (Roncadori, 1968).

It has long been proposed that secondary basidiospore formation is a survival mechanism for spores exposed to improper environmental conditions (Reess, 1870; Weimer, 1917; Hansen and Patton, 1975) or nonhost substrata (Bega, 1960; Van Sickle, 1975). Bauer (1986, 1987) reported that unfavorable positioning on the host, such as adhesion to leaf hairs, can also signal the formation of secondary spores, and demonstrated that one of the nuclei degenerates during this process (Fig. 10).

2.3.2. Influence of Temperature

Many *in vitro* studies have demonstrated that basidiospores can germinate over a wide temperature range. For example, Hansen and Patton (1975) germinated basidiospores of *Cronartium ribicola* between 5 and 24°C, with a maximum observed between 12 and 15°C. . With the same species, Bega (1960) found germination on both water agar and host needles between 1 and 28°C, but with broad maxima between 5 and 24°C and 13 and 20°C, respectively. A similarly broad maximum between 8 and 20°C has been determined *in vitro* for basidiospores of *M. pinitorqua*, a species that may germinate and grow between 2 and 30°C (Desprez-Loustau, 1986). These results and others (Reed and Crabill, 1915; MacLachlan, 1935; Siggers, 1947; Yamada, 1956; Van Sickle, 1975; Ragazzi *et al.*, 1986) show that basidiospores have an enormous germination potential. Thermal death has been determined between 30 and 32°C for a few species (Reed and Crabill, 1915; MacLachlan, 1935; Desprez-Loustau, 1986).

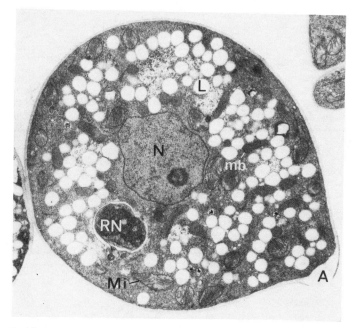

Figure 10. Basidiospore of *Coleosporium asclepiadeum* with intact nucleus (N) and degenerated remnant nucleus (RN) prior to secondary spore formation. A, apiculus; L, lipid body; Mi, mitochondrion; mb, microbody. (11,500×.) (From Bauer and Oberwinkler, 1988.)

2.3.3. Influence of Light

In order to avoid the effects of UV radiation and desiccation by direct sunlight, most basidiospores are formed and released rhythmically at night or early morning (see Gold and Mendgen, 1983). Darkness, however, is not necessary for germination. When other conditions are favorable, light has either no effect (Yamada, 1956; Bega, 1960; Diner and Mott, 1982a; Desprez-Loustau, 1986) or only a marginal negative effect (Gould and Shaw, 1969).

2.3.4. Influence of Host Substances

Fresh leaves of sagebrush inhibit the germination of *Cronartium comandrae* completely (Krebill, 1972). Leaf fragments with abundant glandular hairs (Robinson, 1914) or water-soluble exudates from *Pinus pinaster* inhibit germination in *Puccinia malvacearum* and *M. pinitorqua* (Desprez-Loustau, 1986), respectively. The inhibitory effect on *M. pinitorqua* is stronger with exudates from disease-resistant species. Water-insoluble monoterpenic compounds from epi-cuticular waxes are also inhibitory, although there are no differences between resistant and susceptible host species.

2.3.5. Influence of Massing

Massing of basidiospores has been reported to inhibit germination (Hansen and Patton, 1975). Spaulding and Rathbun-Gravatt (1926) attributed the effect to insufficient oxygen or secretion of toxic substances. Self-inhibitors of basidiospore germination *per se* have not been reported.

3. BASIDIOSPORE GERMLINGS IN VITRO

Physiological studies of germ tube development have been limited by difficulties in obtaining sufficient spore material. Furthermore, the limited life span of basidiospores and their restricted growth pattern in axenic culture have also hindered such research.

3.1. Germ Tube Differentiation

3.1.1. Influence of Environmental Conditions

C. ribicola germ tubes grow best at 12–15°C; however, most intraepidermal vesicles are formed between 19 and 26°C. Incubation at 16°C followed by treatment at 24 or 28°C for 1–2 hr increases the incidence of vesicle formation considerably (Hansen and Patton, 1975), indicating that basidiospore germ tubes can react to temperature shocks as do those of urediniospores (see Hoch and Staples, this volume).

Robinson (1914) studied the tropic effect of light on germ tube growth in *Puccinia malvacearum*. Freshly cast spores were illuminated unilaterally with incident light from a window. In this simple experiment the germ tubes emerged predominantly from the shadowed side of the spores, and when these were subsequently rotated through 90°, exhibited a marked negative phototropic response.

3.1.2. Influence of Artificial Substrata

The induction of infection structures *in vitro* depends greatly on the hardness of the substratum. For example, in water or on 0.2–0.7% water agar, spores of *Gymnosporangium clavariiformae* germinate with a broad germ tube and subsequently form a secondary spore (Fig. 9) (Bauer, 1987). On 1–2% water agar, a thin, unbranched, and undifferentiated nonseptate germ tube is formed (Fig. 8). Agar concentrations around 4–5% induce the formation of short germ tubes with appressoria (Bauer, 1986; Freytag *et al.*, 1988). With *Uromyces viciae-fabae*, *U. vignae*, and *U. appendiculatus*, infection structure development may proceed as far as a vesicle (Fig. 11) or a primary hypha (Freytag *et al.*, 1988).

Artificial membranes on agar, such as nitrocellulose (Figs. 12, 19), collodion (Fig. 13), polyethylene (Fig. 15), and dialysis tubing (Figs. 16, 17), can induce infection structures from basidiospores (Hansen and Patton, 1975; Freytag *et al.*, 1988; Mims and Richardson, 1989). The hard cuticular surface of barley coleoptiles (R. E. Gold, unpublished) and even glass (Fig. 14) (Bauer, 1986; Freytag *et al.*, 1988) induce the formation of appressoria, suggesting that surface recognition in the haploid stage is not very specific. In all these cases the surface hardness and the amount of water available determine whether (1) a germ tube forms, (2) a secondary spore is produced, or (3) germ tube growth continues to produce normal infection structures (see Fig. 7).

Isolated cuticles of *Vicia faba*, *Vigna sinensis*, and *Phaseolus vulgaris* induce the formation of infection structures of *U. viciae-fabae*, *U. vignae*, and *U. appendiculatus*, respectively (Freytag *et al.*, 1988). Cross-inoculations show this response to be nonhost specific (S. Freytag and L. Bruscaglioni, unpublished). On leaf cuticles mounted on 2% agar, basidiospores germinate, form short germ tubes, differentiate appressoria and intraepidermal vesicles either on the surface of the cuticle (Fig. 18) or below the cuticle in the agar (S. Freytag, unpublished). Under natural conditions, such vesicles develop within host epidermal cells (Figs. 22, 23a).

3.2. Surface Carbohydrates

In vitro, basidiospore germ tubes may be copiously covered with an extracellular matrix (Figs. 16, 17) (Mims and Richardson, 1989). Basidiospores and germ tubes of *U. viciae-fabae*

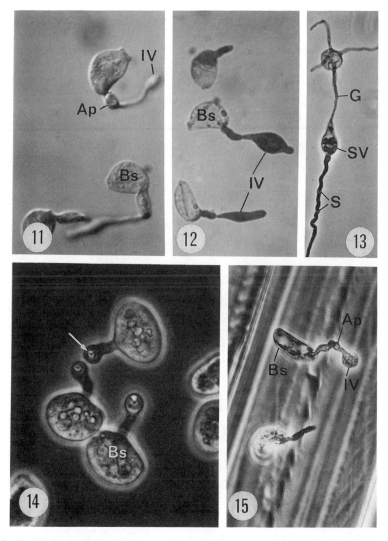

Figure 11. Basidiospore germlings of *Uromyces vignae* on 5% water agar. Appressoria (Ap) of the basidiospores (Bs) are visible on the agar surface. Vesicles (IV, not in focus) form within the agar. (640×.) (From Freytag *et al.*, 1988.)

Figure 12. Germinated basidiospore (Bs) of *Uromyces vignae* on nitrocellulose membrane filter showing vesicle (IV). (640×.) (From Freytag *et al.*, 1988.)

Figure 13. Basidiospore germling of *Cronartium ribicola* on collodion membrane with four germ tubes (G) and pyriform vesicle (SV). Septa (S) are visible proximal to the vesicle. (420×, estimated.) (From Hansen and Patton, 1975.)

Figure 14. Basidiospore germlings of *Gymnosporangium clavariiformae* on glass. The appressorial ring (arrow) is clearly visible. Bs, basidiospore. (1060×.) (Courtesy of R. Bauer, unpublished.)

Figure 15. Basidiospore germlings of *Uromyces appendiculatus* on polyethylene sheet. One basidiospore (Bs) is shown with appressorium (Ap) and vesicle (IV). (730×.)

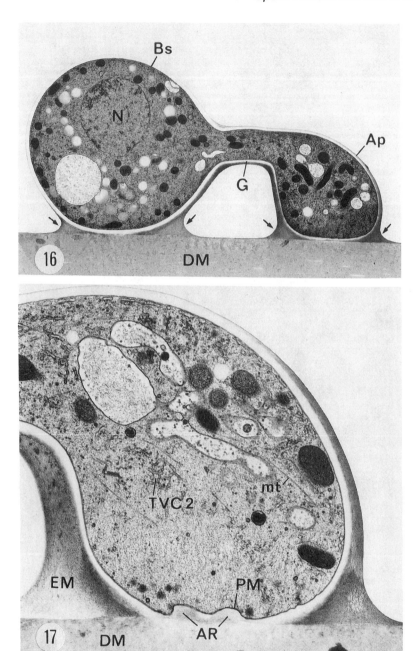

Figures 16 and 17. Basidiospore germlings of *Gymnosporangium juniperi-virginianae* differentiate infection structures *in vitro* on dialysis tubing membrane (DM). *Figure 16.* Longitudinal section of a germinated basidiospore (Bs) showing one of its nuclei (N), germ tube (G), and prominent appressorium (Ap). The electron-dense extracellular matrix (arrows) is abundant and attaches the germling to membrane. (6500×.) (From Mims and Richardson, 1989.) *Figure 17.* Appressorium with appressorial ring (AR) outside plasma membrane (PM) and in close contact with membrane. The extracellular matrix (EM) covers appressorium. Note microtubules (mt) and tubular–vesicular complex (TVC2). (20,000×.) (Courtesy of C. W. Mims and E. A. Richardson, unpublished.)

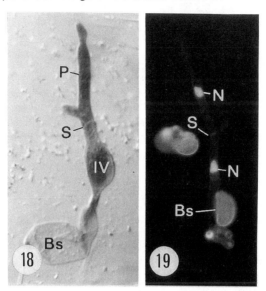

Figure 18. Germinated basidiospore (Bs) of *Uromyces appendiculatus* on isolated host bean leaf cuticle. The vesicle (IV) and septate hypha (P) formed on the surface of the cuticle. S, septum. (960×.) (From Freytag *et al.*, 1988.)

Figure 19. Basidiospore germling of *Uromyces viciae-fabae* on nitrocellulose filter after staining with DAPI/calcofluor. The hypha contains two nuclei (N) separated by a septum (S). Bs, basidiospore. (500×.) (Courtesy of S. Freytag, unpublished.)

strongly bind lectins of *Triticum vulgaris* and *Canavalia ensiformis*, and weakly those of *Lens culinaris*, *Tetragonolobus purpureas*, and *Ricinus communis* (Fig. 20). These results suggest that they are covered with carbohydrates or glycoproteins that consist mainly of chitin, α-D-mannose, α-D-glucose, and some α-L-fucose and β-D-galactose. In comparison to urediniospore-derived infection structures, which appear to cover its chitin with other carbohydrates as soon as it penetrates the plant (Mendgen *et al.*, 1985), the basidiospore-derived infection structure reduces chitin while the affinity to other lectins remains at the same level (Fig. 20).

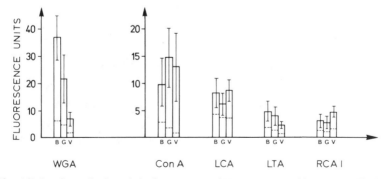

Figure 20. Affinity of some lectins to infection structures of *Uromyces viciae-fabae in vitro*. The lectins were labeled with FITC and the fluorescence was measured with a microscope photometer. B, basidiospore; G, germ tube; V, intraepidermal vesicle. WGA, wheat germ agglutinin; Con A, *Canavalia ensiformis* agglutinin; LCA, *Lens culinaris* agglutinin; LTA, *Tetragonolobus purpureas* agglutinin; RCA, *Ricinus communis* agglutinin. (Courtesy of S. Freytag, unpublished.)

4. BASIDIOSPORE GERMLINGS IN VIVO

4.1. Historical Perspective

Much of the early rust work focused on basidiospore-derived infections of stem rust of cereals and coniferous tree rust diseases. de Bary (1865) demonstrated the relationship between infections of *Puccinia graminis* on wheat and barberry, which stimulated numerous subsequent studies on host penetration and infection (de Bary, 1866; Reess, 1870; Ráthay, 1881; Fischer, 1898, 1901; Eriksson, 1911; Waterhouse, 1921; Allen, 1930, 1932, 1934, 1935; Nusbaum, 1935). These initial studies, often containing accurate light microscopic observations (see Figs. 21, 22), gave way to studies on the epidemiologically more important urediniospore, which dominate the literature today.

4.2. Penetration

The anatomy of basidiospore germination, host penetration, and the development of characteristic inter- and intracellular structures is shown diagrammatically in Figs. 23 and 28. Penetration directly through the epidermis has been observed more often than indirect, stomatal penetration. Indirect penetration is common for species that attack gymnosperms, although exceptions do occur (Figs. 24, 42). Gymnosperms have a thick-walled epidermis covered by a thick cuticle. The extreme rigidity of even young needles presents a tough physical barrier and rusts may have adapted to stomatal penetration as a response to the impenetrable host epidermis (Fischer, 1898; Patton and Johnson, 1970; Grill *et al.*, 1980; Bauer, 1986).

4.2.1. Development on Host Epidermis

Basidiospores of many rusts produce a single, unbranched germ tube (Fig. 33) (Van Sickle, 1975; Kohno *et al.*, 1977b; Jacobi *et al.*, 1982; Metzler, 1982; Gold and Mendgen, 1984a) that generally emerges opposite the apiculus (Fig. 24) or laterally (Fig. 26). In *Cronartium ribicola* and *C. asclepiadeum*, several germ tubes may develop (Figs. 34, 35), which can become highly branched (Bega, 1960; Hansen and Patton, 1975; Bauer, 1986; Ragazzi *et al.*, 1987). Gold and

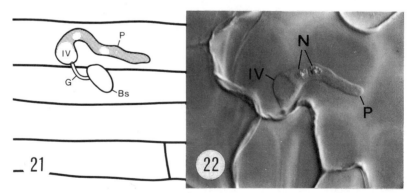

Figures 21 and 22. Epidermal infection structures. *Figure 21.* Diagrammatic representation. Bs, basidiospore; G, germ tube; IV, intraepidermal vesicle; P, primary hypha. (Modified after de Bary, 1863.) *Figure 22.* Ovate intraepidermal vesicle (IV) and primary hypha (P) of *Uromyces appendiculatus* are visible in leaf epidermis. Note the two nuclei (N). (830×.) (From Gold and Mendgen, 1984a.)

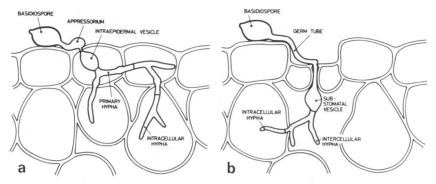

Figure 23. Diagrammatic representation of a direct-penetrating (a) and an indirect-penetrating (b) basidiospore germling.

Littlefield (1979) described branched germ tubes in *Melampsora lini* associated with septum formation.

In contrast to urediniospore germ tubes, which tend to grow at right angles to the epidermal ridges (Hoch and Staples, this volume), basidiospore germ tubes grow more randomly on the epidermal surface (Hansen and Patton, 1977; Gray *et al.*, 1983). The amount of germ tube growth prior to appressorium formation varies considerably and is dependent on the selection of an appropriate penetration site (Gold and Mendgen, 1984a).

The extracellular matrix, a common feature of direct-penetrating rusts, accumulates along

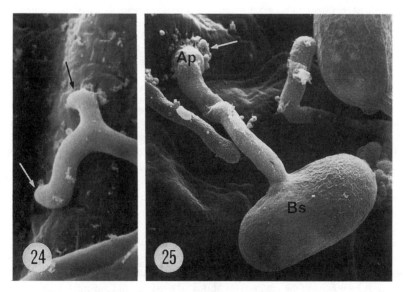

Figure 24. Sugar pine hypocotyl inoculated with vegetative hyphae of an axenic basidiospore subculture of *Cronartium ribicola*. Note apparent direct penetration of epidermal cell by the appressorium-like hyphal tips (arrow) closely appressed to the epidermis. (3450×.) (From Diner and Mott, 1982c.)

Figure 25. Germinated basidiospore (Bs) of *Gymnosporangium fuscum* on pear leaf. Note extracellular matrix material (arrow) near appressorium (Ap) at the penetration site. (2620×.) (From Metzler, 1982.)

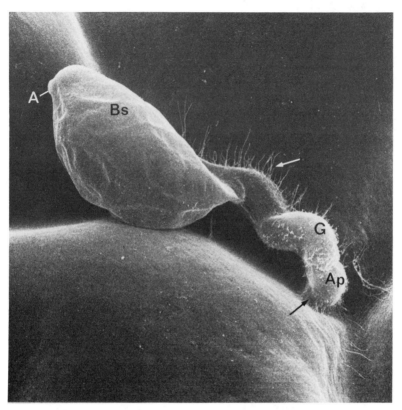

Figure 26. Direct penetration by basidiospore germling of *Uromyces appendiculatus* near the junction of three epidermal cells. Note the fibrouslike extracellular matrix (arrows) along the germ tube (G) and at the base of the appressorium (Ap). A, apiculus; Bs, basidiospore. (4020×.) (From Gold and Mendgen, 1984a.)

all areas of contact between the epidermis and fungus (Figs. 25, 26), especially at the periphery of the appressorium (Fig. 27) (Waterhouse, 1921; Littlefield and Heath, 1979; Metzler, 1982; Gold and Mendgen, 1984a). Jacobi and co-workers (1982) also showed that it occurs at contact points between basidiospores, suggesting a nonspecific contact response. It appears to attach the germ tube and appressorium to the plant surface, protect it from desiccation and UV radiation, and serve as a reservoir for "penetration enzymes" (see Nicholson and Epstein, this volume). The enzymes necessary for penetration may arise from the appressorial ring (Metzler, 1982), a tubular, collarlike zone that flanks the periphery of the penetration pore (Figs. 14, 28, 32) (Gold and Mendgen, 1984a; Mims and Richardson, 1989). The appressorial ring may also be important for attachment.

The formation of a slightly swollen, terminal appressorium (Figs. 24, 25, 32) by direct-penetrating species occurs preferentially at or near epidermal cell junctions (Fischer, 1898; Eriksson, 1911; Melhus *et al.*, 1920; Allen, 1932, 1935; Gold and Mendgen, 1984a). This site preference may be related to the greater availability of nutrients and moisture due to a higher rate of exosmosis in these areas and/or the chemical composition and physical structure of the wax layer that affects both pH and wettability of the leaf surface (Cutler *et al.*, 1982). This characteristic appears to be independent of nuclear condition, as evidenced by direct-penetrating urediniospore germlings (see Koch and Hoppe, 1988). Appressoria do not form in indirect-penetrating species (Figs. 33–35).

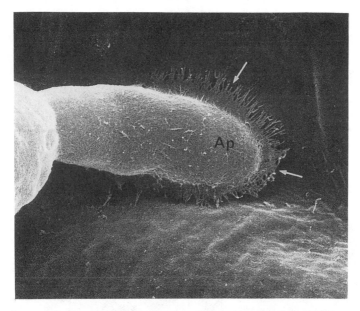

Figure 27. Apparent direct penetration near junction of two epidermal cells by basidiospore germling of *Uromyces appendiculatus*; note the copious deposition of the extracellular matrix (arrows). Ap, appressorium. (12,120×.)

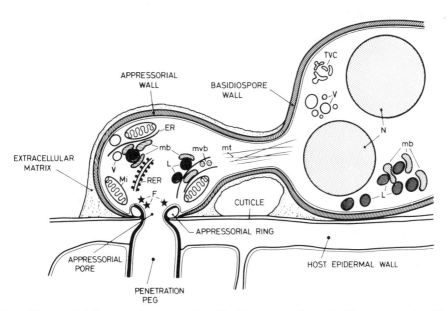

Figure 28. Detailed diagrammatic representation of basidiospore germling and epidermal penetration. (Modified after Metzler, 1982.) ER, endoplasmic reticulum; F, filasomes; L, lipid bodies; mb, microbodies; Mi, mitochondrion; mt, microtubules; mvb, multivesicular bodies; N, nucleus; RER, rough endoplasmic reticulum; TVC, tubular–vesicular complex; V, vacuoles.

4.2.2.   Mode of Ingress

The earliest stages of direct penetration are presumably effected by a narrow (diam. = ca. 0.5 μm) peg that has only been observed by light microscopy (Fig. 29). Once through the outer region, the peg expands in diameter (Metzler, 1982; Gray *et al.*, 1983; Gold and Mendgen, 1984a), breaches the epidermal wall, and invaginates the host plasmalemma. After a brief period of apical growth, it enlarges distally to form an ovate to oblong intraepidermal vesicle (Fig. 30). At this stage the nuclei migrate from the spore to the vesicle (see Figs. 30, 32).

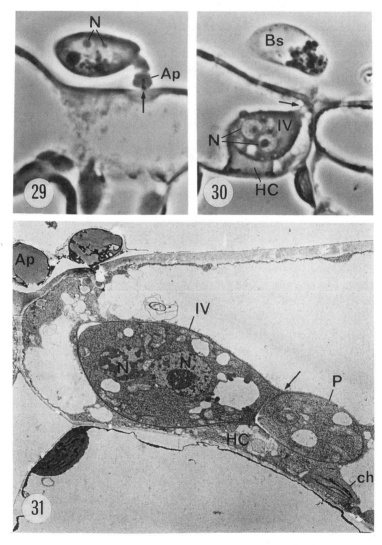

Figures 29–31.   Penetration and infection of bean epidermis by *Uromyces appendiculatus. Figure 29.* Initial stage of penetration. Binucleate basidiospore germling with appressorium (Ap) and phase-dark penetration peg (arrow). N, nucleus. (1500×.) (From Gold and Mendgen, 1984a.) *Figure 30.* The two nuclei (N) and cytoplasm have migrated from the basidiospore (Bs) through the narrow penetration peg (arrow) into the intraepidermal vesicle (IV). Note aggregation of host cytoplasm (HC) in the invaded epidermal cell. (1390×.) (From Gold and Mendgen, 1984a.) *Figure 31.* TEM of penetration site showing appressorium (Ap), intraepidermal vesicle (IV) with two nuclei (N) and primary hypha (P), separated by a septum (arrow). Note aggregation of host cytoplasm (HC) and chloroplast (ch) below primary hypha. (3540×.)

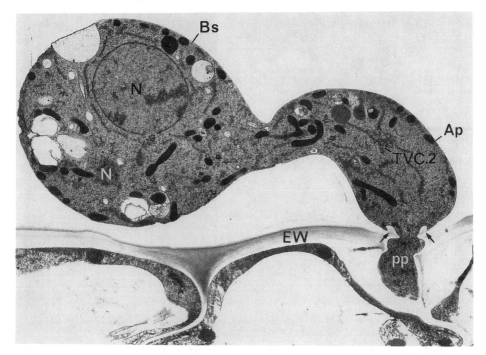

Figure 32. *Gymnosporangium juniperi-virginianae* on susceptible apple leaf prepared for TEM by freeze substitution. Penetration peg (pp) from appressorium (Ap) has grown through epidermal wall (EW). The appressorial ring (arrows) is visible at the point of penetration. Only one of the two nuclei (N) is clearly visible in the basidiospore (Bs). (8000×.) (From Mims and Richardson, 1989.)

de Bary (1884) reported that the penetration pore becomes plugged after migration of the spore protoplast into the intraepidermal vesicle. Ultrastructural evidence of this was presented by Metzler (1982) as a septumlike structure in the neck region of the vesicle and by Gold and Mendgen (1984a) as an electron-dense occlusion of the penetration pore. Normal septum formation has been shown in *Cronartium fusiforme* (Fig. 43) (Gray *et al.*, 1983).

The average minimum time for penetration and formation of a vesicle is 4–6 hr (Weimer, 1917; Waterhouse, 1921; Snow, 1968b; Van Sickle, 1975; Gold and Mendgen, 1984a), although periods of up to 24 hr have also been reported (Allen, 1932, 1935; Nusbaum, 1935; Kohno *et al.*, 1977a). Light has either no influence (Hansen and Patton, 1975; Gold and Mendgen, 1984a) or a small inhibitory effect (Cotter, 1932).

4.3. Differentiation of Infection Structures

In direct-penetrating species, the vesicle forms within an epidermal cell and gives rise via apical growth to the primary hypha (Figs. 31, 37, 43) (Allen, 1983; Gold and Mendgen, 1984a). The primary hypha develops rapidly into a branched, multicellular network of mycelia (Fig. 41). Growth directly into neighboring epidermal cells (Figs. 37, 38, 41) and subtending palisade parenchyma is common (Figs. 39, 40) as is excellular growth into intercellular spaces (Figs. 42, 43). Here, the invaginated plasmalemma of the epidermal cell becomes reunited with the native plasmalemma at points directly adjacent to the exit site (Fig. 43) (Gray *et al.*, 1983). Once the hyphae reach the intercellular spaces, extensive ramification of the tissue occurs (Fig. 44) (Gold *et al.*, 1979; Grill *et al.*, 1980).

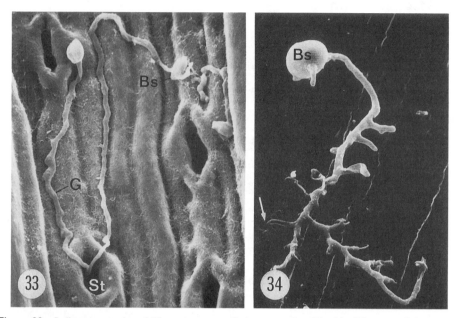

Figure 33.   Indirect penetration of *Pinus pinaster* needle by germ tubes (G) of basidiospores of *Cronartium flaccidum*. Bs, basidiospore. St, stoma. (625×.) (From Ragazzi *et al.*, 1987.)

Figure 34.   Basidiospore germling of *Coleosporium tussilaginis* produces a highly branched germ tube on a needle of Scotch pine. Penetration of host through host stoma (arrow). Bs, basidiospore. (610×.) (From Bauer, 1986.)

In indirect-penetrating species, intercellular hyphae are derived from a vesicle that forms in the substomatal chamber (Fig. 36). Further development occurs via infection structures equivalent to those of direct-penetrating species.

The bi- or multinucleate condition remains until development of the primary hypha and subsequent infection structures. Thereafter, many rusts revert to the mononucleate condition (Eriksson, 1911; Colley, 1918; Allen, 1930, 1935; Ashworth, 1935; Nusbaum, 1935; Metzler, 1980; Freytag *et al.*, 1988), but the binucleate (Lindfors, 1924; Thirumalachar, 1939; Kohno *et al.*, 1977b) or multinucleate (Allen, 1934; B. Müller, unpublished) condition persists in others.

Ultrastructural studies often refer to an amorphous, electron-opaque extracellular matrix between the host and the intercellular hyphae (Fig. 45) (Harder, 1978; Borland and Mims, 1980; Al-Khesraji and Lösel, 1981; Longo *et al.*, 1982; Gray *et al.*, 1983; Hopkin and Reid, 1988). The origin and function of this material are uncertain, but many authors assume it to be of fungal origin and to act as an aid to attachment (Littlefield and Heath, 1979; Harder and Chong, 1984; Hopkin and Reid, 1988). It may also be involved in signal exchange between host and parasite (Mendgen *et al.*, 1988).

Intercellular hyphae grow in contact with mesophyll cells and enter them to form intracellular hyphae (Fig. 48). The morphology of intracellular hyphae has been described in detail recently by Littlefield and Heath (1979), Harder and Chong (1984), and Hopkin and Reid (1988) and thus need not be repeated here. A characteristic of intracellular hyphae is the septum near the site of penetration either just outside (Figs. 46, 48) or inside (Figs. 49, 50) the host cell wall. Septum formation appears associated with a differentiation of intracellular infection structures. A highly conserved feature is the fingerlike or filamentous growth habit (Figs. 46–51). Conventional TEM shows no evidence of a definable neck ring (Harder and Chong, 1984). The extrahyphal

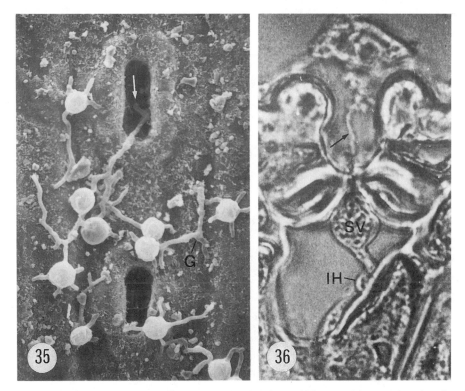

Figures 35 and 36. Development of *Cronartium ribicola* before and after stomatal penetration of needles of eastern white pine. *Figure 35.* Penetration by basidiospore germling through host stoma (arrow). Note branched germ tube (G) and lack of prominent appressorium. (600×.) (From Patton and Spear, 1980.) *Figure 36.* Cross section of stomatal penetration of secondary needle. Germ tube of the basidiospore penetrated between guard cells (arrow) to form substomatal vesicle (SV) and intercellular hypha (IH). (900×, estimated.) (From Patton and Johnson, 1970.)

membrane formed by the host usually stains weakly with PACP (a stain for plasma membranes) and lacks ATPase activity (Woods and Gay, 1987). The extrahyphal matrix is more pronounced around the distal end of intracellular hyphae and stains more intensively for polysaccharides (Chong *et al.*, 1981).

Intracellular hyphae commonly do not terminate within host cells. Several light- (Fig. 52) (Colley, 1918; Allen, 1930; Pady, 1935; Gold and Mendgen, 1984b) and electron-microscopic studies (Figs. 53, 54) (Gold *et al.*, 1979; Littlefield and Heath, 1979; Longo and Naldini Longo, 1982; Gray *et al.*, 1983; Gold and Mendgen, 1984b) have confirmed this. Improved methods of location (Zeyen and Bushnell, 1981) should promote future microscopical analyses of intracellular infection structures.

4.4. Host Response

Many investigators have taken a microscopical approach to study the incompatible interaction. The results of these studies suggest several mechanisms for resistance. On *Pinus sylvestris*, a relatively resistant species, basidiospore germination was reduced and delayed compared to more susceptible species (Ragazzi and Dellavalle Fedi, 1982). Natural occlusion of the stomata of *Pinus*

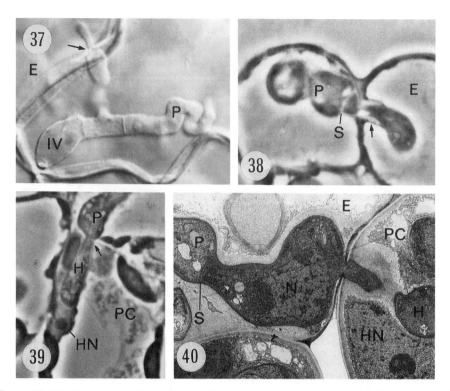

Figures 37–39. Development of *Uromyces appendiculatus* in bean epidermis as seen in whole leaf samples or in cross section. (From Gold and Mendgen, 1984a.) *Figure 37.* A branch of the primary hypha (P) penetrates into an adjacent epidermal cell (E) at arrow. IV, intraepidermal vesicle. (970×.) *Figure 38.* Primary hypha (P) grows into adjacent epidermal cell (E). Note septum (S) proximal to penetration point and the phase-light collarlike wall apposition (arrow) of the anticlinal wall. (1770×.) *Figure 39.* Growth of the primary hypha (P) into a palisade parenchyma cell (PC) at arrow. The nucleus (HN) of the PC is closely associated with apex of the intracellular hypha (H). (1560×.)

Figure 40. Transcellular penetration site of *Gymnosporangium fuscum*. The primary hypha (P) grew from an epidermal cell (E) into a subtending palisade parenchyma cell (PC). Abundant cell wall appositional deposits are associated with the transcellular hypha. Note fungal nucleus (N), narrow penetration site and host nucleus (HN) closely appressed to the intracellular hypha (H). S, septum. (5180×.) (Courtesy of B. Metzler, unpublished.)

strobus seedlings with wax (Fig. 55) seems to reduce the penetration of *Cronartium ribicola* and the number of needle lesions (Patton and Spear, 1980). However, most other reports on this host–pathogen system suggest that resistance is based on hypersensitive reactions (Kinloch and Littlefield, 1977; Diner and Mott, 1982b). Furthermore, Grill and co-workers (1980) reported that wax occlusions do not hinder penetration by *Chrysomyxa abietis*. They suggested that other physiological factors are responsible for the resistance observed. The thickness of the cuticle has been proposed as a factor preventing direct penetration (Melhus *et al.*, 1920; Melander and Craigie, 1927), although this has not been confirmed in other systems (Reed and Crabill, 1915; Ashworth, 1931; Nusbaum, 1935).

In tissue with hypersensitive lesions, necrosis of host cells (Fig. 57) and host cell wall appositions (Fig. 58) have been observed (Robb *et al.*, 1975; Walkinshaw, 1978; Gray and Amerson, 1983). The hypersensitive response may be triggered by attempted penetration of the

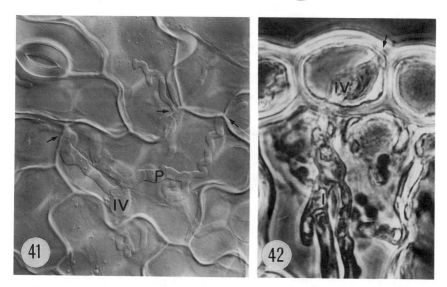

Figure 41. Highly branched, multicellular primary hypha (P) of *Uromyces appendiculatus*. Transcellular penetration sites are shown at arrows. IV, intraepidermal vesicle. (540×.)

Figure 42. Direct penetration (arrow) by *Cronartium fusiforme* in needle of slash pine seedling. An ovate intraepidermal vesicle (IV) developed in the epidermal cell, prior to forming an intercellular mycelium (I). (1150×.) (From Miller *et al.*, 1980.)

host epidermis (Fig. 56) or by dead or encased haustoria (Jonsson *et al.*, 1978; Gray and Amerson, 1983).

As most studies have focused on tree rusts, the practical value of the described resistance mechanisms can only be evaluated after long-term field trials (Kuhlmann, 1988).

5. CONCLUSIONS

Recent studies on basidiospores and their germlings have centered around ultrastructural and morphological descriptions of the penetration and infection process, with particular emphasis on the morphology of intracellular infection structures. In contrast to urediniospore research, less emphasis has been given to biochemical and physiological aspects of germination and differentiation of infection structures. Such information is crucial to the understanding of disease initiation and in the development of improved plant disease control methods.

The fact that two haploid, homokaryotic nuclei in basidiospores versus the heterokaryotic condition in urediniospores can effect significantly different penetration and infection patterns on the same host merits further investigation. Apparently, when complementary nuclei are brought together into the same cytoplasm, different parts of the fungal genome are expressed, as evidenced by the distinct penetration and infection process initiated by urediniospores. Future studies should include parallel experiments with basidiospores and urediniospores of the same species for a better understanding of the nuclear control of differentiation processes.

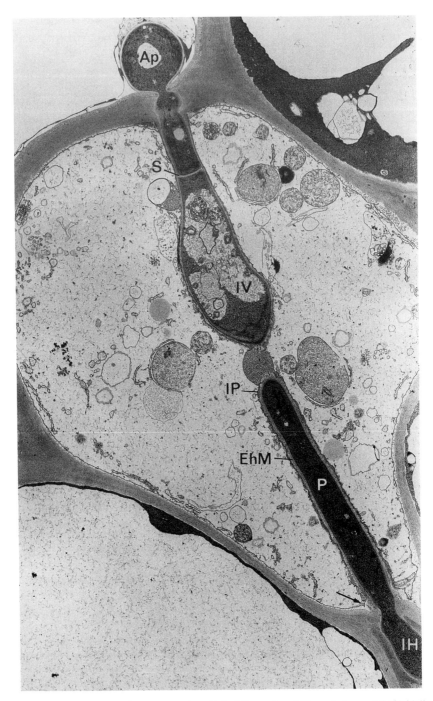

Figure 43. Direct penetration from appressorium (Ap) of *Cronartium fusiforme* in hypocotyl of loblolly pine. The vacuolated intraepidermal vesicle (IV) and primary hypha (P) are bordered by the invaginated plasmalemma (IP) and a prominent extrahyphal matrix (EhM). The IP is fused with noninvaginated plasmalemma (arrow) where P exits the epidermal cell to become an intercellular hypha (IH). Note septate (S) neck region of the vesicle. (7160×.) (From Gray *et al.*, 1983.)

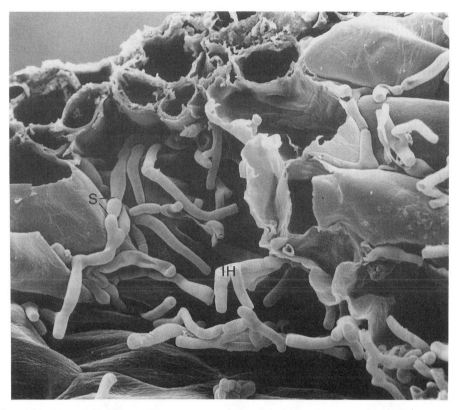

Figure 44.   Intercellular hyphae (IH) of *Chrysomyxa abietis* in spruce needle. Note clearly visible septa (S). (1000×.) (From Grill *et al.*, 1980; Courtesy of Center for Electron Microscopy, University of Graz, Austria.)

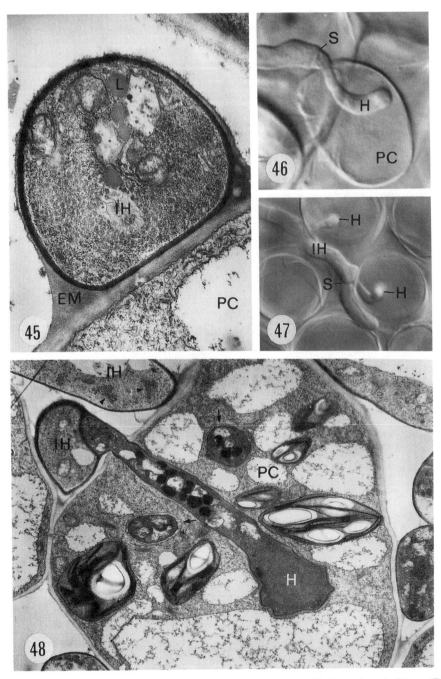

Figures 45–48. Inter- (IH) and intracellular hyphae (H) of *Uromyces appendiculatus* in bean leaf tissue. (From Gold and Mendgen, 1984b.) *Figure 45.* IH in cross section showing extracellular matrix (EM) along its periphery and at points of contact with host cell. L, lipid body; PC, parenchyma cell. (21,260×.) *Figure 46.* Terminal, fingerlike H. Note septum (S) prior to penetration site of parenchyma cell (PC). (1180×.) *Figure 47.* IH with two H that formed as side branches. Note septa (S) near penetration sites (980×.) *Figure 48.* Median longitudinal section through a penetration site into a parenchyma cell (PC). The invaginated plasma membrane and extrahyphal matrix are only clearly visible surrounding adjacent cross-sectioned H (arrows). Patches of glycogenlike particles (arrowheads) are seen in the IH. (6210×.)

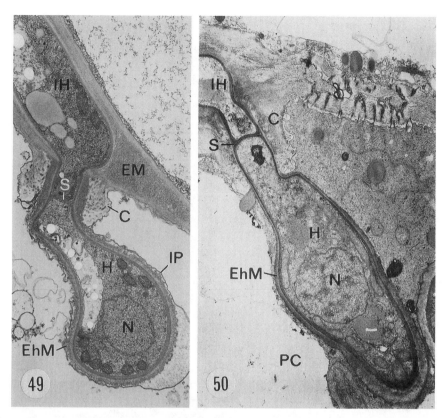

Figure 49. Intracellular hypha (H) of *Cronartium fusiforme* in loblolly pine cell with associated extrahyphal matrix (EhM) and invaginated host plasmalemma (IP). A well-developed collar (C) containing membranous components surrounds the neck region of the H. The EhM appears similar to the extracellular matrix (EM) surrounding the intercellular hyphae (IH). N, nucleus; S, septum. (9960×.) (From Gray *et al.*, 1982.)

Figure 50. Intracellular hypha (H) of *Peridermium pini* in a pine parenchyma cell (PC). Note the collarlike wall apposition (C) and the septum (S). The intercellular hypha (IH) is only slightly constricted at the penetration site. N, nucleus; EhM, extrahyphal matrix. (10,000×.) (From Walles, 1974.)

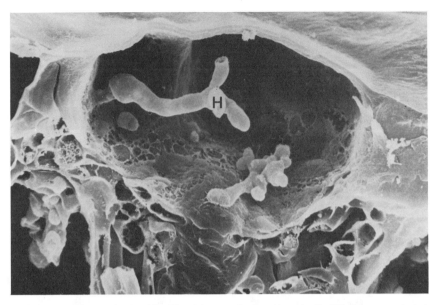

Figure 51. Two intracellular hyphae (H) of *Puccinia recondita* in epidermal cell of *Thalictrum speciosissimum*. Note the branched, hyphalike morphology of the H. (1900×.) (From Gold *et al.*, 1979.)

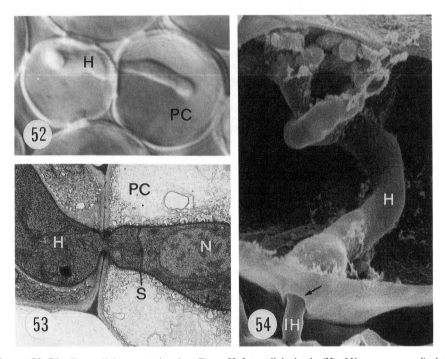

Figures 52–54. Transcellular penetration sites. *Figure 52*. Intracellular hypha (H) of *Uromyces appendiculatus* between two parenchyma cells (PC). (1590×.) (From Gold and Mendgen, 1984b.) *Figure 53*. Intracellular hypha (H) of *Gymnosporangium fuscum* in pear parenchyma cells (PC) with septum (S) prior to penetration site. N, nucleus. (8000×.) (Courtesy of B. Metzler, unpublished.) *Figure 54*. Coiled intracellular hypha (H) of *Puccinia malvacearum* in parenchyma cell of hollyhock. Note the branched. hyphalike nature of the H and exit from the host cell (arrow) to become an intercellular hypha (IH). (5400×.) (From R. E. Gold and L. J. Littlefield, unpublished.)

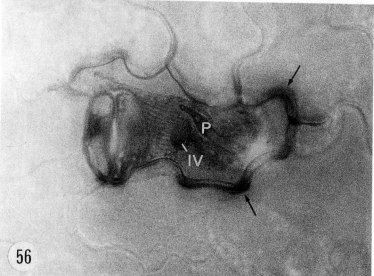

Figure 55. Basidiospore germlings (Bs) of *Cronartium ribicola* on a leaf of a resistant pine cultivar. The germ tubes grew randomly across the wax-covered epidermis without attempting to penetrate the wax-occluded stomatal antechamber (arrow). (500×.) (Courtesy of R. F. Patton, unpublished.)

Figure 56. Attempted penetration by basidiospore germling of *Uromyces vignae* on the nonhost *Vicia faba*. The germling was removed during processing for light microscopy. The incompatible host–parasite interaction resulted in necrosis of the epidermal cell; note rippled appearance of the epidermis and darker-stained anticlinal walls (arrows). The necrotic intraepidermal vesicle (IV) and primary hypha (P) are visible. (500×.) (Courtesy of S. Freytag, unpublished.)

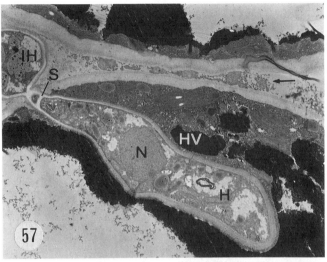

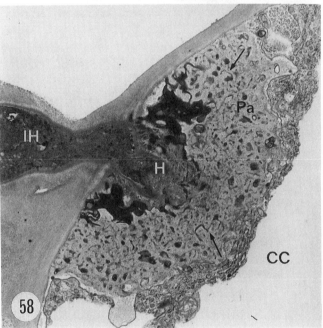

Figure 57. Longitudinal section through a disorganized intracellular hypha (H) of *Cronartium ribicola* in an axenic-cultured pine cell. The H is surrounded by host vacuoles (HV) that contain densely staining tannin deposits and the intercellular space is filled with cellular debris (arrow). IH, intercellular hypha; N, nucleus; S, septum. (4320×.) (From Robb *et al.*, 1975.)

Figure 58. An aborted intracellular hypha (H) of *Cronartium fusiforme* in a cortical cell (CC) of a resistant loblolly pine seed line. A large papillar wall apposition (Pa) formed in response to attempted penetration. Note the heterogeneous appearance of the Pa and abundance of electron-dense membranous material (arrows) in the apposition and in the adjacent host cytoplasm. The intercellular hypha (IH) is also necrotic. (12,000×.) (From Gray and Amerson, 1983.)

ACKNOWLEDGMENTS.   We thank L. Bruscaglioni, S. Freytag, B. Metzler, and R. F. Patton for use of their unpublished micrographs and S. Freytag for putting together her unpublished data in Fig. 20. We thank C. W. Mims and E. A. Richardson for use of their micrographs prior to publication and colleagues who provided prints or negatives of their previously published cytological results. We thank A. Akers and J. B. Speakman for critical reading of the manuscript; especially for their suggestions and discussions. Finally, we gratefully acknowledge H. Fendrich (Universität Konstanz) for preparation of the graphics and the BASF AG for technical support in the photographic laboratory and word processing.

Note added in proof: Since this chapter was written, experiments with basidiospores of *M. pinitorqua* on shoots of *P. sylvestris* have provided information on the effect of the epicuticular wax structure on pre-penetration development of germlings (Desprez-Loustau and Le Menn, 1989). Also, two reports on host response to basidiospore infections have appeared (Heath, 1989; Hopkins and Reid, 1988).

6. REFERENCES

Al-Khesraji, T. O., and Lösel, D. M., 1981, The fine structure of haustoria, intracellular hyphae and intercellular hyphae of *Puccinia poarum*, *Physiol. Plant Pathol.* **19**:301–311.

Allen, R. F., 1930, A cytological study of heterothallism in *Puccinia graminis*, *J. Agric. Res.* **40**:585–614.

Allen, R. F., 1932, A cytological study of heterothallism in *Puccinia coronata*, *J. Agric. Res.* **45**:513–541.

Allen, R. F., 1933, A cytological study of the teliospores, promycelia, and sporidia in *Puccinia malvacearum*, *Phytopathology* **23**:572–586.

Allen, R. F., 1934, A cytological study of heterothallism in flax rust, *J. Agric. Res.* **49**:765–791.

Allen, R. F., 1935, A cytological study of *Puccinia malvacearum* from the sporidium to the teliospore, *J. Agric. Res.* **51**:801–818.

Anikster, Y., 1983, Binucleate basidiospores—A general rule in rust fungi, *Trans. Br. Mycol. Soc.* **81**:624–626.

Anikster, Y., and Wahl, I., 1985, Basidiospore formation and self-fertility in *Puccinia mesnieriana*, *Trans. Br. Mycol. Soc.* **84**:164–167.

Anikster, Y., Moseman, J. G., and Wahl, I., 1980, Development of basidia and basidiospores in *Uromyces* species on wild barley and Liliaceae in Israel, *Trans. Br. Mycol. Soc.* **75**:377–382.

Ashworth, D., 1931, *Puccinia malvacearum* in monosporidial culture, *Trans. Br. Mycol. Soc.* **16**:177–202.

Ashworth, D., 1935, An experimental and cytological study of the life history of *Endophyllum sempervivi*, *Trans. Br. Mycol. Soc.* **19**:240–258.

Bauer, R., 1986, Basidiosporenentwicklung und -keimung bei Heterobasidiomyceten. Teil A: Experimentell-ontogenetische und karyologische Untersuchungen an keimenden Rostpilzbasidiosporen, *Ber. Dtsch. Bot. Ges* **99**:67–81.

Bauer, R., 1987, Uredinales—germination of basidiospores and pycnospores, *Stud. Mycol.* **30**:111–125.

Bauer, R., and Oberwinkler, F., 1988, Nuclear degeneration during ballistospore formation of *Cronartium asclepiadeum* (Uredinales), *Bot. Acta* **101**:272–282.

Bega, R. V., 1960, The effect of environment on germination of sporidia in *Cronartium ribicola*, *Phytopathology* **50**:61–69.

Blank, L. M., and Leathers, C. R., 1963, Environmental and other factors influencing development of southwestern cotton rust (*Puccinia stakmanii*), *Phytopathology* **53**:921–928.

Borland, J., and Mims, C. W., 1980, An ultrastructural comparison of the aecial and telial haustoria of the autoecious rust *Puccinia podophylli*, *Mycologia* **72**:767–774.

Bushnell, W. R., and Roelfs, A. P. (eds.), 1984, *The Cereal Rusts*, Volume I, Academic Press, New York.

Chong, J., Harder, D. E., and Rohringer, R., 1981, Ontogeny of mono- and dikaryotic rust haustoria: cytochemical and ultrastructural studies, *Phytopathology* **71**:975–983.

Colley, R. H., 1918, Parasitism, morphology, and cytology of *Cronartium ribicola*, *J. Agric. Res.* **15**:619–660.

Cotter, R. U., 1932, Factors affecting the development of the aecial stage of *Puccinia graminis*, *U. S. Dep. Agric. Tech. Bull.* **314**:1–37.

Cutler, D. F., Alvin, K. L., and Price, C. E. (eds.), 1982, *The Plant Cuticle*, Academic Press, New York.

de Bary, A., 1863, Recherches sur le développement de quelches champignons parasites, *Ann. Sci. Nat. Bot.* **20**: 5–148.

de Bary, A., 1865, Neue Untersuchungen über die Uredineen, insbesondere die Entwicklung der *Puccinia graminis* und den Zusammenhang derselben mit *Aecidium berberidis*, *Monatsber. K. Akad. Wiss. Berlin* **1865**:15–50.

de Bary, A., 1866, Neue Untersuchungen über Uredineen, *Monatsber. K. Akad. Wiss. Berlin* **1866**:205–215.

de Bary, A., 1884, *Vergleichende Morphologie und Biologie der Pilze Mycetozoen und Bakterien*, Engelmann, Leipzig.

Desprez-Loustau, M.-L., 1986, Physiologie *in vitro* des basidiospores de *Melampsora pinitorqua*: conséquences pour la compréhension des infections, *Eur. J. For. Pathol.* **16**:193–206.

Desprez-Loustau, M.-L. and Le Menn, R., 1989, Epicuticular waxes and *Melampsora pinitorqua* Rostr. preinfection behaviour on maritime pine shoots, *Eur. J. For. Pathol.* **19**:178–188.

Diner, A. M., and Mott, R. L., 1982a, Axenic cultures from basidiospores of *Cronartium ribicola*, *Can. J. Bot.* **60**:1950–1955.

Diner, A. M., and Mott, R. L., 1982b, A rapid axenic assay for hypersensitive resistance of *Pinus lambertiana* to *Cronartium ribicola*, *Phytopathology* **72**:864–865.

Diner, A. M., and Mott, R. L., 1982c, Direct inoculation of five-needle pines with *Cronartium ribicola* in axenic culture, *Phytopathology* **72**:1181–1184.

Eriksson, J., 1911, Der Malvenrost (*Puccinia malvacearum* Mont.), seine Verbreitung, Natur und Entwicklungsgeschichte, *K. Sven. Vetenskapsakad. Handl.* **47**:1–125.

Fischer, E., 1898, Entwicklungsgeschichtliche Untersuchungen über Rostpilze, *Beitr. Kryptogamenflora Schweiz* **1**:1–121.

Fischer, E., 1901, *Aecidium elatinum* Alb. et Schw. der Urheber des Weisstannen-Hexenbesens und seine Uredo- und Teleutosporenform, *Z. Pflanzenkr.* **11**:321–343.

Freytag, S., Bruscaglioni, L., Gold, R. E., and Mendgen, K., 1988, Basidiospores of rust fungi (*Uromyces* species) differentiate infection structures *in vitro*, *Exp. Mycol.* **12**:275–283.

Gasparrini, G., 1848, Osservazioni sulla generazione delle spore nel *Podisoma fuscum*, *Rend. R. Accad. Sci. Napoli* **7**:346–356.

Gold, R. E., and Littlefield, L. J., 1979, Light and scanning electron microscopy of the telial, pycnial, and aecial stages of *Melampsora lini*, *Can. J. Bot.* **57**:629–638.

Gold, R. E., and Mendgen, K., 1983, Activation of teliospore germination in *Uromyces appediculatus* var. *appendiculatus*. II. Light and host volatiles, *Phytopathol. Z.* **108**:281–293.

Gold, R. E., and Mendgen, K., 1984a, Cytology of basidiospore germination, penetration, and early colonization of *Phaseolus vulgaris* by *Uromyces appendiculatus* var. *appendiculatus*, *Can. J. Bot.* **62**:1989–2002.

Gold, R. E., and Mendgen, K., 1984b, Vegetative development of *Uromyces appendiculatus* var. *appendiculatus* in *Phaseolus vulgaris*, *Can. J. Bot.* **62**:2003–2010.

Gold, R. E., and Mendgen, K., 1984c, Cytology of teliospore germination and basidiospore formation in *Uromyces appendiculatus* var. *appendiculatus*, *Protoplasma* **119**:150–155.

Gold, R. E., Littlefield, L. J. and Statler, G. D., 1979, Ultrastructure of the pycnial and aecial stages of *Puccinia recondita*, *Can. J. Bot.* **57**:74–86.

Gopinathan Nair, K. R., 1972, A unique mode of multiplication of basidiospores in *Ravenelia hobsoni* (Uredinales), *Experientia* **28**:604–605.

Gould, C. J., and Shaw, C. G., 1969, Spore germination in *Chyrsomyxa* spp., *Mycologia* **61**:694–717.

Gray, D. J., and Amerson, H. V., 1983, *In vitro* resistance of embryos of *Pinus taeda* to *Cronartium quercuum* f. sp. *fusiforme*: Ultrastructure and histology, *Phytopathology* **73**:1492–1499.

Gray, D. J., Amerson, H. V., and Van Dyke, C. G., 1982, An ultrastructural comparison of monokaryotic and dikaryotic haustoria formed by the fusiform rust fungus *Cronartium quercuum* f. sp. *fusiforme*, *Can. J. Bot.* **60**:2914–2922.

Gray, D. J., Amerson, H. V., and Van Dyke, C. G., 1983, Ultrastructure of the infection and early colonization of *Pinus taeda* by *Cronartium quercuum* formae speciales *fusiforme*, *Mycologia* **75**:117–130.

Grill, D., Hafellner, J., and Waltinger, H., 1980, Rasterelektronenmikroskopische Untersuchungen an *Chrysomyxa abietis* befallenen Fichtennadeln, *Phyton* **20**:279–284.

Hansen, E. M., and Patton, R. F., 1975, Types of germination and differentiation of vesicles by basidiospores of *Cronartium ribicola*, *Phytopathology* **65**:1061–1071.

Hansen, E. M., and Patton, R. F., 1977, Factors important in artificial inoculation of *Pinus strobus* with *Cronartium ribicola*, *Phytopathology* **67**:1108–1112.

Harder, D. E., 1978, Comparative ultrastructure of the haustoria in uredial and pycnial infections of *Puccinia coronata avenae*, *Can. J. Bot.* **56**:214–224.

Harder, D. E., and Chong, J., 1984, Structure and physiology of haustoria, in: *The Cereal Rusts*, Volume I (W. R. Bushnell and A. P. Roelfs, eds.), Academic Press, New York, pp. 431–476.

Heath, M. C., 1989, A comparison of fungal growth and plant responses in cowpea and bean cultivars inoculated with urediospores or basidiospores of the cowpea rust fungus, *Physiol. Molec. Plant Pathol.* **34**:415–426.

Hoffman, A. W. H., 1912, Zur Entwicklungsgeschichte von *Endophyllum sempervivi*, *Centralbl. Bakteriol. Abt. II* **32**:137–158.

Hopkin, A. A., and Reid, J., 1988, Cytological studies of the M-haustorium of *Endocronartium harknessii*: Morphology and ontogeny, *Can. J. Bot.* **66**:974–988.

Hopkin, A. A., and Reid, J., 1988, Host cell responses in susceptible hard pine tissue infected with *Endocronartium harknessii*, *Can. J. Bot.* **66**:2511–2517.

Jacobi, W. R., Amerson, H. V., and Mott, R. L., 1982, Microscopy of cultured loblolly pine seedlings and callus inoculated with *Cronartium fusiforme*, *Phytopathology* **72**:138–143.

Jonsson, A., Holmvall, M., and Walles, B., 1978, Ultrastructural studies of resistance mechanisms in *Pinus sylvestris* L. against *Melampsora pinitorqua* (Braun) Rostr. (pine twisting rust), *Stud. For. Suec.* **145**:1–28.

Kapooria, R. G., 1968, Cytological studies of the germinating teliospores and basidiospores of *Puccinia penniseti*, *Neth. J. Plant Pathol.* **74**:2–7.

Kapooria, R. G., 1971, A cytological study of promycelia and basidiospores and the chromosome number in *Uromyces fabae*, *Neth. J. Plant Pathol.* **77**:91–96.

Kapooria, R. G., and Zadoks, J. C., 1973, Morphology and cytology of the promycelium and the basidiospore of *Puccinia horiana*, *Neth. J. Plant Pathol.* **79**:236–242.

Kinloch, B. B., Jr., and Littlefield, J. L., 1977, White pine blister rust: Hypersensitive resistance in sugar pine, *Can. J. Bot.* **55**:1148–1155.

Knauf, G., and Mendgen, K., 1988, Secretion systems and membrane associated structures in rust fungi after high pressure freezing and freeze fracturing, *Biol. Cell* **64**:351–358.

Koch, E., and Hoppe, H. H., 1988, Development of infection structures by the direct-penetrating soybean rust fungus (*Phakopsora pachyrhizi* Syd.) on artificial membranes, *J. Phytopathol.* (*Berlin*) **122**:232–244.

Kohno, M., Nishimura, T., Ishizaki, H., and Kunoh, H., 1975, Cytological studies on rust fungi. (III) Nuclear behaviors during the process from teliospore stage through sporidial stage in two short-cycled rusts, *Kuehneola japonica* and *Puccinia horiana*, *Bull. Fac. Agric. Mie Univ.* **49**:21–29.

Kohno, M., Ishizaki, H., and Kunoh, H., 1977a, Cytological studies on rust fungi. (VI) Fine structures of infection process of *Kuehneola japonica* (Diet.) Dietel, *Mycopathologia* **61**:35–41.

Kohno, M., Nishimura, T., Noda, M., Ishizaki, H., and Kunoh, H., 1977b, Cytological studies on rust fungi. (VII) The nuclear behavior of *Gymnosporangium asiaticum*, Miyabe et Yamada during the stages from teliospore germination through sporidium germination, *Trans. Mycol. Soc. Jpn.* **18**:211–219.

Krebill, R. G., 1972, Germination of basidiospores of *Cronartium comandrae* on rocks and vegetation, *Phytopathology* **62**:389–390.

Kuhlmann, E. G., 1988, Histology and progression of fusiform rust symptoms on inoculated loblolly pine seedlings, *Plant Dis.* **72**:719–721.

Lindfors, T., 1924, Studien über den Entwicklungsverlauf bei einigen Rostpilzen aus zytologischen und anatomischen Gesichtspunkten, *Sven. Bot. Tidskr.* **18**:1–84.

Littlefield, L. J., and Heath, M. C., 1979, *Ultrastructure of Rust Fungi*, Academic Press, New York.

Longo, N., and Naldini Longo, B., 1982, Behaviour of the monokaryotic haustorium of *Melampsora pinitorqua* (A.Br.) Rostr. Preliminary note, *Caryologia* **35**:471–476.

Longo, N., Moriondo, F., and Naldini Longo, B., 1982, Ultrastructural observations on the host–pathogen interface in infections of *Cronartium flaccidum* on pine, *Caryologia* **35**:307–326.

MacLachlan, J. D., 1935, The dispersal of viable basidiospores of the *Gymnosporangium* rusts, *J. Arnold Arbor. Harv. Univ.* **16**:411–422.

Melander, L. W., and Craigie, J. H., 1927, Nature of resistance of *Berberis* spp. to *Puccinia graminis*, *Phytopathology* **17**:95–114.

Melhus, I. E., Durrell, L. W., and Kirby, R. S., 1920, Relation of the barberry to stem rust in Iowa, *Iowa Agric. Exp. Stn. Res. Bull.* **57**:283–325.

Mendgen, K., Lange, M., and Bretschneider, K., 1985, Quantitative estimation of the surface carbohydrates on the infection structures of rust fungi with enzymes and lectins, *Arch. Microbiol.* **140**:307–311.

Mendgen, K., 1973, Microbodies (glyoxysomes) in infection structures of *Uromyces phaseoli*, *Protoplasma* **78**:477–482.

Mendgen, K., 1984, Development and physiology of teliospores, in: *The Cereal Rusts*, Volume I (W. R. Bushnell and A. P. Roelfs, eds.), Academic Press, New York, pp. 375–398.

Mendgen, K., Schneider, A., Sterk, M., and Fink, W., 1988, The differentiation of infection structures as a result of recognition events between some biotrophic parasites and their hosts, *J. Phytopathol.* (*Berlin*) **123**:259–272.

Metzler, B., 1982, Untersuchungen an Heterobasidiomyceten (23): Basidiosporenkeimung und Infektionsvorgang beim Birnengitterrost, *Phytopathol. Z.* **103**:126–138.

Miller, T., Patton, R. F., and Powers, H. R., Jr., 1980, Mode of infection and early colonization of slash pine seedlings by *Cronartium quercuum* f. sp. *fusiforme*, *Phytopathology* **70**:1206–1208.

Mims, C. W., 1977, Fine structure of basidiospores of the cedar-apple rust fungus *Gymnosporangium juniperi-virginianae*, *Can. J. Bot.* **55**:1057–1063.

Mims, C. W., and Richardson, E. A., 1989, Ultrastructure of appressorium development by basidiospore germlings of the rust fungus *Gymnosporangium juniperi-virginianae*, *Protoplasma* **148**:111–119.

Mims, C. W., Roberson, R. W., and Richardson, E. A., 1988, Ultrastructure of freeze-substituted and chemically fixed basidiospores of *Gymnosporangium juniperi-virginianae*, *Mycologia* **80**:356–364.

Nusbaum, C. J., 1935, A cytological study of the resistance of apple varieties to *Gymnosporangium juniperi-virginianae*, *J. Agric. Res.* **51**:573–596.

Olive, L. S., 1949, A cytological study of typical and atypical basidial development in *Gymnosporangium clavipes*, *Mycologia* **41**:420–426.

Pady, S. M., 1935, The role of intracellular mycelium in systemic infections of *Rubus* with the orange-rust, *Mycologia* **27**:618–637.

Patton, R. F., and Johnson, D. W., 1970, Mode of penetration of needles of eastern white pine by *Cronartium ribicola*, *Phytopathology* **60**:977–982.

Patton, R. F., and Spear, R. N., 1980, Stomatal influences on white pine blister rust infection, *Phytopathol. Mediterr.* **19**:1–7.

Petersen, R. H., 1974, The rust fungus life cycle, *Bot. Rev.* **40**:453–513.

Poirault, G., and Raciborski, M., 1895, Sur les noyaux des Urédinées, *J. Bot.* **9**:381–388.

Ragazzi, A., and Dellavalle Fedi, I., 1982, Observations under fluorescence on progress of basidiospore germination in *Cronartium flaccidum* (Alb. et Schw.) Wint. on the needle surface of certain pine species, *Eur. J. For. Pathol.* **12**:246–251.

Ragazzi, A., Dellavalle Fedi, I., and Mesturino, L., 1986, *Cronartium flaccidum* (Alb. et Schw.) Wint. spores: Temperature requirements for germination, *Phytopathol. Mediterr.* **25**:57–60.

Ragazzi, A., Fagnani, A., Dellavalle Fedi, I., and Mesturino, L., 1987, *Pinus pinaster* e *Pinus sylvestris*: due specie a diverso comportamento verso *Cronartium flaccidum*, *Phytopathol. Mediterr.* **26**:81–84.

Ráthay, E., 1881, Über das Eindringen der Sporidien-Keimschläuche der *Puccinia malvacearum* Mont. in die Epidermiszellen der *Althaea rosea*, *Verh. Zool. Bot. Ges. Wien* **31**:9–10.

Reed, H. S., and Crabill, C. H., 1915, The cedar rust disease of apples caused by *Gymnosporangium juniperi-virginianae* Schw., *Va. Agric. Exp. Stn. Tech. Bull.* **9**:1–106.

Reess, M., 1865, *Chrysomyxa abietis* Unger und die von ihr verursachte Fichtennadelnkrankheit, *Bot. Ztg.* **23**:385–388.

Reess, M., 1870, Die Rostpilzformen der deutschen Coniferen, *Abh. Naturforsch. Ges. Halle* **11**:50–118.

Rijkenberg, F. H. J., and Truter, S. J., 1972, Haustoria and intracellular hyphae in the rusts, *Phytopathology* **62**:281–286.

Robb, J., Harvey, A. E., and Shaw, M., 1975, Ultrastructure of tissue cultures of *Pinus monticola* infected by *Cronartium ribicola*. II. Penetration and post-penetration, *Physiol. Plant Pathol.* **5**:9–18.

Robinson, W., 1914, Some experiments on the effect of external stimuli on the sporidia of *Puccinia malvacearum* (Mont.), *Ann. Bot. (London)* **28**:331–340.

Roncadori, R. W., 1968, The pathogenicity of secondary and tertiary basidiospores of *Cronartium fusiforme*, *Phytopathology* **58**:712–713.

Sappin-Trouffy, P., 1896, Recherches histologiques sur les Urédinées, *Botaniste* **5**:59–244.

Savile, D. B. O., 1939, Nuclear structure and behavior in species of the Uredinales, *Am. J. Bot.* **26**:585–609.

Siggers, P. V., 1947, Temperature requirements for germination of spores of *Cronartium fusiforme*, *Phytopathology* **37**:855–864.

Snow, G. A., 1968a, Basidiospore production by *Cronartium fusiforme* as affected by suboptimal temperatures and preconditioning of teliospores, *Phytopathology* **58**:1541–1546.

Snow, G. A., 1968b, Time required for infection of pine by *Cronartium fusiforme* and effect of field and laboratory exposure after inoculation, *Phytopathology* **58**:1547–1550.

Snow, G. A., and Froelich, R. C., 1968, Daily and seasonal dispersal of basidiospores of *Cronartium fusiforme*, *Phytopathology* **58**:1532–1536.

Spaulding, P., and Rathbun-Gravatt, A., 1926, The influence of physical factors on the viability of sporidia of *Cronartium ribicola* Fischer, *J. Agric. Res.* **33**:397–433.

Thirumalachar, M. J., 1939, Rust on *Jasmium grandiflorum*, *Phytopathology* **29**:783–792.

Thirumalachar, M. J., 1945, Development of spore-forms and the nuclear cycle in the autoecious opsis rust, *Cystopsora oleae*, *Bot. Gaz.* (*Chicago*) **107**:74–86.

Tulasne, L. R., 1854, Second mémoire sur les Urédinées et les Ustilagineées, *Ann. Sci. Nat. Bot. 4 Ser.* **2**:77–196.

Van Sickle, G. A., 1975, Basidiospore production and infection of balsam fir by a needle rust, *Pucciniastrum goeppertianum*, *Can. J. Bot.* **53**:8–17.

Walkinshaw, C. H., 1978, Cell necrosis and fungus content in fusiform rust-infected loblolly, longleaf, and slash pine seedlings, *Phytopathology* **68**:1705–1710.

Walles, B., 1974, Ultrastructure of the rust fungus *Peridermium pini* (Pers.) Lev., *Stud. For. Suec.* **122**:1–30.

Waterhouse, W. L., 1921, Studies in the physiology of parasitism. VII. Infection of *Berberis vulgaris* by sporidia of *Puccinia graminis*, *Ann. Bot.* (*London*) **35**:557–564.

Weimer, J. L., 1917, Three cedar rust fungi, their life histories and diseases they produce, *Cornell Univ. Agric. Exp. Stn. Bull.* **390**:507–549.

Welter, K., Müller, M., Mendgen, K., 1988, The hyphae of *Uromyces appendiculatus* within the leaf tissue after high pressure freezing and freeze substitution, *Protoplasma* **147**:91–99.

Woods, A. M., and Gay, J. L., 1987, The interface between haustoria of *Puccinia poarum* (monokaryon) and *Tussilago farfara*, *Physiol. Mol. Plant Pathol.* **30**:167–185.

Yamada, S., 1956, Experiments on the epidemiology and control of *Chrysanthemum* white rust, caused by *Puccinia horiana* P. Henn., *Ann. Phytopathol. Soc. Japan* **20**:148–154.

Zeyen, R. J., and Bushnell, W. R., 1981, An in-block, light microscope viewing procedure for botanical materials in plastic embedments; with emphasis on location and selection of host cell–microbe encounter sites, *Can. J. Bot.* **59**:397–402.

5

Attachment of Mycopathogens to Cuticle

The Initial Event of Mycoses in Arthropod Hosts

Drion G. Boucias and Jacquelyn C. Pendland

1. INTRODUCTION

Representative members of all major fungal groups have evolved mechanisms required for infection and subsequent development within invertebrates (for review see McCoy *et al.*, 1988). To date, over 700 species within 90 different genera have been determined to be pathogenic to insect and mite hosts (Roberts and Humber, 1981; Samson *et al.*, 1988). In the majority of cases, the spore stage represents the infectious propagule which attaches to and actively gains entry through the exoskeleton by producing penetrant hyphae (Charnley, 1984; see Chapter 6). In nature, the success of an invertebrate mycopathogen will be determined by its ability to produce and deliver infectious propagules to susceptible host systems. In turn, successful delivery will depend on the particular spore-bearing structure and the physical and chemical properties of the infectious propagule. In part, the array of spores and spore structures observed among the pathogenic fungi reflects their adaptation to the diversity of both the target hosts and their respective habitat.

In terrestrial environments, the spore type may range from dry hydrophobic to sticky hydrophilic conidia (Table I). For example, the fungus *Nomuraea rileyi,* a pathogen of lepidopteran larvae, produces extremely hydrophobic conidia. Larvae succumbing to disease become attached to the plant substrate and produce myriads of short external conidiophores over the insect cadaver. Progeny conidia are passively disseminated by wind and contact susceptible larvae feeding on plant foliage (Kish and Allen, 1978). The hyphomycete *Hirsutella thompsonii,* a pathogen of erophyiid mites, produces a network of mycelia extending away from fungus-killed mites on the leaf substrate (McCoy and Kanavel, 1969). At intervals, erect conidiophores bearing single or multiple hydrophilic conidia are produced on these mycelia and provide a means for increasing the contact frequency of the fungus to the mite host. Whether the mucus coat associated with the adhesive conidia is attractive to host insects, as demonstrated with certain endoparasitic nematophagous fungi (Jansson and Nordbring-Hertz, 1983), remains to be tested.

Drion G. Boucias and Jacquelyn C. Pendland • Department of Entomology and Nematology, University of Florida, Gainesville, Florida 32611-0711.

Table I. Characteristics of the Infectious Propagules of Representative Terrestrial Arthropod Mycopathogens

Fungal species	Surface composition of infectious propagules	Relative degree of wettability	References
Nomuraea rileyi (conidia)	Rodlet layer	Hydrophobic	Boucias *et al.* (1988)
Beauveria bassiana (conidia)	Rodlet layer	Hydrophobic	Boucias *et al.* (1988)
Metarhizium anisopliae (conidia)	Rodlet layer	Hydrophobic	Boucias *et al.* (1988)
Paecilomyces fumosoroseus (conidia)	Rodlet layer	Hydrophobic	Boucias (unpublished)
Conidiobolus obscurus (primary discharged spore)	Mucilaginous coat over rodlet layer	Slightly hydrophilic	Brey and Lagey (1986); Latge *et al.* (1986)
Entomophaga aulicae (primary discharged spore)	Mucilaginous coat		Murrin and Nolan (1987)
Entomophthora muscae (primary discharged spore)	Mucilage present between inner and outer wall layers	Hydrophilic	Eilenberg *et al.* (1986)
Entomophthora planchoniana (primary discharged spore)	Mucilage present between inner and outer wall layers	Hydrophilic	Eilenberg *et al.* (1986)
Entomophthora culicis (primary discharged spore)	Mucilage present between inner and outer wall layers	Hydrophilic	Eilenberg *et al.* (1986)
Neozygites floridana (adhesive capilliconidia)	Mucilaginous attachment sac	Hydrophilic	Nemoto *et al.* (1979)
Neozygites fresenii (adhesive capilliconidia)	Mucilaginous attachment sac	Hydrophilic	Brobyn and Wilding (1977)
Ashersonii aleyrodis	Mucilaginous coat	Hydrophilic	Fransen (1987)
Verticillium lecanii (conidia)	Mucilaginous coat	Hydrophilic	Hall (1981)
Hirsutella thompsonii (conidia)	Mucilaginous coat	Hydrophilic	Samson *et al.* (1980)
Culicinomyces clavisporus (conidia)	Mucilaginous coat	Hydrophilic	Sweeney *et al.* (1980)

The coelomycete *Ashersonii aleyrodis*, a pathogen of sessile whiteflies, produces slimy masses of conidia in pycnidia on colonized insects. These conidia are readily splash-dispersed by rainfall and the mucus coat allows the infectious conidia to remain attached to both the leaf substrate and the host cuticle (Fransen, 1987). The entomophthoraleans, mycopathogens associated with a spectrum of insect hosts (King and Humber, 1981), actively discharge primary conidia away from fungus-killed insect. The primary discharge conidia produced by many entomophthoralean genera upon contact with susceptible hosts produce a penetrant germ tube. In other genera, e.g., *Neozygites*, the primary discharge spore will produce adhesive capilliconidia. The erect posture of *Neozygites* capilliconidiophores will enhance the contact frequency between the capilliconidia and the host.

The majority of research on aquatic invertebrate mycopathogens has been directed at the water molds *Aphanomyces*, *Leptolegnia*, *Legenidium*, and *Coelomomyces*. In all cases, motile spores are the infectious propagules which bind to and encyst on host cuticles. The motile zoospores of *Aphanomyces astaci* have been demonstrated to have a chemotactic response to crayfish cuticle exudates, plant exudates, and other nutrient sources (Cerenius and Soderhall, 1984a). Encystment of the *A. astaci* zoospores may be triggered by various chemical and physical stimuli, increasing the probability of encystment on nonhost substrates (Cerenius *et al.*, 1985). This fungus has developed a strategy for repeated zoospore emergence (Cerenius and Soderhall, 1984b, 1985), increasing its chances of binding to a suitable host. The biflagellate zygote of *Coelomomyces psorophorae*, a fungal pathogen of certain mosquito hosts, has been reported to selectively bind to and encyst susceptible hosts (Zebold *et al.*, 1979).

This review focuses on our current knowledge of the processes involved in the attachment of invertebrate pathogenic fungi to their respective hosts. The attachment of the fungal propagule to the host cuticle is mediated by the chemical components present on the outer layer of the spore wall and the epicuticle. With respect to fungi pathogenic to terrestrial invertebrates, the components responsible for spore–cuticle interaction are preexisting and mediate attachment by a potential complex of specific (e.g., glycoprotein, enzyme) and nonspecific (e.g., electrostatic or hydrophobic) recognition systems (Fargues, 1984). Initial attachment may be reinforced further by either the active secretion of adhesive materials or the modification of spore wall material located at the conidia–cuticle interface. Detailed examination of the processes involved in the fungal propagule–cuticle attachment has been limited to only a few fungal–arthropod systems (Travland, 1979; Michel, 1981; Latge *et al.*, 1988a; Boucias *et al.*, 1988). The majority of studies which have dealt with *in vivo* development of fungal invertebrate pathogens have given little or no attention to the initial event of mycosis.

2. STRUCTURAL STUDIES OF THE OUTER WALL LAYER OF FUNGAL PROPAGULES INFECTIOUS TO ARTHROPOD HOSTS

2.1. Hydrophobic Conidia

Carbon replicas have revealed the surface morphology of several arthropod fungal pathogens (Boucias *et al.*, 1988; Fig. 1). The entomopathogenic fungi *Nomuraea rileyi*, *Beauveria bassiana*, *Metarhizium anisopliae*, and *Paecilomyces fumoso-roseus* possess an outer layer on their dry conidia comprised of interwoven fascicles of rodlets (Fig. 1). This rodlet layer appears to be unique to the conidial stage and has not been detected on the vegetative cells (hyphal bodies, mycelium) of these fungi. The relative size and arrangement of the rodlets may vary according to species. For example, the rodlets on *M. anisopliae* conidia are relatively short and are arranged in linear arrays (>50 rodlets/bundle), whereas those on *N. rileyi* vary in length and are arranged in braided bundles containing 4–12 rodlets/bundle. Culture conditions may also influence the

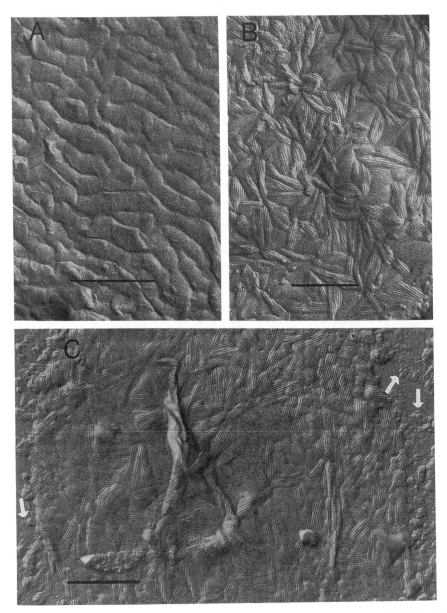

Figure 1. Surface replicas of (A) *Metarhizium anisopliae* and (B, C) *Beauveria bassiana* conidia. In C, rodlets have been partially displaced (arrows) from the cell surface after conidia were suspended in water and vortexed. Samples were shadowed with carbon-platinum at a 45° angle in a vacuum evaporator. Bars = 0.5 μm.

morphological properties of the rodlet layer. The phialoconidia of *B. bassiana* produced on solid media possess a well-organized rodlet layer, whereas conidia produced via the microcycle in broth cultures (Thomas *et al.*, 1987) possess a layer of highly disorganized rodlets.

The presence and stability of the rodlet layer can be correlated to the ability to wet the conidial preparations. Pigmented conidia of *N. rileyi* and *M. anisopliae* possess very resistant rodlet layers. Harsh physical (prolonged sonication) and/or chemical (detergent) treatments are required to suspend these conidia. The resistance of these conidia is believed to be enhanced by the presence of pigments. For example, conidial preparations harvested either from young nonpigmented cultures or from UV-induced white mutant of *N. rileyi* cultures are readily suspended by mild sonication. Similarly, the white conidia of the fungi *B. bassiana* and *P. fumoso-roseus* may be suspended by vortexing in water. The rodlet layer present on these two species appears to be relatively loose and is partly displaced from the conidial surface during suspension (Fig. 1). Again, conidia from young cultures appear to be much easier to suspend than conidia harvested from older cultures. These results suggest that the well-defined rodlet layer detected on the surface of many "dry" conidium types may result from conidial aging and simply reflect a desiccation or hardening process.

2.2. Hydrophilic Conidia

The invertebrate pathogenic fungi producing "hydrophilic" conidia typically lack a well-organized outer rodlet layer. In many cases, the hydrophilic conidia are characterized by the presence of an outer mucilaginous coat produced during spore maturation. For example, the rodlet layer of primary discharge conidia of *Conidiobolus obscurus* is coated with a mucilaginous layer composed of long microfibrils embedded in an amorphous matrix (Latge *et al.*, 1986; Fig. 2). Alternatively, the mucilaginous material may be located between the inner and outer wall layer, as is the case with several species within the genus *Entomophthora* (Eilenberg *et al.*, 1986). Upon impact with a substrate, the mucilaginous coat is released and deposited at the conidium-substrate interface and is believed to serve as a glue (Latge *et al.*, 1988a; Eilenberg *et al.*, 1986; Murrin and Nolan, 1987; Brobyn and Wilding, 1983). Mucilage associated with *Erynia radicans* conidia is believed to interact and modify the epicuticular waxes as indicated by the production of imprints on cuticle (Wraight *et al.*, 1990). With certain entomophthoraleans, e.g., *Neozygites* and *Zoophthora* species, the primary discharge conidia will produce a slender capilliconidiophore. This structure grows in a vertical fashion giving rise to single small adhesive capilliconidia (Nemoto *et al.*, 1979; Brobyn and Wilding, 1977; Glare *et al.*, 1985). These capilliconidia produce an "attaching sac" at their distal ends (Fig. 2) which is believed to mediate attachment to the legs or lower body regions of susceptible mite and/or aphid hosts (Selhime and Muma, 1966; Glare *et al.*, 1985). The hydrophilic conidia of *Hirsutella thompsonii, Verticillium lecanii*, and *Ashersonii aleyrodis* have also been reported to possess a mucus coat which facilitates in both their dispersal by rainfall and their adhesion to host insects (Fransen, 1987; Hall, 1981; McCoy *et al.*, 1988; Latge *et al.*, 1988b; Fig. 3). Carbon replicas of conidia from these fungi failed to reveal the presence of rodlets in the mucilaginous coat (Fig. 3).

2.3. Encystment of Motile Spores

The attachment of the water molds *Lagenidium, Coelomomyces, Leptolegnia*, and *Aphano-myces* spp., to host cuticle is associated with the encystment of the motile spore (Zattau and McInnis, 1987; Travland, 1978; Olson *et al.*, 1984; Brey *et al.*, 1988). For example, the motile zygote of the chytridiomycete *Coelomomyces psorophora*, upon contacting host cuticle, will immediately retract the flagella apparatus (Travland, 1978). Following flagella retraction, the contents of the adhesion vesicles are secreted from small pseudopodia which are in contact with the cuticle. Active secretion of the amorphous material, presumed to contain glycoproteins, firmly binds the naked cyst to the host cuticle. Within 5 min of the onset of encystment, a definite cyst wall is laid down. The primary zoospores of the crayfish mycopathogen, *Aphanomyces astaci*, may be stimulated by various physical and chemical cues to encyst (Cerenius and Soderhall, 1984a). Unlike saprophytic *Aphanomyces* species, the primary cyst of *A. astaci* may

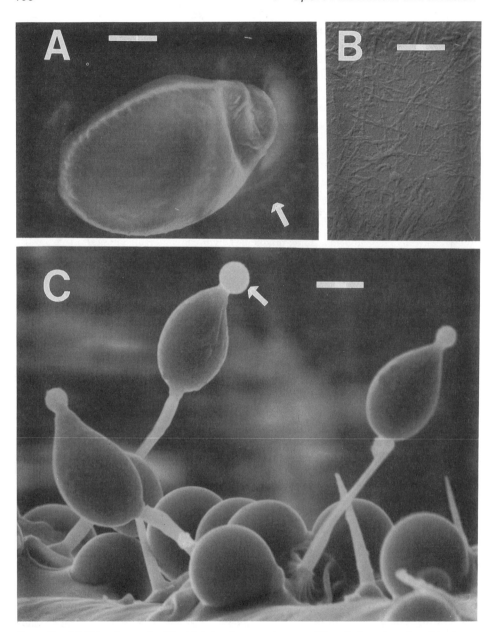

Figure 2. (A) Light micrograph of *Conidiobolus obscurus* primary conidium discharged on glass slide. Arrow indicates mucilaginous halo produced upon impact of conidium with substrate. (B) Replica of microfibrils embedded in the mucilaginous coat of *C. obscurus* conidia. A and B courtesy of J. P. Latge. (C) Scanning electron micrograph of partially freeze-dried capilliconidiophores extending from collapsed primary spores of *Neozygites fresenii* on aphid cuticle. Mucilage drops of some capilliconidia are not fully developed. Arrow indicates mucilage drop. Micrograph by T. M. Butt. Bars = 10 μm (A), 0.5 μm (B), 5 μm (C).

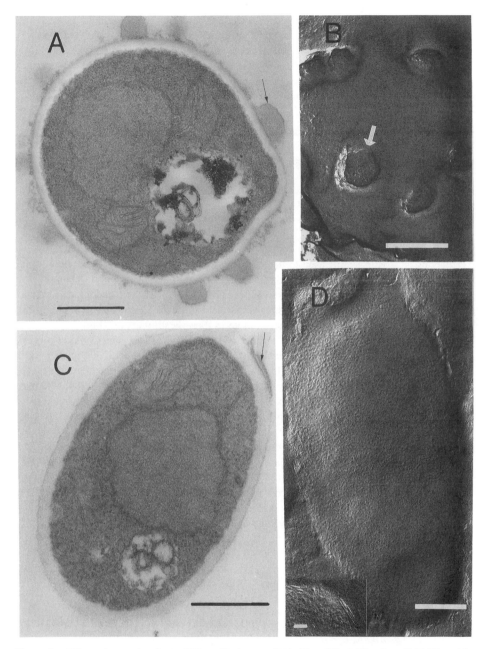

Figure 3. Thin sections and replicas of *Hirsutella thompsonii* (A, B) and *Verticillium lecanii* (C, D) conidia. Mucilaginous material around *H. thompsonii* conidia forms warts (arrows) on the conidial surface. Only a small portion of the mucus coat remains on *V. lecanii* cell wall (arrow, C). Long rodlets are occasionally observed on surface replicas of this fungus (inset, D). Bars = 0.5 μm; inset = 0.1 μm.

produce a germ tube or may release secondary zoospores (Cerenius and Soderhall, 1984b, 1985). Olson *et al.* (1984) reported that the *A. astaci* zoospore encystment is an instantaneous process, flagella dehisce from the spore, peripheral vesicles with granular contents disappear, and a cyst coat is formed around the spore. The outer layer of the mature cyst was speculated to be derived from the expulsion of peripheral vesicles containing flocculent material in a fashion similar to that reported for other Oomycetes (Sing and Bartnicki-Garcia, 1975; Grove and Bracker, 1978). The presence of such a layer strongly suggests that it functions as an adhesive in a manner similar to that reported for plant pathogenic Oomycetes (Nicholson, 1984; see Chapter 1).

2.4. Germ Tubes and Appressoria

Many arthropod pathogenic fungi have been observed to actively secrete mucilage during germ tube and/or appressorial formation (Table II, Fig. 4). For example, *N. rileyi* germ tubes, which do not form appressoria, produce an exocellular sheath which is well stained with ruthenium red, indicating the presence of acidic polysaccharides (Pendland and Boucias, 1984a). At present, the composition of this mucilage is unknown. None of the commercially available FITC-lectins or ferritin-lectins used to detect mannose, galactose, *N*-acetylglucosamine, glucose, and fucose residues on the cell wall of *N. rileyi* bound to the extracellular sheath of germ tubes. Further studies, involving both enzyme treatments ($\beta 1,3$ $\beta 1,6$ exoendoglucanase) and immunocytochemical studies with antilaminaribiosyl antisera and $\beta 1,3$ glucanase colloidal gold, failed to detect the presence of $\beta 1,3$ glucans in germ tube sheath material (Fig. 4).

Mucilage has been detected on germ tubes of the entomophthoraleans *Conidiobolus obscurus* (Brey and Latge, 1986), *Erynia radicans* (Wraight *et al.*, 1990), and *Neozygites fresenii* (Butt, personal communication; Fig. 5). Germ tubes of many arthropod mycopathogens, depending on the host and fungal species, may produce an appressorial cell (Fargues, 1984; Charnley, 1984). In many cases, the secretion of electron-dense mucilage is believed to bind the appressorial cells to the cuticle surface (Zacharuk, 1970b; Schabel, 1978; Travland, 1979; Madelin *et al.*, 1967; Murrin and Nolan, 1987; St. Leger *et al.*, 1989a). The mucilage produced by germ tubes and appressorial structures probably has functions similar to that of preformed mucus associated with hydrophilic conidia. Imprints similar to those observed with mucus-coated conidia of *E. radicans* were also frequently observed in cuticles beneath dislodged germ tubes and appressoria (Wraight *et al.*, 1990). Mucilage produced by penetrant germ tubes or by appressorial cells is hygroscopic and may create a favorable environment for the exocellular enzymes released by these structures (Butt and St. Leger, personal communication).

Table II. Arthropod Mycopathogens Reported to Produce Exocellular Mucilage during Germ Tube or Appressorial Cell Formation

Fungus	Host	Fungal structure	References
Metarhizium anisopliae	Beetle	Appressoria	Zacharuk (1970a,b)
Paecilomyces farinosus	Beetle	Appressoria	Madelin *et al.* (1967)
Entomophaga aulicae	Caterpillar	Germ tube, appressoria	Murrin and Nolan (1987)
Conidiobolus obscurus	Aphid	Germ tube	Brey and Latge (1986)
Neozygites fresenii	Aphid	Germ tube, appressoria	Butt (personal communication)
Coelomomyces psorophorae	Mosquito	Appressoria	Travland (1979)
Erynia radicans	Leafhopper	Germ tube and appressoria	Wraight *et al.* (1990)
Nomuraea rileyi	Caterpillar	Germ tube	Pendland and Boucias (1984a)

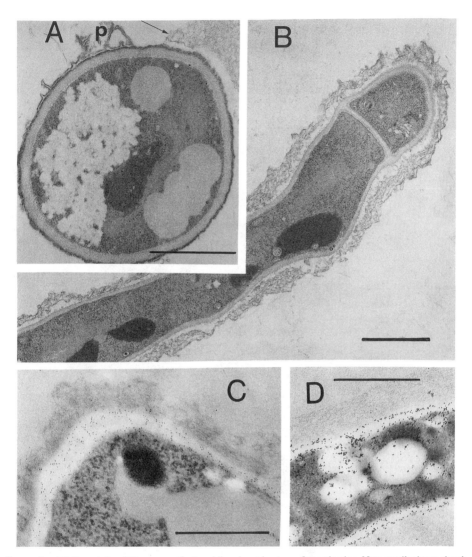

Figure 4. (A) Germ tube of *Nomuraea rileyi* conidium in early stage of germination. Note mucilaginous sheath (arrow) forming at the site of germ tube emergence. The pellicle (P) has become disrupted due to conidial swelling. (B) Germ tube wall surrounded by sheath material. (C) *N*-Acetylglucosamine in germ tube wall labeled by a wheat germ ferritin conjugate. The sheath is not labeled. (D) Inner and outer layers of the germ tube wall labeled by an antilaminaribiosyl (anti-β1,3 glucan) antiserum–colloidal gold probe. Sheath material is not labeled. Bars = 1 μm (A, B, D), 0.5 μm (C).

3. CHEMICAL PROPERTIES OF THE OUTER LAYER OF FUNGAL PROPAGULES INFECTIOUS TO INVERTEBRATES

3.1. Outer Rodlet Layer

The outer rodlet layer of the hydrophobic conidia of *N. rileyi*, *M. anisopliae*, and *B. bassiana* is the structure which initially interfaces with host insect cuticle (Boucias *et al.*, 1988). Attempts to remove this layer by various physiochemical treatments demonstrated the resistant

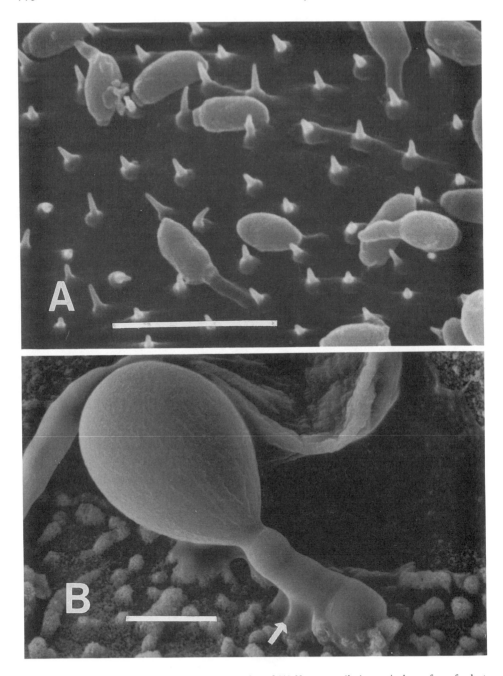

Figure 5. Scanning electron micrographs of germ tubes of (A) *Nomuraea rileyi* on cuticular surface of velvet bean caterpillar and (B) *Neozygites fresenii* on aphid cuticle. Mucilaginous material (arrow) is readily visible on the appressorium in B. B courtesy of T. M. Butt. Bars = 5 μm.

nature of the layer (Table III). Sonication of conidial suspensions at a 70% power setting (300 W Sonic Dismembranator), while removing limited amounts of insoluble and soluble material, produced conidial replicas which were morphologically identical to untreated conidia. Treatment of conidial samples with solvents previously used to extract lipids from fungal samples (Boucias *et al.*, 1984) had no effect on the integrity of the rodlet layer (Fig. 6). Likewise, incubation in various denaturation agents and enzymes (chitinase, protease) or in the sequential chemical and enzymatic treatment used previously for protoplasting vegetative cells (Pendland and Boucias, 1984b), had little or no effect on the integrity of the rodlet layer. Treatment with mild alkali or acid, however, resulted in partial digestion of the rodlet layer (Fig. 6). Analysis of the material removed by the various physiochemical treatments demonstrated that sonication or prolonged incubation in 0.1 M NaOH and 0.1 M HCl removed detectable amounts of sugars (3.0 μg/mg conidia) from *N. rileyi* conidia (Boucias *et al.*, 1988). Incubation in various detergents (1% NP40, Tween 20, Triton X-100, SDS, acetylcetrimide) and solvents inhibited the release of sugars. Mild alkali hydrolysis was the only treatment which released detectable levels of proteins (1.8 μg/mg conidia) from treated *N. rileyi* conidia. Treatment with stronger concentrations of alkali (1.0 M NaOH) effectively removed the rodlet layer. However, such treatment caused significant reductions in both the number and the viability of conidia.

At present, very little is known about the chemical composition of the rodlet layer associated with the hydrophobic conidia of the invertebrate fungal pathogens. Small amounts of rodlet material have been isolated from sonicates of gram quantities of *B. bassiana* conidia (Boucias *et al.*, 1988). Carbon replicas of rodlet preparations extracted by differential centrifugation revealed the presence of fascicles of rodlets embedded in an amorphous matrix. Preparations of dialyzed and lyophilized rodlet preparations contained 40% protein and 10% sugars. Attempts to resolve the structural polypeptides using SDS–PAGE failed to produce detectable bands in Coomassie-stained gels. These results again demonstrated the resistance (insolubility) of the rodlet layer to denaturation agents (Table III). The types of sugars associated with the outer rodlet layer of *N.*

Table III. Chemical Treatments of Conidia and Their Effect on Attachment to Cuticle Ghost Preparations

Physicochemical treatment of conidia[b]	Treatment period	Effect on rodlet layer[a]		
		N. rileyi	*B. bassiana*	*M. anisopliae*
Sonication—setting 30%	1 min	+	+	+
—setting 70%	4,1 min	+	+	+
CHCl$_3$: MEOH (2 : 1) (100°C)	4 hr	+	+	+
Detergents (1% NP40, SDS, Triton X-100, acetylcetrimide)	24 hr	+	+	+
Protease K (37°C)	24 hr	+	+	+
Cell wall lytic enzymes[c]	24 hr	+	+	+
Chitinase (37°C)	24 hr	+	+	+
8 M urea + 1% β-mercaptoethanol	12 hr	+	+	+
0.1 M HCl	24 hr	±	NT[d]	NT
0.1 M NAOH	24 hr	±	±	±
1.0 M NAOH	4 hr	±	−	±
1.0 M NAOH (100°C)	1 hr	−	−	−

[a]The presence (+) or absence (−) of normal intact fascicles of rodlets in carbon replicas of conidial preparations.
[b]Conidial preparations after being incubated with chemicals were washed three times in distilled water prior to being assayed for attachment.
[c]Protocol for using cell wall lytic enzymes was that used for producing protoplasts from vegetative cells of *N. rileyi* (Pendland and Boucias, 1984b).
[d]NT, not tested.

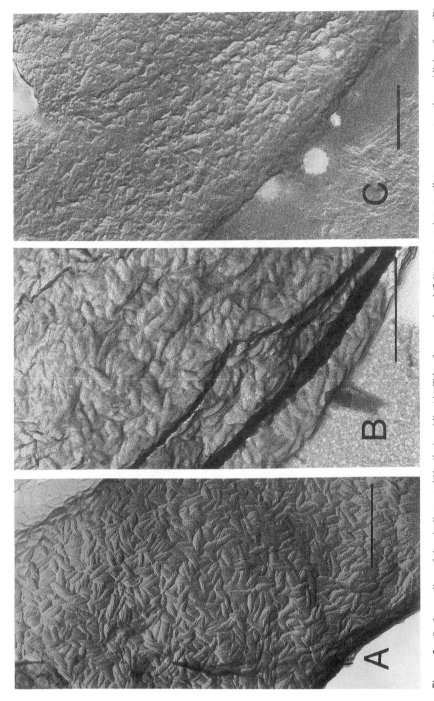

Figure 6. Surface replicas of chemically treated *N. rileyi* conidia. (A) Chloroform–methanol (2:1) treatment: intact rodlets are present on the conidial surface. (B) Urea–mercaptoethanol: rodlets, although not as distinct as in A, are still visible. (C) The outer rodlet layer is seen to be extensively etched by NaOH treatment (0.1 M, 17 hr). Bars = 0.5 μm.

rileyi have been investigated using a series of commercially available FITC- and ferritin-labeled lectins (Pendland and Boucias, 1984c). Both concanavalin A (Con A) and wheat germ agglutinin (WGA) probes faintly stained intact conidia preparations, indicating limited amounts of exposed glucose–mannan and N-acetylglucosamine residues. In areas where rodlets were absent from the conidial surface (onset of germination) the underlying cell wall was intensely stained with both FITC–Con A and FITC–WGA (Pendland and Boucias, 1984c), indicating the presence of exposed N-acetylglucosamine, glucose, and mannose residues.

3.2. Mucus Coat

It has been proposed that the mucus coat associated with hydrophilic conidia may act as an antidesiccant, protect conidia from toxic host polyphenols, contain various enzymatic activities, and act as an adhesive. Very little, if anything, is known of the mucilage associated with the slimy phialoconidia of *V. lecanii* or *H. thompsonii*. However, Latge *et al.* (1984, 1986, 1988) have investigated the chemical composition of the primary discharge spore of *Conidiobolus obscurus*. The spore wall is comprised of an inner electron-translucent thick layer and an outer electron-dense layer comprised of organized rodlets. The entire spore wall is coated in mucilage which was readily stained with horseradish peroxidase and bound to Con A but not WGA. Fibrils associated with the mucus layer were digested with a laminarase solution but were not labeled with antilaminaribiosyl antisera (Latge *et al.*, 1986). The inability of the antilaminaribiosyl antisera to bind to these fibrils may reflect the lack of branching in the 1,3-glucan (Latge *et al.*, 1986) which appears to be required for efficient antiserum binding (Horisberger, personal communication). The fibrils were not tagged with either WGA or peanut agglutinin (PA) and were not susceptible to treatment with papain or chitinase. However, protease solutions readily removed spores attached to a glass substrate, suggesting the presence of proteins or glycoproteins in the mucus layer. Related studies (Latge *et al.*, 1984) have demonstrated the presence of an exocellular carbohydrate-binding protein on *C. obscurus* ballistospores believed to be responsible for the observed binding between mucus fibrils and horseradish peroxidase (Latge *et al.*, 1988a). Underlying the mucus layer is a well-defined rodlet layer which upon impact with host cuticle and subsequent displacement of the mucus layer may contact the aphid cuticle in combination with the amorphous and fibrillar components of the mucus.

3.3. Carbohydrate-Binding Proteins Present on Fungal Propagules

Recognition of carbohydrate moieties located on host cuticle by sugar-binding proteins has been suggested to play a role in the adhesion of arthropod mycopathogens (Grula *et al.*, 1984; Kerwin and Washino, 1986; Latge *et al.*, 1984, 1988a). The best-studied systems where such fungal lectins are involved in recognition are the nematode fungal pathogens *Arthrobotrys oligospora* and *Meria coniospora* (Jansson and Nordbring-Hertz, 1983; Nordbring-Hertz and Mattiasson, 1979; Nordbring-Hertz, 1988). In both of these cases, carbohydrate-binding proteins were detected either on the cell wall of the trapping mechanism or on the conidial wall. Adhesion of fungal propagules to host nematodes was inhibited by incubating the fungal nematode trap with specific hapten sugars or by removing sugar residues with specific carbohydrases. In the case of arthropod mycopathogens, possession of such a recognition system has yet to be clearly defined. The motile zygotes of *Coelomomyces*, which preferentially encyst on susceptible hosts but not on resistant mosquito hosts (Zebold *et al.*, 1979), may possess carbohydrate-binding proteins which do recognize cuticular carbohydrate (Kerwin, 1983a). To date, neither the binding protein nor its cuticular carbohydrate has been identified for this fungus–insect system.

Various investigators have reported the presence of hemagglutinins (HA) associated with conidia of other entomopathogenic fungi (Ishikawa *et al.*, 1979, 1983; Lange *et al.*, 1988a; Grula *et al.*, 1984; Boucias *et al.*, 1988). Latge *et al.* (1988a) detected carbohydrate-binding proteins in the mucus coat of the entomophthoralean *C. obscurus* which had specificity to glucose and *N*-acetylglucosamine residues. Using various neoglycoproteins conjugated to fluorescent dyes and colloidal gold, these sugar-binding proteins were detected throughout the slime layer and on the wall outer layer of *C. obscurus* conidia (Latge *et al.*, 1988a). Similarly, Grula *et al.* (1984) also reported HA on the conidial surface of *B. bassiana* which could be released by mild sonication and could be inhibited by various hapten sugars (i.e., glucose, glucosamine, and *N*-acetyl-*D*-glucosamine). Incubating *B. bassiana* conidial samples with these hapten sugars reduced their affinity to insect larvae. Recently, hemagglutinins were detected from conidial samples of a variety of *B. bassiana*, *M. anisopliae*, and *N. rileyi* isolates (Boucias *et al.*, 1988; Table IV). Untreated conidia, sonicated conidia, and supernates from conidial sonicates all contained detectable HA against two or more vertebrate erythrocyte preparations. Conidial preparations of *M. anisopliae* produced the highest titer against rabbit and human O red blood cells, whereas, *B. bassiana* conidial preparations produced the highest titer against sheep red blood cells. Competitive inhibition with various simple and complex carbohydrates demonstrated the *B. bassiana* HA could be inhibited (one well) by *N*-acetylglucosamine (200 mM), whereas none of the carbohydrates assayed inhibited the *N. rileyi* HA (Boucias *et al.*, 1988). The *M. anisopliae* (strain 5507) HA is heat-sensitive and could be inhibited by glucosyl and galactosyl haptens, as well as Ca^{2+} chelating agents (EDTA). This carbohydrate-binding protein, extracted from supernates of sonicated conidial preparations using a fucose–agarose affinity column, had a molecular mass of 32,000 daltons.

The role that the cell wall carbohydrate-binding proteins play in the conidium–cuticle interaction is unclear. The primary ballistospores of *C. obscurus* are capable of binding to noncuticle substrates (glass slides) as well as cuticle which lacks exposed glucose and *N*-acetylglucosamine residues on the epicuticular surface (Latge *et al.*, 1988a). Second, certain *C. obscurus* strains, lacking detectable sugar-binding proteins on the conidial surface, adhered to pea aphid cuticle as well as the *C. obscurus* strains which possessed detectable sugar-binding proteins. Competitive inhibition studies with various hapten sugars, including those that were reported to be inhibitory for *B. bassiana* (Grula *et al.*, 1984), failed to reduce the binding of *B. bassiana*, *M. anisopliae*, and *N. rileyi* to insect cuticle substrates (Boucias *et al.*, 1988). Furthermore, ferritin and rhodamine conjugates of the extracted *M. anisopliae* fucose lectin did not bind to the epicuticular surface. Possibly, the sugar-binding proteins detected in the conidial

Table IV. Hemagglutination of Various Red Blood Cell Preparations by Conidia and Sonicated Extracts of Conidia

	Hemagglutination titer		
Fungus	Rabbit	Human O	Sheep
Nomuraea rileyi (FL74-6)			
Dry conidia	4	0	0
Sonicated conidia	2–16	2	0
Supernate from sonicate	16	32	0
Beauveria bassiana (#5477)			
Dry conidia	4	0	16
Sonicated conidia	4–8	8	8–16
Supernate from sonicate	2	2	8
Metarhizium anisopliae (#5507)			
Dry conidia	4–64	128	4
Supernates from sonicate (1 hr 60°C)	2048–4096	512–4096	tr[a]
	2	tr[a]	0

[a]tr,trace levels of HA detected.

wall have little or nothing to do with adhesion. Latge *et al*. (1988a) speculated that the primary role of such components may be in the construction of the cell wall matrix.

3.4. Enzyme Activity Associated with Fungal Propagules

In addition to HA, various enzyme activities have been detected in pregerminating conidia as well as in the exocellular mucus produced by germ tubes and appressorial cells (David, 1967; Soderhall *et al*., 1978; Zacharuk, 1970a; Michel, 1981; St. Leger *et al*., 1986, 1988a,b, 1989b). *In situ* histochemical staining revealed the presence of esterase, lipase, and *N*-acetylglucosamini-dase in pregerminating *B. bassiana* conidia deposited on larval cuticle. In screening selected *B. bassiana*, *M. anisopliae*, and *N. rileyi* with commercially available substrates, a spectrum of enzyme activities were detected in conidial preparations (Table V). Fargues (1984) speculated that the enzymes present on the conidial surface may assist in the consolidation of the infectious propagule to the cuticle surface. Enzymes having esterase, protease, and/or lipase activity may be used to digest the epicuticle. Histological studies of pregerminating conidia on cuticle suggest that modification of the epicuticle does occur before germ tube formation (David, 1967). Potentially, the primary function of many of the enzymes associated with the deuteromycete conidia is to hydrolyze the epiculticular wax layer and provide nutrients required for germ tube formation. Various exocellular enzymes have been also reported to be produced by germ tubes and appressorial cells. For example, *A. astaci* cysts upon germ tube emergence and *M. anisopliae* during appressorial cell formation release a complex of endopeptidases (Persson *et al*., 1984; St. Leger *et al*., 1986, 1989a,b). Recent studies by St. Leger *et al*. (1989b) demonstrated that chymoelastase, the major cuticle-degrading endopeptidase of *M. anisopliae*, is the major protein synthesized by appressorial cells. Such exoenzymes, by selectively degrading the cuticle, may further consolidate the appressorial structures to the host cuticle (see Chapter 6).

4. STRUCTURE AND COMPOSITION OF THE EPICUTICLE: THE TARGET SUBSTRATE FOR SPORE ADHESION

The initial host tissue which the majority of mycopathogens contact is the cuticle. In terms of fungal development, the cuticle has a dual function: it is the substrate for conidial attachment,

Table V. Enzyme Activities Detected in Pregerminating Conidia of N. rileyi, M. anisopliae, and B. bassiana

Enzyme systems	Relative levels of enzyme activity[a]		
	N. rileyi	*M. anisopliae*	*B. bassiana*
Alkaline phosphatase	0.5	3.5	0.5
Esterase lipase (C8)	3.0	3.0	2.0
Lipase (C14)	0	0	0.5
Leucine aminopeptidase	2.5	3.0	5.1
Acid phosphatase	3.0	5.0	1.5
Phosphohydrolase	3.0	5.1	1.5
β-Glucosidase	3.5	2.0	1.0
N-Acetyl-β-glucosaminidase	1.0	5.0	1.5
α-Mannosidase	0	2.0	0

[a]Two hundred microliters (10[6] spores) added to wells of APIZYM (Analytab Products) incubated for 1 hr at 37°C and stained. Relative rankings of activity correspond to ≥40 nmoles for 5.0 and to 5 nmoles for 1.0.

and it provides the necessary chemical cues for the production of the penetrant germ tube. Both the surface topography and the chemical properties of the host cuticle are believed to play a role in the attachment of fungal propagules to host cuticle. The integument, composed of the epidermal layer and its secretion product, the cuticle, provides arthropods with a rigid matrix for supporting internal tissues and a protective barrier from both the external environment and invading pathogens (David, 1967). The extremely variable topography of the arthropod cuticle has served as the major morphological determinant in resolving more than 1 million arthropod species. The integument is a highly dynamic tissue and may alter the cuticle topography as it undergoes larval (nymphal), pupal, and adult molts. In addition to altering the surface structure of the cuticle, the molting process may alter the chemical properties of the cuticle.

The epicuticle, the outer layer of the cuticle, represents the substrate on which the fungal propagules initially will contact, adhere, and breach. Unlike the relatively thick procuticle, which is composed of proteins complexed to chitin fibrils, the epicuticle is comprised of proteins, lipoproteins, phenolic compounds, and lipids. Histological studies have revealed the epicuticle to be composed of two to four layers, including the inner layer, cuticulin layer, wax layer, and a cement layer (Filshie, 1982). More recently, Locke (1984) proposed the presence of two functional layers: the outer cuticulin envelope and the intermediate layer. The cuticulin envelope, the initial layer formed at molting, undergoes tanning by the action of cellular phenols and polyphenoloxidases, providing both chemical resistance and hydrophobic properties (Locke, 1984). In electron micrographs, this outer cuticulin layer appears to be very electron-dense (Fig. 7). The lipids and waxes secreted by the epidermal cells through lipid channels will overlay this hydrophobic layer, providing an effective barrier to water loss (Hadley, 1981).

Current understanding of epicuticle chemistry has been hindered by the relative resistance of the epicuticle to chemical treatment and the technical problems associated with obtaining pure epicuticle (Andersen, 1979). Locke (1984), in reference to the chemistry of the cuticulin layer, suggested that it is "the most under researched topic in proportion to its importance in the whole of applied biology." Gross extraction of cuticle or exuviae preparations has provided insight into the complex array of lipids and proteins associated with cuticle (Hackman, 1984; Lockey, 1988). Approximately 50–100 different proteins, comprising 50% of the cuticle's weight, have been detected in cuticle extracts (Hackman, 1984). Using two-dimensional gel analysis of guanidine hydrochloride extracts of cuticle, Cox and Willis (1987) resolved overlapping protein families from cuticle regions having distinct functional properties. In general, flexible cuticle extracts (e.g., intersegmental regions), characterized by the presence of both highly acidic proteins and many glycosylated proteins, contained many more proteins than did rigid cuticle preparations (Cox and Willis, 1985, 1987; Willis, 1987). Mannose and *N*-acetylglucosamine were the main sugars associated with glycosylation of the flexible cuticle proteins. Due to differences in amino acid compositions, the hydrophobicity in rigid (sclerotized) cuticles has been estimated to be significantly greater than in flexible cuticles (Andersen, 1979). Whether or not the differences in amino acid and protein content in flexible versus rigid cuticles reflect variation in the epicuticle or are due entirely to differences in the underlying procuticle layer is unclear. In addition to the structural proteins, a variety of enzyme activities including phenoloxidases, esterase, phosphatase, phosphorylase, and peroxidase have been detected in cuticle preparations (Hackman, 1984).

The epicuticle contains large amounts of bound and free lipids. The major function of the epicuticular lipids is to act as a water barrier. Other functions such as chemical messengers have been assigned to epicuticle hydrocarbons (Howard and Bloomquist, 1982). Very little is known about the bound lipids associated with the epicuticle layer. Hackman (1986), sequentially treating cuticle by boiling in chloroform/methanol, boiling in HCl 24 hr, followed by hot 1 M KOH treatment, was able to isolate the outer epicuticle from *Lucilia cuprina* pupae. Chemical analysis revealed the presence of high-molecular-weight cyclic or cross-linked structures composed essentially of methylene groups with small numbers of aromatic and carbonyl groups (Hackman,

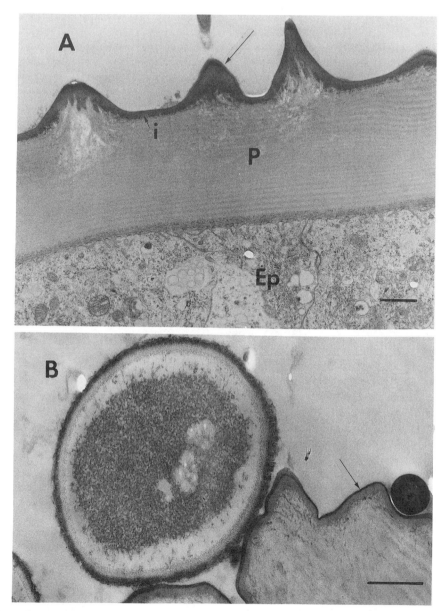

Figure 7. (A) Thin section of cuticle from third-instar velvet bean caterpillar. Arrow indicates cuticulin layer of epicuticle; I = intermediate layer of epicuticle; P = procuticle; EP = epidermis. (B) *N. rileyi* conidium attached to cuticle ghost. Boiling in 1% SDS did not remove conidia from ghost cuticle. Dark-staining cuticulin layer (arrow) is easily discernible in sections of these preparations. Bars = 1 µm (A), 0.5 µm (B).

1986). The majority of epicuticle lipid research to date has been concerned with the "free" lipid fraction which is readily extracted from cuticle preparations with organic solvents. Both intrinsic (host species, age, physiological stage) and extrinsic factors (nutritional status, environmental conditions) may be responsible for the observed qualitative and quantitative differences in extractable cuticular lipids. Representative wax esters, glycerides, free fatty acids, sterols, alcohols, aldehydes, and hydrocarbons have been detected in solvent extracts of cuticle preparations (Lockey, 1985). The hydrocarbons, *n*-alkanes, alkenes, alkadienes, and branched alkanes comprised the most abundant lipid classes associated with cuticle lipids (Hadley, 1981; Lockey, 1985, 1988). Differentiating between lipids specific to the epicuticle and those associated with the whole cuticle requires the adoption of very careful solvent extraction protocols. For example, Brey *et al.* (1985), examining solvent-treated pea aphid cuticle, reported that even gentle extractions, such as a 5-min treatment with chloroform/methanol of aphid cuticle or 4 hr with pentane, influence the integrity of the entire integument. Interestingly, with 5 min chloroform/methanol the hydrocarbons comprised the major lipid class (Brey *et al.*, 1985), whereas with pentane extraction free fatty acids were reported to be the major lipid class (Stransky *et al.*, 1973, cited by Brey *et al.*, 1985).

The "free" lipid fraction, while not always influencing the passive adhesion of spores to cuticle, may affect germination and appressorial formation, thus influencing the consolidation of the infectious propagule to the cuticle. Various workers have reported that solvent extracts of insect cuticle as well as isolated short-chain fatty acids from insect cuticle may either inhibit germination (Koidsuma, 1957; Smith and Grula, 1982) or result in the production of nonpenetrant germ tubes (Kerwin, 1984; Latge *et al.*, 1987). In other cases, these cuticle extracts will stimulate germ tube formation (Boucias and Pendland, 1984; Boucias and Latge, 1988; Latge *et al.*, 1987; St. Leger *et al.*, 1989a). For example, the hydrocarbon and polar lipid classes extracted from host cuticle were very effective in stimulating germination of *N. rileyi*. *N. rileyi*, known to possess lipase activity capable of utilizing *n*-alkanes and branched alkanes as sole carbon source (St. Leger *et al.*, 1988c), is likely to hydrolyze and assimilate hydrocarbons associated with the epicuticle.

In addition to protein and lipid components found in the cuticle, a complex of components derived from either plant, substrate, host fecal material, or from microbial flora present on the cuticle surface may be detected on the epicuticular surface. Brey *et al.* (1985), analyzing aqueous and methanol extracts of the aphid, reported the presence of free amino acids and sugars (glucose, mannose, and acetylglucosamine). These components, originating from the aphid's honeydew secretions, coated the entire epicuticular surface. Similarly, investigations by Woods and Grula (1984) reported the presence of amino acids, peptides, and glucosamine on the surface of corn earworm (*Heliothis zea*) larvae. *In vitro* assays of water or methanol extracts of insect cuticle have demonstrated that these surface components may play a role in both the stimulation and growth of penetrant germ tubes (Woods and Grula, 1984; Latge *et al.*, 1987; St. Leger *et al.*, 1989a). Microbial exudates produced by bacteria and fungi associated with the cuticle may also be present on the epicuticular surface and may have in certain cases acted as antifungal agents and prevented conidial germination (Delmas, 1973; Schabel, 1978).

5. RELATIONSHIP BETWEEN FUNGAL PROPAGULES AND THE HOST CUTICLE

5.1. Physical Association between the Epicuticle and Fungal Propagule

In general, conidia of many arthropod mycopathogens attach in a nondiscriminate fashion to the epicuticle of both host and nonhost arthropods. The hydrophobic conidia of *N. rileyi*, *M.*

anisopliae, and *B. bassiana* attached readily to both lepidopteran and aphid cuticle (Boucias *et al.*, 1988). Similarly, conidia of *Culicinomyces clavisporus* attach over the entire cuticular surface of the mosquito foregut (Sweeney, 1975). The primary discharge spores of the entomophthoraleans *Entomophaga aulicae*, *Conidiobolus obscurus*, *Entomophthora muscae*, and *Erynia radicans* will bind randomly to the epicuticular surface of their respective host insects (Murrin and Nolan, 1987; Brobyn and Wilding, 1983; Brey and Latge, 1986; Wraight *et al.*, 1990). Similarly, the hydrophilic conidia of *Verticillium lecanii* and *Ashersonii aleyrodis* nonspecifically attach to the epicuticle of host and nonhost substrates (Fig. 8).

The topography and general chemical properties of the epicuticle may enhance conidial adhesion and subsequent germ tube orientation. For example, while the hydrophobic conidia of *N. rileyi*, *M. anisopliae*, and *B. bassiana* were capable of attaching to all body regions, a preference to cuticle surfaces containing short cuticular spines has been reported (Boucias *et al.*, 1988). Conidia are frequently found to be trapped by and tightly bound to these spines (Fig. 8). Comparative experiments have also revealed that substantially greater numbers of these conidia bind to neonate larval cuticle than to the less hydrophobic fourth-instar larval cuticle. Similarly, Pekrul and Grula (1979) reported that *B. bassiana* conidia at low inoculum concentrations tended to adhere to the vicinity of setae and noncellular nodules of larval cuticle. At high concentrations, conidia clumped together and attached over the entire surface. McCauley *et al.* (1968) and Zacharuk (1970a) reported that *M. anisopliae* conidia, while capable of binding over the entire cuticle, are easily removed from smooth exposed sclerite epicuticle but remain firmly attached to the epicuticle associated with the protected intersegmental folds. Floating conidia of *M. anisopliae*, when assayed in an aquatic system against mosquito larvae, bound to the epicuticle of the exposed breathing siphon tube (Lacey *et al.*, 1988). The preferential attachment of *Tolypocladium cylindrosporum* to the head region (base of mandibles, maxillae) of mosquito larval hosts (Goettel, 1988) reflects the feeding behavior of the host larvae and probably not conidial recognition of these cuticle regions.

In only a few cases have terrestrial mycopathogens been reported to specifically recognize host cuticle. Conidia of an *M. anisopliae* strain isolated from the scarab *Cetonia aurata* have been reported to attach readily to *C. aurata* cuticle but failed to adhere to larvae of the scarab *Oryctes rhinoceros* (Fargues, 1984). Similarly, McGuire (1985) reported that the *E. radicans* discharge conidia adhered to host *Empoasca fabae* but not to the nonhost leafhoppers *Macrosteles fascifrons* and *Circulifer tenellus*. In the case of the aquatic mycopathogens, host recognition appears to be the rule rather than the exception. The motile zoospores of *Leptolegnia chapmanii* tend to aggregate and encyst on the cervical collar and intersegmental regions of mosquito larvae (Zattau and McInnis, 1987; Lord and Fukuda, 1988). The chytridiomycete *Coelomomyces psorophora* is unique in that it utilizes alternate arthropod hosts, mosquito and copepod, to complete its life cycle (for review, see Couch and Bland, 1985). Motile haploid zoospores and diploid biflagellate zygotes selectively encyst on the intersegmental membranes and head region of their respective copepod and mosquito hosts (Zebold *et al.*, 1979; Travland, 1979). Whether these motile zoospores and zygotes detect the glycoproteins present in the flexible cuticle regions (Cox and Willis, 1987) or are trapped physically by the cuticle–membrane folds is unknown. Regardless, the motile zygotes have been reported to promptly encyst on larvae of "susceptible" mosquito species but rarely attach to and encyst on larvae of resistant mosquito species (Zebold *et al.*, 1979). These results strongly suggest that the initial recognition by motile spores may be a key determinant regulating the host specificity of this mycopathogen.

Attachment of a fungal propagule to a nonhost substrate may elicit a variety of responses in which the conidia may (1) not germinate; (2) produce a nonpenetrant germ tube; (3) form conidiophores producing microconidia, secondary discharge conidia, or adhesive conidia; or (4) convert to thick-walled resting spores. The development of germ tubes and appressorial structures produced by terrestrial arthropod mycopathogens attaching to receptive hosts may be

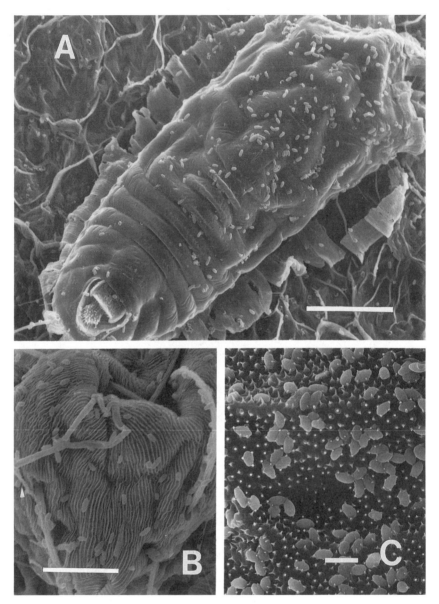

Figure 8. Scanning electron micrographs of *V. lecanii* conidia attaching to (A) host whitefly and (B) nonhost mite cuticle. (C) *N. rileyi* conidia on host velvet bean caterpillar. Note association of conidia with cuticular spines. Bars = 50 μm (A), 25 μm (B), 5 μm (C).

influenced by the epicuticular surface. Ultrastructural studies suggest that conidia landing on heavily sclerotized regions (head capsule) produce germ tubes which grow over the surface until they contact flexible interscleral regions (Pekrul and Grula, 1979; McCauley *et al.*, 1968; Schabel, 1978). Even on the softer abdominal sclerites, germ tubes of *N. rileyi* frequently orientate to and penetrate the membranous region surrounding cuticular spines (Fig. 5). Likewise, *in vitro* studies have shown that germ tubes of *M. anisopliae* on melanized cuticle grow errantly, whereas nonmelanized cuticle germ tubes orientate to and penetrate the cuticle (St. Leger *et al.*, 1988d). The phenols and oxidized phenolic compounds associated with sclerotized cuticle are toxic to fungal growth (Soderhall and Ajaxon, 1982) and reduce the activity of cuticle-degrading enzymes (St. Leger *et al.*, 1988d). Recent studies by Wraight *et al.* (1990) utilizing the fluorescent dye Uvitex BOPT (Butt, 1987) have analyzed the *in situ* attachment and penetration process associated with *Erynia radicans*, Their results demonstrated that the primary discharge spores randomly attach and germinate at equivalent rates over the head, thorax, and abdomen of the leafhopper *Empoasca fabae*. Germ tubes produced either penetrant germ tubes, replicative conidiophores, or capilliconidiophore. The majority of successful penetrations occurred in the abdominal region. Many of the spores located on sclerites produced germ tubes which grew parallel to the long axis of the body. Upon contact with flexible cuticle, germ tubes produced an appressorial cell which gave rise to a penetration tube. The physical and chemical stimuli responsible for this directional tropism are unknown (Wraight *et al.*, 1990).

5.2. Stability of the Conidia–Cuticle Interaction

Fargues (1984), drawing from information on bacterial adhesion, speculated that the conidia–cuticle interaction is a biphasic process involving a combination of long-range nonspecific and short-range specific forces. At present, very little is known about the chemical or physical forces responsible for conidial adhesion to the cuticle substrate. Nonspecific binding of hydrophobic conidia of *M. anisopliae*, *N. rileyi*, and *B. bassiana* has been examined using "insect ghosts" (Boucias *et al.*, 1988). Insect ghosts, produced by boiling larvae in 1% sodium dodecyl sulfate followed by water rinses, are transparent and are highly conducive to light microscopic examination. The detergent treatment, while effectively removing internal tissues and partially disrupting the endocuticle, does leave the epicuticle highly receptive to conidial attachment. Conidia of the previously mentioned deuteromycetes, if allowed to incubate with such ghosts for 5 min, were not displaced from the cuticle substrates with either detergent (ionic, nonionic), mild acid or alkali, solvent, or enzyme treatments (Fig. 7B). Furthermore, conidial preparations killed either by boiling in detergent or solvent or by incubation in fixatives readily attached to the cuticle substrate after being rinsed with water. These results suggest that attachment of hydrophobic conidia is passive and does not require the synthesis or release of adhesive material. The tenacity with which these spores adhere to host cuticle may reflect the prolonged time (12–48 hr) required for germination and subsequent cuticle penetration of host insects. The passive adhesion of many mycopathogens, whether due to the outer rodlet layer or mucilaginous coat, will be consolidated with the production of adhesives by developing germ tubes and appressoria. Very little, if anything, is known about the binding stability of the mucilages produced during the postgermination events. During his ultrastructural studies, Zacharuk (1970b,c) observed that the *M. anisopliae* conidia–cuticle interface could be separated by a weak electron beam, whereas the appressoria remained attached to the cuticle during intense electron beam exposure.

5.3. Binding Forces Associated with Hydrophobic Conidia–Cuticle Interaction

Research on the attachment of *M. anisopliae*, *N. rileyi*, and *B. bassiana* has shown that these fungi do possess sugar-binding proteins associated with spore walls, an overall net negative charge, and are extremely hydrophobic (Boucias *et al.*, 1988). Attempts to competitively inhibit conidial binding to larval cuticle of *A. gemmatalis* with hapten sugars tested alone or in combination failed, suggesting that these compounds are not involved in the recognition process. In addition to tightly adhering to insect cuticle, conidia of these species also bind to chitosan and DEAE resins (Boucias *et al.*, 1988). Binding to these noncuticle substrates, while not affected by incubation with sugars, could be inhibited by poly-L-lysine, suggesting that binding was being mediated by electrostatic attraction between the negatively charged conidia and the positively charged noninsect substrates. Cherbit and Delmas (1979) and Delmas (1983) proposed that such electrostatic forces may play a role in the preferential adhesion of *M. anisopliae* conidia to specific sites on the cuticle surface of *Oryctes rhinoceros* larvae. However, poly-L-lysine or poly-DL-aspartic acid alone or in combination with hapten sugars failed to competitively inhibit the binding of hydrophobic conidia to cuticle substrates (Boucias *et al.*, 1988).

Experimental data suggest that the hydrophobic interactions occurring between the conidia wall and cuticulin layer of epicuticle are responsible for the passive, nonspecific adhesion observed with the hydrophobic conidia. Detergents, solvents, and high-molecular-weight proteins, known to neutralize hydrophobic interactions, reduced adhesion when added to conidial suspensions prior to incubation with cuticle samples (Boucias *et al.*, 1988; Table VI). Alkali and acid, when mixed with conidia cuticle samples at zero time, also interfered with the initial attachment process. The chemicals are known to disrupt both the outer epicuticle layer and the rodlet layer present on the conidial surface. Data from salt aggregation and phase exclusion assays clearly demonstrated that the conidia of *N. rileyi*, *B. bassiana*, and *M. anisopliae* were extremely hydrophobic. For example, at pH 3.0, 6.0, and 9.0, 100% of the *M. anisopliae* conidia were excluded from the aqueous phase (Boucias *et al.*, 1988). Intact rodlet preparations extracted from *B. bassiana* were extremely hydrophobic, as demonstrated by the phase exclusion assay. These rodlet preparations, conjugated to rhodamine, were found to adhere efficiently to the epicuticule surface. Harsh alkali treatment of conidia, while effectively stripping away the rodlet layer (Fig. 6), only reduced the relative hydrophobicity by 4–18% and failed to cause significant reductions in the binding of these spores to the cuticle. These results indicate that while the rodlet layer is hydrophobic and is capable of binding to cuticle, the underlying layers of the conidial walls also contain hydrophobic proteins which are capable of adhering spores to the cuticle surface. The involvement of the hydrophobic cuticulin layer as the target-binding substrate has been demonstrated via selective degradation. The epicuticle, resistant to various denaturation agents, solvents, and chitinase, can be degraded with 1.0 M NaOH or with protease K. Examination of protease K-treated cuticle–conidial preparations revealed that conidia were binding only to the areas of intact epicuticle and not to the underlying endocuticle.

Table VI. Effect of Various Chemicals on the Attachement of Conidia to Fourth-Instar Cuticle Ghost Preparations of Anticarsia gemmatalis[a,b]

Chemical treatment	Relative number of conidia adhering to cuticle preparations		
	N. rileyi	*B. bassiana*	*M. anisopliae*
H$_2$O control	49 (18)	37 (19)	30 (8)
1.0 M HCl	22 (10)	22 (17)	6 (4)
1.0 M NaOH	18 (15)	7 (4)	3 (1)
1% SDS	6 (3)	7 (7)	11 (8)
1% NP40	6 (5)	9 (5)	5 (3)
CHCl$_3$/methanol (2 : 1)	14 (7)	16 (7)	3 (2)
Bovine serum albumin (1 mg/ml)	4 (1)	8 (6)	6 (3)
Ovalbumin (1 mg/ml)	10 (7)	2 (2)	4 (3)

[a]Modified from Boucias *et al.* (1988).
[b]Mean values of attached conidia counted within a microscope field (600×) having an area of 0.013 mm^2 (8–16 fields for each treatment). Values in parentheses represent standard deviation.

6. SUMMARY

Adhesion of the fungal propagule to arthropod cuticle represents the initial stage of mycoses. Many arthropod mycopathogens depend upon passive, nonspecific mechanisms to contact and attach to their respective hosts. At present, very little is known about the physiochemical forces mediating the interaction between the conidial and cuticle surfaces. Passive attachment of the "dry" hydrophobic conidia to cuticle is a result of strong binding forces involving hydrophobic and to a lesser extent electrostatic interactions. Lectins and various enzyme activities detected on the conidial surface are not believed to play a role in the initial attachment event but may be active in the consolidation of conidia, germ tubes, or appressoria to the cuticle surface.

Examination of the surfaces of "dry" conidia produced by various arthropod mycopathogens revealed the presence of a well-organized outer rodlet layer. The rodlet layer present on these mycopathogens possessed chemical and physical properties similar to those reported previously for other fungi (Hashimoto et al., 1976; Beever and Dempsey, 1978; Hallet and Beever, 1981; Hobot and Gull, 1981; Cole and Pope, 1981; Figeuras et al., 1988). In general, the rodlet layer, resistant to various detergents, enzymes, solvents, and denaturation agents, serves to protect fungal spores against biotic and abiotic degradation. The chemical composition of the rodlet layer varies among species with the major component(s) being protein, lipoprotein, glycoprotein, or polysaccharide (Hashimoto et al., 1976, Cole et al., 1983; Hobot and Gull, 1981; Smucker and Pfister, 1978). Whether or not these results represent true differences in rodlet composition of different fungal species or reflect the problem of isolating and analyzing the rodlet layer is unknown. In the case of ascospores of the yeast *Saccharomyces cerevisiae*, the outer layer, having an affinity to lipophilic dyes and being osmophilic, is comprised of protein (Briley et al., 1970) containing large numbers of dityrosine residues (Briza et al., 1986). It is the presence of these sporulation-specific dityrosine residues which provides the ascospore with both hydrophobic properties and resistance to lytic enzymes (Briza et al., 1986). Whether or not such residues exist in the rodlet layer of arthropod mycopathogens is unknown.

In addition to providing chemical stability, the rodlet layer present on fungal propagules has been reported to protect against dehydration and to serve in the aerial dispersal of conidia (Sasson et al., 1967; Hess et al., 1968; Beever and Dempsey, 1978). Conidia of arthropod mycopathogens may also be protected against dehydration and provide a means for dispersal in air currents by the hydrophobic rodlet layer. However, a major role of this layer may be the attachment of conidia to host cuticle. This interaction, while nonspecific, does provide a mechanism for preferential adhesion of conidia to the epicuticle.

Over the past years, research on arthropod mycopathogens has been stimulated in part by the potential of developing these fungi as selective control agents against a spectrum of insect pests (Burges, 1981). At present, emphasis has been placed on developing the "dry" hydrophobic conidia into wettable powder formulation which may be applied with current application technology. Formulators of conidial-based inocula have not considered the importance of the conidia–cuticle relationships reviewed in this chapter. Detailed information on the biochemistry of the surface components of both the fungal propagule and the epicuticle will be required before the adhesion process is fully understood. Knowledge of the polymers associated with adhesion to the cuticle may provide an avenue for developing formulations of fungal inocula and/or toxins which can be preferentially delivered to the host insect.

ACKNOWLEDGMENTS. The authors are grateful to Drs. Tariq Butt and R. St. Leger for providing various unpublished results and photomicrographs, Dr. J. P. Latge for contributing micrographs, and Drs. C. McCoy, J. Latge, and T. Butt for their critical review of the manuscript. This work has been supported in part by NATO Grant 375/84.

7. REFERENCES

Anderson, S. O., 1979, Biochemistry of insect cuticle, *Annu. Rev. Entomol.* **24**:29–61.

Beever, R. E., and Dempsey, G. P., 1978, Function of rodlets on the surface of fungal spores, *Nature* **272**:608–610.

Boucias, D. G., and Latge, J. P., 1988, Nonspecific induction of germination of *Conidiobolus obscurus* and *Nomuraea rileyi* with host and non-host cuticle extracts, *J. Invertebr. Pathol.* **51**:168–171.

Boucias, D. G., and Pendland, J. C., 1984, Nutritional requirements for conidial germination of several host range pathotypes of the entomopathogenic fungus *Nomuraea rileyi*, *J. Invertebr. Pathol.* **43**:288–292.

Boucias, D. G., Brasaemle, D. L., and Nation, J. L., 1984, Lipid composition of the entomopathogenic fungus *Nomuraea rileyi*, *J. Invertebr. Pathol.* **43**:254–258.

Boucias, D. G., Pendland, J. C., and Latge, J. P., 1988, Nonspecific factors involved in the attachment of entomopathogenic deuteromycetes to host insect cuticle, *Appl. Environ. Microbiol.* **54**(7):1795–1805.

Brey, P. T., and Latge, J. P., 1986, Integumental penetration of the pea aphid, *Acyrthosiphon pisum*, by *Conidiobolus obscurus* (Entomophthoraceae), *J. Invertebr. Pathol.* **48**:34–41.

Brey, P. T., Ohayon, H., Lesourd, M., Castex, H., Roucache, J., and Latge, J. P., 1985, Ultrastructure and chemical composition of the outer layers of the cuticle of the pea aphid *Acyrthosiphon pisum* (Harris), *Comp. Biochem. Physiol.* **82A**:401–411.

Brey, P. T., Lebrun, R. A., Papierok, B., Ohayon, H., Vennavalli, S., and Hafez, J., 1988, Defense reactions by larvae of *Aedes aegypti* during infection by the aquatic fungus *Lagenidium giganteum* (Oomycete), *Cell Tissue Res.* **253**:245–250.

Briley, M. S., Illingworth, F. R., Rose, A. H., and Fisher, D. J., 1970, Evidence for a surface protein layer on the *Saccharomyces cerevisiae* ascospore, *J. Bacteriol.* **104**:588–589.

Briza, P., Winkler, G., Kalchhauser, H., and Breitenbach, M., 1986, Dityrosine is a prominent component of the yeast ascospore wall, *J. Biol. Chem.* **261**:4288–4294.

Brobyn, P. J., and Wilding, N., 1977, Invasive and developmental processes of *Entomophthora* species infecting aphids, *Trans. Br. Mycol. Soc.* **69**:349–366.

Brobyn, P. J., and Wilding, N., 1983, Invasive and developmental processes of *Entomophthora muscae* infecting houseflies (*Musca domestica*), *Trans. Br. Mycol. Soc.* **80**:1–8.

Burges, H. D., 1981, *Microbial Control of Pests and Plant Disease 1970–1980*, Academic Press, New York.

Butt, T. M., 1987, A fluorescence microscopy method for the rapid localization of fungal spores and penetration sites on insect cuticle, *J. Invertebr. Pathol.* **50**:72–74.

Cerenius, L., and Soderhall, K., 1984a, Chemotaxis in *Aphanomyces astaci*, an arthropod-parasitic fungus, *J. Invertebr. Pathol.* **43**:278–281.

Cerenius, L., and Soderhall, K., 1984b, Repeated zoospore emergence from isolated spore cysts of *Aphanomyces astaci*, *Exp. Mycol.* **8**:370–377.

Cerenius, L., and Soderhall, K., 1985, Repeated zoospore emergence as a possible adaptation to parasitism in *Aphanomyces astaci*, *Exp. Mycol.* **8**:370–377.

Cerenius, L., 1985, Morphology and physiology of the differentiation process in an aquatic fungus, *Aphanomyces astaci*. Doctoral thesis, University of Uppsala, Institute of Physiological Botany, Uppsala, Sweden, pp. 29.

Charnley, A. K., 1984, Physiological aspects of destructive pathogenesis in insects by fungi: A speculative review, in: *Invertebrate–Microbial Interactions* (J. M. Anderson, A. D. M. Rayner, and D. W. H. Warton, eds.), Cambridge University Press, London, pp. 229–270.

Cherbit, C., and Delmas, J. C., 1979, Potentiels cutanes et points de moindra impedence chez *Oryctes rhinoceros* (Coleoptera: Scarabaeidae), *C.R. Acad. Sci.* **289**:1077–1080.

Cole, G. T., and Pope, L. M., 1981, Surface wall components of *Aspergillus niger* conidia, in: *The Fungal Spore: Morphogenic Controls* (G. Turian and D. R. Hohl, eds.), Academic Press, New York, pp. 192–215.

Cole, G. T., Pope, L. M., Huppert, M., Sun, S. H., and Starr, P., 1983, Ultrastructure and composition of conidial walls fractions of *Coccidiodes immitis*, Exp. Mycol **7**:297–318.

Couch, J. N., and Bland, C. E., 1985, *The Genus Coelomomyces*, Academic Press, New York.

Cox, D. L., and Willis, J. H., 1985, The cuticular proteins of *Hyalophora cecropia* from different anatomical regions and metamorphic stages, *Insect Biochem.* **15**:349–362.

Cox, D. L., and Willis, J. H., 1987, Analysis of the cuticular proteins of *Hyalophora cecropia* with two dimensional electrophoresis, *Insect Biochem.* **17**:457–468.

David, W. A. L., 1967, The physiology of the insect integument in relation to the invasion of pathogens, in: *Insects and Physiology* (S. W. L. Beament and S. E. Treherne, eds.) Edinburg, Oliver and Boyd, pp. 17–35.

Delmas, J. C., 1973, Influence du lieu de contamination tegumentaire sur le developpement de la mycose a *Beauveria tenella* (Delacr.) Siemaszko (Fungi imperfecti) chez les larves du coleoptire *Melolontha melolontha* L., *C. R. Acad. Sci.* **277**:433–435.

Eilenberg, J., Bresciani, J., and Latge, J. P., 1986, Ultrastructural studies of primary spore formation and discharge in the genus *Entomophthora*, *J. Invertebr. Pathol.* **48**:318–324.

Fargues, J., 1984, Adhesion of the fungal spore to the insect cuticle in relation to pathogenicity, in: *Infection Processes of Fungi* (D. W. Roberts and J. R. Aist, eds.), Rockefeller Foundation Conference Report, pp. 90–110.

Figeuras, M. S., Garro, J., and Dijk, F., 1988, Rodlet structure on the surface of *Chaetomium* spores, *Microbios* **53**:101–107.

Filshie, B. K., 1982, Fine structure of the cuticle of insects and other arthropods, in: *Insect Ultrastructure*, Volume 1 (R. C. King and H. Akai, eds.), Plenum Press, New York, pp. 281–312.

Fransen, J. J., 1987, *Aschersonii aleyrodis* as a microbial control agent of greenhouse whitefly, Doctoral thesis, University of Wageningen, Holland.

Glare, T. R., Chilvers, G. A., and Milner, R. J., 1985, Capilliconidia as infective spores on *Zoophthora phalloides* (Entomophthorales), *Trans. Br. Mycol. Soc.* **85**:463–470.

Goettel, M. S., 1988, Pathogenesis of the hyphomycete *Tolypocladium cylindrosporum* in the mosquito *Aedes aegypti*, *J. Invertebr. Pathol.* **51**:259–274.

Grove, S. N., and Bracker, C. E., 1978, Protoplasmic changes during zoospore encystment and cyst germination in *Pythium aphanidermation*, *Exp. Mycol.* **2**:51–98.

Grula, E. A., Woods, S. P., and Russell, H., 1984, Studies utilizing *Beauveria bassiana* as an entomopathogen, in: *Infection Processes of Fungi* (D. W. Roberts and J. R. Aist, eds.), Rockefeller Foundation Conference Report, pp. 147–152.

Hackman, R. H., 1984, Cuticle: Biochemistry, in: *Biology of the Integument*, Volume 1 (J. Bereiter-Hahn, A. G. Mateltsy, and K. S. Richards, eds.), Springer-Verlag, Berlin, pp. 583–610.

Hackman, R. H., 1986, The chemical nature of the outer epicuticle from *Lucilia cuprina* larvae, *Insect Biochem.* **16**:911–916.

Hadley, N. F., 1981, Cuticular lipids of terrestrial plants and arthropods, a comparison of their structure, composition and waterproofing function, *Biol. Rev.* **56**:23–47.

Hall, R. A., 1981, The fungus *Verticillium lecanii* as microbial insecticide against aphids and scales, in: *Microbial Control of Pests and Plant Diseases, 1970–1980* (H. D. Burger, ed.), Academic Press, New York, pp. 483–498.

Hallet, I. C., and Beever, R. E., 1981, Rodlets on the surface of *Neurospora* conidia, *Trans. Br. Mycol. Soc.* **77**:662–665.

Hashimoto, T., Wu-Yuan, C. D., and Bluemthal, H. J., 1976, Isolation and characterization of the rodlet layer of *Trichophyton mentagrophytes* microconidial wall, *J. Bacteriol.* **127**:1543–1549.

Hess, W. M., Sassen, M. M. A., and Remsen, C. C., 1968, Surface characteristics of *Penicillium* conidia, *Mycologia* **60**:290–303.

Hobot, J. A., and Gull, K., 1981, Changes in the organization of surface rodlets during germination of *Syncephalastrum racemosum* sporangiospores, *Protoplasma* **107**:339–343.

Howard, R. W., and Bloomquist, G. J., 1982, Chemical ecology and biochemistry of insect hydrocarbons, *Annu. Rev. Entomol.* **27**:149–172.

Ishikawa, F., Oisha, K., and Aida, K., 1979, Chitin-binding hemagglutinin produced by *Conidiobolus* strains, *Appl. Environ. Microbiol.* **37**:1110.

Ishikawa, F., Oisha, K., and Aida, K., 1983, Purification and immunological properties of chitin-binding hemagglutinin and n-acetylglucosaminidase produced by *Conidiobolus lamprauges*, Agric. Biol. Chem. **47**:149.

Jansson, H., and Nordbring-Hertz, B., 1983, The endoparasitic nematophagous fungus *Meria coniospora* infects nematodes specifically at the chemosensory organs, *J. Gen. Microbiol.* **129**:1121–1126.

Kerwin, J. L., 1983a, Biological aspects of the interaction between *Coelomomyces psorophorae* zygotes and the larvae of *Culiseta inornata*: Host-mediated factors, *J. Invertebr. Pathol.* **41**:224–232.

Kerwin, J. L., 1984, Fatty acid regulation of the germination of *Erynia varabilis* conidia on adults and puparia of the lesser housefly, *Fannia canicularis*, *Can. J. Microbiol.* **30**:158–161.

Kerwin, J. L., and Washino, R. K., 1986, Cuticular regulation of host recognition and spore germination by entomopathogenic fungi, in: *Fundamental and Applied Aspects of Invertebrate Pathology* (R. A. Samson, J. M. Vlak, and D. Peters, eds.), *Found. 4th Colloq. Invertebr. Pathol.* pp. 423–425.

King, D. S., and Humber, R. A., 1981, Identification of the entomophthorales, in: *Microbial Control of Pests and Plant Diseases, 1970–1980* (H. D. Burger, ed.), Academic Press, New York, pp. 107–127.

Kish, L. E., and Allen, G. E., 1978, The biology and ecology of *Nomuraea rileyi* and a program predicting its incidence on *Anticarsia gemmatalis* in soybean, *Fl. Agric. Exp. Stn. Bull.* **795**.

Koidsuma, K., 1957, Antifungal action of cuticular lipids in insects, *J. Insect Physiol.* **1**:40–51.

Lacey, C. M., Lacey, A. L., and Roberts, D. R., 1988, Route of invasion and histopathology of *Metarrhizium anisopliae* in *Culex quinquefasciatus*, *J. Invertebr. Pathol.* **52**:108–118.

Latge, J. P., Monsigny, M., Prevost, M. C., Roche, A. C., Kiedo, C., and Fournet, B., 1984, Carbohydrate binding proteins in the entomogenous fungus, *Conidiobolus obscurus*, *Biol. Cell.* **51**:51–52 (abstract).

Latge, J. P., Cole, G. T., Horisberger, M., and Prevost, M. C., 1986, Ultrastructure and chemical composition of the ballistospore wall of *Conidiobolus obscurus*, *Exp. Mycol.* **10**:99–113.

Latge, J. P., Sampedro, L., Brey, P., and Diaquin, M., 1987, Aggressiveness of *Conidiobolus obscurus* against the pea aphid: Influence of cuticular extracts on ballistospore germination of aggressive and non-aggressive strains, *J. Gen. Microbiol.* **133**:1987–1997.

Latge, J. P., Monsigny, M., and Prevost, M. C., 1988a, Visualization of exocellular lectins in the entomopathogenic fungus *Conidiobolus obscurus*, *J. Histochem. Cytochem.* **36**:1419–1424.

Latge, J. P., Cabrera Cabrera, R. I., and Prevost, M. C., 1988b, Microcycle conidiation in *Hirsutella thompsonii*, *Can. J. Microbiol.* **34**:625–630.

Locke, M., 1984, The structure of the insect cuticle, in: *Infection Processes of Fungi* (D. W. Roberts and J. R. Aist, eds.), Rockefeller Foundation Conference Report, pp. 38–53.

Lockey, K. H., 1985, Insect cuticular lipids, *Comp. Biochem. Physiol.* **81B**:263–273.

Lockey, K. H., 1988, Lipids of the insect cuticle: Origin, composition and function, *Comp. Biochem. Physiol.* **89B**:595–645.

Lord, J. C., and Fukuda, T., 1988, An ultrastructural study of the invasion of *Culex quinquefasciatus* larvae by *Leptolegnia chapmanii* (Oomycetes: Saprolegniales), *Mycopathologia* **104**:67–73.

McCauley, V. J. E., Zacharuk, R. Y., and Tinline, R. D., 1968, Histopathology of green muscardine in larvae of four species of Elateridae (Coleoptera), *J. Invertebr. Pathol.* **12**:444–459.

McCoy, C. W., Samson, R. A., and Boucias, D. G., 1988, Entomogenous fungi, in: *CRC Handbook of Natural Pesticides*, Volume 5, Part A (C. M. Ignoffo, ed.), CRC Press, Boca Raton, pp. 151–236.

McGuire, M. R., 1985, *Erynia radicans*: Studies on its distribution, pathogenicity, and host range in relation to potato leafhopper, *Empoasca fabae*, Ph.D. thesis, University of Illinois, Urbana Champaign.

Madelin, M. F., Robinson, R. F., and Williams, R. S., 1967, Appressorium like structures in insect parasitizing deuteromycetes, *J. Invertebr. Pathol.* **9**:404–412.

Michel, B., 1981, Recherches expermentals sur la penetration des champignons pathogens chez les insects, These 3e cycle, University of Montpellier.

Murrin, F., and Nolan, R. A., 1987, Ultrastructure of the infection of spruce budworm larvae by the fungus *Entomophaga aulicae*, *Can. J. Bot.* **65**:1694–1706.

Nemoto, H., Kobayashi, M., and Takizawa, Y., 1979, Scanning electron microscopy of *Entomophthora* (Triplosporium) *floridana* (Zygomycetes: Entomophthorales) attacking the sugi spider mite, *Oligonychus hondoensis* (Acarina: Tetranychidae), *Appl. Entomol. Zool.* **14**:376–382.

Nicholson, R. L., 1984, Adhesion of fungi to the plant cuticle, in: *Infection Processes of Fungi* (D. W. Roberts and J. R. Aist, eds.), Rockefeller Foundation Conference Report, pp. 74–90.

Nordbring-Hertz, B., 1988, Nematophagous fungi: Strategies for nematode exploitation and for survival, *Microbio. Sci.* **5**:108–116.

Nordbring-Hertz, B., and Mattiason, B., 1979, Action of a nematode-trapping fungus shows lectin mediated host–microorganism interaction, *Nature* **281**:477–479.

Olson, L. W., Cerenius, L., Lange, L., and Soderhall, K., 1984, The primary and secondary spore cyst of *Aphanomyces* (Oomycetes, Saprolegniales), *Nord. J. Bot.* **4**:681–696.

Pekrul, S., and Grula, E. A., 1979, Mode of infection of the corn earworm (*Heliothis zea*) by *Beauveria bassiana* as revealed by scanning electron microscopy, *J. Invertebr. Pathol.* **34**:238–247.

Pendland, J. C., and Boucias, D. G., 1984a, Ultrastructural studies of conidial germination in the entomopathogenic hyphomycete *Nomuraea rileyi*, J. Invertebr. Pathol. **43**:432–434.

Pendland, J. C., and Boucias, D. G., 1984b, Production and regeneration of protoplasts in the entomogenous hyphomycete, *Nomuraea rileyi*, *J. Invertebr. Pathol.* **43**:285–287.

Pendland, J. C., and Boucias, D. G., 1984c, Use of labeled lectins to investigate cell wall surfaces of the entomogenous hyphomycete *Nomuraea rileyi*, *Mycopathologia* **81**:141–148.

Persson, M., Hall, L., and Soderhall, K., 1984, Comparison of peptidase activity in some fungi pathogenic to arthropods, *J. Invertebr. Pathol.* **44**:342–348.

Roberts, D. W., and Humber, R. A., 1981, Entomogenous fungi, in: *Biology of Conidial Fungi*, Volume 2 (G. T. Cole and B. Kendrick, eds.), Academic Press, New York, pp. 201–236.

St. Leger, R. J., Charnley, A. K., and Cooper, R. M., 1986, Cuticle-degrading enzymes of entomopathogenic fungi: Mechanisms of interaction between pathogen enzymes and insect cuticle, *J. Invertebr. Pathol.* **47**:295–302.

St. Leger, R. J., Durrands, P. K., Cooper, R. M., and Charnley, A. K., 1988a, Regulation of production of proteolytic enzymes by the entomopathogenic fungus *Metarhizium anisopliae*, *Arch. Microbiol.* **150**:413–416.

St. Leger, R. J., Durrands, P. K., Charnley, A. K., and Cooper, R. M., 1988b, Role of extracellular chymoelastase in the virulence of *Metarhizium anisopliae* for *Manduca sexta*, *J. Invertebr. Pathol.* **52**:285–293.

St. Leger, R. J., Cooper, R. M., and Charnley, A. K., 1988c, Utilization of alkanes by entomopathogenic fungi, *J. Invertebr. Pathol.* **52**:356–359.

St. Leger, R. J., Cooper, R. M., and Charnley, A. K., 1988d, The effect of melanization of *Manduca sexta* cuticle on growth and infection by *Metarhizium anisopliae*, *J. Invertebr. Pathol.* **52**:459–470.

St. Leger, R. J., Butt, T. M., Goettel, M. S., Staples, R. C., and Roberts, D. W., 1989a, Production in vitro of appressoria by the entomopathogenic fungus *Metarhizium anisopliae*, *Exp. Mycol.* **13**:274–288.

St. Leger, R. J., Butt, T. M., Staples, R. C., and Roberts, D. W., 1989b, Synthesis of proteins including a cuticle-degrading protease during differentiation of the entomopathogenic fungi *Metarhhizium anisopliae*, *Exp. Mycol.* **13**:253–262.

Samson, R. A., McCoy, C. W., and O'Donnell, K. L., 1980, Taxonomy of the acarine parasite, *Hirsutella thompsonii*, *Mycologia* **72**:359.

Samson, R. A., Evans, H. C., and Latge, J. P., 1988, *Atlas of Entomopathogenic Fungi*, Springer-Verlag, Berlin.

Sassan, M. M., Remsen, C. C., and Hess, W. M., 1967, Fine structure of *Penicillum megasporum* conidiospores, *Protoplasma* **64**:75–88.

Schabel, H. G., 1978, Percutaneous infection of *Hylobius pales* by *Metarrhizium anisopliae*, *J. Invertebr. Pathol.* **31**:180–187.

Selhime, A. G., and Muma, M. H., 1966, Biology of *Entomophthora floridana* attacking *Eutetranychus banksi*, *Fla. Entomol.* **49**:161–168.

Sing, V. O., and Bartnicki-Garcia, S., 1975, Adhesion of *Phytophora* zoospores: Detection and ultrastructural visualization of concanavalin A receptor sites appearing during encystment, *J. Cell Sci.* **19**:11–20.

Smith, R. J., and Grula, E. A., 1982, Toxic components on the larval surface of the corn earworm (*Heliothis zea*) and their effects on germination and growth of *Beauveria bassiana*, *J. Invertebr. Pathol.* **39**:15–22.

Smucker, R. A., and Pfister, R. M., 1978, Characteristics of *Streptomyces coelicolor* aerial spore rodlet mosaic, *Can. J. Microbiol.* **24**:397–408.

Soderhall, K., and Ajaxon, R., 1982, Effect of quinones and melanin on mycelial growth of *Aphanomyces* spp. and extracellular protease of *Aphanmyces astaci*, a parasite on crayfish, *J. Invertebr. Pathol.* **39**:105–109.

Soderhall, K., Svensson, E., and Unestam, T., 1978, Chitinase and protease activities in germinating zoospore cysts of a parasitic fungus, *Aphanomyces astaci* (Oomycetes), *Mycopathologia* **64**:9–11.

Stransky, K., Ubik, K., Holman, J., and Streibl, M., 1973, Chemical composition of compounds produced by the pea aphid, *Acyrthosiphon pissum* (Harris): Pentane extract of surface lipids, *Collect. Czech. Chem. Commun. Engl. Ed.* **38**:770–780.

Sweeney, A. W., 1975, The mode of infection of the insect pathogenic fungus *Culicinomyces* in larvae of the mosquito *Culex fatigans*, *Aust. J. Zool.* **23**:49–57.

Sweeney, A. W., Wright, R. C., and vanderLubbe, L., 1980, Ultrastructural observations on the invasion of mosquito larve by the fungus *Culicinomyces*, *Micron* **11**:487.

Thomas, K. C., Khachatourians, G. G., and Ingledew, W. M., 1987, Production and properties of *Beauveria bassiana* conidia cultivated in submerged culture, *Can. J. Microbiol.* **33**:12–20.

Travland, L. B., 1978, Structures of the motile cells of *Coelomomyces psorophorae* and function of the zygote in encystment on a host, *Can. J. Bot.* **57**:1021–1035.

Travland, L. B., 1979, Initiation of infection of mosquito larvae (*Culiseta inornata*) by *Coelomomyces psorophorae*, *J. Invertebr. Pathol.* **33**:95–105.

Willis, J. H., 1987, Cuticular proteins: The neglected component, *Arch. Insect Biochem. Physiol.* **6**:203–215.

Woods, S. P., and Grula, E. A., 1984, Utilizable surface nutrients on *Heliothis zea* available for growth of *Beauveria bassiana*, *J. Invertebr. Pathol.* **43**:259–269.

Wraight, S. P., Butt, T. M., Galini-Wright, S., Allee, L. L., and Roberts, D. W., 1990, Germination and infection processes of the entomophthoralean fungus, *Erynia radicans* on the potato leafhopper, *Empoasca fabae*, *J. Invertebr. Pathol.* (in press).

Zacharuk, R. Y., 1970a, Fine structure of the fungus *Metarrhizium anisopliae* infecting three species of larval Elateridae (Coleoptera). I. Dormant and germinating conidia, *J. Invertebr. Pathol.* **15**:63–80.

Zacharuk, R. Y., 1970b, Fine structure of the fungus *Metarrhizium anisopliae* infecting three species of larval Elateridae (Coleoptera). II. Conidial germ tubes and appressoria, *J. Invertebr. Pathol.* **15**:81–91.

Zackaruk, R. Y., 1970c, Fine structure of the fungus *Metarrhizium anisopliae* infecting three species of larval Elateridae (Coleoptera). III. Penetration of the host integument, *J. Invertebr. Pathol.* **15**:372–396.

Zattau, W. C., and McInnis, T., Jr., 1987, Life cycle and mode of infection of *Leptolegnia chapmanii* (Oomycetes) parasitizing *Aedes aegypti*, *J. Invertebr. Pathol.* **50**:134–145.

Zebold, S. L., Whisler, H. C., Schemanchuk, J. A., and Travland, L. B., 1979, Host specificity and penetration in the mosquito pathogen *Coelomomyces psorophorae*, *Can. J. Bot.* **57**:2766–2770.

6

The Fate of Fungal Spores in the Insect Gut

R. J. Dillon and A. K. Charnley

1. INTRODUCTION

Suctorial insects in the main are not exposed to a large enteric microbial inoculum. In contrast, mandibulate insects ingest a wide range of microorganisms with their food. Enforced association between insects and their microbes particularly during feeding has led to the evolution of a wide variety of formal, and informal relationships which are amply illustrated by insect–fungus interactions. Fungi inhabit the gut as commensals, e.g., Trichomycetes (Moss, 1979); symbionts, e.g., yeasts provide essential nutrients (Pant and Fraenkel, 1950; Chararas *et al.*, 1983); commuters, e.g., plant pathogens in transit (Ingold, 1971); exoparasites, e.g., Laboulbeniales (Bell, 1974); or potential endoparasites (pathogens; Roberts and Humber, 1981). Exploitation of fungi by insects occurred early in evolution as feeding on dead or decaying material is probably the primitive condition among the class Insecta (Southwood, 1972). Direct mycophagy is also an important feeding habit (Martin, 1979; Kukor and Martin, 1987). Symbiotic associations have developed between insects and fungi with possibly the pinnacle being the fungal gardens of leaf-cutting ants (Weber, 1966; Boyd and Martin, 1975a,b).

The primary role of the alimentary canal is the breakdown of ingested food and as such it is a potentially hostile environment to many biodegradable materials. Thus, the ability of fungal propagules to withstand conditions within the gut may limit or bias the relationship between the fungus and the mandibulate insect.

The failure of most pathogenic fungi to invade via the gut, preferring rather the external cuticle, underlines the inhospitable nature of the gut. The object of the present contribution is to consider the reasons for the relatively low incidence of gut invasion. This information is relevant to the use of fungal pathogens for pest control and on a wider level it highlights problems facing fungi in developing relationships with insects. Particular reference will be made to the well-studied interaction between the entomopathogen *Metarhizium anisopliae* and the desert locust *Schistocerca gregaria* but other insect–fungus relations will be considered where they are relevant to the colonization and invasion processes of fungal pathogens.

R. J. Dillon and A. K. Charnley • School of Biological Sciences, University of Bath, Bath, Avon BA2 7AY, England.

2. THE STRUCTURE OF THE INSECT GUT AND THE ASSOCIATED MICROBIOTIA

The gut of the desert locust possesses all the features of the alimentary canal which in more specialized insects becomes highly modified. The gut is divided into three sections (viz., the foregut, midgut, and hindgut) based on function and morphology (Fig. 1). Ingested food is mixed in the foregut with salivary enzymes and regurgitated midgut enzymes. The foregut acts as a storage organ, allowing controlled release of the bolus into the midgut where digestion is completed. Enzyme secretion and absorption of the products of digestion occur primarily in the six ceca which are located at the interface between the foregut and midgut. The mid/hindgut junction is the site of attachment of the Malpighian tubules; these are part of the secretory system of the insect and they secrete an isosmotic or slightly hyperosmotic filtrate of the hemolymph ("primary urine") containing waste nitrogenous products as well as detoxified xenobiotics (Bradley, 1985).

The principal function of the hindgut is the absorption of water and salts from the feces and the primary urine (Chapman, 1985). Amino acids and other low-molecular-weight metabolites from the primary urine may also be absorbed through the hindgut cuticle of *Schistocerca gregaria*

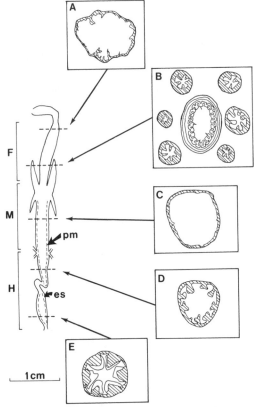

Figure 1. Diagrammatic representation of digestive tract of *S. gregaria.* F, foregut; M, midgut; H, hindgut; pm, peritrophic membrane; es, ectoperitrophic space. (A) Section through esophagus; (B) section through crop with anterior ceca, note ridged cuticle with spines; (C) section through the anterior midgut epithelium, cuticle absent; (D) section through ileum of hindgut showing lobed epithelium and cuticle; (E) section through the rectal pads showing subcuticular space.

(Phillips, 1980), but this may only be of importance in insects which harbor a fermentative hindgut biotia (Breznak, 1984).

The foregut and hindgut are of ectodermal origin and thus are lined with cuticle. There is little information available on the composition of the cuticle but it is thought to be similar in composition to that of the external cuticle apart from the absence of the outer wax layer consistent with a requirement for permeability (Bignell, 1984). Most of the gut cuticle remains unhardened due to the absence of phenolic cross-links between the protein molecules (Bignell, 1984). The foregut cuticle and parts of the hindgut cuticle are equipped with backward-directed spines thought to prevent the forward movement of the food bolus during peristalsis (Chapman, 1985). The cuticle is absent in the midgut (Bertram and Bird, 1961). But mechanical protection of the midgut epithelium is afforded by the peritrophic membrane (pm) in the absence of mucus-producing cells. In *S. gregaria* the membrane is secreted in layers by the midgut epithelium; approximately every 15 min (Bernays, 1981), it coats the food bolus as it moves down the gut, and is voided in the feces. The composition of the pm is not known in detail but is thought to comprise a chitin fibril network set in a protein–carbohydrate complex (Richards and Richards, 1977). In addition to protecting the gut wall from abrasion, the pm may also promote the conservation of enzymes in the ectoperitrophic space (see Fig. 1) (Terra *et al.*, 1979) and provide some protection against pathogenic microorganisms (Brandt *et al.*, 1978). Interestingly, glucose increases permeability of the pm in *Manduca sexta* to polyethylene glycol 6000 (4-nm diameter) while low doses of δ-endotoxin from *Bacillus thuringiensis* have the opposite effect (Rupp, 1986). It is not yet known whether permeability to microorganisms is similarly affected by these treatments. There are no pores or other discontinuities in the pm of the larvae of *Orygia pseudootsugata* (Brandt *et al.*, 1978). However, in the locust there is a break in the pm in the colon (Goodhue, 1963); this provides access to the cuticular surface of the ileum and rectum for bacteria and cysts of the sporozoan *Malamoeba locustae* (Hunt and Charnley, 1981) and for fungal propagules (Dillon and Charnley, 1986b). In many insects the pm is broken up in the hindgut by cuticular spines (Becker, 1978) and thus does not prevent the colonization of the gut cuticle by bacteria.

The desert locust possesses an abundant gut bacterial biotia (Hunt and Charnley, 1981), consisting predominantly of Enterobacteriaceae, which increase in numbers posteriorly through the gut. Streptococci are also present but at a concentration 10–100 times less than that of the Enterobacteriaceae. The bacterial population increased during fasting of the insect. No strict anaerobes were isolated and the composition of the gut microbiotia appeared to be reflective of the flora on the food (Hunt, 1982). In many insect species where there is no obligate microbial association the composition of the gut microbiotia is thought to be due to chance contamination from the external environment (Brooks, 1963; Lysenko, 1985). Newly hatched locusts rapidly acquire a gut microbiotia which is maintained throughout the life of the insect. However, the gut biotia is not a stable community as without additional inocula the population declines during larval life (Dillon and Charnley, unpublished).

The microbiotia inhabiting the insect guts may be commensals, pathogens, or symbionts, though the distinction between these types is blurred (Lysenko, 1985). It is often difficult to determine the degree to which either insect or microbiotia derives any physiological benefit. In some cases microbially derived nutrients may be essential to the insect whereas in others commensals may provide low levels of nutrients not essential to the insect (Brooks, 1963). The contribution of the locust microbiotia to the host nutrition appears to be at best nonessential as germfree insects are in most respects comparable to conventional parasite-free insects (Charnley *et al.*, 1985).

In common with the situation in most insects, bacteria in the anterior part of the locust gut are restricted mainly to the lumen and inside the pm. Colonization of the midgut surface is unusual (see, e.g., Bracke *et al.*, 1979; Klug and Kotarski, 1980) possibly because of the rapid turnover of the epithelial cells. The hindgut cuticle is the main site for the resident biotia (Hunt and Charnley, 1981). Presumably the considerable macerating activity in the foregut is not conducive to the establishment of a microbial population, in contrast to the hindgut, where the folds of the ileum and crypts of the rectum provide a protected niche away from the mainstream of the gut. In some insects there are additional features to promote the gut microbiotia, e.g., the hindgut of the house cricket *Acheta domestica* is lined with cuticular bristles which are colonized by the bacteria (Ulrich *et al.*, 1981). Further modifications of the mid- and hindgut include paunches, diverticula, and mycetomas which provide specialized chambers to accommodate an obligate biotia (Buchner, 1965).

Studies on the role of the gut bacteria have tended to rely on the use of broad-spectrum antibiotics to reduce or eliminate the bacteria, but such methods have serious drawbacks. Antibiotics may directly affect the insect's physiology (Eutick *et al.*, 1978), fail to achieve complete elimination of the microbes (Bracke *et al.*, 1978), or promote the development of resistant strains of bacteria (Henry, 1964).

Axenic locusts were reared in a plastic isolator system similar to that developed for studies on gnotobiotic animals (Fuller, 1978). Locusts harboring a conventional microbiotia were also reared free from the protozoan parasites *Malamoeba locustae* which are endemic in laboratory locust colonies (Tobe and Pratt, 1975). The axenic insects were found to be physiologically comparable to parasite-free insects and were able to reproduce normally (Charnley *et al.*, 1985). A comparative study of the immunology of the three types of locust—conventional, parasite-free, and axenic—showed that the absence of the gut microbiotia did not compromise the hemocytic response of the insect to injection of fungal conidia (Dillon *et al.*, 1987). Axenic and conventional insects were equally susceptible to a topical dose of *M. anisopliae* conidia (Dillon, 1984).

3. RETENTION OF SPORES IN THE GUT

A prerequisite for fungal invasion via the gut is that the fungal propagule be present for a sufficient length of time for germination and penetration to occur. On the external surface of the insect, failure of the spore to adhere precludes germination. In the alimentary canal, germination can potentially occur prior to adhesion and host penetration; this may provide the fungus with an opportunity for an increase in the inoculum potential.

However, the gut transit time of a meal in a locust is rapid, ca. 2.2 hr (Dillon and Charnley, 1986b). Since a large proportion of a conidial inoculum of *M. anisopliae* passes through the locust gut with the test meal, most conidia are flushed out before they have had time to germinate, with or without prior adhesion (Dillon and Charnley, 1986b). Some conidia appeared in fecal pellets within as little as 1 hr of completion of the test meal, having traveled down the gut suspended in the gut fluid in advance of the food bolus, while at the other extreme 0.8% of the initial dose was retained for 24 hr (long enough for germination and penetration to have occurred). Other studies have also shown long-term retention of spores in insects with short food transit times. *M. anisopliae* conidia were retained in the gut of *S. gregaria* for > 24 hr (Veen, 1968) and in the gut of *Hylobius pales* for 74 hr (Schabel, 1976). Allee *et al.* (unpublished) located conidia of *Beauveria bassiana* 32 hr postinoculation in the gut of the potato Colorado beetle.

The pm may play an important role in prolonging the retention of spores in the gut. This may be particularly so in those insects (e.g., *Aedes aegypti*) which possess a type II pm (produced by a discrete group of cells at the anterior end of the midgut; Richards and Richards, 1977) and as a consequence, a pronounced ectoperitrophic space which provides shelter from the "tidal" flow cf. trypanosomes in mosquito larvae (Stohler, 1961). This option may not be available in locusts which have a type I pm, i.e., membrane is the product of all cells in the midgut and there is no distinct ectoperitrophic space in the midgut.

The presence of a midgut countercurrent flow as proposed by Dow (1981) may contribute to the retention of spores in the locust gut. During feeding the locust crop empties and flow is unidirectional down the midgut. But in insects deprived of food for > 4 hr, fluid from the Malpighian tubules moves forward through the food in the midgut into the anterior midgut where water is absorbed in the ceca. The countercurrent may also account for the fact that an inert dye (amaranth) was completely eliminated from the gut in less than 12 hr in continuously feeding locusts but was retained in the gut for at least 60 hr in insects which were starved after intake of the test meal, though the solid food material was voided in less than 12 hr (Baines *et al.*, 1973).

Dillon and Charnley (1986b) isolated *M. anisopliae* conidia from all three regions of the gut (viz., foregut, midgut, and hindgut) 24 hr after inoculation. Presumably the conidia had lodged under the spines in the foregut and hindgut, in the folds of the ileum, and between the rectal pads after rupture of the pm in the colon allowed access to the ectoperitrophic space in the hindgut. Hasan (1982) found conidia of the phytopathogen *Colletotrichum graminicola* in the posterior ceca of *S. gregaria* 48 hr after ingestion; this site may also provide a refuge for *M. anisopliae* conidia.

4. INFLUENCE ON SPORE GERMINATION OF PASSAGE THROUGH THE GUT

In an experiment to test viability of gut-passaged conidia, Dillon and Charnley (1986a) allowed locusts to feed on a grass meal containing conidia. Fecal pellets produced 3–4 hr postingestion were homogenized and placed on a medium selective for *M. anisopliae*, 26.6 ± 4.7% of the conidia germinated compared to 78.3 ± 4.5% germination for control conidia. Hasan (1982) found a similar reduction in the germination of *C. graminicola* after passage through the gut of the desert locust. Fifty percent germination occurred in fecal pellets produced just after feeding and viability declined to 10–15% in conidia retained for 8 hr prior to ejection in the feces. Thus, a further decline in germination of *M. anisopliae* conidia may be predicted when they are retained in the gut for periods of time sufficient for germination to occur (> 12 hr). The detrimental effects of gut passage on *M. anisopliae* conidia were mirrored by the failure of retained conidia to show signs of swelling or germ tube formation up to 30 hr after ingestion.

Ten other species of entomopathogens and plant pathogens also failed to germinate in nutrient-supplemented gut fluid. Interestingly, three species of *Penicillium* were not affected by the gut fluid. Conidia of *M. anisopliae* were inhibited from germinating in the gut fluid (nutrient enriched) from seven other diverse species of Orthoptera: *Schistocerca cancellata*, *S. picifrons picifrons*, *Locusta migratoria migratorioides*, *Ephippiger perforatus*, *Poecilimon ornatus*, *Pholidoptera griseoptera*, and *Chorthippus brunneus*; thus, the phenomenon may be widespread among the order.

The few other published studies on the effects of gut passage and gut fluid on the viability of entomopathogenic fungi suggest a wide variation in the properties of the gut among other insect orders. Thus, Schabel (1976) observed 95% germination of *M. anisopliae* conidia after 20 hr in gut fluid extracted from *H. pales* and little effect of passage through the gut, though conidia failed to germinate after a period of 3 days in the gut. Gabriel (1959) also found that *M. anisopliae* germinated in the gut contents of *Bombyx mori* and *Tenebrio molitor*. Broome *et al.* (1976) noted a reduction to 37% germination of *B. bassiana* in the gut of a fire ant *Solenopsis richteri*. Allee *et al.* (unpublished) recorded germination of *B. bassiana* in the gut of the Colorado potato beetle fed on an artificial diet, though germination was rarely seen in guts of insects fed on potato leaves.

Mosquito larvae as filter feeders are particularly prone to the rapid acquisition of large numbers of conidia, to the extent that the gut becomes packed with conidia (see, e.g., Agudelo-Silva and Wassink, 1984; Lacey *et al.*, 1988). Thus, it is not surprising that they appear to have evolved a potent gut antifungal defense. Crisan (1971) reported that 89% of conidia of *M. anisopliae* were disrupted or digested in the gut of *Culex pipiens*. Lacey *et al.* (1988) reported apparent digestion of *M. anisopliae* in the gut of *C. quinquefasciatus*. Ravallec *et al.* (1989) observed disruption of ingested spores of *Tolypocladium cylindrosporium* in the gut of *Aedes albopictus*; in contrast, the viability of this fungus was not reduced after a 2.5 hr passage through the gut of the mosquito *A. aegypti* (Goettel, 1988a,b).

Other published work on the viability of gut-passaged spores of entomopathogens, phytopathogens, and saprophytes involves the use of qualitative observations (e.g., Veen, 1968; Price, 1976). The demonstration of fungal colonies on agar plates spread with insect fecal material yields no information on potential inhibitory effects of gut transit.

An interesting example of a beneficial effect of gut passage on fungal spores is given by Nuss (1982). Proteospores of *Ganoderma* formed at the beginning of the sporulation period germinated readily whereas other spores which developed later on the same fruit body must first pass through the gut of a fly larva before they will germinate; thus, the species is adapted to dispersal by the insect.

5. CONDITIONS IMPOSED BY THE INSECT AND GUT MICROBIOTIA WHICH INFLUENCE SPORE ADHESION AND GERMINATION

The gut is a complex nonuniform environment which is in a constant state of flux, and its ecophysiology can change markedly from one region to another. Thus, it seems likely that inhibition of fungal germination may be caused by a number of interacting factors which change in space, time, and between species. One of the potential difficulties in assessing the significance of factors affecting fungal viability is that even modest effects may be critical in preventing colonization.

The activities of the gut microbiotia can make a substantial contribution to the chemical and physical conditions prevailing in the gut. It is well established for vertebrates that the normal microbiotia helps to prevent the establishment of alien or allochthonous microorganisms including fungi (see, e.g., reviews by Tannock, 1984; Pivnik and Nurmi, 1982; Lee, 1985; Hill *et al.*, 1986). If the introduced organism is a potential pathogen, then the normal biotia is effectively contributing to the host defense. Van der Waaij *et al.* (1971) coined the term *colonization resistance* (CR) to describe the phenomenon. Oral streptococci were implicated in the colonization resistance to the potential pathogen *Candida albicans* in germfree mice (Liljemark and Gibbons, 1973). The intactness of the anaerobic biotia is of key importance in preventing colonization by *C. albicans* (Van der Waaij, 1982). *C. albicans* was nonpathogenic to chickens harboring a conventional gut microbiotia or a monobiotia of *E. coli* but caused lesions in germfree chicks (Balish and Phillips, 1966a,b). Some authors have discussed the potential for the normal insect gut microbiotia to exert CR against bacterial pathogens (e.g., Smirnov, 1978; Greenberg *et al.*, 1970; Jarosz, 1979a; Veivers *et al.*, 1982; Lysenko, 1981). The possibility of CR against fungal entomopathogens has until recently received little attention.

5.1. pH

Most conidial fungi can grow between pH4 and 9 (Cochrane, 1958). This range encompasses the pH found in many insect guts (Bignell, 1984). However, some lepidopterans possess a midgut pH in excess of 10 (Waterhouse, 1940); similarly, a pH of > 10 has been recorded in species of higher termite (Bignell, 1984). Conversely, Greenberg *et al.* (1968) recorded an acid pH of 3 in some dipteran larvae. It is interesting to note that many of the species in which extreme pH conditions are known to occur harbor a large microbial population which presumably are physiologically adapted to the extreme conditions. High gut pH may account for the lack of gut infection in the lepidopteran *Heliothis zea* by *B. bassiana* (Pekrul and Grula, 1979).

The pH of the gut contents may be adversely affected by the activities of the microbiotia. Streptococci increased the acidity of the gut of *Lymantria dispar* by the production of lactic acid (Kodama and Nakasuji, 1971). The inclusion of the nonpathogenic fermentative *Bacillus* with a pathogenic chitinolytic bacterium resulted in increased mortality of *L. dispar*. This was thought to be due to the drop in pH which facilitated the activity of the bacterial chitinase (Daoust and Gunner, 1979).

Dillon and Charnley (1986a) recorded germination and growth of *M. anisopliae* throughout the pH range of 3–10 with an optimal range of 5–7. The pH range of the gut of *S. gregaria*—foregut pH 5.7, hindgut pH 7.0 (Evans and Payne, 1964)—is therefore suitable for growth of *M. anisopliae*.

5.2. Oxygen Availability

Germination of fungal conidia may be limited by oxygen availability (Hall, 1981). There are little data published on the oxygen status of insect guts. Where oxygen is restricted, it is usually inferred that this provides the conditions for fermentation of recalcitrant dietary materials (Bignell, 1984).

Veivers *et al.* (1980) examined the guts of representatives from all termite families. In all cases they found that the fore- and midgut were aerobic but that the hindgut (the location of the symbiotic microbiotia) was anaerobic. Similarly, cockroaches possess a complex anaerobic microbiotia in the hindgut (Bracke *et al.*, 1979). In these insects it may be postulated that anaerobiosis provides a barrier to fungal colonization of the hindgut.

Most insects investigated by Bignell (1984) had a high gut redox potential, indicative of aerobiosis. Similarly, no oxygen deficit was found in the gut of *S. gregaria* (Dillon and Charnley, 1986a), though this does not preclude the possibility that small anaerobic pockets may be created by the large populations of facultative anaerobes in the hindgut (Hunt and Charnley, 1981). Such microniches are known to exist in predominantly aerobic environments such as the oral cavity of mammals (Bowden *et al.*, 1979).

Enhanced levels of CO_2 or bicarbonate ions in the gut of *Apis mellifera* were thought to activate the ascospores of *Ascosphaera apis*, the causative agent of chalkbrood disease (Heath and Gaze, 1987).

5.3. Nutrient Availability

The availability of nutrients may determine the ability of the fungal spore to germinate in the gut as many of the entomopathogenic Deuteromycotina require an external nutrient (carbon and nitrogen) source (Charnley, 1984). The nutrient availability will depend on the region of the gut in which the fungus is located. Thus, spores in the foregut or midgut may have access to ingested low-molecular-weight nutrients and products of digestion which are absent in the hindgut. In most insects the hindgut is the site of the indigenous microbiotia and pathogenic fungi will have to compete with the resident bacterial population for nonabsorbed nutrients and waste products secreted by the Malpighian tubules. It is conceivable that a large gut microbiotia may act as a nutrient sink (Hunt and Charnley, 1981). Lepesme (1938) proposed that the inability of *Asper-gillus flavus* to infect the locust via the gut was due to a lack of nutrients. A nutrient supplement, however, did not promote the germination of *M. anisopliae* in gut fluid or filter sterilized gut fluid (gut bacteria excluded) from *S. gregaria* (Dillon and Charnley, 1986a).

Provision of micronutrients, particularly iron, is potentially important in the germination of fungal conidia. The production of siderophores by bacteria enables them to scavenge available iron (Griffiths, 1986). The pathogenicity of bacteria and fungi is known to depend in part on their successful development in an iron-restricted environment on plants (Griffiths, 1983). Thus, iron assimilation by the host and gut microbiotia could result in a condition of fungistasis due to iron deprivation in the gut lumen.

5.4. Osmotic Pressure

The maceration of food material in the gut releases large amounts of osmotically active solutes into the gut lumen and the absorption of water from the gut lumen by the insect will further increase the osmotic pressure within the gut. High osmotic pressures can be detrimental to fungal spore germination (Gottlieb, 1978). However, the osmotic pressure in the foregut fluid of locusts was not at an inhibitory level (Dillon and Charnley, 1986a).

5.5. Enzymes

The presence of fungal cell wall-degrading enzymes in the digestive tract of insects would have important implications for the fate of fungal spores. Although conidial wall chemistry of any fungus has not been studied in detail (Cole, 1986), it is thought that it contains chitin (polymer of N-acetylglucosamine), $\beta1,3$- and $\beta1,6$-glucans as major polysaccharide wall components, and significant quantities of protein (e.g., *Coccidioides immitis*, Cole, 1986; *Conidiobolus obscurus*, Latge *et al.*, 1986; *Aspergillus nidulans*, Claverie-Martin *et al.*, 1988).

Mycophagy is widespread in insects and presupposes the production of fungal cell wall-degrading chitinases and $\beta1,3$- and $\beta1,6$-glucanases (Kukor and Martin, 1987; Martin, 1979). Some insects may not produce the enzymes themselves but acquire the requisite enzymes from the fungal biomass during maceration of the fungal tissue in the foregut. Alternatively, extracellular enzymes may be obtained from the substrate surrounding the fungus (Martin, 1979). In addition to mycophagous insects, chitinase has been located in cockroach guts (Waterhouse and McKellar, 1961), tenebrioid beetle (Saxena and Sarin, 1972), and ants (Martin *et al.*, 1976).

Dillon *et al.* (1986) investigated the possibility that the fungitoxicity of gut fluid from conventional locusts was due to the presence of fungal cell wall-degrading enzymes. Since fungitoxicity was confined to insects with a bacterial microbiota, they reasoned that any candidate enzymes must be of microbial origin. Morgan (1976) concluded that some of the enzyme activity within the locust gut may be of microbial origin. Interestingly, Charnley *et al.* (1985) found that α-glucosidase activity in desert locust hindguts did relate to bacterial concentration in the order conventional locusts > parasite-free > axenic. However, Charnley *et al.* (1985) and Dillon *et al.* (1986) found no differences in the activities of a wide range of carbohydrases including $\beta1,3$-glucanase between dialyzed gut extracts from conventional, parasite-free, and axenic insects. Endochitinase was absent from axenic and conventional insects (Dillon and Charnley, 1986a). It seems unlikely, therefore, that the toxic properties of the locust gut fluid are due to enzymatic disruption of the conidial wall.

5.6. Antimicrobial Compounds

5.6.1. Production from the Insect and Diet

Antifungal agents play an important part in plant protection against disease (Deverall, 1976; Schönbeck and Grunewaldt-Stöcker, 1986) and some of these preformed compounds may play an additional role in preventing the proliferation of fungi in the guts of phytophagous insects.

The saponins, the largest group of plant compounds which possess specifically fungitoxic properties, bind to sterols in the fungal cell membrane (Schönbeck and Grunewaldt-Stöcker, 1986). These compounds are found, for example, in the leaves of *Avena sativa*, and Solanaceae. It is interesting that Allee *et al.* (unpublished) rarely found germinating conidia of *B. bassiana* in the gut of potato Colorado beetle fed on potato leaves; in contrast, high conidial germination was found when the insects were given an artificial diet. Other potential antifungal compounds include terpenoids, e.g., avenacin (Burkhardt *et al.*, 1964), and alkaloids, e.g., berberine (Greathouse and Watkins, 1938). Phenols are universal in plants (Harborne, 1984) and many are known to be inhibitors of fungal spore germination (Walker and Stahmann, 1955; Friend, 1981; Macko, 1981; Boonchird and Flegel, 1982). The polyphenolic tannins are potent precipitating agents of protein and enzyme inhibitors (Goldstein and Swain, 1965). Preformed compounds stored in an inactive state in the plant, e.g., as glycosides, may be converted to toxic forms by plant enzymes released during maceration of the plant in the gut. Tissue damage of leaves of *Brassica* species, for example, results in the release of a volatile fungitoxin, allyl isothiocyanate, formed by enzymatic hydrolysis of the glucoside sinigrin (Greenhalgh and Watkins, 1976). Obviously, some compounds may be rendered inoperative under the conditions in the gut. Alternatively, metabolism in the gut may cause the production of additional antifungal agents, e.g., transformation of chlorogenic acid in mulberry leaves into caffeic acid and associated quinones in the gut of the silkworm *Bombyx mori* (Koike *et al.*, 1979). In the latter case, caffeic acid was thought to be directly responsible for suppressing the growth of bacterial pathogens (*Streptococcus* spp.). However, this is unlikely because the level of caffeic acid fell rapidly after a meal even though the antibacterial activity was maintained. The digestive fluid in *B. mori* is alkaline (pH 9.9) and the caffeic acid was probably oxidized over several hours to the corresponding antibacterial quinones.

In vitro assays have demonstrated activity against entomogenous bacteria by substances in plants by many taxa (e.g., Kushner and Harvey, 1962; Maksymiuk, 1970). Presumably, many of these compounds would be active against fungi, but few investigations have addressed the potential for antifungal activity of plant constituents in the guts of insects. Ramoska and Todd (1985) found that the plant diet of the chinch bug *Blissus leucopterus leucopterus* was important in determining susceptibility to the fungus *B. bassiana*. But a suboptimal diet rather than the absence of the antifungal compounds may predispose the insect to disease in this case.

The significance of plant antifungal compounds in insect–fungus associations is seen in the behavior of the leaf-cutting *Atta cephalotes* (Hubbell *et al.*, 1983). The ants appear to avoid many potential plants as substrates for the mutualistic fungus. Extractions from the nonhost plant *Hymenaea courbaril* revealed the presence of caryophyllene epoxide, a compound which displayed both insect deterrent and fungitoxic activities at the *in vivo* concentrations. The production of the fungicide myrmicacin (β-hydroxydecanoic acid) by the leaf-cutting ant *Atta sexdens* to prevent germination of fungal contaminants in the fungal "garden" (Schildknecht, 1971) is perhaps one of the most refined uses of antifungal compounds by an insect species.

Some insects appropriate toxic plant secondary metabolites for use as repellents against predators and parasites (Eisner, 1970). The compounds are discharged from the gut or a specialized gland in response to traumatic stimuli (Blum, 1981). These secretions may play an additional role in preventing gut infections by fungal pathogens.

Fuhrer and Willers (1986) found that insect-derived acid mucopolysaccharides and nitrogenous compounds, discharged from the hindgut of the endoparasitic larva *Pimpula turionellae*, inhibited the germination of *B. bassiana*.

5.6.2. Production by the Microbiotia

Antagonism between bacteria and entomopathogenic fungi is well established. Fungal antibiotics may suppress bacterial growth in insect cadavers (Kodaira, 1961; Veen, 1968; Ferron, 1978). Conversely, Fargues *et al.* (1983) found that blastospores of *B. bassiana* were lysed by soil bacteria causing a reduction in the intracytoplasmic reserves and eventual autolysis. Latge *et al.*

(1987) noted that the cuticular microbiotia of the pea aphid, *Acyrethosiphon pisum*, inhibited germination of the entomopathogen *Conidiobolus obscurus*.

The influence of the gut microbiotia on the production of antifungal or indeed antibacterial compounds has not received much attention. It is possible that some of the antimicrobials in insect guts, thought to be derived from the diet, may require the presence of the microbiotia.

Jarosz (1979a,b, 1983) showed that *Enterococcus faecium* suppressed *Pseudomonas aeruginosa* and the yeasts *Candida guilliermondi*, *C. krusei*, and *Geototrichum candida*. The bacterium, *E. faecium*, usually present as a monobiotia in the gut of the wax moth *Galleria mellonella*, was found to produce a bacteriocin and lysozyme in the insect intestinal fluid.

The fermentation activities of the gut microbiotia in some insects lead to the production of significant quantities of volatile fatty acids [e.g., termites (Odelson and Breznak, 1983), cockroaches (Cruden and Markovetz, 1987), *Oryctes nasicornis* (Bayon, 1980)]. Short-chain fatty acids are thought to contribute to colonization resistance in guts of vertebrates (Rolfe, 1984; Walden and Hentges, 1975) and to inhibit germination of entomopathogens (Smith and Grula, 1982) and may therefore help to prevent colonization and infection by fungi in insect guts.

Brood food of bees and royal jelly possess antimicrobial activity and this may account for the failure to isolate microbes from the gut of the queen bee and the young larvae. The conversion of pollen to bee bread is the result of a lactic acid fermentation by lactobacilli and yeasts which leads to the reduction of the contamination of bee bread in the cell (Foote, 1957; Haydak, 1958).

Gilliam *et al.* (1988) observed that some bee colonies were resistant to the fungus *Ascosphaera apis*, the causative agent of chalkbrood disease. Although good hygienic behavior was important, resistance was also attributed to the production of antimycotic agents by the microbiotia. Eight-two fungal and bacterial saprophytes found in the gut of foraging and nurse bees and in bee bread produced antimycotic substances active against *A. apis*. The largest degree of inhibition was caused by *Rhizopus* spp. and unidentified Mucorales.

The gut bacteria of the desert locust are implicated in the production of antifungal toxins which help to prevent invasion by the fungus *M. anisopliae* (Dillon and Charnley, 1986b). Germination of *M. anisopliae* in gut fluid from conventional insects and *in vivo* in gut fluid was reduced compared to controls. In contrast, germination in the gut lumen and gut fluid of axenic insects was not inhibited. The reduced ability of *M. anisopliae* to germinate in gut fluid of conventional insects was not due to the presence of protozoan parasites since gut fluid from parasite-free insects was also inhibitory to germination. Furthermore, adverse pH, osmotic pressure, dietary chemicals, oxygen or nutrient deficiency, fungal cell wall-degrading enzymes, and competition for nutrients could not account for the inhibitory properties in the conventional locust gut (see above). Since there was no apparent difference in gut or whole animal physiology between axenic and conventional (nonparasitized) locusts (Charnley *et al.*, 1985), it was hypothesized that an antifungal toxin, produced by the gut bacteria, was responsible (Dillon and Charnley, 1986a).

Resuspension of inhibited conidia in fresh nutrient medium devoid of the gut fluid did not result in germination. In addition, a few conidia incubated with gut fluid fluoresced when treated with the vital stain fluorescein diacetate (Dillon and Charnley, 1988). Thus, the gut fluid is fungitoxic rather than fungistatic.

Conidia suspended in distilled water for 20 hr prior to exposure to neat gut fluid were able to swell and form germ tubes (Dillon and Charnley, 1988). Since water without nutrients will only support a preswelling phase of germination (Dillon and Charnley, 1990), the fungitoxicity must be directed at the early stage of germination.

Dialysis and ultrafiltration experiments revealed that the toxic component of the gut fluid was of a low molecular weight (< 1000). The toxin bound to polyvinylpyrrolidone and in ultrafiltered preparations was not affected by heat (100°C for 5 min). The action of the toxin was pH dependent; inhibition of germination occurred at pH 4 and 5.5 but not at pH 8. Further purification of the toxin was carried out using extracts from fecal pellets of conventional insects. This enabled larger quantities of toxin to be purified than was possible with gut fluid. The toxic activity in the pellets was similar to that found in the gut fluid with regard to heat stability, activity in acid conditions, and size.

A purification process was established using an anion exchange resin followed by fractionation on polyacrylamide gel. Germination assays with *M. anisopliae* showed that the toxic component of the pellets was adsorbed to the resin thus enabling separation from basic and neutral elements. Bio-gel P-2 (fraction range 100–1800) was used to separate the remaining low-molecular-weight components of the extract. The fractions obtained by passage through the gel were assayed for the presence of germination-inhibiting activity. Fungitoxic activity was located in a group of low-molecular-weight (≈ 200) UV-absorbing, Folin-positive fractions designated as peak 4 (Dillon and Charnley, 1988). This group of inhibitory fractions was recovered in Bio-Gel elution profiles of extracts from conventional and parasite-free locusts (fed axenic diet; see Charnley *et al.*, 1985) but was absent from extracts from axenic insects. Preliminary analysis of the fractions established the presence of phenols. Fractions coincident with peak 4 from conventional, parasite-free, and axenic insects were dried, silylated, and injected into gas chromatography (GC) and GC–MS systems. No compounds were detected in fractions from axenic insects (Dillon and Charnley, 1988). However, hydroquinone, 3,4-dihydroxybenzoic acid, and 3,5-dihydroxybenzoic acid were identified in derivatized extracts from peak 4 fractions of conventional fecal extracts by comparison of mass spectra with authentic compounds and by gas cochromatography (Dillon and Charnley, 1988). A provisional structure was assigned to two other compounds, viz. 3,4-dihydroxyphenylketo acid and 3,4-dihydroxyethanol. A solution of the authentic phenols inhibited germination of *M. anisopliae* at concentrations estimated to occur in the fecal extract. At least ten other compounds were observed in the GC profile of the inhibitory fraction from the conventional extract, but some of the compounds may be partially silylated derivatives of the identified compounds.

The phenols were located in the extracts from the conventional and parasite-free insects and not from axenic insects. Since axenic insects are, in most respects studied, physiologically comparable with parasite-free insects (Charnley *et al.*, 1985), and are fed the same diet, it is unlikely that the phenols are released directly from the food by the insects' own digestive processes. Instead, their production relates directly to the presence of the gut bacteria.

When *Enterobacter agglomerans*, a prominent member of the locust gut biotia (Hunt and Charnley, 1981), was introduced into axenic insects, the gut fluid displayed significant antifungal activity. The level of inhibition was lower than that found in conventional insects; this may be because of the lower numbers of bacteria present in the gut lumen of the monoassociated insect or the absence of other members of the gut microbiotia (Dillon and Charnley, unpublished).

Extracts from fecal pellets of monoassociated insects contained two inhibitory phenols, viz., *p*-hydroxybenzoic acid and 3,4-dihydroxybenzoic acid, in concentrations sufficient to cause inhibition of conidial germination. One of the two phenols was found in the feces from conventional insects; the other (*p*-hydroxybenzoic acid) is a closely related compound.

There are several mechanisms by which the gut bacteria could produce the antifungal phenols: (1) release of the phenols from the diet of the insect. The antifungal phenols have all been isolated from plants where they are associated with cell walls conjugated as esters or glycosides (Harborne, 1984). (2) Bacterial transformation of the waste products from the host insect which enter the gut via the Malpighian tubules. 3,4-Dihydroxybenzoic acid, 3,4-dihydroxyphenyl-ethanol, and 3,4-dihydroxyketo acid are also present in locust cuticle (Andersen, 1980). Waste phenols in insects are detoxified by conjugation (e.g., to glucose, aromatic amino acids, or sulfate) and then excreted via the Malpighian tubules into the gut (Dauterman, 1985; Wilkinson, 1986). Thus, plant- or insect-derived phenolic conjugates could be hydrolyzed by members of the Enterobacteriaceae in the gut (Soleim and Scheline, 1972; Scheline, 1968). (3) The phenols are produced *de novo* by the bacteria. Synthesis of phenols from tyrosine (Enei *et al.*, 1972) or nonaromatic cyclic compounds (Scheline, 1973) has been recorded for species of the Enterobacteriaceae and low-molecular-weight phenols are formed from proteins by human gut bacteria (Allison and Macfarlane, 1988). Duffey and Blum (1977) isolated a bacterium that produced phenol from tyrosine in the gut of the millipede *Oxidus gracilis*. Vanillic acid was detected in plant homogenates and all locust fecal extracts. This compound is demethylated to dihydroxybenzoic acid by the microbiotia in the gut of rats (Scheline, 1966). A comparison of the metabolism of flavenoids by conventional and germfree rats showed that phenolic acids detected in the urine of conventional rats were produced as a result of microbial action on flavenoids in the gut (Griffiths and Barrow, 1972). Nolte *et al.* (1973) also identified a microbially derived phenol (the phase pheromone, ethylmethoxyphenol) from locust guts. The pheromone was absent in antibiotic-treated insects and was thought to be produced by the action of the gut bacteria on plant cell walls. Thus, either or several of alternative mechanisms could be implicated in the production of the phenols.

Antifungal activity was found in locusts fed different diets and since the profile of bacterial species may alter with diet it is likely that more than one species is involved in the production of phenols (Dillon and Charnley, 1986a). The extracts from monoassociated locusts also showed that different phenols may be produced by changes in the gut microbiotia.

The antifungal activity of phenolic compounds is well known (see previous section). Antifungal phenols feature in the protection of insects against fungal infection in the cuticle (Schildknecht, 1971; St. Leger *et al.*, 1988) and in the blood (Söderhall and Ajaxon, 1982). The mechanism of toxicity of the phenols has not been established for phenols in blood, cuticle, or gut though the activity of gut phenols is directed primarily toward the initial swelling phase of germination. The capacity of phenols to form hydrogen bonds is probably important in their toxic effects (Singleton, 1981); hence, the phenols may cause general coagulation at high concentrations or inhibit active membrane transport (Hugo, 1983) by binding to components of the plasma membrane or they may uncouple oxidative phosphorylation (Cotmore *et al.*, 1979).

The antifungal activity and concentration of the phenolic compounds may be influenced by surfactants in the insect gut lumen. *S. gregaria* possesses a variety of surfactants thought to be important in the prevention of complex formation between polyphenolic tannins and proteins (Martin and Martin, 1984; Martin *et al.*, 1987). The action of surfactants on low molecular weight phenols may result in enhanced levels of the compounds within the gut lumen.

Antifungal activity was found in the gut fluid of seven other species of Orthopteran (Section 4). Fecal pellets were collected from two of these species, *Pholidoptera griseoptera* and *Chorthippus brunneus*. GC profiles of the low-molecular-weight fractions of the *C. brunneus* extract revealed the presence of *p*-hydroxybenzoic acid and 3,5-dihydroxybenzoic acid (Dillon and Charnley, unpublished). No inhibitory phenols were detected in the corresponding fractions from extracts of *P. griseoptera*.

The detection of the antifungal phenols in the feces of conventional insects allied to their absence in axenically produced feces correlates with the ability of *M. anisopliae* to germinate in the axenic gut and invade through the hindgut (Dillon and Charnley, 1986b; see Section 6). Similar compounds were detected in *C. brunneus*, thus suggested that these low-molecular-weight phenols may be important, at least in some Orthoptera, in preventing fungal infections via the gut.

5.7. Adhesion

The ability of the fungal conidia to adhere to the wall of the digestive tract may be decisive in the attempt of the fungus to invade via the gut. Many of the Trichomycete fungi are found as obligate endocommensals in the guts of some insect species. The Trichomycetes have adapted to the requirement for rapid attachment to overcome the fast passage of food through the gut by the rapid production of thalli which are attached to, but do not penetrate, the cuticular lining of their hosts (Moss, 1979). Trichospore germination and thallus attachment of *Smittium* spp. were observed in the hindgut of *Aedes aegypti* within 1.5 hr after ingestion (Williams and Lichtwardt, 1972). During germination the protoplast projected rapidly (2–10 sec) through the apical end of the spore. The subsequent process of holdfast formation occurred in a period of 10–30 sec during which part of the protoplasm appeared to pass through the base of the thallus to form a small adhesive pad.

Many entomopathogenic fungal conidia possess characteristics which enable successful adhesion and germination on the external insect cuticle (Boucias *et al.*, 1988). Deuteromycete pathogens like *M. anisopliae* and *Nomuraea rileyi* rely on hydrophobic forces for attachment; the apparent absence of the outer wax layer on gut epicuticle may interfere with this bonding. However, not enough is known about the chemistry of the cuticle to draw any firm conclusions.

The application of an excess of nutrients to cuticle represses appressorium formation and penetration by *M. anisopliae* on *Manduca sexta* (St. Leger *et al.*, 1988a). Thus, penetration may be similarly reduced in the foregut where nutrients are available in excess.

Fungal penetration by entomopathogens has not been observed in the midgut of insects (apart from *Ascosphaera apis*; see Section 6); indeed, this site is not commonly colonized by bacteria. The rapid turnover of the midgut epithelial cells (Chapman, 1985) means that there would be insufficient time for penetration by most fungi. In vertebrates, such a niche is colonized by bacteria which possess the ability to rapidly transfer their holdfast to the young underlying cells (Chase and Erlandson, 1976).

Gut peristalsis is thought to deter adhesion of invading microbes in mammalian guts (see, e.g., Savage, 1977) and could also provide a barrier to gut infection in insects (Gabriel, 1959; Rozsypal, 1930). Infection by *M. anisopliae* in the germfree locust gut only occurred in insects deprived of food; this condition is thought to result in reduced gut peristalsis (Bignell, 1984). Interestingly, peristalsis is stimulated by the conventional microbiotia in the mammalian digestive tract and is thought to reduce infection by bacterial pathogens (Abrams and Bishop, 1966).

Colonization of the gut wall by bacteria may impede adhesion of fungi. The resident microbiotia in the stomach of the mouse successfully competed for adhesion sites required by yeasts (Savage, 1969).

The best sites of adhesion and penetration for fungi in the axenic locust gut were the folds of the hindgut and between the rectal pads. These sites, which provide a protected niche away from the mainstream of the gut, were also the most extensively colonized by bacteria in the conventional insect (Dillon and Charnley, 1986b).

6. GUT INFECTIONS IN INSECTS

The yeast *M. unicuspidata* is thought to invade its host *Dasyhelea obscura* exclusively through the intestine. The yeast forms club-shaped asci each containing a single needlelike spore, within the body of the dying host. The asci are ingested by healthy *D. obscura* when they feed on the dead and diseased insects. Once ingested, the asci break up and release the needlelike spores which perforate the gut wall and thus enter the body cavity.

The normal route of infection of *Culicinomyces clavisporus* in the host *Culex fatigans* is also through the digestive tract (Sweeney, 1975, 1979). The ingested conidia of *C. clavisporus* adhered firmly to the foregut and hindgut cuticle (Fig. 2). This adhesion may be assisted by the mucilage (possibly mucopolysaccharide) which encases the mature conidia. The adhesion was strong enough to resist displacement due to the abrasion by food particles and vigorous peristaltic movements of the gut. After emergence the germ tube apex became tightly bound to the

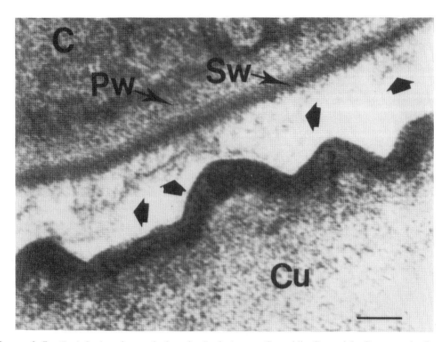

Figures 2–5. Gut infection of mosquito larva by *C. clavisporus*. C, conidia; Cu, cuticle; Gt, germ tube; Lu, gut lumen; Pw, primary wall; Sw, secondary wall; Ph, penetrant hypha; Hy, hypodermis; mc, melanized cuticle. (From Sweeney *et al.*, 1983; reprinted by permission of Academic Press.)

Figure 2. Conidium of *C. clavisporus* adhering to the foregut cuticle of mosquito larva. Note mucoid coating (arrowed) responsible for adhesion to cuticle. Bar = 0.1 μm.

epicuticle (Fig. 3). A penetrant hypha, slightly narrower than the germ tube, was directed through the procuticle (Fig. 4). In some cases a defense reaction by the insect resulted in the production of melanized areas in the pathway of the penetrant hyphae. The melanized area deflected hyphae and forced apart the cuticular layers by mechanical pressure (Fig. 5). Penetration of the cuticle was usually completed 6–18 hr after exposure to the conidia. In this case the high incidence of gut infection may be promoted by the nonselective filter feeding of the larvae which tend to increase the contact between large numbers of spores and the gut wall. The fungus is capable of adhering to the external cuticle and in some cases may penetrate through the anal papillae. However, the latter only occurs in the presence of large concentrations of spores (10⁶/ ml). Al-Aidroos and Roberts (1978) also noted gut invasion of the aquatic mosquito larva *A. aegypti* with *M. anisopliae*, confirming that mosquitoes are particularly vulnerable to this mode of attack. The importance of the digestive tract, however, as a site of infection for the mosquito pathogen *Tolypocladium cylindrosporum* appears to vary with the strain. The New Zealand strain of *T. cylindrosporum* invades through the foregut and hindgut (see Sweeney, 1981). But the digestive tract is a secondary site of infection for a Californian strain of the fungus (Sweeney, 1981). Goettel (1988a) found that infection via the gut of *A. aegypti* was rare and only occurred when the insect was molting. *T. cylindrosporum* does not infect via the gut of *A. albopictus* (Ravallec *et al.*, 1989). There have been a number of reports of insect mortality due to the toxic activities of large numbers of ingested but ungerminated conidia in the filter-feeding mosquito larvae (e.g., Crisan, 1971).

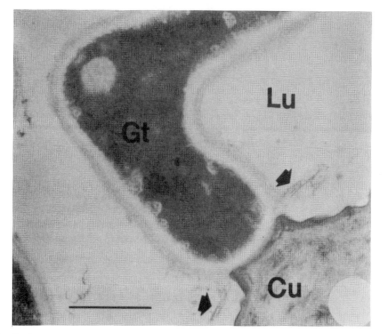

Figure 3.   *C. clavisporus* germ tube apex adhering to foregut cuticle of mosquito larva showing mucoid coating sloughed off at point of contact. Bar = 0.5 μm.

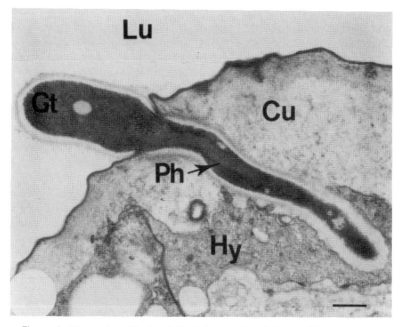

Figure 4.   Penetration of hypha of *C. clavisporus* through host cuticle. Bar = 1 μm.

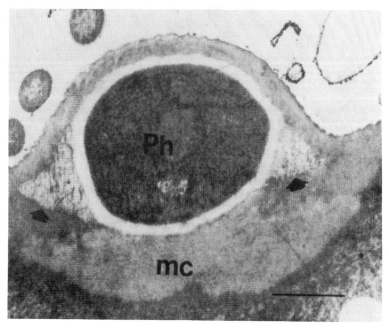

Figure 5.   Transverse section of penetrant hypha of *C. clavisporus* in foregut cuticle of mosquito larva. Note zone of melanization and evidence of physical rupture of cuticle. Bar = 0.25 μm.

Fungi from the genus *Ascosphaera*, the causative agent of chalkbrood disease of bees, may also be primarily invasive via the gut of their insect hosts (Vandenberg and Stephen, 1983; Bamford and Heath, 1990). The disease only infects the larvae (Gilliam *et al.*, 1988) which display an occlusion between the midgut and hindgut. This adaptation, which prevents fouling of the larval cell, may be important in promoting gut infection since it greatly reduces the movement of food through the gut. Activated spores of *Ascosphaera apis* were observed in the gut of *Apis mellifera* 18 hr postinoculation and germination occurred after 24 hr (Bamford and Heath, 1990). The pm failed to act as a barrier to the hyphae; histological sections showed penetration through the gut epithelium after 48 hr.

Other references to gut infection concern fungi which normally penetrate via the external cuticle. In many instances it is difficult to evaluate the importance of the alimentary canal as a mode of entry for these pathogens. The reasons for this are threefold: (1) Surface sterilization is often not carried out, in which case integumental infection cannot be discounted; e.g., Kish and Allen (1978) fed microdrops of *Nomuraea rileyi* conidial suspension to *Anticarsia gemmitalis* larvae but could not confirm gut infection. (2) Histological confirmation of gut infection is not sought; e.g., Agudelo and Falcon (1983) claimed that hyphal bodies of *Paecilomyces farinosus* were infective *per os* in surface-sterilized *Spodoptera exigua*. (3) Large doses of conidia may be used which are unlikely to have any biological significance; e.g., doses of 10^9 conidia of *N. rileyi* were required by Bell and Hamalle (1980) to give over 50% mortality in *Heliothis zea* and similarly, Kramm and West (1982) attributed the death of *Reticulitermes* spp. to the ingestion of a large quantity of ingested *M. anisopliae* conidia.

An additional problem when attempting to determine whether gut infection has occurred is when the gut is the primary target after penetration of the fungus through the external cuticle. Cheung and Grula (1982) observed preferential invasion of the gut wall of *H. zea* by *B. bassiana*.

Attempts have been made to initiate infection through the gut without risk of surface contamination by introducing a conidial inoculum directly into the alimentary tract via an anal or oral catheter. Lefebvre (1934) injected conidia of *B. bassiana* into the hindgut of the European corn borer using a fine pipette. The insects were surface sterilized with alcohol. The inoculated larvae were infected by the fungus and Lefebvre suggested that this was good evidence of infection via the gut. In the absence of histological evidence, however, penetration via the anal aperture could not be discounted. Conidia of *M. anisopliae* or *B. bassiana* were microfed to *Tenebrio molitor*, *Bombyx mori*, and *Galleria mellonella* by Gabriel (1959). The external cuticle of the larvae was disinfected with merthiolate tincture. The microfed *M. anisopliae* caused lower mortality than when conidia were applied externally. With *B. bassiana* there was also a trend toward a lower incidence of infection in microfed insects. However, the gut wall was injured during some of the microfeeding, facilitating invasion by bacteria and fungi. No histological study was done to confirm the incidence of infection via the gut.

Gut infection can only be detected with certainty using histological methods (Madelin, 1963). Veen (1966) observed penetration by *M. anisopliae* in the mouthparts of surface-sterilized *S. gregaria* but not in any other part of the body. Schabel (1976) microfed *M. anisopliae* conidia to immobilized *Hylobius pales*, relying on the immobility of the insect to prevent external integumental infection. Histological sections revealed invasion of the fungus via the buccal cavity but no penetration was observed in the intestinal tract. Yendol and Paschke (1965) noted infections by *Entomophthora* spp. of the preoral cavity, pharynx, and esophagus of *Reticulitermes flavipes*; penetration of the crop, midgut, and hindgut was not observed. Allee *et al.* (unpublished) inoculated surface-sterilized *Leptinotarsa decemlineata per os* with *B. bassiana*; histological studies of fed and starved larvae did not reveal any infections via the gut apart from one individual starved larva. One of the few unequivocal reports of gut invasion was given by Broome *et al.* (1976) working on the interaction between *B. bassiana* and the larvae of *Solenopsis richteri*. Their method of surface sterilization accompanied by appropriate controls ensured that *per os*-inoculated insects could not have been infected through the external cuticle. There was 79% mortality of *per os*-inoculated surface-sterilized larvae and fungal hyphae were found penetrating the gut wall. Table I lists the references where gut infection of insects by fungi was recorded.

No gut infections were observed in conventional or axenic *S. gregaria* when insects were allowed access to food after inoculation (via a cannula) with conidia of *M. anisopliae* (Dillon and Charnley, 1986b). However, 41% of axenic and 26% of conventional insects showed symptoms of mycosis, viz., hyphal bodies in the hemolymph, penetration in the hindgut epithelium, and tetanic paralysis. The infection was thought to have developed after fungal penetration via the mouthparts (Dillon and Charnley, 1986b). Veen (1966) also noted that infection via the buccal cavity occurred in *per os*-inoculated insects. Although a cannula was used to introduce the conidia into the gut, the conidia appeared to be regurgitated when the locusts started to feed. This mode of infection did not occur with "starved" locusts in which antiperistalsis was reduced (Anstee and Charnley, 1977).

When locusts were denied access to food after *per os* inoculation with *M. anisopliae*, a different pattern of infection emerged (Dillon and Charnley, 1986b). None of the conventional locusts exhibited signs of infection. In contrast, 82% of the axenic locusts had hyphal fragments or blastospores in the gut and 44% showed symptoms of mycosis.

Gut infection by *M. anisopliae* in axenic locusts was confined to the hindgut. There was extensive growth of the fungus in the gut lumen adjacent to the hindgut cuticle before penetration occurred (Fig. 6). The sites of penetration were characterized by deposition of melanin, disruption of cells, and swelling of the gut wall. Infection sites were located in the colon end of the ileum (Figs. 7 and 8) and the posterior portion of the rectal sac (Fig. 9). Growth of the fungus was observed to be confined to the gut lumen and epithelium before any blastospores were observed in the hemolymph and it was concluded that infection via the gut wall rather than via the external

Table I. References to Gut Infection in Insects by Fungal Pathogens

Insect host	Fungal species	Reference	Comments
	Infection thought to have occurred via the gut		
Dasyhelea obscura	Monosporella unicuspidata	Keilin (1920)	Needle-shaped ascospore
Ostrinia nubilalis	B. bassiana	Lefebvre (1934)	Surface sterilized, no histology
Bombyx mori	Aspergillus flavus	Sussman (1952)	Noted as hindgut penetration, no histology
Leptinotarsa decemlineata	B. bassiana	Schaerffenberg (1957)	
Bombyx mori	M. anisopliae	Masera (1957)	Route of infection not examined
Galleria mellonella	M. anisopliae	Gabriel (1959)	Possible accidental gut puncture
Bombyx mori	M. anisopliae	Gabriel (1959)	
Tenebrio molitor	M. anisopliae	Gabriel (1959)	
Reticuliternes flavipes	Entomophthora coronata	Yendol and Paschke (1965)	
Reticuliternes flavipes	B. bassiana	Bao and Yendol (1971)	
Reticuliternes sp.	B. bassiana	Kramm and West (1982)	Foregut infection
Solenopsis richteri	B. bassiana	Broome et al. (1976)	Midgut penetration
Aedes aegypti	M. anisopliae	Al-Aidroos and Roberts (1978)	Thought to be primary mode of infection
Culex fatigans	Culicinomyces clavisporus	Sweeney (1975)	Primary route of infection
		Sweeney et al. (1983)	
Megachile rotundata	Ascophaera apis	Vandenberg and Stephens (1983)	
Aedes aegypti	Tolypocladium cylindrosporum	Goettel (1988a)	Infection during molting
Schistocerca gregaria	M. anisopliae	Dillon and Charnley (1986b)	Infection only observed in gut of axenic insects
Apis mellifera	Ascophaera apis	Bamford and Heath (1990)	Midgut/hindgut occlusion may promote infection
Leptinotarsa decemlineata	B. bassiana	Allee et al. (unpublished)	Only one starved insect infected via gut
	Infection thought to have occurred via the buccal cavity		
Schistocerca gregaria	M. anisopliae	Veen (1966)	No intestinal penetration
Hylobius pales	M. anisopliae	Schabel (1976)	
Spodoptera exigua	Paecilomyces farinosus	Agudelo and Falcon (1983)	
	Death attributed to the release of toxins in the gut		
Apis mellifera	Aspergillus flavus	Toumanoff (1931)	
Apis mellifera	Aspergillus flavus	Burnside (1930)	
Culex pipiens	M. anisopliae	Crisan (1971)	Large numbers of conidia
Culex quinquefasciatus	M. anisopliae	Lacey et al. (1988)	Conidia filled gut

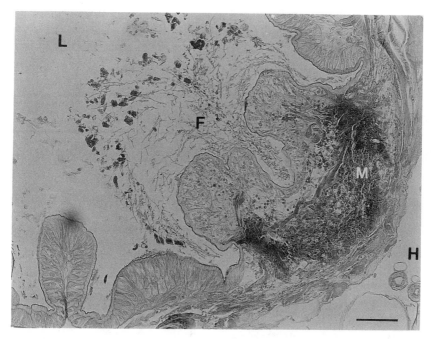

Figures 6–9. Fungal infection by *M. anisopliae* via the hindgut cuticle of an axenic locust. (From Dillon and Charnley, 1986b; copyright 1986 by Martinus Nijhoff Publishers and reprinted by permission of Kluwer Academic Publishers.) L, gut lumen; H, hemocoel; F, fungus; M, melanin deposition; mt, Malpighian tubules; C, hindgut cuticle; S, subintimal space.

Figure 6. Extensive growth of fungus in gut lumen between folds of the ileum. Bar = 100 μm.

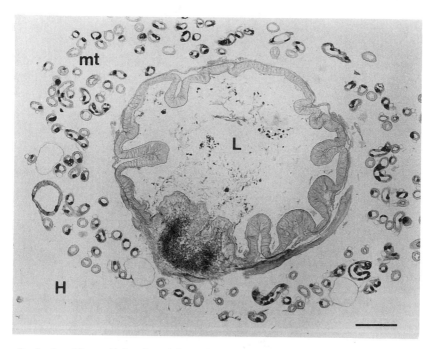

Figure 7. Section of ileum of infected axenic locust showing swelling and melanin deposition in gut wall. Bar = 200 μm.

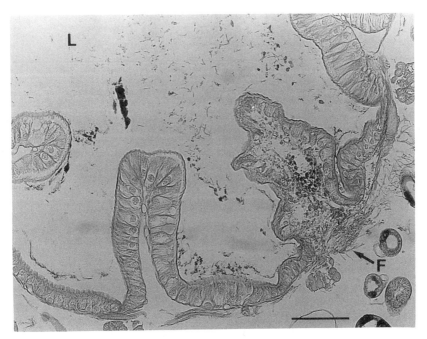

Figure 8. Penetration of hyphae into hemocoel of axenic locust. Note disruption of epidermal cells. Bar =
100 μm.

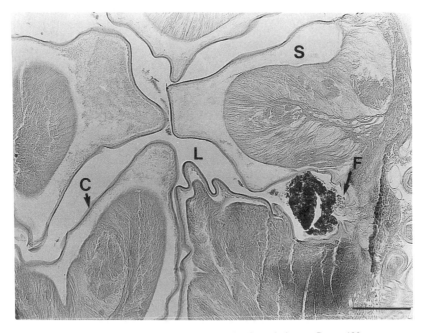

Figure 9. Infection between the rectal pads of axenic locust. Bar = 100 μm.

cuticle had occurred. In contrast, topical infections resulted in the initial growth of the fungus in the hemocoel, there was no growth in the gut lumen until the hyphae had ramified throughout the epithelium (Fig. 10).

7. CONCLUSIONS

It is clear from this review that there is a paucity of information on the effects of passage through the guts of insects on fungal propagules. In some insects, entomopathogenic fungi are unharmed by gut transit. Any resistance to the conditions in the insect gut is probably an adaptation to promote dispersal and maintain inoculum potential rather than a prelude to invasion. The gut is seldom used as a primary route of infection. Peristalsis and a rapid food throughput apparently prevent establishment of insect pathogenic fungi which are otherwise unaffected by their host's gut. Specific holdfast structures as present in the commensalistic Trichomycetes have not evolved among fungal pathogens.

In other insect–entomopathogenic fungus interactions, fungistasis or fungitoxicity may ensue during gut passage of the fungal propagules. Such phenomena may provide significant additions to insect physical defenses against pathogenic fungi.

There is a vast array of potentially fungitoxic secondary chemicals in the diet of phytophagous insects. There is also ample opportunity for further transformation within the gut (with or without the participation of the microbiotia). Thus, it seems surprising that the desert locust and related Orthoptera provide the only examples to date of the involvement of diet or microbially

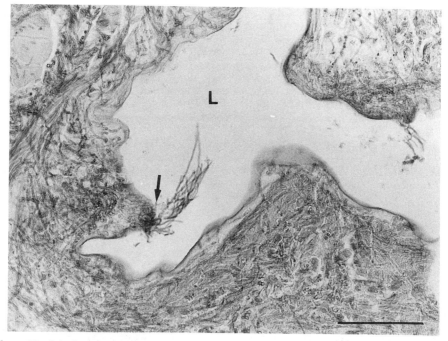

Figure 10.   Infection by *M. anisopliae* after application of conidia to external cuticle of axenic locust. In contrast to infection via the gut, the fungal hyphae have penetrated throughout the gut epidermis before penetration into the gut lumen (arrow). Bar = 100 μm.

derived antifungal compounds in the host defense against pathogenic fungi. In part, the omission probably reflects the lack of work in this area. However, the increasing use of artificial diet for rearing experimental insects may reduce the gut microbiotia as well as deprive the insect of secondary plant chemicals.

Interesting comparisons can be made between vertebrates and insects in the interactions of fungi and yeasts with the host alimentary canal. In both systems there may be restrictions on the growth of fungi due to pH, degree of peristalsis, movement of the food bolus, oxygen tension, enzymes, diet, and epithelial turnover, and infection may be promoted by a disturbance in the ecophysiology, e.g., after the application of antibiotics. A vertebrate gut, however, is a bewilderingly complex environment. Fed by a comparatively varied diet, the vertebrate gut may contain a diverse microbiotia of up to 500 species (Tannock, 1984). In contrast, many phytophagous insects are restricted to one or a few related host plants and harbor only a few resident bacterial species. Thus, such insects offer a simple model system with which to explore the nature of animal gut–fungal interactions.

ACKNOWLEDGMENT.   The authors thank Leslie Allee of Boyce Thompson Institute for a preview of an unpublished manuscript.

8. REFERENCES

Abrams, G. D., and Bishop, J. E., 1966, Effect of the normal microbial flora on the resistance of the small intestine to infection, *J. Bacteriol.* **92**:1604–1608.

Agudelo, F., and Falcon, L. A., 1983, Mass production, infectivity and field application studies with the entomogenous fungus *Paecilomyces farinosus*, *J. Invertebr. Pathol.* **42**:124–132.

Agudelo-Silva, F., and Wassink, H., 1984, Infectivity of a Venezuelan strain of *Metarhizium anisopliae* to *Aedes aegypti* larvae, *J. Invertebr. Pathol.* **43**:435–436.

Al-Aidroos, K., and Roberts, D. W., 1978, Mutants of *Metahizium anisopliae* with increased virulence towards mosquito larvae. *Can. J. Genet. Cytol.* **20**:211–219.

Allee, L. L., Goettel, M. S., Golberg, A., Whitney, H. S., and Roberts, D. W., 1990, Infection by *Beauveria bassiana* of *per os* inoculated *Leptinotarsa decemlineata* larvae occurs following fecal contamination of the integument, unpublished.

Allison, C., and Macfarlane, G. T., 1988, Production of volatile phenols by human intestinal bacteria, *J. Appl Bacteriol.* **65**(6):xviii.

Andersen, S. O., 1980, Cuticular sclerotization, in: *Cuticle Techniques in Arthropods* (T. A. Miller, ed.), Springer-Verlag, Berlin, pp. 185–215.

Anstee, J. H., and Charnley, A. K., 1977, Effects of frontal ganglion removal and starvation on activity and distribution of six gut enzymes in *Locusta*, *J. Insect Physiol.* **23**:965–974.

Baines, D. M., Bernays, E. A., and Leather, E. M., 1973, Movement of food through the gut of fifth instar males of *Locusta migratoria migratorioides* (R and F), *Acrida* **2**(4):319–332.

Balish, E., and Phillips, A. W., 1966a, Growth, morphogenesis, and virulence of *Candida albicans* after oral inoculation in the germ free and conventional chick, *J. Bacteriol.* **91**(5):1736–1743.

Balish, E., and Phillips, A. W., 1966b, Growth and virulence of *Candida albicans* after oral inoculation in the chick with a monoflora of either *Escherichia coli* or *Streptococcus faecalis*, *J. Bacteriol.* **91**(5):1744–1749.

Bamford, S., and Heath, L. A. F., 1990, The infection of *Apis mellifera* larvae by *Ascosphaera apis*, *J. Apic. Res.* (in press).

Bao, C., and Yendol, W. G., 1971, Infection of the eastern subterranean termite, *Reticulitermes flavipes* with the fungus *Beauveria bassiana*, *Entomophaga* **16**:343–352.

Bayon, C., 1980, Volatile fatty acids and methane production in relation to anaerobic carbohydrate fermentation in *Oryctes nasicornis* larvae (Coleoptera: Scarabaeidae), *J. Insect Physiol.* **26**:819–828.

Becker, B., 1978, Determination of the formation rate of peritrophic membranes in some Diptera, *J. Insect Physiol.* **24**:529–533.

Bell, J. V., 1974, Mycoses, in: *Insect Diseases*, Volume 1 (G. E. Cantwell, ed.), Dekker, New York, pp. 185–236.

Bell, J. V., and Hamalle, R. J., 1980, *Heliothis zea* larval mortality time from topical and *per os* dosages of *Nomuraea rileyi* conidia, *J. Invertebr. Pathol.* **35**:182–185.

Bernays, E. A., 1981, A specialized region of the gastric caeca in the locust, *Schistocerca gregaria*, *Physiol. Entomol.* **6**:1–6.

Bertram, D. S., and Bird, R. G., 1961, Studies on mosquito borne viruses in their vectors. I. The normal fine structure of the midgut epithelium of the female *Aedes aegypti* and the functional significance of its modifications following a blood meal, *Trans. R. Soc. Trop. Med. Hyg.* **55**:404–423.

Bignell, D. E., 1984, The arthropod gut as an environment for microorganisms, in: *Invertebrate, Microbial Interactions, British Mycological Symposium*, Volume 6 (J. M. Anderson, A. D. M. Rayner, and D. Walton, eds.), Cambridge University Press, London, pp. 205–227.

Blum, M. S. (ed.), 1981, *Chemical Defenses of Arthropods*, Academic Press, New York.

Boonchird, C., and Flegel, T. W., 1982, *In vitro* antifungal activity of eugenol and vanillin against *Candida albicans* and *Cryptococcus neoformans*, *Can. J. Microbiol.* **28**:1235–1241.

Boucias, D. G., Pendland, J. C., and Latge, J. P., 1988, Nonspecific factors involved in attachment of entomopathogenic Deuteromycetes to host insect cuticle, *Appl. Environ. Microbiol.* **54**:1795–1805.

Bowden, G. H. W., Elwood, D. C., and Hamilton, I. R., 1979, Microbial ecology of the oral cavity, in: *Advances in Microbial Ecology*, Volume 3 (M. Alexander, ed.), Plenum Press, New York.

Boyd, N. D., and Martin, M. M., 1975a, Faecal proteinases of the fungus growing ant, *Atta texana*: Properties, significance and possible origin, *Insect Biochem.* **5**:619–635.

Boyd, N. D., and Martin, M. M., 1975b, Faecal proteinases of the fungus growing ant, *Atta texana*: Their fungal origin and ecological significance, *J. Insect Physiol.* **21**:1815–1820.

Bracke, J. W., Cruden, D. L., and Markovetz, A. J., 1978, Effect of metronidazole on the intestinal microflora of the American cockroach, *Periplaneta americana* L. *Antimicrob. Agents Chemother.* **13**(1):115–120.

Bracke, J. W., Cruden, D. L., and Markovetz, A. J., 1979, Intestinal microbial flora of the American cockroach, *Periplaneta americana* L., *Appl. Environ. Microbiol.* **38**:945–955.

Bradley, T. J., 1985, The excretory system: Structure and physiology, in: *Comprehensive Insect Physiology, Biochemistry and Pharmacology*, Volume 4 (G. A. Kerkut and L. I. Gilbert, eds.), Pergamon Press, New York, pp. 421–506.

Brandt, C. R., Adang, M. J., and Spence, K. D., 1978, The peritrophic membrane: Ultrastructural analysis and function as a mechanical barrier to microbial infection in *Orygia pseudotsugata*, *J. Invertebr. Pathol.* **32**: 12–24.

Breznak, J. A., 1984, Biochemical aspects of symbiosis between termites and their intestinal microbiotia, in: *Invertebrate, Microbial Interactions, British Mycological Symposium*, Volume 6 (J. M. Anderson, A. D. M. Rayner, and D. Walton, eds.), Cambridge University Press, London, pp. 173–203.

Brooks, M. A., 1963, The microorganisms of healthy insects, in: *Insect Pathology*, Volume 1 (E. A. Steinhaus, ed.), Academic Press, New York, pp. 215–250.

Broome, J. R., Sikorowski, P. P., and Norment, B. R., 1976, A mechanism of pathogenicity of *Beauveria bassiana* on larvae of the imported fire ant, *Solenopsis richteri*, *J. Invertebr. Pathol.* **28**:87–91.

Buchner, P. (ed.), 1965, *Endosymbiosis of Animals with Plant Microorganisms*, Interscience, New York.

Burkhardt, H. J., Maizel, J. V., and Mitchell, H. K., 1964, Avenacin, and antimicrobial substance isolated from *Avena sativa*. II. Structure, *Biochemistry* **3**:426–431.

Burnside, C. E., 1930, Fungous diseases of the honey bee, *U.S. Dept. Agric. Tech. Bull.* **149**:43.

Chapman, R. F., 1985, Structure of the digestive system, in: *Comprehensive Insect Physiology, Biochemistry, and Pharmacology*, Volume 4 (G. A. Kerkut and L. I. Gilbert, eds.), Pergamon Press, New York, pp. 165–211.

Chararas, C., Pignal, M. C., Vodjdani, G., and Bourgeay-Causse, 1983, Glycosidases and B group vitamins produced by six yeast strains from the digestive tract of *Phoracantha semipunctata* larvae and their role in insect development, *Mycopathologia* **83**:9–15.

Charnley, A. K., 1984, Physiological aspects of pathogenesis in insects by fungi: A speculative review, in: *Invertebrate, Microbial Interactions, British Mycological Symposium*, Volume 6 (J. M. Anderson, A. D. M. Rayner, and D. Walton, eds.), Cambridge University Press, London, pp. 229–271.

Charnley, A. K., Hunt, J., and Dillon, R. J., 1985, The germ free culture of desert locusts *Schistocerca gregaria*, *J. Insect Physiol.* **31**:477–485.

Chase, D. G., and Erlandsen, S. L., 1976, Evidence for a complex life cycle and endospore formation, in attached, filamentous, segmented bacterium from murine ileum, *J. Bacteriol.* **127**:572–583.

Cheung, P. Y. K., and Grula, E. A., 1982, *In vivo* events associated with entomopathology of *Beauveria bassiana* for the corn earworm *Heliothis zea*, *J. Invertebr. Pathol.* **39**:303–313.

Claverie-Martin, F., Diaz-Torres, M. R., and Geoghegan, M. J., 1988, Chemical composition and ultrastructure of wild type and white mutant *Aspergillus nidulans* conidial walls, *Curr. Microbio.* **16**:281–287.

Cochrane, V. W. (ed.), 1958, *Physiology of Fungi*, Wiley, New York.

Cole, G. T., 1986, Models of cell differentiation in conidial fungi, *Microbiol. Rev.* **50**:95–132.

Cotmore, J. M., Burke, A., Lee, N. H., and Shapiro, I. M., 1979, Respiratory inhibition of isolated rat liver mitochondria by eugenol, *Arch. Oral Biol.* **34**:565–568.

Crisan, E. V., 1971, Mechanism responsible for release of toxin by *Metarhizium* spores in mosquito larvae, *J. Invertebr. Pathol.* **17**:260–264.

Cruden, D. L., and Markovetz, A. J., 1987, Microbial ecology of the cockroach gut, *Annu. Rev. Microbiol.* **41**: 617–643.

Daoust, R. A., and Gunner, H. G., 1979, Microbial synergists pathogenic to *Lymantria dispar*: Chitinolytic and fermentative bacterial interactions, *J. Invertebr. Pathol.* **33**:368–377.

Dauterman, W. C., 1985, Insect metabolism: Extramicrosomal, in: *Comprehensive Insect Physiology, Biochemistry and Pharmacology*, Volume 4 (G. A. Kerkut and L. I. Gilbert, eds.), Pergamon Press, New York, pp. 713–730.

Deverall, B. J., 1976, Current perspectives in research on phytoalexins, in: *Biochemical Aspects of Plant Parasite Relationships*, (J. Friend and D. R. Threlfall, eds.), Academic Press, New York, pp. 207–223.

Dillon, R. J., 1984, Studies on the pathogenicity of *Metarhizium anisopliae* (Metch.) Sorok. to *Schistocerca gregaria* (Forsk.) with particular reference to the gut, Ph.D. thesis, University of Bath, England.

Dillon, R. J., and Charnley, A. K., 1986a, Inhibition of *Metarhizium anisopliae* by the gut bacterial flora of the desert locust, *Schistocerca gregaria*: Evidence for an antifungal toxin, *J. Invertebr. Pathol.* **47**:350–360.

Dillon, R. J., and Charnley, A. K., 1986b, Invasion of the pathogenic fungus *Metarhizium anisopliae* through the guts of germ-free desert locusts, *Schistocerca gregaria*, *Mycopathologia* **96**:59–66.

Dillon, R. J., and Charnley, A. K., 1988, Inhibition of *Metarhizium anisopliae* by the gut bacterial flora of the desert locust: characterisation of antifungal toxins, *Can. J. Microbiol.* **34**:1075–1082.

Dillon, R. J., and Charnley, A. K., 1990, Initiation of germination in conidia of the entomopathogenic fungus, *Metarhizium anisopliae*, *Mycol. Res.* **94**(3):299–304.

Dillon, R. J., Charnley, A. K., and Hunt, J., 1986, Contribution of the gut bacterial flora to the physiology of the desert locust, *Schistocerca gregaria*, *J. Appl. Bacteriol.* **61**(6):xiv.

Dillon, R. J., Charnley, A. K., and Ratcliffe, N. A., 1987, The immunology of germ-free desert locusts, *Schistocerca gregaria*, *20th Annual Meeting of Society of Invertebrate Pathology*, Abstract, p. 28.

Dow, J. A. T., 1981, Countercurrent flows, water movements and nutrient absorption in the locust midgut, *J. Insect Physiol.* **27**:579–585.

Duffey, S. S., and Blum, M. S., 1977, Phenol and guaiacol: Biosynthesis, detoxication, and function in a polydesmid millipede, *Oxidus gracilis*, *Insect Biochem.* **7**:57–65.

Eisner, T., 1970, Chemical defense against predation in arthropods, in: *Chemical Ecology* (E. Sondheimer and J. B. Simeone, eds.), Academic Press, New York, pp. 157–217.

Enei, H., Matsui, H., Yamashita, K., Okumura, S., and Yamada, H., 1972, Distribution of tyrosine phenol lyase in microorganisms, *Agric. Biol. Chem.* **36**:1861–1868.

Eutick, M. L., O'Brien, R. W., and Slaytor, M., 1978, Bacteria from the gut of australian termites, *Appl. Environ. Microbiol.* **35**:823–828.

Evans, W. A. L., and Payne, D. W., 1964, Carbohydrases of the alimentary tract of the desert locust *Schistocerca gregaria* Forsk., *J. Insect Physiol.* **10**:657–674.

Fargues, J., Reisinger, O., Robert, P. H., and Aubart, C., 1983, Biodegradation of entomopathogenic Hyphomycetes: Influence of clay coating on *Beauveria bassiana* blastospore survival in the soil, *J. Invertebr. Pathol.* **41**:131–142.

Ferron, P., 1978, Biological control of insect pests by entomogenous fungi, *Annu. Rev. Entomol.* **23**:409–442.

Foote, H. L., 1957, Possible use of microorganisms in synthetic bee bread production, *Am. Bee J.* **98**:476–478.

Friend, J., 1981, Plant phenolics, lignification and plant disease, in *Progress in Phytochemistry*, Volume 7 (L. Reinhold, J. B., Harborne, and T. Swain, eds.), Pergamon Press, New York, pp. 197–261.

Fuhrer, E., and Willers, D., 1986, The anal secretion of the endoparasitic larva *Pimpula turionellae*: Sites of production and effects, *J. Insect Physiol.* **32**:361–367.

Fuller, R., 1978, Epithelial attachment and other factors controlling the colonization of the intestine of the gnotobiotic chicken by lactobacilli, *J. Appl. Bacteriol.* **45**:389–395.

Gabriel, B. P., 1959, Fungus infection of insects via the alimentary tract, *J. Insect Pathol.* **1**:319–330.

Gilliam, G., Taber, III, S., Lorenz, B. J., and Prest, D. B., 1988, Factors affecting development of chalkbrood disease in colonies of honey bees *Apis mellifera*, fed pollen contaminated with *Ascosphaera apis*, *J. Invertebr. Pathol.* **52**:314–325.

Goettel, M. S., 1988a, Pathogenesis of the hyphomycete *Tolypocladium cylindrosporum* in the mosquito *Aedes aegypti*, *J. Invertebr. Pathol.* **51**:259–274.

Goettel, M. S., 1988b, Viability of *Tolypocladium cylindrosporum* conidia following ingestion and excretion by larval *Aedes aegypti*, *J. Invertebr. Pathol.* **51**:275–277.

Goldstein, J. L., and Swain, T., 1965, The inhibition of enzymes by tannins, *Phytochemistry* **4**:185–192.

Goodhue, D., 1963, Some differences in the passage of food through the intestines of the desert and migratory locusts, *Nature* **200**:288–289.

Gottlieb, D. (ed.), 1978, *The Germination of Fungus Spores*, Meadowfield Press, Durham.

Greathouse, G. A., and Watkins, G. M., 1938, Berberine as a factor in the resistance of *Mahonia trifoliata* and *M. swaseyi* to *Phymatotrichum* root rot, *Am. J. Bot.* **25**:743–748.

Greenberg, B., Kowalski, J. A., and Klowden, M. J., 1968, Model for destruction of bacteria in the midgut of blowfly maggots, *J. Med. Entomol.* **5**:31–38.

Greenberg, B., Kowalski, J. A., and Klowden, M. J., 1970, Factors affecting the transmission of *Salmonella* by flies: Natural resistance to colonization and bacterial interference, *Infect. Immun.* **2**:800–809.

Greenhalgh, J. R., and Watkins, N. D., 1976, The involvement of flavour volatiles in the resistance to downy mildew of wild and cultivated forms of *Brassica oleracea*, *New Phytol.* **77**:391–398.

Griffiths, E., 1983, Availability of iron and survival of bacteria in infection, in: *Medical Microbiology*, Volume 3 (C. S. F. Easmon, M. Brown, and P. A. Lambert, eds.), Academic Press, New York, pp. 150–177.

Griffiths, E., 1986, Iron and biological defense mechanisms, in: *Natural Antimicrobial Systems, FEMS Symposium No. 35* (G. W. Gould, M. E. Rhodes-Roberts, A. K. Charnley, R. M. Cooper, and R. G. Board, eds.), Bath University Press, England, pp. 56–71.

Griffiths, L. A., and Barrow, A., 1972, Metabolism of flavenoid compounds in germ free rats, *Biochem. J.* **130**:1161–1162.

Hall, R., 1981, Physiology of conidial fungi, in: *Biology of Conidial Fungi*, Volume 2 (G. T. Cole, and B. Kendrick, eds.), Academic Press, New York, pp. 417–457.

Harborne, J. B. (ed.), 1984, *Phytochemical Methods*, Chapman & Hall, London.

Hasan, S., 1982, The possible role of two species of Orthoptera in the dissemination of a plant pathogenic fungus, *Ann. Appl. Biol.* **101**:205–209.

Haydak, M. H., 1958, Pollen, pollen substitutes, bee bread, *Am. Bee J.* **98**:145–146.

Heath, L. A. F., and Gaze, B. M., 1987, Carbon dioxide activation of spores of the chalkbrood fungus *Ascosphaera apis*, *J. Apic. Res.* **26**(4):243–246.

Henry, S. M., 1964, Intestinal microorganisms of the cockroach with and without intracellular symbionts, *XIIth Int. Congr. Entomol. Proc.*, Section III, p. 748.

Hill, M. J., Fadden, K., Fernandez, F., and Roberts, A. K., 1986, Biochemical basis for microbial antagonism in the intestine, in: *Natural Antimicrobial Systems, FEMS Symposium No. 35* (G. W. Gould, M. E. Rhodes-Roberts, A. K. Charnley, R. M. Cooper, and R. G. Board, eds.), Bath University Press, England, pp. 29–39.

Hubbell, S. P., Wiemer, D. F., and Adejare, A., 1983, An antifungal terpenoid defends a neotropical tree (*Hymenaea*) against attack by fungus growing ants (*Atta*), *Oecologia (Berlin)* **60**:321–327.

Hugo, W. B., 1983, Mode of action of non-antibiotic antimicrobial agents, in: *Pharmaceutical Microbiology* (W. B. Hugo and A. D. Russell, eds.), Blackwell, Oxford, pp. 258–264.

Hunt, J., 1982, Studies on the intestinal microflora of the desert locust *Schistocerca gregaria* (Forsk), Ph.D. thesis, University of Bath, England.

Hunt, J., and Charnley, A. K., 1981, Abundance and distribution of the gut flora of the desert locust, *Schistocerca gregaria*, *J. Invertebr. Pathol.* **38**:378–385.

Ingold, C. T. (ed.), 1971, *Fungal Spores: Their Liberation and Dispersal*, Oxford University Press, London.

Jarosz, J., 1979a, Gut flora of *Galleria mellonella* suppressing ingested bacteria, *J. Invertebr. Pathol.* **34**:192–198.

Jarosz, J., 1979b, Yeast-like fungi from greater wax moth larvae (*Galleria mellonella*) fed antibiotics, *J. Invertebr. Pathol.* **34**:257–262.

Jarosz, J., 1983, *Streptococcus faecium* in the intestine of the greater wax moth, *Galleria mellonella*, *Microbios Lett.* **23**:125–128.

Keilin, D., 1920, On a new saccharomycete *Monosporella unicuspidata* Gen. N. Nom., N.sp., parasitic in the body cavity of a dipterous larva (*Dasyhelea obscura* Winnertz). *Parasitology* **12**:83–91.

Kish, L. P., and Allen, G. E., 1978, The biology and ecology of *Nomuraea rileyi* and a program of predicting its incidence on *Anticarsia gemmatalis* in soybean, *Fl. Agric. Exp. Stn. Bull.* **795**:1–47.

Klug, M. J., and Kotarski, S., 1980, Bacteria associated with the gut tract of larval stages of the aquatic cranefly *Tipula abdominalis* (Diptera; Tipulidae), *Appl. Environ. Microbiol.* **40**:408–416.

Kodaira, Y., 1961, Biochemical studies on the muscardine fungi in the silkworms, *Bombyx mori*, *J. Fac. Text. Sci. Technol. Shinshu Univ.* **29**:1–68.

Kodama, R., and Nakasuji, Y., 1971, Further studies on the pathogenic mechanism of bacterial diseases in gnotobiotic silkworm larvae, *IFO Res. Commun.* **5**:1–9.

Koike, S., Izuka, T. I., and Mizutani, J., 1979, Determination of caffeic acid in the digestive juice of silkworm larvae and its antibacterial activity against pathogenic *Streptococcus faecalis* AD-4, *Agric. Biol. Chem.* **43**(8):1727–1731.

Kramm, K. R., and West, D. F., 1982, Termite pathogens: Effects of ingested *Metarhizium, Beauveria,* and *Gliocladium* conidia on worker termites (*Reticulitermes* sp.), *J. Invertebr. Pathol.* **40**:7–11.

Kukor, J. J., and Martin, M. M., 1987, Nutritional ecology of fungus feeding arthropods, in: *The Nutritional Ecology of Insects, Mites, and Spiders* (F. Slansky, Jr., and J. G. Rodriguez, eds.), Wiley, New York, pp. 791–814.

Kushner, D. J., and Harvey, G. T., 1962, Antibacterial substances in leaves: Their possible role in insect resistance to disease, *J. Insect Pathol.* **4**:155–184.

Lacey, M. C., Lacey, L. A., and Roberts, D. W., 1988, Route of invasion of *Metarhizium anisopliae* in *Culex quinquefasciatus, J. Invertebr. Pathol.* **52**:108–118.

Latge, J. P., Cole, G. T., Horisberger, M., and Prevost, M. C., 1986, Ultrastructure of chemical composition of the ballistospore of *Conidiobolus obscurus, Exp. Mycol.* **10**:99–113.

Latge, J. P., Sampedro, L., Brey, P., and Diaquin, M., 1987, Aggressiveness of *Conidiobolus obscurus* against the pea aphid: Influence of cuticular extracts on ballistospore germination of aggressive and non-aggressive strains, *J. Gen. Microbiol.* **133**:1987–1997.

Lee, A., 1985, Neglected niches: The microbiol ecology of the gastrointestinal tract, in: *Advances in Microbial Ecology,* Volume 8 (K. C. Marshall, ed.), Plenum Press, New York, pp. 115–162.

Lefebvre, C. L., 1934, Penetration and development of the fungus *Beauveria bassiana,* in the tissues of the corn borer, *Ann. Bot.* **48**:441–452.

Lepesme, P., 1938, Recherches sur une aspergillose des Acridiens, *Bull. Soc. Hist. Nat. Afr. Nord.* **29**:372–381.

Liljemark, W. F., and Gibbons, R. J., 1973, Suppression of *Candida albicans* by human oral streptococci in gnotobiotic mice *Infect. Immun.* **8**:846–849.

Lysenko, O., 1981, Principles of pathogenesis of insect bacterial diseases as exemplifed by the nonsporeforming bacteria, in: *Pathogenesis of Invertebrate Microbial Diseases* (E. W. Davidson, ed.), Allanheld, Osmun, Montclair, N. J., pp. 163–188.

Lysenko, O., 1985, Non-sporeforming bacteria pathogenic to insects: Incidence and mechanisms, *Annu. Rev. Microbiol.* **39**:673–695.

Macko, V., 1981, Inhibitors and stimulants of spore germination and infection structure formation in fungi, in: *The Fungal Spore: Morphogenetic Controls* (G. Turian and H. R. Hohl, eds.), Academic Press, New York, pp. 565–584.

Madelin, M. F., 1963, Diseases caused by hyphomycetous fungi, in: *Insect Pathology: An Advanced Treatise* (E. A. Steinhaus, ed.), Academic Press, New York, pp. 233–271.

Maksymiuk, B., 1970, Occurrence and nature of antibacterial substances in plants affecting *Bacillus thuringiensis* and other entomogenous bacteria *J. Invertebr. Pathol.* **15**:356–371.

Martin, J. S., Martin, M. M., and Bernays, E. A., 1987, Failure of tannic acid to inhibit digestion or reduce digestibility of plant protein in gut fluids of insect herbivores: implications for theories of plant defense, *J. Chem. Ecol.* **13**:605–621.

Martin, M. M., 1979, Biochemical implications of insect mycophagy, *Biol. Rev.* **54**:1–21.

Martin, M. M., and Martin, J. S., 1984, Surfactants: Their role in preventing the precipitation of proteins by tannins in insect guts, *Oecologia* **61**:342–345.

Martin, M. M., Gieselmann, M. J., and Martin, J. S., 1976, The presence of chitinase in the digestive fluids of ants, *Comp. Biochem. Physiol.* **53A**:331–332.

Masera, E., 1957, *Metarhizium anisopliae* (Metchnikoff) Sorokin, parassita del baco da seta, *Ann. Sper. Agrar.* **11**:281–298.

Morgan, M. R. J., 1976, Gut carbohydrases in locusts and grasshoppers, *Acrida* **5**:45–58.

Moss, S. T., 1979, Commensalism of the Trichomycetes, in: *Insect–Fungus Symbiosis* (L. R. Batra, ed.), Allanheld, Osmun, Monclair, N.J., pp. 175–227.

Nolte, D. J., Eggers, S. H., and May, I. R., 1973, A locust pheromone: locustol, *J. Insect Physiol.* **19**:1547–1554.

Nuss, R., 1982, Die Bedeutugn der Proterosporen: Schlußfolgerungen aus Untersuchungen an *Ganoderma* (Basidiomycetes), *Plant Syst. Evol.* **141**:53–79.

Odelson, D. A., and Breznak, J. A., 1983, Volatile fatty acid production by the hindgut microbiota of xylophagous termites, *Appl. Environ. microbiol.* **45**:1602–1613.

Pant, N. C., and Fraenkel, G., 1950, The function of the symbiotic yeasts of 2 insect species *Lasioderma serricorne* F. and *Stegobium (Sitodrepa) paniceum* L., *Science* **112**:498.

Pekrul, S., and Grula, E. A., 1979, Mode of infection of the corn earworm (*Heliothis zea*) by *Beauveria bassiana* as revealed by scanning electron microscopy, *J. Invertebr. Pathol.* **34**:238–247.

Phillips, J. E., 1980, Epithelial transport and control in recta of terrestrial insects, in: *Insect Biology in the Future,* *'VBW 80'* (M. Locke and D. S. Smith, eds.), Academic Press, New York, pp. 145–177.

Pivnik, H., and Nurmi, E., 1982, The Nurmi concept and its role in the control of salmonellae in poultry, in: *Developments in Food Microbiology*, Volume 1 (R. Davies, ed.), Applied Science Publishers, London, pp. 41–70.

Price, D. W., 1976, Passage of *Verticillium albo-atrum* through the alimentary canal of the bulb mite, *Phytopathology* **66**:46–50.

Ramoska, W. A., and Todd, T., 1985, Variation in efficacy and viability of *Beauveria bassiana* in the chinch bug (Hemiptera: Lygaeidae) as a result of feeding activity on selected host plants, *Environ. Entomol.* **14**:146–148.

Ravallec, M., Vey, A., and Riba, G., 1989, Infection of *Aedes albopictus* by *Tolypocladium cylindrosporum*, *J. Invertebr. Pathol.* **53**:7–11.

Richards, A. G., and Richards, A. G., 1977, The peritrophic membranes of insects, *Annu. Rev. Entomol.* **22**:219–240.

Roberts, D. W., and Humber, R. A., 1981, Entomogenous fungi, in: *Biology of Conidial Fungi*, Volume 2 (G. T. Cole and B. Kendrick, eds.), Academic Press, New York, pp. 201–236.

Rolfe, R. D., 1984, Role of volatile fatty acids in colonisation resistance to *Clostridium difficile*, *Infect. Immun.* **45**:185–191.

Rozsypal, J., 1930, The sugar-beet pest, *Bothynoderes punctiventria* Germ., and its natural enemies [in Czech.] *Sb. Chir. Pohyb. Ustroji* C16; 1931, *Rev. Appl. Entomol.* **A19**:427–429 (abstr.).

Rupp, R. A., 1986, The role of the peritrophic membrane of *Manduca sexta* in insect defense, Ph.D. thesis, Washington State University.

St. Leger, R. J., Butt, T. M., Goettel, M. S., Staples, R. C., and Roberts, D. W., 1988a, Production *in vitro* of appressoria by the entomopathogenic fungus *Metarhizium anisopliae*, *Exp. Mycol.* **13**:274–288.

St. Leger, R. J., Cooper, R. M., and Charnley, A. K., 1988b, The effect of melanization of *Manduca sexta* cuticle on growth and infection by *Metarhizium anisopliae*, *J. Invertebr. Pathol.* **52**:459–470.

Savage, D. C., 1969, Microbial interference between indigenous yeast and lactobacilli in the rodent stomach, *J. Bacteriol.* **98**:1278–1283.

Savage, D. C., 1977, Microbial ecology of the gastrointestinal tract, *Annu. Rev. Microbiol.* **31**:107–133.

Saxena, S. C., and Sarin, K., 1972, Chitinase in the alimentary tract of the lesser mealworm, *Alphitobius diaperinus* (Panzer) (Coleoptera: Tenebrionidae), *Appl. Entomol. Zool.* **7**:94.

Schabel, H. G., 1976, Oral infection of *Hylobius pales* by *Metarhizium anisopliae*, *J. Invertebr. Pathol.* **27**:377–383.

Schaerffenberg, B., 1957, *Beauveria bassiana* (Vuill.) Link als parasit des kartoffelkaffers (*Leptinotarsa decemlineata* Say), *Anz. Schaedlingsk.* **30**:69–74.

Scheline, R. R., 1966, Decarboxylation and demethylation of some phenolic benzoic acid derivatives of rat caecal contents, *J. Pharm. Pharmacol.* **18**:664–669.

Scheline, R. R., 1968, The metabolism of drugs and other organic compounds by the intestinal microflora, *Acta Pharmacol. Toxicol.* **26**:332–342.

Scheline, R. R., 1973, Metabolism of foreign compounds by gastrointestinal microorganisms, *Pharmacol. Rev.* **25**:451–523.

Schildknecht, H., 1971, Evolutionary peaks in the defensive chemistry of insects, *Endeavour* **30**:136–232.

Schönbeck, F., and Grunewaldt-Stöcker, G., 1986, Preformed antimicrobial compounds in relation to disease resistance, in: *Natural Antimicrobial Systems, FEMS Symposium No. 35* (G. W. Gould, M. E. Rhodes-Roberts, A. K. Charnley, R. M. Cooper and R. G. Board, eds.), Bath University Press, England, pp. 176–190.

Schroth, M. N., and Hancock, J. G., 1982, Disease suppressive soil and root colonising bacteria, *Science* **216**:1376–1381.

Singleton, V. L., 1981, Phenolic substances as food toxicants, *Adv. Food Res.* **27**:149–242.

Smirnov, O. V., 1978, Interrelationships between entomopathogens in the same insect, *Entomol. Obozr.* **57**(3):473–476.

Smith, R. J., and Grula, E. A., 1982, Toxic components on the larval surface of the corn earworm (*Heliothis zea*) and their effects on germination and growth of *Beauveria bassiana*, *J. Invertebr. Pathol.* **39**:15–22.

Söderhall, K., and Ajaxon, R. A., 1982, Effect of quinones and melanin on mycelial growth of *Aphanomyces* spp. and extracellular protease of *Aphanomyces astaci*, a parasite on crayfish, *J. Invertebr. Pathol.* **39**:105–109.

Soleim, H. A., and Scheline, R. R., 1972, Metabolism of xenobiotics by strains of intestinal bacteria, *Acta Pharmacol. Toxicol.* **31**:471–480.

Southwood, T. R. E., 1972, The insect/plant relationship—An evolutionary perspective, in: *Insect/Plant Relationships, Symposia of the Royal Society of London*, No. 6, Blackwell, Oxford, pp. 3–30.

Stohler, H., 1961, The peritrophic membrane of blood sucking Diptera in relation to their role as vectors of blood parasites, *Acta Trop.* **18**:263–266.

Sussman, A. S., 1952, Studies of an insect mycosis. V. Color changes accompanying parasitism in *Platysamia cecropia*, *Ann. Entomol. Soc. Am.* **45**:223–245.

Sweeney, A. W., 1975, The mode of infection of the insect pathogenic fungus *Culicinomyces* in larvae of the mosquito *Culex fatigans*, *Aust. J. Zool.* **23**:49–57.

Sweeney, A. W., 1979, Infection of mosquito larvae by *Culicinomyces* sp. through anal papillae, *J. Invertebr. Pathol.* **33**:249–251.

Sweeney, A. W., 1981, Fungal pathogens of mosquito larvae, in: *Pathogenesis of Invertebrate Microbial Diseases* (E. W. Davidson, ed.), Allanheld, Osmun, Montclair, N.J., pp. 403–424.

Sweeney, A. W., Inman, A. O., Bland, C. E., and Wright, R. G., 1983, The fine structure of *Culicinomyces clavisporus* invading mosquito larvae, *J. Invertebr. Pathol.* **42**:224–243.

Tannock, G. W., 1984, Control of gastrointestinal pathogens by normal flora, in: *Microbial Ecology: Current Perspectives in Microbial Ecology* (M. J. Klug and C. A. Reddy, eds.), American Society for Microbiology, Washington, D.C., pp. 374–382.

Terra, W. R., Ferreira, C., and De Bianchi, A. G., 1979, Distribution of enzymes among the endo- and ectoperitrophic spaces and midgut cells of *Rhynchosciara* and its physiological significance, *J. Insect Physiol* **25**:487–494.

Tobe, S. S., and Pratt, G. E., 1975, Corpus allatum activity *in vitro* during ovarian maturation in the desert locust, *Schistocerca gregaria*, *J. Exp. Biol.* **62**:611–627.

Toumanoff, C., 1931, Actions des champignons entomophytes sur les abeilles, *Ann. Parasitol. Hum. Comp.* **9**:462–482.

Ulrich, R. G., Buthala, D. A., and Klug, M. J., 1981, Microbiotia associated with the gastrointestinal tract of the common house cricket, *Acheta domestica*, *Appl. Environ. Microbiol.* **42**:246–254.

Vandenberg, J. D., and Stephens, W. P., 1983, Pathogenesis of chalkbrood in the alfalfa leafcutting bee, *Megachile rotundata*, *Apidologie* **14**(4):333–341.

Van der Waaij, D., 1982, Gut resistance to colonization: Clinical usefulness of selective use of orally administered antimicrobial and antifungal drugs, in: *Infections in Cancer Patients* (J. Klastersky, ed.), Raven Press, New York, pp. 73–85.

Van der Waaij, D., Berghuis-de Vries, J. M., and Lekkerkerk-van der Wees, J. E. C., 1971, Colonization resistance of the digestive tract in conventional and antibiotic-treated mice, *J. Hyg.* **69**:405–413.

Veen, K. H., 1966, Oral infection of second instar nymphs of *Schistocerca gregaria* by *Metarhizium anisopliae*, *J. Invertebr. Pathol.* **8**:254–256.

Veen, K. H., 1968, Recherches sur la maladie, due a *Metarhizium anisopliae* chez le criquet pelerin, *Meded Landbouwhoquesch. Wageningen* **68**.

Veivers, P. C., O'Briend, R. W., and Slaytor, M., 1980, The redox state of the gut of termites, *J. Insect Physiol.* **26**:75–77.

Veivers, P. C., O'Briend, R. W., and Slaytor, M., 1982, Role of bacteria in maintaining the redox potential in the hindgut of termites and preventing entry of foreign bacteria, *J. Insect Physiol.* **28**:947–951.

Walden, W. C., and Hentges, D. J., 1975, Differential effects of oxygen and oxidation reduction potential on the multiplication of three species of anaerobic intestinal bacteria, *Appl. Microbiol.* **30**:781–785.

Walker, J. C., and Stahmann, M. A., 1955, Chemical nature and disease resistance in plants, *Annu. Rev. Plant Physiol.* **6**:351–366.

Waterhouse, D. F., 1940, Studies on the physiology and toxicology of blowflies. 5. The hydrogen ion concentration in the alimentary canal, *Pamphl. Counc. Sci. Ind. Res. Aust.* **102**:7–27.

Waterhouse, D. F., and McKellar, J. W., 1961, The distribution of chitinase activity in the body of the American cockroach, *J. Insect Physiol.* **6**:185–195.

Weber, N. A., 1966, Fungus growing ants, *Science* **153**:587–604.

Wilkinson, C. F, 1986, Xenobiotic conjugation in insects, in: *Xenobiotic Conjugation Chemistry* (G. D. Paulson, J. Caldwell, D. H. Hutson, and J. J. Menn, eds.), American Chemical Society, Washington, D.C., pp. 48–61.

Williams, M. C., and Lichtwardt, R. W., 1972, Infection of *Aedes aegypti* by axenic cultures of the fungal genus *Smittium* (Trichomycetes), *Am. J. Bot.* **59**:189–193.

Yendol, W. G., and Paschke, J.D., 1965, Pathology of an *Entomophthora* infection in the eastern subterranean termite *Reticulitermes flavipes* (Kollar), *J. Invertebr. Pathol.* **7**:414–422.

7

Candida Blastospore Adhesion, Association, and Invasion of the Gastrointestinal Tract of Vertebrates

Michael J. Kennedy

1. INTRODUCTION

Survival, implantation, and dissemination from the alimentary tract by *Candida* species, and *Candida albicans* in particular, may play an important role in human and animal health. Recurrent vaginitis and systemic candidosis, for instance, have both been linked to colonization of the gastrointestinal (GI) tract by *C. albicans* (Krause *et al.*, 1969; Stone, 1974; Miles *et al.*, 1977; Myerowitz *et al.*, 1977; Nystatin Multicenter Study Group, 1986), and the alimentary tract is now regarded by some as the proximate source of infection in these and other types of infections caused by *Candida* species (Odds, 1988). The "overgrowth" of *C. albicans* in the GI tract, and its subsequent passage through the gut mucosa into the host bloodstream is thus believed to be the primary mechanism leading to systemic candidosis in severely compromised patients (Stone *et al.*, 1974). This type of "autoinoculation" or "self-inoculation" appears to be particularly prevalent among individuals with acute leukemia (Myerowitz *et al.*, 1977). Similarly, colonization of the GI tract by *C. albicans* and other *Candida* species may also lead to involvement in a number of other disease syndromes that include esophageal, gastric, and intestinal thrush, gastric and intestinal ulceration, GI bleeding, diarrhea, peritonitis, perianal itch, napkin dermatitis, chronic "irritable bowel" syndrome, autobrewery syndrome, and growth depression in food-producing animals (Odds, 1988; Rippon, 1988).

Although the determinants of colonization and dissemination from the GI tract by *C. albicans* have not been completely defined, findings from experimental animal models and patients suggest that association with GI mucosal surfaces plays an important role in initiating both colonization and infection (Kennedy *et al.*, 1987). Colonization of the small intestine from which dissemination of *Candida* is thought to occur (Umenai *et al.*, 1980), probably could not take place in the face of the rapid passage of material through the small intestine due to peristalsis unless these organisms could associate with the mucosa. Considering further the sloughing of

---

Michael J. Kennedy • The Upjohn Company, Kalamazoo, Michigan 49001.

epithelial cells and the bathing actions of fluids over host mucosal surfaces suggests that not only does "attachment" play an important role in virulence and colonization, but that mucosal association by *C. albicans* is also likely to be a very dynamic event. Thus, although the importance of *Candida*–mucosal association and gut colonization to human and animal health has clearly been established, at present neither of these aspects can be controlled or regulated such that certain beneficial effects of yeast colonization are retained and the detrimental effects are eliminated (Kennedy, 1989). This will only be possible when the determinants of mucosal association and gut colonization are more thoroughly understood. To that end, this chapter will present an overview of the factors influencing intestinal colonization and infection by *C. albicans* and will emphasize early events in these processes.

2. SEQUENTIAL STEPS IN MUCOSAL ASSOCIATION AND INVASION

Mucosal association and infection by *C. albicans* are complex, multifactoral events that cannot be described using a single descriptor term such as *adhesion* or *invasion*. Several steps appear to take place in sequential fashion before an active lesion can develop or even before colonization can occur (Kennedy and Volz, 1985a; Kennedy *et al.*, 1987). According to Freter and Jones (1983), in order for any microorganism to colonize a mucosal surface it must progress sequentially through the following steps: (1) it must make contact with the thick "mucus gel" or "mucus blanket" covering the mucosal epithelium, (2) it must penetrate (actively or passively) the mucus gel, (3) it must attach to the epithelial cell surface, and (4) it must multiply either on the epithelial surface or intracellularly once penetration of the epithelial layer has occurred. The complexity of mucosal association and invasion appears to be further complicated by the likelihood that each individual step in this process may also occur in phases, some of which may proceed rapidly and others more slowly. As one example, "permanent" adhesion to epithelial cells is preceded by nonspecific adhesion, which, in turn, may also occur in phases (Kennedy, 1988). This will be discussed in further detail below but each of the steps listed may be necessary to resist physical removal from the mucosa to some demonstrable degree.

The interaction of *C. albicans*, *C. tropicalis*, or any other species of *Candida* that gains entry to the GI tract, with the intestinal mucosa may consist of at least three consecutive but overlapping phases. The first phase consists of the accumulation or adsorption of the fungal cells to the mucosal surface. This probably occurs randomly as these species of fungi are not motile and are thus dependent on the action of GI mixing to come into contact with the mucosa. It has been shown that intestinal contents are mixed rather rapidly in the gut and that an inoculum delivered to the stomach is passed into the small intestine within a relatively short period of time (Freter *et al.*, 1983; Kennedy and Volz, 1985a). Thus, the random collision of yeast cells to the mucus gel in any one area of the GI tract may occur rapidly. Much of the complexity of *Candida*–host interactions in the GI tract, nevertheless, is probably due to the various effects of the mucus gel since the mucus gel represents a definite but imperfect barrier to the passive penetration of particles in the size range of yeast cells (Wells *et al.*, 1988). In instances other than those where large numbers of *C. albicans* are ingested (Krause *et al.*, 1969) or injected into experimental animals (Pope *et al.*, 1979; Field *et al.*, 1981; Kennedy and Volz, 1983), then, only a small fraction of the original inoculum may reach the epithelium.

After contact with the mucus gel, *C. albicans* may penetrate through the gel until it reaches the epithelial surface proper. This may occur rather rapidly due to progressive digestion of the mucus gel by proteolytic activity (Cole *et al.*, 1988), or it may occur more slowly via passage of yeast cells through channels or pores in the viscous gel (Freter *et al.*, 1981) due to surface tension. Thereafter, *Candida* cells enter into the third and final phase of mucosal association before invasion can take place (Kennedy *et al.*, 1987). This phase consists of the adhesion of yeast cells to the epithelial surface. The extent of adhesion found may be a strain- and species-dependent property (Kennedy, 1988). This stage probably also occurs in phases, i.e., nonspecific or reversible adhesion followed by specific or irreversible adhesion, and may be complicated by the suggestion that *C. albicans* may attach to receptors in the mucus gel itself (Kennedy and Volz, 1985a). The apparent interplay between adhesion to the mucus gel and the epithelial surface proper remains to be fully characterized. Nevertheless, both adsorption and adhesion to the mucus gel and epithelium are probably necessary prerequisites to invasion (Kennedy, 1989). At this point, it is worth noting that all *Candida* species do not necessarily need to carry out all of these steps with equal efficiency and intensity. Also, at each of the above-mentioned steps the progress of the invading fungus may be accelerated or retarded by a number of modifying factors, and association with the mucosa is required for colonization only in those regions of the GI tract where the rate of physical removal exceeds the rate of fungal multiplication. This will be discussed in more detail at the end of the chapter.

3. PHYSICOCHEMICAL ASPECTS OF MUCOSAL ASSOCIATION

The association of *C. albicans* with the GI mucosa *in vivo* may vary considerably depending on host species and physiology, cell phenotype, and tissue involved (Kennedy, 1988). Likewise, a number of long-range, short-range, and hydrodynamic forces may also influence these processes (Marshall, 1976; Rutter and Vincent, 1984; Loeb, 1985). To date, however, few studies have addressed the role these factors play in adhesion and invasion of mucosal surfaces by *C. albicans*. Relatively little is known, for instance, about the binding sites of *C. albicans* adhesins and even less is known about *Candida*–mucosal binding kinetics. A complete physicochemical description of the association and invasion of *C. albicans* with the gut mucosa, therefore, is not yet possible. The following general discussion, nevertheless, will present the fundamental physicochemical principles involved in the early phases of *Candida*–mucosal interactions, and will serve as a springboard for future experiments.

In general, the cell surfaces of *C. albicans* and the GI mucosa have an overall negative surface charge (Jones, 1977). These negative potentials result from the ionization of various chemical groups (e.g., sialic acid carboxyl groups of epithelial glycocalyces, acidic amino acid side chains, glycolipids, and phospholipids) of the cell surface, and can be modified by the ionic strength of the surrounding milieu (Marshall, 1976). These charged surface groups loosely attract oppositely charged ions, termed *gegen-* or *counterions*, from the surrounding vicinity to form a diffuse double layer of ions (Marshall, 1976). The resulting electric double layer is considered to be a part of the cell surface, and can, therefore, modify attractive and repulsive forces (Jones, 1977) because repulsion will occur when layers from two surfaces overlap (Paerl, 1985). It is likely that differences in pH between the stomach and other portions of the GI tract may have very different binding kinetics for *C. albicans*. The acidic nature of the stomach probably renders a surface charge to *Candida* cells that is quite different from those yeast cells that have entered the large intestine. Thus, adhesion and association by *C. albicans* in these two sites may be quite different. Not surprisingly, several studies have shown that *C. albicans* adhesion to mucosal cells varied significantly according to selection of assay medium (Kennedy, 1988).

The most useful descriptions of adhesive interactions between two negatively charged surfaces, which theoretically should repel one another, are those provided by the now classic lyophobic colloid theory of Derjaguin and Landau (1941) and of Verwey and Overbeek (1948) (DLVO theory) and its many derivations (Jones and Isaacson, 1983). Briefly, the DLVO theory proposes that the forces of repulsion (electrostatic interactions in the overlapping double layers) and attraction (London–van der Waals forces) between two similarly charged surfaces are additive, but vary independently with the distance of separation between the bodies. There are two separation distances between cells at which attraction is greater than repulsion. These are called the primary minimum (at small distances of separation of < 1 nm) where attraction forces are strong, and the secondary minimum (at relatively large distances of separation of > 10 nm) where attraction forces are weaker and are easily reversible, for example, by mild fluid shear (Jones and Isaacson, 1983). Interposed between these two positions is a point at which repulsive forces predominate and the potential energy is maximized (Jones and Isaacson, 1983). The overall charge and shape of the bodies are important and contribute significantly to the net interaction (Isaacson, 1985). For example, repulsion forces decrease with bodies of decreasing radius of curvature (Jones and Isaacson, 1983; Isaacson, 1985). Likewise, as the ionic strength of an environment increases, the energy maximum repulsion barrier decreases, at high electolyte concentrations, the repulsion energy barrier may be eliminated altogether (Jones and Isaacson, 1983).

The DLVO theory describes long-range adhesive interactions (i.e., those occurring at the secondary minimum) adequately where the bodies are held, more or less, in a state of mutual attraction. This model is inadequate, however, to describe close-range interactions, which are regulated by adhesin–receptor binding, because bacteria and yeast simply do not possess sufficient kinetic energy to overcome the repulsion barrier (Rutter and Vincent, 1984). Thus, the DVLO theory remains valid to describe initial, weak adhesive interactions (Jones and Isaacson, 1983; Rutter and Vincent, 1980), whereas the relatively irreversible adhesion (i.e., binding at the primary minimum) is the result of adhesive appendages (i.e., adhesins) that bridge the gap between cell surfaces to bind their complementary receptor (Jones and Isaacson, 1983; Marshall, 1984). The role of fungal adhesins may simply be to extend the points of contact between the fungal cell through the surrounding electrostatically charged layers to the mucosal surface. Both types of interactions may be necessary for adhesion to and colonization of GI mucosal surfaces.

Similarly, curved bodies come closer together at the secondary minimum and require less kinetic energy to reach the primary minimum than do planar bodies (Isaacson, 1985). Adhesion of yeast cells to cells with microvilli is thus energetically more favorable compared with other tissue of planar configuration, although the effect of curvature may not decrease the repulsion suffi- ciently to permit C. albicans to reach the primary minimum (i.e., to establish firm adhesion). Adhesive appendages would also offer structures of very small radii of curvature relative to the yeast cell that would further favor contact with the mucosal surface. Individual fibrillar adhesins, as those described below, with considerably smaller radii of curvature than the yeast cell could overcome repulsion due to the potential energy maximum required to bridge the gap between C. albicans and the epithelium. In the small intestine, the effect of fluid shear due to peristalsis and the bathing actions of mucosal secretions should dislodge Candida cells, resulting in their removal from the tissue surface, if only nonspecific adhesion was involved. Firm adhesion of C. albicans to small intestinal microvilli, as has been observed by Pope and Cole (1981, 1982), may thus be sufficient to explain the persistence of C. albicans in the small intestine (Kennedy et al., 1987).

Two types of Candida–epithelial cell interaction that preceded cell invasion have been noted by Marrie and Costerton (1981). The first was a "loose" adhesion apparently mediated by a ruthenium red-positive matrix, followed by a "tight" adhesion where no space could be seen between host and yeast cell. By introducing into an assay various substances that might block C. albicans adhesion, it was also found that C. albicans may bind reversibly for up to 20 min, but thereafter the cells bind relatively irreversibly (Sandin, 1987). This further suggests that mucosal association after the initial yeast–mucus or yeast–epithelium contact may be due to nonspecific adhesion, which is followed by specific adhesion. Thus, if the initial attachment is the result of nonspecific adhesion, for example, "loose" binding, additional adhesion or association mecha- nisms may be required for irreversible or "tight" binding (Kennedy, 1988). Modification of both host and yeast cells has been observed once C. albicans attaches (Howlett and Squier, 1980;

Tronchin *et al.*, 1984). It may be, then, that as *Candida* cells "bump" into epithelial cells and bind reversibly, physiologic changes occur in both host and fungus that strengthen the adhesion. These changes could modify the epithelium to the extent that more or different receptors are exposed that, if bound to *Candida* cells, would stabilize and strengthen adhesion.

Likewise, modifications of the cell wall of *C. albicans*, possibly causing an increase or concentration of adhesins, or an unmasking of different or more adhesins, have been noted at yeast–epithelial binding sites. Alternatively, the deposition of new cell wall material could occur after initial attachment. Tronchin *et al.* (1984) have noted the reorganization and proliferation of an external cell wall layer of *C. albicans* during adhesion. Moreover, they noted an abundant extracellular material with numerous binding sites for concanavalin A that appeared to be released from the yeast surface, leaving underlying cell wall layers exposed. Several descriptions of phases in *Candida* adhesion to epithelial cells exist (Kennedy, 1988), and the ultrastructural appearance of the cell wall of *C. albicans* has been observed to change within the first few hours after adhesion and invasion of GI mucosa (Cole *et al.*, 1988). Further studies will be necessary, however, to characterize in greater detail the affinity and number of binding sites that are available to *C. albicans* in various regions of the gut and to present a more complete and accurate description of *Candida*–mucosal interactions. One may speculate from what is presently known about *Candida*–mucosal interactions, however, that once long-range electrostatic forces are overcome, specific adhesin-receptor interactions become the important determinants of mucosal association and adhesion. Studies with bacteria are consistent with the view, and suggest that specific adhesion predominates over nonspecific adhesion (Jones, 1984). Therefore, a complex adhesive system may be involved in the attachment of *C. albicans* to various mucosal surfaces and other microorganisms, and may be mediated by two or more distinct adhesive entities in which nonspecific interactions (e.g., hydrophobicity) may play a secondary or stabilizing role (Kennedy and Volz, 1985a; Kennedy, 1987).

At the shorter distances, several other interactions are also important in facilitating and stabilizing adhesion. Such interactions include: dipole–dipole (Keesom) interactions, dipole-induced dipole (Debye) interactions, ion–dipole interactions, chemical bonds (e.g., electrostatic, covalent, and hydrogen), and hydrophobic interactions (Rutter and Vincent, 1984; Tadros, 1980). A number of these types of interactions have been shown to be important in bacterial adhesion (Marshall, 1984), and are thus likely to be important in fungal adhesion (Kennedy, 1988). It should be noted, however, that short-range effects may be repulsive or attractive depending on the nature of the surfaces involved (Rutter and Vincent, 1980; Tadros, 1980), and are particularly important in aqueous systems (Marshall, 1976; Rutter and Vincent, 1984). The role most of these short-range forces may play in the adhesion of *C. albicans* to GI mucosal surfaces has not been examined. Nevertheless, the variability of the surface of *C. albicans* (e.g., in charge, shape, appendages) may lead to various types of synergistic interactions that take place simultaneously. These and other factors that may influence the surface chemistry of *C. albicans*, and therefore, may influence adhesion and invasion, have been described in more detail elsewhere (Kennedy, 1988) and are summarized in Table I.

4. MECHANISMS OF ADHESION AND ASSOCIATION WITH GASTROINTESTINAL MUCOSA

The association of *C. albicans* with GI mucosal surfaces is likely to play an important role in colonization and pathogenesis of GI and systemic *Candida* infections. This has been alluded to above and can be further illustrated from recent studies on the dynamics of luminal and wall-associated *Candida* populations (Kennedy *et al.*, 1987, 1990). Although the kinetics of population shifts of *C. albicans* in the gut contents versus those associated with the mucosa have not

Table I. Factors Influencing Mucosal Association
and Invasion by Candida albicans in Vivo[a]

Factor	Examples
Host factors	Antimicrobic therapy
	Body site
	Diet
	Environmental factors (e.g., pH)
	Hormonal status
	Immunologic status
	Number and type of receptors available
	Physiologic status
Indigenous microflora	Alteration of physicochemical nature of environment
	Colonization of adhesion sites
	Modification of substrate
	Production of inhibitors
Fungal factors	"Cell type" or "phenotype'"
	Concentration and type of adhesins synthesized
	Enzyme production (e.g., proteinases)
	Germination
	Morphology
	Surface properties

[a]Modified from Kennedy (1989).

been defined in detail, it was found that in the large intestine less than 10% of the total *Candida* population was associated with the gut wall early after challenge (Kennedy *et al.*, 1987). Thereafter, as the intestinal microflora reestablished, mucosal *Candida* populations predominated (Kennedy *et al.*, 1990). It seems likely, therefore, that mucosal association is important to the ecology and pathogenesis of infection in that it prevents *Candida* cells from being "washed out" of the GI tract due to antagonism by intestinal bacteria, and the passage of material due to peristalsis. Considering further that the metabolic activity of the indigenous microflora causes *C. albicans* and other fungi to have a doubling time that can be significantly longer than this dilution rate and that a significant lag phase is imparted to fungi upon entering the GI tract (discussed below), suggests that mucosal association would not just simply facilitate survival in this harsh ecosystem but that it may actually be necessary for long-term implantation to occur (Kennedy, 1987). Likewise, passage of viable *Candida* organisms through the small intestinal mucosa to initiate systemic infection by the hematogenous route (Myerowitz *et al.*, 1977) probably could not take place in the face of the rapid passage of material through the small intestine (Freter *et al.*, 1983), unless these organisms could associate rather irreversibly with the mucosa (Kennedy, 1989). Several other arguments could also be given to support the role of mucosal association in colonization and dissemination from the GI tract. Nevertheless, the above list does implicate the ecologic and pathologic importance of mucosal association and colonization. The discussion in this section focuses on the adhesion and association mechanisms of *C. albicans*, particularly as they relate to mucosal colonization and invasion. The mechanisms that may allow *C. albicans* to pass through or invade the intestinal mucosa will be considered separately in this chapter, as will certain interactions between *Candida* and mucosal bacteria.

Although studies using animal models have yielded the majority of information regarding the association of *C. albicans* with GI mucosal surfaces, these studies have confirmed clinical findings that *Candida* can associate with the mucosa of all regions of the GI tract (Kennedy, 1988, 1989). Animal models used to study colonization of the GI tract have been described in detail elsewhere (Guentzel *et al.*, 1985; Kennedy, 1989) and have included neonatal and adult conventional, specific pathogen-free, germfree, antibiotic-treated, and/or athymic gnotobiotic mice, hamsters, chickens, and rats. Most of these studies have involved removing infected tissues from experimental animals infected with *C. albicans* and viewing the resulting association by scanning (SEM) or transmission (TEM) electron microscopy. Other studies have included examination of stained histologic sections, and some have included quantitative cultures of tissue collected from patients or infected laboratory rodents to determine population levels of attached yeast. Tissue slices and isolated mucus gel have also been used to study mucosal association *in vitro* (Kennedy and Volz, 1985a).

The association of *C. albicans* with the stomach mucosa has been examined in both conventional infant and germfree athymic infant and adult mice (Helstrom and Balish, 1979; Pope and Cole, 1981, 1982; Balish *et al.*, 1984). It is of interest to note that different preferential sites of colonization have been observed between these and other animal models. The stomach was the primary site of colonization in infant mice (Field *et al.*, 1981), whereas the cecum was most heavily colonized by *C. albicans* in antibiotic-treated or germfree adult mice (Kennedy and Volz, 1985a,b; Balish *et al.*, 1984). Similarly, the GI tract of neonatal mice was not colonized as quickly with *C. albicans* as were adult mice (Balish *et al.*, 1984). Nevertheless, these studies have revealed that at early times after oral inoculation with *C. albicans*, yeast cells were attached directly to both keratinized squamous epithelial and columnar secreting epithelial surfaces (Fig. 1). At later times, yeast cells were also observed attached to the secreting epithelium surrounded by and embedded in mucus. *Candida* cells were also observed in association with and attached to lactobacilli at the junction of the keratinized and secreting epithelium (Pope and Cole, 1981, 1982; Balish *et al.*, 1984).

Hyphal invasion of the keratinized region of the stomach has also been observed in both infant and adult mice (Pope and Cole, 1981, 1982; Balish *et al.*, 1984). This site appears to be the preferential mucosal location in the stomach in some experimental animals (Guentzel *et al.*,

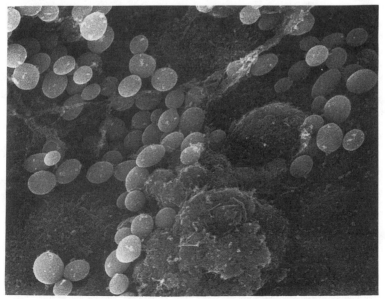

Figure 1. Yeast cells associated with the secreting epithelium of the stomach from an infant mouse 30 min after oral–intragastric inoculation with *Candida albicans* CA30. ×1500. (From Pope and Cole, 1981.)

1985), and in some studies this site was observed to be the sole colonization and invasion site by *C. albicans* (Helstrom and Balish, 1979; Pope and Cole, 1981, 1982; Balish *et al.*, 1984). The reason why *C. albicans* colonizes and associates with this region of the stomach preferentially is not clear, but it may be that the cardial–atrium ridge contains unique or an abundant number of receptors that bind *C. albicans*. It appears, nevertheless, that *C. albicans* can associate with the stomach mucosa by a number of distinct adhesion and association mechanisms. These include: (1) direct adhesion to the epithelium, (2) indirect attachment to the epithelium by coadhering to attached microorganisms, and (3) penetration or invasion of the epithelium. At present, no attempt has been made to identify *Candida* adhesins or mucosal receptors that might be involved in gastric colonization by *C. albicans*. Thus, whether more than one adhesin is involved in mucosal association by *C. albicans*, or whether the adhesive activities observed for *C. albicans* result from a single multifactorial adhesive cell wall component, is presently not known.

The association of *Candida* with small intestinal mucosal surfaces has similarly been studied in infected animals. These studies have demonstrated, like those described above, that *C. albicans* can associate with small intestinal mucosa by a redundancy of mechanisms (Field *et al.*, 1981; Pope and Cole, 1981, 1982). Using an infant mouse model, Pope and Cole (Pope *et al.*, 1979; Pope and Cole, 1981, 1982) visualized large numbers of *C. albicans* in association with the surface of villi (Fig. 2) in the small intestine between 1 and 6 hr after inoculation, indicating that active adhesion was involved. Ultrathin sections from small intestinal mucosa (Fig. 3) showed that yeast cells probably attached to microvilli by binding to the epithelial glycocalyx (Pope and Cole, 1982). Further examination revealed that yeast cells adhered to all areas of the epithelium and that cells of *C. albicans* were frequently associated with mucus. Numerous yeast cells appeared to be attached to, embedded in, and covered by the thick mucus blanket (Field *et al.*,

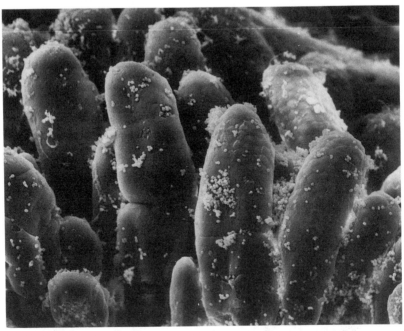

Figure 2. Scanning electron micrograph of the small intestine of an infant mouse 6 hr after oral–intragastric inoculation with *Candida albicans* CA30. A large number of yeast cells can be seen attached to intestinal villi. ×550. (From Pope and Cole, 1982.)

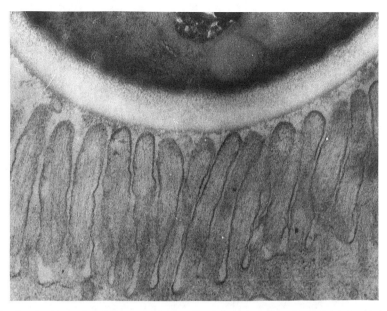

Figure 3. An ultrathin section of the upper intestine of an infant mouse 3 hr after oral–intragastric inoculation with *Candida albicans* NS33. A yeast cell has attached to the villus surface without apparent damage to the microvilli. Note the dark fibrous material, which is probably epithelial glycocalyx, at the yeast–microvillus junction. ×40,000. (From Pope and Cole, 1981.)

1981; Pope and Cole, 1981, 1982). As is shown in Fig. 4, histologic sections from infant mice at 3 and 6 hr postchallenge also revealed yeast cells embedded in the villus surface, and that some yeast cells were associating indirectly with the mucosa by attaching to adherent yeast (Pope and Cole, 1982).

Examination of intestinal mucosal surfaces from the large intestines of adult hamsters and mice revealed that *C. albicans* associated with these surfaces by similar mechanisms (Kennedy and Volz, 1985a; Kennedy *et al.*, 1987). However, in adult conventional animals this was true only when the animals were given antibiotics to disrupt the ecology of the indigenous bacterial flora (Kennedy and Volz, 1983, 1985b). SEM studies revealed that large numbers of *C. albicans* were present on the surface of the epithelium and in mucus in antibiotic-treated animals challenged with *C. albicans* (Kennedy and Volz, 1985a; Kennedy *et al.*, 1987), whereas no yeast were observed associating with host mucosal surfaces of untreated animals similarly challenged. Large numbers of indigenous bacteria were observed colonizing the mucosa of these animals (Kennedy *et al.*, 1987). In mice or hamsters given antibiotics, *C. albicans* could associate with the mucosa of all areas of the GI tract, but the cecum consistently maintained the highest population levels (Kennedy and Volz, 1985b).

Further examination of the cecal mucosa of antibiotic-treated animals challenged with *C. albicans* revealed that *Candida* cells also apparently attached to the epithelium by attaching to the epithelial glycocalyx (Kennedy *et al.*, 1987). This is shown in Fig 5. Yeast cells were also observed to attach directly to or were seen embedded in mucus material and it was found that *C. albicans* could associate with the cecal mucosal surface indirectly by attaching to other adherent organisms. *Candida* cells were observed attached to both adherent yeast cells and adherent bacteria (Kennedy and Volz, 1985a; Kennedy *et al.*, 1987). The possible importance of these interactions is apparent in the finding that microcolonies (Fig. 6) of *Candida* were evident in the

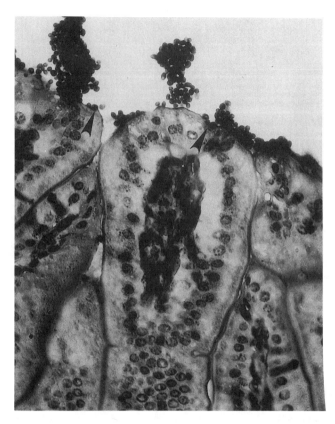

Figure 4. Histologic section of tissue from the upper portion of the small intestine of a neonatal mouse 3 hr postchallenge with *Candida albicans* NS33. Numerous yeast cells are attached to the epithelium en masse apparently due to coadhesion. Note individual yeast cells adherent to the villus tips associated with shallow depressions (arrows). ×1100. (From Pope and Cole, 1982.)

cecal mucosa as early as 72 hr after oral challenge (Kennedy and Volz, 1985a; Kennedy *et al.*, 1987). In such microcolonies, some yeast cells were associated with the epithelium proper, were attached to mucus material, and were coattached to neighboring yeast cells. The finding that microcolonies can be formed in intestinal mucosa within a relatively short period of time after implantation may explain why isolated plaques are observed on endoscopic examination of intestinal mucosa in patients infected with *C. albicans* (Eras *et al.*, 1972; Joshi *et al.*, 1981). Microcolony formation has been suggested to be of ecologic and pathologic importance in colonization of mucosal surfaces by pathogenic bacteria and the indigenous bacterial flora (Cheng *et al.*, 1981; Costerton *et al.*, 1985). Depressions in the epithelium were also observed under *Candida* cells, and were presumably due to enzymatic lysis of host tissue (Kennedy *et al.*, 1987). Digestion of the mucosa may expose underlying receptors and thus stabilize *C. albicans* adhesion to the epithelium after the initial attachment. It is interesting to note that germ tubes or hyphae, which are frequently observed in infected oral and stomach mucosa (Kennedy, 1988), were not observed to penetrate the mucosa of the small or large intestine in any of the studies described above.

   Although *C. albicans* has been observed to associate with and attach to mucus, mucus is not maintained intact and appears somewhat amorphous in gut sections processed for electron microscopy (Savage, 1983). To determine more conclusively whether *C. albicans* could associate

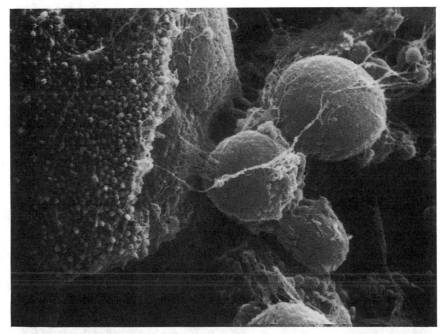

Figure 5. Scanning electron micrograph of cecal mucosa from an adult penicillin-treated mouse 72 hr after oral–intragastric inoculation with *Candida albicans* CA34 showing yeast associated with the epithelium, possibly by adhesion to the epithelial glycocalyx (arrow). ×10,000. (From Kennedy *et al.*, 1987.)

with and attach to the mucus gel proper, and to ensure that the apparent attachment to mucus was not due to specimen processing (which can cause mucus to shrink), *Candida* adhesion to mucus gel was studied *in vitro* (Kennedy and Volz, 1985a). Large numbers of *C. albicans* were observed in association with intestinal mucus very rapidly (within 5 min) when mixed together *in vitro*, and *Candida* cells were attached to and embedded in the gel matrix.

The precise nature of all the adhesive events between *C. albicans* and the intestinal mucosa described above, as well as between *Candida* and other adherent microorganisms, remains to be determined. It seems likely, however, that a complex set of interactions, involving both specific and nonspecific adhesion mechanisms, regulates the adhesion and association of *C. albicans* with GI mucosal surfaces (Kennedy, 1987). Table II summarizes the various mechanisms of adhesion and association that have been observed for *C. albicans* with gastric and intestinal mucosa.

5. MECHANISMS OF MUCOSAL INVASION

It is apparent from the above discussion that in the GI tract it is possible to define three different ecologic sites for *C. albicans* and other *Candida* species. These include the intestinal contents, the mucus gel, and the epithelium proper; which itself may contain three microsites (cells indirectly attached to the epithelium, cells directly attached to the epithelium, and cells penetrating into the epithelium) (Kennedy *et al.*, 1987). As noted, the ecology of *C. albicans* gut colonization and the dynamics of luminal and mucosa-associated *Candida* populations have not been completely defined. Nevertheless, the data presented above, together with reports on mucosal association by bacteria (Freter, 1982), emphasize the importance of distinct steps in the

Figure 6. Scanning electron micrograph of cecal mucosa from an adult conventional mouse treated with penicillin 72 hr after oral–intragastric challenge with *Candida albicans* CA34 showing a microcolony on the mucosal surface. Note that some yeast cells are attached to the surface, while others are attached to adherent yeast. ×3000. (From Kennedy *et al.*, 1987.)

association, invasion, and passage of *C. albicans* and other fungi through intestinal mucosa: (1) contact with the surface of the mucus gel; (2) adhesion to and penetration or trapping in the mucus gel; and (3) adhesion to and penetration of the epithelial surface. These, of course, are influenced by how well the organism can actively multiply on or in the mucosa.

The initial association with the viscous mucus gel by *C. albicans* and other fungi is probably entirely random, unlike bacteria in which motility and chemotaxis may greatly facilitate mucosal association and penetration of the mucus gel (Kennedy, 1987). The latter steps in mucosal

Table II. Adhesion and Association Mechanisms of Candida albicans[a]

	Nature of mechanism		
Mechanism	Active or passive	Specific or nonspecific	Direct or indirect
Adhesin–receptor interaction	A	S	D, I
Nonspecific adhesion	A, P	N	D, I
Coadhesion to adherent organisms	A	S, N	I
Entrapment in mucus or tissue	P	N	D
Germ tube penetration	A	N, S	D, I
Enzymatic digestion	A	N, S	D, I

[a]Modified from Kennedy (1988).

association and invasion by *C. albicans* appear not to be random, in that they may involve the interaction of specific adhesin–receptor binding and active penetration due to enzymatic modification of the mucus gel and epithelial surface (Kennedy *et al.*, 1987). Recent ultrastructural examination of mucosal invasion in the GI tract of infant mice verified this view and showed that yeast phase cells of *C. albicans* were capable of progressive extracellular digestion of the intestinal mucus gel and microvillus layer (Cole *et al.*, 1988). This was followed by invasion of columnar epithelial cells and the appearance of intraepithelial yeast cells in "vacuoles." Yeast cells were evident within the cytoplasm of the host cells within 3 hr and no continuous membrane was noted at the interface between this vacuolelike space and host cytoplasm, thus indicating that invasion was due to tissue degradation and not a phagocytic mechanism of uptake (Cole *et al.*, 1988). Figure 7 shows this process in the jejunum of an infant mouse 3 hr after intragastric inoculation with *C. albicans*. Germ tubes or hyphae were not observed in association with or invading the mucosa of the small intestine in this study. Thus, the mechanism(s) operating in this instance may well represent the primary method of invasion in the GI tract at regions below the stomach, especially when low population levels of *C. albicans* exist on the mucosa. This type of invasion mechanism appears to be similar to that observed for penetration of *C. albicans* into endothelial tissue (Klotz *et al.*, 1983).

Penetration of stomach mucosa by germ tubes may also play an important role in the systemic spread from the GI tract. Hyphal invasion of the keratinized region of the stomach, and the cardial–atrium ridge in particular, has been observed in both infant and adult animals (Helstrom and Balish, 1979; Pope and Cole, 1981, 1982; Balish *et al.*, 1984), and gastric ulceration due to *C. albicans* in certain patient populations (Odds, 1988). Although invasion of stomach mucosa by *C. albicans* has not been examined in the same detail as invasion in the small intestine, ultrastructural characterization indicates that the process may be similar to that for penetration of oral mucosa (Kennedy, 1988). Contact between *C. albicans* germ tubes and the epithelium in the latter case has shown that no alteration of the epithelial cell surface was noted either at the point of entry or as hyphae grew along the surface (Kennedy, 1988). These findings indicate that if enzymatic lysis is associated with the invasive process of hyphae, it is probably localized to the hyphal tips (Pugh and Cawson, 1975). Other reports support the role of hydrolytic enzyme activity in cell invasion of *Candida* blastospores and germ tubes or hyphae, and a number of proteinases and phospholipases have been implicated (Kennedy, 1988; Odds, this volume).

Neither of the aforementioned invasive mechanisms, however, can account for the rapid "persorption" of *Candida* from the GI tract that has been observed in man, dogs, monkeys, and infant mice (Kennedy, 1989). In all of these studies, *Candida* cells were recovered from the blood or systemic organs in large numbers in less that 2 hr. It seems unlikely, therefore, that large numbers of *Candida* cells could be cleared from the stomach, associate with the jejunal mucosa, degrade the mucus gel and microvillus layer to invade and pass through epithelial cells in this length of time. More unlikely is the possibility that *Candida* cells could germinate, penetrate the epithelium, and generate large numbers of blastospores that would reach extraintestinal sites. This suggests that at least one additional mechanism may be responsible for the rapid passage of *C. albicans* through the gut mucosa as was noted in the studies mentioned above. The pericellular passage of yeast cells that is apparently more frequent in the desquamation zones of the intestinal epithelium is one such possibility (Guentzel *et al.*, 1985). As has been noted recently by Wells *et al.* (1988), the literature abounds with reports of the invasion of the intestinal mucosa by inert, nonmotile particles such as starch granules, latex beads, and nonpathogenic yeasts. Experiments with latex beads also suggest that phagocytic cells may participate, at least in some phase of the trip, with deep penetration of the mucosa and systemic spread to extraintestinal sites (Wells *et al.*, 1988).

Apparently, this type of dissemination is regulated by the population level of yeast cells in the GI tract (Kennedy and Volz, 1983). For instance, significantly lower numbers of viable yeast cells were found in the GI tracts of animals that did not show signs of systemic spread to visceral

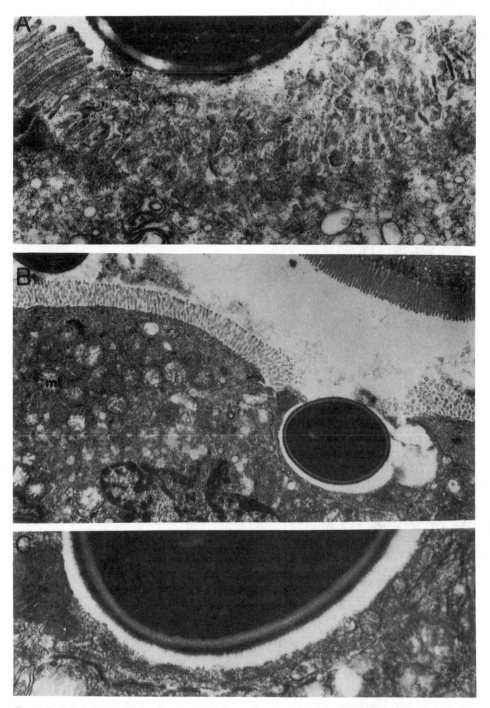

Figure 7. Transmission electron micrographs of thin sections of jejunal mucosa showing morphologic aspects of mucosal invasion by *Candida albicans* in the infant mouse. Apparent digestion of microvillus layer (A), and invasion of cytoplasm of epithelial cells (B, C). The micrographs represent invasion in the jejunum of infant mice at 1 hr (A) and 2 hr (B, C) after oral–intragastric challenge with *C. albicans* CA30. (From Cole *et al.*, 1988.)

organs compared to animals that did (Kennedy and Volz, 1983, 1985a,b). Thus, a "threshold" yeast population level appears to be a determinant for "passive" fungal spread from the gut (Kennedy *et al.*, 1982; Kennedy and Volz, 1983, 1985a,b). In long-term colonization of the GI tract, then, either germ tube penetration or invasion by blastospores as the result of the activity of extracellular, hydrolytic enzymes may be the predominant forms of systemic spread. Odds (1988) has noted antibodies to *Candida* proteinase(s) in both healthy individuals harboring *Candida* in their GI tract and patients with systemic candidosis. It may be that *C. albicans* regularly and continually invades the bloodstream by one or both of these mechanisms, from several sites along the alimentary tract (including the mouth and esophagus), similar to that of indigenous bacteria (see Freter, 1983). This possibility lends support to the importance of bacterial antagonism in controlling *Candida* populations in the gut; which would keep yeasts that spread systemically at numbers that are manageable by both local and systemic immunity (Rogers and Balish, 1980). This is consistent with the findings of Wingard *et al.* (1980), who found that treatment of mice with a combination of antibiotics and cytarabine significantly increased the incidence of systemic spread by *C. albicans* and *C. tropicalis*.

6. CANDIDA CELL SURFACE COMPOSITION, ULTRASTRUCTURE, AND POSSIBLE ADHESINS

The cell wall of *C. albicans* is a complex structure approximately 100 to 400 nm thick in which at least five to eight distinct layers have been identified (Kennedy, 1988). Relatively little is known at present regarding the biosynthesis and architecture of the cell wall of *C. albicans*, although the chemical composition and ultrastructure of the cell wall of *C. albicans* have been studied in great detail (Cabib *et al.*, 1982; Poulain *et al.*, 1985). Several investigators, for instance, have examined the ultrastructure of the cell wall of *C. albicans* using a variety of procedures to reveal the precise location of the principal wall components (Poulain *et al.*, 1985). Several cell wall layers have been identified for *C. albicans* (Kennedy, 1988) and these layers have been illustrated schematically by various investigators (Poulain *et al.*, 1985). Cytochemical staining (Djaczenko and Cassone, 1971) has revealed that each layer contains a dominant polysaccharide (Poulain *et al.*, 1985), and that the gross appearance of the cell wall as viewed by electron microscopy is that of a sandwich structure (Kennedy, 1988). The ultrastructural appearance of the cell wall of *C. albicans* appears to consist of an outermost set of layers of high electron density, an inner layer of high electron density, and an intermediate set of layers of lower electron density (Poulain *et al.*, 1985).

It should be stressed that the gross appearance of the cell wall has been shown to be directly influenced by the growth parameters used for preparation of the test cells (Cassone *et al.*, 1973, 1979; Kennedy and Sandin, 1988). Poulain and co-workers (1985), for instance, found that both the growth medium and the age of the cells affected the number and appearance of layers in the cell wall. Recent studies have shown further that other growth parameters (e.g., temperature) can also have a significant effect on cell wall ultrastructure (McCourtie and Douglas, 1981; Kennedy and Sandin, 1988), and that some of these changes may significantly alter the adhesiveness of *C. albicans* (McCourtie and Douglas, 1984; Kennedy and Sandin, 1988).

Ultrastructural studies on the adhesion of *C. albicans* suggest that at least three general morphologic classes of *Candida* adhesins may exist (Kennedy, 1988; Kennedy and Sandin, 1988). The first, or so-called floccular adhesins, are present as a thick (approximately 100 to 400 nm) "fuzzy" cell wall coat and appear to be somewhat amorphous (McCourtie and Douglas, 1981; Tronchin *et al.*, 1984; Kennedy and Sandin, 1988). This material may have an ordered alignment around the cell wall (Tronchin *et al.*, 1984), or it may be unevenly distributed on the cell surface (Kennedy and Sandin, 1988) or localized only at an adhesive junction (Howlett and Squier, 1980;

Marrie and Costerton, 1981). Ultrastructural studies of the adhesion of *C. albicans* showed that floccular structures are present on the yeast cell surface in infected tissues, and that they may indeed mediate *Candida* adhesion (Howlett and Squier, 1980). *Candida* blastospores harvested from media with high concentrations of certain carbohydrates have been found to be highly adhesive, and contained an outermost floccular layer that was absent from blastospores that were significantly less adhesive which were cultivated in media without these carbohydrates (McCourtie and Douglas, 1981).

The second class of adhesins exist as fibrillar material [called elsewhere fimbriae (Gardiner *et al.*, 1982)], which can be seen on the cell surface as thin filamentous structures arranged perpendicularly to the cell surface and evenly distributed around the entire cell (Barnes *et al.*, 1983; Lee and King, 1983a). The diameters of these structures have not been measured, but appear to be within the size range (2 to 10 nm) of fimbrial adhesins of bacteria (Jones and Isaacson, 1983). Likewise, the length of the fibrillar adhesins has not been accurately measured, but these appear to be less than 0.5 μm long (Lee and King, 1983a), considerably shorter than bacterial fimbriae (Jones and Isaacson, 1983). Fibrillae have been observed to mediate the adhesion of *C. albicans* to epithelial and endothelial surfaces *in vivo* (Marrie and Costerton, 1981; Rotrosen *et al.*, 1985). Both types of adhesins stain with ruthenium red (Marrie and Costerton, 1981), which has an affinity for anionic polymers such as polysaccharides (Luft, 1971). A third general or morphologic type of adhesin that exists for *C. albicans* is what might be termed a nonfloccular or nonfibrillar adhesin. Thus, adhesion is mediated by structures that are not discernible by electron microscopy. Cells of this type have been observed attached to and invading intestinal mucosal surfaces and buccal epithelial cells (Kennedy and Sandin, 1988; Cole *et al.*, 1988). Aside from no extra surface layers, electron-dense outer layers, or obvious adhesive appendages, the cells appeared to have a relatively thin cell wall. In the former instance, however, there were also cells that have an outer electron-dense layer. Thus, *C. albicans* may bind transiently by one mechanism which then gives way to another, or may simply suspend the synthesis of an adhesin during various stages of invasion.

The precise chemical and organizational nature of the floccular or fibrillar structures is not known. These structures may represent distinct adhesins or they may simply be the same adhesive moiety that has been assembled or modified differently due to environmental pressures. There is some evidence to suggest that *C. albicans* does produce more than one adhesin (McCourtie and Douglas, 1985; Sandin, 1987; Kennedy and Sandin, 1988). It should be noted that the presence of either floccular or fibrillar structures on the cell surface of *C. albicans* does not, by itself, indicate that either structure plays a role in adhesion (Kennedy and Sandin, 1988). *Candida* cells devoid of floccular or fibrillar structures have been shown to be highly adhesive to epithelial cells (Kennedy and Sandin, 1988). Furthermore, not all known microbial adhesins are identifiable as discrete surface structures (Jones and Isaacson, 1983).

Studies involving the chemical composition of the cell wall of *C. albicans* have revealed that two polysaccharides, α-mannan and β-glucan, represent about 75–85% of the dry matter of the cell wall, and that the remainder is composed of lesser amounts of chitin, protein, and lipid similar to that of *Saccharomyces cerevisiae* (Ballou, 1976; Poulain *et al.*, 1985; Kennedy, 1988). Detailed chemical analyses of the cell wall composition of *C. albicans* suggest that mannan makes up nearly 35–40% of the total dry weight of the cell wall (Bartnicki-Garcia, 1973), and that glucan appears to be the most abundant polymer (making up some 40–50% of the cell wall) (Bishop *et al.*, 1960). *Candida* cell wall mannan is a highly branched polysaccharide that consists of a backbone of α1,6-linked mannose residues to which side chains attach via α1,2 and rare α1,3 bonds. Glucan, which is considered to be the essential component of the "microfibrillar skeleton" (Poulain *et al.*, 1985), is a highly branched polysaccharide that consists of a backbone of β1,6-linked (and possibly β1,3-linked) glucose residues to which side chains of β1,3-linked glucose residues are attached (Yu *et al.*, 1967).

Various polysaccharide–protein complexes (e.g., glucoprotein, mannoprotein, and gluco-mannoprotein) have also been isolated from the cell wall of *C. albicans* (Kessler and Nickerson, 1959). These molecules appear to be more or less tightly associated with one another and with glucan and mannan (Kolarova *et al.*, 1973), and may serve as necessary adhesive links between glucan and mannan (San-Blas, 1982). The proteins that are complexed with mannan and glucan, at least in yeast phase cells, are thought to be present as water-soluble proteins and water-insoluble glycopeptides (Kolarova *et al.*, 1982). Mannan–protein complexes are thought to be linked through an *N*-acetylglucosamine residue and side chain amino groups (Odds, 1988). Glucan is also thought to be complexed to chitin (polymer of *N*-acetylglucosamine joined by $\beta1,4$ linkages) adjacent to the plasma membrane (Gopal *et al.*, 1984). Although chitin helps to anchor glucan to the cell membrane, it comprises only a small part ($< 1\%$) of the total cell wall composition by dry weight (Poulain *et al.*, 1985). Most (90%) of the chitin has been shown to be associated with bud scars of yeast-phase cells (Bacon *et al.*, 1966), and the remainder forms the link between the cell membrane and the innermost insoluble glucan possibly by a mixed $\beta1,3$–$\beta1,6$ linkage (Gopal *et al.*, 1984).

The protein content of the cell wall of *C. albicans* has been studied less extensively than the polysaccharide content though protein may, in some instances, make up to about 30% of the cell wall content (Chaffin and Stocco, 1983). Historically, it has been believed that nearly all of the proteins found in the outer layers of the cell wall of *C. albicans* were associated with large mannan molecules (Poulain *et al.*, 1985). Recent studies from several investigators, however, do not support this view and indicate that many cell wall proteins of *C. albicans* may actually have a negligible mannose content. Ponton and Jones (1986a,b) have suggested that there may actually be a complex latticework of proteins situated in the outer layers of the cell wall of both yeast- and mycelial-phase *C. albicans*, and that inter- and intra-disulfide bonds in these proteins are important in maintaining the structure of this latticework. This, coupled with the fact that these or similar proteins were secreted from and were located in the outer layers of the cell wall (Ponton and Jones, 1986a), lends support to the view that cell surface protein may serve as adhesins (Kennedy, 1988). Evidence that protein alone may serve as an adhesin, or that the protein portion of the mannoprotein complex is more important than the carbohydrate moiety in mediating attachment, is evident from several recent studies. Adhesion of *C. albicans* to epithelial cells has been shown to be decreased after exposure to heat or various proteolytic enzymes (Lee and King, 1983b; Sobel *et al.*, 1981). Pretreatment of *C. albicans* with trypsin, chymotrypsin, or pronase reduced adhesion to fibrin–platelet clots (Lee and King, 1983b), as did pretreatment of an extracellular polymer (thought to serve as an adhesin) from *C. albicans* with heat, dithiothreitol, or certain proteases, but not sodium metaperiodate or α-mannosidase (Critchley and Douglas, 1987). Similarly, a protein component from *C. albicans* pseudohyphae also reduced the binding affinity of pseudohyphae for neutrophils in competitive binding assays (Diamond and Krzesicki, 1978). Studies on bacterial adhesion are consistent with the proteinaceous nature of microbial adhesins, as virtually all have shown that most of the well-characterized bacterial adhesins (with only two exceptions) are proteins (Jones, 1977; Jones and Isaacson, 1983).

Analogous studies have also suggested that a mannose-containing moiety, probably a mannoprotein (McCourtie and Douglas, 1985), on the surface of *C. albicans* is the *Candida* adhesin (Sandin *et al.*, 1982). Cell wall chitin and lipid have similarly been implicated to play a role in *C. albicans* adhesion (Segal *et al.*, 1982; Ghannoum *et al.*, 1986). It is important to note that there have been major differences in various experimental parameters used by investigators who have suggested a role for these various cell wall components in adhesion (Kennedy, 1988). The majority of reports using lectin and carbohydrate blocking experiments, nevertheless, have indicated that a mannoprotein on the surface of *C. albicans* probably serves as the adhesin in most of the adhesive reactions that have been studied to date (Kennedy, 1988). Table III summarizes the possible cell surface moieties suggested to serve as *Candida* adhesins.

Table III. Possible Cell Surface Components Suggested
to Serve as Candida Adhesins[a]

Possible adhesin moieties	Examples of inhibitors	References
Chitin	Glucosamine	Segal *et al.* (1982)
	Chitin-soluble extract	Segal *et al.* (1982)
	N-Acetylglucosamine	Segal *et al.* (1982)
	Mannosamine	Segal *et al.* (1982)
Mannan/mannoprotein	Concanavalin A	Sandin *et al.* (1982)
	Mannose	Sandin *et al.* (1982)
	α-Mannosidase	Sandin *et al.* (1982)
	D-Methylmannopyranoside	Sandin *et al.* (1982)
	α-Methylmannosidase	Lee and King (1983b)
	Tunicamycin	Douglas and McCourtie (1983)
Glucose	Glucose	
	Glucan	
Protein	Chymotrypsin	Lee and King (1983a), Sobel *et al.* (1981)
	Trypsin	Sobel *et al.* (1981)
	Papain	Lee and King (1983b)
	Pepsin	Lee and King (1983b)
	Pronase	Lee and King (1983b)
Lipid	Sterols	Ghannoum *et al.* (1986)

[a]Modified from Kennedy (1988).

7. FACTORS INFLUENCING ADHESION, INVASION, AND COLONIZATION

Colonization of host tissue is a hazardous event for many pathogens including *C. albicans*. Association with GI mucosal surfaces, for instance, indicates that *Candida* cells will be confronted by competition and antagonism by the indigenous microflora and by local immunity (e.g., secretory IgA). In at least some instances, moreover, several of these host defense mechanisms may act synergistically (Shedlofsky and Freter, 1974) to deter survival of *Candida* at mucosal surfaces (Kennedy and Volz, 1985a). Indeed, for significant adhesion, colonization, and invasion of *C. albicans* with the GI mucosa to occur, it has been found that the ecology of the indigenous microbiotia must first be disrupted (e.g., by antibiotic treatment) (Kennedy and Volz, 1983, 1985a,b), or the *Candida* must become implanted in the gut before the mucosal flora has been established (Pope *et al.*, 1979; Field *et al.*, 1981).

The GI tract and its contents make up a dynamic ecosystem that exists in the climax stage. The GI tract, therefore, contains many interdependent phenomena, each of which may directly or indirectly influence the association, invasion, and colonization of GI mucosal surfaces (Kennedy, 1989). Recent studies clearly demonstrate the importance of the anaerobic intestinal microflora in the suppression of intestinal mucosal association, colonization and subsequent invasion and dissemination by *C. albicans* (Kennedy and Volz, 1985a,b), and that an intact and functioning microflora is more important than local immunity in suppressing *C. albicans* gut colonization (Helstrom and Balish, 1979). The suppression of *C. albicans* is likely to be due to a complex series of interactions that is regulated by an equally complex intestinal microflora (Kennedy, 1990), whereby several distinct mechanisms act collectively to regulate the survival, implantation, and dissemination of *C. albicans* and other fungi from the GI tract (Kennedy *et al.*, 1989). These mechanisms have been summarized in Table IV and include two general mechanisms: (1) antagonism of *C. albicans* growth and (2) inhibition of the association of *Candida* cells with the intestinal mucosa (Kennedy, 1989).

Table IV. Mechanisms by Which the Intestinal Microflora Inhibits Candida albicans Colonization and Dissemination from the Gastrointestinal Tract[a]

Type	Mechanism
I	Inhibition of mucosal association:
	—formation of thick layers of bacteria that inhibit penetration into the mucosa
	—competition for adhesion sites
	—production of inhibitory substances (e.g., SCFA and DCBA)
	—increased rate of dissociation
	—modification of metabolic activity of *C. albicans* so that adhesins or digestive enzymes cannot be produced
II	Regulation of the *Candida* population size:
	—anaerobic condition of the gut
	—production of inhibitory substances (e.g., SCFA, DCBA, and H_2S)
	—competition for limiting nutrients
	—restricting substrate availability
	—prolonging the lag phase and doubling time

[a]Modified from Kennedy (1989).

Several breakdown products from the large number of primary dietary or host-derived substrates in the gut may serve to antagonize *Candida* species, as might a number of substances synthesized by the microflora itself (Kennedy *et al.*, 1990). The most likely inhibitory substances have been suggested to be those that are produced by predominate anaerobic species of bacteria that colonize the gut and, therefore, are those substrates that are produced in the highest concentrations. Examples of substances that may antagonize *C. albicans* include short-chain fatty acids (SCFA), hydrogen sulfide, and deconjugated bile acids (DCBA) (Kennedy, 1989). These are produced in relatively high concentrations in the gut due to the metabolic activity of intestinal anaerobes (Argenzio, 1981) and have been shown to antagonize *C. albicans* growth by prolonging both its lag phase and doubling times (Kennedy *et al.*, 1990).

Although it is likely that several host-derived or anaerobe-synthesized substrates may antagonize the growth of *Candida* in the gut, Freter (1983) has suggested that it is theoretically impossible to account for the total suppression of a sensitive microorganism in the GI tract solely on the basis of the production of growth inhibitors. If this were the case, *C. albicans* would be eliminated if too much inhibitor were present, whereas if too little inhibitor were present the *Candida* population would increase until it became limited by some other mechanism (Freter, 1983). Studies using a continuous-flow culture model of the ecology of the intestinal microflora (Kennedy *et al.*, 1990) have shown that *C. albicans* can adapt itself and become resistant or tolerant to these "toxic" substrates when it is implanted before or simultaneously with the intestinal microflora.

The ability of *C. albicans* to become resistant to certain mechanisms of antagonism alone, however, was not sufficient to allow implantation of this fungus in the continuous-flow culture model or in the gut (Kennedy *et al.*, 1990). The presence of any one of these inhibitory substances probably imparts a prolonged lag phase and doubling time to *C. albicans*, and stresses the importance of mucosal association in the initiation of colonization and infection (Kennedy *et al.*, 1990). Studies employing antibiotic-treated mice confirmed this view and showed that *C. albicans* populations predominated initially in the lumen but that mucosa-associated populations predominated after the intestinal microflora reestablished (Kennedy *et al.*, 1990). Thus, the adhesion and association of *Candida* with the GI mucosa may allow *C. albicans* to remain in the gut even at a severely depressed growth rate (Kennedy, 1989). The ability of the indigenous microflora to inhibit implantation and infection by *C. albicans* in the GI tract, therefore, may key on the ability of the microflora to inhibit mucosal association (Kennedy and Volz, 1985a).

As has been noted, the first step in mucosal association must be the penetration of the mucus gel (Freter, 1982). In infant mice, which lack a complete bacterial flora including the dense bacterial populations in the mucus gel (Schaedler *et al.*, 1965; Davis *et al.*, 1973), *C. albicans* can readily associate with and pass through the mucosa to initiate systemic infection (Pope *et al.*, 1979; Field *et al.*, 1981). Similarly, in adult mice treated with certain antibiotics to eliminate wall-associated populations of indigenous bacteria, *C. albicans* was found to associate with and invade the GI mucosa and spread systemically to visceral organs (Kennedy and Volz, 1983, 1985b). In contrast, it was observed that when intestinal tissues possessed an indigenous wall-associated microflora there was significant inhibition of mucosal association (*in vitro* and *in vivo*) and dissemination of *C. albicans* from the GI tract (Kennedy and Volz, 1985a).

Mucosal association by *C. albicans*, like growth in the intestinal contents, appears also to be inhibited by a redundancy of mechanisms that may act synergistically to inhibit *Candida* attachment and invasion (Kennedy and Volz, 1985a). For example, mucosal association by *C. albicans* was retarded by the indigenous microflora by outcompeting the yeast cells for adhesion sites and physically blocking the larger yeast cells from penetrating deeply into the mucosa (Kennedy and Volz, 1985a). Moreover, the rate of dissociation from intestinal tissues by *C. albicans* from antibiotic-treated animals was nearly three times that of untreated animals (Kennedy and Volz, 1985a). Thus, when an intact and functioning mucosal flora is present, *Candida* cells may be more localized to the surface of the mucosa because they may not be able to penetrate as deeply into the mucus gel (Kennedy and Volz, 1985a). This, apparently, may further provoke mucosa-associated yeast cells to dissociate from the tissue to a faster and much greater extent than occurs in the mucosa of antibiotic-treated animals (Kennedy and Volz, 1985a).

Finally, certain chemical factors produced in the gut by the metabolic activity of the normal flora also appear to inhibit the association of *C. albicans* with intestinal mucosa (Kennedy and Volz, 1985a). SCFA and DCBA, for instance, were found to significantly inhibit the association of *Candida* with the mucosa in an *in vitro* adhesion assay using intestinal tissue slices from both antibiotic-treated and untreated adult hamsters (Table V). Although the exact nature of inhibition of mucosal association by SCFA and DCBA has not been elucidated, a correlation between the

Table V. Association of Candida albicans with Cecal Slices: Antagonism by Intestinal Bacteria[a]

Assay System	Assay Solution[b]	Source of cecal slice[c]	Log_{10} mean no. of *C. albicans* per slice	Association index[d]
1	IC	AB-TR	2.5	0.08
2	IF	AB-TR	4.0	1.82
3	PBS	AB-TR	4.8	13.04
4	IC	UN-TR	2.4	0.06
5	IF	UN-TR	3.4	0.25
6	PBS	UN-TR	3.7	0.73
7	PBS + BA	AB-TR	4.3	4.19
8	PBS + VFA	AB-TR	4.1	1.86

[a]From Kennedy and Colz (1985a).
[b]IC = intestinal contents in PBS; IF = intestinal filtrates in PBS; PBS + BA = PBS containing bile acids (lithocholic acid, 3.0 mM; deoxycholic acid, 2.6 mM); PBS + VFA = PBS containing volatile fatty acids (valeric acid, 1.3 mM; isovaleric acid, 2.2 mM; isobutyric acid, 1.4 mM; propionic acid, 20.1 mM; acetic acid, 49.3 mm).
[c]AB-TR = hamsters that were given antibiotics in the drinking water for 3 days; UN-TR = untreated control hamsters.
[d]The association index = $(t/[t + K]) \times 100$, where t is the number of *C. albicans* associating with intestinal slices after rinsing, and k is the number of viable *C. albicans* per milliliter of assay solution after 2 hr of incubation.

drop in the concentration of SCFA and DCBA in the intestinal tract and an increase in mucosal association by *C. albicans* following antibiotic treatment has clearly been shown (Kennedy and Volz, 1985a). It may be that bacterial substrates produced or altered in the intestinal tract (probably as a result of the metabolic activity by indigenous anaerobes) act to reduce the ability of *Candida* cells to attach to certain mucosal receptors by modifying *Candida* adhesins or mucosal receptors (Kennedy and Volz, 1985a).

8. REFERENCES

Argenzio, R. A., 1981, Short-chain fatty acids and the colon, *Dig. Dis. Sci.* **26**:97–99.

Bacon, J. S. D., Davidson, E. D., Jones, D., and Taylor, I. F., 1966, The location of chitin in the yeast cell wall, *J. Biochem.* **101**:36C–38C.

Balish, E., Balish, M. J., Salkowski, C. A., Lee, K. W. and Bartizal, K. F., 1984, Colonization of congenitally athymic gnotobiotic mice by *Candida albicans*, *Appl. Environ. Microbiol.* **47**:647–652.

Ballou, C., 1976, Structure and biosynthesis of the mannan component of yeast cell envelope, *Adv. Microb. Physiol.* **14**:93–158.

Barnes, J. L., Osgood, W., Lee, J. C., King, R. D., and Stein, J. H., 1983, Host–parasite interactions in the pathogenesis of experimental renal candidiasis, *Lab. Invest.* **49**:460–467.

Bartnicki-Garcia, S., 1973, Fungal cell wall composition, in: *Handbook of Microbiology*, Volume II (A. I. Laskin and H. A. Lechevalier, eds.), CRC Press, Cleveland, pp. 201–214.

Bishop, C. T., Blank, F., and Gardner, P., 1960, The cell wall polysaccharides of *Candida albicans*, glucan, mannan and chitin, *Can. J. Chem.* **38**:869–881.

Cabib, E., Roberts, R., and Bowers, B., 1982, Synthesis of yeast cell wall and its regulation, *Annu. Rev. Biochem.* **51**:763–793.

Cassone, A., Simonette, N., and Strippoli, V., 1973, Ultrastructural changes in the wall during the germ-tube formation from blastospores of *C. albicans*, *J. Gen. Microbiol.* **77**:417–426.

Cassone, A., Kerridge, D., and Gale, E. F., 1979, Ultrastructural changes in the wall of *Candida albicans* following cessation of growth and their possible relationship to the development of polyene resistance, *J. Gen. Microbiol.* **110**:339–349.

Chaffin, L. W., and Stocco, D. M., 1983, Cell wall proteins of *Candida albicans*, *Can. J. Microbiol.* **29**:1438–1444.

Cheng, K. J., Irvin, R. T., and Costerton, J. W., 1981, Autochthonous and pathogenic colonization of animal tissues by bacteria, *Can. J. Microbiol.* **27**:461–490.

Cole, G. T., Seshan, K. R., Pope, L. M., and Yancey, R. J., 1988, Morphological aspects of gastrointestinal tract invasion by *Candida albicans* in the infant mouse, *J. Med. Vet. Mycol.* **26**:173–185.

Costerton, J. W., Marrie, T. J., and Cheng, K. J., 1985, Phenomena of bacterial adhesion, in: *Bacterial Adhesion: Mechanisms and Physiological Significance.* (D. C. Savage and M. Fletcher, eds.), Plenum Press, New York, pp. 3–43.

Critchley, I. A., and Douglas, L. J., 1987, Isolation and partial characterization of an adhesin from *Candida albicans*, *J. Gen. Microbiol.* **133**:629–636.

Davis, C. P., McAllister, J. S., and Savage, D. C., 1973, Microbial colonization of the intestinal epithelium in suckling mice, *Infect. Immun.* **7**:666–672.

Derjaguin, B. V. and Landau, L., 1941, Theory of the stability of strongly charged lyophobic sols and of the adhesion of strongly charged particles in solutions of electrolytes, *Acta Physiochim. USSR* **14**:633–662.

Diamond, R. D., and Krzesicki, R., 1978, Mechanisms of attachment of neutrophils to *Candida albicans* pseudohyphae in the absence of serum, and of subsequent damage to pseudohyphae by microbial processes of neutrophils *in vitro*, *J. Clin. Invest.* **61**:360–369.

Djaczenko, W., and Cassone, A., 1971, Visualization of new ultrastructural components in the cell wall of *Candida albicans* with fixatives containing tapo, *J. Cell Biol.* **52**:186–190.

Douglas, L. J., and McCourtie, J., 1983, Effect of tunicamycin treatment on the adherence of *Candida albicans* to human buccal epithelial cells, *FEMS Microbiol. Letters* **16**:199–202.

Eras, P., Goldstein, M. J., and Sherlock, P., 1972, *Candida* infection of the gastrointestinal tract, *Medicine (Baltimore)* **51**:367–379.

Field, L. H., Pope, L. M., Cole, G. T., Guentzel, M. N., and Berry, L. J., 1981, Persistence and spread of *Candida albicans* after intragastric inoculation of infant mice, *Infect. Immun.* **31**:783–791.

Freter, R., 1982, Bacterial association with the mucus gel system of the gut, in: *Microbiology*. (D. Schlessinger, ed.), American Society for Microbiology, Washington, D.C., pp. 278–281.

Freter, R., 1983, Mechanisms that control the microflora in the large intestine, in: *Human Intestinal Microflora in Health and Disease*. (D. J. Hentges, ed.), Academic Press, New York, pp. 33–54.

Freter, R., and Jones, G. W., 1983, Models for studying the role of bacterial attachment in virulence and pathogenesis, *Rev. Infect. Dis.* **5**:5647–5658.

Freter, R., O'Brien, P. C. M., and Macsai, M. S., 1981, Role of chemotaxis in the association of motile bacteria with intestinal mucosa: *in vitro* studies, *Infect. Immun.* **34**:241–249.

Freter, R., Brickner, H., Fekete, J., Vickerman, M. M., and Carey, K. E., 1983, Survival and implantation of *Escherichia coli* in the intestinal tract, *Infect. Immun.* **39**:686–703.

Gardiner, R., Podgorski, C., and Day, A. W., 1982, Serological studies on the fimbriae of yeasts and yeast-like species, *Bot. Gaz.* **143**:534–541.

Ghannoum, M. A., Burns, G. R., Elteen, K., and Radwin, S. S., 1986, Experimental evidence for the role of lipids in adherence of *Candida* spp. to human buccal epithelial cells, *Infect. Immun.* **54**:189–193.

Gopal, P., Sullivan, P. A., and Shepard, M. G., 1984, Isolation and structure of glucan from regenerating spheroplasts of *Candida albicans*, *J. Gen. Microbiol.* **130**:1217–1225.

Guentzel, M. N., Cole, G. T., and Pope, L. M., 1985, Animal models for candidiasis, in: *Current Topics in Medical Mycology*, Volume 1 (M. R. McGinnis, ed.), Springer-Verlag, Berlin, pp. 57–116.

Helstrom, P. B., and Balish, E., 1979, Effect of oral tetracycline, the microbial flora and the athymic state on gastrointestinal colonization and infection of BALB/c mice with *Candida albicans*, *Infect. Immun.* **23**:764–774.

Howlett, J. A., and Squier, C. A., 1980, *Candida albicans* ultrastructure: Colonization and invasion of oral epithelium, *Infect Immun.* **29**:252–260.

Isaacson, R. E., 1985, Pilus adhesins, in: *Bacterial Adhesion: Mechanisms and Physiological Significance* (D. C. Savage and M. Fletcher, eds.), Plenum Press, New York, pp. 307–336.

Jones, G. W., 1977, The attachment of bacteria to the surfaces of animal cells, in: *Microbial Interactions: Receptors and Recognition*, Series B, Volume 3 (J. L. Reissig, ed.), Chapman & Hall, London, pp. 139–176.

Jones, G. W., 1984, Adhesion to animal surfaces, in: *Microbial Adhesion and Aggregation* (K. C. Marshall, ed.) Springer-Verlag, Berlin, pp. 71–84.

Jones, G. W., and Isaacson, R. E., 1983, Proteinaceous bacterial adhesins and their receptors, *CRC Crit. Rev. Microbiol.* **10**:229–260.

Joshi, S. N., Garvin, P. J., and Sunwoo, Y. C., 1981, Candidiasis of the duodenum and jejunum, *Gastroenterology* **80**:829–833.

Kennedy, M. J., 1987, Role of motility, chemotaxis and adhesion in microbial ecology, *Ann. N. Y. Acad. Sci.* **506**:260–273.

Kennedy, M. J., 1988, Adhesion and association mechanisms of *Candida albicans*, in: *Current Topics in Medical Mycology*, Volume 2 (M. R. McGinnis, ed.) Springer-Verlag, Berlin, pp. 73–169.

Kennedy, M. J., 1989, Regulation of *Candida albicans* populations in the gastrointestinal tract: Mechanisms and significance in GI and systemic candidosis, in: *Current Topics in Medical Mycology*, Volume 3 (M. R. McGinnis and M. Borgers, eds.), Springer-Verlag, Berlin, pp. 315–402.

Kennedy, M. J., and Sandin, R. L., 1988, Influence of growth conditions on *Candida albicans* adhesion, hydrophobicity and cell wall ultrastructure, *J. Med. Vet. Mycol.* **26**:79–92.

Kennedy, M. J., and Volz, P. A., 1983, Dissemination of yeasts after gastrointestinal inoculation in antibiotic-treated mice, *Sabouraudia* **21**:27–33.

Kennedy, M. J., and Volz, P. A., 1985a, Ecology of *Candida albicans* gut colonization: Inhibition of *Candida* adhesion, colonization, and dissemination from the gastrointestinal tract by bacterial antagonism, *Infect. Immun.* **49**:654–663.

Kennedy, M. J., and Volz, P. A., 1985b, Effect of various antibiotics on gastrointestinal colonization and dissemination by *Candida albicans*, *Sabouraudia* **23**:265–274.

Kennedy, M. J., Bajwa, P. S., and Volz, P. A., 1982, Gastrointestinal inoculation of *Sporothrix schenckii* in mice, *Mycopathologia* **78**:141–143.

Kennedy, M. J., Volz, P. A., Edwards, C. A., and Yancey, R. J., 1987, Mechanisms of association of *Candida albicans* with intestinal mucosa, *J. Med. Microbiol.* **24**:333–341.

Kennedy, M. J., Rogers, A. L., and Yancey, R. J., 1990, Mechanisms that control *Candida albicans* populations in continuous-flow culture models of intestinal microflora and in the large intestine, (submitted for publication).

Kessler, G., and Nickerson, W. J., 1959, Glucomannan–protein complexes from cell walls of yeast, *J. Biol. Chem.* **234**:2281–2285.

Klotz, S. A., Drutz, D. J., Harrison, J. L., and Huppert, M., 1983, Adherence and penetration of vascular endothelium by *Candida* yeasts, *Infect. Immun.* **42**:374–384.

Kolarova, N., Masler, L., and Sikl, D., 1973, Cell wall glycopeptides of *Candida albicans* serotypes A and B, *Biochim. Biophys. Acta* **328**:221–227.

Krause, W., Matheis, H., and Wulf, K., 1969, Fungaemia and funguria after oral administration of *Candida albicans*, *Lancet* **1**:598–599.

Lee, J. C., and King, R. D., 1983a, Adherence mechanisms of *Candida albicans*, in: *Microbiology—1983* (D. Schlessinger, ed.), American Society for Microbiology, Washington, D.C., pp. 269–272.

Lee, J. C., and King, R. D., 1983b, Characterization of *Candida albicans* adherence to human vaginal epithelial cells in vitro, *Infect. Immun.* **41**:1024–1030.

Loeb, G. I., 1985, The properties of nonbiological surfaces and their characterization, in: *Bacterial Adhesion: Mechanisms and Physiological Significance* (D. C. Savage and M. Fletcher, eds.) Plenum Press, New York, pp. 111–129.

Luft, J. H., 1971, Ruthenium red and ruthenium violet. II. Fine structural purification, methods for use for electron microscopy and localization in animal tissues, *Anat. Rec.* **171**:369–416.

McCourtie, J., and Douglas, L. J., 1981, Relationship between cell surface composition of *Candida albicans* and adherence to acrylic after growth on different carbon sources, *Infect. Immun.* **32**:1234–1241.

McCourtie, J., and Douglas, L. J., 1984, Relationship between cell surface composition, adherence and virulence of *Candida albicans*, *Infect. Immun.* **45**:6–12.

McCourtie, J., and Douglas, L. J., 1985, Extracellular polymer of *Candida albicans*: Isolation, analysis and role in adhesion, *J. Gen. Microbiol.* **131**:495–503.

Marrie, T. J., and Costerton, J. W., 1981, The ultrastructure of *Candida albicans* infections, *Can. J. Microbiol.* **27**:1156–1164.

Marshall, K. C., 1976, *Interfaces in Microbial Ecology*, Harvard University Press, Cambridge, Mass.

Marshall, K. C. (ed.), 1984, *Microbial Adhesion and Aggregation*, Springer-Verlag, Berlin.

Miles, M. R., Olsen, L., and Rogers, A., 1977, Recurrent vaginal candidiasis: Importance of an intestinal reservoir, *J. Am. Med. Assoc.* **238**:1836–1837.

Myerowitz, R. L., Pazin, G. J., and Allen, C. M., 1977, Disseminated candidiasis: Changes in incidence, underlying diseases, and pathology, *Am. J. Clin. Pathol.* **68**:29–38.

Nystatin Multicenter Study Group, 1986, Therapy of candidal vaginitis: The effect of eliminating intestinal *Candida*, *Am. J. Obstet. Gynecol.* **155**:651–655.

Odds, F. C., 1988, *Candida and Candidosis*, University Park Press, Baltimore.

Paerl, H. W., 1985, Influence of attachment on microbial metabolism and growth in aquatic ecosystems, in: *Bacterial Adhesion: Mechanisms and Physiological Significance* (D. C. Savage and M. Fletcher, eds.), Plenum Press, New York, pp. 363–400.

Ponton, J., and Jones, J. M., 1986a, Analysis of cell wall extracts of *Candida albicans* by sodium dodecyl sulfate–polyacrylamide gel electrophoresis and Western blot techniques, *Infect. Immun.* **53**:565–572.

Ponton, J., and Jones, J. M., 1986b, Identification of two germ-tube-specific cell wall antigens of *Candida albicans*, *Infect. Immun.* **54**:864–868.

Pope, L. M., and Cole, G. T., 1981, SEM studies of adherence of *Candida albicans* to the gastrointestinal tract of infant mice, *Scanning Electron Microsc.* **3**:73–80.

Pope, L. M., and Cole, G. T., 1982, Comparative studies of gastrointestinal colonization and systemic spread by *Candida albicans* and nonlethal yeast in the infant mouse, *Scanning Electron Microsc.* **4**:1667–1676.

Pope, L. M., Cole, G. T., Guentzel, M. N., and Berry, L. J., 1979, Systemic and gastrointestinal candidiasis of infant mice after intragastric challenge, *Infect Immun.* **25**:702–707.

Poulain, D., Hopwood, V., and Vernes, A., 1985, Antigen variability of *Candida albicans*, *Crit. Rev. Microbiol.* **12**:223–271.

Pugh, D., and Cawson, R. A., 1975, The cytochemical localization of phospholipase and lysophospholipase in *Candida albicans*, *Sabouraudia* **13**:110–115.

Rippon, J. W., 1988, *Medical Mycology: The Pathogenic Fungi and the Pathogenic Actinomycetes*, Saunders, Philadelphia.

Rogers, T. J., and Balish, E., 1980, Immunity to *Candida albicans*, *Microb. Rev.* **44**:660–682.

Rotrosen, D., Edwards, J. E, Gibson, T. R., Moore, J. C., Cohen, A. H., and Green, I., 1985, Adherence of *Candida* to cultured vascular endothelial cells: Mechanisms of attachment and endothelial cell penetration, *J. Infect. Dis.* **152**:1264–1274.

Rutter, P. R., and Vincent, B., 1980, The adhesion of microorganisms to surfaces: Physiochemcial aspects, in: *Microbial Adhesion to Surfaces*. (R. C. W. Berkley, J. M. Lynch, P. R. Rutter, and B. Vincent, eds.), Ellis Horwood, London, pp. 79–92.

Rutter, P. R., and Vincent, B., 1984, Physicochemical interactions of the substratum, microorganisms, and the fluid phase, in: *Microbial Adhesion and Aggregation* (K. C. Marshall, ed.), Springer-Verlag, Berlin, pp. 21–38.

San-Blas, G., 1982, The cell wall of fungal human pathogens: Its possible role in host–parasite relationships, *Mycopathologia* **79**:159–184.

Sandin, R. L., 1987, The attachment to human buccal epithelial cells by *Candida albicans*: An in vitro kinetic study using concanavalin A, *Mycopathologia* **98**:179–184.

Sandin, R. L., Rogers, A. L., Patterson, R. J., and Beneke, E. S., 1982, Evidence for mannose-mediated adherence of *Candida albicans* to human buccal epithelial cells in vitro, *Infect Immun.* **35**:79–85.

Savage, D. C., 1983, Associations of indigenous microorganisms with gastrointestinal epithelial surfaces, in: *Human Intestinal Microflora in Health and Disease* (D. J. Hentges, ed.), Academic Press, New York, pp. 55–78.

Schaedler, R. W., Dubos, R., and Costello, R., 1965, The development of the bacterial flora in the gastrointestinal tract of mice, *J. Exp. Med.* **122**:59–66.

Segal, E., Lehrer, N., and Ofek, I., 1982, Adherence of *Candida albicans* to human vaginal epithelial cells: Inhibition by amino sugars, *Exp. Cell Biol.* **50**:13–17.

Shedlofsky, S., and Freter, R., 1974, Synergism between ecological and immunological control mechanisms of intestinal flora, *J. Infect. Dis.* **129**:296–303.

Sobel, J. D., Myers, P. G., Kaye, D., and Levison, M. E., 1981, Adherence of *Candida albicans* to human vaginal and buccal epithelial cells, *J. Infect. Dis.* **143**:76–82.

Stone, H. H., 1974, Studies in the pathogenesis, diagnosis and treatment of *Candida* sepsis in children, *J. Pediatr. Surg.* **9**:127–133.

Stone, H. H., Kolb, L. D., Currie, C. H., Gebeher, C. E., and Cuzzell, J. F., 1974, *Candida* sepsis: Pathogenesis and principles of treatment, *Ann. Surg.* **179**:697–711.

Tadros, T. F., 1980, Particle–surface adhesion, in: *Microbial Adhesion to Surfaces* (R. C. W. Berkley, J. M. Lynch, P. R. Rutter, and B. Vincent (eds.), Ellis Horwood, London, pp. 79–92.

Tronchin, G., Poulain, D., and Vernes, A., 1984, Cytochemical and ultrastructural studies of *Candida albicans*. III. Evidence for modifications of the cell wall coat during adherence to human buccal epithelial cells, *Arch. Microbiol.* **139**:221–224.

Umenai, T., Konno, S., Yamauchi, A., Iimura, Y., and Fujimoto, H., 1980, Growth of *Candida* in the upper intestinal tract as a possible source of systemic candidiasis in mice, *Tohoku J. Exp. Med.* **130**:101–102.

Verwey, E. J. W., and Overbeek, J. T. G., 1948, *Theory of the Stability of Lyophobic Colloids*, Elsevier, Amsterdam.

Wells, C. L., Maddaus, M. A., and Simmions, R. L., 1988, Proposed mechanisms for the bacterial translocation of intestinal bacteria, *Rev. Infect. Dis.* **10**:958–979.

Wingard, J. R., Dick, J. D., Merz, W. G., Sanford, G. R., Saral, R., and Burns, W. H., 1980, Pathogenicity of *Candida tropicalis* and *Candida albicans* after gastrointestinal inoculation in mice, *Infect. Immun.* **29**:808–813.

Yu, R., Bishop, C., Cooper, F., and Hasenclever, H., 1967, Glucans from *Candida albicans* (serotype B) and from *Candida parapsilosis*, *Can. J. Chem.* **45**:2264–2267.

8

Infectious Propagules of Dermatophytes

Tadayo Hashimoto

1. INTRODUCTION

1.1. Definition, Etiological Agents, and Classification

Dermatophytosis is a mycotic infection that affects the superficial keratinized tissues (cornified epidermis, hair, horn, nail, and feathers) of humans and other animals. The disease is caused by a taxonomically related group of fungi, the dermatophytes. Currently, approximately 40 species of the dermatophytes have been identified and grouped into three genera—*Trichophyton*, *Microsporum* and *Epidermophyton*. Of these, 11 species are implicated in human infections (Rippon, 1988).

Although most dermatophytes are classified as imperfect fungi (Deuteromycetes), the existence of a sexual life cycle has been demonstrated in some species. All *Microsporum* species with the demonstrated perfect stage are now reclassified in the genus *Nannizzia*, and the perfect *Trichophyton* species are the members of the genus *Arthroderma*.

The dermatophytes are also categorized on the basis of the environment in which they prefer to inhabit. Geophilic strains can grow and strive in soil. They are also isolated from the hair of small, apparently healthy, mammals and some birds. Zoophilic dermatophytes normally parasitize on animals other than humans. However, all of the known zoophilic species occasionally produce infections in humans. Anthropophilic strains are primarily parasitic on humans. Strict anthropophilic species depend entirely on humans for their survival and reproduction. While this type of strict host dependency is occasionally found in pathogenic bacteria, it is extremely unusual for fungal agents. Certain strict anthropophilic dermatophytes such as *T. rubrum* cannot infect lower animals and they cannot be isolated from saprophytic sources other than humans.

1.2. Incidence and Clinical Features

Dermatophytoses are among the most prevalent mycotic infections in the world (Rippon, 1988; Mitchell, 1988). The incidence is highest in hot, humid climates or in crowded environments where a high standard of hygiene is difficult to maintain (Kane *et al.*, 1988). Due to its close

Tadayo Hashimoto • Department of Microbiology, Loyola University Stritch School of Medicine, Maywood, Illinois 60153.

association with certain professions, dermatophytosis is regarded by some as an occupational mycosis (Kane and Krajden, 1983). According to a recent survey, many species of the dermatophytes have been isolated from human patients at various clinics throughout the United States (Sinski and Kelley, 1987).

Clinically, dermatophytoses are known as tinea or "ringworm" infections. The many species of dermatophytes elicit a variety of well-defined clinical syndromes (Rippon, 1988). Dermatophytoses are, however, not debilitating or life-threatening infections, although serious invasive granulomatous forms involving visceral organs have been reported in patients whose defense mechanisms are immunologically compromised (Ahmed, 1982; Hironaga et al., 1983; Novick et al., 1987). In dermatophytosis, spontaneous resolution of the lesions is not uncommon; clinical symptoms often disappear even left untreated. However, the etiological fungi may survive in the infected loci for years, creating a state of "healthy" carrier. Exposure of the clinically healed lesion to trauma or warm humid stimuli often triggers the relapse of the disease (Rippon, 1988). Treatment failures and relapses occur with all currently available therapeutic agents. Certain forms of ringworm infections, especially those caused by T. rubrum, are especially recalcitrant to chemotherapeutic treatment.

1.3. Communicability and Epidemiological Problems

One of the characteristics that distinguishes dermatophytosis from the rest of mycotic infections is the communicable nature of the disease. The infection is transmissible from person to person directly by means of contact or indirectly via fomites contaminated with infected skin scales or hairs. Numerous outbreaks of dermatophytosis have been reported in schools or orphanages, dormitories, chronic-care institutions, and other public and military installations. In many countries, dermatophytosis still remains a major public health problem.

Dermatophytosis is also acquired by humans from infected animals and other saprophytic sources. Initially, animals were most likely infected by geophilic dermatophytes proliferating in soil and evolved first to zoophilic species. Dermatophytosis in humans can be acquired by direct exposure to soil infested with dermatophytes. Numerous cases of severe infections due to geophilic or zoophilic dermatophytes were reported among the U.S. military personnel during the Vietnam war (Rippon, 1988).

1.4. Mediators of Disease Transmission

As mentioned earlier, what makes dermatophytosis unique among mycotic infections is the communicable nature of the infection. Then, what are the mediators of disease transmission in dermatophytosis? The answer to this important epidemiological question seems to be found in the unique ability of the dermatophytes to produce two distinctly different types of infective conidia depending on whether they are inhabiting under saprophytic or parasitic environments. In other words, disease transmission in dermatophytosis is mediated by two distinctly different types of propagules depending on the source of infection. From saprophytic sources such as soil or shed animal skin scales, the disease is transmitted to humans via saprophytic conidia, namely macro- or microconidia. From human sources, the disease is transmitted to other individuals by parasitic conidia.

In this chapter, I shall describe the salient biological characteristics of these infectious conidia that the dermatophytes produce under saprophytic and parasitic environments and discuss their roles in disease transmission and disease initiation. Although references will be made to other dermatophytic species or strains, the discussion will be focused mainly on the infectious propagules of T. mentagrophytes which have been extensively investigated in my laboratory.

2. INFECTIVE CONIDIA PRODUCED BY THE DERMATOPHYTES

2.1. Conidial Type and Definition

Depending on the environment in which they grow, the dermatophytes produce two types of asexual propagules: saprophytic conidia or parasitic conidia. Under nonparasitic environments, they produce saprophytic conidia which include two morphologically different types; macroconidia and microconidia. Within infected tissues, however, dermatophytes exclusively produce arthroconidia or parasitic conidia. The sporadic formation of arthroconidia in cultured dermatophytes was noted in the earlier literature (Emmons, 1934; Chin and Knight, 1957). All these conidia possess the universal properties of dormancy and resistance usually ascribed to microbial spores. In the dermatophytes, these conidia are the important propagules of the species and the mediators of infection. The asexual life cycles of *T. mentagrophytes* involving either saprophytic or parasitic types of conidia are illustrated in Fig. 1.

Some dermatophytes produce chlamydospores and other resistant forms of cells under saprophytic conditions (Rebell and Taplin, 1970). The perfect dermatophytes produce ascospores under certain conditions. However, in this chapter, the discussion will be limited only to the three types of conidia—macroconidia, microconidia, and arthroconidia—because these represent by far the important infectious propagules of the dermatophytes.

2.2. Saprophytic Conidia: Macroconidia and Microconidia

Saprophytically, most dermatophytes produce two types of conidia: a small single-celled microconidium and a large multicellular macroconidium (Fig. 1). The type and the microscopic characteristics of saprophytic conidia are the most important aid in the identification of dermatophytes (Emmons, 1934; Rebell and Taplin, 1970; Ajello and Padhye, 1974).

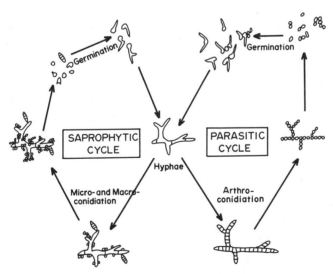

Figure 1. The asexual life cycle of the dermatophyte *T. mentagrophytes*. When fertile hyphae enter the saprophytic cycle, they proliferate by apical elongation and lateral branching forming macro- or microconidiogenous hyphae. Swelling of the apical tips of the conidiogenous hyphae is the initial event that leads to macro- or microconidium formation. Under parasitic conditions, fertile hyphae are repeatedly segmented and eventually fragmented to form arthroconidia. For the physiological conditions conducive to each cycle, see the text. (Original figure by Emyanitoff and Hashimoto.)

For a given species of the dermatophytes, either microconidia or macroconidia are usually the predominant type of conidia produced under saprophytic conditions. Occasionally, however, both microconidia and macroconidia are produced in the same species. For example, in *T. mentagrophytes*, microconidia are the predominant form of saprophytic conidia but some macroconidia are also produced. Among pathogenic species, *E. floccosum* is unique in that although macroconidia are abundant, microconidia are never produced.

2.3. Evolutionary Change in Conidiation Patterns

A natural evolution from geophilic dermatophytes to zoophilic or anthropophilic strains apparently occurred by association with and invasion of keratinous tissues in living animals and humans. The animals which acted initially as hosts were probably the soil-inhabiting rodents. Saprophytic conidia (micro- or macroconidia) produced in the soil dermatophytes are the most likely mediator of such infections. In establishing a primary parasitic existence with the host, the fungus underwent evolutionary changes. As a result, many strict anthropophilic dermatophytes such as *M. audouinii* and *T. rubrum* no longer produce abundant saprophytic conidia in culture and they no longer depend on saprophytic conidia for their survival and dissemination (Rippon, 1988). In such dermatophytes, parasitic conidia or arthroconidia become the sole means of survival and propagation.

In most dermatophytes, the ability to form saprophytic conidia tends to diminish or disappear during repeated transfers on artificial media. The asporogenic variants emerging as a result of such subculturing are referred to as downy variants contrasting to the granular parent strains. Sometimes, passages of such downy variants through susceptible animals or culturing under increased carbon dioxide concentration restores their ability to produce saprophytic conidia (Chin and Knight, 1957). This phenomenon is known as pleomorphism in dermatophytes. The precise mechanism involved in the development of pleomorphism in dermatophytes is not known. The fact that certain asporogenic downy forms may revert back to the sporogenic granular form by changing the growth environment seems to argue against mutation. Pleomorphism in dermatophytes presents a problem in both maintenance and the identification of these fungi. It is our experience that the development of downy form variants from granular strains can be minimized only when several individual conidia from well-isolated granular colonies are inoculated on each plate and are transferred at weekly intervals.

It is important to recognize that no saprophytic conidia (macro- and microconidia) are produced by any dermatophyte under circumstances of active parasitization. Macro- and microconidia may be produced by certain dermatophytes when growing in epilated or fallen hairs, but not in growing hairs. Apparently, the mechanism controlling saprophytic conidiogenesis (macro- or microconidium formation) in the dermatophytes is suppressed under the state of parasitization. The fungi seem to activate an alternative conidiation system in order to survive and reproduce under adverse parasitic circumstances.

Clearly, macro- and microconidia are the means of survival and propagation of the dermatophytes under saprophytic conditions, and they are the most likely mediators of disease transmission from saprophytic sources such as soil to humans and animals.

2.4. Parasitic Conidia: Arthroconidia

As previously stated, arthroconidia are the only infective propagules produced under parasitic conditions (Fig. 1). They are rarely formed by the saprophytically grown dermatophytes (Emmons, 1934). Agents of tinea capitis such as *M. audouinii* produce arthroconidia only when invading hairs that are actively growing. Agents of tinea cruris and tinea pedis such as *E. floccosum* and *T. rubrum* produce them in the keratinized layer of the dermis and the hair follicle of infected humans (Rippon, 1988). Because of the morphological resemblance, arthroconidia of *T. rubrum* were referred to as yeastlike spherules (Rippon and Scherr, 1959; Weigl and Hejtmánek, 1984) or spherical cells (Miyaji and Nishimura, 1971). Weigl and Hejtmánek (1984) felt that the spherules observed in *T. rubrum* were abnormally transformed sporogenic cells.

There seems to be a positive correlation between the abilities of the dermatophytes to produce saprophytic conidia and parasitic conidia. In *T. mentagrophytes*, only the granular variants which produce microconidia saprophytically can undergo arthroconidiogenesis when placed under parasitic conditions. The "fluffy" variant (downy variant) that is unable to microconidiate saprophytically is incapable of forming arthroconidia parasitically (Weigl and Hejtmánková, 1984). One must be cautious in generalizing this correlation because certain anthropophilic strains, which are aconidiogenic *in vitro*, can form arthroconidia parasitically. At present, very little is known about the control mechanism that regulates conidiogenesis in the dermatophytes.

Since arthroconidia are frequently seen in the biopsy specimens of tinea lesions (Fig. 2), their close association with the parasitic environment and their possible role in disease transmission have long been suspected (Chin and Knight, 1957; Miyaji and Nishimura, 1971). However, only in recent years has their infectivity been clearly established. These arthroconidia are now believed to be the natural pathogenic elements of the dermatophytes and the major mediator of disease transmission in dermatophytosis (Hashimoto and Blumenthal, 1977; Allen and King, 1978; Kane and Krajden, 1983; Weigl and Hejtmánková, 1984; Fujita and Matsuyama, 1987; Rippon, 1988).

3. Formation of Conidia

Currently, two different modes of conidial development are recognized in the dermatophytes (Cole and Samson, 1979). The saprophytic conidia (macro- and microconidia) are formed by the holothallic mode: conversion of a terminal or intercalary segment of a fertile hypha into a propagule surrounded by multilayered walls. The parasitic arthroconidia are formed by consecutive segmentation and fragmentation of an existing hypha.

3.1. Ontogeny and Morphogenesis

3.1.1. Saprophytic Conidia: Macroconidia and Microconidia

The holothallic mode of macroconidiogenesis in *M. gypseum* is described in detail by Cole and Samson (1979). It is initiated by the swelling of the apical segment of fertile hyphae. Some extension growth occurs as the swelling of the hyphal tip continues due to intussusception occurring in association with thickening of the conidial wall. The distal segment of the fertile hypha is converted to a conidium leaving the hyphal wall layer intact. A transverse septum is formed at the base of the thallic conidium delimiting the developing macroconidium from the distal cylindrical cells. After the macroconidium has become septate, it is released from the hypha. Macroconidia of some dermatophytes develop an additional layer of ornament on the surface of the conidial wall.

It is believed that microconidia of the dermatophytes are also formed by the holothallic mode (Cole and Samson, 1979). In *T. mentagrophytes*, it is initiated by the swelling of the apical portion of the conidiogenous hypha which has extended either terminally or laterally. When the apical portion of the hypha is converted to a spherical or oval-shaped microconidium, the transverse septum is formed at the base of the thallic conidium. Upon maturation of the microconidium, it is released from the distal portion of the fertile hypha. In *T. mentagrophytes*, the surface of the

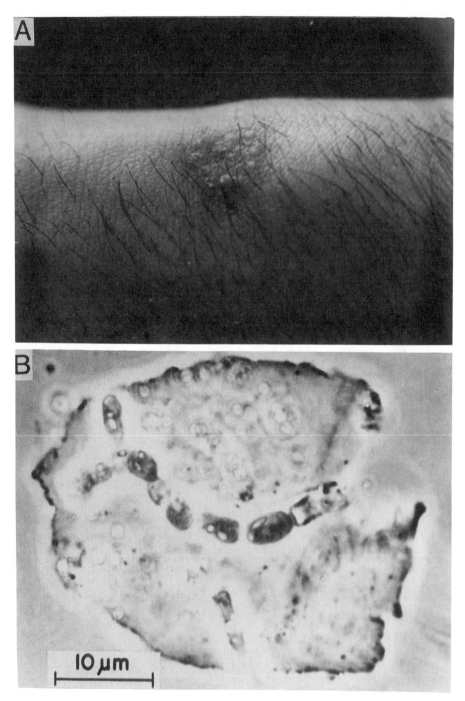

Figure 2. (A) Human dermatophytosis caused by *T. mentagrophytes* and (B) parasitic arthroconidia found in the skin specimen taken from the lesion shown in A. The infection was accidentally caused by *T. mentagrophytes* microconidia. The specimen was viewed under a phase-contrast microscope using an oil immersion lens. (Original photographs by Hashimoto, unpublished.)

conidiogenous hypha and the microconidium is coated with an additional layer of rodlets suggesting that the outer hyphal layer has extended over the conidial cell surface. An alternative blastic mode of microconidium development has been described in *T. mentagrophytes* when cultured at 37 instead of 28°C (Vannini *et al.*, 1981).

3.1.2. Parasitic Conidia: Arthroconidia

Because arthroconidia of the dermatophytes were produced only under parasitic conditions, systematic studies of their ontogeny was not possible before an *in vitro* arthroconidiation system was developed for *T. mentagrophytes* (Bibel *et al.*, 1977; Hashimoto and Blumenthal, 1977).

In the dermatophytes, arthroconidia are formed by segmentation and subsequent fragmentation of a fertile hypha (Fig. 3). Time-lapse photomicrography of *T. mentagrophytes* undergoing arthroconidiation revealed that the new septa were not inserted in any discernibly regular order along the length of hypha; septation did not proceed from tip to base or from base to tip (Hashimoto *et al.*, 1984). Each new septum appeared to divide a preexisting compartment cell approximately in half. The number of septa continuously increased until the entire hypha was segmented into square or rectangular immature conidia. The segmented cells became highly refractile, probably due to the thickening of the wall, and gradually transformed into spherical shape. Finally, the chains of conidia began to disarticulate into shorter chains. This time-lapse micrographic study clearly demonstrated that the tip of an arthroconidiating hypha continued to elongate while multiple septum formation started to divide the distal portion of the hypha into smaller compartments. In other words, the complete cessation of terminal hyphal tip elongation was not required for the initiation or even progression of arthroconidiation (Hashimoto *et al.*, 1984).

The arthroconidiation process of *T. mentagrophytes* can be divided into six stages: vegetative growth (stage 0); appearance of a distinct inner wall layer (stage I); formation of nonperforate double-walled sporulation septa (stage II); wall thickening, during which the inner wall layer increased in thickness while the outer wall layer became thinner (stage III); arthroconidium rounding (stage IV); and secession of the conidia into single cell units (stage V). Any segments which failed to receive at least one nucleus appeared to disintegrate during maturation. The sequence of arthroconidium ontogeny based on our electron microscopic study of *T. mentagrophytes* (Hashimoto *et al.*, 1984) is diagrammatically illustrated in Fig. 4.

From the morphogenic studies, it can be concluded that the mode of arthroconidiation in *T. mentagrophytes* is enteroarthric (Hashimoto *et al.*, 1984). This conclusion appears to be supported by Barrera (1983) who investigated the formation and ultrastructure of *Mucor rouxii* arthrospores and compared its mode of arthroconidiation with that of the dermatophytes. Cole and Samson (1979), however, thought that arthroconidiogenesis of *T. mentagrophytes* occurs by the holoarthric mode. Their conclusion was based on the electron micrographs published earlier by Bibel *et al.*, (1977). Their micrographs seem to support the conclusion of Cole and Samson (1979) although the wall of the arthroconidiating hyphae was not well resolved. Clearly, more work is needed in this area in order to establish general patterns of conidial development in dermatophytes. Once the arthroconidiation process is induced, the rest of the process appears to be completed endotrophically.

3.2. Conditions Affecting Conidium Formation

3.2.1. Saprophytic Conidia

As long as the development of pleomorphic variants is prevented, most dermatophytes produce characteristic macro- and microconidia on routine mycological media such as Sabou-

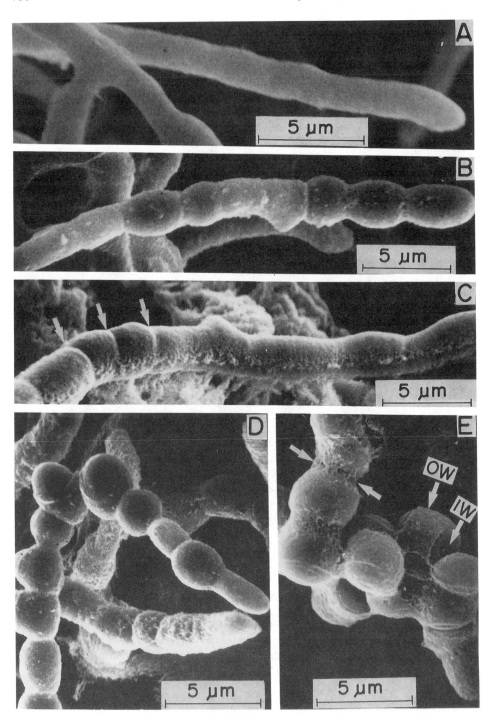

Figure 3. *T. mentagrophytes* undergoing arthroconidiation *in vitro*. Vegetative hyphae (A) show numerous septations, segmentation, and constrictions (B, C) transforming the hyphae into chains of arthroconidia (D) and eventually individual mature arthroconidia (E). Note that the surface of the mature arthroconidia is covered with the remnants of the hyphal wall (OW and double arrows in E). The *in vitro* conditions employed in this experiment are similar to those reported by Hashimoto *et al.* (1984). (Original scanning electron micrographs by Mock and Hashimoto, unpublished.)

Stage

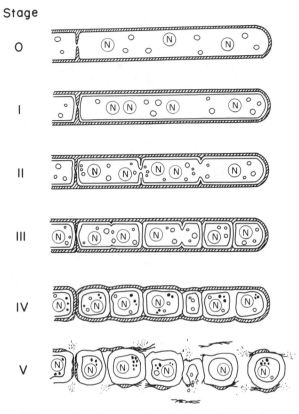

Figure 4. Diagrammatic interpretation of arthroconidium development in *T. mentagrophytes*. The main conidial wall (inner wall) is clearly formed *de novo* during arthroconidiogenesis. The original hyphal wall partially disintegrates during arthroconidiogenesis and only its remnant remains on the surface of the mature arthroconidia. The process of arthroconidium formation can be divided into six stages (Emyanitoff, 1978). See the text for the description of each stage.

raud dextrose agar. Generally, the formation of these saprophytic conidia occurs in *T. mentagrophytes* at temperatures between 20 and 28°C. Temperatures above 32°C are usually suppressive for the formation of saprophytic conidia in *T. mentagrophytes* (Vannini *et al.*, 1981). Aerobic conditions are essential for saprophytic conidiation for all dermatophytes. Although carbon dioxide stimulates microconidiation of *T. mentagrophytes* and *T. rubrum* (Chin and Knight, 1957), the gas is inhibitory for macroconidiation of *M. gypseum*.

3.2.2. Parasitic Conidia

Very little is known about the precise nutritional or biophysical factors responsible for arthroconidium formation in the infected host. The fact that the production of arthroconidia by *T. rubrum* occurs only in particular hosts—"*T. rubrum* people"—may suggest that certain unique metabolites or tissue components are required for its formation (Rippon, 1988).

Although arthroconidia were seen in the dermatophyte cultures grown *in vitro* (Emmons, 1934; Miyaji and Nishimura, 1971), it was generally assumed that the formation of parasitic arthroconidia requires a living host. Recent evidence clearly indicates that this is not the case. *T. mentagrophytes* hyphae incubated at 37°C on the isolated human stratum corneum under high

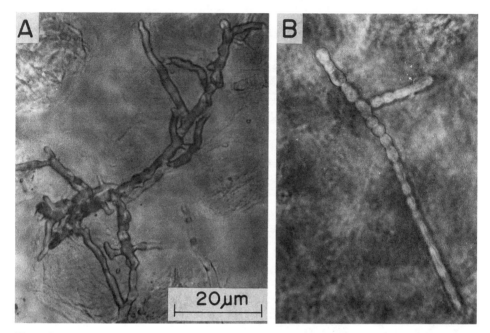

Figure 5. *T. mentagrophytes* arthroconidia formed on the human stratum corneum incubated *in vitro* using the method described by Knight (1973). The vegetative hyphae (A) applied on the sterile adhesive tape (Steri-Drape, 3M Co.), which had been applied repeatedly to the upper arm until it was nonadhesive, were incubated in a moisture chamber at 37°C for 9 days. Most hyphae became arthroconidia in several days (B). (Emyanitoff and Hashimoto, unpublished.)

humidity form abundant arthroconidia in several days (Fig. 6). Environmental factors that affect their induction and formation were investigated extensively in *T. mentagrophytes* (Chin and Knight, 1957; Miyaji and Nishimura, 1971; Bibel *et al.*, 1977; Hashimoto and Blumenthal, 1977; Weigl and Hejtmánek, 1980; Wright *et al.*, 1984; Fujita *et al.*, 1986). The *in vitro* arthroconidiation system developed in our laboratory has enabled us to investigate various properties of the purified arthroconidia of *T. mentagrophytes* (Hashimoto and Blumenthal, 1977, 1978). The system was subsequently modified by inoculating cells on a cellulose dialysis membrane, placed on the surface of Sabouraud dextrose medium instead of directly on the agar (Hashimoto and Blumenthal, 1978). Fujita *et al.* (1986) adopted a similar method to prepare arthroconidia of *T. mentagrophytes* which they used for testing their infectivity.

By far, the most critical factor affecting arthroconidiogenesis in the dermatophytes appears to be the temperature of incubation. Abundant arthroconidium formation was seen when *T. rubrum* was grown at 37°C, but not at 25°C, in brain heart infusion glucose liquid medium (Miyaji and Nishimura, 1971). Temperatures between 32 and 39°C (37°C optimal) also strongly favor the arthroconidiation of *T. mentagrophytes* in Sabouraud dextrose medium (Emyanitoff and Hashimoto, 1979; Vannini *et al.*, 1981). Although good mycelial growth occurs in the same medium at temperatures between 25 and 30°C, little or no arthroconidium formation occurs. Interestingly, the optimal temperature for the arthroconidiation in *T. mentagrophytes* slightly drops when protein-rich media are used. This may partially explain why good arthroconidiation of the dermatophytes occurs optimally within the skin or hairs. The relatively high temperatures seem to be required for the induction, rather than the progression, of arthroconidiation in the

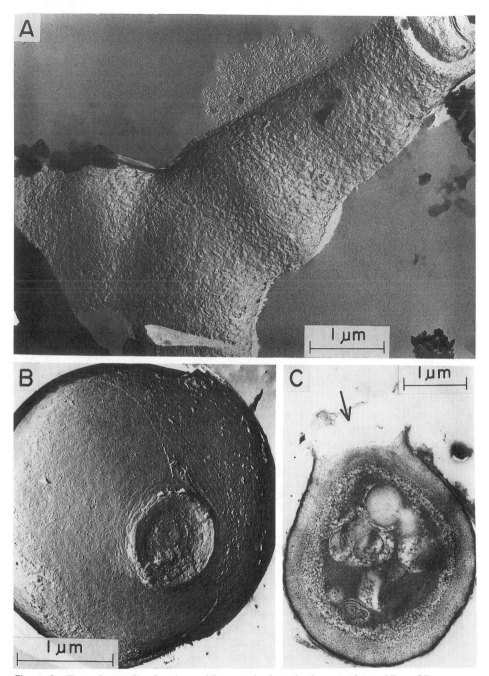

Figure 6. The surface profile of a microconidiogenous hypha and a dormant microconidium of *T. mentagrophytes*. Note the resistant rodlet patches cover the surface of the conidiogenous hypha (A) and the microconidium (B) except the abscission scar area, where concentrically arranged chitin fibrils are seen. The absence of the rodlet layer at the abscission scar is more clearly seen in the thin-sectioned microconidium (arrow in C). (B from Hashimoto *et al.*, 1976; A and C from Hashimoto, unpublished.)

dermatophytes. Once induced, shifting down to lower temperatures (30°C) does not seem to prevent arthroconidiation from completion (Emyanitoff and Hashimoto, 1979).

The pH of the medium appears to be vitally important for the arthroconidiation of certain dermatophytes. While both *T. mentagrophytes* and *T. tonsurans* undergo complete arthroconidiation at 37°C in the regular Sabouraud dextrose medium (pH 4.5), essentially no arthroconidia of *T. rubrum* are formed in the same medium. However, when the pH of the medium is raised slightly (pH 7.0), *T. rubrum* produces abundant arthroconidia in the same medium (Mock and Hashimoto, unpublished). It is possible that the pH of the local environment within the skin is an important factor in arthroconidium formation in *T. rubrum*.

In contrast to saprophytic conidium formation, high humidity is required for arthroconidiation in the dermatophytes. Miyaji *et al.* (1971a) observed that parasitic arthroconidia were more frequently found in the moist lesions of *T. rubrum* infection than in the dry lesions. For the production of *T. mentagrophytes* arthroconidia *in vitro*, high humidity must be maintained during the entire course of incubation at 37°C (Hashimoto and Blumenthal, 1977; Wright *et al.*, 1984).

Arthroconidium formation in certain strains of *T. mentagrophytes* is reported to be stimulated under increased carbon dioxide concentration (Chin and Knight, 1957; Bibel *et al.*, 1977; Emyanitoff and Hashimoto, 1979; Weigl and Hejtmánek, 1979, 1980; Wright *et al.*, 1984). It has been postulated that arthroconidium formation in the lesions may be stimulated by the diffusion of carbon dioxide through the skin (King *et al.*, 1976). Whether the stimulated arthroconidiogenesis is due to the presence of carbon dioxide itself, or reduced oxygen tension has been the subject of some controversy (Barrera, 1986). We believe that exogenous carbon dioxide is not the absolute requirement. The stimulatory effect of carbon dioxide more pronounced when arthroconidiogenesis is allowed to occur at suboptimal temperatures. *T. mentagrophytes* hyphae growing on Sabouraud dextrose agar at 37°C were able to complete arthroconidiation even when residual carbon dioxide was removed by the addition of NaOH pellets to the incubation jar (Emyanitoff and Hashimoto, 1979). Weigl and Hejtmánek (1980) found that the 8–10% concentration of carbon dioxide in the atmosphere is not necessarily a sufficient stimulus to arthroconidium formation for all arthroconidiogenic strains of *T. mentagrophytes*. Several strains that failed to form arthroconidia on an agar medium under increased carbon dioxide tension were found to do so on guinea pig skin.

Certain tissue components and metabolites are known to stimulate arthroconidium formation in certain dermatophytes at 37°C. These include cysteine for *T. rubrum* and *M. audouinii* (Rippon and Scherr, 1959), and 0.85% NaCl and 1% ornithine for *T. rubrum* (Miyaji *et al.*, 1971b). However, these compounds showed no stimulatory activity for the arthroconidiation of *T. mentagrophytes* at all concentrations tested (Emyanitoff and Hashimoto, 1979). At 37°C, *T. mentagrophytes* arthroconidia can be formed in a minimal salts medium containing a single amino acid if a suitable carbon source such as glucose or acetate is provided (Emyanitoff and Hashimoto, unpublished). Of 15 amino acids (25 mM) tested in this defined medium, 7(L-Ala, gly, L-leu, L-Ile, L-Ser, L-Val, and L-Arg) supported arthroconidium formation in *T. mentagrophytes*. No inorganic nitrogen compounds such as ammonium chloride or ammonium sulfate were effective. It could be significant that only certain amino acids supported arthroconidiogenesis in the dermatophytes (Emyanitoff, 1978). Probably, amino acids present in the skin may be utilized during arthroconidium formation *in situ*.

Interestingly, griseofulvin (0.5–1.0 μg/ml), clotrimazole (0.5–1.0 μg/ml), and amphotericin B (5.0–10.0 μg/ml), therapeutic agents for dermatophytosis, are stimulatory for arthroconidium formation in *T. mentagrophytes* when tested at 30°C (Emyanitoff and Hashimoto, 1979). Similar stimulatory effect on arthroconidiation was also observed with sublethal concentrations of amphotericin B and cycloheximide. This observation may be clinically significant because such antimycotics may promote the rapid conversion of some dermatophytes to arthroconidia if sufficiently fungicidal concentrations are not attained in some areas of the lesion. In such cases, relapse or exacerbation is expected to occur as long as the arthroconidia remain viable in the apparently healed lesions. It is conceivable that elevated temperatures, high humidity, high carbon dioxide tension, and some stimulatory metabolites may all contribute to the arthroconidiation of the dermatophytes in the skin.

4. BIOLOGICAL CHARACTERISTICS

Fungal conidia are a means of species reproduction and survival under adverse conditions. Some remain at their place of origin, preserving the species over an unfavorable period, while others are dispersed and initiate new vegetative hyphae to propagate the species. Conidia possess the universal properties of dormancy and resistance ascribed to spores. By controlling two distinctly different modes of conidiation, the dermatophytes seem to increase the chance of their survival and propagation in all environments in which they are able to inhabit.

4.1. Properties Contributing to Resistance

Several structural and physiological properties of dermatophytic conidia contribute to their resistance to harmful environmental stresses. Such properties of the dermatophytic conidia include: metabolic inactivity (Hashimoto et al., 1972), presence of the unique resistant layers in the conidial walls (Hashimoto et al., 1976; Pollack et al., 1983), elevated resistance to various physical, chemical, and biological agents (Hashimoto et al., 1972; Wu-Yuan and Hashimoto, 1977; Hashimoto and Blumenthal, 1978), and the specific requirements for germination (Hashimoto et al., 1972; Hashimoto and Blumenthal, 1977).

As reported by Hashimoto et al., (1976), the surface of the microconidia of T. mentagrophytes is coated with a layer consisting of patches of rodlets (Fig. 6B). Similar rodlet structures have been found on the surface of other fungal and bacterial spores. Similar rodlet structures are absent in the young vegetative hyphae and appear to be formed during microconidiogenesis of this dermatophyte (Fig. 6A). Interestingly, the rodlet patches are absent at the abscission scar area (Fig. 6B, C). This structural defect, as will be discussed later, is responsible for their susceptibility to lytic enzymes elaborated by soil microorganisms. In the case of T. mentagrophytes microconidia, the rodlet layer is predominantly made of a protein unusually resistant to various chemicals and hydrolytic enzymes (Hashimoto et al., 1976). Additionally, the rodlet layer contains some glucomannan, lipids, and a melaninlike pigment. The hydrophobicity of T. mentagrophytes microconidia is attributed to this rodlet layer. No similar rodlet structures are found in either macroconidia or arthroconidia of this fungus.

In contrast to the saprophytic conidia, T. mentagrophytes arthroconidia do not have the rodlet layer on their surface. Instead, they possess another type of resistant layer in the wall (Pollack et al., 1983). The mature arthroconidia of T. mentagrophytes are surrounded by the thick inner wall, which is synthesized de novo during arthroconidiogenesis, and by the outer wall, the remnant of the hyphal wall (Hashimoto et al., 1984). This inner wall is usually very thick (0.5μm in some parts). Chemically, two thirds of the isolated arthroconidial wall is chitinous in nature and is made of two distinct forms of chitin: chitinase-sensitive microfibrillar form and chitinase-resistant nonfibrillar form (Pollack et al., 1983). From the adsorption capacity of chitin and its analogues, this unique chitin-rich layer may be responsible for the resistance of the arthroconidia to antifungal agents (Hashimoto and Blumenthal, 1978).

The arthroconidia of T. mentagrophytes contain characteristic pigments not found in the vegetative hyphae or in the saprophytic conidia. They contain several types of carotenoids in their intracellular granules (Hashimoto et al., 1978). Although some dermatophytes produce naphthoquinone or its derivatives such as xanthomegnin during their saprophytic growth, carotenoids have not been reported in the dermatophytes. *T. mentagrophytes* var. interdigitale arthroconidia recently produced *in vitro* by Fujita *et al*. (1986) contain orange-colored pigments, although their chemical identity has not been elucidated. We also found that *T. tonsurans* produces carotenoids during arthroconidiogenesis (Mock and Hashimoto, unpublished). The occurrence of carotenoid pigments in the arthroconidia has raised an interesting possibility that they may be involved in the UV resistance of the arthroconidia. The evidence so far obtained suggests that this is the case. The function of carotenoid pigments in the arthroconidia of *T. mentagrophytes* remains unknown. *T. rubrum* arthroconidia produced *in vitro* do not seem to contain any carotenoid pigments (Mock and Hashimoto, unpublished). *T. mentagrophytes* produces perfectly normal arthroconidia under blue light irradiation but very little, if any, carotenoid accumulation occurred in such conidia (Mock and Hashimoto, 1985). Thus, carotenogenesis and arthroconidiogenesis appear to be two completely dissociable events in *T. mentagrophytes*. It is not known whether arthroconidia of the dermatophytes produce any carotenoid pigments during parasitization.

4.2. Resistance to Physical, Chemical, and Biological Agents

The dormant microconidia of *T. mentagrophytes* can remain viable for at least 2 years at 4°C under desiccated conditions. They are resistant to moderate heat and antifungal agents (Hashimoto *et al.*, 1972). Most microconidia are resistant to 50°C but start to lose their viability rather rapidly at 55°C (90% in 8 min). In contrast to microconidia, the germinated microconidia or hyphae lose 99% of their viability in 2 min. The microconidia of *T. mentagrophytes* are no more sensitive to UV irradiation than the hyphae. The resistance of the dormant microconidia to chemicals or lytic enzymes may be attributed to the hydrophobic nature of the spore surface. When hydrated, however, *T. mentagrophytes* microconidia are surprisingly sensitive to lytic enzymes (chitinase or β1,3-glucanase) produced by soil microorganisms. The susceptibility of the mature microconidia to the lytic enzymes appears to be due to the absence of the resistant rodlet patches at the abscission scar areas (Fig. 5B, C). It is very unlikely that microconidia of this dermatophyte can survive long in the soil where microbial lytic enzymes are abundant.

Dermatophytes parasitizing in skin scales or clinical specimens can tolerate a moderate heat treatment (Lorincz and Sun, 1963; Sinski *et al.*, 1979, 1980). The arthroconidia of *T. mentagrophytes* produced on the laboratory media are less resistant to heat when compared with the saprophytic conidia. More than 90% of dormant arthroconidia become nonviable within 5 min of exposure to 50°C. Even at 48°C, approximately 50% of the arthroconidia are inactivated within 10 min (Hashimoto and Blumenthal, 1978). This heat resistance is significantly lower than the range reported for the dormant cells surviving in the tissue fragments. *T. mentagrophytes* infecting guinea pig skin scales can remain viable even exposed to 60°C for 4 hr (Sinski *et al.*, 1980). The most plausible explanation for this discrepancy is that the keratinous tissue protein surrounding the infective fungal elements serves as a protective shield. It may be pertinent to recall that the coats of bacterial spores, which are highly resistant to adverse environmental stresses, are made of proteins similar to keratin.

In contrast to lack of thermoresistance, the arthroconidia of *T. mentagrophytes* are remarkably resistant to chilling and freezing. At low temperatures (4–10°C), more than 90% of the arthroconidia can survive for more than 2 months. At −20°C, more than 95% of the arthroconidia remain viable for 1 year, and approximately 65% for 2 years (Hashimoto and Blumenthal, 1978).

The pure arthroconidia of *T. mentagrophytes* are fairly susceptible to desiccation (Hashimoto and Blumenthal, 1978). The air-dried or lyophilized arthroconidia lose their viability rapidly regardless of the temperature employed during storage (Hashimoto and Blumenthal, 1978). However, the resistance dramatically increases when the arthroconidia are first desiccated in the presence of proteins such as casein or albumin and then stored at various temperatures

(Hashimoto and Blumenthal, 1978). Again, in this case, the protective effect of the protein coating the conidial surface is evident. This may account for the high degree of resistance to desiccation that characterizes the parasitic conidia present in the skin scales (Lorincz and Sun, 1963; Dvořák *et al.*, 1968; Sinski *et al.*, 1979, 1980).

Dormant arthroconidia are resistant to common antimycotics such as clotrimazole, griseofulvin, miconazole nitrate, and nystatin (Hashimoto and Blumenthal, 1978). This resistance is speculated to be due to the relatively impervious nature of the arthroconidial wall. The resistance to antimycotics is completely lost upon germination of the arthroconidia.

4.3. Germination Requirements

When exposed to a favorable environment, conidia undergo a change from the dormant and resistant state to the metabolically active form. This process is broadly defined as germination. For fungal conidia, the accepted criterion of germination is an emergence of recognizable germ tubes or short hyphae. During germination, various properties uniquely associated with the dormant conidia are lost and their level of resistance drops to that of the vegetative hyphae. Conidia of each fungal species usually require a specific compound(s) for the initiation of germination. These compounds are known as germination inducers or simply germinants. They are relatively simple organic compounds often found in nature. Amino acids, sugars, and nucleosides are some of the most common germination inducers for conidia. Sometimes, simple inorganic salts or even water can induce germination in certain conidia.

Germination of *T. mentagrophytes* microconidia is triggered by several amino acids or dipeptides, most effectively by L-leucine (Hashimoto *et al.*, 1972). Amino acid-induced germination of the arthroconidia occurs optimally at 35°C at pH 6–6.5. No sugars, organic acids, purines, pyrimidines, nucleosides, or inorganic compounds are effective. No germination of *T. mentagrophytes* microconidia occurs in buffers or water, even when incubated at 37°C for an extended period of time. This germination requirement is important when one considers the infectivity of the microconidia. While the fastidious germination requirement may aid the survival of the conidia in nature, their infectivity would be greatly reduced especially when the required germinants are not readily available at the site of infection. This may account for the difficulty in producing an experimental dermatophytosis with the microconidia of the dermatophytes. If the sweat, serous exudates, or normal or abnormal skin metabolites provide the required germinant, the conidia may germinate and initiate infection.

Freshly harvested arthroconidia of *T. mentagrophytes* germinate only in a rich complex medium such as Sabouraud dextrose broth or vitamin-free casamino acids. When activated, however, they are able to germinate rapidly over a wide range of pH (5.5–8.0, optimal 6.5) and temperature (20–39°C, optimal 37°C) in the presence of single amino acids or oligopeptides (Hashimoto and Blumenthal, 1977). The activation of *T. mentagrophytes* arthroconidia is achieved by storing in distilled water at 24°C for 24 hr or by a brief exposure to sublethal doses of heat (45°C for 10 min). The fully activated arthroconidia can germinate spontaneously in water at 37°C, requiring no exogenous nutrients. This may indicate that the activated *T. mentagrophytes* arthroconidia can be highly infectious for the normal hydrated skin. The significantly higher infectivity of *T. mentagrophytes* arthroconidia, compared with that of the microconidia, for experimental animals (Fujita and Matsuyama, 1987) may well be due to this difference in the germination requirement.

Germination of *T. mentagrophytes* arthroconidia is inhibited by imidazole antifungal agents (Scott *et al.*, 1984). In most instances, higher concentrations are required to inhibit arthroconidium germination than are obtained in *in vitro* MIC tests. Ketoconazole, even at higher concentrations shows little effect on the germination process of arthroconidia (Scott *et al.*, 1984).

5. INFECTIVITY AND PATHOGENICITY

There is little doubt that the transmission of dermatophytosis is mediated by both sapro-phytic and parasitic conidia. Although no well documented evidence is available, saprophytic conidia are thought to act only occasionally as an inoculum for man and animals (Austwick, 1966). Most epidemiological problems associated with dermatophytosis are, however, caused by infective arthroconidia.

5.1. Infections Mediated by Saprophytic Conidia

The infectious nature of saprophytic conidia has been demonstrated experimentally on numerous occasions. These saprophytic conidia have been used to produce experimental derma-tophytosis in animals (Chittasobhon and Smith, 1979; Poulain *et al.*, 1980; Tagami *et al.*, 1982; Okuwa and Horio, 1986; Hay *et al.*, 1988). Chittasobhon and Smith (1979), for example, reported the successful production of experimental tinea lesions in guinea pigs using microconidia or macroconidia of *T. mentagrophytes, T. rubrum, T. tonsurans, M. canis, M. gypseum, M. persicolar*, and *E. floccosum*. The number of saprophytic conidia used to produce experimental dermatophytoses ranged from 1×10^4 to 15×10^6. Almost without exception, the experimental dermatophytoses reported in the literature were produced by inoculating a mixture of both hyphae and conidia, rather than pure conidia.

For successful production of experimental dermatophytosis in animals, superficial skin abrasion is usually required prior to application of the spores (Fig. 7). Such scarification appears

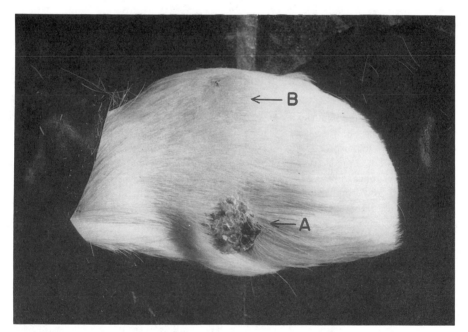

Figure 7.   Experimental dermatophytosis produced in a guinea pig by inoculating 1×10^5 dormant microconidia of *T. metagrophytes*. Active infection developed only when an abrasive pretreatment was given to the skin prior to inoculation (A). No infection resulted if inoculated nonabrasively (B). This abrasion requirement may not be necessary if high humidity is maintained at the infected site for the first 24–48 hr. by occlusive dressing (Watanabe, 1987). (Original photograph by Hashimoto, unpublished.)

to serve two purposes: to provide the microconidia with enough moisture and nutrients required for germination, and to allow the microconidia to become wedged deep in the stratum corneum where they normally parasitize. If sufficient moisture is kept at the inoculated site using an occlusive dressing for the first 24 hr, it is possible to incite dermatophytosis nonabrasively on the depiliated skin (Watanabe, 1987).

Experimental dermatophytosis has been produced in human volunteers using primarily microconidia (Reinhardt et al., 1974). Although rare, dermatophytosis occurs among laboratory personnel due to the exposure to *T. mentagrophytes* microconidia. Generalized dermatophytosis can be experimentally produced in animals by the systemic administration of dermatophytes (Van Cutsem and Janssen, 1984; Watanabe, 1987).

5.2. Infections Mediated by Parasitic Conidia

It has long been suspected that the transmission of dermatophytosis from one individual to another is mediated by the arthroconidia surviving in fallen hairs and loosened squames (Klingman, 1952; Gotz, 1959; Georg, 1960). This suspicion has gained further support from the following two observations: (1) arthroconidia are the only resistant form of cells produced by the dermatophytes during parasitization, and (2) viable dermatophytes can be recovered from clinical specimens (skin, hair, and nails) stored for extended periods of time under adverse conditions (Sinski et al., 1979, 1980). Many surveys have repeatedly shown that dermatophytes can be isolated from combs, brushes, the back of theater seats, and caps. They are also frequently isolated from bed linens, towels, undergarments, and jockstraps. Locker-room and shower floors trod over by "athlete feet" are often contaminated with skin debris containing infective arthroconidia.

Airborne transmission of infection with many zoophilic tineas is common (English, 1972). The dissemination of dermatophytosis via arthroconidia has been suspected among experimentally infected animals as well as in dermatological clinics (Midgley and Clayton, 1972). Although the infectivity of the parasitically formed arthroconidia has been demonstrated in experimental animals and human volunteers (Klingman, 1952; Smith, 1982, cited by Watanabe, 1987), it is only recently that the arthroconidia prepared *in vitro* have been proven to be infective (Fujita and Matsuyama, 1987; Watanabe, 1987). Using the arthroconidia of *T. mentagrophytes* prepared in the laboratory, Fujita and Matsuyama (1987) have produced in guinea pigs an experimental tinea pedis clinically and histopathologically similar to human infections. They fixed a paper disk containing arthroconidia on the plantar part of the hind foot with an adhesive elastic tape for 7 days. They also determined 50% minimal foot infection dose ($MFID_{50}$) for both the anthropophilic and zoophilic strains of *T. mentagrophytes* they employed. They found that 280 arthroconidia are sufficient to infect 50% of inoculated feet by the anthropophilic strain, and 80 by the zoophilic strain. This figure is significantly lower than the number of microconidia usually required to produce experimental infections in animals.

To my knowledge, this is the first convincing evidence that dermatophytic arthroconidia are indeed infectious and are capable of transmitting infection. The tinea pedis produced in guinea pigs by Fujita and Matsuyama (1987) is significant in three respects. First, unlike most previous experimental dermatophytoses, the infection was produced in the animal nonabrasively, although the possible presence of submicroscopic wounds in the skin could not be precluded. Second, the infections that developed in the guinea pig feet were quite similar to human tinea pedia histologically, clinically, and in terms of the chronicity. Most experimental dermatophytoses produced in animals usually undergo spontaneous cure in several weeks. Third, the infection could be incited with relatively small numbers of arthroconidia proving that arthroconidia are most infectious propagules of the dermatophytes and are the major mediators of disease transmission.

It is pertinent to recall that the aged or activated *T. mentagrophytes* arthroconidia require no special nutrients for germination (Hashimoto and Blumenthal, 1977). They are fully capable of transforming to infective hyphae when only proper temperatures and high humidity are provided. The macerated skin appears to provide an excellent environment for germination of the infective arthroconidia. These germination characteristics of *T. mentagrophytes* arthroconidia may explain why the parasitic conidia of the dermatophytosis are so infectious and relapses are so frequent in dermatophytosis.

The nonabrasive animal model developed by Fujita and Matsuyama (1987) is a useful experimental system to investigate the mechanism of penetration of the normal skin barrier by infective arthroconidia in order to establish active infections. It also appears to be a useful model to test prospective new antimycotics or therapeutic methods which cannot be accurately assessed in the conventional models where rapid spontaneous cures usually occur at high frequencies.

6. MECHANISMS OF DISEASE INITIATION IN DERMATOPHYTOSIS

It is clear from the preceding discussion that dermatophytosis is a communicable disease. It is transmitted to humans by infective saprophytic or parasitic conidia. In either case, the transmission of infection is by direct contact of human and animal via contaminated fomites, so that the infective conidia are passed on in large numbers within the protective sheaths of keratin. In order for an active infection to occur, these conidia lodging on the skin surface should be able to penetrate the protective surface layer. The resistance of the conidia against adverse environmental stresses would certainly increase the chance of their survival on the host's skin surface.

The next phase of infection is the penetration of the surface layer of the skin by invasive germlings. For this, infective conidia require high humidity and germinants for germination. If the conidia have fastidious germination requirements, their change for successful tissue invasion is considerably reduced. The surface layer of the skin contains a variety of chemical agents derived from sweat, sebaceous glands, and the keratinization process. There are also many metabolites produced by surface bacteria utilizing skin lipids and proteins. These agents or metabolites may sufficiently induce germination of the conidia. On the other hand, certain cutaneous lipids or fatty acids may act as inhibitors for conidial germination. If the conidia require no special compounds for germination, their infectivity will be greatly enhanced.

To establish an active infection in the normal skin, the germinated conidia must penetrate the keratin layer and reach the ultimate site of parasitization. This step of infection may be bypassed if the conidia lodge in small orifices or abrasions present in the skin. If this occurs, the conidia may enter within the keratinous layer, directly initiating the parasitic activity almost immediately. The microconidia of *T. quinckeanum* infecting the scarified skin of mice first adhere specifically to keratinocytes before germination takes place (Hay *et al.*, 1988). Hay *et al.* (1988) estimated that germination of microconidia occurs with 24 hr.

Usually the emerging hyphae are capable, either mechanically or enzymatically, of penetrating the intact keratin. Only those germinated conidia which penetrate the intact keratin can initiate the full parasitic activity. In *T. quinckeanum*, which produces dermatophytosis in mice, emerging hyphae tend to enter hair follicles without hair shaft penetration (Hay *et al.*, 1988), in contrast to *Microsporum* species which almost always invade hairs in the early stage of infection.

The initial host reaction to the invasion of germinated microconidia includes infiltration of neutrophils and the epidermal edema. Neutrophils and, to a lesser extent, monocytes accumulate in the epidermal filtrates probably in response to the chemotactic components of the dermatophytic cells (Tagami *et al.*, 1982; Hay *et al.*, 1988). Although little attention has been paid previously, these phagocytes accumulating in the epidermal filtrates are believed to play an important role in the initial host defense to the invasion of dermatophyte germlings. It has been suggested that the dermatophytes are damaged by the generation by neutrophils of toxic oxidative products (Hay *et al.*, 1988). Whether a similar phagocytic cellular response to dermatophytic conidia occurs in the early phase of nonabrasively produced dermatophytosis remains to be seen. Relatively early in the course of infection, increased epidermal proliferation, not dependent on T cell activation, occurs. This increased epidermal proliferation is also implicated in host defense against dermatophytes (Hay *et al.*, 1988).

The ultimate establishment of dermatophytosis and the subsequent course of the infection are influenced by other host factors such as immune state, previous exposure to dermatophytes, or inherent genetic predisposition (Ahmed, 1982; Hay *et al.*, 1988). Antibody, however, appears to have no protective role in dermatophytosis.

Some dermatophytes produce specialized hyphae that facilitate their invasion into the keratinous tissues. For example, some geophilic and zoophilic dermatophytes produce "infection structures" utilized for hair penetration by invading hyphae (English, 1968; Rebell and Taplin, 1970; Kanbe and Tanaka, 1982). *M. gypseum* produces specialized flattened fronds of hyphae or eroding mycelium at the penetration site, and a perforating organ from these fronds to penetrate the hair (Kanbe and Tanaka, 1982). Lomasomes abundantly present in the perforating organ are implicated in the secretion of keratinase (Kanbe and Tanaka, 1982). So far, these specialized infection structures have been demonstrated only in the dermatophytes infecting the detached hair *in vitro* but not in natural diseases (Kanbe and Tanaka, 1981; Rippon, 1988). Other species erode the hair cuticle but may not form perforating structures (Rebell and Taplin, 1970). To my knowledge, no studies have investigated the invasive process of the host tissue by germinating arthroconidia.

Dermatophytes are known to produce various hydrolytic enzymes extracellularly. These include proteases, lipases, phosphatases, nucleases, and glucosidases (Calvo *et al.*, 1986). Among these enzymes, proteolytic enzymes (keratinase, collagenase, and elastase) are most frequently implicated in the pathogenesis of dermatophytes (Rippon, 1967, 1968; Grappel and Blank, 1972; Kunert, 1972; Meevootisom and Niederpruem, 1979; Calvo *et al.*, 1985; Asahi *et al.*, 1985; Skorepova and Hauck, 1987; Apodaca and McKerrow, 1989; Tsuboi *et al.*, 1989). Some of these enzymes are involved in (1) facilitating the penetration of germ tubes or hyphae into keratinous epidermal tissues, (2) securing nutrients for fungal growth, and (3) eliciting local inflammatory immune response. The ability of the dermatophytes to produce these hydrolytic enzymes *in vivo* may be an important virulence factor of the dermatophytes. The majority of these studies have been performed by using either vegetative hyphae or hyphal culture filtrates. It is not clear at this time whether these hydrolytic enzymes play any significant role in disease initiation in dermatophytosis. The precise tissue penetration mechanism by the infective dermatophytic conidia remains to be elucidated.

Conidia are widely known to play an important role in the dissemination of mycotic infections in both animals (Garrett, 1966) and plants (Austwick, 1966). The process of tissue invasion by fungi has been investigated by a number of animal and plant mycologists. Some plant pathogens are known to develop specialized infection structures by which they penetrate and colonize host tissue (Staples and Macko, 1980; Staples and Hoch, 1987). These structures consist of an appressorium and infectious hyphae differentiating from developing germ tubes. Apparently, certain plant pathogenic fungi utilize appressoria for sensing possible points of entry and associated infectious hyphae for actual penetration. Similar mechanisms of tissue penetration have not been detected in animal pathogens.

Since extracellular lytic enzymes have been postulated to be the main mechanism that facilitates tissue penetration, many attempts have been made to develop a sensitive method that determines the ability of fungal spores to produce various degrading enzymes at various times during germination. Recently, Hagerman *et al.* (1985) reported a sensitive plate assay method for lytic enzymes (protease, cellulase, pectinase) of germinating conidia. By applying this method, they could demonstrate that cellulase and protease were present before any conidia of *Botrytis cinerea* germinated (Hagerman *et al.*, 1985). Pectin lyase was detected only after the conidia had germinated. This type of *in vitro* assay system is believed to be extremely useful for elucidating the disease-initiating mechanisms of human pathogens mediated by infective conidia. In dermatophytosis, the availability of pure saprophytic and parasitic conidia and the animal model that clinically resembles human infection have made the study of the pathogenic mechanism more feasible and worth challenging.

ACKNOWLEDGMENT.   I thank Dr. Jordan Pollack for helpful discussions during the preparation of the manuscript.

8. REFERENCES

Ajello, L., and Padhye, A. A., 1974, Dermatophytes and agents of superficial mycoses, in *Manual of Clinical Microbiology* 2nd ed (E. H. Lennette, E. H. Spaulding, and J. O. Truant, eds.) American Society for Microbiology, Washington, D.C., pp. 469–481.

Allen, A. M., and King, R. D., 1978, Occlusion, carbon dioxide and fungal skin infections, *Lancet* vol. 1 for 1978 (No. 8060): 360–362.

Ahmed, A. R., 1982, Immunology of human dermatophyte infections, *Arch Dermatol.* **118**:521–525.

Apodaca, G., and McKerrow, J. H., 1989, Regulation of *Trichophyton rubrum* proteolytic activity, *Infect. Immun.* **57**:3081–3090.

Asahi, M., Lindquist, R., Fukuyama, K., Apodaca, G., Epstein, W., and McKerrow, J. H., 1985, Purification and characterization of major extracellular proteinases from *Trichophyton rubrum*, *Biochem. J.* **232**:139–144.

Austwick, P. K., 1966, The role of spores in the allergies and mycoses of man and animals, in: *The Fungus Spore* (M. F. Madelin, ed.), Butterworths, London, pp. 321–333.

Barrera, C. R., 1983, Formation and ultrastructure of *Mucor rouxii* arthrospores, *J. Bacteriol* **155**:886–895.

Barrera, C. R., 1986, Formation and germination of fungal arthroconidia, *CRC Critical Rev. Microbiol.* **12**: 271–292.

Bibel, D. J., Crumrine, D. A., Yee, K., and King, R. D, 1977, Development of arthrospores of *Trichophyton mentagrophytes*, *Infect. Immun.* **15**:958–971.

Calvo, M. A., Bruguera, T., Cabanes, F. J., Calvo, R. M., Trape, J., and Abarca, L., 1985, Brief communication: Extracellular enzymatic activities of dermatophytes, *Mycopathologia* **92**:19–22.

Calvo, M. A., Trape, J., Abarca, L., Cabanes, F. J., Calvo, R. M., and Bruguera, T., 1986, Variability of biochemical characteristics in strains of *Trichophyton mentagrophytes*, *Mycopathologia* **93**:137–139.

Chin, B., and Knight, S. G., 1957, Growth of *Trichophyton rubrum* in increased carbon dioxide tensions, *J. Gen. Microbiol.* **16**:642–646.

Chittasobhon, N., and Smith, M. B., 1979, The production of experimental dermatophyte lesions in guinea pigs, *J. Invest. Dermatol.* **73**:198–201.

Cole, G. T., and Samson, R. S., 1979, *Patterns of Development in Conidial Fungi*, Pitman, London, pp. 106–110.

Dvořák, J., Hubalekz, Z., and Otčenášek, M., 1968, Survival of dermatophytes in human skin scales, *Arch. Dermatol.* **98**:540–542.

Emmons, C. W., 1934, Dermatophytes: Natural grouping based on the form of the spores and accessory organs, *Arch. Dermatol. Syphilol.* **30**:337–362.

Emyanitoff, R. G., 1978, Cytology and physiology of arthrospore formation in the dermatophyte *Trichophyton mentagrophytes*, Ph.D. dissertation, Loyola University of Chicago.

Emyanitoff, R. G. and Hashimoto, T., 1979, The effects of temperature, incubation atmosphere, and medium composition on arthrospore formation in the fungus *Trichophyton mentagrophytes*, *Can. J. Microbiol.* **25**:362–366.

English, M. P., 1968, The developmental morphology of the perforating organs and eroding mycelium of dermatophytes, *Sabouraudia* **6**:218–229.

English, M. P., 1972, The epidemiology of animal ringworm in man, *Br. J. Dermatol.* **86**(Suppl. 8):78–87.

Fujita, S., and Matsuyama, T., 1987, Experimental tinea pedia induced by non-abrasive inoculation of *Trichophyton mentagrophytes* arthrospores on the plantar part of a guinea pig foot, *J. Med. Vet. Mycol.* **25**: 203–213.

Fujita, S., Matsuyama, T., and Sato, Y., 1986, A simple and reliable culturing method for production of arthrospores by dermatophytes, *Jpn. J. Med. Mycol.* **27**:175–181 (in Japanese).

Garrett, S. D., 1966, Spores as propagules of disease, in: *The Fungus Spore* (M. F. Madelin, ed.), Butterworths, London, pp. 309–318.

Georg, L. K., 1960, Epidemiology of the dermatomycoses, sources of infection, modes of transmission and epidemicity, *Ann. N.Y. Acad. Sci.* **89**:69–77.

Gotz, H., 1959, Zur Morphologie der Pilzelemente im stratum corneum bei tinea (Epidermophytia) pedis, manus et inguinalis, *Mycopathol. Mycol. Appl.* **12**:124–140.

Grappel, S. F., and Blank, F., 1972, Role of keratinases in dermatophytosis: I. Immune response of guinea pigs infected with *Trichophyton mentagrophytes* and guinea pigs immunized with keratinases, *Dermatologia* **145**:245–255.

Hagerman, A. E., Blau, D. M., and McClure, A. L., 1985, Plate assay for determining the time of production of protease, cellulase, and pectinases by germinating fungal spores, *Anal. Biochem.* **151**:334–342.

Hashimoto, T., and Blumenthal, H. J., 1977, Factors affecting germination of *Trichophyton mentagrophytes* arthrospores, *Infect Immun.* **18**:470–486.

Hashimoto, T., and Blumenthal, H. J., 1978, Survival and resistance of *Trichophyton mentagrophytes* arthrospores, *Appl. Environ. Microbiol.* **35**:274–277.

Hashimoto, T., Wu, C. D. R., and Blumenthal, H. J., 1972, Characterization of L-leucine-induced germination of *Trichophyton mentagrophytes* microconidia, *J. Bacteriol.* **112**:967–978.

Hashimoto, T., Wu-Yuan, C. D., and Blumenthal, H. J., 1976, Isolation and characterization of the rodlet layer of *Trichophyton mentagrophytes* microconidial wall, *J. Bacteriol.* **127**:1543–1549.

Hashimoto, T., Pollack, J. H., and Blumenthal, H. J., 1978, Carotenogenesis associated with arthrosporulation of *Trichophyton mentagrophytes*, *J. Bacteriol.* **136**:1120–1126.

Hashimoto, T., Emyanitoff, R. G., Mock, R. C., and Pollack, J. H., 1984, Morphogenesis of arthroconidiation in the dermatophyte *Trichophyton mentagrophytes* with special reference to wall ontogeny, *Can. J. Microbiol.* **30**:1415–1421.

Hay, R. J., Calderon, R. A., and Mackenzie, C. D., 1988, Experimental dermatophytosis in mice: Correlation between light and electron microscopic changes in primary, secondary and chronic infections, *Br. J. Exp. Pathol.* **69**:703–715.

Hironaga, M., Okazaki, N., Saito, K., and Watanabe, S., 1983, *Trichophyton mentagrophytes* granulomas: Unique systemic dissemination to lymph node, testes, vertebrae, and brain, *Arch. Dermatol.* **119**:482–490.

Kanbe, T., and Tanaka, K., 1982, Ultrastructure of the invasion of human hair in vitro by the keratinophilic fungus *Microsporum gypseum*, *Infect. Immun.* **38**:706–715.

Kane, J., and Krajden, S., 1983, Dermatophytosis, in: *Occupational Mycoses* (A. F. DiSalvo, ed.), Lea & Febiger, Philadelphia, pp. 143–182.

Kane, J., Leavitt, E., Summerbell, R. C., Krajden, S., and Kasatiya, S. S., 1988, An outbreak of *Trichophyton tonsurans* dermatophytosis in a chronic care institution for the elderly, *Eur. J. Epidemiol.* **4**:144–149.

King, R. D., Dillavou, C. L., Greenberg, J. H., Jeppsen, J. C., and Jaegar, J. S., 1976, Identification of carbon dioxide as a dermatophyte inhibitory factor produced by *Candida albicans*, *Can. J. Microbiol.* **22**:1720–1727.

Klingman, A. M., 1952, The pathogenesis of tinea capitis due to *Microsporum audounii* and *Microsporum canis*. I. Gross observations following the inoculation of humans, *J. Invest. Dermatol.* **18**:231–246.

Knight, A. G., 1973, Culture of dermatophytes upon stratum corneum, *J. Invest. Dermatol.* **59**:427–431.

Kunert, J., 1972, Keratin decomposition by dermatophytes: Evidence of the sulphitolysis of the protein, *Experientia* **28**:1025.

Lorincz, A. L., and Sun, S. H., 1963, Dermatophyte viability at modestly raised temperatures, *Arch. Dermatol.* **88**:393–402.

Meevootisom, V., and Niederpruem, D. J., 1979, Control of exocellular proteases in dermatophytes and especially *Trichophyton rubrum*, *Sabouraudia* **17**:91–106.

Midgley, G., and Clayton, Y. M., 1972, Distribution of dermatophytes and *Candida* spores in the environment, *Br. J. Dermatol.* **86**(Suppl. 8):69–77.

Mitchell, T. G., 1988, Dermatophytosis and other cutaneous mycoses, in: *Zinsser Microbiology* (W. K. Joklik, H. P. Willett, D. B. Amos, and C. M. Wilfert, eds.), Appleton & Lange, Norwalk, Conn., pp. 922–929.

Miyaji, M., and Nishimura, K., 1971, Studies on arthrospore of *Trichophyton rubrum* (I), *Jpn. J. Med. Mycol.* **12**:18–23.

Miyaji, M., Nishimura, K., and Kariya, H., 1971a, Studies on arthrospore of *Trichophyton rubrum* (II): Relationship between the types of eruption and parasitic forms of *Trichophyton rubrum*, *Jpn. J. Med. Mycol.* **12**:81–85 (in Japanese).

Miyaji, M., Fujiwara, K., and Nishimura, K., 1971b, Studies on arthrospore of *Trichophyton rubrum* (III): Effects of amino acids, salt and pH on the spherical cell formation, *Jpn. J. Med. Mycol.* **12**:200–205 (in Japanese).

Mock, R. C., and Hashimoto, T., 1985, Effect of visible light on carotenogenesis in arthroconidiating *Trichophyton mentagrophytes*, *Arch. Microbiol.* **140**:271–275.

Novick, N. L., Tapia, L., and Bottone, E. J., 1987, Invasive *Trichophyton rubrum* infection in an immuno-compromised host: Case report and review of the literature, *Am. J. Med.* **82**:321–325.

Okuwa, T., and Horio, T., 1986, The inhibitory effect of PUVA on the immunity of experimental dermatophytosis in guinea pigs, *Arch. Dermatol. Res.* **278**:320–323.

Pollack, J. H., Lange, C. F., and Hashimoto, T., 1983, Nonfibrillar chitin associated with walls and septa of *Trichophyton mentagrophytes* arthrospores, *J. Bacteriol.* **154**:965–975.

Poulain, D., Tronchin, G., Vernes, A., Delabre, M., and Biguet, J., 1980, Experimental study of resistance to infection by *Trichophyton mentagrophytes*: Demonstration of memory skin cells, *J. Invest. Dermatol.* **74**:205–209.

Rebell, G., and Taplin, D., 1970, *Dermatophytes: Their Recognition and Identification*, rev. ed., University of Miami Press, Coral Gables.

Reinhardt, J. H., Allen, A. M., Gunnison, D., and Akers, W. A., 1974, Experimental human *Trichophyton mentagrophytes* infections, *J. Invest. Dermatol.* **63**:419–422.

Rippon, J. W., 1967, Elastase: Production by ringworm fungi, *Science* **157**:947.

Rippon, J. W., 1968, Extracellular collagenase from *Trichophyton schoenleinii*, *J. Bacteriol.* **95**:43–46.

Rippon, J. W., 1988, *Medical Mycology*, 3rd ed., Saunders, Philadelphia.

Rippon, J. W., and Scherr, G. H., 1959, Induced dimorphism in dermatophytes, *Mycologia* **51**:902–914.

Scott, E. M., Gorman, S. P., and Wright, L. R., 1984, The effect of imidazoles on germination of arthrospores and microconidia of *Trichophyton mentagrophytes*, *J. Antimicrob. Chemother.* **13**:101–110.

Sinski, J. T., and Kelley, L. M., 1987, A survey of dermatophytes isolated from human patients in the United States from 1982 to 1984, *Mycopathologia* **98**:35–40.

Sinski, J. T., Wallis, B. M., and Kelley, L. M., 1979, Effect of storage temperature on viability of *Trichophyton mentagrophytes* in infected guinea pig skin scales, *J. Clin. Microbiol.* **10**:841–843.

Sinski, J. T., Moore, T. M., and Kelley, L. M., 1980, Effect of moderately elevated temperatures on dermatophyte survival in clinical and laboratory-infected specimens, *Mycopathologia* **71**:31–35.

Skorepova, M., and Hauck, H., 1987, Extracellular proteinases of *Trichophyton rubrum* and the clinical picture of tinea, *Mykosen* **30**:25–27.

Staples, R. C., and Hoch, H. C., 1987, Infection structures—form and function, *Exp. Mycol.* **11**:163–169.

Staples, R. C., and Macko, V., 1980, Formation of infection structures as a recognition response in fungi, *Exp. Mycol.* **4**:2–16.

Tagami, H., Natsume, N., Aoshima, T., Inoue, F., Suehira, S., and Yamada, M., 1982, Analysis of transepidermal leukocyte chemotaxis in experimental dermatophytosis in guinea pigs, *Arch. Dermatol. Res.* **273**:205–217.

Tsuboi, R., Ko, I., Takamori, K., and Ogawa, H., 1989, Isolation of a keratinolytic proteinase from *Trichophyton mentagrophytes* with enzymatic activity of acidic pH, *Infect. Immun.* **57**:3479–3483.

Van Cutsem, J., and Janssen, P. A. J., 1984, Experimental systemic dermatophytosis, *J. Invest. Dermatol.* **83**:26–31.

Vannini, G. L., Dall'olio, G., and Scannerini, S., 1981, Effects of termperature on microconidium ontogeny in *Trichophyton mentagrophytes*: A tentative interpretation based on ultrastructural aspects, *Microbiologia* **4**:141–151.

Watanabe, S., 1987, Animal models of cutaneous and subcutaneous mycoses, in: *Animal Models in Medical Mycology* (M. Miyaji, ed.), CRC Press, Boca Raton, pp. 53–78.

Weigi, E., and Hejtmánek, M., 1979, Differentiation of *Trichophyton mentagrophytes* arthrospores controlled by physical factors, *Mykosen* **22**:167–172.

Weigl, E., and Hejtmánek, M., 1980, Arthrosporogenesis in *Trichophyton mentagrophytes* on agar medium and in guinea pig skin, *Mykosen* **23**:486–493.

Weigl, E., and Hejtmánek, M., 1984, Attempts of inducing dimorphism in *Trichophyton rubrum*, *Acta Univ. Palacki. Olomuc. Fac. Med.* **107**:31–42.

Weigl, E., and Hejtmánková, N., 1984, Vegetative segregation and arthrosporogenesis in *Trichophyton mentagro-phytes*, *Acta Univ. Palacki. Olomuc. Fac. Med.* **107**:23–30.

Wright, L. R., Scott, E. M., and Gorman, S. P., 1984, Spore differentiation in a clinical strain of *Trichophyton mentagrophytes*, *Microbios* **39**:87–93.

Wu-Yuan, C. D., and Hashimoto, T., 1977, Architecture and chemistry of microconidial walls of *Trichophyton mentagrophytes*, *J. Bacteriol.* **129**:1584–1592.

II

Fungal Spore Products and Pathogenesis

9

Melanin Biosynthesis

Prerequisite for Successful Invasion of the Plant Host by Appressoria of Colletotrichum and Pyricularia

Yasuyuki Kubo and Iwao Furusawa

1. INTRODUCTION

A prerequisite for invasion of host plants by some plant pathogenic fungi is a differentiation of nonpathogenic spores into germinating spores that are accompanied by formation of the appressorium, which functions as an infection structure.

Recent studies on appressorial penetration by *Colletotrichum* species indicated that appressorium formation is accompanied by metabolic changes that are essential for penetration of the host plant. Appressorial melanization of *Colletotrichum lagenarium* and *C. lindemuthianum* (Kubo *et al.*, 1982a, 1983; Wolkow *et al.*, 1983), cellulase synthesis by *C. lagenarium* (Suzuki *et al.*, 1981, 1982a), cutinase synthesis by *C. gloeosporioides* (Dickman and Patil, 1986), secretion of cellulase and penetration peg protrusion of *C. lagenarium* (Katoh *et al.*, 1988) were essential factors for successful penetration of the host by appressoria.

The significance of appressorial melanization has been poorly understood; however, recent studies with albino mutants of *C. lagenarium* in which colorless appressoria were unable to penetrate the host plant have enhanced our understanding of melanization and the infection process (Kubo *et al.*, 1982a). Appressorial melanization has been assumed to contribute to an increased endurance under adverse environmental conditions (Emmett and Parbery, 1975). The use of several melanin-deficient mutants of *C. lagenarium*, genetically blocked at different steps in the melanin biosynthetic pathway, helped clarify the essential role of melanin for appressorial penetration (Kubo, 1986; Kubo and Furusawa, 1986; Kubo *et al.*, 1982a, 1983, 1985, 1987; Suzuki *et al.*, 1982b).

Rice blast disease, caused by *Pyricularia oryzae*, has been controlled with compounds that act through mechanisms other than those relating to conventional fungicidal activities. One of the most attractive groups of such compounds are inhibitors of melanin biosynthesis such as tricyclazole, pyroquilon, PP389, and tetrachlorophthalide (Woloshuk and Sisler, 1982). These

Yasuyuki Kubo and Iwao Furusawa • Laboratory of Plant Pathology, Faculty of Agriculture, Kyoto University, Kyoto 606, Japan.

chemicals are highly effective in controlling rice blast at concentrations well below those that inhibit hyphal growth (Woloshuk *et al.*, 1981). The mechanism of action of these chemicals on the pathogen was investigated in studies of infection by *Pyricularia* (Chida and Sisler, 1987a,b; Woloshuk *et al.*, 1983; Yamaguchi *et al.*, 1983a,b). In *C. lagenarium* the use of melanin-deficient mutants and melanin-inhibiting fungicides, afforded clear evaluation of the effect of these chemicals on melanin biosynthesis and penetration by appressoria (Kubo *et al.*, 1985).

This review is concerned with the significance of melanin biosynthesis for appressorial penetration of *Colletotrichum* and *Pyricularia*.

2. MODE OF APPRESSORIAL PENETRATION

Both *Colletotrichum* and *Pyricularia* form well-developed melanized appressoria. *C. lagenarium* strain 104-T (laboratory strain of Kyoto University) forms abundant spores that differentiate synchronously to form appressoria. The infection process is divided into spore germination, appressorium development and melanization, infection peg formation, and penetration hypha formation. The latter structure is involved in penetration of the host. Cellulose membranes, simulating plant cell walls, have been used as a substrate allowing observation of the infection process (Araki and Miyagi, 1977; Suzuki *et al.*, 1981, 1982a, 1983). Spores of *C. lagenarium* germinate and form melanized appressoria. Infection hyphae develop from the appressoria and penetrate the cellulose membranes. When the membranes are treated with a $ZnCl_2/KI/I_2$ solution, clear zones are observed around penetration sites of the appressoria (Fig. 1A) (Suzuki *et al.*, 1981, 1983). Such clear zones resulted from degradation of the cellulose by enzymes secreted from the appressoria. Appressorium formation and penetration peg formation were shown not to be dependent upon protein synthesis, including cellulase synthesis. When *C. lagenarium* germlings were incubated in cycloheximide, a protein synthesis inhibitor, appressorium and infection structure formation proceeded normally (Suzuki *et al.*, 1983; Katoh *et al.*, 1988). Cellulase is considered to be essential for acquiring nutrients needed for the elongation of penetration hyphae (Suzuki *et al.*, 1983). From these observations, penetration by appressoria up to penetration peg formation seems to occur mechanically.

In *C. gloeosporioides*, bleaching of host plant cell walls by cutinase is essential for appressorial penetration (Dickman and Patil, 1986).

Melanin biosynthesis is also a prerequisite for infection peg protrusion from appressoria in plant pathogenic fungi such as *Colletotrichum* and *Pyricularia*.

3. PENETRATION ABILITY OF MELANIN-DEFICIENT MUTANTS

The importance of appressorial melanization in penetration was first noted in *C. lagenarium* albino mutants (Kubo *et al.*, 1982a). Pigmented appressoria of the wild-type strain 104-T formed penetration hyphae and penetrated cellulose membranes (Fig. 1A), while albino mutants formed colorless appressoria that germinated laterally to form secondary appressoria (Fig. 1B). The mutant appressoria rarely penetrated the membranes. Furthermore, it has been shown that the albino mutants are nonpathogenic on their host cucumber leaves (Kubo *et al.*, 1982a).

The role of appressorial melanization in penetration was clarified by elucidating the biosynthetic pathway of melanin. A color mutant, strain 8015, was isolated which secreted a reddish pigment into the medium. On agar culture, melanization of albino mutants was restored by dual culturing with mutant 8015. A metabolite from mutant 8015 with melanin precursor activity was isolated and identified as scytalone (Kubo *et al.*, 1983). Since scytalone is a melanin precursor derived from pentaketide synthesis in *Verticillium dahliae* (Bell *et al.*, 1976a,b; Stipanovic and Bell, 1976, 1977; Tokousbalides and Sisler, 1979), *Thielaviopsis basicola* (Stipa-

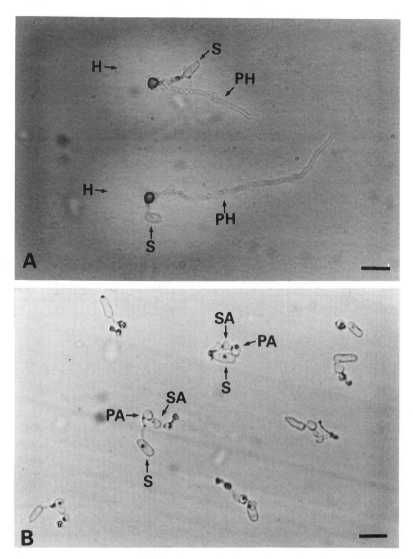

Figure 1. Penetration of cellulose membranes by appressoria of (A) parent strain 104-T and (B) albino mutant 79215 of *Colletotrichum lagenarium*. S, spores; PH, penetration hyphae; H, halos; PA, primary appressoria; SA, secondary appressoria. Bars = 10 μm.

novic and Wheeler, 1980), and *P. oryzae* (Woloshuk *et al.*, 1980), the pathway of *C. lagenarium* melanin was considered to be similar (Fig. 2) (Kubo *et al.*, 1985). Mutant 8015 is considered to be defective in the enzyme involved in the conversion of scytalone to 1,3,8-trihydroxynaphthalene (1,3,3-THN). Therefore, scytalone accumulates in the medium. Since albino mutants did not accumulate detectable levels of melanin intermediates, the deficient site in albino mutants is thought to be at or before pentaketide cyclization.

Scytalone also restored appressorial melanization of albino mutants, indistinguishable from that of the parent strain. Restored appressoria of the albino mutant effectively penetrated both cellulose membranes and plant cell walls. In addition, they did not germinate laterally as did the nonmelanized appressoria (Kubo *et al.*, 1983). Thus it should be clear that appressorial melaniza-

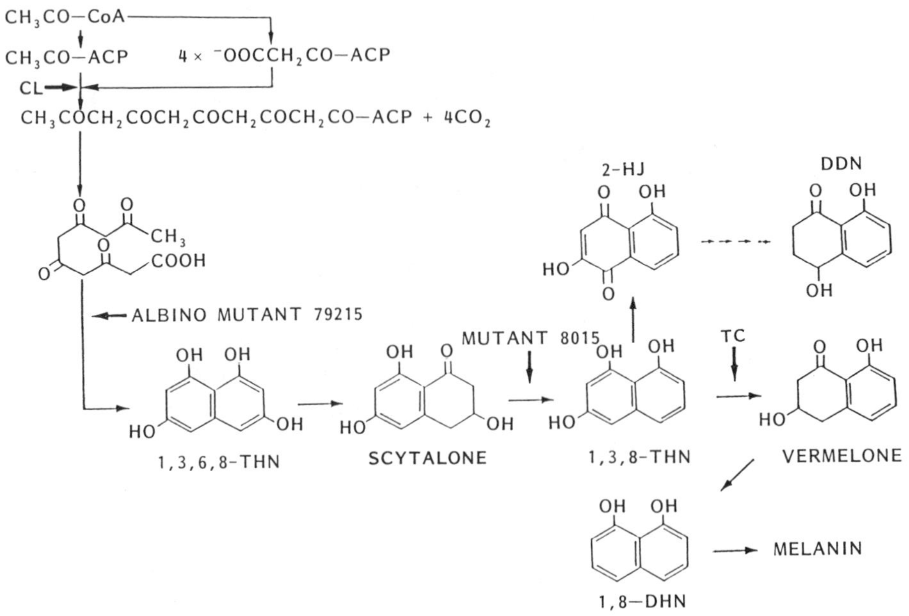

Figure 2. Biosynthetic pathway for melanin in *Colletotrichum lagenarium*, with deficient sites of color mutants and inhibition sites of tricyclazole (TC) and cerulenin (CL).

tion is essential for penetration by appressoria. Since melanin is resistant to cell wall-degrading enzymes (Potgieter and Alexander, 1966; Bloomfield and Alexander, 1967; Kuo and Alexander, 1967; Bull, 1970a,b), a possible role for melanin is inhibition of lateral germination through suppression of appressorium-derived wall-splitting enzymes. It may be that such enzymes induce vertical germination from the pore, a nonmelanized area, formed at the base of appressoria normally in contact with the substratum.

In *P. oryzae*, Woloshuk and co-workers (1980) reported that several melanin-deficient mutants whose colony color on agar media differed from that of the wild-type strain, were nonpathogenic to their host rice plants. However, a correlation between appressorial melanization and penetration was not demonstrated. Appressorial melanization and penetration ability in melanin-deficient mutants of *P. oryzae* have been reported (Okuno *et al.*, 1983; Bustamam and Sisler, 1987; Chida and Sisler, 1987a,b). The appressoria of melanin-deficient mutants were colorless and had little penetration ability. But unlike *Colletotrichum*, the colorless appressoria rarely germinated laterally. Penetration ability of *P. oryzae* was restored by application of the melanin precursors, scytalone or 1,8-dihydroxynaphthalene (1,8-DHN) (Okuno *et al.*, 1983; Bustamam and Sisler, 1987; Chida and Sisler, 1987a,b). In *P. oryzae*, it is thought that melanin has a role in providing strength and rigidity to the wall so that the force for penetration can be confined and focused (Woloshuk *et al.*, 1983). Thus, the role of melanin in appressorial penetration may differ in *Colletotrichum* and *Pyricularia*.

4. LOCATION AND FORM OF MELANIN

The ultrastructural location of electron-dense material in appressoria of *Colletotrichum* species has been suggested to be representative of melanin (Ishida and Akai, 1968; Griffiths and

Campbell, 1973; Politis and Wheeler, 1973; Mercer *et al.*, 1975; Kozar and Netlitzky, 1978). The location of melanin in relation to its role in penetration by appressoria was investigated in *C. lagenarium* (Kubo and Furusawa, 1986), *C. lindemuthianum* (Wolkow *et al.*, 1983), and *P. oryzae* (Woloshuk *et al.*, 1983). In these fungi, the location of melanin was verified by comparing the ultrastructure of appressoria of wild-type strains with that from a melanin-deficient albino mutant or from a melanin inhibitor-treated wild-type strain. In *C. lagenarium*, the site of melanization was determined from observations of melanized appressoria of normally albino mutants formed in the presence of scytalone. In the *C. lagenarium* parent strain 104-T, there is a smooth, thin, electron-dense (ED) layer just outside of the plasmalemma (Fig. 3A), whereas the layer is absent in the appressorial wall of an albino mutant (Fig. 3B). In appressoria of the albino mutant treated with scytalone, an ED layer is present in the same location and pattern as in the parent strain (Fig. 3C) (Kubo and Furusawa, 1986). A similar melanin layer is also observed in appressoria of *C. lindemuthianum* (Wolkow *et al.*, 1983). The melanin layer extends around the appressorium except in the immediate region of the pore (Fig. 4A,B) (Ishida and Akai, 1968; Politis and Wheeler, 1973; Mercer *et al.*, 1975; Seifers and Ammon, 1980).

From the above discussion, a role for melanin in appressorial penetration can be proposed. During hyphal growth, cell wall-degrading enzymes appear to be involved in directing the incipient sites at hyphal branching (Mitchel and Sabar, 1966; Gull and Trinci, 1974; Polacheck and Rosenberger, 1977; Reyes and Lahoz, 1977). Furthermore, it is known that melanin is resistant to these same cell wall-degrading enzymes (Potgieter and Alexander, 1966; Bloomfield and Alexander, 1967; Kuo and Alexander, 1967; Bull, 1970a,b). It seems possible that lateral germination (branching) of appressoria of *C. lagenarium* is prevented by melanin because it is resistant to lytic enzymes involved in wall softening. Vertical germination (penetration) thus ensues through the pore, which is the only nonmelanized region of the appressorium. In the case of *C. lindemuthianum*, melanin is thought to provide the physical rigidity necessary to focus the turgor pressure of the protoplasm onto a small area of the host cuticle to facilitate penetration (Wolkow *et al.*, 1983). Similar observations were made with *P. oryzae* (Woloshuk *et al.*, 1983). The authors proposed that wall rigidity, provided by melanin, was necessary for focusing turgor forces in the vertical direction. When appressoria were not melanized, the wall was flexible, resulting in the basal edge being displaced laterally, leaving a relatively large area of the plant surface depressed. Such appressoria were not able to penetrate the host surface.

The melanins reported for spores or resting structures of other fungi appear granular or amorphous when examined ultrastructurally (Durrell, 1964; Tsao and Tsao, 1970; Griffiths and Swart, 1974; Ellis and Griffiths, 1974, 1975; Wheeler *et al.*, 1976, 1978; Wheeler and Stipanovic, 1979). Those melanins are thought to contribute to the protection of the cells from adverse environments (Lockwood, 1960; Durrell, 1964; Sussman, 1968). The thin ultrastructural appearance of smooth melanin in appressoria should be distinguished from the spore melanin in both appearance and function.

5. ACTION MECHANISM OF MELANIN-INHIBITING FUNGICIDES

One of the most potent groups of agents that control rice blast disease caused by *P. oryzae* are inhibitors of melanin biosynthesis. This group includes compounds such as tricyclazole [5-methyl-1,2,4-triazolo-(3,4-b)benzothiazole], pyroquilon [1,2,5,6-tetrahydropyrrolo-(3,2,1-i,j) quinolin-4-one], PP389 [4,5-dihydro-4-methyltetrazolo-(1,5-a)quinazolin-5-one], chlobenthiazone [4-chloro-3-methyl-2(3H)benzothiazolone], and fthalide (4,5,6,7-tetrachlorophthalide) (Woloshuk *et al.*, 1981; Woloshuk and Sisler, 1982; Inoue and Kato, 1983; 1984a,b; Inoue *et al.*, 1985). Tricyclazole, pyroquilon, and fthalide are currently used commercially for control of rice blast disease. These chemicals are only weakly toxic to mycelial growth of *P. oryzae*, but possess high protectant activity (Tokousbalides and Sisler, 1978; Woloshuk *et al.*, 1981). Furthermore,

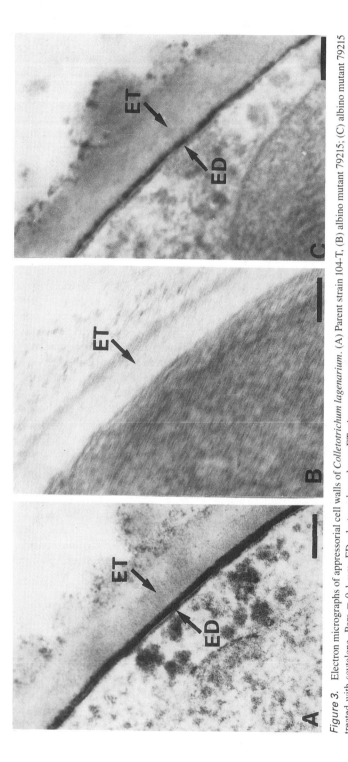

Figure 3. Electron micrographs of appressorial cell walls of *Colletotrichum lagenarium*. (A) Parent strain 104-T, (B) albino mutant 79215; (C) albino mutant 79215 treated with scytalone. Bars = 0.1 μm. ED, electron-dense layer; ET, electron-transparent layer.

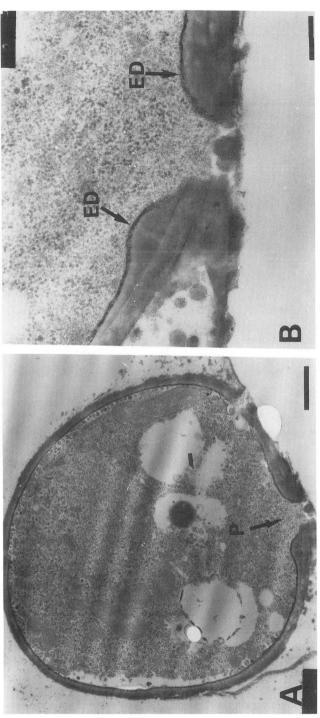

Figure 4. Electron micrographs of albino mutant 79215 appressorium of *Colletotrichum lagenarium* treated with scytalone. (A) Transverse section through the appressorium. Bar = 1 μm. P, pore. (B) Enlargement of the pore area of A. Bar = 0.1 μm. ED, electron-dense layer.

these inhibitors also exhibit high selectivity for rice blast disease, and under experimental conditions they have been shown to effectively control *C. lagenarium* (Kubo *et al.*, 1982b) and *C. lindemuthianum* (Wolkow *et al.*, 1983).

A direct relationship between inhibition of melanin biosynthesis and disease control is suggested by the fact that melanin-deficient mutants of *P. oryzae*, genetically blocked at the inhibition site by tricyclazole, are nonpathogenic (Woloshuk *et al.*, 1980). The importance of melanization for penetration has already been implicated from studies using albino mutants of *C. lagenarium* (Kubo *et al.*, 1982a). Treatment of wild-type *C. lagenarium* appressoria with tricyclazole inhibited melanization and resulted in colorless appressoria. Such appressoria germinated laterally to form secondary appressoria, similar to those of the albino mutants, and exhibited poor penetration efficiency (Kubo *et al.*, 1982b).

The inhibition site of tricyclazole in the melanin biosynthetic pathway of *P. oryzae* was clarified by identifying the shunt metabolites. The primary site of action of tricyclazole lies between 1,3,8-THN and vermelone (Woloshuk *et al.*, 1980). The biosynthetic pathway of melanin is thought to be derived from pentaketide synthesis and melanin is formed from 1,8-DHN subunits via oxidative polymerization (Woloshuk *et al.*, 1980). The pathway is similar or identical to that previously proposed in *V. dahliae* or *T. basicola* (Stipanovic and Bell, 1977; Tokousbalides and Sisler, 1979; Wheeler and Stipanovic, 1979; Stipanovic and Wheeler, 1980).

The DHN melanin biosynthesis was determined for *Colletotrichum* (Kubo *et al.*, 1982b, 1985; Wolkow *et al.*, 1983), *Cochliobolus* (Kubo *et al.*, 1989; Tajima *et al.*, 1989), *Wangiella* (Geis *et al.*, 1984), *Alternaria* (Lazarovits and Stoessl, 1988), and *Botrytis* (Zeun and Buchenauer, 1985) by studying the intermediate products that resulted through tricyclazole inhibition or secretion of melanin intermediates to 1,8-DHN melanin from melanin-deficient mutants.

Dehydratases and reductases are involved in the melanin biosynthetic pathway (Wheeler, 1982). Reductase and dehydratase activities can be estimated in cell-free homogenates by supplying various intermediates of melanin biosynthesis as substrates and by determining subsequent products. Measurement of reductase and dehydratase activities in cell-free homogenates was used to demonstrate DHN melanin in 36 species of fungi (Wheeler, 1983; Wheeler and Stipanovic, 1985; Bell and Wheeler, 1986).

Cell-free homogenates were used to determine the inhibition site of the various inhibitors (Wheeler, 1982; Wheeler and Greenblatt, 1988). The results of *in vitro* reactions with extracts of *P. oryzae* in the presence of several inhibitors (e.g., tricyclazole, fthalide, and pyroquilon) indicated that a direct relationship exists between the effect of inhibitors on *in vitro* reductase activity and their effect on melanin biosynthesis and penetration of host plant leaves (Wheeler and Greenblatt, 1988).

In *C. lagenarium*, the action mechanism of melanin-inhibiting fungicides was investigated by elucidating the inhibition site of the chemicals in the melanin biosynthetic pathway. A comparison was made of metabolites occurring in tricyclazole-treated cultures of the parent strain 104-T, mutant 8015 which is defective in conversion of scytalone to 1,3,8-THN, and albino mutant 79215. A shunt metabolite of melanin biosynthesis was isolated from culture media of the parent strain, and was identified as 3,4-dihydro-4,8-dihydroxy-1(2*H*)naphthalenone (DDN). DDN is the last shunt product derived from 1,3,8-THN through inhibition of its reduction to vermelone (Tokousbalides and Sisler, 1979; Woloshuk *et al.*, 1980). Since mutant 8015 lacks the enzyme involved in conversion of scytalone to 1,3,8-THN (Kubo *et al.*, 1983), and also the metabolite DDN, which was found in the parent strain 104-T treated with tricyclazole, it seems probable that DDN was derived from 1,3,8-THN. From these results, it became clear that tricyclazole inhibits the conversion of 1,3,8-THN to vermelone in melanin biosynthesis of *C. lagenarium* (Fig. 2) (Kubo *et al.*, 1985).

Melanin-inhibiting fungicides, including tricyclazole, PP389, and pyroquilon, inhibited appressorial pigmentation and host penetration by *C. lagenarium*. The application of various melanin precursors to cells in which melanin synthesis had been inhibited has been used to elucidate the site of inhibition. Vermelone, a melanin precursor, restored appressorial pigmentation and subsequent host penetration by the parent strain 104-T in the presence of tricyclazole, PP389, or pyroquilon at concentrations sufficient to normally inhibit pigmentation and penetration (Fig. 5) (Table I) (Kubo *et al.*, 1985). To the contrary, scytalone restored neither appressorial pigmentation nor penetration. These results indicated that tricyclazole inhibited conversion of 1,3,8-THN to vermelone and that lack of melanin in the appressoria is the major cause of penetration failure. While colorless appressoria of *P. oryzae* treated with melanin inhibitors or those of melanin-deficient mutants cannot penetrate host substrates, application of 1,8-DHN to the appressoria partially restored penetration ability (Woloshuk *et al.*, 1983; Okuno *et al.*, 1983; Yamaguchi *et al.*, 1983a,b). The restoration efficiency of *P. oryzae* was occasionally low compared with that of *C. lagenarium*. Yamaguchi and co-workers (1982, 1983a,b) reported that melanin biosynthesis intermediates such as scytalone or 2-hydroxyjuglone (2-HJ) accumulated in the appressoria from inhibition of the melanin biosynthetic pathway and that the metabolites were cytotoxic for appressorial penetration. It is evident, however, that melanization of appressoria is of primary importance, since albino mutants do not accumulate cytotoxic intermediates, and are also nonpathogenic (Kubo *et al.*, 1982a, 1983; Bustamam and Sisler, 1987; Chida and Sisler, 1987a,b).

Further evidence that melanization is essential for penetration was provided by experiments using the antibiotic cerulenin [(2S)(3R)2,3-epoxy-4-oxo-7,10-dodecadienoylamide]. Cerulenin inhibits synthesis of compounds *via* the polyketide synthetic pathway by inhibiting the malonyl-ACP: acyl-ACP condensation reaction as well as that of fatty acid synthesis (Ōmura, 1976). Since melanin of *C. lagenarium* is synthesized from the polyketide pathway, the effect of cerulenin on appressorial melanization was investigated (Kubo *et al.*, 1986). Cerulenin inhibited appressorial melanization effectively, and on cellulose membranes the colorless appressoria germinated laterally similar to those of the albino mutants. Such cerulenin-treated appressoria rarely

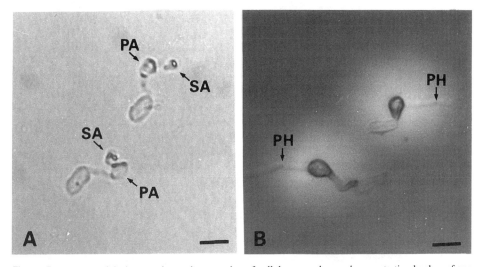

Figure 5. Appressorial pigmentation and penetration of cellulose membranes by penetration hyphae of vermelone-treated appressoria of the parent strain 104-T of *Colletotrichum lagenarium* in the presence of PP389. (A) No treatment; (B) treatment with vermelone. Bars = 10 μm. PA, primary appressoria; SA, secondary appressoria; PH, penetration hyphae.

Table I. Effect of Melanin Precursors on Appressorial Pigmentation
and Penetration by Appressoria in the Presence of Melanin-Inhibiting
Chemical in Colletotrichum lagnearium[a]

Chemicals	Concentration (μM)		Appressorial pigmentation[b]	Percentage of appressoria with		
	Scylatone	Vermelone		Penetration hyphae	Halos	Lateral germination
Tricyclazole (100 μM)	0	0	−	0.9	1.5	82.5
	100	0	−	0.0	0.0	99.2
	500	0	−	0.0	0.0	98.0
	1000	0	±	2.9	3.6	79.0
	0	1	−	3.1	2.1	66.7
	0	5	±	20.3	19.0	31.6
	0	10	+	48.3	40.4	14.2
pp389 (500 μM)	0	0	−	0.0	0.0	96.2
	100	0	−	0.5	0.5	97.5
	500	0	−	0.7	5.6	91.8
	1000	0	±	5.1	4.5	89.8
	0	1	−	0.6	0.6	96.8
	0	5	±	26.7	24.8	47.6
	0	10	+	48.4	36.3	8.2
Pyroquilon (500 μM)	0	0	−	0.0	0.0	96.8
	100	0	−	0.0	0.0	99.5
	500	0	−	0.0	0.0	98.0
	1000	0	±	0.0	0.0	96.5
	0	1	−	20.5	18.2	78.2
	0	5	±	29.4	25.2	63.8
	0	10	+	50.5	46.2	8.1
None	0	0	+	75.7	74.0	9.7

[a]Spores were incubated on nitrocellulose membranes for 72 hr at 24°C.
[b]Appressoria were darkly pigmented (+), light brown (±), or colorless (−).

penetrated the membranes. When scytalone, a melanin precursor to cerulenin, was applied in the presence of cerulenin, appressorial melanization was restored along with the ability of the appressoria to penetrate cellulose membranes (Kubo et al., 1986). It is evident that cerulenin inhibited pentaketide synthesis which normally leads to melanin biosynthesis of C. lagenarium (Fig. 2). Since inhibition of melanin biosynthesis early in the pathway effectively brought about penetration failure without the accumulation of cytotoxic polyketide melanin intermediates, further proof for the essential role of appressorial melanization in penetration was established.

In contrast to C. lagenarium, it may be necessary to consider the cytotoxic effect of the intermediates in addition to the inhibition of appressorial melanization in P. oryzae. Chida and Sisler (1987a,b) reported that the penetration ability of appressoria treated with low concentrations of fthalide or tricyclazole was restored by 1,8-DHN, but was not restored well in appressoria treated with high concentrations of the chemicals unless cerulenin was also present. Similar results were also obtained with an untreated buff mutant P-2 m-1. In the former case, inhibition of appressorial melanization accounted exclusively for failure of penetration. In the latter case, accumulation of cytotoxic intermediates, which operates concurrently with the first, is also the cause of interference of appressorial penetration.

6. REGULATION OF MELANIN BIOSYNTHESIS DURING APPRESSORIUM FORMATION

During cellular differentiation of *C. lagenarium*, from spore germination to penetration hypha formation, melanization is confined to the appressoria; spores, germ tubes, and penetration hyphae are never melanized, indicating that melanin biosynthesis is a developmentally regulated system. The relation of appressorium differentiation in *C. lagenarium* to the metabolic development of melanin biosynthesis was investigated using an albino mutant and a melanin precursor, scytalone (Kubo *et al.*, 1984).

Germinating spores were labeled with [^{14}C]scytalone for 2-hr periods after the onset of incubation. Incorporation of [^{14}C]scytalone into the melanin fraction was coincident with the period of appressorial swelling and pigmentation. Appressorial melanization was not inhibited by the protein synthesis inhibitor cycloheximide when the inhibitor was applied 1 hr after the onset of spore incubation. These results indicate that reductases and dehydratases involved in melanin biosynthesis subsequent to scytalone synthesis are considered to be preexisting spore enzymes, or they are synthesized as inactive forms during the first hour of incubation (Kubo *et al.*, 1984).

Temperature sensitivity of enzymes involved in melanin biosynthesis was also investigated. Structurally mature colorless appressoria of *C. lagenarium* strain 104-T were formed after incubation at 24°C for 5.5 hr. When similar cells were postincubated at 32°C for an additional 6.5 hr, appressorial melanization did not occur. When postincubation was performed in the presence of scytalone, however, appressoria became melanized. From these results, we interpreted that the melanin biosynthetic pathway in appressoria of *C. lagenarium* is divided into two steps: the processes leading to pentaketide cyclization are temperature sensitive, while the processes subsequent to the cyclization mediated by reductase and dehydratase are not sensitive (Kubo *et al.*, 1984).

7. CONCLUSION

For some plant pathogenic fungi such as *Colletotrichum* and *Pyricularia*, melanin plays an essential role in appressorial penetration. In plant pathogenic fungi that do not form appressoria or those which do not form melanized appressoria, melanin biosynthesis is not needed for infection of the host. For species of fungi that require melanization for pathogenicity, the significance of melanin biosynthesis for pathogenicity should be determined. A precise role for melanin in appressorial penetration will be clarified from molecular, enzymatic and immuno-cytologic studies.

Most of the chemicals that are not inhibitory to fungal growth *in vitro* but which interfere with pathogenicity have been developed empirically. Elucidation of metabolisms that are essential to the expression of pathogenicity should provide unique targets for development of new fungicides. Scytalone dehydratase of melanin biosynthesis seems to be one of those potential targets, since the enzyme is unique in 1,8-DHN melanin synthesis and is apparently involved in pathogenicity of *Colletotrichum* and *Pyricularia*.

Progress in research on fungal pathogenicity will lead to a clearer understanding about the mode of action of these and other fungicides.

8. REFERENCES

Araki, F., and Miyagi, Y., 1977, Effects of fungicides on penetration by *Pyricularia oryzae* as evaluated by improved cellphane method, *J. Pestic. Sci.* **2**:457–461.

Bell, A. A., and Wheeler, M. H., 1986, Biosynthesis and formation of fungal melanins, *Annu. Rev. Phytopathol.* **24**:411–451.

Bell, A. A., Puhalla, J. E., Tolmsoff, W. J., and Stipanovic, R. D., 1976a, Use of mutants to establish (+)-scytalone as an intermediate in melanin biosynthesis by *Verticillium dahliae, Can. J. Microbiol.* **22**:787–799.

Bell, A. A., Stipanovic, R. D., and Puhalla, J. E., 1976b, Pentaketide metabolites of *Verticillium dahliae*: Identification of (+)-scytalone as a natural precursor to melanin, *Tetrahedron* **32**:1353–1356.

Bloomfield, B. J., and Alexander, M., 1967, Melanins and resistance of fungi to lysis, *J. Bacteriol.* **93**:1276–1280.

Bull, A. T., 1970a, Chemical composition of wild and mutant *Aspergillus nidulans* cell walls. The nature of polysaccharides and melanin constituents, *J. Gen. Microbiol.* **63**:75–94.

Bull, A. T., 1970b, Inhibition of polysaccharases by melanin: Enzyme inhibition in relation to mycolysis, *Arch. Biochem. Biophys.* **137**:345–356.

Bustamam, M., and Sisler, H. D., 1987, Effect of pentachloronitrobenzene, pentachloroaniline, and albinism on epidermal penetration by appressoria of *Pyricularia, Pestic. Biochem. Physiol.* **28**:29–37.

Chida, T., and Sisler, H. D., 1987a, Restoration of appressorial penetration ability by melanin precursors in *Pyricularia oryzae* treated with anti-penetrants and in melanin-deficient mutants, *J. Pestic. Sci.* **12**:49–55.

Chida, T., and Sisler, H. D., 1987b, Effect of inhibitors of melanin biosynthesis on appressorial penetration and reductive reactions in *Pyricularia oryzae* and *Pyricularia grisea, Pestic. Biochem. Physiol.* **29**:244–251.

Dickman, M. B., and Patil, S. S., 1986, Cutinase deficient mutants of *Colletotrichum gloeosporioides* are nonpathogenic to papaya fruit, *Physiol. Mol. Plant Pathol.* **28**:235–242.

Durrell, L. W., 1964, The composition and structure of walls of dark fungus spores, *Mycopathol. Mycol. Appl.* **23**:339–345.

Ellis, D. H., and Griffiths, D. A., 1974, The location and analysis of melanins in the cell walls of some soil fungi, *Can. J. Microbiol.* **20**:1379–1386.

Ellis, D. H., and Griffiths, D. A., 1975, Melanin deposition in the hyphae of a species of *Phomopsis, Can. J. Microbiol.* **21**:442–452.

Emmett, R. W., and Parbery, D. G., 1975, Appressoria, *Annu. Rev. Phytopathol.* **13**:147–167.

Geis, P. A., Wheeler, M. H., and Szaniszlo, P. J., 1984, Pentaketide metabolites of melanin synthesis in the dematiaceous fungus *Wangiella dermatitidis, Arch. Microbiol.* **137**:324–328.

Griffiths, D. A., and Campbell, W. P., 1973, Fine structure of conidial germination and appressorial development in *Colletotrichum atramentarium, Trans. Br. Mycol. Soc.* **61**:529–536.

Griffiths, D. A., and Swart, H. J., 1974, Conidial structure in two species of *Pestalotiopsis, Trans. Br. Mycol. Soc.* **62**:295–304.

Gull, K., and Trinci, A. P. J., 1974, Detection of areas of wall differentiation in fungi using fluorescent staining, *Arch. Microbiol.* **96**:53–57.

Inoue, S., and Kato, T., 1983, Mode of rice blast control by chrobenthiazone, *J. Pestic. Sci.* **8**:333–338.

Inoue, S., Maeda, K., Uematsu, T., and Kato, T., 1984a, Comparison of tetrachlorophthalide and pentachloro-benzyl alcohol with chrobenthiazone and other melanin inhibitors in the mechanism of rice blast control, *J. Pestic. Sci.* **9**:731–736.

Inoue, S., Uematsu, T., and Kato, T., 1984b, Effects of chrobenthiazone on the infection process by *Pyricularia oryzae, J. Pestic. Sci.* **9**:689–695.

Inoue, S., Uematsu, T., Kato, T., and Ueda, K., 1985, New melanin biosynthesis inhibitors and their structural similarities, *J. Pestic. Sci.* **16**:589–598.

Ishida, N., and Akai, S., 1968, Electron microscopic observation of cell wall structure during appressorium formation in *Colletotrichum lagenarium, Mycopathol. Mycol. Appl.* **35**:68–74.

Katoh, M., Hirose, I., Kubo, Y., Hikichi, Y., Kunoh, H., Furusawa, I., and Shishiyama, J., 1988, Use of mutants to indicate factors prerequisite for penetration of *Colletotrichum lagenarium* by appressoria, *Physiol. Mol. Plant Pathol.* **32**:177–184.

Kozar, F., and Netlitzky, H. J., 1978, Studies on hyphal development and appressorium formation of *Colletotrichum graminicola, Can. J. Bot.* **56**:2234–2242.

Kubo, Y., 1986, Melanin biosynthesis in fungi, *Trans. Mycol. Soc. Jpn.* **27**:487–500.

Kubo, Y., and Furusawa, I., 1986, Localization of melanin in appressoria of *Colletotrichum lagenarium, Can. J. Microbiol.* **32**:280–282.

Kubo, Y., Suzuki, K., Furusawa, I., Ishida, N., and Yamamoto, M., 1982a, Relation of appressorium pigmentation and penetration of nitrocellulose membranes by appressoria of *Colletotrichum lagenarium, Phytopathology* **72**:498–501.

Kubo, Y., Suzuki, K., Furusawa, I., and Yamamoto, M., 1982b, Effect of tricyclazole on appressorial pigmentation and penetration from appressoria of *Colletotrichum lagenarium, Phytopathology* **72**:1198–1200.

Kubo, Y., Suzuki, K., Furusawa, I., and Yamamoto, M., 1983, Scytalone as a natural intermediate of melanin biosynthesis in appressoria of *Colletotrichum lagenarium*, *Exp. Mycol.* **7**:208–215.

Kubo, Y., Furusawa, I., and Yamamoto, M., 1984, Regulation of melanin biosynthesis during appressorium formation in *Colletotrichum lagenarium*, *Exp. Mycol.* **8**:364–369.

Kubo, Y., Suzuki, K., Furusawa, I., and Yamamoto, M., 1985, Melanin biosynthesis as a prerequisite for penetration by appressoria of *Colletotrichum lagenarium*: Site of inhibition by melanin-inhibiting fungicides and their action on appressoria, *Pestic. Biochem. Physiol.* **23**:47–55.

Kubo, Y., Katoh, M., Furusawa, I., and Shishiyama, J., 1986, Inhibition of melanin biosynthesis by cerulenin in appressoria of *Colletotrichum lagenarium*, *Exp. Mycol.* **10**:301–306.

Kubo, Y., Furusawa, I., and Shishiyama, J., 1987, Relation between pigment intensity and penetrating ability in appressoria of *Colletotrichum lagenarium*, *Can. J. Microbiol.* **33**:871–873.

Kubo, Y., Tsuda, M., Furusawa, I., and Shishiyama, J., 1989, Genetic analysis of genes involved in melanin biosythesis of *Cochliobolus miyabeanus*, *Exp. Mycol.* **13**:77–89.

Kuo, M. J., and Alexander, M., 1967, Inhibition of the lysis of fungi by melanins, *J. Bacteriol.* **94**:624–629.

Lazarovits, G., and Stoessl, A., 1988, Tricyclazole induces melanin shunt products and inhibits altersolanol A accumulation by *Alternaria solani*, *Pestic. Biochem. Physiol.* **31**:36–45.

Lockwood, J. L., 1960, Lysis of mycelium of plant pathogenic fungi by natural soil, *Phytopathology* **50**:787–789.

Mercer, P. C., Wood, R. K. S., and Greenwood, A. D., 1975, Ultrastructure of the parasitism of *Phaseolus vulgaris* by *Colletotrichum lindemuthianum*, *Physiol. Plant Pathol.* **5**:203–214.

Mitchel, R., and Sabar, N., 1966, Autolytic enzymes in fungal cell walls, *J. Gen. Microbiol.* **42**:39–42.

Okuno, T., Matsuura, K., and Furusawa, I., 1983, Recovery of appressorial penetration by some melanin precursors in *Pyricularia oryzae* treated with tricyclazole and in a melanin deficient mutant, *J. Pestic. Sci.* **8**:357–360.

Ōmura, S., 1976, The antibiotic cerulenin, a novel tool for biochemistry as an inhibitor of fatty acid synthesis, *Bacteriol. Rev.* **40**:681–697.

Polacheck, I., and Rosenberger, R. F., 1977, *Aspergillus nidulans* mutant lacking α-(1,3)-glucan, melanin, and cleistothecia, *J. Bacteriol.* **132**:650–656.

Politis, D. J., and Wheeler, H., 1973, Ultrastructural study of penetration of maize leaves by *Colletotrichum graminicola*, *Physiol. Plant Pathol.* **3**:465–471.

Potgieter, H. J., and Alexander, M., 1966, Susceptibility and resistance of several fungi to microbial lysis, *J. Bacteriol.* **91**:1526–1532.

Reyes, F., and Lahoz, R., 1977, Variations in lysis of walls of *Sclerotinia fructigena* with age of culture, *J. Gen. Microbiol.* **98**:607–610.

Seifers, D., and Ammon, V., 1980, Mode of penetration of sycamore leaves by *Gloeosporium platani*, *Phytopathology* **70**:1050–1055.

Stipanovic, R. D., and Bell, A. A., 1976, Pentaketide metabolites of *Verticillium dahliae*. 3. Identification of (−)-3,4-dihydro-3,8-dihydroxy-1(2H)napthalenone[(−)-vermelone] as a precursor to melanin, *J. Org. Chem.* **41**:2468–2469.

Stipanovic, R. D., and Bell, A. A., 1977, Pentaketide metabolites of *Verticillium dahliae*. II. Accumulation of naphthol derivatives by the aberrant-melanin mutant brm-2, *Mycologia* **69**:164–172.

Stipanovic, R. D., and Wheeler, M. H., 1980, Accumulation of 3,3'-biflaviolin, a melanin shunt product, by tricyclazole treated *Thielaviopsis basicola*, *Pestic. Biochem. Physiol.* **13**:198–201.

Sussman, A. S., 1968, Longevity and survivability of fungi, in: *The Fungi*, Volume III (G. C. Ainsworth and A. S. Sussman, eds.), Academic Press, New York, pp. 447–486.

Suzuki, K., Furusawa, I., Ishida, N., and Yamamoto, M., 1981, Protein synthesis during germination and appressorium formation of *Colletotrichum lagenarium* spores, *J. Gen. Microbiol.* **124**:61–69.

Suzuki, K., Furusawa, I., Ishida, N., and Yamamoto, M., 1982a, Chemical dissolution of cellulose membranes as a prerequisite for penetration from appressoria of *Colletotrichum lagenarium*, *J. Gen. Microbiol.* **128**:1035–1039.

Suzuki, K., Kubo, Y., Furusawa, I., Ishida N. and Yamamoto, M., 1982b, Behavior of colorless appressoria in an albino mutant of *Colletotrichum lagenarium*, *Can. J. Microbiol.* **28**:1210–1213.

Suzuki, K., Furusawa, I., and Yamamoto, M., 1983, Role of chemical dissolution of cellolose membranes in the appressorial penetration by *Colletotrichum lagenarium*, *Ann. Phytopathol. Soc. Jpn.* **49**:481–487.

Tajima, S., Kubo, Y., Furusawa, I., and Shishiyama, J., 1989, Purification of a melanin biosynthetic enzyme converting scytalone to 1,3,8-trihydroxynapthalene from *Cochliobolus miyabeanus*, *Exp. Mycol.* **13**:69–76.

Tokousbalides, M. C., and Sisler, H. D., 1978, Effect of tricyclazole on growth and secondary metabolism in *Pyricularia oryzae*, *Pestic. Biochem. Physiol.* **8**:26–32.

Tokousbalides, M. C., and Sisler, H. D., 1979, Site of inhibition by tricyclazole in the melanin biosynthetic pathway of *Verticillium dahliae*, *Pestic. Biochem. Physiol.* **11**:64–73.

Tsao, P. W., and Tsao, P. H., 1970, Electron microscopic observation on the spore wall and operculum formation in chlamydospores of *Thielaviopsis basicola*, *Phytopathology* **60**:613–616.

Wheeler, M. H., 1982, Melanin biosynthesis in *Verticillium dahliae*: Dehydration and reduction reactions in cell free homogenates, *Exp. Mycol.* **6**:171–179.

Wheeler, M. H., 1983, Comparison of fungal melanin biosynthesis in ascomycetes, imperfect and basidiomycetous fungi, *Trans. Br. Mycol. Soc.* **81**:29–36.

Wheeler, M. H., and Greenblatt, G. A., 1988, The inhibition of melanin biosynthetic reactions in *Pyricularia oryzae* by compounds that prevent rice blast disease, *Exp. Mycol.* **12**:151–160.

Wheeler, M. H., and Stipanovic, R. D., 1979, Melanin biosynthesis in *Thielaviopsis basicola*, *Exp. Mycol.* **3**: 340–350.

Wheeler, M. H., and Stipanovic, R. D., 1985, Melanin biosynthesis and the metabolism of flaviolin and 2-hydroxyjuglone in *Wangiella dermatitidis*, *Arch Microbiol.* **142**:234–241.

Wheeler, M. H., Tolmsoff, W. J., and Meola, S., 1976, Ultrastructure of melanin formation in *Verticillium dahliae* with (+)-scytalone as a biosynthetic intermediate, *Can. J. Microbiol.* **22**:702–711.

Wheeler, M. H., Tolmsoff, W. J., and Bell, A. A. 1978, Ultrastructural and chemical distinction of melanins formed by *Verticillium dahliae* from (+)-scytalone, 1,8-dihydroxynaphthalene, and L-3,4-dihydroxynaphthalene, *Can. J. Microbiol.* **24**:289–297.

Wolkow, P. M., Sisler, H. D., and Vigil, E. L., 1983, Effect of inhibitors of melanin biosynthesis on structure and function of appressoria of *Colletotrichum lindemuthianum*, *Physiol. Plant Pathol.* **23**:55–72.

Woloshuk, C. P., and Sisler, H. D., 1982, Tricyclazole, pyroquilon, tetrachlorophthalide, PCBA, coumarine and related compounds inhibit melanization and epidermal penetration by *Pyricularia oryzae*, *J. Pestic. Sci.* **7**:161–166.

Woloshuk, C. P., Sisler, H. D., Tokousbalides, M. C., and Dutky, S. R., 1980, Melanin biosynthesis in *Pyricularia oryzae*: Site of tricyclazole inhibition and pathogenicity of melanin-deficient mutants, *Pestic. Biochem. Physiol.* **14**:256–264.

Woloshuk, C. P., Wolkow, P. M., and Sisler, H. D., 1981, The effect of three fungicides, specific for the control of rice blast disease, on the growth and melanin biosynthesis by *Pyricularia oryzae* Cav., *Pestic. Sci.* **12**:86–90.

Woloshuk, C. P., Sisler, H. D., and Vigil, E. L., 1983, Action of antipenetrant, tricyclazole on appressoria of *Pyricularia oryzae*, *Physiol. Plant Pathol.* **22**:245–259.

Yamaguchi, I., Sekido, S., and Misato, T., 1982, The effect of non-fungicidal antiblast chemicals on the melanin biosynthesis and infection by *Pyricularia oryzae*, *J. Pestic. Sci.* **7**:523–529.

Yamaguchi, I., Sekido, S., and Misato, T., 1983a, Inhibition of appressorial melanization in *Pyricularia oryzae* by non-fungicidal anti-blast chemicals, *J. Pestic. Sci.* **8**:229–232.

Yamaguchi, I., Sekido, S., Seto, H., and Misato, T., 1983b, Cytotoxic effect of 2-hydroxyjuglone, a metabolite in the branched pathway of melanin biosynthesis in *Pyricularia oryzae*, *J. Pestic. Sci.* **8**:545–550.

Zeun, R., and Buchenauer, H., 1985, Effect of tricyclazole on production and melanin contents of sclerotia of *Botrytis cinerea*, *Phytopathol. Z.* **112**:259–267.

10

The Plant Cuticle

A Barrier to Be Overcome by Fungal Plant Pathogens

Wolfram Köller

1. INTRODUCTION

Most plant-pathogenic fungi gain access into their host by penetration of unwounded tissue. Some pathogens such as rusts invade the host via stomata (Hoch and Staples, 1987 and Chapter 2), whereas others penetrate the intact leaf surface without the requirement of natural openings (Aist, 1976; Emmett, 1975; Kunoh, 1984). The latter type of direct penetration encounters the plant cuticle, a noncellular hydrophobic structure covering the layer of epidermal cells. The cuticle thus serves as the first surface barrier that the pathogen has to breach. There is little evidence for the mere physical strength of the plant cuticle as a major factor in plant defense against pathogens. In some cases, the thickness of plant cuticles has been correlated with an increased passive resistance against fungal attack. This correlation, however, appears to be an exception rather than the rule (Martin, 1964). Furthermore, the cuticle has not been considered to play a major role in the active defense mechanisms of disease-resistant cultivars. There is good evidence that the cuticle is penetrated by the attacking pathogen before the sequential steps of disease development are halted by the active defense reactions of the challenged plant. Recent examples for this lack of cuticle involvement in cultivar resistance are the host–pathogen interactions of *Venturia inaequalis*–apple (Valsangiacomo and Gessler, 1988) or *Phytophthora infestans*–potato (Gees and Hohl, 1987). The breaching of the cuticle can also be accomplished in many interactions of pathogens with nonhost plants (Heath, 1987).

Regardless of the restricted importance of the cuticle as a factor in cultivar resistance or plant immunity, this barrier has to be breached in order for fungal pathogens to establish a parasitic relationship. The mechanism by which the cuticle is penetrated has been debated for nearly a century. Two conflicting views are that cuticle penetration is aided by the mechanical force exerted by the infection structure, or that the penetration step would require the enzymatic dissolution of the cuticle. The early literature was generally in favor of the "mechanical hypothesis," as summarized by Wood (1960): "At present, therefore, the evidence is that

Wolfram Köller • Department of Plant Pathology, Cornell University, New York State Agricultural Experiment Station, Geneva, New York 14456.

penetration of the cuticle is mechanical and does not depend on substances produced by the hyphae of invading organisms." However, the same author highlighted one of the major obstacles in this field of research: "No one has demonstrated that plant pathogens are able to degrade the cuticle chemically, but apparently plant pathologists have not made serious attempts to obtain fungal preparations able to do this." Indeed, most of the evidence for both models was based on ultrastructural observations.

The first serious attempts to obtain fungal preparations able to hydrolyze plant cuticles were reported by Heinen and co-workers in the early 1960's. This initial work on cutinolytic enzymes from fungi was summarized in the first review article dedicated to cutinase (Van den Ende and Linskens, 1974). Tremendous progress has been made since then, initiated mainly by P.E. Kolattukudy and his group. The development and refinement of an "enzymatic hypothesis" have been summarized in several recent reviews (Kolattukudy, 1984a,b, 1985; Kolattukudy and Köller, 1983; Kolattukudy and Soliday, 1985; Kolattukudy and Crawford, 1987; Kolattukudy et al., 1985, 1987a,b,c).

The purpose of this chapter is to summarize and discuss the current evidence for the involvement of cutinase in the invasion of plants by fungal pathogens. Emphasis will be placed on the description of host–pathogen systems that irrefutably involve cutinase as a factor of pathogenicity. Unfortunately, only a few cases have been investigated by direct means, and most of the additional evidence either supporting or disputing the importance of cutinase in plant infection remains circumstantial. Consequently, one of the additional goals of this chapter will be to identify and describe some of the currently unresolved questions.

2. THE PLANT CUTICLE

2.1. Composition and Structure

The outer walls of epidermal plant cells are covered by a continuous layer of lipid material, the cuticular membrane or, in a broader sense, the plant cuticle. The cuticle is attached to the cell wall by an intermediate layer enriched in pectin. The thickness and the morphology of the cuticle are variable, depending on the plant species and cultivar, on the specific organ of a given plant (e.g., leaf, stem, fruit), on the developmental stage of the plant tissue, or on the environmental conditions. This aspect of cuticle diversity has been summarized in great detail (Holloway, 1982a; Martin and Juniper, 1970).

The general features of plant cuticles are most appropriately characterized in chemical rather than structural terms. Both the chemistry and the biochemistry of the cuticle have been described in several comprehensive reviews (Holloway, 1982b; E. Baker, 1982; Kolattukudy, 1980a,b, 1981, 1984c, 1987; Kolattukudy and Espelie, 1985; Kolattukudy et al., 1981a) and will only be summarized. Plant cuticles can be chemically or enzymatically separated from cell walls. The isolated material consists of two major lipid components, a soluble complex mixture of long-chain aliphatic compounds collectively called waxes, and an insoluble polymeric material called cutin. The most common constituents of the wax fraction are hydrocarbons (C_{29}, C_{31}), primary alcohols (C_{26}, C_{28}, alcanoic acids (C_{24}, C_{26}, C_{28}), and monoesters (C_{44}–C_{50}). The monomers of depolymerized cutin consist primarily of a C_{16} and a C_{18} family of fatty acids (Fig. 1). The characteristic feature of cutin fatty acids is a hydroxyl group in ω-, and often a second and third hydroxyl group in a midchain position. The C_{16} family derives from palmitic acid via subsequent direct hydroxylations in ω- and 9- or 10-position, most likely mediated by a cytochrome P-450 monooxygenase. The C_{18} family originates predominantly from oleic or linoleic acid. The bioconversion involves a hydroxylation in the ω-position, the epoxidation of a double bond, and a subsequent hydration of the epoxy group generating a trihydroxy fatty acid. Numerous minor

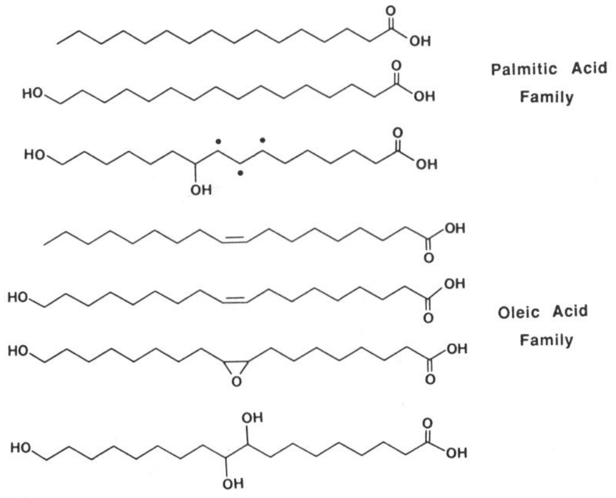

Figure 1. Structures of major cutin monomers. Positional isomers of dihydroxypalmitic acid are indicated in the structure.

components of cutin have been identified, but the function of these constituents remains largely unexplored.

In the cutin polymer, the fatty acid monomers are held together by ester linkages (Fig. 2). Most of the primary hydroxyl groups are esterified, resulting in a predominantly linear polymer. Crosslinking between polyester chains is achieved by esterification of some but not all midchain hydroxyl groups. According to the structure and chemical composition, the plant cuticle is best described as a polyester that provides a network for the waxes embedded in and protruding from the cutin.

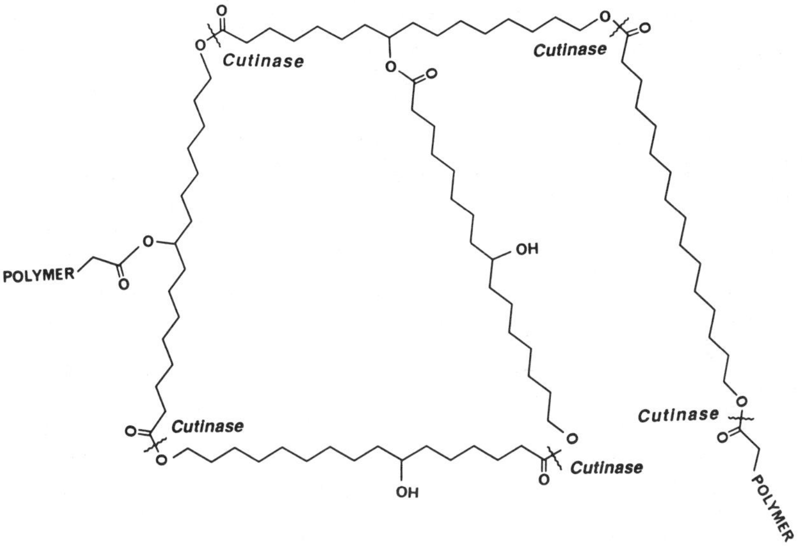

Figure 2. A structural model representing the polymer cutin. The sites of ester hydrolysis by cutinase are indicated.

2.2. Function

The primary functions of the hydrophobic cuticle are to act as a barrier against the diffusion of water and nutrients from the plant, and to protect the plant from the adverse effects of chemicals in the environment (Schönherr, 1982). Interestingly, this protective nature of the cuticle is sometimes an undesirable feature. For example, many agrichemicals targeted to the inside of plant cells have to diffuse through the cuticle, a requirement not easily met (Price, 1982). Experimental work with isolated cuticles has demonstrated that the wax fraction and not the polymer cutin provides the major diffusion barrier (Schönherr, 1982).

Wax components have also been related to the chemical communication between plants and other organisms, including pathogenic fungi. There is some evidence that cuticular components are involved in the control of fungal differentiation on leaf surfaces (Macko, 1981; Trione, 1981). In many cases, plant waxes were also shown to contain fungitoxic components (Blakeman and Atkinson, 1981). However, a decisive role for these toxic substances in the protection of plants from fungal infection is often not clear (e.g., Hargreaves et al., 1982). The intriguing aspect of wax components acting as chemical signals in the interaction of plants with microorganisms is probably the least understood function of plant cuticles.

Besides the function as a diffusive barrier against molecules, the cuticle serves as a protective layer against microorganisms. In general, the protective quality of the cuticle is highly effective, and a great variety of phylloplane organisms are not capable of overcoming this layer (Blakeman, 1981). However, some pathogenic microorganisms have developed means to breach the cuticle, and this step is regarded as one of the crucial factors of pathogenicity. The most efficient "cuticle breachers" are fungal pathogens, many of which are known to invade plants through the intact plant surface. In this case, the polymeric structure of the cutin, and not the monomeric wax fraction, is thought to provide the major contribution to the physical strength of the cuticle.

3. CUTINASE FROM FUNGAL PATHOGENS

The deposition and attachment of fungal propagules to plant surfaces comprise the initial steps in the infection of plants by fungi. This aspect is described in detail by Nicholson and Epstein (Chapter 1). Propagule germination is followed by growth of germ tubes, which might be directed, and the formation of specialized infection structures (see Chapters 1 and 2). All of these first steps have to be envisioned as prepenetration events and, thus, as the preparation for the crucial step of plant invasion. The first surface barrier to be breached during disease initiation is the plant cuticle, and any enzymatic dissolution of this barrier by fungal cutinases would become crucial at this particular step of fungal attack. Cutinases from a number of plant-pathogenic fungi have been purified and characterized; the enzymatic and structural properties of these cutinases will be described in the following sections.

3.1 Identification of Cutinases

Any biochemical investigation of enzymes requires a suitable enzyme substrate that provides both ease of measuring and a clear indication for the enzyme reaction under question. These requirements were not easily met with cutin as the natural substrate for putative cutinases. First, cutin is an amorphous and insoluble polymer, and second, the chemical composition and structure of cutin remained largely unexplored when the first studies on fungal cutinases were initiated. However, it had been long known that hydroxylated fatty acids linked by ester bonds were a major structural component (Martin, 1964). The first studies on cutinolytic activity from fungi were accomplished with *Penicillium spinulosum*, a soil saprophyte isolated from rotting leaves (Van den Ende and Linskens, 1974). Based on the principles of the cutin structure, the release of fatty acids from polymeric cutin had been taken as a measure of cutinolytic activity, and a simple titration method was used to determine the acidification of the test medium. The results of these early studies indicated the existence of a "cutinase complex." An esterase was thought to hydrolyze ester bonds and to generate "carboxycutin" still linked by peroxide bridges. These peroxide bridges would then be cleaved by a "carboxycutin peroxidase" (Van den Ende and Linskens, 1974). The abundant existence of ether or peroxide bridges in cutin has not been confirmed in subsequent chemical studies (Holloway, 1982b; Kolattukudy, 1980a,b, 1981; Kolattukudy *et al.*, 1981b), and some uncertainties with regard to the identity of the "cutinase complex" remain.

A more elaborate and sensitive means of assaying cutinase activity was the utilization of radioactive cutin. Several different methods for the preparation of radioactive cutin substrates have been described. Fatty acid monomers of cutin were labeled biochemically by applying radioactive precursors, such as acetate (Kolattukudy *et al.*, 1981b) or palmitic acid (Salinas *et al.*, 1986), to the epidermal cell layer of rapidly expanding apple or tomato fruits. This procedure led to a ^{14}C-labeled enzyme substrate suitable for cutinase studies. ^{3}H-labeled cutin as an alternative was prepared by the reduction of double bonds in apple cutin with ^{3}H$_2$ gas (Kolattukudy *et al.*, 1981b), or by the reduction of carbonyl groups present in grapefruit cutin with NaB^{3}H$_4$ (Espelie *et al.*, 1983; Köller *et al.*, 1982a.). The activity of cutinase was assayed by measuring the radioactivity released from the polymeric cutin. It is advisable to extract the acidified cutin acids into ethyl ether whenever ^{3}H-labeled cutin is used as a substrate (Baker and Bateman, 1978; Köller *et al.*, 1982a). This step reduces the relatively high background radioactivity. The identity of the hydrolysis products liberated by the enzyme as cutin monomers should be verified, whenever a "new" cutinase is being investigated.

All enzyme assays with radioactive cutin are relatively expensive and laborious, and the esterase substrate *p*-nitrophenyl butyrate has been widely applied as a suitable model substrate for spectrophotometrical assays (Kolattukudy *et al.*, 1981b). The molar extinction coefficient of *p*-nitrophenyl is strongly pH-dependent. This dependency must be taken into account, whenever enzyme units are calculated (Köller and Parker, 1989). However, it should be pointed out that *p*-nitrophenyl butyrate used as the substrate indicates only the presence of nonspecific esterase activity. The conclusive proof for the identity of cutinase activity has to be obtained by analysis of cutin hydrolysis and should involve the identification of hydrolysis products.

The presence of cutinase activity in culture fluids of fungal plant pathogens was reported as early as 1963 (Linskens and Haage, 1963). The pathogens investigated since then are summarized in Table I. The great number of fungi shown to be cutinase producers suggests that the ability of plant pathogens to excrete this enzyme is widespread. Unfortunately, the stringent criteria for the enzyme assays of cutinases are not always met, and the cutinase nature of the reported enzymes is not entirely certain in all cases. Thus, some skepticism is in order, especially when the relatively crude titration method had been used for enzyme assays, or when very small amounts of radioactivity were released from radioactively labeled cutin. Some of the cutinolytic activities described in these earlier reports might have been derived from hydrolases other than cutinase. For example, cutin was shown to be depolymerized by rat pancreatic enzymes (Brown and Kolattukudy, 1978). Considerable progress has been made since the first extracellular cutinase was purified from the culture fluid of *Fusarium solani* f.sp. *pisi* grown on cutin polymer as the sole carbon source (Purdy and Kolattukudy, 1975a). Since then, cutinases from several fungal pathogens have been purified and characterized (Table I), and it is mainly this set of cutinases whose identity has been clarified.

Table I. Cutinases Detected in Cultures of Fungal Plant Pathogens

Pathogen	Enzyme assay[a]	Hydrolysis products identified?	Purified?	Reference[b]
Alternaria alternata	RC	No	Yes	a
Botrytis cinerea	TI, RC	Yes	No	b–e
Cladosporium cucumerinum	TI, RC	No	No	b, c
Colletotrichum gloeosporioides	RC	No	Yes	f
Colletotrichum graminicola	RC	No	No	c
Gloeocercospora sorghi	RC	No	Yes	c
Helminthosporium carbonum	RC	No	No	c
Helminthosporium maydis	RC	No	No	c
Helminthosporium sativum	RC	No	Yes	g
Fusarium solani f.sp. *phaseoli*	RC	No	No	c
Fusarium solani f.sp. *pisi*	RC	Yes	Yes	h, i
Fusarium roseum culmorum	RC	No	Yes	j
Fusarium roseum sambucinum	RC	No	Yes	g
Pythium aphanidermatum	RC	No	No	c
Pythium arrhenomanes	RC	No	No	c
Pythium ultimum	RC	No	No	c
Rhizoctonia solani	TI, RC	No	No	b, c
Stempyllium loti	RC	No	No	c
Sclerotium rolfsii	RC	No	No	c
Ulocladium consortiale	RC	No	Yes	g
Venturia inaequalis	RC	Yes	Yes	k

[a]TI, titration of fatty acids; RC, assayed with radioactive cutin.

[b]a, Tanabe *et al.* (1988a); b, Linskens and Haage (1963); c, C. Baker and Bateman (1978); d, Shishiyama *et al.* (1970); e, Salinas *et al.* (1986); f, Dickman *et al.* (1982); g, Lin and Kolattukudy (1980b); h, Purdy and Kolattukudy (1975a); i, Purdy and Kolattukudy (1975b); j, Soliday and Kolattukudy (1976); k, Köller and Parker (1989).

3.2. Structural Elements of Cutinases

The purification procedures for cutinases from various sources were summarized by Kolattukudy *et al.* (1981b). Since then, one major modification to the general purification scheme has been introduced, namely the application of hydrophobic chromatography on octyl- (Kolattukudy, 1985) or phenyl-Sepharose (Köller and Parker, 1989). All purified cutinases are relatively small enzymes composed of a single polypeptide chain. The molecular weights range from 22,000 for *F. solani* (Purdy and Kolattukudy, 1975a) to 32,000 for *Alternaria alternata* (Tanabe *et al.*, 1988a). Some enzymes appear as two fragments of approximately equal sizes after reduction of a disulfide bridge and electrophoresis under denaturing conditions (Lin and Kolattukudy, 1980b). This cleavage, best explained by a proteolytic nick of the native protein, did not affect the enzyme activities.

The amino acid composition of 14 different cutinases was recently summarized by Kolattukudy (1985). Their overall composition is quite similar. The majority of fungal cutinases contain only one tryptophan, one methionine, one or two histidines, and two or four cysteines involved in disulfide bridges. An unusually high content of glycine was recently identified for the enzyme isolated from *Venturia inaequalis*, the first cutinase to be purified from a leaf pathogen (Köller and Parker, 1989). The complete amino acid sequence of cutinases from *F. solani* f.sp. *pisi* (Soliday *et al.*, 1984), *Colletotrichum gloeosporioides*, and *C. capsici* (Ettinger *et al.*, 1987) were deduced from the sequences of cloned cDNAs. The overall sequence homology between the *Fusarium* and the *Colletotrichum* enzymes was approximately 40%, with the longest homologous sequences positioned around the active serine residue and one of the cysteines involved in a crucial disulfide bond. This relatively low degree of relatedness is reflected by a low degree of cross-reactivity of antibodies prepared against cutinases (Soliday and Kolattukudy, 1976; Lin and Kolattukudy, 1980b). Also interesting is the remarkable variability of cutinases purified from different isolates of *F. solani* f.sp *pisi*. Six enzymes with slight differences in amino acid composition have been reported (Kolattukudy, 1985). Possible functional reasons for this remarkable structural diversity remain to be explored. Cutinases from *F. solani* and some other fungi were found to contain an unusual N-terminus in an amide linkage with glucuronic acid (Lin and Kolattukudy, 1977, 1980a,b). Both the function and the biosynthesis of this N-terminal amide remain unknown.

Fungal cutinases thus far characterized are glycoproteins containing 3–16% carbohydrate (Lin and Kolattukudy, 1980b; Dickman *et al.*, 1982; Köller and Parker, 1989). Chemical linkage analysis conducted with some of the enzymes revealed that most of the carbohydrate was bound O-glycosidically involving serine, threonine, and sometimes β-hydrophenyalanine and β-hydroxytyrosine, the latter two amino acids being novel structures in proteins (Lin and Kolattukudy, 1976, 1980a,b.). The function of the glycoprotein nature has not been investigated. However, O-linked carbohydrates might be a requirement for the secretion of the enzyme, as suggested for extracellular endoglucanases from *Trichoderma reesei* (Kubicek *et al.*, 1987).

3.3. Enzyme Catalysis

The catalytic properties of cutinase were investigated in detail with the purified enzyme from *F. solani* (Purdy and Kolattukudy, 1975b). This enzyme acted as both an endo- and an exohydrolase. Short-term incubation with radioactive cutin released monomeric but also oligomeric products. These oligomers were converted to monomers after prolonged incubation. A fraction larger than cutin monomers were isolated even after longer periods of enzyme action and were thought to represent esters of secondary alcohols derived from cross-linked regions within the cutin polymer. This result suggested a strong preference of cutinase for primary alcohol esters, the major structural element found in cutin. This preference was clearly demonstrated with several model esters (Purdy and Kolattukudy, 1973, 1975b). A less pronounced substrate specificity existed for an esterase coexcreted with cutinase in cultures of *F. solani* grown on cutin (Purdy and Kolattukudy, 1975a,b). However, this nonspecific esterase, which also was present in cultures from *F. roseum, Ulocladium consortiale*, and *Helminthosporium sativum* (Soliday and Kolattukudy, 1976; Lin and Kolattukudy, 1980b), but not from *Colletotrichum gloeosporioides* (Dickman *et al.*, 1982) or *Venturia inaequalis* (Köller and Parker, 1989), lacked cutin-hydrolyzing activity. A function for this nonspecific esterase has not been identified. The series of *p*-nitrophenyl esters was shown to be readily hydrolyzed by cutinases (Kolattukudy *et al.*, 1981b). For cutinase from *F. solani*, a fatty acid chain of C_4 was the optimal substrate. The maximum velocity of hydrolysis dropped sharply with fatty acids $> C_6$ (Purdy and Kolattukudy, 1975b). This strong dependency on the chain length was not observed with cutinase purified from *C. gloeosporioides* (Dickman *et al.*, 1982).

Differences between cutinases exist with respect to the optimal pH for cutin hydrolysis. The optimum for cutinases isolated from *F. solani* f.sp *pisi* (Purdy and Kolattukudy, 1975b) and *F. roseum culmorum* (Soliday and Kolattukudy, 1976) was the alkaline pH of 10. To the contrary, purified cutinase from *V. inaequalis* (Köller and Parker, 1989) and a crude cutinase preparation from *Botrytis cinerea* (Salinas *et al.*, 1986) were reported to hydrolyze cutin optimally at a slightly acidic pH of 6.5. This pH meets the environmental conditions on plant surfaces more closely than does pH 10. Interestingly, both cutinases with an acidic pH optimum were derived from pathogens infecting aerial tissues of plants, whereas many *Fusarium* spp. including *F. solani* are known as damping-off pathogens and, thus, attack the upper root system or the stem base (Nelson *et al.*, 1981). Since most of the purified enzymes have been assayed only at an alkaline pH, a general correlation between the pH-optima of fungal cutinases and the tissue specificity of the respective pathogens cannot be drawn. First direct evidence was derived from a set of pathogens with distinctly different tissue specificities (Trail and Köller, 1990). *Rhizoctonia solani* (AG 2-2), a stem base-specific pathogen of beans excreted a cutinase with an alkaline pH-optimum, whereas the cutinase produced by *Cochliobulos heterostrophus*, a leaf-specific pathogen of corn, exhibited a slightly acidic optimum. Both types of cutinase could be separated from culture fluids of *Alternaria brassicicola*, a pathogen infecting both leaves and stems of *Brassica oleracea*. Furthermore, a stem-specific isolate of *R. solani* nonpathogenic on leaves penetrated the bean leaf surface and produced disease symptoms when the inoculum was amended with cutinase derived from a leaf pathogen. To the contrary, inoculum amended with cutinase from a stem-base pathogen remained nonpathogenic. The general validity of this relationship between cutinase properties and tissue specificity, and the role of different catalytic properties in this host-pathogen specificity remain unknown.

Inhibitor studies on the first cutinase purified to homogeneity suggested that the enzyme belonged to the group of esterases characterized by a serine residue involved in enzyme catalysis (Purdy and Kolattukudy, 1975b). A more detailed study revealed that the phosphorylation of one active serine with diisopropyl fluorophosphate was sufficient to inhibit both esterase and cutinase activity (Köller and Kolattukudy, 1982). In addition, the modification of one histidine with diethylpyrocarbonate was required to inhibit enzyme activity. This histidine labeling was only possible after the deactivation of cutinase with small amounts of SDS, which was easily reversed with Triton X-100 (Köller and Kolattukudy, 1982). A histidine modification could not be accomplished prior to this conformational change, since the modifying agent was shown to be readily hydrolyzed and deactivated by active cutinase (Foster and Kolattukudy, 1987). A similar structural change with SDS was required to chemically modify one essential carboxylic group (Köller and Kolattukudy, 1982). The results of the chemical modification study indicated that one serine, one histidine, and one carboxylic group were involved in the catalysis of ester bond hydrolysis. This catalytic triad is a well-characterized feature of serine proteases like trypsin or elastase (Kraut, 1977; Polgár, 1987). Active serines are also known to exist in different classes of esterases. Some examples of the sequences surrounding the active serine residues of various hydrolases are shown in Table II. All serine hydrolases have conserved the common sequence Gly-X-Ser-X-Gly. The homologies between enzymes that catalyze similar reactions (e.g., cutinases or proteases) are usually more pronounced (Table II).

Several other amino acids in addition to the catalytic triad were shown to be crucial elements for the function of cutinase from *F. solani* (Kolattukudy, 1984d; Köller and Kolattukudy, unpublished results). The reductive cleavage of one disulfide bridge resulted in complete deactivation of enzyme activity and in pronounced changes of the enzyme structure. Thus, the intact disulfide bridge appears to be of crucial importance for the structural integrity of the extracellular enzyme. Two arginine residues are most likely essential for the binding of cutinase from *F. solani* to the natural substrate cutin. Chemical modification of these amino acids strongly inhibited the hydrolysis of radioactive cutin, but had no effect on the esterase activity with *p*-nitrophenol butyrate as the substrate. Arginine-modified cutinase was also shown to have lost the binding capacity to polymeric cutin.

3.4. Induction and Biosynthesis of Cutinases

The induction of cutinase synthesis in mycelia of *F. solani* f.sp. *pisi* was shown to exhibit the features of a classical catabolite repression. No cutinase activity was present in cultures grown on

Table II. Amino Acid Sequence Surrounding the Active Serine Residue Present in Serine Hydrolases

Enzyme	Source	Sequence	Reference[a]
Cutinase	*Fusarium solani*	Ile-Ala-Gly-**Gly**-Tyr-**Ser**-Gln-**Gly**-Ala-Ala	a
	Colletotrichum gloeosporioides	Val-Ser-Gly-**Gly**-Tyr-**Ser**-Gln-**Gly**-Thr-Ala	b
	Colletotrichum capsici	Val-Ala-Gly-**Gly**-Tyr-**Ser**-Gln-**Gly**-Thr-Ala	b
Thioesterase	Rat	Arg-Val-Ala-**Gly**-Tyr-**Ser**-Phe-**Gly**-Ala-Cys	c
	Rat	Ala-Phe-Phe-**Gly**-His-**Ser**-Phe-**Gly**-Tyr-Ile	d
	Duck	Ala-Leu-Phe-**Gly**-His-**Ser**-Phe-**Gly**-Ser-Phe	e
Carboxylic ester hydrolase	*Drosophila melanogaster*	Leu-Leu-Val-**Gly**-His-**Ser**-Ala-**Gly**-Gly-Ala	f
Acetylcholinesterase	*Torpedo californica*	Thr-Ile-Phe-**Gly**-Glu-**Ser**-Ala-**Gly**-Gly-Ala	g
Cholinesterase	Human	Thr-Leu-Phe-**Gly**-Glu-**Ser**-Ala-**Gly**-Ala-Ala	h
Trypsin	Bovine	Ser-Cys-Gln-**Gly**-Asp-**Ser**-Gly-**Gly**-Pro-Val	h
Elastase	Pig	Gly-Cys-Pro-**Gly**-Asp-**Ser**-Gly-**Gly**-Pro-Leu	h
Protease A	*Streptomyces griseus*	Ala-Glu-Pro-**Gly**-Asp-**Ser**-Gly-**Gly**-Ser-Leu	h

[a]a, Soliday *et al* (1984); b, Ettinger *et al.* (1987); c, Naggert *et al.* (1988); d, Randhawa and Smith (1987); e, Poulose *et al.* (1985); f, Oakeshott *et al.* (1987); g, Schumacher *et al.* (1986); h, Daydoff (1972).

glucose. Cutinase production could be induced by low levels of chemically prepared cutin hydrolysate, but only after the glucose pool had been fully depleted. Consequently, glucose acted as an efficient repressor of cutinase production when added along with cutin monomers to glucose-depleted cultures. The structural requirement for a cutinase inducer was shown to be a primary hydroxyl group attached to an aliphatic chain with an optimal chain length around 16, a characteristic structural feature of cutin monomers (Lin and Kolattukudy, 1978). In summary, cutinase production in fungal cultures growing saprophytically is induced by small amounts of cutin monomers, but remains repressed as long as more suitable carbon sources such as glucose are present. A similar induction of cutinase in germinating spores has been demonstrated more recently and will be discussed later.

The *in vitro*-translation product of cutinase mRNA isolated from induced cultures was shown to be larger than the native enzyme (Flurkey and Kolattukudy, 1981). This indicated that cutinase was first synthesized as a precursor protein with an N-terminal leader peptide. The sequence of this leader peptide was recently shown to consist of 36 amino acids (Soliday *et al.*, 1984). Posttranslational processes such as the cleavage of the leader peptide, the introduction of a modified N-terminus, glycosylation, the secretory pathway within the cell, and the final secretion have not been investigated in detail.

4. CUTINASE AS A FACTOR IN PATHOGENICITY

The basic strategy to demonstrate the unequivocal existence of cutinases in cultures of fungal plant pathogens grown on cutin as the sole carbon source, and the characterization of the enzyme properties have been highly successful and informative. However, the mere presence of cutinase in the culture fluid of fungi grown saprophytically on cutin cannot be regarded as conclusive proof for a role in the infection of plants. This role has been investigated by more direct means with the three pathogen–host systems *F. solani*–pea, *C. gloeosporioides*–papaya, and, more recently, *A. alternata*–pear.

4.1. Immunocytology

A major requirement for the proof that cutinase is involved in the penetration step of plant pathogens into their hosts would be the presence of the enzyme at the penetration site. This presence has been demonstrated for pea stems in the region where germinating spores of *F. solani* were breaching the cuticle. Transmission electron microscopy demonstrated the presence of ferritin-labeled antibodies, when cutinase-specific antibodies were applied to the area where the intact cuticle was inoculated with the pathogen (Shaykh *et al.*, 1977). Thus, the infection structures of *F. solani* indeed excreted the same cutinase identified in cultures grown saprophytically on the cutin polymer.

4.2. Cutinase-Deficient Mutants

Several isolates of *F. solani* f.sp. *pisi* were compared with respect to their pathogenicity on intact or mechanically wounded pea stem segments (Köller *et al.*, 1982a). Four representative isolates are described in Table III. The T-8 isolate showed a high infection rate independent of the integrity of the cuticle. In contrast, the T-30 isolate was highly pathogenic on wounded but not intact surfaces. The K1-17 isolate was intermediate, and T-204 was only weakly pathogenic on either surface. The pathogenicity of the latter isolate might be greatly impaired by the failure to detoxify pisatin, the phytoalexin produced by peas (VanEtten *et al.*, 1987). The differences in pathogenicity on intact plant surfaces between T-8, K1-17, and T-30 were reflected by the amount of cutinase released from spores germinating in water (Table III). The release of cutinase preceded spore germination. The final level of cutinase was almost reached after 6 hr, when only 10–17% of the spores had germinated (Köller *et al.*, 1982a).

The correlation between early cutinase release and pathogenicity on intact surfaces suggested that the ability to release sufficient amounts of cutinase was involved in the pathogenicity of respective isolates. The strategy to investigate this hypothesis was to supplement the spores of isolate T-30 with exogenous cutinase. The infection rate was indeed increased by this treatment. However, only when pectinase, cellulase, and pectinesterase were added as an additional supplement did the infection rate increase to the high level expressed on wounded surfaces (Köller *et al.*, 1982a). Thus, mechanical wounding could be fully substituted by a set of "penetration enzymes," including cutinase. One of the additional penetration enzymes, a pectate lyase, was recently purified from isolate T-8. Antibodies prepared for the homogenous enzyme were shown to inhibit both the enzyme activity and the infection of pea stems (Crawford and Kolattukudy, 1987). A mutant of the isolate T-8 with a greatly decreased ability to produce cutinase was shown to exhibit a considerably reduced virulence on intact surfaces of pea stems. Again, virulence

Table III. Infection Rate and Cutinase Activity Released into the Extracellular Fluid of Germinating Spores of F. solani Isolates[a]

Isolate	Infection rate (%)		2 hr germination		6 hr germination	
	Wounded	Intact	Cutinase activity[b]	Germination (%)	Cutinase activity[b]	Germination (%)
T-8	90	90	50,500	3	2,620,000	12
K1-17	88	27	8,100	2	12,500	12
T-30	82	6	2,800	2	3,100	10
T-204	4	4	8,500	1	19,300	17

[a]Data from Köller *et al.* (1982a).
[b]Amount of radioactivity (cpm) released from radioactive cutin per hr per 10^{10} spores.

could be restored with exogenous cutinase (Dantzig *et al.*, 1986). Moreover, there is an indication that T-8 contains two copies of the cutinase gene, whereas the low cutinase producer T-30 has only one (Kolattukudy and Soliday, 1985). The function of this difference remains unknown.

Similar experiments were reported for cutinase-deficient laboratory mutants of *C. gloeosporioides* (Dickman and Patil, 1986). The mutants were highly pathogenic on wounded surfaces of papaya fruits, but disease did not occur when intact surfaces were inoculated. Pathogenicity could be fully restored with exogenous cutinase. Additional cell wall-degrading enzymes were not required in this case. The difference between *F. solani* and *C. gloeosporioides* is interesting with regard to their different ways of infecting their hosts. *Fusarium* spp. penetrate the cuticle and enter the plant cells immediately after germination, while *C. gloeosporioides* penetrates the cuticle of immature papaya fruits, and remains subcuticular in a latent stage until after the fruit matures (Dickman and Alvarez, 1983). This difference would require a synchronous induction of cutinase and cell wall-degrading enzymes in the first case, but a stepwise induction of penetration enzymes in the latter, starting with cutinase. Additional evidence for the requirement of cutinase in pathogenesis was derived from experiments with cutinase-deficient mutants of *A. alternata* (Tanabe *et al.*, 1988b). Three different mutants were shown to be greatly impaired pathogens on pear leaves. Interestingly, pectic enzyme-deficient mutants of the same fungus retained full pathogenicity, and the authors concluded that these cell wall-degrading enzymes were not essential for the expression of pathogenicity of *A. alternata*. Further support for the necessity of cutinase in plant infection was derived from *Mycosphaerella*, a wound parasite of papaya fruits. This fungus became pathogenic on intact fruits when the inoculation fluid was supplemented with cutinase (Dickman *et al.*, 1982). This result suggested that only the deficiency in cutinase production was restricting this wound pathogen from full pathogenicity.

4.3. Inhibitors of Cutinase

If exogenous cutinase can restore the pathogenicity of cutinase-deficient mutants, the specific deactivation of cutinase activity should be expected to restrict greatly the pathogenicity of fungal pathogens. This experimental approach led to conclusive results with *F. solani*, *C. gloeosporioides*, and *A. alternata*. Cutinase-specific antibodies were shown to protect peas (Maity and Kolattukudy, 1979; Köller *et al.*, 1982a) and papaya fruits (Dickman *et al.*, 1982) from disease. Moreover, phosphoroganic esters, well-established inhibitors of acetylcholinesterase in insects, were also powerful inhibitors of cutinase (Dickman *et al.*, 1983; Köller *et al.*, 1982b; Tanabe *et al.*, 1988a). These cutinase inhibitors were shown to protect peas, papaya fruits, and pear leaves from disease at concentrations not inhibitory to fungal growth. However, the protective activity was not expressed when wounded surfaces were inoculated. The results strongly suggest that the deactivation of cutinase activity is sufficient to prevent the penetration step and, consequently, the development of plant disease.

4.4. Genetic Transformation Studies

Considerable progress has been made by applying the techniques of molecular genetics to questions regarding the function of fungal cutinases. Using the cutinase cDNA as a probe, the transcriptional regulation of the cutinase gene was studied in an *in vitro* system with nuclei isolated from *F. solani* f.sp. *pisi* (Podila *et al.*, 1988). As indicated in an earlier study (Lin and Kolattukudy, 1978), typical cutin monomers were specific inducers of the *in vitro* gene transcription when combined with a proteinaceous factor present in the cytosol. This factor might function as both a receptor and transcription factor, similar to steroid hormone receptors (Godowski and Picard, 1989).

The detailed structure of the cutinase gene from *F. solani* f.sp. *pisi* was recently described by Soliday *et al.* (1989). The sequencing of the a 2,800 base pair genomic DNA fragment revealed an open reading frame of 690 base pairs identical to the cDNA sequence, and a 51-base pair intron. In addition to this coding region, a 360 base pair promoter region immediately 5′ to the cutinase initiation codon could be identified. The cutinase gene from *F. solani* f.sp. *pisi*, surrounded by extensive portions of the 5′ and 3′ flanking region, was transferred into a *Mycosphaerella* spp. (Podila *et al.*, 1989), a wound-pathogen previously shown to become pathogenic on intact plant surfaces after amendment with exogenous cutinase. After transformation with the cutinase gene, the transformants excreted *F. solani* cutinase induced by cutin monomers and acquired the capacity to infect intact papaya fruits. Moreover, the production of cutinase and the virulence of transformants was dependent on the number of cutinase gene copies. This result is compelling evidence for the crucial role of cutinase in fungal pathogenicity.

Similar results were obtained when a cutinase-less mutant of *F. solani* f.sp. *pisi* previously shown to be non-pathogenic on intact surfaces (Köller *et al.*, 1982a) was transformed with the cutinase gene (Kolattukudy *et al.*, 1989). Again, several transformants became pathogenic on intact pea stems, and the virulence ranking correlated with the level of cutinase produced by the genetically altered strains. The results with both the transformed *Mycoshperella* spp. and *F. solani* f.sp *pisi* greatly substantiate the conclusions drawn from previous 'cutinase amendment' experiments (Dickman *et al.*, 1982, Köller *et al.*, 1982a).

4.5. The Cutinase Concept

The three lines of evidence described above—the presence of cutinase at the site of penetration, the greatly restricted capability of cutinase-deficient mutants to penetrate into their hosts, and the antipenetrant nature of various cutinase inhibitors—provide conclusive evidence for the crucial role of cutinase in the establishment of disease and for the enzymatic dissolution of the plant cuticle during the penetration of plant surfaces. One critical question, however, remained open: was the amount of preformed cutinase released by germinating spores (Köller *et al.*, 1982a) sufficient for cuticle penetration, or was additional cutinase production and excretion required for the successful penetration of the pathogen? This question has been investigated more recently by Woloshuk and Kolattukudy (1986). Germinating spores of *F. solani* f.sp. *pisi* were shown to synthesize cutinase *de novo* only when cutin hydrolysate was added to the spore suspension. The typical structural element of cutin monomers, an ω- and a midchain hydroxyl group, was a requirement for efficient cutinase induction. Cutinase production could also be induced with polymeric cutin, but the cutinase level was greatly enhanced after the addition of small amounts of purified cutinase. Cutinase production was regulated at the transcriptional level, and mRNA could be detected within 15 min after treatment with inducers.

These results provide the basis for a relatively refined cutinase concept. The fungal spore attached to the plant surface releases a small quantity of preformed cutinase at the onset of germination. The function of this "sensing" cutinase is the liberation of small amounts of cutinase monomers, which act as transcriptional inducers of the bulk production of cutinase required for the enzymatic penetration of the cuticle. The cutinase level detected with spores of *F. solani* germinating in water (Köller *et al.*, 1982a) represents the "sensing" cutinase responsible for cutinase production during cuticle penetration. Small amounts of esterase activity were also extracted from the mucilaginous matrix surrounding spores of *C. graminicola* (Nicholson and Moraes, 1980), or from germinating conidia of *Erysiphe graminis* (Nicholson *et al.*, 1988). These enzymes were suggested to represent cutinases with a direct function in cutin hydrolysis, but they might well resemble a "sensing" cutinase similar to the enzyme released by spores of *F. solani*.

Cutinase induction by cutin monomers was demonstrated earlier for saprophytically growing mycelia of *F. solani* (Lin and Kolattukudy, 1978). In this case, glucose was shown to act as an efficient repressor of cutinase production. The presence of a similar catabolite repression has not been investigated for germinating spores. This question is of particular interest, since carbohydrates (including glucose) might well be present on leaf surfaces. Cutin monomers acting as inducers and glucose acting as a repressor of cutinase production would suggest that the presence of glucose on the plant surface should have an adverse effect on disease development. Within some experimental limitations, the pathogenesis of *F. solani* f.sp. *phaseoli* was indeed delayed when glucose was present in the infection droplet. To the contrary, disease development was enhanced by arginine, but also by a mixture of peptone and glucose (Toussoun *et al.*, 1960). Clearly, the influence of free nutrient sources present on plant surfaces on the induction of the penetration enzyme cutinase needs further investigation.

The principles of the cutinase concept comprise one of the best understood systems in molecular plant pathology. However, some interesting questions have not been addressed. For example, a critical evaluation exploring the impact of the cutin composition and structure on cutinase action and induction is largely lacking. These compositional changes of cuticles can be very pronounced (e.g., Riederer and Schönherr, 1988), and it appears feasible to expect differences in cutinase accessibility for different cuticle structures. More importantly, even the mere presence of cutinase during the penetration of plant surfaces is far from clear for the vast majority of plant diseases.

5. MODES OF PENETRATION AND INVOLVEMENT OF CUTINASE

The time required for penetration into the host tissue varies considerably, with extremes ranging from 2 hr for *Peronospora tabacina* invading tobacco leaves (McKeen and Svircev, 1981) to 4 days for *C. gloeosporioides* infecting papaya fruits (Chau and Alvarez, 1983). However, in most host–pathogen interactions the penetration step is completed between 12 and 24 hr after the onset of spore germination. Most often, the penetration event is preceded by the formation of specialized infection structures such as appressoria or infection cushions. This cell differentiation, however, is not a stringent requirement (Aist, 1976; Emmett and Parbery, 1975; Kunoh, 1984). Penetration can occur through stomata (Hoch and Staples, 1987) or directly through the cuticle (Aist, 1976; Emmett and Parbery, 1975; Kunoh, 1984). The sites of cuticle penetrations, however, are not always random. It has long been noted that cell junctures are often the favored sites for appressorium formation by fungal pathogens (Preece *et al.*, 1967).

Numerous ultrastructural studies have been dedicated to the interaction of pathogens with their hosts, many of which emphasize the early steps in disease development. Surprisingly, the cell wall and not the cuticle was often envisioned as the first and exclusive penetration barrier. This situation changed after the biochemical work on the role of cutinase purified from *F. solani* and *C. gloeosporioides* received more attention among plant pathologists, and the ultrastructure of cuticle penetration has been increasingly investigated and discussed. Because conclusions based solely on cytological methods have their experimental limitations, it is not surprising that the original controversy surrounding the mode of cuticle penetration is still not entirely clarified. In particular, research with inhibitors of melanin biosynthesis and melanin-deficient mutants supported a mechanical rupturing of the cuticle in those host–pathogen interactions, where the melanization of appressoria is a requirement for successful penetration (see Chapter 9).

In most cases the penetration of the cuticle is preceded by the formation of specialized infection structures. The morphological appearance of these structures varies considerably, and in some cases the plant surface is even breached by the tip of the growing germ tube (Aist, 1976). More common, however, is the formation of appressoria prior to the penetration step. Penetration is often accomplished by small infection pegs emerging from the base of the appressoria. These infection pegs are sometimes "naked"—a cell wall is either missing or very thin. Three major types of appressoria have been described: (1) Hyaline appressoria that emerge from germ tubes. They may or may not be delimited by a septum. (2) Compound appressoria, most commonly found on hyphae growing on the plant surface, can be multicellular. They often become infection cushions formed from the aggregation of several hyphae. Many pathogens such as *Botrytis* spp. or *Sclerotinia* spp. form both hyaline and compound appressoria. (3) Dark appressoria are characterized by heavily melanized cell walls. They are described in detail by Kubo and Furusawa (Chapter 9).

The mode of penetration as typified by the morphology of the infection structures might be the most appropriate way to describe and discuss the current state of knowledge regarding the role of cutinase in cuticle penetration. It is beyond the scope of this chapter to present a comprehensive coverage of this subject, and only some more recently investigated cases with model character will be discussed.

5.1. Direct Penetration

Fusarium spp. cause two types of disease on a wide variety of crops: cortical rots and vascular wilts (Nelson *et al.*, 1981). Intact surfaces of roots and stems are penetrated by the growing tip of the germ tube without the formation of appressoria (e.g., Parry and Pegg, 1985). Direct penetration of pea tissue has also been reported for *F. solani* f.sp. *pisi* (Shaykh *et al.*, 1977). Cutinase and its role in the infection process have been most extensively investigated for this particular host–pathogen system. As described in great detail above, cutinase has to be regarded as a crucial pathogenicity factor in this particular case. *Pythium acanthicum* (Aist, 1976) and *Cladosporium cucumerinum* (Paus and Raa, 1973) are other examples for the direct mode of penetration. *Pythium* spp. and *C. cucumerinum* were reported to release cutinase into the medium when grown on cutin as a carbon source (Table I), but direct evidence for the involvement of cutinases in infection is lacking.

5.2. Penetration from Hyaline Appressoria

V. inaequalis, the causal agent of apple scab, is an excellent example to illustrate the development of "penetration concepts." Infection structures originating from ascospores or conidia penetrate the intact cuticle after which the formation of subcuticular stroma is initiated (Gessler and Stumm, 1984; Smereka *et al.*, 1987, 1988). Cuticle penetration from appressoria of *V. inaequalis* is illustrated in Fig. 3. Whether the breaching of the first barrier of apple leaves is achieved by enzymatic degradation of the cuticle or by mechanical force has been debated for over 70 years. Wiltshire (1915) reported that scab hyphae grew between the cuticle and the cell wall, and that the cuticle above these hyphae was always thinner. This indication for an enzymatic degradation of the cuticle was not confirmed by Nusbaum and Keitt (1938). Furthermore, the fungitoxicity of cutin monomers was taken as evidence against an enzymatic degradation of the cuticle. It was speculated that the fatty acids, once liberated by the action of the enzyme, would exert a suicidal effect on the invading fungus (Martin, 1964). However, a transitory esterase activity detected during the early stage of infection (Nicholson *et al.*, 1972), and the presence of cutin monomers in a cutin-based culture medium (Sparapano and Graniti, 1977) provided the first circumstantial evidence for the involvement of cutinase in the penetration of apple leaves. Cutinase from *V. inaequalis* has been purified only recently (Köller and Parker, 1989).

Alternaria brassicae is the incitant of blackspot disease of oilseed rape. The pathogen enters leaves of the host either through stomata or directly through the cuticle after formation of simple appressoria (Tsuneda and Skoropad, 1978). Whenever direct penetration was observed, cuticle penetration was followed by a brief period of subcuticular hyphal growth. The cuticle was not

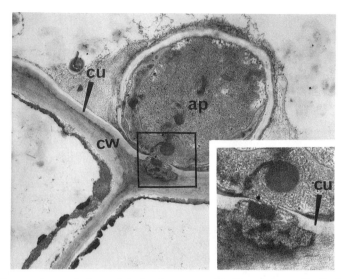

Figure 3.   Electron micrograph of an apple leaf surface penetrated by *Venturia inaequalis*. The insert represents a magnification of the indicated section. The cuticle is breached at the base of an appressorium, and a subcuticular stroma is being formed. The cuticle at the penetration site appears to be degraded. The photograph represents the developmental stage 36 hr after inoculation. ap, appressorium; cu, cuticle; cw, cell wall. (Micrograph kindly provided by M. Yepes, Department of Plant Pathology, Cornell University, NYS Agricultural Experiment Station, Geneva.)

deformed or indented at the site of penetration, and the involvement of cutinase rather than a mechanical mode of penetration was discussed (Tewari, 1986). Conclusive evidence for cutinase was provided for the host–pathogen system of pear leaves and *A. alternata* (Tanabe *et al.*, 1988a,b). Cutinase was purified and characterized, the pathogenicity of cutinase-deficient mutants was greatly reduced, and cutinase inhibitors acted as effective antipenetrants.

Penetration of wheat leaves by *Septoria nodorum* (leaf spot of wheat) involves the formation of appressoria and infection pegs, from which the cuticle is penetrated (Karjalainen and Lounatmaa, 1986). Direct penetration by germ tubes without the prior formation of appressoria has also been reported (Zinkernagel *et al.*, 1988). Ultrastructural evidence for enzymatic dissolution of the cuticle during penetration was evident in the latter case. This observation is interesting with respect to the similar mode of penetration by *F. solani*, in which cutinase was shown to be involved. Cuticle penetration of *S. nodorum* is frequently followed by extensive subcuticular mycelial growth before the epidermal cells are entered (Zinkernagel *et al.*, 1988).

Germinating conidia of *Pyrenophora teres* (net blotch of barley) form appressoria on leaves of barley, from which infection hyphae penetrate the cuticle (Keon and Hargreaves, 1983). The cuticle was physically deformed around the penetration site, although the penetration holes beneath detached appressoria were round with smooth edges. Keon and Hargreaves (1983) interpreted their results as evidence for a combined enzymatic dissolution and mechanical mechanism for cuticle penetration. The presence of cutinase activity has not been reported from *P. teres*, but from a similar pathogen, *Helminthosporium sativum* (Table I).

Erysiphe graminis (powdery mildew of barley and wheat) penetrates the host surface from a small infection peg that emerges at the base of an appressorium. Involvement of cutinase in cuticle penetration has been suspected (Kunoh, 1981). McKeen and Rimmer (1973) reported that the tip of the penetration peg was blunt and lacked a well-developed cell wall. The authors concluded that a mechanical penetration of the host cell wall was unlikely without the rigidity of a cell wall. However, the cuticle was reported to be buckled inward and conformed to the tip of the penetration peg, an indication that the stressed cuticle was mechanically ruptured (Sargent and Gay, 1977). Alternatively, the regular appearance of penetration holes on leaves of barley and cucumber inoculated with *E. graminis* and *E. cichoracearum* was discussed as evidence for an enzymatic dissolution of the cuticle (Staub *et al.*, 1974). Furthermore, conidia of *E. graminis* were shown to release small amounts of esterase activity, possible a cutinase, as an immediate response to a contact stimulus (Nicholson *et al.*, 1988). Heintz (1986) investigated the mode of cuticle penetration with a similar pathogen, *Uncinula necator* (grapevine powdery mildew), and concluded that cuticle penetration was mechanical. Infection pegs were able to perforate thin synthetic membranes made of methacrylate polymer. Interestingly, the diameter of the penetration hole in the artificial membranes was eight times larger than that observed in leaf cuticles. It might be possible that the weakening of the synthetic membrane and the subsequent perforation was caused by "detergents" released from the germinating hyphae and appressoria. Impressions into the surface wax of barley and cucumber leaves have been reported after conidial hyphae and appressoria of *E. graminis* and *E. cichoracearum* were removed (Staub *et al.*, 1974). This interesting aspect of wax dissolution prior to penetration is of more general importance with respect to cutinase action, but has rarely been considered and discussed. Waxes comprise the outer layer of the cuticle, and it appears logical that these hydrophobic and solid waxes have to be dissolved before cutinase can diffuse to the polymeric cutin substrate. Whether this wax removal is aided by enzymes or merely by a detergent characteristic of the extracellular adhesive material remains unexplored.

The initial stages of infection by *Phytophthora infestans* (tomato late blight) occur after the formation and germination of zoospore cysts (Coffey and Wilson, 1983). Appressoria are formed with preference for the "stomatal complex" consisting of guard cells and adjacent epidermal cells (Gees and Hohl, 1987), and the plant leaf surface is penetrated at both of these sites. The mode of cuticle penetration has drawn little attention and is only rarely discussed (e.g., Coffey and Wilson, 1983). However, an enzyme with esterase activity similar to cutinase was reported to be actively secreted from germinating zoospore cysts of this fungus (Moreau and Seibles, 1985).

Botrytis species infecting healthy plant tissues commonly form simple appressoria at the tip of conidial germ tubes (Verhoeff, 1980). Infection can also occur from compound appressoria and infection cushions (Backhouse and Willetts, 1987). The mechanism of cuticle penetration by *B. cinerea* has been controversial. The earlier literature was clearly in favor of a mechanical rupture as the mode of cuticle penetration (Verhoeff, 1980). Involvement of cutinase in the penetration process was first discussed for the interaction of *B. cinerea* with *Vicia faba* (McKeen, 1974). The author observed round holes with sharp edges at the sites of penetration. He further demonstrated histochemically the presence of a transitory esterase activity during the early stage of infection. Since the fungal cell wall around the penetration peg was absent, any mechanical force necessary for penetration would have to be generated by the protoplast alone. The ultrastructural data suggesting an enzymatic dissolution of the cuticle during the penetration of intact plant surfaces by *B. cinerea* were strongly supported by Rijkenberg and co-workers (1980). The presence of cutinase in culture fluids of this fungus and the preliminary characterization of the enzyme (Salinas *et al.*, 1986) are further evidence for the possible involvement of cutinase.

5.3. Penetration from Compound Appressoria and Infection Cushions

Sclerotinia sclerotiorum is a necrotrophic pathogen with a broad host range. The mechanism of cuticle penetration (enzymatic versus mechanical) has long been debated, starting with de Bary in 1887, who suggested that a "ferment" was involved in cuticle penetration. This early literature is summarized by Tarique and Jeffries (1986) in their study on the ultrastructure of the infection of *Phaseolus* spp. Cuticle penetration occurred by a narrow infection peg emerging from appressoria that differentiated from aerial mycelia. Frequently, the mycelia aggregated to form infection cushions. The zone where the infection peg emerged was characterized by vesicular material outside the fungal plasma membrane. The contents of the vesicles were discharged into the cuticular adhesion zone after the dissolution of the fungal cell wall. The wall-less infection peg penetrated through the cuticle without causing any physical depressions. This mode of penetration is indicative of the involvement of cutinase, a suggestion later substantiated by the cytochemical demonstration of lipolytic activity during cuticle penetration (Tarique and Jeffries, 1987).

Rhizoctonia solani is a widely distributed plant pathogen and attacks a tremendous range of plants, causing damping-off, root rots, fruit decays, and foliar diseases (Parmeter, 1970). The host range and tissue specificity are partly determined by the particular anastomosis group (Ogoshi, 1987). Different modes of penetration have been reported, but the formation of infection cushions prior to penetration is considered the typical infection structure for this group of pathogens (Dodman and Flentje, 1970; Armentrout and Downer, 1987). Epiphytic hyphae produce numerous side branches that continue to proliferate within localized areas on the plant surface, eventually forming infection cushions. Penetration occurs from simple appressoria originating at the base of the cushions. An involvement of cutinase in cuticle penetration has been suggested (Linskens and Haage, 1963; Baker and Bateman, 1978). The penetration stage has recently been studied cytologically for the interaction of *R. solani* (AG 1) with rice (Matsuura, 1986). The penetration pores visible in the epidermis had smooth edges, although it was not clear whether cutinase action was involved. A cutinase from *R. solani* has been purified (Trail and Köller, unpublished), and studies are under way to investigate whether or not the penetration is accompanied by cutin dissolution.

5.4. Penetration from Melanized Appressoria

Pyricularia oryzae, the causal agent of rice blast disease, forms dark melanized appressoria. Penetration of the cuticle and the cell wall is accomplished by a small infection peg that emerges at the base of the appressorium (see Chapter 9). In an early ultrastructural study, the partial dissolution of the rice cuticle beneath the appressoria was thought to be an indication of a chemical rather than a physical penetration (Hashioka *et al.*, 1967). This concept changed when inhibitors of melanin biosynthesis were shown to act at the level of penetration (Kubo and Furusava, Chapter 9; Sisler, 1986). The mechanical force exerted by the emerging penetration peg was considered to comprise the primary means by which the cuticle was penetrated; melanin was considered essential in providing the strength and rigidity in anchoring appressoria to the plant surface. Under normal conditions, the mechanical force produced by the emerging infection peg was focused on a small cuticular area. Failure of nonmelanized appressoria to develop infection pegs was explained as an insufficient anchorage of appressoria to the leaf surface and, therefore, a dissipation of the mechanical force to a relatively large area of the cuticle. The "mechanical" concept was supported by the observation that diisopropyl fluorophosphate, a cutinase and penetration inhibitor in the *F. solani*–peas interaction (Maiti and Kolattukudy, 1979), lacked activity as an antipenetrant (Woloshuk *et al.*, 1983). A refinement of the model in support of a mechanical penetration was presented by Howard and Ferrari (1989). Using wild-type and melanin-deficient strains the authors concluded that melanin had a different permeability to water and solute and allowed the built-up of a high internal pressure when appressoria were 'glued' to surfaces. This pressure was sufficiently high for the penetration of plastic surfaces and presumably plant surfaces. Although the evidence against a role for cutinase is strong, the question deserves more direct investigation and more conclusive data. Mucilage responsible for the attachment of conidia to the hydrophobic rice leaf surface was shown to be expelled from the conidial apex before germ tube emergence (Hamer *et al.*, 1988). Appressoria are similarly attached by mucilage (Woloshuk *et al.*, 1983), and such attachment was shown to be strongly inhibited by melanin biosynthesis inhibitors (Inoue *et al.*, 1987). Thus, the melanin biosynthesis inhibitors appear to interfere with secretion of extracellular mucilage, and this inhibitory action might also involve cutinase secretion.

Colletotrichum spp. form melanized appressoria similar to those of *P. oryzae* (see Chapter 9). Cutinase from *C. gloeosporioides* has been purified and characterized (Dickman *et al.*, 1982; Ettinger *et al.*, 1987). It was noted that the specific activity of cutin hydrolysis was 100-fold lower than that determined for cutinase from *F. solani* (Dickman *et al.*, 1982). This difference is remarkable and might relate to the particular mode of penetration via melanized appressoria. *C. gloeosporioides* causes fruit rot on a wide variety of plants such as avocado and papaya. Although disease symptoms do not appear until harvest, infections occur early in the season and remain latent until fruit ripening. Latency is established either after the formation of dark appressoria (Binyamini and Schiffmann-Nadel, 1972) or after cuticle penetration and formation of subcuticular hyphae (Chau and Alvarez, 1983). The crucial role of cutinase was proven for the infection of papaya fruits (Dickman and Patil, 1986). Here, a relatively thick cuticle of a fruit has to be penetrated, and it is possible that the need for cutinase is restricted to these particular cases. The involvement of cutinases in the penetration of thin cuticles on young leaf and stem tissue by *Colletotrichum* spp. is less well understood. This question has been indirectly investigated for *C. lindemuthianum* and bean (Wolkow *et al.*, 1983). Penetration of the plant surface was completely blocked by inhibitors of melanin biosynthesis. The cutinase inhibitor diisopropyl fluorophosphate was not active as an antipenetrant. However, the penetration of bean stems was slowed by the inhibitor, indicating a potential role for cutinase in addition to the melanization of appressoria. A cytological study of this particular host–pathogen system did not conclusively answer the question of cutinase involvement (O'Connell *et al.*, 1985). Cutinase activity was detected in culture filtrates of *C. graminicola* (Table I). In addition, the mucilaginous matrix surrounding spores of this pathogen has been shown to contain small amounts of esterase activity (Nicholson and Moraes, 1980). An esterase with cutinolytic activity was recently purified from *Colletotrichum lagenarium* (Bonnen and Hammerschmidt, 1989a). In contrast to the cutinase characterized from *C. gloeosporioides* with its molecular weight in the range of fungal cutinases (Dickman *et al.*, 1982), the enzyme from *C. lagenarium* had a molecular weight of 60,000 and, thus, resembled the nonspecific esterase purified from *F. solani* grown on cutin as carbon source (Purdy and Kolattukudy, 1975a). No evidence was found for a role of cutinolytic enzymes in the infection of cucumber by *C. lagenarium* (Bonnen and Hammerschmidt, 1989b). In summary, while there is substantial evidence for the "physical" mode of cuticle penetration by pathogens forming melanized appressoria, an additional involvement of cutinase cannot be excluded.

5.5. Penetration through Stomata

Some pathogens, such as *rust fungi*, penetrate the plant exclusively via stomata (Hoch and Staples, 1987, and Chapter 2). This route of invasion represents the most complex type of penetration, and the involvement of cutinase is less likely though not entirely excluded. A requirement for cutinase would largely depend on the question of whether the substomatal cavity is covered by a thin cuticular layer. This question is controversial and remains unresolved (Kolattukudy, 1985). Interestingly, the penetration route of rusts depends on the sexual stage of the spores (Littlefield and Heath, 1979). For example, urediniospores of *Uromyces appendiculatus* penetrate exclusively through stomata (see Chapter 2), whereas the basidiospores form appressoria on the intact surface and penetrate the cuticle directly (see Chapter 4). It is intriguing to speculate that cutinase is involved in the latter type of plant infection, and progress regarding the presence and function of cutinase genes in rust fungi will most likely result from work with basidiospores.

5.6. Conclusion

The examples described above clearly demonstrate that our knowledge of the general role of cutinase in penetration is still limited. Most of our current understanding is based on electron microscopic studies and, consequently, on the physical appearance of the cuticle in contact with the invading pathogen. Round and smooth penetration holes have been used as a criterion for the enzymatic dissolution of the cuticle. A mechanical mode of penetration would be expected to leave signs of tearing and irregularities around the penetration hole. As frequently criticized, such irregularities might have been removed during sample preparation. The fact that many infection structures are cemented to the cuticle, that infection pegs exert sufficient physical pressure to penetrate through artificial membranes, and that a strengthening of appressoria by melanin is sometimes a prerequisite for successful penetration are often taken as evidence for a mechanical rupture of the cuticle. The histochemical demonstration of transient esterase activity during the early penetration stage (McKeen, 1974; Nicholson *et al.*, 1972) provides more direct and conclusive evidence for the presence of cutinase at the penetration site, but even in these cases the cutinase nature of the esterase activity remains to be clarified by biochemical and immuno-cytological studies. Conclusive proof for the crucial involvement of cutinase during penetration, would require several lines of evidence: For example (1) cutinase must be present at the site of penetration, (2) cutinase-deficient mutants must be avirulent, (3) the transfer of the cutinase gene into these mutants should restore pathogenicity, and (4) specific cutinase inhibitors should block the penetration step and protect plants from disease. Clearly, the demonstration of cutinase activity *in vitro* and, most ideally, the purification of the enzyme will be a requirement for major achievements. This task is especially difficult for the group of strictly obligate pathogens such as the rust or powdery mildew fungi. In these cases, the modern techniques of molecular genetics can be expected to provide the tools for future progress.

6. CUTINASE AS A TARGET FOR PLANT DISEASE CONTROL

Soon after the first cutinase had been purified from *F. solani*, the value of this enzyme as a target for plant disease control was recognized (Shaykh *et al.*, 1977). The deactivation of cutinase at the plant surface would prevent penetration and, thus, would protect plants from fungal disease. The principles of this concept were soon investigated and shown to be feasible. Both cutinase-specific antibodies and diisopropyl fluorophosphate, a known inhibitor of serine esterases, protected pea stems from infection by *F. solani* (Maiti and Kolattukudy, 1979).

The need for a biorational design of chemicals to control plant diseases has been frequently discussed (e.g., Schwinn and Geissbühler, 1986). Cutinase might be an excellent target enzyme to test the feasibility of this approach. However, a major shortcoming still exists. As described above, our knowledge is almost exclusively based on cutinases from *F. solani* and *C. gloeosporioides*, both of which are relatively specialized pathogens. *F. solani* infects upper roots and lower stem bases with cuticle penetration occurring directly from undifferentiated germ tubes. *C. gloeosporioides* penetrates the relatively thick cuticle of immature fruits and then becomes latent. A major concern still is that the involvement of cutinase might be restricted to a relatively narrow spectrum of diseases. The preceding section clearly demonstrates the lack of a clear understanding of a substantial role for cutinase in a broad spectrum of economically important diseases. This represents a question of great practical importance with respect to the development of agrichemicals. Nevertheless, investigations dedicated to this practical aspect of cutinase inhibition have been initiated. Work so far has been primarily done to test the ground for the biorational design of chemical inhibitors.

6.1. Inhibition of Enzyme Action

The identification of the enzyme mechanism of cutinase as a typical serine hydrolase (Köller and Kolattukudy, 1982) presented the opportunity for elucidating the cutinase-inhibitory effects of organophosphorus pesticides with their known mode of action as serine esterase inhibitors (Köller et al., 1982b). The potency of these compounds as cutinase inhibitors varied greatly and followed structural rules established for the inhibition of their insecticidal target acetylcholinesterase. All pesticides prevented the infection of intact pea stems by F. solani under controlled bioassy conditions. Infection, however, was not prevented on wounded surfaces. The effective inhibitor concentrations reflected the corresponding inhibitor values for cutinase inhibition, and no direct effect on mycelial growth could be observed. The data suggested that the pesticides prevented penetration and infection by inhibiting cutinase action. The principles of this study were successfully expanded to the system C. gloeosporioides–papaya (Dickman et al., 1983). One of the inhibitors, diethyl trichloropyridyl, was reported to protect corn from C. graminicola and apples from V. inaequalis under greenhouse conditions (Kolattukudy, 1984b). The structure of this compound was successfully optimized according to the special requirements of the active site of cutinase (Maurer et al., 1986). Cutinase from F. solani was also inhibited by the fungicide benomyl (Köller et al., 1982c). This irreversible inhibition was shown to be caused by n-butylisocyanate, a breakdown product of benomyl. This compound was shown to be responsible for the protective activity of benomyl in laboratory assays. Apparently, this protective mode of action of benomyl is less pronounced under field conditions. The development of field resistance to benomyl was not different from other benzimidazole fungicides that lack the activity of cutinase inhibition (Davidse, 1987). This example demonstrates, however, the possibility of combining more than one mode of action in a chemical structure. The "design" of potential cutinase inhibitors can be based on the analogies of cutinase and various other serine hydrolases. Numerous mechanistically different inhibitors of this class of enzymes have been reported in the past, including isatoic anhydrides (Moorman and Abeles, 1982), trifluoromethylketones (Gelb et al., 1985), chloroisocoumarines (Harper and Powers, 1985), or isobenzofuranones (Hemmi et al., 1985), to name only a few examples. The strategy would be to adopt the inhibitors to the special requirements of the cutinase binding site.

Alternatives to chemical inhibitors of cutinase, although highly speculative, might emerge in the future. For example, antibodies against cutinase were shown to prevent infection very similar to chemical inhibitors (Köller et al., 1982a; Maiti and Kolattukudy, 1979). It might become feasible to utilize and modify the gene coding for the production of inhibitory monoclonal antibodies. These antibodies or, more likely, active fragments thereof might then be produced by transformed organisms inhabiting the phylloplane of plants, or they might be applied directly after fermentative production. Much progress would still have to be made in this area of research, but promising developments are proceeding rapidly (e.g., Bird et al., 1988).

6.2. Inhibition of Synthesis

There are several opportunities during the infection process to block cutinase action, in addition to direct chemical interference with the cutin hydrolytic activity. Vulnerable steps are both synthesis and excretion, and the rationale of how to interfere with either event might become clearer in the future. For example, the flanking regions of the cutinase gene have been identified (Ettinger et al., 1987). An in vitro system for studies on the molecular mechanism of the cutinase gene activation has been developed and described (Podila et al., 1988), and promoter activity in the 5'-flanking region of the cutinase gene has been identified (Soliday et al., 1989). The understanding of the molecular events of cutinase induction might lead to an opportunity to block the step of gene activation. All posttranslational steps involved in cutinase processing and excretion such as the cleavage of the leader sequence, the introduction of a unique N-terminus, the glycosylation, and the final excretion step (Kolattukudy, 1985) might serve equally well as targets for control agents. Unfortunately, little progress has been made in this field of cutinase research, and much work needs to be done to evaluate these possibilities.

7. ECOLOGY OF CUTINASE

It is generally accepted that nonpathogenic and epiphytic microorganisms inhabiting the plant surface are utilizing plant leachates as a nutrient base (Morris and Rouse, 1985). There is evidence, however, that at least some of the phylloplane microflora might also utilize components of the cuticle. Cuticle degradation might even be more pronounced in the soil. Since all cuticular components are recycled in the litter and soil after leaf fall, extensive degradation must take place. It is likely that soil saprophytes participate in this cuticle decomposition. Surprisingly little is known about this ecological aspect.

7.1. Cutinases from Phylloplane Organisms

The possibility that phylloplane organisms might utilize cuticle components as a carbon source was first suggested by Ruinen (1963). A more detailed study on this subject was published later (Ruinen, 1966). Two yeasts (Cryptococcus laurentii and Rhodotorula glutinis and two nitrogen-fixing bacteria (Azotobacter chroococcum and Beijerinckia flumigensis) were shown to erode the surfaces of isolated plant cuticles. Cryptococcus laurentii released titratable fatty acids into the growth medium when cultivated with isolated cuticles. Since the cuticle preparations used in this study were not dewaxed, it is not clear whether these fatty acids were cleaved from the polymer cutin or were generated by the dissolution of waxes. Leaf waxes were suggested to be utilized by the phylloplane yeast Sporobolomyces roseus (McBride, 1972). But again, the conclusions were based on the physical appearance of the cuticle in the vicinity of the yeast cells, and the erosion of the cuticle surface might have been caused by the detergent nature of adhesive material excreted by the organisms. Cuticular wax was subsequently shown to stimulate the growth of S. roseus under suitable growth conditions and, thus, might indeed serve as a nutrient source (MacNamara and Dickinson, 1981).

The presence of bacterial cutinase has been demonstrated for Pseudomonas putida inhabiting the phylloplane (Sebastian et al., 1987). In contrast to fungi, cutinase production could be induced only by polymeric cutin and not by cutin monomers. P. putida was closely associated with a nitrogen-fixing Corynebacterium sp., and detailed studies on the nutritional requirements suggested that the two microorganisms coexist in a symbiotic relationship. P. putida was shown to provide the carbon source in the form of cutin monomers liberated by the action of cutinase, and the Corynebacterium provided the reduced nitrogen essential for growth and proliferation.

In general, bacteria penetrate the plants via natural openings or wounds (Huang, 1986), and unlike fungi, not directly through intact plant surfaces. Nevertheless, a role for a bacterial cutinase in the penetration of plant cuticles was suggested for Pseudomonas syringae pv. tomato (Bashan et al., 1985). Cutinase could be extracted from tomato leaves infected with the pathogen, and enzyme activity increased for the first 48 hr of disease development. These increases were postulated to originate from bacteria multiplying within the stomatal cavity. Cutinase released by the growing organism might lead to the dissolution of the thin cuticular penetration barrier possibly present in the cavity. The work of Bashan and co-workers (1985) represents the first indication that bacteria can produce cutinase as a penetration enzyme. Unfortunately, the cutinase assay used in this study was the relatively insensitive method of fatty acid titration. Furthermore, the hydrolysis products released from cutin were reported to comigrate with oleic acid when analyzed by thin-layer chromatography. Fatty acids are only minor cutin components; major components are hydroxylated fatty acids, which are less hydrophobic and thus migrate more slowly in chromatography systems similar to the one used in the study (Woloshuk and Kolattu-kudy, 1983). Consequently, the cutinase nature of the isolated enzyme remains questionable.

7.2. Cutinases from Soil Organisms

The biodegradation of plant cuticles in soil must be considerable, since large amounts of this complex lipid material have to be recycled annually after leaf fall. This aspect was first investigated by de Vries and co-workers (1967) employing light microscopy. Dependent on the season, cuticle strips buried in soil lost their integrity after 3–9 months. Interestingly, the biochemistry of cutin hydrolysis and the initiation of research on cutinases were initiated by Heinen and co-workers with their investigations on the soil saprophyte *Penicillium spinolosum* (Heinen, 1960; Van den Ende and Linskens, 1974). The biochemical work on cutinases from plant pathogenic fungi originated later (Linskens and Haage, 1963). Although there is little doubt that *P. spinolosum* and other soil saprophytic fungi can utilize cutin as a carbon source (Heinen and de Vries, 1966), cutinase from a saprophytic fungus has not yet been purified.

Little information is available for cutin degradation by soil bacteria. One out of five unidentified bacteria isolated from soil was shown to release substantial amounts of fatty acids into the medium when grown on cutin (Heinen and de Vries, 1966). Similarly, a *Pseudomonas* sp. isolated from orchard soil was capable of utilizing purified apple cutin as the sole carbon source, and liberated fatty acids identical with the monomers present in the polymeric cutin (Hankin and Kolattukudy, 1971). In another study, *Micrococcus* isolated from soil reportedly degraded cuticular wax (Hankin and Kolattukudy, 1968).

In summary, little is known about the ecological aspects of cuticle degradation and utilization by saprophytic microorganisms. In particular, the potential of cuticular components to provide a nutrient source for organisms inhabiting the outer surfaces of plants deserves more attention. This aspect has become increasingly important since progress has been made in the biocontrol of plant diseases. One of the problems in this area is the competent and competitive establishment of biocontrol microorganisms in close proximity to the plant surface. It may be feasible to expect that the trait of microorganisms to liberate and utilize wax or cutin components as nutrient source could be beneficial in the establishment of significant populations of biocontrol agents on plant surfaces.

8. REFERENCES

Aist, J. R., 1976, Cytology of penetration and infection, in: *Physiological Plant Pathology* (R. Heitefuss and P. H. Williams, eds.), Springer-Verlag, Berlin, pp. 197–221.

Armentrout, V. N., and Downer, A. J., 1987, Infection cushion development by *Rhizoctonia solani* on cotton, *Phytopathology* **77**:619–623.

Backhouse, D., and Willetts, H.J., 1987, Development and structure of infection cushions of *Botrytis cinerea*, *Trans. Br. Mycol. Soc.* **89**:89–95.

Baker, C. J., and Bateman, D. F., 1978, Cutin degradation by plant pathogenic fungi, *Phytopathology* **68**:1577–1584.

Baker, E. A., 1982, Chemistry and morphology of plant epicuticular waxes, in: *The Plant Cuticle* (D. F. Cutler, K.L. Alvin, and C. E. Price, eds.), Academic Press, New York, pp. 139–166.

Bashan, Y., Okon, Y., and Henis, Y., 1985, Detection of cutinases and pectic enzymes during infection of tomato by *Pseudomonas syringae* pv. *tomato*, *Phytopathology* **75**:940–945.

Binyamini, N., and Schiffmann-Nadel, P., 1972, Latent infection in avocado fruit due to *Colletotrichum gloeosporioides*, *Phytopathology* **62**:592–594.

Bird, R. E., Hardman, K.D., Jacobson, J. W., Johnson, S., Kaufman, B. M., Lee, S.-M., Lee, T., Pope, S. H., Riordan, G. S., and Whitlow, M., 1988, Single-chain antigen-binding proteins, *Science* **242**:423–426.

Blakeman, J. P. (ed.), 1981, *Microbial Ecology of the Phylloplane*, Academic Press, New York.

Blakeman, J. P., and Atkinson, P., 1981, Antimicrobial substances associated with the aerial surface of plants, in: *Microbial Ecology of the Phylloplane* (J. P. Blakeman, ed.), Academic Press, New York, pp. 245–264.

Bonnen, A. M., and Hammerschmidt, R., 1989a, Cutinolytic enzymes from *Colletotrichum lagenarium*, *Physiol. Molec. Plant Pathol.* **35**:463–474.

Bonnen, A. M., and Hammerschmidt, R., 1989b, Role of cutinolytic enzymes in infection of cucumber by *Colletotrichum lagenarium*, *Physiol. Molec. Plant Pathol.* **35**:475–481.

Brown, A., and Kolattukudy, P. E., 1978, Mammalian utilization of cutin, the cuticular polyester of plants, *J. Agric. Food Chem.* **26**:1263–1266.

Chau, K. F., and Alvarez, A. M., 1983, A histological study of anthracnose on *Carica papaya*, *Phytopathology* **73**:1113–1116.

Coffey, M. D., and Wilson, U. E., 1983, Histology and cytology of infection and disease caused by *Phytophthora* in *Phytophthora: Its Biology, Taxonomy, Ecology, and Pathology* (D. C. Erwin, S. Bartnicki-Garcia, and P. H. Tsao, eds), American Phytopathological Society, St. Paul, pp. 289–301.

Crawford, M. S., and Kolattukudy, P. E., 1987, Pectate lyase from *Fusarium solani* f.sp. *pisi*. Purification, characterization, *in vitro* translation of rthe mRNA, and the involvement in pathogenicity, *Arch. Biochem. Biophys.* **258**:196–205.

Dantzig, A. H., Zuckerman, S. H., and Andonov-Roland, M. M., 1986, Isolation of a *Fusarium solani* mutant reduced in cutinase activity and virulence, *J. Bacteriol.* **168**:911–916.

Davidse, L., 1987, Biochemical aspects of benzimidazole fungicides—Action and resistance, in: *Modern Selective Fungicides* (H. Lyr, ed.), Wiley, New York, pp. 245–257.

Dayhoff, M. O., 1972, *Atlas of Protein Sequence and Structure*, Volume 5, National Biomedical Research Foundation, Washington, D.C.

de Vries, H., Bredemeijer, G., and Heinen, W., 1967, The decay of cutin and cuticular components by soil microorganisms in their natural environment, *Acta Bot. Neerl.* **16**:102–110.

Dickman, M. B., Podila, G. K., and Kolattukudy, P. E., 1989, Insertion of cutinase gene into a wound pathogen enables it to infect intact host, *Nature* **342**:446–448.

Dickman, M. B., and Alvarez, A. M., 1983, Latent infection of papaya caused by *Colletotrichum gloeosporioides*, *Plant Des.* **67**:748–750.

Dickman, M. B., and Patil, S. S., 1986, Cutinase deficient mutants of *Colletotrichum gloeosporioides* are nonpathogenic to papaya fruit, *Physiol. Mol. Plant Pathol.* **28**:235–242.

Dickman, M. B., Patil, S. S., and Kolattukudy, P. E., 1982, Purification, characterization and role in infection of an extracellular cutinolytic enzyme from *Colletotrichum gloeosporioides* Penz. on *Carica papaya* L., *Physiol. Plant Pathol.* **20**:333–347.

Dickman, M. B., Patil, S. S., and Kolattukudy, P. E., 1983, Effects of organophosphorus pesticides on cutinase activity and infection of papayas by *Colletotrichum gloeosporioides*, *Phytopathology* **73**:1209–1214.

Dodman, R. L., and Flentje, N. T., 1970, The mechanism and physiology of plant penetration by *Rhizoctonia solani*, in: *Rhizoctonia Solani: Biology and Pathology* (J. R. Parmeter, ed.), University of California Press, Berkeley, pp. 149–160.

Emmett, R. W., and Parbery, D. G., 1975, Appressoria, *Annu. Rev. Phytopathol.* **13**:147–167.

Espelie, K. E., Köller, W., and Kolattukudy, P. E., 1983, 9,16-Dihydroxy-10-oxo-hexadecanoic acid, a novel component in citrus cutin, *Chem. Phys. Lipids* **32**:13–26.

Ettinger, W. F., Thukral, S. K., and Kolattukudy, P. E., 1987, Structure of cutinase gene, cDNA, and the derived amino acid sequence from phytopathogenic fungi, *Biochemistry* **26**:7883–7892.

Flurkey, W. H., and Kolattukudy, P. E., 1981, *In vitro* translation of cutinase mRNA: Evidence for a precursor form of an extracellular fungal enzyme, *Arch. Biochem. Biophys.* **212**:154–161.

Foster, R. J., and Kolattukudy, P. E., 1987, Enzymatic hydrolysis of diethylpyrocarbonate, a commonly used histidine modifying agent, by esterases, *Int. J. Biochem.* **19**:391–394.

Gees, R., and Hohl, H. R., 1987, Cytological comparison of specific (*R3*) and general resistance to late blight in potato leaf tissue, *Phytopathology* **78**:350–357.

Gelb, M. H., Svaren, J. P., and Abeles, R. H., 1985, Fluoro ketone inhibitors of hydrolytic enzymes, *Biochemistry* **24**:1813–1817.

Gessler, C., and Stumm, D., 1984, Infection and stroma formation by *Venturia inaequalis* on apple leaves with different degrees os susceptibility to scab, *Phytopathol. Z.* **110**:119–126.

Godowski, P. J., and Picard, D., 1989, Steroid receptors: How to be both a receptor and a transcription factor, *Biochem. Pharmacol.* **38**:3135–3143.

Hamer, J. E., Howard, R. J., Chumley, F. G., and Valent, B., 1988, A mechanism for surface attachment in spores of a plant pathogenic fungus, *Science* **239**:288–290.

Hankin, L., and Kolattukudy, P.E., 1968, Metabolism of a plant wax paraffin (n-nonacosane) by a soil bacterium (*Micrococcus cerificans*), *J. Gen. Microbiol.* **51**:457–463.

Hankin, L., and Kolattukudy, P. E., 1971, Utilization of cutin by a pseudomonad isolated from soil, *Plant Soil* **34**:525–529.

Hargreaves, J. A., Brown, G. A., and Holloway, P. J., 1982, The structural and chemical characteristics of the leaf surface of *Lupinus albus* L. in relation to the distribution of antifungal compounds, in: *The Plant Cuticle* (D. F. Cutler, K. L. Alvin, and G. E. Price, eds.), Academic Press, New York, pp. 331–340.

Harper, J. W., and Powers, J. C., 1985, Reaction of serine proteases with substituted 3-alkoxy-4-chloro-isocoumarines and 3-alkoxy-7-amino-4-chloroisocoumarines: New reactive mechanism based inhibitors, *Biochemistry* **24**:7200–7213.

Hashioka, Y., Ikegami, H., Horino, O., and Kamei, T., 1967, Fine structure of the rice blast. II. Electron micrographs of the initial infection, *Res. Bull. Fac. Agric. Gifu Univ.* **24**:78–90.

Heath, M. C., 1987, Host vs. nonhost resistance, in: *Molecular Strategies for Crop Protection* (C.J. Arntzen and C. Ryan, eds.), Liss, New York, pp. 25–34.

Heinen, W., 1960, Über den enzymatischen Cutin Abbau. I. Mitteilung: Nachweis eines "Cutinase" Systems, *Acta Bot. Neerl.* **9**:167–190.

Heinen, W., and de Vries, H., 1966, Stages during the breakdown of plant cutin by soil microorganisms, *Arch. Microbiol.* **54**:331–338.

Heintz, C., 1986, Infection mechanisms of grapevine powdery mildew (*Oidium tuckeri*): Comparative studies of the penetration process on artificial membranes and leaf epidermis, *Vitis* **25**:215–225.

Hemmi, K., Harper, J., and Powers, J. C., 1985, Inhibition of human leukocyte elastase, cathepsin G, chymotrypsin A, and porcine pancreatic elastase with substituted isobenzofuranones and benzopyrandiones, *Biochemistry* **24**:1841–1848.

Hoch, H. C., and Staples, R. C., 1987, Structural and chemical changes among the rust fungi during appressorium development, *Annu. Rev. Phytopathol.* **25**:231–247.

Holloway, P. J., 1982a, Structure and histochemistry of plant cuticular membranes: An overview, in: *The Plant Cuticle* (D. F. Cutler, K. L. Alvin, and C. E. Price, eds.), Academic Press, New York, pp. 1–32.

Holloway, P. J., 1982b, The chemical constitution of plant cuticles, in: *The Plant Cuticle* (D. F. Cutler, K. L. Alvin, and C. E. Price, eds.), Academic Press, New York, pp. 45–86.

Howard, R. J., and Ferrari, M. A., 1989, Role of melanin in appressorium function, *Exp. Mycol.* **13**:403–418.

Huang, J. S., 1986, Ultrastructure of bacterial penetrations in plants, *Annu. Rev. Phytopathol.* **24**:141–157.

Inoue, S., Kato, T., Jordan, V. W. L., and Brent, K. J., 1987, Inhibition of appressorial adhesion of *Pyricularia oryzae* to barley leaves by fungicides, *Pestic. Sci.* **19**:145–152.

Karjalainen, R., and Lounatmaa, K., 1986, Ultrastructure of penetration and colonization of wheat leaves by *Septoria nodorum*, *Physiol. Mol. Plant Pathol.* **29**:263–270.

Keon, J. P. R., and Hargreaves, J. A., 1983, A cytological study of the net blotch disease of barley caused by *Pyrenophora teres*, *Physiol. Plant Pathol.* **22**:321–329.

Kolattukudy, P. E., 1980a, Biopolyester membranes of plants, *Science* **208**:990–1000.

Kolattukudy, P. E., 1980b, Cutin, suberin and waxes, in: *The Biochemistry of Plants*, Volume 4 (P. K. Stumpf, ed.), Academic Press, New York, pp. 571–645.

Kolattukudy, P. E., 1981, Structure, biosynthesis, and biodegradation of cutin and suberin, *Annu. Rev. Plant Physiol.* **32**:539–567.

Kolattukudy, P. E., 1984a, How do pathogenic fungi break the plant cuticular barrier? in: *Infection Processes of Fungi* (D. W. Roberts and J. R. Aist, eds.), The Rockefeller Foundation, New York, pp. 31–37.

Kolattukudy, P. E., 1984b, Fungal penetration of defensive barriers of plants, in: *Structure, Function, and Biosynthesis of Plant Cell Walls* (W. M. Dugger and S. Bartnicki-Garcia, eds.), American Society of Plant Physiologists, Rockville, pp. 31–37.

Kolattukudy, P. E., 1984c, Biochemistry and function of cutin and suberin, *Can. J. Bot.* **62**:2918–2933.

Kolattukudy, P. E., 1984d, Cutinases from fungi and pollen, in: *Lipases* (B. Borgström and H. L. Brockman, eds.), Elsevier, Amsterdam, pp. 471–504.

Kolattukudy, P. E., 1985, Enzymatic penetration of the plant cuticle by fungal pathogens, *Annu. Rev. Phytopathol.* **23**:223–250.

Kolattukudy, P. E., 1987, Lipid-derived defensive polymers and waxes and their role in plant–microbe interaction, in: *The Biochemistry of Plants*, Volume 9 (P. K. Stumpf, ed.), Academic Press, New York, pp. 291–314.

Kolattukudy, P. E., and Crawford, M. S., 1987, The role of polymer degrading enzymes in fungal pathogenesis, in: *Molecular Determinants of Plant Diseases* (S. Nishimura, C. P. Vance, and N. Doke, eds.), Springer–Verlag, Berlin, pp. 75–96.

Kolattukudy, P. E., and Espelie, K. E., 1985, Biosynthesis of cutin, suberin and associated waxes, in : *Biosynthesis and Biodegradation of Wood Components* (T. Higuchi, ed.), Academic Press, New York, pp. 162–208.

Kolattukudy, P. E., and Köller, W., 1983, Fungal penetration of the first line defensive barriers of plants, in: *Biochemical Plant Pathology* (J. A. Callow, ed.), Wiley, New York, pp. 79–100.

Kolattukudy, P. E., and Soliday, C. L., 1985, Effects of stress on the defensive barriers of plants, in: *Cellular and Molecular Biology of Plant Stress* (J. L. Key and T. Kosuge, eds.), Liss, New York, pp. 381–400.

Kolattukudy, P. E., Espelie, K. E., and Soliday, C. L., 1981a, Hydrophobic layers attached to cell walls. Cutin, suberin and associated waxes, in: *Plant Carbohydrates II—Extracellular Carbohydrates* (W. Tanner and F. A. Loewus, eds.), Springer-Verlag, Berlin, pp. 225–254.

Kolattukudy, P. E., Purdy, R. E., and Maiti, I. B., 1981b, Cutinases from fungi and pollen, *Methods Enzymol.* **71**:652–664.

Kolattukudy, P. E., Ettinger, W. F., and Sebastian, J., 1987a, Cuticular lipids in plant microbe interaction, in: *The Metabolism, Structure, and Function of Plant Lipids* (P. K. Stumpf, J. B. Mudd and W. D. Nes, eds.), Plenum Press, New York, pp. 473–480.

Kolattukudy, P. E., Soliday, C. L., Woloshuk, C. P., and Crawford, M., 1985, Molecular biology of the early events in the fungal penetration into plants, in: *Molecular Genetics of Filamentous Fungi* (W. E. Timberlake, ed.), Liss, New York, pp. 421–438.

Kolattukudy, P. E., Sebastian, J., Ettinger, W. F., and Crawford, M. S., 1987b, Cutinase and pectinase in host–pathogen and plant–bacterial interaction, in: *Molecular Genetics of Plant-Microbe Interaction* (D. P. S. Verma and N. Brisson, eds.), Nijhoff, The Hague, pp. 43–50.

Kolattukudy, P. E., Crawford, M. S., Woloshuk, C. P., Ettinger, W. F., and Soliday, C. L., 1987c, The role of cutin, the plant cuticular hydroxy fatty acid polymer, in fungal interaction with plants, in: *Ecology and Metabolism of Plant Lipids* (G. Fuller and W. D. Nes, eds.), American Chemical Society, Washington, D. C., pp. 152–175.

Kolattukudy, P. E., Podila, G. K., Roberts, E., and Dickman, M. D., 1989, Gene expression resulting from the early signals in plant-fungus interaction, in: *Molecular Biology of Plant-Pathogen Interactions* (B. Staskawicz, B. Ahlquist and O. Yoder, eds.), Liss, New York, pp. 87–102.

Köller, W., and Kolattukudy, P. E., 1982, Mechanism of action of cutinase: Chemical modification of the catalytic triad characteristic for serine hydrolases, *Biochemistry* **21**:3083–3090.

Köller, W., and Parker, D. M., 1989, Purification and characterization of cutinase from *Venturia inaequalis*, *Phytopathology* **79**:278–283.

Köller, W., Allan, C. R., and Kolattukudy, P. E., 1982a, Role of cutinase and cell wall degrading enzymes in infection of *Pisum sativum* by *Fusarium solani* f.sp. *pisi*. *Physiol. Plant Pathol.* **20**:47–60.

Köller, W., Allan, C. R., and Kolattukudy, P. E., 1982b, Protection of *Pisum sativum* from *Fusarium solani* f.sp *pisi* by inhibition of cutinase with organophosphorous pesticides, *Phytopathology* **72**:1425–1430.

Köller, W., Allan, C. R., and Kolattukudy, P. E., 1982c, Inhibition of cutinase and prevention of fungal penetration into plants by benomyl—A possible protective mode of action, *Pestic. Biochem. Physiol.* **18**:15–25.

Kraut, J., 1977, Serine proteases: Structure and mechanism of catalysis, *Annu. Rev. Biochem.* **46**:331–358.

Kubicek, C. P., Panda, T., Schreferl-Kunar, G., Gruber, F., and Messner, R., 1987, O-linked but not N-linked glycosylation is necessary for the secretion of endoglucanases I and II by *Trichoderma reesei*, *Can. J. Microbiol.* **33**:698–703.

Kunoh, H., 1981, Early stages of infection process of *Erysiphe graminis* on barley and wheat, in: *Microbial Ecology of the Phylloplane* (J. P. Blakeman, ed.), Academic Press, New York, pp. 85–101.

Kunoh, H., 1984, Cytological aspects of penetration of plant epidermis by fungi, in: *Infection Processes of Fungi* (D. W. Roberts and J. R. Aist, eds.), The Rockefeller Foundation, New York, pp. 137–146.

Lin, T. S., and Kolattukudy, P. E., 1976, Evidence for novel linkages in a glycoprotein involving β-hydrophenyl-alanine and β-hydroxytyrosine, *Biochem. Biophys. Res. Commun.* **72**:243–250.

Lin, T. S., and Kolattukudy, P. E., 1977, Glucoronyl glycin, a novel N-terminus in a glycoprotein, *Biochem. Biophys. Res. Commun.* **75**:87–93.

Lin, T. S., and Kolattukudy, P. E., 1978, Induction of a polyester hydrolase (cutinase) by low levels of cutin monomers in *Fusarium solani* f.sp. *pisi*, *J. Bacteriol.* **133**:942–951.

Lin, T. S., and Kolattukudy, P. E., 1980a, Structural studies on cutinase, a glycoprotein containing novel amino acids and glucoronic acid amide at the N-terminus, *Eur. J. Biochem.* **106**:341–351.

Lin, T. S., and Kolattukudy, P. E., 1980b, Isolation and characterization of a cuticular polyester (cutin) hydrolyzing enzyme from phytopathogenic fungi, *Physiol. Plant Pathol.* **17**:1–15.

Linskens, H. F., and Haage, P., 1963, Cutinase-Nachwies in phytopathogenen Pilzen, *Phytopathol. Z.* **48**:306–311.

Littlefield, L. J., and Heath, M. C., 1979, *Ultrastructure of Rust Fungi*, Academic Press, New York.

McBride, R. P., 1972, Larch leaf waxes utilized by *Sporobolomyces roseus* in situ, *Trans. Br. Mycol. Soc.* **58**: 329–331.

McKeen, W. E., 1974, Mode of penetration of epidermal cell walls of *Vicia faba* by *Botrytis cinerea*, *Phytopathology* **64**:461–467.

McKeen, W. E., and Rimmer, S. R., 1973, Initial penetration process in powdery mildew infection of susceptible barley leaves, *Phytopathology* **63**:1049–1053.

McKeen, W. E., and Svircev, A. M., 1981, Early development of *Peronospora tabacina* in the *Nicotiana tabacum* leaf, *Can. J. Plant Pathol.* **3**:145–158.

Macko, V., 1981, Inhibitors and stimulants of spore germination and infection structure formation in fungi, in: *The Fungal Spore: Morphogenetic Controls* (G. Turian and H. R. Hohl, eds.), Academic Press, New York, pp. 565–584.

MacNamara, O. C., and Dickinson, C. H., 1981, Microbial degradation of plant cuticle, in: *Microbial Ecology of the Phylloplane* (J. P. Blakeman, ed.), Academic Press, New York, ppg. 455–473.

Maiti, I. B., and Kolattukudy, P. E., 1979, Prevention of fungal infection of plants by specific inhibitors of cutinase, *Science* **205**:507–508.

Martin, T. J., 1964, Role of the cuticle in the defense against plant diseases, *Annu. Rev. Phytopathol.* **2**:81–100.

Martin, T. J., and Juniper, B. E., 1970, *The Cuticles of Plants*, St. Martin's Press, New York.

Matsuura, K. 1986, Scanning electron microscopy of the infection process of *Rhizoctonia solani* in leaf sheaths of rice plants, *Phytopathology* **76**:811–814.

Maurer, F., Sommer, H., Köller, W., Brandes, W., and Reinecke, P., 1986, Preparation of pyridyl phosphate esters as agrochemical fungicides, *Chem. Abstr.* **107**:231426n.

Moorman, A. R., and Abeles, R. H., 1982, A new class of serine protease inactivators based on isatoic anhydride, *J. Am. Chem. Soc.* **104**:6785–6786.

Moreau, R. A., and Seibles, T. S., 1985, Production of extracellular enzymes by germinating cysts of *Phytophthora infestans*, *Can. J. Bot.* **63**:1811–1816.

Morris, C. E., and Rouse, D. I. 1985, Role of nutrients in regulating epiphytic bacterial populations, in: *Biological Control on the Phylloplane* (C. E. Windels and S. E. Lindow, eds.), American Phytopathological Society, St. Paul, pp. 63–82.

Naggert, J., Witkowski, A., Mikkelsen, J., and Smith, S., 1988, Molecular cloning and sequencing of a cDNA encoding the thioesterase domain of the rat fatty acid synthetase, *J. Biol. Chem.* **263**:1146–1150.

Nelson, P. E., Toussoun, T. A. and Cook, R. J., 1981, *Fusarium: Diseases, Biology, and Taxonomy*, The Pennsylvania State University Press, University Park.

Nicholson, R. L., and Moraes, B. C., 1980, Survival of *Colletotrichum graminicola*: Importance of the spore matrix, *Phytopathology,* **70**:255–261.

Nicholson, R.L., Kuc, J., and Williams, E. B., 1972, Histochemical demonstration of transitory esterase activity in *Venturia inaequalis*, *Phytopathology* **62**:1242–1247.

Nicholson, R. L., Yoshioka, H., Yamaoka, N., and Kunoh, H., 1988, Preparation of infection court by *Erysiphe graminis*. II. Release of esterase enzyme from conidia in response to a contact stimulus, *Exp. Mycol.* **12**: 336–349.

Nusbaum, C. J., and Keitt, G. W., 1938, A cytological study of host–parasite relations of *Venturia inaequalis* on apple leaves, *J. Agric. Res.* **66**:595–618.

Oakeshott, J. G., Collet, C., Phillis, R. W., Nielsen, K. M., Russell, R. J., Chambers, G. K., Ross, V., and Richmond, R. C., 1987, Molecular cloning and characterization of esterase-6, a serine hydrolase from *Drosophila*, *Proc. Natl. Acad. Sci. USA* **84**:3359–3363.

O'Connell, R. J. O., Bailey, J. A., and Richmond, D. V., 1985, Cytology and physiology of infection of *Phaseolus vulgaris* by *Colletotrichum lindemuthianum*, *Physiol. Plant Pathol.* **27**:75–98.

Ogoshi, A., 1987, Ecology and pathogenicity of anastomosis and intraspecific groups of *Rhizoctonia solani* Kühn, *Annu. Rev. Phytopathol.* **25**:125–143.

Parmeter, J. R. (ed.), 1970, *Rhizoctonia solani: Biology and Pathology*, University of California Press, Berkeley.

Parry, D. W., and Pegg, G. F., 1985, Surface colonization, penetration and growth of three *Fusarium* species in lucerne, *Trans. Br. Mycol. Soc.* **85**:495–500.

Paus, F., and Raa, J., 1973, An electron microscope study of infection and disease development in cucumber hypocotyls inoculated with *Cladosporium cucumerinum*, *Physiol. Plant Pathol.* **3**:461–464.

Podila, G. K., Dickman, M. B., and Kolattukudy, P. E., 1988, Transcriptional activation of a cutinase gene in isolated fungal nuclei by plant cutin monomers, *Science* **242**:922–925.

Polgár, L., 1987, Structure and function of serine proteases, in: *Hydrolytic Enzymes* (A. Neuberger and K. Brocklehurst, eds.), Elsevier, Amsterdam, pp. 159–200.

Poulose, A. J., Rogers, L., Cheesebrough, T. M., and Kolattukudy, P. E., 1985, Cloning and sequencing of the cDNA for S-acyl fatty acid synthase thioesterase from the uropygial gland of mallard duck, *J. Biol. Chem.* **260**:15953–15958.

Preece, T. F., Barnes, G., and Baylay, J. W., 1967, Junctions between epidermal cells as sites for appressorium formation by plant pathogenic fungi, *Plant Pathol.* **16**:117–118.

Price, C. E., 1982, A review of the factors influencing the penetration of pesticides through plant leaves, in: *The Plant Cuticle* (D. F. Cutler, K. L. Alvin, and G. E. Price, eds.), Academic Press, New York, pp. 237–252.

Purdy, R. E., and Kolattukudy, P. E., 1973, Depolymerization of a hydroxy fatty acid biopolymer, cutin, by an extracellular enzyne from *Fusarium solani* f.sp. *pisi*: Isolation and some properties of the enzyme, *Arch. Biochem. Biophys.* **159**:61–69.

Purdy, R. E., and Kolattukudy, P. E., 1975a, Hydrolysis of plant cuticle by plant pathogens. Purification, amino acid composition, and molecular weight of two isozymes of cutinase and a nonspecific esterase from *Fusarium solani* f. *pisi*, *Biochemistry* **14**:2824–2831.

Purdy, R. E., and Kolattukudy, P. E., 1975b, Hydrolysis of plant cuticle by plant pathogens. Properties of cutinase I, cutinase II, and a nonspecific esterase isolated from *Fusarium solani* f. *pisi*, *Biochemistry* **14**:2832–2840.

Randhawa, Z. I., and Smith, S., 1987, Complete amino acid sequence of the medium-chain S-acyl fatty acid synthase thio ester hydrolase from rat mammary gland, *Biochemistry* **26**:1365–1373.

Riederer, M., and Schönherr, J., 1988, Development of plant cuticles: Fine structure and cutin composition of *Clivia miniata* Reg. leaves, *Planta* **174**:127–138.

Rijkenberg, F. H. J., De Leeuw, G. T. N., and Verhoeff, K., 1980, Light and electron microscopy studies on the infection of tomato fruits by *Botrytis cinerea*, *Can. J. Bot.* **58**:1394–1404.

Ruinen, J. 1963, The phyllosphere. II. Yeasts from the phyllosphere of tropical foliage, *Antonie van Leeuwenhoek*, *J. Microbiol. Serol.* **29**:425–438.

Ruinen, J., 1966, The phyllosphere. IV. Cuticle decomposition by microorganisms in the phyllosphere, *Ann. Inst. Pasteur. Paris* **111**:342–346.

Salinas, J., Warnaar, F., and Verhoeff, K., 1986, Production of cutin hydrolyzing enzymes by *Botrytis cinerea* in vitro, *J. Phytopathol.* **116**:299–307.

Sargent, C., and Gay, J. L., 1977, Barley epidermal apoplast structure and modification by powdery mildew contact. *Physiol. Plant Pathol.* **11**:195–205.

Schönherr, J., 1982, Resistance of plant surfaces to water loss, in: *Physiological Plant Ecology II–Water Relations and Carbon Assimilation* (O. L. Lange, P. S. Nobel, C. B. Osmond, and H. Ziegler, eds.), Springer-Verlag, Berlin, pp. 153–179.

Schumacher, M., Camp, S., Maulet, Y., Newton, M., MacPhee-Quigley, K., Taylor, S. S., Friedmann, T., and Taylor, P., 1986, Primary structure of *Torpedo californica* acetylcholinesterase deduced from its cDNA sequence, *Nature* **319**:407–409.

Schwinn, F., and Geissbühler, H., 1986, Towards a more rational approach to fungicide design, *Crop Pro.* **5**:33–40.

Sebastian, J., Chandra, A. K., and Kolattukudy, P. E., 1987, Discovery of a cutinase-producing *Pseudomonas* sp. cohabiting with an apparently nitrogen-fixing *Corynebacterium* sp. in the phyllosphere, *J. Bacteriol.* **169**:131–136.

Shaykh, M., Soliday, C., and Kolattukudy, P. E., 1977, Proof for the production of cutinase by *Fusarium solani* f. *pisi* during the penetration into its host, *Pisum sativum*, *Plant Physiol.* **60**:170–172.

Shishiyama, J., Araki, F., and Akai, S., 1970, Studies on cutin esterase II. Characteristics of cutin-esterase from *Botrytis cinerea* and its activity on tomato-cutin, *Plant Cell Physiol.* **11**:937–945.

Sisler, H. D., 1986, Control of fungal diseases by compounds acting as antipenetrants, *Crop Prot.* **5**: 306–313.

Smereka, K. J., MacHardy, W. E., and Kausch, A. P., 1987, Cellular differentiation in *Venturia inaequalis* ascospores during germination and penetration of apple leaves, *Can. J. Bot.* **65**:2549–2561.

Smereka, K. J., Kausch, A. P., and MacHardy, W. E., 1988, Intracellular junctional structures in germinating ascospores of *Venturia inaequalis*, *Protoplasma* **142**:1–4.

Soliday, C. L., and Kolattukudy, P. E., 1976, Isolation and characterization of a cutinase from *Fusarium roseum culmorum* and its immunological comparison with cutinases from *F. solani pisi*, *Arch. Biochem. Biophys.* **176**:334–343.

Soliday, C. L., Flurkey, W. H., Okita, T. W., and Kolattukudy, P. E., 1984, Cloning and structure determination of cDNA for cutinase, an enzyme involved in fungal penetration of plants, *Proc. Natl. Acad. Sci. USA* **81**:3939–3943.

Soliday, C. L., Dickman, M. B., and Kolattukudy, P. E., 1989, Structure of the cutinase gene and detection of promoter activity in the 5'-flanking region by fungal transformation, *J. Bacteriol.* **171**:1942–1951.

Sparapano, L., and Graniti, A., 1977, Cutin degradation by two scab fungi, *Spilocaea olegania* (Cast.) Hugh. and *Venturia inaequalis*, in: *Current Topics in Plant Pathology* (Z. Kiraly, ed.), Akademiai Kiado, Budapest, ppg. 117–138.

Staub, T., Dahmen, H., and Schwinn, F. J., 1974, Light- and scanning electron microscopy of cucumber and barley powdery mildew on host and nonhost plants, *Phytopathology* **64**:264–272.

Tanabe, K., Nishimura, S., and Kohmoto, K., 1988a, Cutinase production by *Alternaria alternata* Japanese pear pathotype and its role in pathogenicity, *Annu. Phytopathol. Soc. Jpn.* **54**:483–492.

Tanabe, K., Nishimura, S., and Kohmoto, K., 1988b, Pathogenicity of cutinase- and pectic enzymes-deficient mutants of *Alternaria alternata* Japanese pear pathotype, *Annu. Phytopathol. Soc. Jpn.* **54**:552–555.

Tarique, V.-N., and Jeffries, P., 1986, Ultrastructure of penetration of *Phaseolus* spp. by *Sclerotinia sclerotiorum*, *Can. J. Bot.* **64**:2909–2915.

Tarique, V.-N., and Jeffries, P., 1987, Cytochemical localization of lipolytic enzyme activity during the penetration of host tissues by *Sclerotinia sclerotiorum*, *Physiol. Mol. Plant Pathol.* **30**:77–91.

Tewari, J. P., 1986, Subcuticular growth of *Alternaria brassicae* in rapeseed, *Can. J. Bot.* **64**:1227–1231.

Toussoun, T. A., Nash, S. M., and Snyder, W. C., 1960, The effect of nitrogen sources and glucose on the pathogenesis of *Fusarium solani* f. *phaseoli*, *Phytopathology* **50**:137–140.

Trail, F., and Köller, W., 1990, Diversity of cutinases from plant pathogenic fungi: Evidence for a relationship between enzyme properties and tissue specificity, *Physiol. Molec. Plant Pathol.* (in press).

Trione, E. J. 1981, Natural regulators of fungal development, in: *Plant Disease Control* (R. C. Staples and G. H. Toenissen, eds.), Wiley, New York, pp. 85–102.

Tsuneda, A., and Skoropad, W. P., 1978, Behavior of *Alternaria brassicae* and its mycoparasite *Nectria inventa* on intact and on excised leaves of rapeseed, *Can. J. Bot.* **56**:1333–1340.

Valsangiacomo, C., and Gessler, C., 1988, Role of the cuticular membrane in ontogenic and Vf-resistance of apple leaves against *Venturia inaequalis*, *Phytopathology* **78**:1066–1068.

Van den Ende, G., and Linskens, H. F., 1974, Cutinolytic enzymes in relation to pathogenesis, *Annu. Rev. Phytopathol.* **12**:247–258.

VanEtten, H. D., Matthews, D. E., and Mackintosh, S. F., 1987, Adaption of pathogenic fungi to toxic barriers of plants, in: *Molecular Strategies for Crop Protection* (C. J. Arntzen and C. Ryan, eds.), Liss, New York, pp. 59–70.

Verhoeff, K., 1980, The infection process and host–pathogen interactions, in: *The Biology of Botrytis* (J. R. Coley-Smith, K. Verhoeff, and W. R. Javaris, eds.), Academic Press, New York, pp. 153–179.

Wiltshire, S. P., 1915, Infection and immunity studies on the apple and pear scab fungi, *Ann. Appl. Biol.* **1**:335–350.

Wolkow, P. M., Sisler, H. D., and Vigil, E. L., 1983, Effect of inhibitors of melanin biosynthesis on structure and function of appressoria of *Collectotrichum lindemuthianum*, *Physiol. Plant Pathol.* **23**:55–71.

Woloshuk, C. P., and Kolattukudy, P. E., 1986, Mechanism by which contact with plant cuticle triggers cutinase gene expression in the spores of *Fusarium solani* f.sp. *pisi*, *Proc. Natl. Acad. Sci. USA* **83**:1704–1708.

Woloshuk, C. P., Sisler, H. D., and Vigil, E. L., 1983, Action of the antipenetrant, tricyclazole, on appressoria of *Pyricularia oryzae*, *Physiol. Plant Pathol.* **22**:245–259.

Wood, R. K. S., 1960, Chemical ability to breach the host barriers, in: *Plant Pathology—An Advanced Treatise*, Volume 2 (J. G. Horsfall and A. E. Dimond, eds.), Academic Press, New York, pp. 232–272.

Zinkernagel, V., Riess, F., Wendland, M., and Bartscherer, H.-C., 1988, Infektionsstrukturen von *Septoria nodorum* in Blättern anfälliger Weizensorten, *Z. Pflanzenkr. Pflanzenschutz* **95**:169–175.

11

Appearance of Pathogen-Related Proteins in Plant Hosts

Relationships Between Compatible and Incompatible Interactions

Ingrid M. J. Scholtens-Toma, Matthieu H. A. J. Joosten, and Pierre J. G. M. De Wit

1. PLANT–PATHOGEN INTERACTIONS

1.1. Introduction

There are generally four relationships between plants and microorganisms: (1) no obvious interaction or a zero relationship, (2) basic incompatibility or nonhost resistance (plants are resistant to the majority of pathogens), (3) a beneficial interaction (nitrogen-fixing bacteria–legumes, mycorrhizae–various plants), and (4) a pathogenic relationship (basic compatibility). At the cultivar level, compatibility may be restricted to a few genotypes as a result of the presence of avirulence genes in the pathogen and resistance genes in the host.

Generally, nonhost resistance is based on a number of preformed (passive) plant defense factors, such as physical barriers and antimicrobial compounds. Also, a number of inducible resistance mechanisms exist: elicitation of cell death [the hypersensitive response (HR)], lignification and accumulation of hydroxyproline-rich glycoproteins (HRGP) in plant cell walls, accumulation of phytoalexins, inhibitors of fungal enzymes, and accumulation of glucanases, chitinases, and other pathogenesis-related (PR) proteins. Furthermore, thionins, cell wall polypeptides with *in vitro* antifungal activity, may play a role in active resistance against microbial infection (Bohlmann *et al.*, 1988). Preformed and inducible resistance mechanisms have been reviewed extensively in recent years (e.g., Dunkle, 1984; Fraser, 1985).

It should be realized that most of the host defense responses are not elicited by pathogens exclusively, but by abiotic stress as well. Moreover, defense responses are not specific for a particular pathogen, but rather their induction in the plant host is usually a specific event.

Ingrid M. J. Scholtens-Toma, Matthieu H. A. J. Joosten, and Pierre J. G. M. De Wit • Department of Phytopathology, Wageningen Agricultural University, 6700 EE Wageningen, The Netherlands.

A widely studied defense response is HR. An array of structures has been described that can act as elicitors of HR (De Wit, 1986). Such a diversity of structures suggests that plants have different receptors which after interaction with elicitors, trigger defense-related genes by signal-transducing mechanisms. Most elicitors that have been studied are elicitors of phytoalexin accumulation.

Only a relatively small number of microorganisms have gained basic compatibility and are able to infect a host plant. In order to establish infection, the pathogen must breach or avoid plant defense mechanisms. In a compatible interaction the pathogen seems to have all the tools needed for successful colonization of the host. It is therefore not likely for a pathogen to gain access to new host plants through a few mutations, since basic compatibility is thought to be controlled by many genes encoding pathogenicity factors. A few of these pathogenicity factors are briefly mentioned here and will be discussed in more detail later.

Necrotrophic pathogens produce toxins to kill the host cells before they can elicit a defense reaction (Knoche and Duvick, 1987). Some fungi are able to degrade preformed antimicrobial compounds (Mansfield, 1983; Keen, 1986). Many actively penetrating pathogens have cutinase for degrading or altering the structure of the cuticle. The role of cutinase in penetration of the cuticle has been studied extensively (Kolattukudy, 1985) and is discussed in Chapter 10. Degradation of the underlying plant cell wall, a heterogeneous and complex physical barrier, requires the action of many specific cell wall-degrading enzymes (CWDE). An array of pectic enzymes act to modify or depolymerize the pectic substrates and render other cell wall polymers more susceptible to enzymatic attack (Dunkle, 1984).

Obligate and biotrophic parasites usually colonize their host without obvious macroscopic responses of the plant in the early stages of the infection. Ultrastructural studies suggest highly localized action by CWDE. CWDE produced by these parasites are presumably under strict regulatory control and cell-wall bound (Cooper, 1983; Keon et al., 1987). Defense responses are possibly prevented or suppressed. Among these parasites the phenomenon of physiological specialization has often been observed. Such specialization is based on a gene-for-gene interaction between physiological races of the fungus and plant cultivars with different genes for resistance.

This chapter will deal mainly with the role of proteins in compatible and incompatible fungus–plant relationships with emphasis on the *Cladosporium fulvum*–tomato interaction. These include proteinaceous pathogenicity factors, protein and glycoprotein elicitors of plant defense responses, and proteinaceous plant defense factors.

1.2. The Gene-for-Gene System

Single and often dominant resistance genes have been introduced into susceptible host plants by plant breeders, but unfortunately new fungal races appeared which could easily overcome these genes (race-specific resistance). The phenomenon of race-specific resistance is frequently observed among biotrophic and obligate parasites and is based on a gene-for-gene concept first coined by Flor (1942) who studied the relationship between flax and flax rust. Presently this hypothesis is thought to hold true for many host–pathogen interactions (Crute, 1985). In natural pathosystems the phenomenon of physiological specialization has also been described (Dinoor and Eshed, 1984). The gene-for-gene concept implies that for each gene conditioning avirulence in the pathogen, there is a corresponding gene conditioning resistance in the host plant. Usually resistance genes are dominant, but recessive resistance genes have also been described, albeit less frequently.

Two arrangements of gene-for-gene systems have frequently been studied: the "quadratic check" (Table I) and the "reciprocal check" (Table II). In the quadratic check three different genotypical combinations (A_1/r_1, a_1/r_1, a_1/R_1) all lead to a compatible interaction, while only one

Table I. The "Quadratic Check"[a]

Genotypes of pathogen isolates	Genotypes of host lines	
	R_1R_1	r_1r_1
A_1	I	C
A_2	C	C

[a]Possible combinations between two host lines and two haploid parasite isolates differing in resistance and parasite capability, respectively. R_1 is an allele for resistance and A_1 an allele for avirulence; r_1 and a_1 are the opposite alleles for susceptibility and virulence. C, compatible/susceptible; I, incompatible/resistant.

combination (A_1/R_1) results in an incompatible interaction. This indicates that the specificity of the interaction is controlled by the interaction of the gene products of the dominant A and R alleles. In the three other combinations, there is most likely no interaction between the gene products eventually leading to compatibility. The reciprocal check is an extension of the quadratic check with two dominant resistance genes and two dominant avirulence genes. Incompatibility is thought to be epistatic to compatibility.

1.3. A Few Model Gene-for-Gene Systems

The gene-for-gene system of lettuce, *Lactuca sativa* L., and the biotrophic, haustorium-forming parasite *Bremia lactucae* has been thoroughly characterized (Michelmore *et al.*, 1988). At least 18 dominant genes for resistance to Downy mildew (Dm) are matched by 18 complementary avirulence gene products in *B. lactucae*. All resistance genes map to one of four linkage groups. As a rule, avirulence is inherited in a dominant manner, but in certain crosses expression of avirulence was suppressed by dominant inhibitor alleles (I_1, I_4, I_5). A dominant allele at these inhibitor loci will lead to a virulent phenotype. Such inhibitor genes are thought to be independent and epistatic to avirulence genes.

In this interaction the specific recognition event is closely linked to the initiation of HR (Mansfield *et al.*, 1988). The possible existence of suppressors was investigated in experiments with mixed inocula, but no evidence was found for their existence. This indicates a direct interaction between avirulence gene products and resistance gene products. Apoplastic fluids from compatible interactions, however, did not elicit HR in any race–cultivar interaction. This suggests that, in contrast to the *C. fulvum*–tomato interaction, there are no diffusible elicitors.

Table II. The "Reciprocal Check"[a]

Genotypes of pathogen isolates	Genotypes of host lines	
	$R_1R_1r_2r_2$	$r_1r_1R_2R_2$
A_1a_2	I	C
a_1A_2	C	I

[a]Possible combinations between two host lines and two haploid parasite isolates differing in resistance and parasite capability, respectively. A_1 and A_2 are different alleles for avirulence, R_1 and R_2 are alleles for resistance; r_1, r_2 and a_1, a_2 are the opposite alleles for susceptibility and virulence, respectively. C, compatible/susceptible; I, incompatible/resistant.

Also, when tissue of an infected susceptible cultivar was transplanted onto a resistant cultivar, widespread necrosis did not occur (Crucefix *et al.*, 1984). It was concluded that the recognition event is highly localized, probably at the plant cell wall–fungal wall interface. Also, the responses characteristic of race-specific resistance were expressed by lettuce suspension cells but not by protoplasts prior to cell wall regeneration. Crucefix and co-workers (1987) have tried to isolate race-specific elicitors of the HR from fungal infection structures, but none have been identified.

A gene-for-gene system has been proposed for the interaction between *Phytophthora infestans* and potato (Shaw, 1988). Potato lines with one dominant resistance gene are available. Avirulence seems to be governed by single dominant avirulence genes although results are not conclusive. Also, gene dosage effects may be involved (Shaw, 1988). In this interaction, there are indications for race-specific suppressors. All races produce nonspecific elicitors, most notably the unsaturated lipids arachidonic acid and eicosapentaenoic acid (for review see Kuc *et al.*, 1984). Virulent races have been shown to contain mycolaminarins, which suppress the defense reactions elicited by avirulent races of the fungus (Doke and Tomiyama, 1980). Similar glucans from avirulent races were found that did not suppress phytoalexin accumulation. In addition, glucans have been found that acted as "enhancers" of the activity of the fatty acid elicitors (Maniara *et al.*, 1984; Preisig and Kuc, 1985).

The hemibiotrophic fungus *Colletotrichum lindemuthianum* is the causal agent of anthracnose on bean and exists as several physiological races which can be differentiated by their interaction with various cultivars of bean (Krüger *et al.*, 1977). Nonspecific glycoprotein elicitors of *C. lindemuthianum* of which the glyco moiety consisted mainly of mannose, galactose and glucose, have been found in cell walls of the fungus and in culture filtrates. They are heterogeneous with respect to molecular weight and could be separated as both neutral and anionic molecules (Hamdan and Dixon, 1986). Tepper and Anderson (1986), however, found galactoglucomannans and glycoproteins from the α race of *C. lindemuthianum* exhibiting differential elicitor activity on bean cultivars. One α race elicitor, consisting mainly of carbohydrate, has been purified and exhibited high elicitor activity on the resistant cultivar Dark Red Kidney, but not on the susceptible cultivar Great Northern. No extracellular compounds of the β race were found having activity on the susceptible cultivars Dark Red Kidney and Great Northern. In this interaction there is no evidence for the presence of suppressors.

The interaction between the biotrophic fungus *C. fulvum* and tomato is thought to conform to a gene-for-gene system (Day, 1956). A number of physiological races of the fungus have been isolated. Near-isogenic tomato lines with resistance genes giving clear-cut resistance to *C. fulvum* are available (Table III). The interaction between the fungus and its host is confined to the apoplast. Apoplastic fluids can easily be obtained and analyzed for the presence of different classes of compounds involved in basic compatibility and (a)virulence. Apoplastic fluids of compatible interactions between *C. fulvum* and tomato contain race-specific proteinaceous elicitors. The specificity of these elicitors is determined by the race of *C. fulvum* and not by the resistance genes present in the tomato cultivar. The putative product of avirulence gene *A*9 of the fungus has been purified and the amino acid sequence of the peptide has been determined (Scholtens-Toma and De Wit, 1988). Further details of this are discussed in Section 3.1.

Although the emphasis is on fungal–plant gene-for-gene systems, we will briefly review a few bacterial–plant gene-for-gene interactions (for detailed reviews see Keen and Staskawicz, 1988; Daniels *et al.*, 1988). A large number of plant pathogenic bacteria have races which only attack a set of cultivars of certain host plants carrying different genes for resistance. The most extensively studied plant pathogenic bacteria are *Pseudomonas syringae* pv. *glycinea* (*Psg*), *P. syringae* pv. *tomato* (*Pst*), *P. syringae* pv. *phaseolicola*, *P. syringae* pv. *syringae*, *P. syringae* pv. *tabaci*, *P. solanacearum*, *Xanthomonas campestris* pv. *campestris*, *X. campestris* pv. *vesicatoria*, and *X. campestris* pv. *malvacearum*. Genetic studies with these bacterial pathogens have

Table III. Differential Interactions Between Various Races of Cladosporium fulvum and Tomato Cultivars with Different Genes for Resistance to C. fulvum

Cultivars	Resistance genes	Races of C. fulvum						
		0	2	4	5	2.4	2.4.5	2.4.5.9
		$A_2A_4A_5A_9$	$a_2A_4A_5A_9$	$A_2a_4A_5A_9$	$A_2A_4a_5A_9$	$a_2a_4A_5A_9$	$a_2a_4a_5A_9$	$a_2a_4a_5a_9$
Moneymaker	None	C[a]	C	C	C	C	C	C
Near-isogenic line Cf_2 of Moneymaker	Cf_2	I[b]	C	I	I	C	C	C
Near-isogenic line Cf_4 of Moneymaker	Cf_4	I	I	C	I	C	C	C
Near-isogenic line Cf_5 of Moneymaker	Cf_5	I	I	I	C	I	C	C
Near-isogenic line Cf_9 of Moneymaker	Cf_9	I	I	I	I	I	I	C
Sonato	Cf_2Cf_4	I	C	C	C	C	C	C
Sonatine	$Cf_2Cf_4Cf_9$	I	I	I	I	I	I	C

[a] C, compatible/susceptible.
[b] I, incompatible/resistant.

identified a number of pathogenicity genes of which the hypersensitive reaction pathogenicity (hrp) genes are very interesting. Some of these genes are induced when the bacteria are inoculated into plants. Clusters of these hrp genes appear to have great homology between *X. campestris* pv. *campestris* and *P. solanacearum*. The hrp genes from *P. syringae*, however, are not homologous with those of the above-mentioned genera. The hrp genes from *X. campestris* pv. *campestris* seem to be involved in secretion of pectic lyases, while the hrp genes of the other genera seem to be involved in pathogenicity as well as in the induction of HR in resistant cultivars, as both functions can be abolished by mutations in this cluster.

In recent years a number of bacterial avirulence (avr) genes have been identified by "shotgun" cloning. From *Psg*, *avrA*, *avrB*, and *avrC* have been cloned and sequenced. These genes encode 100-, 36- and 39-kDa proteins, respectively, when expressed in *Escherichia coli*. The proteins, however, did not induce HR when injected into the respective resistant cultivars, indicating that in addition other regulatory proteins are needed for induction of HR. From *Pst*, *avrD* encoding a 34-kDa protein has been cloned and sequenced. This gene seems to be plant inducible and is highly homologous to *avrD* from *Psg* which acts there presumably as a recessive allele as it does not induce HR in the *Psg* background. The 34-kDa protein isolated form *E. coli* overexpressing the *avrD* gene did not induce HR in the appropriate cultivars. However, from these overproducing cells a low-molecular-weight race-specific elicitor could be isolated (N. T. Keen, personal communication). In addition to the avr genes from *Psg*, several other avr genes have been cloned from the plant pathogenic bacteria mentioned above. One interesting finding is that the *avrA* from *Psg* seems to be present in all races of *Pst* tested so far, indicating that avr genes are not unique to a single pathovar. The same *avrA* gene introduced into *P. syringae* pv. *tabaci* elicited HR on three different tobacco species. Similarly, an avr gene from *X. campestris* pv. *vesicatoria* that causes disease on tomato did induce HR on pepper. Inactivation of this avr gene by gene replacement extended the host range of the tomato pathogen to include pepper. This is the first proof that avr genes may be responsible for host range specificities above the race-cultivar level.

2. THE COMPATIBLE INTERACTION

2.1. Fungal Proteins Specific for the Compatible Interaction

Toxins can be important pathogenicity factors for necrotrophic fungi. Most toxins are low-molecular-weight metabolites synthesized along different well-known pathways, but some, which are of interest in the context of this chapter, are (glyco)peptides or (glyco)proteins. The low-molecular-weight, often cyclic peptide toxins are not synthesized by ribosomes but by a synthetase enzyme mechanism, which exhibits the amino acid activation, racemization, trans-peptidation, and cyclization. The peptide and protein toxins, however, require the ribosomal protein-synthesizing system for their formation. Examples of the latter case are the proteinaceous toxin cerato-ulmin (128 amino acid residues) produced by *Ophiostoma ulmi* (Scheffer *et al.*, 1987; Richards and Takai, 1988) and possibly the peptide toxins (11 amino acid residues) produced by race 1 and 2 of *Verticillium dahliae* (Nachmias *et al.*, 1987).

Cerato-ulmin was generally produced in larger quantities by aggressive isolates than by nonaggressive isolates of *O. ulmi*. However, often a large variation in production was observed, which made the production of cerato-ulmin an uncertain criterion for assessing the aggressiveness of an individual isolate.

The peptide toxins from race 1 and 2 of *V. dahliae* were isolated from culture filtrates. The two peptides were found to differ in amino acid composition and toxicity to tomato plants with (*Ve*) and without (*ve*) resistance genes to *V. dahliae*. The race 1 peptide caused symptoms on *ve* plants but not on *Ve* plants, while the race 2 peptide caused symptoms on both cultivars. So it seems that *Ve* confers tolerance to the phytotoxic effects of the race 1 peptide, while the pathogenicity of race 2 on *Ve* cultivars may be due to its production of an altered peptide.

One of the most studied fungal enzymes is cutinase (see Chapter 10). The major structural surface component of the plant cell wall is cutin, a biopolyester composed of hydroxy and hydroxyepoxy fatty acids derived from the common cellular fatty acids. To penetrate the underlying plant cells the biopolymer has to be degraded or broken by the fungal extracellular enzyme cutinase. Cutinase most probably plays an important role in the interaction between *Fusarium solani* f.sp. *pisi* and pea. Cutinase-deficient mutants were nonpathogenic on pea. Pathogenicity was restored by addition of purified cutinase. Also, cutinase antibodies and cutinase inhibitors eliminated pathogenicity (Kolattukudy, 1985). cDNA clones for the *Fusarium* cutinase have been isolated and sequenced (Soliday *et al.*, 1984). Southern hybridizations with genomic DNA and cutinase cDNA probes indicate that in high-cutinase-producing strains there are two copies of the gene and in low-cutinase-producing strains there is only one. The gene has been cloned and sequenced (Kolattukudy *et al.*, 1985). Cutinase production is triggered by cutin hydrolysates. These hydrolysates may be generated by low amounts of constitutive cutinase already present in the spores. The increase in cutinase is regulated at the transcriptional level (Kolattukudy *et al.*, 1985).

After penetration of the cuticle or after penetration through natural openings, a pathogen encounters the complex cell wall of which the wall polymer pectin is the first layer to be attacked. The CWDE are produced in sequence often in the order: pectic enzymes, hemicellulases, and cellulases, respectively. The gradual appearance of different CWDE presumably reflects the physicochemical susceptibilities of corresponding wall polymers. Often, synthesis of CWDE is sensitive to catabolite repression, which is probably part of the regulation of its synthesis *in vivo*. The fungi for which the role of CWDE has been studied most intensively are *C. lindemuthianum*, *Verticillium albo-atrum*, and *Fusarium oxysporum* f.sp. *lycopersici* (Pegg, 1985; Cooper, 1983; Keon *et al.*, 1987).

Induction of mutants in a common genetic background is a powerful approach in which the role of certain CWDE can be studied (Durrands and Cooper, 1988a,b). These authors investigated the role of pectinase in the vascular wilt pathogen *V. albo-atrum*. Three pectinase-deficient mutants were produced and tested on pectin, host cell walls, galacturonic acid, and glucose to determine changes in production and regulation of pectin lyase, polygalacturonases, cellulase (C_x), β-D-galactosidase, β-D-glucosidase, and L-leucine arylamidase. One isolate, c23, was defective in secretion of all enzymes and pectin lyase accumulated in the mycelium. Isolate 34i was deficient in production of inducible polygalacturonase and pectin lyase, as it was unable to utilize galacturonides. The third mutant, 24d, produced reduced levels of five of the six pectin lyase isozymes. The three pectinase-deficient mutants produced no symptoms or symptoms were less severe or appeared later. Isolate 34i and 24d, but not the secretory mutant c23, colonized plants to levels comparable to that of the wild type. These results indicate that pectinases, particularly endo-pectin lyase, are virulence factors, but probably not crucial determinants of pathogenicity, because the host plants can also be colonized in their absence.

The second wave of CWDE which often appears *in vivo* are the hemicellulases. Although these enzymes do not cause severe maceration like the endo-polygalacturonases do, the role of hemicellulases as pathogenicity factors may be well underestimated. The hemicellulases include enzymes such as (endo)galactanases, arabinosidases, xylanases, and galactosidases (Cooper, 1983; Keon *et al.*, 1987). Cellulase is often one of the enzymes which appear only during later stages of infection, but in a few diseases it may be active during early stages (Kelman and Cowling, 1965).

Cell walls of plants also contain proteins such as extensin which, along with wall polysaccharides, can form a potential barrier against invading microorganisms. Microbial proteases may function as pathogenicity factors in order to break down plant cell wall proteins or inactivate plant enzymes or proteinaceous enzyme inhibitors. The fungi which have been studied for the production of proteases *in vitro* and *in vivo* are *Cladosporium cucumerinum*, *Colletotrichum lagenarium*, *Colletotrichum lindemuthianum*, *Monilia fructigena*, and *Ustilago maydis*. Robertson (1987) studied the pathogenicity of a *Cladosporium cucumerinum* wild type and an alkaline protease-mutant and found that the mutant was as pathogenic as the wild type, indicating that protease was not an important pathogenicity factor for this fungus. For the bacterial plant pathogen *X. campestris* pv. *campestris*, similar results were obtained by Daniels *et al.*, (1988).

Proteinaceous inhibitors of CWDE and proteases have been isolated from many species and have frequently been proposed as factors being involved in limitation of cell wall breakdown (Brown and Adikaram, 1983; Ryan and An, 1988). However, their biological relevance *in vivo* seems to be questionable as it is often not known whether these proteins are apoplastic or cytoplastic (Turner and Hoffman, 1985). To be effective, these proteins should be cell wall-bound and this has often not been clearly established.

The ability to break down phytoalexins is thought to be an important pathogenicity factor for necrotrophic fungi. The pea pathogen *Nectria haematococca* is able to detoxify pisatin to a less toxic metabolite, called DMDP (3, 6α-dihydroxy-8, 9-methylenedioxypterocarpan) by the enzyme pisatin demethylase, a cytochrome P-450 monooxygenase. A survey of the *in vitro* pisatin demethylase activity (PDA) of field isolates of *N. haematococca* revealed Pda⁻ (no ability to demethylate), Pdaⁱ (inducible PDA), and Pdaⁿ (noninducible PDA) isolates (VanEtten *et al.*, 1987). PDA was correlated with virulence. By classical genetic analyses, several unlinked genes were detected. A genomic library of Pdaⁱ isolate was constructed in a vector for *Aspergillus nidulans*. The library was screened for transformants able to demethylate pisatin and a gene determining pisatin demethylating activity was isolated (Weltring *et al.*, 1988).

Also, suppressors of defense reactions can be pathogenicity factors (Heath, 1980). *Mycosphaerella pinodes*, a pea pathogen, produces both elicitors and suppressors of pisatin accumulation (Oku *et al.*, 1987). The suppressors appeared to be low-molecular-weight, nonphytotoxic peptides (600–1000 kDa). A carbohydrate elicitor isolated from this fungus was suppressed in its activity by these compounds. Addition of suppressor to pea leaves allowed infection by several nonpathogens. Host specificity of *M. pinodes* was correlated with the biological activity of the suppressor. From *M. meloni* and *M. ligulicola*, two similar specific suppressors were isolated. The mode of action of one of the *M. pinodes* suppressors was investigated. Not only did it suppress the elicitation of the resistance response, but it also inhibited some enzymes of the pisatin biosynthetic pathway in pea and other host plants. In the nonhosts bean and soybean, the suppressor acted as an elicitor. The induction of phenylalanine ammonia lyase (PAL) and chalcone synthase (CHS) in bean was delayed at the transcriptional level (Yamada *et al.*, 1988). There was a 3 hr delay in PAL and CHS mRNA activity, 6 hr in PAL enzymatic activity, and 6–9 hr in pisatin accumulation. The suppressor was soon inactivated in the plant, probably by proteolytic enzymes of the host.

The proteins accumulating in the intercellular spaces of several compatible and incompatible interactions between *C. fulvum* and tomato plants have been compared by a number of PAGE techniques and Western blotting (De Wit *et al.*, 1986). Several proteins accumulate that are specific for the compatible interaction. So far only one (14 kDa) protein has been isolated which is most probably of fungal origin (Fig. 1). It was present in the apoplastic fluids of all compatible interactions (Joosten and De Wit, 1988). The protein was not detectable in *in vitro* cultures of the fungus nor did it accumulate after infection with a tomato strain of *P. infestans* or after abiotic stress (Joosten and De Wit, 1988). The protein was purified to homogeneity and part of the amino acid sequence has been determined. In apoplastic fluids from compatible interactions between *C. fulvum* and tomato, the 14-kDa protein was detected 6–8 days after inoculation. No serological relationship with other fungal or plant proteins could be detected on Western blots. The biological function of the protein is not known. It may play a role in obtaining or maintaining basic

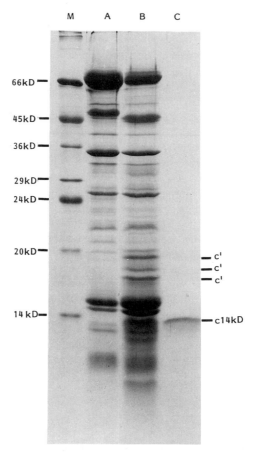

Figure 1. SDS–PAGE profiles of apoplastic fluids obtained 14 days after inoculation from an incompatible interaction (Cf4/race 5; lane A), a compatible interaction (Cf5/race 5; lane B), the purified compatible-interaction-specific 14-kDa protein (lane C). Lane M contains molecular weight markers. C′ refers to not yet purified compatible-interaction-specific proteins present in lane B.

compatibility. A partly degenerated synthetic DNA probe derived from the amino acid sequence gave clear signals on Northern blots with poly(A) RNA from all compatible interactions. No positive signals were obtained with poly(A) RNA from the healthy plant or fungus grown *in vitro*, indicating that the gene encoding the 14-kDa protein is only expressed when the fungus colonizes the tomato plant (De Wit *et al.*, unpublished).

Also in the interaction between *C. fulvum* and tomato there is evidence for the existence of suppressors (Peever and Higgins, 1989b). These suppressors act against nonspecific glycoproteinaceous elicitors (De Wit and Kodde, 1981). The intercellular fluids of all compatible interactions contained suppressor activity. When coinjected with nonspecific elicitor, necrosis induction and callose deposition, but not electrolyte leakage, were suppressed. The suppressor is most probably of host origin since intercellular fluids from uninoculated plants also possessed suppressor activity. The suppressors showed no race-cultivar specificity and their activity might be explained by enzymatic breakdown of the nonspecific elicitors. The ability of intercellular fluids to suppress the activity of the nonspecific elicitors suggests that this type of elicitor may not exist for long *in vivo*.

2.2.   Plant Proteins Specific for the Compatible Interaction

Basic compatibility may be realized by positive functions (pathogenicity factors) from the pathogen, but it might well be that the pathogen needs some support from the host in the form of proteinaceous factors. In plant–fungus interactions, one can only speculate about the existence of these factors, but for a number of plant–virus and plant–bacterium interactions such plant proteins have been shown to exist.

Plants must supply a wide variety of proteins for the replication of viruses in systemic plant–virus interactions. Virus replication is likely to be carried out by some virus–encoded proteins in association with host-encoded proteins (Hohn and Schell, 1987). It is not known whether the specific activation of host genes is required for virus reproduction. Among the plant pathogenic bacteria, *Agrobacterium tumefaciens* and *A. rhizogenes* insert a piece of bacterial plasmid DNA (T-DNA) into the plant genome and direct their host plants to synthesize a whole array of proteins involved in opine metabolism and catabolism as well as the production of the plant growth hormones auxin and cytokinin (Hohn and Schell, 1987). In the course of a *Rhizobium*–legume interaction, a large number of plant proteins, the so-called nodulins, are induced which seem to be indispensable for nodulation and nitrogen fixation (Vance *et al.*, 1988).

For obligate and biotrophic fungi which form intercellular hyphae and haustoria, undoubtedly a number of yet unknown plant proteins will be required in the tissue around the functional intercellular hyphae and haustoria. It is unclear, however, whether these proteins are already present at the time of infection, or either have to be synthesized *de novo* as a result of the interaction with the fungus. No data on such proteins are available. However, it is striking that until now, only data on plant proteins involved in compatible plant–microbe interactions are available for those interactions which show clear-cut phenotypes (e.g., *Agrobacterium*, *Rhizobium*).

3.   THE INCOMPATIBLE INTERACTION

3.1.   Protein and Glycoprotein Elicitors of Defense Reactions

In an incompatible interaction, elicitors produced by the pathogen are recognized by the plant and as a result plant defense responses are elicited.

Cruickshank and Perrin (1968) isolated a protein, monilicolin A, from mycelium of *Monilinia fructicola*, that induced the accumulation of phaseolin in bean (host-specific elicitor).

Three ~ 10-kDa proteins were isolated from culture filtrates of *Phytophthora cryptogea*, *P. cinnamomi*, and *P. capsici* (Billard *et al.*, 1988). They elicited necrosis and accumulation of PR proteins in tobacco plants and induced resistance against *P. nicotianae*. These proteins were called cryptogein, cinnamomin, and capsicein, respectively. The complete amino acid sequences of cryptogein and capsicein have been determined. They are novel proteins with about 85% amino acid sequence homology. Cryptogein and cinnamomin are basic peptides while capsicein is acidic, according to their behavior on ion-exchange chromatography. The necrosis-inducing activity of the basic peptides was about 100-fold higher than that of the acidic capsicein.

A *P. megasperma* f.sp. *glycinea* glucomannan released by a β-1, 3-endoglucanase from soybean had race-specific elicitor activity in soybean hypocotyls (Keen *et al.*, 1983). A higher level of glyceollin accumulation occurred in the incompatible interaction. However, in the interaction between *P. megasperma* f.sp. *glycinea* and soybean, race-specific suppressors have been found (Ziegler and Pontzen, 1982). Mannan-glycoproteins inhibited the accumulation of the phytoalexin glyceollin in soybean cotyledons treated with a *P. megasperma* f.sp. *glycinea* cell wall glucan elicitor. The suppressor from race 1 inhibited glyceollin accumulation in the susceptible cultivars Harosoy and Wayne, but not in the resistant cultivar Harosoy 63. The suppressor isolated from race 3 inhibited glyceollin accumulation in all three, susceptible, cultivars. The carbohydrate part of the glycoprotein was responsible for the race-specific suppressor activity.

Crude cell wall elicitors from *P. megasperma* f.sp. *glycinea* acted as elicitors of phytoalexin accumulation in parsley and soybean (Parker *et al.*, 1988). Proteinaceous constituents of these elicitors were active in parsley but not in soybean. Proteinase-treated elicitor and a defined heptaglucan elicited phytoalexin synthesis in soybean but not in parsley. These results indicate that soybean and parsley cells perceive different signals from *P. megasperma* f.sp. *glycinea* cell walls.

In germ tube walls of *Puccinia graminis* f.sp. *tritici*, Con A-binding glycoproteins were detected which were inducers of lignification and PAL activity in wheat leaves (Kogel *et al.*, 1988). One glycoprotein elicitor (67 kDa) was isolated and the carbohydrate portion consisted mainly of mannose (50%) and galactose (47%). The activity was not destroyed after treatment with pronase or trypsin, indicating that the carbohydrate part is the active portion of the molecule. Intercellular fluids of a compatible interaction between *Puccinia graminis* f.sp. *tritici* and wheat contained an elicitor with identical molecular mass and Con A binding properties. This indicates that the elicitor is released from fungal cell walls during infection. The elicitor did not show cultivar specificity.

The intercellular fluids isolated from the leaves of different compatible *C. fulvum*–tomato interactions contain race-specific proteinaceous elicitors (De Wit and Spikman, 1982). One elicitor, the putative product of avirulence gene *A9*, has been purified and its amino acid sequence determined (Scholtens-Toma and De Wit, 1988). It was isolated from apoplastic fluids of compatible interactions with races having the *A9* avirulence gene. The peptide elicitor (3 kDa) was not detectable in compatible interactions involving races such as 2.4.5.9 which do not have the *A9* avirulence gene. The peptide could not be detected in incompatible interactions, presumably because it is only present in very low amount and it possibly binds to receptor sites as soon as it is produced. The peptide contains 27 amino acids of which 6 are cysteine residues (Fig. 2). It is likely that the peptide is processed from a precursor protein. The purified peptide caused severe necrosis when injected into the intercellular spaces of tomato leaves of cultivars with the Cf9 resistance gene. The necrosis-inducing activity of the elicitor was destroyed after reduction with β-mercaptoethanol or dithiothreitol. However, the chemically synthesized peptide was not biologically active. Several attempts to oxidize the chemically synthesized peptide under different conditions in order to obtain an active molecule have failed. A partly degenerated synthetic DNA probe derived from the amino acid sequence gave clear signals on Northern blots with poly(A) RNA from compatible *C. fulvum*–tomato interactions involving races carrying avirulence gene *A9*, but not on those involving races without the *A9* gene, nor from the healthy tomato or the fungus grown *in vitro*. These results clearly indicate that the avirulence gene is only expressed *in vivo* when the fungus colonizes the tomato plant (De Wit *et al.*, unpublished).

```
AMINO ACID SEQUENCE OF THE NECROSIS-INDUCING PEPTIDE(A9-RELATED)

  1   2   3   4   5   6   7   8   9  10  11  12  13  14  15  16
TYR CYS ASN SER SER CYS THR ARG ALA PHE ASP CYS LEU GLY GLN CYS

 17  18  19  20  21  22  23  24  25  26  27
GLY ARG CYS ASP PHE HIS LYS LEU GLN CYS VAL
```

Figure 2. The amino acid sequence of the 3-kDa necrosis–inducing peptide present in apoplastic fluids of compatible interactions involving races carrying avirulance gene *A9*.

Peever and Higgins (1989a) compared electrolyte leakage, lipoxygenase activity, and lipid peroxidation induced in response to nonspecific glycoprotein elicitor and specific elicitor (the previously described peptide) from *C. fulvum*. Both elicitors induced electrolyte leakage, lipoxygenase activity and lipid peroxidation, which could be inhibited by nonsteroidal anti-inflammatory drugs. In the dark, lipoxygenase and lipid peroxidation were enhanced as under normal light conditions. Necrosis induction, however, was light dependent and did not occur when plants were incubated in the dark.

3.2. Synthesis of Enzymes Involved in Defense Responses

Plants respond actively to infection with avirulent races of a pathogen or after treatment with elicitors by synthesizing phytoalexins and defense-related proteins. Phytoalexins are low-molecular-weight antimicrobial compounds synthesized by the pathogen from remote precursors via biosynthetic pathways. The enzymes involved in these reactions are produced through *de novo* gene expression. Extensive research on the key enzymes involved in the synthesis of phytoalexins has been carried out in the host–fungus interactions *C. lindemuthianum*–French bean and *P. megasperma* f.sp. *glycinea*–soybean and the fungus–nonhost interaction *P. megasperma* f.sp. *glycinea*–parsley.

In elicitor-treated bean cell cultures, transcriptional activation of defense genes commences for some genes less than 5 min after addition of elicitor, indicating that only a few biochemical steps exist between elicitor interaction and initiation of defense gene transcription. The enzymes studied here were phenylalanine ammonia lyase (PAL), cinnamyl alcohol dehydrogenase (CAD), chalcone synthase (CHS), chalcone isomerase (CHI), cinnamic acid 4-hydroxylase (4CH), chitinase, and other PR proteins.

However, the information obtained from suspension cell cultures cannot be translated directly to the intact plant–fungus interaction. In incompatible interactions between *C. lindemuthianum* and bean, there was indeed a strong increase in the synthesis of mRNAs for PAL and CHS commencing approximately 40 hr after inoculation and reaching a maximum at about 70 hr which is during the initial contact between the fungal penetration peg and the plant cell and significantly before the appearance of HR (Bell *et al.*, 1984, 1986). In compatible interactions, these defense genes were not induced until approximately 125 hr after inoculation, during which time considerable biotrophic growth of the fungus has occurred. The strong induction of these genes in the later stages of the compatible interaction correlates well with the switch by the fungus from biotrophic to necrotrophic growth. For a recent review on defense gene activation in the interaction *C. lindemuthianum*–bean, the reader is referred to Templeton and Lamb (1988).

Most enzymes involved in the biosynthesis of isoflavonoid phytoalexins in the *C. lindemuthianum*–bean interaction are also induced very rapidly in the *P. megasperma* f.sp. *glycinea*–soybean interaction. In the latter interaction, stimulation of enzyme activities occurs more rapidly *in vivo* than in the *C. lindemuthianum*–bean interaction. This is a reflection of the penetration and growth pattern of the fungus in this interaction. In incompatible interactions of soybean and *P. megasperma* f.sp. *glycinea*–soybean, the stimulation of enzyme activities of the isoflavonoid pathway begins 2–4 hr after inoculation, which correlates with the onset of phytoalexin (glyceollin) accumulation and the occurrence of HR. In contrast to the incompatible interaction, in the compatible interaction the enzyme activities remain the same as, or are only slightly higher than, in the controls over the period of investigation (2–8 hr) (Ebel, 1986; Bonhoff *et al.*, 1986a,b).

Parsley suspension cells treated with an elicitor preparation from the nonpathogen, *P. megasperma* f.sp. *glycinea*, showed increased levels of a number of mRNAs and enzymes involved in the furanocoumarin pathway as well as PR proteins (Kombrink *et al.*, 1986). This model system has provided much information on the mechanisms underlying the induction of defense responses, but it is difficult to translate it directly to the *in vivo* situation.

3.3. Pathogenesis-Related (PR) Proteins

PR proteins are low-molecular-weight polypeptides (10–40 kDa) which accumulate extracellularly, are relatively protease resistant, and often have extreme isoelectric points (Van Loon, 1985). They accumulate in several plant species after infection by viroids, viruses, bacteria, and fungi (Van Loon, 1985). They also can be induced by a variety of chemicals (Van Loon, 1983) and even accumulate in large amounts in flowering plants (Fraser, 1981).

PR proteins were first detected in tobacco reacting hypersensitively to tobacco mosaic virus (Van Loon and Van Kammen, 1970). The major acidic PR proteins could be resolved on native gels and were named PR-la, -1b, -1c, -2, -N, -O, -P, -Q, -R, and -S in order of decreasing mobility (Pierpoint, 1986).

In the interaction between C. fulvum and tomato, several PR proteins accumulated in the intercellular fluid. Such proteins were observed on low-pH nondenaturing polyacrylamide gels. In incompatible interactions a number of PR proteins accumulated 4–6 days earlier than in compatible ones (De Wit and Van der Meer, 1986; De Wit et al., 1986; Joosten and De Wit, 1988). One of the most distinct PR proteins in tomato is P14. It accumulates not only after inoculation with C. fulvum but also after inoculation of tomato with viroids (Lucas et al., 1985). The entire amino acid sequence of P14 has been determined (Lucas et al., 1985) and exhibits close homology (about 60%) to the tobacco PR-la protein. However, these PR proteins are not related to other proteins whose sequences are known.

A serological relationship was found between the PR-1 proteins of tobacco and PR proteins of tomato and cowpea (Nassuth and Sänger, 1986). Serological homologies exist between PR proteins of a number of other plant species (White, 1983; White et al., 1987). There is 65% amino acid homology between the tobacco PR-S protein and thaumatin, a sweet protein of Thaumatococcus danielli (Cornelissen et al., 1986), and a bifunctional α-amylase/trypsin inhibitor protein from maize, which is believed to function as a protease inhibitor directed against herbivorous insects (Richardson et al., 1987). These data indicate that the genes for the PR proteins are highly conserved in plants.

3.4. The Role of PR Proteins in Active Plant Defense

Only recently has the biochemical function of some of the PR proteins been discovered. Kauffmann and co-workers (1987) isolated three acidic 1,3-β-glucanases (PR-2, -N, -O) and one basic 1,3-β-glucanase from tobacco. Legrand and co-workers (1987) isolated two acidic chitinases (PR-P and PR-Q) and two additional basic chitinases.

Kombrink and co-workers (1988) isolated two 1,3-β-glucanases (36 and 36.2 kDa, pI 9.6 and 9.8) and six chitinases (38.7, 38.0, 34.3, 33.2, 33.2, and 32.6 kDa, all pI's > 7 from apoplastic fluids and homogenates of potato after infection with P. infestans or after treatment with fungal elicitor.

Two predominant PR proteins which accumulated in the interaction between C. fulvum and tomato were purified (Joosten and De Wit, 1989). One protein (35 kDa, pI 6.4) showed endo-1,3-β-glucanase activity, while the other one (26 kDa, pI 6.1) showed endochitinase activity. One additional 1,3-β-glucanase (33 kDa) and three additional chitinases (27, 30, and 32 kDa) were identified on Western blots of C. fulvum-infected tomato leaf homogenates using antisera raised against the purified enzymes from tomato and chitinases and 1,3-β-glucanases isolated from other plant species. The chitinase and 1,3-β-glucanase increased much faster in incompatible than in compatible interactions of C. fulvum and tomato (Fig. 3). Sparse fungal growth occurs in incompatible interactions, so the accumulation of hydrolases is not a result of nonspecific stress caused by colonization of intercellular spaces. This fast response may be caused by the recognition of a race-specific elicitor produced by avirulent races of the fungus.

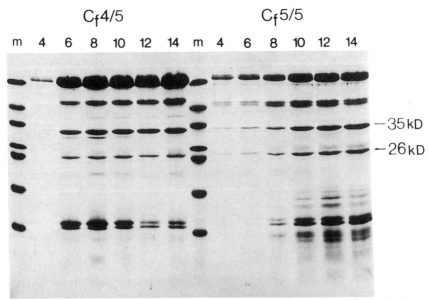

Figure 3.   SDS–PAGE profiles of apoplastic fluids from an incompatible interaction (Cf4/race 5) and a compatible interaction (Cf5/race 5). Samples were taken at different times after inoculation (days). The 1,3-glucanase (35 kDa) and the chitinase (26 kDa) are indicated. Lanes marked "m" are molecular weight markers (see Fig. 1 for details on molecular weight markers).

Hedrick and co-workers (1988) showed that in bean suspension cells, treated with *C. lindemuthianum* cell wall elicitor, chitinase activity was stimulated. cDNAs encoding chitinase were isolated and sequenced. Elicitor treatment caused a tenfold increase in chitinase activity within 30 min, leading to a rapid and transient accumulation of chitinase transcripts with maximum levels after 2 hr, concomitant with the phase of rapid enzyme synthesis. There was also a marked accumulation of chitinase transcripts in wounded and infected hypocotyls.

The chitinases and 1,3-β-glucanases may be involved in degradation of the fungal cell wall. Chitin and 1,3-β-glucan are exposed near the hyphal tip (Boller, 1987). The concentration of these enzymes is possible very high at or around the penetration sites. Many plant chitinases also possess lysozyme activity and may therefore be involved in protection against plant pathogenic bacteria.

Chitinase and 1,3-β-glucanase isolated from pea pods had *in vitro* synergistic degradative activity against several fungi (Mauch *et al.*, 1988a,b). Chitinase or 1,3-β-glucanase alone had no effect on most fungi tested. *Fusarium solani* f.sp. *pisi* and *F. solani* f.sp. *phaseoli* were both inhibited, indicating that the differential pathogenicity of these two fungi is not due to different sensitivity to the pea hydrolytic enzymes. Rather, lysis of the hyphal tips was the cause of the growth inhibition.

A correlation between PR protein synthesis and resistance in various plant species to alfalfa mosaic virus has been reported (Hooft van Huijsduijnen *et al.*, 1986). The role, however, of chitinases and 1,3-β-glucanases in resistance of plants against viruses is uncertain. Probably other intracellular proteins are needed for expression of viral resistance.

Vera and Conejero (1988) isolated a 69-kDa protein, exhibiting protease activity, from tomato infected with citrus exocortis viroid. The protease had a pH optimum of 8.5–9 and a pI of 9 as determined by isoelectric focusing. This protease may be involved in degenerative processes in the chloroplast. The activity also increased in aging healthy plants. A protein of similar molecular weight also accumulated after infection of tomato with *C. fulvum* (De Wit *et al.*, 1986). Similarly, after infection of tomato plants with tomato macho viroid (TPMV) or potato spindle tuber viroid (PSTV), a protein of similar molecular weight was induced (Galindo *et al.*, 1984). If these two proteins are identical to the protease found by Vera and Conejero (1988), the enzyme may be involved in plant defense.

4. CONCLUDING REMARKS

It is difficult to translate the data presented in this chapter to the intact plant–pathogen interaction. While progress has been made in understanding mechanisms that lead to a compatible or incompatible outcome of an interaction, a clear picture has not yet emerged. Attempts to correlate elicitor activity with species or race-cultivar specificity were generally not successful. Elicitors prepared from different races of a fungus grown *in vitro* stimulated phytoalexin production in both resistant and susceptible cultivars. Possibly, determinants of basic compatibility and race-specificity are predominantly expressed *in planta* as has been found in the interaction between *C. fulvum* and tomato (Scholtens-Toma and De Wit, 1988; Joosten and De Wit, 1988) and in plant pathogenic bacteria–plant interactions (Daniels *et al.*, 1988; Keen and Staskawicz, 1988; Osbourn *et al.*, 1987). The nature of basic compatibility and race-specificity, however, will only be understood properly by cloning and studying the genes involved.

Cloning of pathogenicity genes and avirulence genes has been carried out for a number of plant pathogenic bacteria (Daniels *et al.*, 1988; Keen and Staskawicz, 1988), but has only just begun for plant pathogenic fungi (VanEtten *et al.*, 1987; Weltring *et al.*, 1988). Transformation of *C. fulvum* has also become possible (Oliver *et al.*, 1987). Applying molecular biological approaches and exploiting recent data obtained with comparative biochemical studies would give a strong impulse in the research concerning host–fungal interactions and the interaction *C. fulvum*–tomato in particular.

ACKNOWLEDGMENTS. Part of this research was carried out in the framework of contract BAP-0074-NL of the Biotechnology Action Programme of the Commission of the European Comunities.

5. REFERENCES

Bell, J. N., Dixon, R. A., Bailey, J. A., Rowell, P. M., and Lamb, C. J., 1984, Differential accumulation of chalcone synthase mRNA activity at the onset of phytoalexin accumulation in compatible and incompatible plant pathogen interactions, *Proc. Natl. Acad. Sci. USA* **81**:3384–3388.

Bell, J. N., Ryder, T. B., Wingate, V. P. M., Bailey, J. A., and Lamb, C. J., 1986, Differential accumulation of plant defense gene transcripts in a compatible and incompatible plant–pathogen interaction, *Mol. Cell. Biol.* **6**:1615–1623.

Billard, V., Bruneteau, M., Bonnet, P., Ricci, P., Pernollet, J. C, Huet, J. C., Vergne, A., Richard, C., and Michel, G., 1988, Chromatographic purification and characterization of elicitors of necrosis on tobacco produced by incompatible *Phytophthora* species, *J. Chromatogr.* **44**:87–94.

Bohlmann, H., Clausen, S., Behnke, H. G., Giese, H., Hiller, C., Reimann-Philipp, U., Schrader, G., Barkholt, V., and Apel, K., 1988. Leaf-specific thionins of barley—a novel class of cell wall proteins toxic to plant-pathogenic fungi and possibly involved in the defense mechanism of plants, *EMBO J.* **7**:1559–1565.

Boller, T., 1987, Hydrolytic enzymes in plant disease resistance, in: *Plant–Microbe Interactions: Molecular and Genetic Perspectives*, Volume 2 (T. Kosuge and E. W. Nester, eds.), Macmillan Co., New York, pp. 385–413.

Bonhoff, A., Loyal, R., Ebel, J., and Grisebach, H., 1986a, Race: cultivar-specific induction of enzymes related to phytoalexin biosynthesis in soybean roots following infection with *Phytophthora megasperma* f.sp. *glycinea*, *Arch. Biochem. Biophys.* **246**:149–154.

Bonhoff, A., Loyal, R., Feller, K., Ebel, J., and Grisebach, H., 1986b, Further investigation of race: cultivar-specific induction of enzymes related to phytoalexin biosynthesis in soybean roots following infection with *Phytophthora megasperma* f.sp. *glycinea*, *Biol. Chem. Hoppe-Seyler* **367**:797–802.

Brown, A. E., and Adikaram, N. K. B., 1983, A role for pectinase and protease inhibitors in fungal rot development in tomato fruits, *Phytopathol. Z.* **106**:239–251.

Cooper, R. M., 1983, The mechanisms and significance of enzymatic degradation of host cell walls by parasites, in: *Biochemical Plant Pathology* (J. A. Callow, ed.), Wiley, New York, pp. 101–135.

Cornelissen, B. J. C., Hooft van Huijsduijnen, R. A. M., and Bol, J. H., 1986, A tobacco mosaic virus-induced tobacco protein is homologous to the sweet tasting protein thaumatin, *Nature* **321**:531–532.

Crucefix, D. N., Mansfield, J. W., and Wade, M., 1984, Evidence that determinants of race specificity in lettuce downy mildew disease are highly localized, *Physiol. Plant Pathol.* **24**:93–106.

Crucefix, D. N., Rowell, P. M., Street, P. F. S., and Mansfield, J. W., 1987, A search for elicitors of the hypersensitive reaction in lettuce downy mildew disease, *Physiol. Mol. Plant Pathol.* **30**:39–54.

Cruickshank, I. A. M., and Perrin, D. R., 1968, The isolation and partial characterization of monilicolin A, a polypeptide with phaseolin inducing activity from *Monilinia fructicola*, *Life Sci.* **7**:449–458.

Crute, I. R., 1985, The genetic basis of relationships between microbial parasites and their hosts, in: *Mechanisms of Resistance to Plant Diseases*, (R. S. S. Fraser, ed.), Nijhoff, The Hague, pp. 80–142.

Daniels, M. J., Dow, J. M., and Osbourn, A. E., 1988, Molecular genetics of pathogenicity in phytopathogenic bacteria, *Annu. Rev. Phytopathol.* **26**:285–312.

Day, P. R., 1956, Race names of *Cladosporium fulvum*, *Report of the Tomato Genetics Cooperative* **6**:13–14.

De Wit, P. J. G. M., 1986, Elicitation of active resistance mechanisms, in: *Biology and Molecular Biology of Plant–Pathogen Interactions*, NATO ASI Series Volume H1 (J. Bailey, ed.), Springer-Verlag, Berlin, pp. 149–169.

De Wit, P. J. G. M., and Kodde, E., 1981, Induction of polyacetylenic phytoalexins in *Lycopersicon esculentum* after inoculation with *Cladosporium fulvum*, *Physiol. Plant Pathol.* **18**:143–148.

De Wit, P. J. G. M., and Spikman, G., 1982, Evidence for the occurrence of race- and cultivar-specific elicitors of necrosis in intercellular fluids of compatible interactions between *Cladosporium fulvum* and tomato, *Physiol. Plant Pathol.* **21**:1–11.

De Wit, P. J. G. M., and Van der Meer, F. E., 1986, Accumulation of the pathogenesis-related tomato leaf protein P14 as an early indicator of incompatibility in the interaction between *Cladosporium fulvum* (syn. *Fulvia fulva*) and tomato, *Physiol. Plant Pathol.* **28**:203–214.

De Wit, P. J. G. M., Buurlage, M. B., and Hammond, K. E., 1986, The occurrence of host-, pathogen- and interaction-specific proteins in the apoplast of *Cladosporium fulvum* (syn. *Fulvia fulva*)-infected tomato leaves, *Physiol. Mol. Plant Pathol.* **29**:154–172.

Dinoor, A., and Eshed, N., 1984, The role and importance of pathogens in natural plant communities, *Annu. Rev. Plant Pathol.* **22**:443–466.

Doke, N., and Tomiyama, K., 1980, Suppression of the hypersensitive response of potato tuber protoplasts to hyphal wall components by water soluble glucans isolated from *Phytophthora infestans*, *Physiol. Plant Pathol.* **16**:177–186.

Dunkle, D. C., 1984, Factors in pathogenesis, in: *Plant–Microbe Interactions: Molecular and Genetic Perspectives*, Volume 1 (T. Kosuge and E. W. Nester, eds.), Macmillan Co., New York, pp. 19–41.

Durrands, P. K., and Cooper, R. M., 1988a, Selection and characterization of pectinase-deficient mutants of the vascular wilt pathogen *Verticillium albo-atrum*, *Physiol. Mol. Plant Pathol.* **32**:343–362.

Durrands, P. K., and Cooper, R. M., 1988b, The role of pectinase in vascular wilt disease as determined by defined mutants of *Verticillium albo-atrum*, *Physiol. Mol. Plant Pathol.* **32**:363–371.

Ebel, J., 1986, Phytoalexin synthesis: The biochemical analysis of the induction process, *Annu. Rev. Phytopathol.* **24**:235–264.

Flor, H. H., 1942, Inheritance of pathogenicity in *Melampsora lini*, *Phytopathology* **32**:653–669.

Fraser, R. S. S., 1981, Evidence for the occurrence of the 'pathogenesis-related' proteins in leaves of healthy tobacco plants during flowering, *Physiol. Plant Pathol.* **19**:69–76.

Fraser, R. S. S., (ed.), 1985, *Mechanisms of Resistance to Plant Disease*, Nijhoff, The Hague.

Galindo, J. A., Smith, D. R., and Diener, T. O., 1984, A disease-associated host protein in viroid-infected tomato, *Physiol. Plant Pathol.* **24**:257–275.

Hamdan, M. A. M. S., and Dixon, R. A., 1986, Differential biochemical effects of elicitor preparations from *Colletotrichum lindemuthianum*, *Physiol. Mol. Plant Pathol.* **28**:329–344.

Heath, M. C., 1980, The absence of active defence mechanisms, in: *Active Defence Mechanisms in Plants* (R. K. S. Wood, ed.), Plenum Press, New York, pp. 143–156.

Hedrick, S. A., Bell, J. N., Boller, T., and Lamb, C. J., 1988, Chitinase complementary DNA cloning and messenger RNA induction by fungal elicitor, wounding and infection, *Plant Physiol.* **86**:182–186.

Hohn, T., and Schell, J., (eds.), 1987, *Plant DNA Infectious Agents*, Springer-Verlag, Berlin.

Hooft van Huijsduijnen, R. A. M., Alblas, S. W., De Rijk, R. H., and Bol, J. F., 1986, Induction by salicylic acid of pathogenesis-related proteins and resistance to alfalfa mosaic virus infection in various plant species, *J. Gen. Virol.* **67**:2135–2143.

Joosten, M. H. A. J., and De Wit, P. J. G. M., 1988, Isolation, purification and preliminary characterization of a protein specific for compatible *Cladosporium fulvum* (syn. *Fulvia fulva*)–tomato interactions, *Physiol. Mol. Plant Pathol.* **33**:241–253.

Joosten, M. H. A. J., and De Wit, P. J. G. M., 1989, The identification of several pathogenesis-related (PR) proteins in tomato leaves inoculated with *Cladosporium fulvum* (syn. *Fulvia fulva*) as 1,3-β-glucanases and chitinases, *Plant Physiol.* **89**:945–951.

Kauffmann, S., Legrand, M., Geoffroy, P., and Fritig, B., 1987, Biological function of 'pathogenesis-related' proteins: Four PR proteins of tobacco have 1,3-β-glucanase activity, *EMBO J.* **6**:3209–3212.

Keen, N. T., 1986, Pathogenic strategies of fungi, in: *Recognition in Microbe–Plant Symbiotic and Pathogenic Interactions*, NATO ASI Series Volume H4 (B. Lugtenberg, ed.), Springer-Verlag, Berlin, pp. 171–188.

Keen, N. T., and Staskawicz, B., 1988, Host range determinants in plant pathogens and symbionts, *Annu. Rev. Microbiol.* **42**:421–440.

Keen, N. T., Yoshikawa, M., and Wang, M. C., 1983, Phytoalexin elicitor activity of carbohydrates from *Phytophthora infestans* f.sp. *glycinea* and other sources, *Plant Physiol.* **71**:466–471.

Kelman, A., and Cowling, E. B., 1965, Cellulases of *Pseudomonas solanacearum* in relation to pathogenesis, *Phytopathology* **55**:148–155.

Keon, J. P. R., Byrde, R. J. W., and Cooper, R. M., 1987, Some aspects of fungal enzymes that degrade plant cell walls, in: *Fungal Infection of Plants* (G. F. Pegg and P. G. Ayres, eds.), Cambridge University Press, London, pp. 133–157.

Knoche, H. W., and Duvick, J. P., 1987, The role of fungal toxins in plant disease, in: *Fungal Infection of Plants* (G. F. Pegg and P. G. Ayres, eds.), Cambridge University Press, London, pp. 158–192.

Kogel, G., Beissmann, B., Reisener, H. J., and Kogel, K. H., 1988, A single glycoprotein from *Puccinia graminis* f.sp. *tritici* cell walls elicits the hypersensitive lignification response in wheat, *Physiol. Mol. Plant Pathol.* **33**:173–186.

Kolattukudy, P. E., 1985, Enzymatic penetration of the plant cuticle by fungal pathogens, *Annu. Rev. Phytopathol.* **23**:223–250.

Kolattukudy, P. E., Soliday, C. L., Woloshuk, C. P., and Crawford, M., 1985, Molecular biology of the early events in the fungal penetration into plants, in: *Molecular Genetics of Filamentous Fungi*, Volume 34 (W. E. Timberlake, ed.), Liss, New York, pp. 421–438.

Kombrink, E., Bollmann, J., Hauffe, K. D., Knogge, D. Schell, E., Schmelzer, I., Somssich, I., and Hahlbrock, K., 1986, Biochemical responses of nonhost plant cells to fungi and fungal elicitors, in: *Biology and Molecular Biology of Plant–Pathogen Interactions*, NATO ASI Series Volume H1 (J. A. Bailey, ed.), Springer-Verlag, Berlin, pp. 253–262.

Kombrink, E., Schröder, M., and Hahlbrock, K., 1988, Several 'pathogenesis-related' proteins in potato are 1,3-β-glucanases and chitinases, *Proc. Natl. Acad. Sci. USA* **85**:782–786.

Krüger, J., Hoffman, C. M., and Hubbeling, N., 1977, The kappa race of *Colletotrichum lindemuthianum* and sources of resistance to anthracnose in *Phaseolus* beans, *Euphytica* **26**:23–25.

Kuc, J., Tjamos, E., and Bostock, R., 1984, Metabolic regulation of terpenoid accumulation and disease resistance in potato, in: *Isopentenoids in Plant Biochemistry and Function* (W. S. Nes, G. Fuller, and L.-S. Tsai, eds.), Dekker, New York, pp. 103–126.

Legrand, M., Kauffmann, S., Geoffroy, P., and Fritig, B., 1987, Biological function of pathogenesis-related proteins: Four tobacco pathogenesis-related proteins are chitinases, *Proc. Natl. Acad. Sci. USA* **84**:6750–6754.

Lucas, J., Camacho Henriquez, A., Lottspeich, F., Henschen, A., and Sänger, H. L., 1985, Amino acid sequence of the 'pathogenesis-related' leaf protein P14 from viroid infected tomato reveals a new type of structurally unfamiliar proteins, *EMBO J.* **4**:2745–2749.

Maniara, G., Laine, R., and Kuc, J., 1984, Oligosaccharides from *Phytophthora infestans* enhance the elicitation of sesquiterpenoid stress metabolites by arachidonic acid in potato, *Physiol. Plant Pathol.* **24**:177–186.

Mansfield, J. W., 1983, Antimicrobial compounds, in *Biochemical Plant Pathology* (J. A. Callow, ed.), Wiley, New York, pp. 237–265.

Mansfield, J. W., Wood, A. M., Street, P. F. S., and Rowell, P. M., 1988, Recognition processes in lettuce downy mildew disease, in: *Eukaryote Cell Recognition Concepts and Model Systems* (G. P. Chapman, C. C. Ainsworth, and C. J. Chatham, eds.), Cambridge University Press, London, pp. 241–256.

Mauch, F., Hadwicher, L. A., and Boller, T., 1988a, Antifungal hydrolases in pea tissue. I. Purification and characterization of two chitinases and two β-1,3-glucanases differentially regulated during development and in response to fungal infection, *Plant Physiol.* **87**:325–333.

Mauch, F. C., Mauch-Mani, B., and Boller, T., 1988b, Antifungal hydrolases in pea tissue. II. Inhibition of fungal growth by combinations of chitinase and β-1,3-glucanase, *Plant Physiol.* **88**:936–942.

Michelmore, R. W., Ilott, T., Hulbert, S. H., and Farrara, B., 1988, The downy mildews, in: *Genetics of Plant Pathogenic Fungi* (G. S. Sidhu, ed.), Academic Press, New York, pp. 53–79.

Nachmias, A., Buchner, V., Tsror, L., Burstein, Y., and Keen, N., 1987, Differential phytotoxicity of peptides from culture fluids of *Verticillium dahliae* races 1 and 2 and their relationship to pathogenicity of the fungi on tomato, *Phytopathology* **77**:506–510.

Nassuth, A., and Sänger, H. L., 1986, Immunological relationship between 'pathogenesis-related' leaf proteins from tomato, tobacco and cowpea, *Virus Res.* **4**:229–242.

Oku, H., Shiraishi, T., and Ouchi, S., 1987, Role of specific suppressors in pathogenesis of *Mycospaerella* species, in: *Molecular Determinants of Plant Diseases* (S. Nishimura, C. P. Vance, and N. Doke, eds.), Springer-Verlag, Berlin, pp. 145–156.

Oliver, R. P., Roberts, I. N., Harling, R., Kenyon, L., Punt, P. J., Dingemanse, M. A., and Van den Hondel, C. A. M. J. J., 1987, Transformation of *Fulvia fulva*, a fungal pathogen of tomato, to hygromycin β resistance, *Curr. Genet.* **12**:231–233.

Osbourn, A. E., Barber, C. E., and Daniels, M. J., 1987, Identification of plant induced genes of the bacterial pathogen *Xanthomonas campestris* pathovar *campestris* using a promoter-probe plasmid, *EMBO J.* **6**:23–28.

Parker, J. E., Hahlbrock, K., and Scheel, D., 1988, Different cell wall components from *Phytophthora megasperma* f.sp. *glycinea* elicit phytoalexin production in soybean and parsley, *Planta* **176**:75–82.

Peever, T. L., and Higgins, V. J., 1989a, Electrolyte leakage, lipoxygenase and lipid peroxidation induced in tomato leaf tissue by specific and nonspecific elicitors from *Cladosporium fulvum*, *Plant Physiol.* **90**:867–875.

Peever, T. L., and Higgins, V. J., 1989b, Suppression of the activity of nonspecific elicitor from *Cladosporium fulvum* by intercellular fluids from tomato leaves, *Physiol. Mol. Plant Pathol.* **34**:471–482.

Pegg, G. F., 1985, Life in a black hole—The micro-environment of the vascular pathogen, *Trans. Br. Mycol. Soc.* **85**:1–20.

Pierpoint, W. S., 1986, The pathogenesis-related proteins of tobacco leaves, *Phytochemistry* **25**:1595–1601.

Preisig, C. L., and Kuc, J. A., 1985, Arachidonic acid-related elicitors of the hypersensitive response in potato and enhancement of their activities by glucans from *Phytophthora infestans* (mont.) deBary, *Arch. Biochem. Biophys.* **236**:379–389.

Richards, W. C., and Takai, S., 1988, Production of cerato-ulmin in white elm following artificial inoculation with *Ceratocystis ulmi*, *Physiol. Mol. Plant Pathol.* **33**:279–285.

Richardson, M., Valdes-Rodriguez, S., and Blanco-Labra, A., 1987, A possible role for thaumatin and a TMV-induced protein suggested by homology to a maize inhibitor, *Nature* **327**:432–434.

Robertson, B., 1987, Endo-polygalacturonase from *Cladosporium cucumerinum* elicits lignification in cucumber hypocotyls, *Physiol. Mol. Plant Pathol.* **31**:361–374.

Ryan, C. A., and An, G., 1988, Molecular biology of wound-inducible proteinase inhibitors in plants, *Plant Cell Environ.* **11**:345–349.

Scheffer, R. J., Liem, J. I., and Elgersma, D. M., 1987, Production *in vitro* of phytotoxic compounds by non-aggressive isolates of *Ophiostoma ulmi* the Dutch elm disease pathogen, *Physiol. Mol. Plant Pathol.* **30**:321–336.

Scholtens-Toma, I. M. J., and De Wit, P. J. G. M., 1988, Purification and primary structure of a necrosis-inducing peptide from the apoplastic fluids of tomato infected with *Cladosporium fulvum* (syn. *Fulvia fulva*), *Physiol. Mol. Plant Pathol.* **33**:59–67.

Shaw, D. S., 1988, The *Phytophthora* species, in: *Genetics of Plant Pathogenic Fungi* (G. S. Sidhu, ed.), Academic Press, New York, pp. 27–51.

Soliday, C. L., Flurkey, W. H., Okita, T. W., and Kolattukudy, P., 1984, Cloning and structure determination of cDNA for cutinase, an enzyme involved in fungal penetration of plants, *Proc. Natl. Acad. Sci. USA* **81**:3939–3943.

Templeton, M. D., and Lamb, C. J., 1988, Elicitors and defense gene activation, *Plant Cell Environ.* **11**:395–401.

Tepper, C. S., and Anderson, A. J., 1986, Two cultivars of bean display a differential response to extracellular components from *Colletotrichum lindemuthianum*, *Physiol. Mol. Plant Pathol.* **29**:411–420.

Turner, J. G., and Hoffman, R. M., 1985, Effect on the polygalacturonase inhibitor from pea on the hydrolysis of pea cell walls by endopolygalacturonase from *Ascochyta pisi*, *Plant Pathol.* **34**:54–60.

Vance, C. P., Egli, M. A., Griffith, S. M., and Miller, S. S., 1988, Plant regulated aspects of nodulation and N₂ fixation, *Plant Cell Environ.* **11**:413–427.

VanEtten, H. D., Matthews, D. E., and Mackintosh, S. F., 1987, Adaption of pathogenic fungi to toxic chemical barriers in plants: The pisatin demethylase of *Nectria haematococca* as an example, in: *Molecular Strategies for Crop Protection* (C. Arntsen and C. Ryan, eds.), Liss, New York, pp. 59–70.

Van Loon, L. C., 1983, The induction of pathogenesis-related proteins by pathogens and specific chemicals, *Neth. J. Plant Pathol.* **89**:265–273.

Van Loon, L. C., 1985, Pathogenesis-related proteins, *Plant Mol. Biol.* **4**:111–116.

Van Loon, L. C., and Van Kammen, A., 1970, Polyacrylamide disc electrophoresis of the soluble leaf proteins from *Nicotiana tabacum* var "Samsun" and "Samsun NN". II. Changes in protein constitution after infection with tobacco mosaic virus, *Virology* **40**:199–211.

Vera, P., and Conejero, V., 1988. Pathogenesis-related proteins of tomato, P-69 as an alkaline endoprotease, *Plant Physiol.* **87**:58–63.

Weltring, K. M., Turgeon, B. G., Yoder, O. C., and VanEtten, H. D., 1988, Isolation of a phytoalexin-detoxification gene from the plant pathogenic fungus *Nectria haematococca* by detecting its expression in *Aspergillus nidulans*, *Gene* **68**:335–344.

White, R. F., 1983, Serological detection of pathogenesis-related proteins, *Neth. J. Plant Pathol.* **89**:311.

White, R. F., Rybicki, E. P., Von Wechmar, M. B., Dekker, J. L., and Antoniw, J. F., 1987, Detection of PR1 type proteins in *Amaranthaceae*, *Chenopodiaceae*, *Gramineae* and *Solanaceae* by immunoelectroblotting, *J. Gen. Virol.* **68**:2043–2048.

Yamada, T., Hashimoto, H., Shiraishi, T., and Oku, H., 1989, Suppression of pisating phenylalanine ammonia-lyase mRNA and chalcone sythase mRNA accumulation by putative pathogenicity factor from the fungus *Mycosphaerella pinodes*. *Molecular Plant–Microbe Interactions* **2**:256–261.

Ziegler, E., and Pontzen, R., 1982, Specific inhibition of glucan elicited glyceollin accumulation in soybeans by an extracellular mannan-glycoprotein of *Phytophthora megasperma* f.sp. *glycinea*, *Physiol. Mol. Plant Pathol.* **20**:321–331.

12

The Role of Cuticle-Degrading Enzymes in Fungal Pathogenesis in Insects

A. K. Charnley and R. J. St. Leger

1. INTRODUCTION

Insects are members of the Arthropoda. Among the characteristics of this phyllum is the presence of an external skeleton or cuticle. Because of its location, the cuticle serves a variety of functions in addition to the skeletal roles of support and muscle anchorage. The defensive capability of the cuticle is clear since only one group of entomopathogens, the fungi, have acquired the ability to invade insects actively via this route. The other major groups of disease-causing microorganisms, the viruses and bacteria, are restricted primarily to the alimentary canal, where the midgut provides an exposed mucosal surface.

Fungal penetration of intact insect cuticle appears to be by a combination of mechanical force and enzymatic degradation, the relative contribution of the two components depending on the structure and composition of the cuticle encountered (Charnley, 1984). Most fungi are nonentomopathogenic because they are unable to degrade cuticle or overcome cuticle-based host defenses. Consequently, normally innocuous fungi can prove lethal if the cuticle is wounded or bypassed by direct injection of spores (Madelin, 1968).

Entomopathogenic fungi are present in all classes of the fungi though possibly the most widespread occur in the Deuteromycetes and Entomophthorales (Gillespie, 1988). Epizootics of insect fungus diseases are relatively common in nature and fungi can be key regulating factors in pest insect populations. Consequently, there have been many attempts to use entomopathogenic fungi for pest control (Gillespie, 1988). However, the realization of the full potential of insect pathogenic fungi for pest control is hampered by a number of factors including ignorance of the determinants of fungal pathogenicity. A prominent exception to this is recent progress in our understanding of how fungi invade insects. The object of this contribution is to review our knowledge of the cuticle-degrading enzymes produced by entomopathogenic fungi and to evaluate the role of these enzymes in pathogenicity.

A. K. Charnley • School of Biological Sciences, University of Bath, Bath, Avon BA2 7AY England.   *R. J. St. Leger* • Boyce Thompson Institute, Cornell University, Ithaca, New York 14853.

2. THE INVASION PROCESS

Adhesion of a spore to the surface of host cuticle is the first step in the establishment of mycosis; the mechanisms involved are reviewed elsewhere in this volume (Chapter 1). *In vitro* germination of deuteromycete entomopathogens with broad host ranges—e.g., *Metarhizium anisopliae* and *Beauveria bassiana*—occurs in response to nonspecific sources of carbon and/or nitrogen (Smith and Grula, 1981; St. Leger *et al.*, 1986a). However, for many such fungi, the ability to utilize the lipids that make up the outer layer of the cuticle (epicuticle; see Fig. 1) may be fundamental to pathogenesis (Charnley, 1984; St. Leger, 1990). Entomopathogenic fungi with restricted host ranges appear to have more specific requirements for germination. *Nomuraea rileyi*, which primarily infects lepidopterans, responds to diacyglycerols and polar lipids (Boucias and Pendland, 1984). Similarly, *Erynia variabilis* is restricted to small dipterans in part by a requirement for oleic acid to induce germination (Kerwin, 1984).

Successful germination presupposes a tolerance of potentially toxic substances in the outer layers of the cuticle. Short-chain fatty acids have been highlighted but evidence of a defensive role for these compounds is incomplete (for review see St. Leger, 1990). Antagonism from saprophytic flora is a further hazard to a pathogenic fungus on insect cuticle (see Charnley, 1989, and Chapter 6).

The invasion of cuticle often requires the formation of specialized infection structures such as appressoria and penetration pegs (see Fig. 2). In contrast to germination, differentiation of *M. anisopliae* has specific nutrient requirements. Low levels of complex nitrogenous compounds induce appressorial formation against hard hydrophobic surfaces whereas high levels suppress infection-related morphogenesis and cuticle penetration. Instead, germ tubes continue to elon-

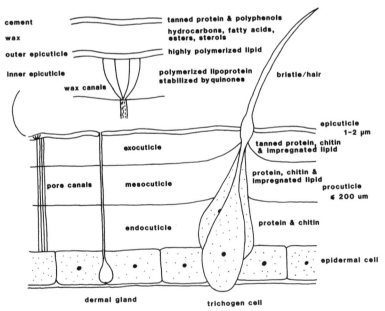

Figure 1.   Structure of insect cuticle. Protein is the major component of insect cuticle. Tanning, or sclerotization, is the cross-linking of the protein by aromatic compounds such as *N*-acetyldopamine; it hardens the cuticle.

In soft-bodied insects (e.g., lepidopteran larvae), the exocuticle is thin and indistinguishable from the epicuticle. The arthrodial membrane at joints and between segments has no exocuticle.

Chitin is a β-1, 4-linked hexose polymer made up of *N*-acetylglucosamine units.

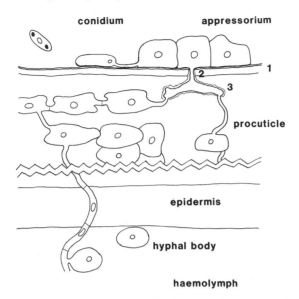

Figure 2. Cuticular penetration by entomopathogenic fungi. 1, epicuticle; 2, penetration peg; 3, penetration plate.

gate and branch forming a hyphal mat (St. Leger *et al.*, 1989a). As with plant pathogenic fungi (see Chapter 2), specific recognition factors may also play a part in either attachment or formation of infection structures. Some preliminary studies have implicated carbohydrates and specific carbohydrate-binding proteins (lectins) in host recognition (e.g., Grula *et al.*, 1984; Kerwin and Washino, 1986), but evidence is far from complete. Entomopathogens with a broad host range are likely to respond to nonspecific stimuli on their hosts. Indeed, *in vitro* emerging germ tubes and appressoria of *M. anisopliae* will adhere to any hydrophobic surface via an amorphous mucilage. However, in addition to nutrition, surface topography influences appressorial formation in *M. anisopliae*, occurring close to the spore on the comparatively fat cuticles of *Calliphora vomitoria*, *Schistocerca gregaria*, and late instar larvae of *Manduca sexta*, but only after extensive growth on the highly convoluted cuticle (microfolds) of young *M. sexta* larvae (St. Leger *et al.*, 1989b). In the case of the last named, appressorial formation occurs preferentially on hair sockets which are covered by thin cuticle and thus constitute zones of weakness (David, 1967).

Most pathogenic fungi need to enter the insect body to obtain nutrients for their growth and reproduction. The only exceptions are the exoparasitic Laboulbeniales (Ascomycetes) (Zacharuk, 1981) and the largely commensal Trichomycetes which inhabit dipteran guts (see Chapter 6). Penetration can occur via wounds, sense organs, and most commonly by direct penetration of intact cuticle, particularly arthrodial membranes at joints and between segments (see Charnley, 1984, 1989; St. Leger, 1990). There is little evidence for penetration via natural channels such as pore channels (David, 1967) while the alimentary canal is often an inhospitable environment for fungi (see Chapter 6).

3. CUTICLE AS A BARRIER TO FUNGAL INFECTION

The epicuticle is the first barrier encountered (Fig. 1). Penetration of this layer is either by infection pegs produced from the underside of appressoria or by direct entry by germ tubes (Fig.

2) (Zacharuk, 1981; Pekrul and Grula, 1979). The disappearance of the wax layer beneath appressoria of *M. anisopliae* on wireworm cuticle indicates enzyme activity (Zacharuk, 1970a) as does the presence of circular holes around germ tubes of *B. bassiana* at the point of entry into larvae of *Heliothis zea* (Pekrul and Grula, 1979). Physical penetration is prominent, however, in host invasion by some Entomophthoralean pathogens where characteristic triradiate and tetra-radiate fissures appear in the epicuticle (Brobyn and Wilding, 1983).

The bending and fracture properties of the epicuticle suggest that it is pliant but not extensible, stronger in compression than in tension (Hepburn, 1985). This may be important in providing resistance to fungal penetration as compressive forces would be expected at the growing point of a penetration peg. The epicuticle is multilayered and each layer has its own properties. The outer epicuticle appears in most insects to be fragile (Hackman, 1984) and may be penetrated by a weak force (St. Leger, 1990). Its resistance to degradation and impermeability (Locke, 1984) suggest that until physically disrupted, it could prevent passage of fungal enzymes (St. Leger, 1990). The inner epicuticle is thought to consist of polymerized lipoprotein stabilized by quinones (Dennell, 1946). This composition implies physical toughness, but enzymes produced by entomopathogens may be equal to the task (see later). The chemical complexity and resilience of the epicuticle may make it of particular importance in deterring fungi, but the remainder of the cuticle, the procuticle (see Fig. 1), which is several hundred times thicker than the epicuticle, presents a significant barrier to infection. The procuticle comprises chitin fibrils embedded in a protein matrix, together with lipids and quinones (Neville, 1984). The mechanical properties of different cuticles depend on the proportions of the two main constituents, the nature and extent of hydration of the proteins, and the degree of sclerotization (cross-linking of the proteins by quinones) (Hillerton, 1984). Being a composite material, the procuticle has impact resistance, tensile, flexural, and compressive strength (Neville, 1984).

After breaching the epicuticle, penetrant structures often expand laterally in the outer layers of the procuticle, producing penetration plates (Fig. 2; Zacharuk, 1970b; Brobyn and Wilding, 1983). These expansions can cause fractures which favor penetration (Brey *et al.*, 1986). Passage of the fungus across the procuticle may be more or less direct or involve a degree of lateral extension parallel to the lamellae before or during vertical penetration. In an extreme case this may involve a stepwise progression of hyphae toward the epidermis (Fig. 2; Zacharuk, 1981). Thickness of the procuticle correlates with disease resistance in that young larvae (with thin cuticles) are more susceptible than old larvae (with thick cuticles) (David, 1967). However, additional changes in cuticular composition occur with advancing age that may influence pathogenesis. Thus, increased resistance to *N. rileyi* by fifth-instar compared with first- and second-instar *Anticarsia gemmatalis* larvae, in addition to cuticle thickness, correlates with a reduced ability of conidia to adhere to a less hydrophobic cuticle (Boucias *et al.*, 1988) and the appearance of a protease inhibitor (see later; Boucias and Pendland, 1987).

The degree of cuticular sclerotization also has a strong influence on penetrability. In the main, insects with heavily sclerotized body segments are invaded via arthrodial membranes or spiracles (see Charnley, 1984; St. Leger, 1990). Resistance to compressive force by sclerotized cuticle (exocuticle) suggests that it presents a stronger physical barrier than nonsclerotized endo- and mesocuticle (Fig. 1) (St. Leger, 1990). Although exocuticle from *S. gregaria* supports growth and cuticle-degrading enzyme activity by *M. anisopliae in vitro* (see later), the integrity of the cuticle remains intact. This comparative resistance of exocuticle to degradation has been noted also with endogenous enzymes in the molting fluid (Hepburn, 1985). St. Leger (1990) suggested that since water extrusion accompanies sclerotization, water availability in addition to the phenolic cross-links may be a factor limiting penetration in exocuticle. Ultrastructural evidence that hyphae tend to grow between the cuticular lamellae of exocuticle mechanically cleaving their way along lines of least resistance underlines the importance of mechanical pressure during penetration of this region of hard cuticles (Zacharuk, 1970b). In addition, vertical penetrant hyphae may appear thin and constricted within the outer layers of the procuticle (exocuticle) whereas the growing tip swells in the more plaint, more easily degraded inner layers (endocuticle) (Robinson, 1966).

The strength of composite materials tends to be highly directional; in most insect cuticles this is avoided by laying down chitin fibrils parallel to the surface with the orientation changing in successive layers. Growth of soft-bodied larvae, e.g., Lepidoptera (where sclerotization is largely restricted to the head), between molts is facilitated by intersusception of new chitin into the old cuticle as it stretches, appositional deposition, and the inclusion of vertical cuticular columns (Figs. 3 and 4), where the chitin fibrils are laid down at right angles to the surface (Wolfgang and Riddiford, 1981; Hassan and Charnley, 1987). Chitin exists in highly crystalline forms in at least some areas of insect cuticle. This combined with the protection afforded by the protein matrix, may interfere with the action of chitinolytic enzymes (Kramer and Kogan, 1986).

Enzymatic degradation of procuticle is suggested by changes in the staining reactions of cuticles around penetrant hyphae (see Charnley, 1984). Brey and co-workers (1986) reported wide zones of complete hydrolysis in cuticle beneath penetrant hyphae of *Conidiobolus obscurus* in *Acyrthosiphon pisum*. However, this appears to be the exception rather than the rule (Charnley, 1984). Absence of obvious zones of histolysis where mechanical damage is reduced or nonexistent suggests that fungal enzymes are active but limited to the vicinity of fungal structures. Interestingly, in recent ultrastructural studies Hassan and Charnley (1989) and Goettel and colleagues (1989) noted clearing of the lamellar pattern but not complete histolysis around hyphae of *M. anisopliae* in the cuticle of *M. sexta* (Fig. 5). This is associated with the dispersal of a pathogen protease (detected by an immunogold technique) in the inner layers of the cuticle.

4. CUTICLE-DEGRADING ENZYMES

When *M. anisopliae*, *B. bassiana*, and *Verticillium lecanii* (deuteromycete) are grown on pulverized cuticle in liquid medium, they produce a range of extracellular cuticle-degrading enzymes corresponding to the major components of insect cuticles, viz. protein, chitin, and lipid (St. Leger *et al.*, 1986a). Although marked variations occurred in enzyme levels between isolates and species, endoproteases were consistently produced in large amounts by all isolates.

Enzymes appeared sequentially. Esterase and proteolytic enzymes (endoprotease, aminopeptidase, and carboxypeptidase) were produced first ($\leq$ 24 hr) followed by N-acetylglucosaminidase (NAGase). Chitinase and lipase were produced 3–5 days later.

The order of appearance of the enzymes is supported by the sequence of cuticle constituents solubilized into the culture medium, where a rapid increase in amino sugars followed early release of amino acids. Since chitinase is an inducible enzyme (St. Leger *et al.*, 1986b; Smith and Grula, 1983) (see later), and cuticular chitin is masked by protein (St. Leger *et al.*, 1986c), the late appearance of chitinase is presumably a result of induction as chitin eventually becomes available after degradation of encasing cuticle proteins. The late detection of lipase appears to be due to the fact that the enzyme is largely cell bound in young cultures (St. Leger, Charnley, and Cooper, unpublished).

Testing purified enzymes against locust cuticle *in vitro*, St. Leger and co-workers (1986c) showed that pretreatment or combined treatment with endoprotease (Pr1; see later) was necessary for high chitinase activity. Samsinakova and co-workers (1971) and Smith and co-workers (1981) also concluded that cuticular chitin is shielded by protein from studies using semipure commercial enzyme preparations against cuticles from *Galleria mellonella* larvae and *Heliothis zea* larvae, respectively. When locust exuviae (nondigested remains of old cuticle shed at ecdysis; exocuticle only) were used as substrate for purified pathogen enzymes instead of cuticle from larval sclerites (exo- and endocuticle), comparatively little hydrolysis occurred (St. Leger *et al.*,

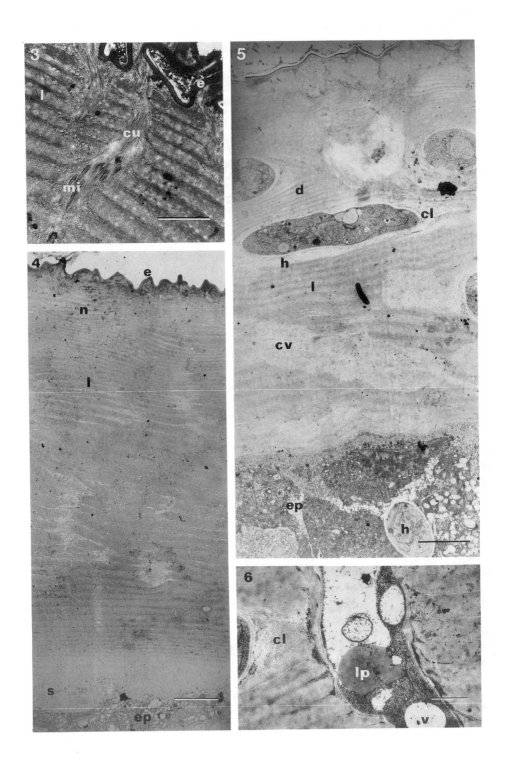

1986c). Similarly, while unsclerotized cuticle from fledgling locusts was rapidly degraded by fungal proteases, the cross-linking of cuticle proteins with glutaraldehyde (as a model for sclerotization) substantially reduced their susceptibility to proteolytic attack (St. Leger, Charnley, and Cooper, unpublished).

5. PROTEOLYTIC ENZYMES

5.1. Properties of Enzymes Purified from Culture Filtrates

Entomopathogenic fungi produce a variety of endo- and exo-acting proteolytic enzymes in culture. Proteases so far characterized are the collagenases produced by *Entomophthora coronata* (zygomycete) (Hurion *et al.*, 1977) and *Lagenidium giganteum* (oomycete) (Hurion *et al.*, 1979), chymoelastases and trypsins produced by a range of hyphomycetes (*M. anisopliae, V. lecanii, B. bassiana, N. rileyi*, and *A. aleyrodis*) (St. Leger *et al.*, 1987b,c), and chymotrypsins produced by *Erynia* spp. (Entomophthorales; zygomycete) (Samuels *et al.*, 1990). A variety of assays have been used to detect protease production by entomopathogens in culture (St. Leger *et al.*, 1987b; Huron *et al.*, 1979) and a protocol has been published for the purification of pathogen proteases by isoelectric focusing and affinity chromatography (St. Leger *et al.*, 1987b). The proteolytic enzymes of *M. anisopliae* are reviewed in detail because they provide the best understood model and their role in pathogenesis has been established.

5.1.1. Endoproteases

Extracellular activity in culture filtrates of *M. anisopliae* can be resolved into three components (versus Hide protein azure), two with alkaline pH optima (Pr1 and Pr2) and a third with an acid pH optimum (Pr3) (St. Leger *et al.*, 1987b). Inhibition studies demonstrated that both Pr1 and Pr2 are serine endoproteases and also contain essential histidine residues in the active site (St. Leger *et al.*, 1987b).

Pr1 occurs as a single very alkaline (pI > 10) hydrolase. It possesses a broad primary specificity for amino acids with a hydrophobic side group at the second carbon atom (e.g.,

---

Figure 3. Electron micrograph showing a transverse section through abdominal tergite cuticle of a newly ecdysed fourth-instar *Manduca sexta*. Note the vertical cuticular column. Bar = 1.5 μm. Abbreviations: a, appressoria; cl, absence of lamellae in association with hyphae; cu, cuticular column; cv, absence of lamellae in advance of hyphae; d, deformation of lamellae; dr, cell debris; e, epicuticle; ep, epidermis; h, hypha; he, hemocyte; l, lamella; lp, lipid droplet; m, melaninlike material; mi, long epidermal microvilli which give rise to the cuticular column; n, newly ecdysed cuticle; po, postecdysial cuticle; pr, preecdysial cuticle; s, subcuticle; v, vacuole. (From Hassan and Charnley, 1989, with permission of Academic Press.)

Figure 4. Fine structure of cuticle of 48-hr-old fourth-instar *Manduca sexta*. Note the increased thickness of the cuticle, the thinning (through stretching) of the original newly ecdysed cuticle (n), and the unfolding of the epicuticle (e) of Fig. 3. Bar = 3.3 μm. Abbreviation as in Fig. 3. (From Hassan and Charnley, 1989, with permission of Academic Press.)

Figure 5. Fine structure of the cuticle of fourth-instar *Manduca sexta*, 48 hr after inoculation with *Metarhizium anisopliae*. Note the deformation of the lamellae around the hyphae and the absence of the lamellar pattern in areas of the cuticle in association with or in advance of a penetrant hypha. Bar = 3.1 μm. Abbreviations as in Fig. 3. (From Hassan and Charnley, 1989, with permission of Academic Press.)

Figure 6. Fine structure of the cuticle of fourth-instar *Manduca sexta*, 48 hr after inoculation with *Metarhizium anisopliae*. The micrograph shows a penetrant hypha growing down a cuticular column. Bar = 1.35 μm. Abbreviations as in Fig. 3. (From Hassan and Charnley, 1989, with permission of Academic Press.)

phenylalanine, methionine, and alanine) but also possesses a secondary specificity for extended hydrophobic peptide chains with the active site recognizing at least five subsite residues. This comparative nonspecificity (as cf Pr2) accounts for it being a good general protease with activity against a range of proteins (casein, elastin, bovine serum albumin, collagen) and insect cuticle (St. Leger et al., 1987b). Analogous peptidases with alkaline, neutral, or acid IEF points have been resolved in culture filtrates from B. bassiana, V. lecanii, N. rileyi, and A. aleyrodis (St. Leger, et al., 1987c). The acidic enzymes occur as multiple isozymes (St. Leger, Charnley, and Cooper, unpublished) which resemble alkaline Pr1 in their primary specificity but do not degrade elastin. The negative charge of the elastin molecule precludes adsorption of the enzyme to elastin which is a prerequisite for activity (St. Leger et al., 1987b). Likewise, essential binding of M. anisopliae (ME1) Pr1 to negatively charged cuticle groups is dictated by its basic nature (St. Leger et al., 1986d). Only following adsorption does the active site come into contact with susceptible peptide bonds; solubilized peptides are further degraded until a chain length of about 5 is obtained (St. Leger et al., 1986c).

M. anisopliae (ME1) Pr2 occurs as multiple isozymes (ca. pI 4–4.5) with little or no activity versus insect cuticle or elastin but high activity versus casein. It has a primary specificity for arginine and lysine residues comparable to that of bovine trypsin and is sensitive also to trypsin inhibitors (e.g., leupeptin, tosyl-lysine-chloroketone, soybean trypsin inhibitor). However, like Pr1, the catalytic efficiency of Pr2 can be influenced by subsite residues at a distance from the cleaved site (St. Leger et al., 1987b). This effect is most pronounced in analogous peptidases produced by isolates of V. lecanii, N. rileyi, and A. aleyrodis. These are specific for a Phe-Val-Arg group and demonstrate less sensitivity to trypsin inhibitors (St. Leger et al., 1987c). Pr2 enzymes may be involved in cellular control mechanisms, catalyzing specific proteolytic inactivation and activation processes (St. Leger et al., 1987c). In this context it is interesting that inhibition of M. anisopliae (ME1) Pr2 with tosyl-lysine-chloroketone selectivity repressed formation by germlings of infection structures, implying a role for Pr2 in control of differentiation (St. Leger, unpublished).

5.1.2. Exoproteases

Two classes of aminopeptidase were isolated from cuticle-grown cultures of M. anisopliae and classified as an aminopeptidase M of broad specificity and a post-proline dipeptidyl aminopeptidase IV (St. Leger, Charnley, and Cooper, unpublished). The aminopeptidase M (pH optimum 7–8, 33kDa) exists as multiple isozymes (pI 5–6) with optimal activity for alanine residues and side activities versus other apolar and hydrophobic amino acids. The enzyme is sensitive to typical inhibitors of metalloenzymes (e.g., EDTA, 1,10, phenanthroline). The dipeptidyl aminopeptidase (pH optimum 8, 74kDa) exists as two isozymes (pI ca. 4.6) and removes X-prolyl groups from polypeptides (X = an apolar amino acid). The enzyme is inhibited by DFP (but not PMSF), indicating that it is a serine hydrolase. Neither peptidase alone hydrolyzed intact insect cuticle. However, when combined with Pr1 they effected enhanced release of amino acids. They presumably function in situ to break down Pr1-derived peptides and thus provide nutrition for the fungus.

5.2. Regulation of Production

Regulation of Pr1 and Pr2 are not identical, although both are produced rapidly (< 2 hr) in culture by carbon and nitrogen derepression alone. In minimal medium, the soluble protein BSA represses production of Pr1, while it allows enhanced synthesis of Pr2. In addition, while the extracellular Pr1 level in minimal growth medium exceeds that of Pr2, the reverse is true of endocellular activities. An endocellular role for Pr2, e.g., catalyzing specific proteolytic activation processes, could account for it being less subject than Pr1 to catabolite repression (CR). Extracellular levels of both Pr1 and Pr2 were enhanced in cultures supplied with insect cuticle or other insoluble polymers (e.g., cellulose) that were insufficient to produce CR. Addition of more readily utilized metabolites (e.g., glucose or alanine) repressed extracellular protease production, confirming that production is constitutive but repressible (St. Leger *et al.*, 1988a). Likewise, Pr1 synthesis on infection structures (appressoria) produced on an artificial surface or during growth on host cuticle is overridden by addition of readily utilized nutrients (St. Leger *et al.*, 1989b). Studies utilizing [^{35}S] methionine incorporation and "Western" blot analysis demonstrated that Pr1 is the major protein product when *M. anisopliae* produces appressoria against polystyrene or *in situ* during penetration of host (*M. sexta*) cuticle (Fig. 7) (St. Leger *et al.*, 1989b). Synthesis during maturation of appressoria and production of penetration pegs far exceeded synthesis of other proteins (Fig. 7). Such rapid protease synthesis is only possible in

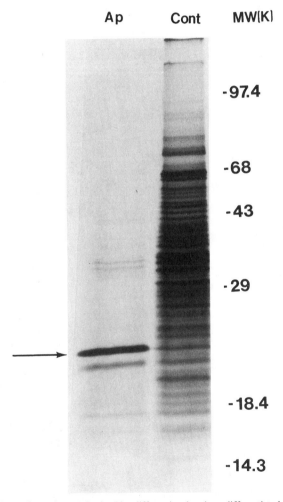

Figure 7. Comparison of proteins synthesized by differentiated and nondifferentiated germlings of *M. anisopliae*. Differentiated (Ap) and nondifferentiated (Cont) germlings were pulse-labeled with ^{35}S-methionine between 22 and 35 hr postinoculation (maturation of appressoria). Pr1 is indicated by the arrow. (From St. Leger *et al.*, 1989b.)

host tissues where the concentration of readily metabolizable compounds is low. This is the case with insect cuticles as the components are largely insoluble until released by cuticle-degrading enzymes (St. Leger *et al.*, 1986c, 1987b,c). However, repression could operate if ever the release from cuticle of degradation products exceeded fungal requirements. This was confirmed by addition to inocula of readily utilized nutrients, e.g., alanine, which resulted in extensive growth on the host cuticle but repressed penetration and synthesis of Pr1 (St. Leger *et al.*, 1989b). Thus, the pathogenic process involving infection-related morphogenesis and enzyme production occurs only when it is necessary for the pathogen to establish a nutritional relationship with the host.

Studies with inhibitors of some phase of protein, RNA, DNA synthesis in *M. anisopliae* indicated that control of enzyme production is at the level of transcription as it is sensitive to actinomycin D and 8-azaguanine. *In vitro* translation of poly(A) RNA isolated from appressoria using a rabbit reticulocyte system confirmed *de novo* synthesis of Pr1-specific mRNA during differentiation (Fig. 8).

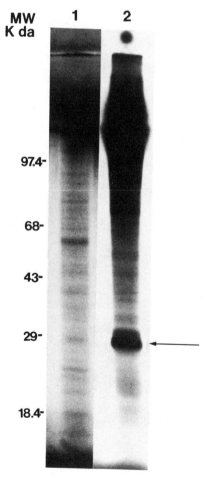

Figure 8. SDS-PAGE of *in vitro* translation products of poly(A) RNA. Poly (A) RNA isolated from differentiating germlings 10 hr (1) and 24 hr (2) after inoculation into polystyrene plates in yeast extract medium (0.0125%) was translated by using a cell-free rabbit reticulocyte system. Pr1 is indicated by the arrow. (From St. Leger, unpublished.)

The second messenger system that mediates the effects of starvation and regulates transcription has not been determined. In other systems, cyclic AMP has been found to overcome CR. However, evidence is accumulating that cyclic AMP is not involved as a secondary messenger in regulating Pr1 in *M. anisopliae* (St. Leger *et al.*, 1988a, 1989c). Ca^{2+} also regulates a diverse array of cellular functions by its ability to activate several protein kinases (Cohen, 1985). Secretion but not synthesis of Pr1 in *M. anisopliae* is inhibited by agents which depress calcium-dependent protein phosphorylation. Thus, depleting cellular calcium with a combination of the calcium-selective ionophore Br-A23187 and EGTA or antagonism of plasmalemma Ca^{2+} with lanthanum ions massively repressed net phosphorylation at serine and threonine residues and secretion of Pr1. However, antagonists of calmodulin (a sensor and effector of many Ca^{2+}-dependent messages) that are potent inhibitors of protein synthesis and phosphorylation during germination do not affect Pr1 synthesis or secretion by mycelia (St. Leger *et al.*, 1989c). While there is strong evidence for stimulus transduction pathways in *M. anisopliae* similar to those in plant and animal systems, additional research will be required to elucidate the nature of the protein kinases, their substrates, and their function in regulating transcription in the fungal cell.

5.3. Evidence for Involvement in Penetration

St. Leger and co-workers (1987a) extracted protease and aminopeptidase from wings of *C. vomitoria* and abdominal cuticle of fifth-instar larvae of *M. sexta* about 16 hours after inoculation with *M. anisopliae*. Endoprotease activity was separated into two compounds which closely resembled Pr1 and Pr2 in pI, substrate specificity, and inhibitor spectrum. The third protease (Pr3) produced in culture was not detected *in vivo* (< 40 hr postinoculation). Purified and nonpurified extracts of infected blowfly wings tested by Ouchterlony gel diffusion against specific antiserum to Pr1 gave a single precipitin line identical to that given by the pure enzyme, confirming the presence of Pr1 during infection.

The translucent wings of *C. vomitoria* have also been used to histochemically locate proteolytic enzymes during penetration (St. Leger *et al.*, 1987a). Substrates and inhibitors specific for Pr1 and Pr2 established the production of these enzymes on appressoria, which developed 10–24 hr after inoculation. Aminopeptidase differed from endoprotease in that it was not present on immature appressoria, and the activity extended into the mucilage surrounding mature appressoria and appressorial plates. An immunogold technique has recently been used to locate Pr1 in cuticle during penetration of larvae of *M. sexta* (Goettel *et al.*, 1989).

Pr1 appears to be a pathogenicity determinant by virtue of its considerable ability to degrade cuticle (St. Leger *et al.*, 1987b) and its production at high levels by the pathogen *in situ* during infection (St. Leger *et al.*, 1987a). Simultaneous application of turkey egg white inhibitor (TEI) and conidia significantly delayed mortality of *Manduca* larvae compared with larvae inoculated with conidia, supporting the importance of Pr1 in penetration (St. Leger *et al.*, 1988b). The inhibitor also reduced melanization of cuticle (a host response to infection, see next section) and invasion of the hemolymph as well as maintaining the host's growth rate. TEI or antibodies raised against Pr1 delayed penetration of the cuticle but did not affect spore viability or prevent growth and formation of appressoria on the cuticle surface. This suggests that inhibition of Pr1 reduced infection by limiting fungal penetration of the insect cuticle. *In vitro* studies using TEI showed that accumulation of protein degradation products from the cuticle, including ammonia, was dependent on active Pr1. This confirms its major part in solubilizing cuticle proteins and making them available for nutrition. Confirmation of the involvement of Pr1 in pathogenesis, however, awaits molecular analysis, viz. restoring pathogenicity to Pr1-deficient mutants by transforming them with cloned Pr1 genes.

6. CHITINOLYTIC ENZYMES

6.1. Properties of Enzymes Purified from Culture Filtrates

6.1.1. Chitinase

St. Leger and co-workers (1986c) purified chitinase from culture filtrates of *M. anisopliae* grown on 1% ground chitin, using ammonium sulfate precipitation, Sephadex G100, and adsorption onto colloidal chitin. The purified enzyme failed to hydrolyze arylglycosides or cellobiose (*N*-acetylglucosamine dimer), showed only trace activity against chitotriose (trimer), but rapidly degraded chitotetraose (tetramer) (St. Leger, Charnley, and Cooper, unpublished). Colloidal chitosan (deacetylated form of chitin) and crystalline chitin were less amenable to degradation than colloidal chitin, but activity against them was still substantial. The chitinase had many similarities to those produced by other microorganisms (e.g., Jeauniaux, 1963; Stirling *et al.*, 1979). These properties include a pH optimum of 5.3, a molecular weight of 33,000, and the lack of any requirement for a cofactor.

Hydrolysis of crystalline chitin produced only one low-molecular-weight reaction product within 24 hr, viz. *N*-acetylglucosamine (NAG). The absence of intermediary oligomers among chitin breakdown products probably means that NAG is released directly from insoluble chitin. Either the chitinase has an exo-acting component or alternatively the reaction proceeds by a single-chain processive mechanism as described for some other endo-acting polysaccharidases (Cooper *et al.*, 1978). This involves the random cleaving of bonds followed by release of monomers or dimers from exposed ends so that a single macromolecule is completely degraded before a new one is attacked. Such a mechanism, especially if it involved simultaneous digestion of several parallel chains, could result in the rapid degradation of chitin fibrils and in addition produce monomers for nutrition and induction for further enzyme synthesis.

6.1.2. N-Acetylglucosaminidase

N-Acetylglucosaminidase activity has been partially purified from culture filtrates of *M. anisopliae* grown on 1% ground chitin using the method employed for chitin (see above), without the affinity step (St. Leger *et al.*, 1986c). The enzyme had substantial activity against *p*-nitrophenol acetylglucosamine, as well as chitobiose, chitotriose, and chitotetraose, the major product in each case being NAG, showing that the enzyme is a true NAGase rather than a chitobiase (St. Leger, Charnley, and Cooper, unpublished). The enzyme had little activity against colloidal or crystalline chitins. Its size, 110–120kDa, is within the range for similar enzymes from other sources (e.g., Reyes and Byrde, 1973).

6.2. Regulation of Production

Chitinase synthesis is regulated in *M. anisopliae* (St. Leger *et al.*, 1986b) and *B. bassiana* (Smith and Grula, 1983) by products of chitin degradation through an inducer-repressor mechanism. The system is best understood for *M. anisopliae*. High chitinase activity was found only in cultures supplied with chitin, but not with other polymers such as pectin, xylan, and cellulose. Slow feeding of cultures with sugars or alanine from diffusion capsules (Pirt, 1971) in a carbon-deficient medium demonstrated that the most effective inducer of chitinase was NAG. Glucosamine also allowed production, possibly an adaptation to the fact that chitin from natural sources (including insect cuticle) appears to be partially deacetylated (Hackman and Goldberg, 1965, 1974). It is unlikely that chitobiose could function as a major inducer of chitinase in *M. anisopliae* as e.g. cellobiose does for cellulase (Cooper and Wood, 1975) because NAGase would degrade it to NAG and the major product of chitinase activity is NAG. Interestingly, NAGase was produced constitutively and was little affected by catabolite repression.

6.3. Evidence for Involvement in Penetration

Chitin constitutes 17–50% of the dry weight of insect cuticle; more pliant cuticles have a higher chitin content than stiff cuticles (Hillerton, 1984). In the main, chitin fibrils are laid down parallel to the cuticular surface and as such present a potential barrier to penetration by entomopathogenic fungi. Preferential penetration of *M. anisopliae* down to the vertical cuticular columns in *M. sexta* cuticle (where the chitin fibrils are perpendicular to the plane of the cuticle) supports this view (Hassan and Charnley, 1989) (Figs. 3 and 6).

In light of the above, it is interesting that St. Leger and co-workers (1987a) failed to find evidence of the production of chitinase during the first critical 40 hr after inoculation of *C. vomitoria* wings or abdominal cuticle of *M. sexta* larvae with *M. anisopliae*. NAGase activity was extracted from infected cuticle, but this enzyme has only trace activity against polymeric chitin (see earlier). The apparent absence of chitinase from infected cuticle could be due to inadequate extraction or inhibitors in the cuticle [chitinase *in vitro* binds tightly to locust chitin in a nonionic manner (St. Leger *et al.*, 1986d)]. Nevertheless, failure to detect the products of chitin hydrolysis in comminuted infected cuticle indicates that the activity of chitinase, if present, is negligible compared to that of protease.

The slow appearance of chitinase *in vivo* (St. Leger *et al.*, 1987a) is consistent with *in vitro* results (St. Leger *et al.*, 1986a) and is probably due to the fact that chitinase is an inducible enzyme (St. Leger *et al.*, 1986b) and cuticular chitin is masked by protein (St. Leger *et al.*, 1986c). It seems likely that chitinase functions largely to provide nutrients during the saprophytic phase of fungal growth in cuticle of moribund insect hosts.

The importance of chitin as a mechanical barrier to penetration and as a stabilizer of the cuticular protein matrix, in the absence of fungal chitinase, is evident from a recent study using the insecticide Dimilin, a specific inhibitor of chitin synthesis in insects. Hassan and Charnley (1983) showed that dual application of Dimilin and *M. anisopliae* had a synergistic effect against larvae of *M. sexta*. Ultrastructural observations demonstrated that fungal penetration through Dimilin-treated cuticle was dramatically enhanced (Hassan and Charnley, 1989). Postecdysial Dimilin-treated cuticle (without chitin) was almost completely destroyed (Figs. 9–11) in contrast to preecdysial cuticle (laid down prior to insecticide treatment) where hydrolysis was apparently selective (protein only?) and restricted to the vicinity of the fungal hyphae (Fig. 5). Consistent with these ultrastructural observations, pharate fifth-instar *Manduca* cuticle, produced during treatment with Dimilin and thus completely disrupted by the insecticide (Hassan and Charnley, 1987), was considerably more susceptible to Pr1 than control cuticle (St. Leger, Charnley, and Cooper, unpublished).

7. LIPOLYTIC AND ESTEROLYTIC ENZYMES

7.1. Properties of Enzymes Purified from Culture Filtrates

In general, esterases may be differentiated from lipases because short-chain fatty acids (C_2–C_4) are preferentially hydrolyzed by the former and long-chain esters $> C_8$) by the latter (Shnitka, 1974). Esterase activity produced by *M. anisopliae* in young cultures (3 days) was greatest against short- and intermediate-length *p*-nitrophenol esters with only trace activity occurring above C_{10} (St. Leger, Charnley, and Cooper, unpublished), suggesting that lipase is not produced extracellularly by young mycelia. Activity versus C_{14} rose in older cultures (7–14 days). Late arrival of

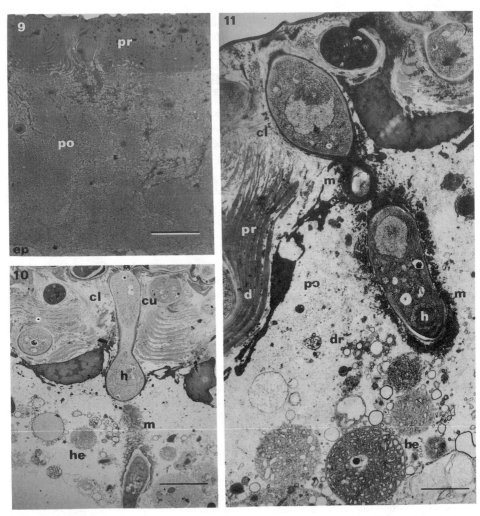

Figure 9. Fine structure of cuticle of 48-hr-old fourth-instar *Manduca sexta* fed on a diet containing the insect chitin synthesis inhibitor Dimilin. Note the disorganized nonlamellate postecdysial cuticle (po). Bar = 8 μm. Abbreviations as in Fig. 3. (From Hassan and Charnley, 1989, with permission of Academic Press.)

Figures 10 and 11. Fine structure of the cuticle of fourth-instar larvae 48 hr after inoculation with *Metarhizium anisopliae* and fed on Dimilin-treated diet. *Figure 10.* Postecdysial cuticle has been virtually completely destroyed. Note the hypha penetrating down a cuticular column. Bar = 4.3 μm. Abbreviations as in Fig. 3. *Figure 11.* Preecdysial cuticle provides limited resistance to penetration. Melaninlike material deposited around penetrant hyphae, particularly in the remains of the Dimilin-affected cuticle. Bar = 2.4 μm. Abbreviations as in Fig. 3. (Both from Hassan and Charnley, 1989, with permission of Academic Press.)

extracellular lipase *in vitro* was confirmed using the "true" lipase substrate olive oil (St. Leger *et al.*, 1986a).

The major esterase peak eluted from a Sephadex G100 column with a profile very similar to that obtained for endoprotease Pr1. As this protease also degrades *p*-nitrophenol esters, it is probably a major contributor to total esterase activity. Flat-bed IEF, however, revealed 25 distinct esterases (isozymes) from culture filtrates of *M. anisopliae* grown on locust cuticle. On the basis of their reactions with naphthyl esters, the isozymes appeared to have different substrate specificities. However, all the bands were inhibited by PMSF, indicating that they are serine carboxyesterases (esterase B) and not arylesterases (esterase A) that are inhibited by N-ethyl-maleimide.

Esterases catalyze many enzymatic reactions though preferentially hydrolyzing aliphatic or aromatic esters and amides (Shnitka, 1974; Heymann, 1980). The considerable heterogeneity of esterases could account for their collective lack of specificity. Multiple enzyme strategies are believed to play an important role in the ability of an organism to adapt to different environments (Moon, 1975; Somero, 1975), presumably including that provided by an insect host.

7.2. Evidence for Involvement in Penetration

St. Leger and co-workers (1987a) detected esterase on pregerminating and germinating conidia and appressoria of *M. anisopliae* on wings of *C. vomitoria* with α-naphthyl acetate (C_2) and α-naphthyl proprionate (C_3) as substrate.

It is very difficult to differentiate histochemically between esterase and lipase (Pearse, 1972). Although many workers have identified lipase produced by entomopathogenic fungi solely on the basis of activity against Tweens (e.g., Michel, 1981), such substrates are degraded by nonspecific esterases (Pearse, 1972). Most microbial lipases are serine enzymes (Brockerhoff and Jensen, 1974) and as such would not be distinguished from the nonspecific esterases produced by *M. anisopliae*. The different sites of activity against Tween 80 (localized on appressoria) and naphthyl-AS-nonanoate (localized on conidia) suggest, however, that at least for *M. anisopliae* the enzyme-degrading Tween is distinguishable from at least one medium-chain-length non-specific esterase (St. Leger *et al.*, 1987a).

Doubts may be cast as to the involvement of lipases in penetration because St. Leger and co-workers (1987a) failed to extract true lipase (with activity against olive oil) from cuticles of *M. sexta* and *C. vomitoria* infected with *M. anisopliae*. Obviously, as with chitinase, failure to detect may reflect numerous factors such as binding of lipase to fungal cell walls and host cuticles. However, notwithstanding this the possible biological role of lipase in pathogenesis is not as obvious as many have assumed. Ultrastructural studies which have demonstrated the early histolysis of the "wax" layer on cuticles underneath infection structures might implicate lipase/esterase activity (e.g., Zacharuk, 1970b). However, in most insects "extractable" cuticular lipids are composed mainly of a complex mixture of alkanes and alkenes with triacylglycerols and wax esters (potential substrates for lipases and esterases) being comparatively minor components (Hepburn, 1985). Lipids could only have a major role if the "bound" surface lipids and "nonextractable" lipid fraction contain esters. Unfortunately, however, very little is known of the chemistry of bound cuticular lipids or their esters (Blomquist, 1984). Wigglesworth (1970) has tentatively suggested that the outer epicuticle (Fig. 1) is a multiple polyester cross-linked by ester bonds. If this is correct, then an esterase with characteristics somewhat similar to cutinase (see Chapter 10) would be required for its hydrolysis. However, no such enzyme is produced by *M. anisopliae* or *B. bassiana* (St. Leger, Charnley, and Cooper, unpublished). Lipoprotein lipases (active against olive oil emulsion activated with human blood plasma) are secreted. This enzyme perhaps in conjunction with Pr1 would aid penetration of the inner epicuticle (polymerized lipoprotein) (see Fig. 1).

8. HOST DEFENSE AGAINST CUTICLE-DEGRADING ENZYMES

Factors that interfere with the cuticle-degrading activity of a fungal pathogen may be preformed or induced (St. Leger, 1990).

8.1. Preformed Defenses

Restriction of enzyme activity to the vicinity of hyphae may be brought about by a combination of the binding of charged enzymes to nonsubstrates and restriction of enzymes because of small intermolecular spaces (though epicuticular filaments, pore canals, and dermal glands could help in distribution). Localized cuticle degradation may, however, be beneficial for the fungus by preventing premature dehydration of the host, reducing the likelihood of invasion by competing microorganisms and facilitating uptake by restricting products of cuticle degradation. Hardening of the outer layers of the cuticle (exocuticle) by phenolic cross-linking of the protein clearly has structural advantages (Hillerton, 1984). However, sclerotized cuticle is comparatively resistant to pathogen enzymes (see earlier). Host-produced protease inhibitors may further restrict enzyme activity of pathogens. Such compounds have been isolated from the cuticles of *Bombyx mori* larvae (Eguchi *et al.*, 1988) and the crayfish *Aphanomyces astaci* (Häll and Söderhäll, 1983) and hemolymph of *Anticarsia gemmatalis* (Boucias and Pendland, 1987).

8.2. Induced Defenses

Induced defenses appear to include the production of phenols which may provide a measure of resistance to fungi, though the case is far from proven (Charnley, 1984; St. Leger, 1990). The oxidation and polymerization of phenolic compounds, typified by the production of brown or black melanic pigments, is a common feature of the response of many insects to infection (Butt *et al.*, (1988). In most cases, penetration of the epicuticle represents the first interaction that results in melanization. The melanic response subsequently extends into the procuticle, appearing as a sheath surrounding the penetrant hyphae or a dark patch covering the whole infection site (St. Leger, 1990).

St. Leger and co-workers (1988c) found that growth and production of cuticle-degrading proteases by *M. anisopliae* were greater on unpigmented cuticles than on melanized-cuticle complexes (prepared by reacting L-dopa or catechol with *Manduca* cuticle), suggesting that phenolic oxidation products are fungitoxic. Although phenols can inhibit microbial enzymes (Bull, 1970), activity of Pr1 and Pr2, in contrast to bovine chymotrypsin, was not inhibited by incubation in a melanizing mixture of *Manduca sexta* tyrosinase and L-dopa, possibly reflecting enzyme adaptation by this pathogen (St. Leger *et al.*, 1988c). Melanized cuticle may be relatively resistant to enzyme attack, though Pr1 still releases melanin from melanized cuticle by hydrolyzing associated proteins (St. Leger *et al.*, 1988c). Physical constraints imposed on the fungus by the melanotic sheath may prove limiting. Apart from restricting growth, the sheath may restrict diffusion of fungal enzymes and toxins and the availability of host nutrients, water, and oxygen (St. Leger, 1990).

9. CUTICLE-DEGRADING ENZYMES AS VIRULENCE DETERMINANTS

The possibility that virulence among isolates may be correlated (at least in part) with cuticle-degrading enzyme activity has stimulated several studies, with conflicting results (see Charnley, 1984). Comparisons between isolates for pathogenicity and production of enzymes may only reveal the great variability within a species for numerous factors, many of which may influence but be unrelated to cuticle-degrading enzyme activity. Induction of mutants within a common genetic background is an alternative approach which has also been exploited. Paris and Ferron (1979) found a link (which is ambiguous; see earlier) between virulence and lipase in mutants of *B. brogniartii*. However, mutants deficient in other respects were also avirulent, suggesting that, as one might expect, pathogenicity was a function of many attributes. Similarly, Pekrul and Grula

(1979) found that low pathogenicity of *B. bassiana* mutants against *H. zea* larvae was not simply a consequence of the lack of a suitable enzyme cocktail. All mutants possessed varying levels of the three major enzyme activities regardless of pathogenicity and some mutants contained very high levels of certain activities and yet were poor pathogens.

The third approach to unraveling the role of cuticle-degrading enzymes in host specificity and virulence is to look at the properties of the enzymes concerned. As is clear from this review, at present only endoproteases for deuteromycete entomopathogens have been studied in sufficient detail to draw any significant conclusions. Cuticle-degrading endoproteases, with similar modes of action, are produced in quantity by all species studied (St. Leger *et al.*, 1987c), and it seems unlikely that they contribute to host specificity or virulence, though the occurrence of two or three types of extracellular protease in five genera of entomopathogen implies an indispensable function for these enzymes (St. Leger *et al.*, 1987c). Cuticle-degrading chymotrypsin/ chymoelastases produced by *M. anisopliae*, *B. bassiana*, *V. lecanii*, *N. rileyi*, and *A. aleyrodis* differed only in charge (IEF). However, this has potential significance for their specificity as binding to the cuticle is a prerequisite for activity. Interestingly, the acidic enzyme from *V. lecanii* (V11) has similar activity to the basic Pr1 from *M. anisopliae* against locust (*Schistocerca gregaria*) cuticle (St. Leger, Charnley, and Cooper, unpublished); thus, they must bind to different regions of the cuticle. Cuticles of *Hyalophora cecropia* (Cox and Willis, 1985) and *Locusta migratoria* (Hojrup *et al.*, 1986) have a nonuniform distribution of charge, with negatively charged and positively charged proteins predominating in flexible cuticle (e.g., arthrodial membranes) and rigid body-wall cuticle, respectively. Thus, it is possible that regions of the cuticle may be favorable or unfavorable to binding (and thus degradation) by individual enzymes, with consequences for the parts of the insect body which can be invaded by enzymatic action. This may influence speed of penetration and thus virulence.

10. REFERENCES

Blomquist, G. J., 1984, Cuticular lipids of insects, in: *Infection Processes of Fungi* (D. W. Roberts, and J. R. Aist, eds.), Conference Report, Rockefeller Foundation, pp. 54–60.

Boucias, D. G., and Pendland, J. C., 1984, Nutritional requirements for conidial germination of several host range pathotypes of the entomopathogenic fungus *Nomuraea rileyi*, *J. Invertebr. Pathol.* **43**:288–293.

Boucias, D. G., and Pendland, J. C., 1987, Detection of protease inhibitors in the hemolymph of resistant *Anticarsia gemmatalis* which are inhibitory to the entomopathogenic fungus *Nomuraea rileyi*, *Experientia* **43**:336–339.

Boucias, D. G., Pendland, J. C., and Latgé, J.P., 1988, Nonspecific factors involved in attachment of entomo- pathogenic deuteromycetes to host insect cuticle, *Appl. Environ. Microbiol.* **54**:1795–1805.

Brey, P. T., Latgé, J. P., and Prevost, M. C., 1986, Integumental penetration of the pea aphid *Acyrthosiphon pisum* by *Conidiobolus obscurus* (Entomophthoraceae), *J. Invertebr. Pathol.* **48**:34–41.

Brobyn, P. J., and Wilding, N., 1983, Invasive and developmental processes of *Entomophthora muscae* infecting houseflies (*Musca domestica*), *Trans. Br. Mycol. Soc.* **1**:1–18.

Brockerhoff, H., and Jensen, R. G., 1974, *Lipolytic Enzymes*, Academic Press, New York.

Bull, A. T., 1970, Inhibition of polysaccharases by melanin in relation to mycolysis, *Arch. Biochem. Biophys.* **137**:345–356.

Butt, T. M., Wraight, S. P., Galani-Wright, S., Humber, R. A., Roberts, D. W., and Soper, R. S., 1988, Humoral encapsulation of the fungus *Erynia radicans* (Entomophthorales) by the potato leafhopper *Empoasca fabae*, *J. Invertebr. Pathol.* **52**:49–57.

Charnley, A. K., 1984, Physiological aspects of destructive pathogenesis in insects by fungi: A speculative review, in: *Invertebrate-Microbial Interactions* (J. M. Anderson, D. M. Rayner, and D. W. A. Walton, eds.), Cambridge University Press, London. pp. 229–270.

Charnley, A. K., 1989, Mechanisms of fungal pathogenesis in insects, in *The Biotechnology of Fungi for Improving Plant Growth* (Whipps, J. M., and R. D. Lumsden, eds.), Cambridge University, London. pp. 85–125.

Cohen, P., 1985, The role of protein phosphorylation in the hormonal control of enzyme activity, *Eur. J. Biochem.* **151**:439–448.

Cooper, R. M., and Wood, R. K. S., 1975, Regulation of synthesis of cell wall-degrading enzymes by *Verticillium albo-atrum* and *Fusarium oxysporum* f. sp. *lycopersici*, *Physiol. Plant Pathol.* **5**:135–156.

Cooper, R. M., Rankin, B., and Wood, R. K. S., 1978, Cell wall-degrading enzymes of vascular wilt fungi. II. Properties and modes of action of polysaccharidases of *Verticillium albo-atrum* and *Fusarium oxysporum* f. sp. *lycopersici*, *Physiol. Plant Pathol.* **13**:101–134.

Cox, D. L., and Willis, J. H., 1985, The cuticular proteins of *Hyalophora cecropia* from different anatomical regions and metamorphic stages, *Insect Biochem.* **15**:349–362.

David, W. A. L., 1967, The physiology of the insect integument in relation to the invasion of pathogens, in: *Insects and Physiology* (J. W. L. Beament, and J. E. Treherne, eds.), Oliver & Boyd, London, pp. 17–35.

Dennell, R., 1946, A study of an insect cuticle: The larval cuticle of *Sarcophaga faculata* Pand. (Diptera), *Proc. R. Soc. London Ser. B* **133**:348–373.

Eguchi, M., Yamashita, M., and Yoshida, S., 1988, Protein inhibitors from the integument and hemolymph of the silkworm, *Bombyx mori* (Lepidoptera: Bombycidae) against fungal proteases, *Proc. XVIII Int. Congr. Entomol. Abstr.* p. 130.

Gillespie, A. T., 1988, Use of fungi to control pests of agricultural importance, in: *Fungi in Biological Control Systems*, (M. N. Burge, ed.), University Press, Manchester, pp. 37–60.

Goettel, M. S., St. Leger, R. J., Rizzo, N. W., Staples, R. C., and Roberts, D. W., 1989, Ultrastructural localization of a cuticle degrading protease produced by the entomopathogenic fungus *Metarhizium anisopliae* during penetration of host cuticle, *J. Gen. Microbiol.* **135**: 2223–2239.

Grula, E. A., Woods, S. P., and Russell, H., 1984, Studies utilizing *Beauveria bassiana* as an entomopathogen, in: *Infection Processes of Fungi* (D. W. Roberts, and J. R. Aist, eds.), Conference Report, Rockefeller Foundation, pp. 147–152.

Hackman, R. H., 1984, Cuticle: Biochemistry, in: *Biology of the Integument. I. Invertebrates* (J. Bereiter-Hahn, A. G. Matolts, and K. S. Richards, eds.), Springer-Verlag, Berlin, pp. 626–637.

Hackman, R. H., and Goldberg, M., 1965, Studies on chitin. 6. The nature of α- and γ-chitin, *Aust. J. Biol. Sci.* **18**:935–946.

Hackman, R. H., and Goldberg, M., 1974, Light scattering and infra-red spectrophotometric studies of chitin and chitin derivatives, *Carbohydr. Res.* **38**:35–45.

Häll, L., and Söderhäll, K., 1983, Isolation and properties of a protease inhibitor in crayfish (*Astacus astacus*) cuticle, *Comp. Biochem. Physiol.* **76B**:699–702.

Hassan, A. E. M., and Charnley, A. K., 1983, Combined effects of diflubenzuron and the entomopathogenic fungus *Metarhizium anisopliae* on the tobacco hornworm *Manduca sexta*, *Proc. 10th Int. Congr. Plant Prot.* **3**:790.

Hassan, A. E. M., and Charnley, A. K., 1987, The effect of Dimilin on the ultrastructure of the integument of *Manduca sexta*, *J. Insect Physiol.* **33**:669–676.

Hassan, A. E. M., and Charnley, A. K., 1989, Ultrastructural study of the penetration by *Metarhizium anisopliae* through Dimilin-affected cuticle of *Manduca sexta*, *J. Invertebr. Pathol.* **54**:117–124.

Hepburn, H. R., 1985, Structure of the integument, in: *Comprehensive Insect Physiology, Biochemistry and Pharmacology*, Volume 3 (G. A. Kerkut and L. I. Gilbert, eds.), Pergamon Press, Elmsford, New York.

Heymann, E., 1980, Carboxyesterases and amidases, in: *Enzymatic Basis of Detoxification*, Volume 1 (W. B. Jakoby, ed.), Academic Press, New York, pp. 291–323.

Hillerton, J. E., 1984, Cuticle: Mechanical properties, in: *Biology of the Integument. I. Invertebrates* (J. Bereiter-Hahn, A. G. Matoltsy, and K. S. Richards, eds.), Springer-Verlag, Berlin, pp. 583–625.

Hojrup, P., Andersen, S. O., and Roepstorff, P., 1986, Isolation, characterisation and N-terminal sequence studies of cuticular proteins from the migratory locust, *Locusta migratoria*, *Eur. J. Biochem.* **154**:153–159.

Hurion, N., Fromentin, H., and Keil, B., 1977, Proteolytic enzymes of *Entomophthora coronata*: Characterization of a collagenase, *Comp. Biochem. Physiol.* **56B**:259–264.

Hurion, N., Fromentin, H., and Keil, B., 1979, Specificity of the collagenolytic enzyme from the fungus *Entomophthora coronata*: Comparison with the bacteria collagenase from *Achromobacter iophagus*, *Arch. Biochem. Biophys.* **192**:438–445.

Jeauniaux, C. H., 1963, *Chitine et Chitinolyse, un chapitre de la Biologique Moleculaire*, Masson, Paris.

Kerwin, J. L., 1984, Fatty acid regulation of the germination of *Erynia variabilis* conidia on adults and puparia of the lesser housefly, *Fannia canicularis*, *Can. Microbiol.* **30**:158–161.

Kerwin, J. L., and Washino, R. K., 1986, Cuticular regulation of host regulation and spore germination by entomopathogenic fungi, in: *Fundamental and Applied Aspects of Invertebrate Pathology* (R. A. Samson,

J. M. Vlak, and D. Peters, eds.), The Foundation of the Fourth International Colloquium of Invertebrate Pathology, Wageningen, pp. 423–425.

Kramer, K. J., and Kogan, D., 1986, Insect chitin, *Insect Biochem.* **16**:851–877.

Locke, M., 1984, The structure of insect cuticle, in: *Infection Processes of Fungi* (D. W. Roberts, and J. R. Aist, eds.), Conference Report, Rockefeller Foundation, pp. 38–53.

Madelin, M. F., 1968, Fungal parasites of invertebrates: 1. Entomogenous fungi, Volume 3 (G. C. Ainsworth, and A. S. Sussman, eds.), Academic Press, New York.

Michel, B., 1981, *Recherches experimentales sur la penetration des chapignons pathogens chez les insectes*, These 3e cycle, University of Montpellier. p. 170.

Moon, T. W., 1975, Temperature adaptation, isozymic function and the maintenance of heterogeneity, in: *Isozymes. II. Physiological Function*, (A. P. Marbert, ed.), Academic Press, New York.

Neville, A. C., 1984, Cuticle: Organisation, in: *Biology of the Integument. I. Invertebrates* (J. Bereiter-Hahn, A. G. Matoltsy, and K. S. Richards, eds.), Springer-Verlag, Berlin, pp. 611–625.

Paris, S., and Ferron, P., 1979, Study of the virulence of some mutants of *Beauveria brogniartii, J. Invertebr. Pathol.* **34**:71–77.

Pearse, A. G. E., 1972, *Histochemistry, Theoretical and Applied*, Volume 2, 3rd ed., Churchill Livingstone, Edinburgh.

Pekrul, S., and Grula, E. A., 1979, Mode of infection of the corn earworm (*Heliothis zea*) by *Beauveria bassiana* as revealed by scanning electron microscopy, *J. Invertebr. Pathol.* **34**:238–247.

Pirt, S. J., 1971, The diffusion capsule, a novel devise for the addition of a solute at a constant rate to a liquid medium, *Biochem. J.* **121**:293–297.

Reyes, F., and Byrde, R. J. W., 1973, Partial purification and properties of N-acetylglucosaminidase from the fungus *Sclerotina fructigena, Biochem. J.* **131**:381–388.

Robinson, R. K., 1966, Studies on penetration of insect integument by fungi, *P.A.N.S.* **12**:131–142.

St. Leger, R. J., 1990, The integument as a barrier to microbial infections, in: *The Physiology of Insect Epidermis* (A. Retnakaran, and K. Binnington, eds.), Inkata Press (in press).

St. Leger, R. J., Charnley, A. K., and Cooper, R. M., 1986a, Cuticle-degrading enzymes of entomopathogenic fungi: Synthesis in culture on cuticle, *J. Invertebr. Pathol.* **48**:85–95.

St. Leger, R. J., Cooper, R. M., and Charnley, A. K., 1986b, Cuticle-degrading enzymes of entomopathogenic fungi: Regulation of production of chitinolytic enzymes, *J. Gen. Microbiol.* **132**:1509–1517.

St. Leger, R. J., Cooper, R. M., and Charnley, A. K., 1986c, Cuticle-degrading enzymes of entomopathogenic fungi: Cuticle degradation *in vitro* by enzymes from entomopathogens, *J. Invertebr. Pathol.* **47**:167–177.

St. Leger, R. J., Charnley, A. K., and Cooper, R. M., 1986d, Cuticle-degrading enzymes of entomopathogenic fungi: Mechanisms of interaction between pathogen enzymes and insect cuticle, *J. Invertebr. Pathol.* **47**: 295–302.

St. Leger, R. J., Cooper, R. M., and Charnley, A. K., 1987a, Production of cuticle-degrading enzymes by the entomopathogen *Metarhizium anisopliae* during infection of cuticles from *Calliphora vomitoria* and *Manduca sexta, J. Gen. Microbiol.* **133**:1371–1382.

St. Leger, R. J., Charnley, A. K., and Cooper, R. M., 1987b, Characterization of cuticle-degrading proteases produced by the entomopathogen *Metarhizium anisopliae, Arch. Biochem. Biophys.* **253**:221–232.

St. Leger, R. J., Cooper, R. M., and Charnley, A. K., 1987c, Distribution of chymoelastases and trypsin-like enzymes in five species of entomopathogenic Deuteromycetes, *Arch. Biochem. Biophys.* **258**:121–131.

St. Leger, R. J., Durrands, P. K., Charnley, A. K., and Cooper, R. M., 1988a, Regulation of production of proteolytic enzymes by the entomopathogen *Metarhizium anisopliae, Arch. Microbiol.* **150**:413–416.

St. Leger, R. J., Durrands, P. K., Charnley, A. K., and Cooper, R. M., 1988b, The role of extracellular chymo-elastase in the virulence of *Metarhizium anisopliae* for *Manduca sexta, J. Invertebr. Pathol.* **52**:285–294.

St. Leger, R. J., Cooper, R. M., and Charnley, A. K., 1988c, The effect of melanization of *Manduca sexta* cuticle on growth and infection of *Metarhizium anisopliae, J. Invertebr. Pathol.* **52**:459–471.

St. Leger, R. J., Butt, T. M., Goettel, M. S., Roberts, D. W., and Staples, R. C., 1989a, Production *In vitro* of appressoria by the entomopathogenic fungus *Metarhizium anisopliae, Exp. Mycol.* **13**:274–288.

St. Leger, R. J., Butt, T. M., Roberts, D. W., and Staples, R. C., 1989b, Synthesis of proteins including a cuticle-degrading protease during differentiation of the entomopathogenic fungus *Metarhizium anisopliae, Exp. Mycol.* **13**:253–262.

St. Leger, R. J., Roberts, D. W., and Staples, R. C., 1989c, Calcium- and calmodulin-mediated protein synthesis and protein phosphorylation during germination, growth and protease production by *Metarhizium aniso-pliae, J. Gen. Microbiol.* **135**:2141–2154.

Samsinakova, A., Misikova, S., and Leopold, J., 1971, Action of enzymatic system of *Beauveria bassiana* on the cuticle of the greater wax moth larvae *Galleria mellonella*, *J. Invertebr. Pathol*. **18**:322–330.

Samuels, R. I., Charnley, A. K., and St. Leger, R. J., 1990, The partial characterization of trypsin-like enzymes and aminopeptidases from three species of entomopathogenic entomophthorales and two species of deuteromycetes, *Mycopathology* **110**:145–152.

Shnitka, T. K., 1974, Esterases-nonspecific esterases, in: *Electron Microscopy of Enzymes*, Volume 3 (M. A. Hayat, ed.), Van Nostrand-Reinhold, Princeton, N. J., pp. 1–53.

Smith, R. J., and Grula, E. A., 1981, Nutritional requirements for conidial germination and hyphal growth of *Beauveria bassiana*, *J. Invertebr. Pathol*. **37**:222–230.

Smith, R. J., and Grula, E. A., 1983, Chitinase is an inducible enzyme in *Beauveria bassiana*, *J. Invertebr. Pathol*. **42**:319–326.

Smith, R. J., Pekrul, S., and Grula, E. A., 1981, Requirements for sequential enzymatic activities for penetration of the integument of the cornworm (*Heliothis zea*), *J. Invertebr. Pathol*. **38**:335–344.

Somero, I., 1975, The role of isoenzymes in adaptation to varying temperatures, in: *Isoenzymes. II. Physiological Function* (A. P. Marbert, ed.), Academic Press, New York.

Stirling, J. L., Cook, G. A., and Pope, A. M. S., 1979, Chitin and its degradation, in: *Fungal Walls and Hyphal Growth*, (J. A. Burnett, and C. Trinci, eds.), Cambridge University Press, London, pp. 169–188.

Wigglesworth, V. B., 1970, Structural lipids in the insect cuticle and the function of oenocytes, *Tissue Cell* **2**:155–179.

Wolfgang, W. J., and Riddiford, L. M., 1981, Cuticular morphogenesis during continuous growth of the final instar of a moth, *Tissue Cell* **13**:757–772.

Zacharuk, R. Y., 1970a, Fine structure of the fungus *Metarhizium anisopliae* infecting three species of larval Elateridae (Coleoptera). II. Conidial germ tubes and appressoria, *J. Invertebr. Pathol*. **15**:81–91.

Zacharuk, R. Y., 1970b, Fine structure of the fungus *Metarhizium anisopliae* infecting three species of larval Elateridae (Coleoptera). III. Penetration of the host integument, *J. Invertebr. Pathol*. **15**:372–396.

Zacharuk, R. Y., 1981, Fungal disease of terrestrial insect, in: *Pathogenesis of Invertebrate Microbial Diseases* (E. W. Davidson, ed.), Allanheld, Osmun, Montclair, N. J., pp. 376–402.

13

Potential for Penetration of Passive Barriers to Fungal Invasion in Humans

Frank C. Odds

The principal passive barriers to fungal invasion in humans are epithelial surfaces. For any fungal propagule that has successfully reached and attached itself to epithelia, penetration of the epithelial surface is the necessary next stage of the pathological process. The extent to which a fungus can penetrate epithelia solely by exerting mechanical pressure on the host cells is unknown: it seems likely that most, if not all, instances of fungal penetration involve enzymatic degradation of host surface macromolecules. The focus of this chapter will therefore be on the elaboration by fungi of enzymes that aid in the penetration of the host's passive barriers.

1. FUNGI WITH POTENTIAL FOR PENETRATION OF EPITHELIAL SURFACES

Epithelial cells are of two general types—keratinized and nonkeratinized—and their vulnerability to fungal invasion differs. Only dermatophytes can break down the intact, dry, keratinized epidermis: other fungal pathogens rely on their ability to penetrate epithelia that are not rich in keratin or that have been weakened by occlusion or maceration.

Some fungi initiate infection in humans by accidental skin penetration, in which the host's passive barriers are breached traumatically. Such a process is beyond the scope of the present discussion. For the remainder of fungi there are three major routes of entry—via ingestion, inhalation, or deposition on the skin. For fungi that are ingested or that reside normally in the gastrointestinal tract, the mechanism of persorption (Volkheimer *et al.*, 1968) may result in gratuitous uptake of infective propagules into the bloodstream. This process has been noted, for example, in the case of *Candida albicans* (Krause *et al.*, 1969) which commonly penetrates gastrointestinal epithelia (Odds, 1988). Otherwise, it is inevitable that a fungus must actively invade the local epithelial surface if infection is to progress beyond the stage of epithelial attachment.

Aspergillus spp., *Cryptococcus neoformans*, zygomycetes such as *Mucor* and *Rhizopus*

Frank C. Odds • Department of Bacteriology and Mycology, Janssen Research Foundation, B-2340 Beerse, Belgium.

spp., and the dimorphic fungi *Blastomyces dermatitidis*, *Coccidioides immitis*, *Histoplasma capsulatum*, and *Paracoccidioides brasiliensis* all normally gain access to the host via inhalation and all possess the potential for invasion beyond the respiratory tract. For these fungi the mechanism of penetration of respiratory tract epithelia is largely a matter of speculation. Such evidence as is available tends to implicate hydrolytic enzyme activity—particularly proteolysis—as the mechanism for degradation and penetration of the epithelial surface. For example, extracellular proteolysis has been demonstrated for *A. fumigatus* (McQuade, 1964; Miyaji and Nishimura, 1977; Kothary *et al.*, 1984; Staib, 1985), *Conidiobolus coronata* (Gabriel, 1968), and *Coccidioides immitis* (Rippon and Varadi, 1968; Rogozhkina *et al.*, 1977; Yuan and Cole, 1987; Lupan and Nziramasanga, 1986). However, similar enzyme activities can also be found in "nonpathogenic" *Aspergillus* spp. (Cohen, 1977) and in saprophytic fungi related to *Coccidioides immitis* (Lupan and Nziramasanga, 1985) so that mere expression of proteolytic activity by a fungus is not of itself evidence of a role for proteolysis in epithelial penetration. More elaborate experimentation that may prove or disprove the significance of proteolysis for epithelial penetration by fungal pathogens of the respiratory tract has not yet been reported.

Candida spp. are the most significant fungi known to invade epithelia of the gastrointestinal tract. For *C. albicans*, the principal pathogenic species, many experimental approaches have been used to study mechanisms of gastrointestinal adhesion and colonization (these are reviewed by Guentzel *et al.*, 1985, and Kennedy, 1988) but only one has specifically mentioned active methods of initial epithelial penetration. Kennedy and colleagues (1987) noted depressions in intestinal epithelial cells associated with adherent *C. albicans* yeast cells and considered that these might be the result of fungal enzyme activities.

The three main groups of fungi that cause skin pathology are the dermatophytes, *Candida* spp., and *Malassezia furfur*. The ability of dermatophytes to lyse keratin has been the subject of considerable experimental attention and there can be little remaining doubt that proteolysis by dermatophytes is, at the very least, an adjunct to their potential for epidermal invasion. Recent evidence similarly implicates proteolysis by *Candida* in epidermal invasion, and dermatophytes and *Candida* are discussed in detail below. For *Malassezia furfur* there is no experimental evidence for specific epidermal invasion mechanisms, but it is clear that this species is able to effect extensive penetration of the outermost keratinized layer (Rippon, 1988).

2. PROTEOLYSIS (AND KERATINOLYSIS) AS A MECHANISM FOR EPITHELIAL PENETRATION

2.1. Dermatophyte Proteinases

The production of proteinase enzymes by pathogenic fungi has been extensively studied only for the dermatophytes: proteinases in *Candida* spp. have also been receiving attention in relatively recent years. Evidence that dermatophytes can secrete proteinases dates back to 1896 when Macfadyen showed that *Trichophyton tonsurans* could liquefy gelatin with a diffusible enzyme. The many publications relating to degradation of keratinaceous substrates by dermatophytes *in vitro* have served to create a general impression that all dermatophytes are capable of enzymatic keratinolysis. However, the biochemical evidence for true keratinolysis by any dermatophyte species is relatively weak.

Keratin is a complex, highly insoluble fibrous protein that is characterized by unusually extensive cross-linking with disulfide bridges. *In vivo*, keratin is never produced in a pure form: the keratin molecules are arranged with various other proteins and cementing substances in organized structures such as hair, wool, and feathers. Materials such as nails, hooves, horns, and the stratum corneum of skin are keratin-rich, but less so than hair and wool (Bradbury, 1973). To prove enzymatic degradation of the keratin molecule itself, it is really necessary to show that peptide material released from natural keratinaceous substrates has not come from the other components within them.

Table I summarizes the main literature that provides biochemical evidence for degradation of keratinaceous substrates by various dermatophyte species *in vitro*. All the papers listed show that the dermatophyte species studied are able to release into their culture media substances that can be characterized as likely to have a protein source. Amino acids, peptides, sulfhydryl groups, and protein labels are all released very slowly, over a period of many days, when dermatophytes or their culture filtrates are incubated with keratinaceous substrates. These observations strongly suggest proteolytic activity by the dermatophytes but they do not necessarily prove degradation of the keratin itself. (Proteolysis is indeed confirmed by the several studies cited in Table I in which nonkeratin proteins such as collagen, casein, and albumin are degraded by dermatophytes, and dermatophyte elastase activity was demonstrated by Rippon in 1967 in agar plate tests.)

There are two main reasons why evidence for proteolytic degradation of wool, hair, and so forth is not specific evidence of keratinolysis. The first is that procedures used to sterilize keratin sources before they are coincubated with dermatophytes are likely to degrade the structure of native keratin molecules and render them more susceptible to proteolytic attack. The second is that the material released from keratinaceous substrates *in vitro* may have come from cementing substances rather than from the keratin component itself.

These problems have been recognized for many years. Stahl and co-workers (1950a) showed that a single passage of wool through a fiber homogenizing mill led to a 10% decrease in its cystine content and made it much more susceptible to digestion by *Microsporum canis* and trypsin. Barlow and Chattaway (1955) showed that chemical breakage of disulfide bonds in hair rendered the hair more vulnerable to degradation by dermatophytes. McQuade (1964) and Ragot (1969) recommended sterilization protocols for wool that involved sequential treatments with organic solvents to avoid the denaturing action of methods such as autoclaving. Noval and Nickerson (1959) found that ethylene oxide sterilization of wool led to no changes in the content of sulfur, nitrogen, or cystine. This observation was confirmed by Hose (1976) who overcame the problem of sterilization of keratin by the use of hair taken aseptically from germ-free guinea pigs. This hair could be used unchanged as a substrate for dermatophyte growth and proteolysis.

The importance of distinguishing between true keratinolysis and degradation of nonkeratin components in natural substrates was quantified by Noval and Nickerson (1959): they showed that nonspecific proteinases may solubilize as little as 10% of the dry weight of wool and yet lead to total disintegration of the fibers. It is because of this phenomenon that evidence for keratinolysis by dermatophytes based solely on microscopic observation of fiber degradation must be considered invalid.

Several lines of evidence suggest that the primary attack of dermatophytes on keratin-rich material is not on keratin *per se*. Stahl and co-workers (1950b) showed by elemental analysis of the residues that the first wool proteins degraded by *M. gypseum* are the nonsulfur-containing molecules. This observation was confirmed by Hose (1976) who showed that release of ^{35}S from guinea pig hair previously radiolabeled *in vivo* occurred after the maximum release of peptide material assayed by chemical methods. Mandels and co-workers (1948) showed microscopically that *M. gypseum in vitro* grows in wool parallel to the axes of keratin fibers but between the cortical (largely keratinous) cells, and Raubitschek (1961) found that *Trichophyton mentagrophytes* grew readily on aqueous extracts of nail and hair, proving that the fungus is able to assimilate nonkeratin material from these substrates. Finally, Chattaway and co-workers (1963) showed that several dermatophyte species had no lytic effect on nail and hair keratin predigested with trypsin unless the disulfide bonds in the keratin were first reduced by thioglycollate.

The evidence discussed so far tends to favor the view that dermatophytes may be proteolytic, but they have limited, if any, ability to degrade keratin *per se*. However, it is clear from other work

Table I. Papers Demonstrating Liberation of Peptides, Amino Acids, or Protein Labels
from Keratins and Other Protein Substrates by Dermatophytes In Vitro

	Substrate			
Dermatophyte sp.	Wool	Hair	Other keratin source[a]	Other proteins
M. canis	Daniels (1953) Evolceanu and Lazar (1960) Weary et al. (1965)	Evolceanu and Lazar (1960) Chattaway et al. (1963) Takiuchi et al. (1981, 1982) Lee et al. (1987)	Evolceanu and Lazar (1960) Chattaway et al. (1963) Takiuchi et al. (1981)	O'Sullivan and Mathison (1971)
M. gypseum	Stahl et al. (1949, 1950b) Evolceanu and Lazar (1960) Weary et al. (1965)	Evolceanu and Lazar (1960) Kunert (1972a) Meevootisom and Niederpruem (1979) Deshmukh and Agrawal (1982)	Evolceanu and Lazar (1960)	
T. ajelloi	Chesters and Mathison (1963) McQuade (1964) Ragot (1967, 1969)	Deshmukh and Agrawal (1982)		Ragot (1966)
T. mentagrophytes	Evolceanu and Lazar (1960)	Evolceanu and Lazar (1960) Chattaway et al. (1963) Yu et al. (1968, 1969a,b) Meevootisom and Niederpruem (1979)	Evolceanu and Lazar (1960) Chattaway et al. (1963)	Day et al. (1968) Yu et al. (1969b)
T. rubrum	Weary and Canby (1967, 1969)	Chattaway et al. (1963) Meevootisom and Niederpruem (1979) Das and Banerjee (1982) Deshmukh and Agrawal (1985)	Chattaway et al. (1963) Mikx and De Jong (1987)	Meevootisom and Niederpruem (1979) Sanyal et al. (1985)
T. schoenleinii	Weary and Canby (1967, 1969)		Chattaway et al. (1963) Wawrzkiewicz et al. (1987)	Rippon (1968)
Other species		Deshmukh and Agrawal (1985)		

[a]Hoof, horn, nail, feathers, callus.

that the lytic capability of dermatophytes for keratin substrates goes beyond that of other proteinases. Chesters and Mathison (1963) showed that *T. ajelloi* could release amino acids from animal wool whereas trypsin had almost no activity in this assay. Stahl and co-workers (1949) found that *M. gypseum* could release cystine from wool when the substrate was exposed to the dermatophyte for several weeks, and a number of studies have implicated sulfitolysis as a mechanism by which dermatophytes can reduce disulfide bonds in keratin (Ragot, 1966; Kunert, 1972b, 1975; Ruffin *et al.*, 1976). Wawrzkiewicz and co-workers (1987) could grow *Trichophyton gallinae* on agar media containing chicken feathers, but the fungus could not utilize human or guinea pig hair—a finding that provides both empirical evidence of a molecular basis for the host specificity of this dermatophyte and a strong indication of particular affinities between dermatophyte species and keratinaceous substrates.

Dermatophyte enzymes able to liberate amino acids from keratin sources have been purified both from culture filtrates and from disrupted fungal cells. The extracellular enzymes have similar pH optima (of the order of 7 to 8), sizes in the range 30–50 kDa, and inhibitor profiles suggesting they are all serine proteinases (Yu *et al.*, 1968; Meevootisom and Niederpruem, 1979; Takiuchi *et al.*, 1982, 1983, 1984; Lee *et al.*, 1987). Rather similar properties have been described for dermatophyte exoenzymes that degrade various proteins but whose activity against keratin substrates was not determined (Roberts and Doetsch, 1967; Day *et al.*, 1968; O'Sullivan and Mathison, 1971; Sanyal *et al.*, 1985). However, the frequent occurrence of multiple pH optima and of enzyme sizes greater than 50 kDa indicates that the dermatophyte proteinases are a heterogeneous group, with multiple isozymes produced even by a single strain (Day *et al.*, 1968; O'Sullivan and Mathison, 1971; Meevootisom and Niederpruem, 1979). Takiuchi and co-workers (1983) found serological crossreactivity between exoproteinases from four dermatophyte species. Viewed overall, the literature so far suggests that dermatophytes secrete a variety of extracellular proteinases with different biochemical properties but some antigenic relatedness.

Intracellular proteinases with broad substrate specificities were purified by Yu and co-workers (1969b, 1971). These enzymes had much greater sizes than the exoproteinases and differed in their biochemical properties.

While there is much known about the proteolytic activities of dermatophytes, comparatively few studies have examined their significance in penetration of host barriers *in vivo*. Cruickshank and Trotter (1956) showed that proteolytic *T. mentagrophytes* culture filtrates, like trypsin, could cause separation of guinea pig epidermis from the dermis *in vitro*. Skořepová and Hauck (1987) found a small degree of correlation between clinical severity of lesions caused by *T. rubrum* and proteolytic activity *in vitro* of the strains isolated from the lesions. Purified intracellular proteinases from *T. mentagrophytes* inhibit migration of guinea pig peritoneal exudate cells *in vitro* (Eleuterio *et al.*, 1973). The skin of guinea pigs infected with *T. mentagrophytes* bears evidence of activity of this species' intracellular proteinase *in vivo* since sections of the skin are positive in indirect fluorescent antibody tests with antiserum raised against the *T. mentagrophytes* enzyme (Collins *et al.*, 1973), and the same enzyme induced delayed-type hypersensitivity reactions in the skin of infected animals (Grappel and Blank, 1972). The importance of proteolysis or keratinolysis in dermatophyte invasion of the epidermis is likely to be established directly in the future by experimentation with proteinase-negative dermatophyte mutants *in vivo*.

2.2. Candida albicans

An extracellular, inducible proteinase of *C. albicans* has been recognized and studied since the 1960s: it is a carboxyl proteinase with a low pH optimum and a size of 40 to 46 kDa (Odds, 1985, 1988). Studies with proteinase-deficient mutants of *C. albicans* have strongly implicated the enzyme as a virulence factor in the pathogenesis of deep-seated *Candida* infections (Macdonald and Odds, 1983; Kwon-Chung *et al.*, 1985) and recent publications indicate that the enzyme may be responsible for the initial invasion of the epidermis in cutaneous infections. The *C. albicans* proteinase readily liberates amino acids from samples of human stratum corneum; it is marginally degradative for hoof keratin and inactive against hair (Hattori *et al.*, 1984; Negi *et al.*, 1984). The most impressive evidence for the penetrative potential of the *C. albicans* proteinase comes from Ray and Payne (1988) who showed by means of scanning electron microscopy that *C. albicans* yeasts adherent to mouse skin formed depressions in the epidermal cells, but that this action was inhibited by pepstatin (the major inhibitor of the *C. albicans* proteinase). Similar observations were made by Borg and Rüchel (1988) with biopsies of nonkeratinized oral epithelia. These two papers probably represent the best scientific evidence so far published associating fungal proteolysis with invasion of passive host barriers to infection.

3. OTHER HYDROLYTIC ENZYMES AS AGENTS FOR EPITHELIAL PENETRATION

Phospholipase activities have been demonstrated in cell envelopes or extracellularly in *C. albicans* (Pugh and Cawson, 1975; Price and Cawson, 1977) and *T. rubrum* (Das and Banerjee, 1978). Lipases have been detected in several *Candida* spp. (Odds, 1988) and dermatophytes (Nobre and Viegas, 1972; Das and Banerjee, 1978). These enzymes possess the ability in theory to facilitate degradation of epithelial cell membranes; however, only minimal evidence is available to support such a supposition. Barrett-Bee *et al.* (1985) found a correlation between adherence of *C. albicans* isolates to exfoliated buccal epithelia and their phospholipase activity; they did not examine penetration of the fungi into the epithelial cells.

4. CONCLUSIONS

It is self-evident that many fungi actively breach the passive barriers offered by human epithelial tissues. To date, there is evidence associating proteolysis in some species with epithelial penetration, but the molecular basis of fungal penetration is otherwise almost entirely unexplored. It is to be hoped that, in the future, research techniques such as specifically engineered mutagenesis of genes coding for known lytic enzymes in pathogenic fungi can be used in conjunction with suitable animal models to determine fully the early mechanisms of host invasion.

5. REFERENCES

Barlow, A. J. E., and Chattaway, F. W., 1955, Attack of chemically modified keratin by certain dermatophytes, *J. Invest. Dermatol.* **24**:65–74.

Barrett-Bee, K., Hayes, Y., Wilson, R. G., and Ryley, J. F., 1985, A comparison of phospholipase activity, cellular adherence and pathogenicity of yeasts, *J. Gen. Microbiol.* **131**:1217–1221.

Borg, M., and Rüchel, R., 1988, Expression of extracellular acid proteinase by proteolytic *Candida* spp. during experimental infection of oral mucosa, *Infect. Immun.* **56**:626–631.

Bradbury, J. H., 1973, The structure and chemistry of keratin fibers, *Adv. Protein Chem.* **27**:111–211.

Chattaway, F. W., Ellis, D. A., and Barlow, A. J. E., 1963, Peptidases of dermatophytes, *J. Invest. Dermatol.* **41**:31–37.

Chesters, C. G. C., and Mathison, G. E., 1963, The decomposition of wool keratin by *Keratinomyces ajelloi*, *Sabouraudia* **2**:225–237.

Cohen, B. L., 1977, The proteases of aspergilli, in: *Genetics and Physiology of Aspergillus* (J. E. Smith and J. A. Pateman, eds.), Academic Press, New York, pp. 281–292.

Collins, J. P., Grappel, S. F., and Blank, F., 1973, Role of keratinases in dermatophytosis. II. Fluorescent antibody studies with keratinase II of *Trichophyton mentagrophytes*, *Dermatologica* **146**:95–100.

Cruickshank, C. N. D., and Trotter, M. D., 1956, Separation of epidermis from dermis by filtrates of *Trichophyton mentagrophytes*, *Nature* **177**:1085–1086.

Daniels, G., 1953, The digestion of human hair by *Microsporum canis* Bodin, *J. Gen. Microbiol.* **8**:289–294.

Das, S. K., and Banerjee, A. B., 1978, Lipolytic enzymes of *Trichophyton rubrum*, *Sabouraudia* **15**:313–323.

Das, S. K., and Banerjee, A. B., 1982, Effect of undecanoic acid on the products of exocellular lipolytic and keratinolytic enzymes by undecanoic acid-sensitive and -resistant strains of *Trichophyton rubrum*, *Sabouraudia* **20**:179–184.

Day, W. C., Toncic, P., Stratman, S. L., Leeman, U., and Harmon, S. R., 1968, Isolation and properties of an extracellular protease of *Trichophyton granulosum*, *Biochim. Biophys. Acta* **167**:597–606.

Deshmukh, S. K., and Agrawal, S. C., 1982, In vitro degradation of human hair by some keratinophilic fungi, *Mykosen* **25**:454–458.

Deshmukh, S. K., and Agrawal, S. C., 1985, Degradation of human hair by some dermatophytes and other keratinophilic fungi, *Mykosen* **28**:463–466.

Eleuterio, M. K., Grappel, S. F., Caustic, C. A., and Blank F., 1973, Role of keratinases in dermatophytosis. III. Demonstration of delayed hypersensitivity to keratinases by the capillary tube migration test, *Dermatologica* **147**:255–260.

Evolceanu, R., and Lazar, M., 1960, Le Problème de l'existence et de la mise en èvidence des ferments kératinolytique chez les dermatophytes, *Mycopathol. Mycol. Appl.* **12**:216–222.

Gabriel, B. P., 1968, Histochemical study of the insect cuticle infected by the fungus *Entomophthora coronata*, *J. Invertebr. Pathol.* **11**:82–89.

Grappel, S. F., and Blank, F., 1972, Role of keratinase in dermatophytosis. I. Immune responses of guinea-pigs infected with *Trichophyton mentagrophytes* and guinea-pigs immunized with keratinases, *Dermatologica* **145**:245–255.

Guentzel, M. N., Cole, G. T., and Pope, L. M., 1985, Animal models for candidiasis, in: *Current Topics in Medical Mycology*, Volume 1 (M. R. McGinnis, ed.), Springer-Verlag, Berlin, pp. 57–115.

Hattori, M., Yoshiura, K., Negi, M., and Ogawa, H., 1984, Keratinolytic proteinase produced by *Candida albicans*, *Sabouraudia J. Med. Vet. Mycol.* **22**:175–183.

Hose, H., 1976, Decomposition of native keratin by dermatophyte fungi, Ph.D. thesis, Department of Microbiology, University of Leeds, Leeds, U.K.

Kennedy, M. J., 1988, Adhesion and association mechanisms of *Candida albicans*, in: *Current Topics in Medical Mycology*, Volume 2 (M. R. McGinnis, ed.), Springer-Verlag, Berlin, pp. 73–169.

Kennedy, M., Volz, P. A., Edwards, C. A., and Yancey, R. J., 1987, Mechanisms of association of *Candida albicans* with intestinal mucosa, *J. Med. Microbiol.* **24**:333–342.

Kothary, M. H., Chase, T., Jr., and MacMillan, J. D., 1984, Correlation of elastase production by some strains of *Aspergillus fumigatus* with ability to cause pulmonary invasive aspergillosis in mice, *Infect. Immun.* **43**:320–327.

Krause, W., Matheis, H., and Wulf, K., 1969, Fungaemia and funguria after oral administration of *Candida albicans*, *Lancet* **1**:598–599.

Kunert, J., 1972a, The digestion of human hair by the dermatophyte *Microsporum gypseum* in a submerged culture, *Mykosen* **15**:59–71.

Kunert, J., 1972b, Thiosulphate esters in keratin attacked by dermatophytes, *Sabouraudia* **10**:6–13.

Kunert, J., 1975, Formation of sulphate, sulphite and S-sulphocysteine by the fungus *Microsporum gypseum* during growth on cystine, *Folia Microbiol.* **20**:142–151.

Kwon-Chung, K. J., Lehman, D., Good, C., and Magee, P. T., 1985, Genetic evidence for role of extracellular proteinase in virulence of *Candida albicans*, **49**:571–575.

Lee, K. H., Park K. K., Park, S. K., and Lee, J. B., 1987, Isolation, purification and characterization of keratinolytic proteinase from *Microsporum canis*, *Yonsei Med. J.* **28**:131–138.

Lupan, D. M., and Nziramasanga, P., 1985, Elastase activity of fungi with anamorphs similar to *Coccidioides immitis*, *Mycopathologia* **92**:169–171.

Lupan, D. M., and Nziramasanga, P., 1986, Collagenolytic activity of *Coccidioides immitis*, *Infect. Immun.* **51**:360–361.

Macdonald, F., and Odds, F. C., 1983, Virulence for mice of a proteinase- secreting strain of *Candida albicans* and a proteinase-deficient mutant, *J. Gen. Microbiol.* **129**:431–438.

Macfadyen, A., 1896, A contribution to the biology of the ringworm organism, *J. Pathol. Bacteriol.* **3**:176–183.

McQuade, A. B., 1964, Microbiological degradation of wool, *Dermatologica* **128**:249–266.

Mandels, G. R., Stahl, W. H., and Levinson, H. S., 1948, Structural changes in wool degraded by the ringworm fungus *Microsporum gypseum* and other organisms, *Text. Res. J.* **18**:224–231.

Meevootisom, V., and Niederpruem, D. J., 1979, Control of exocellular proteases in dermatophytes and especially *Trichophyton rubrum*, *Sabouraudia* **17**:91–106.

Mikx, F. H., and DeJong, M. H., 1987, Keratinolytic activity of cutaneous and oral bacteria, *Infect. Immun.* **55**:621–625.

Miyaji, M., and Nishimura, K., 1977, Relationship between proteolytic activity of *Aspergillus fumigatus* and the fungal invasiveness of mouse brain, *Mycopathologia* **62**:161–166.

Negi, M., Tsuboi, R., Matsui, T., and Ogawa, H., 1984, Isolation and characterization of proteinase from *Candida albicans*, *J. Invest. Dermatol.* **83**:32–36.

Nobre, G., and Viegas, M. P., 1972, Lipolytic activity of dermatophytes, *Mycopathol. Mycol. Appl.* **46**:319–323.

Noval, J. J., and Nickerson, W. J., 1959, Decomposition of native keratin by *Streptomyces fradiae*, *J. Bacteriol.* **77**:251–263.

Odds, F. C., 1985, *Candida albicans* proteinase as a virulence factor in the pathogenesis of *Candida* infections, *Zentralbl. Bakteriol. Hyg.[A]* **260**:539–542.

Odds, F. C., 1988, *Candida and Candidosis*, 2nd ed., Baillière Tindall, London.

O'Sullivan, J., and Mathison, G. E., 1971, The localization and secretion of a proteolytic enzyme complex by the dermatophytic fungus *Microsporum canis*, *J. Gen. Microbiol.* **68**:319–326.

Price, M. F., and Cawson, R. A., 1977, Phospholipase activity in *Candida albicans*, *Sabouraudia* **15**:179–185.

Pugh, D., and Cawson, R. A., 1975, The cytochemical localization of phospholipase a and lysophospholipase in *Candida albicans*, *Sabouraudia* **13**:110–115.

Ragot, J., 1966, Activité kératinolytique du *Keratinomyces ajelloi* vanbr. sur la laine non dénaturée, *C. R. Acad. Sci. Ser. D* **262**:412–415.

Ragot, J., 1967, Activité protéolytique du *Keratinomyces ajelloi* vanbr. en présence de kératine, *C. R. Acad. Sci. Ser. D* **265**:493–496.

Ragot, J., 1969, Influence de certains agents de sterilisation sur la kératine, *Mycopathol. Mycol. Appl.* **39**:177–186.

Raubitschek, F., 1961, Mechanical versus chemical keratinolysis by certain dermatophytes, *Sabouraudia* **1**:87–90.

Ray, T. L., and Payne, C. D., 1988, Scanning electron microscopy of epidermal adherence and cavitation in murine candidiasis: A role for *Candida* acid proteinase, *Infect. Immun.* **56**:1942–1949.

Rippon, J. W., 1967, Elastase: Production by ringworm fungi, *Science* **157**:947.

Rippon, J. W., 1968, Extracellular collagenase from *Trichophyton schoenleinii*, *J. Bacteriol.* **95**:43–46.

Rippon, J. W., 1988, *Medical Mycology*, 3rd ed., Saunders, Philadelphia, p. 158.

Rippon, J. W., and Varadi, D. P., 1968, The elastases of pathogenic fungi and actinomycetes, *J. Invest. Dermatol.* **50**:54–58.

Roberts, F. F., and Doetsch, R. N., 1967, Purification of a highly active protease from a *Microsporum* species, *Antonie van Leeuwenhoek J. Microbiol. Serol.* **33**:145–152.

Rogozhkina, N. M., Klimova, I. M., and Zelenskaya, L. I., 1977, Izuchenia nekotorykh uslovii abrazovaniya proteazy *Coccidioides immitis*, *Zh. Mikrobiol. Epidemiol. Immunobiol.* **8**:87–90.

Ruffin, P., Andrieu, S., Biserte, G., and Biguet, J., 1976, Sulphitolysis in keratinolysis. Biochemical proof, *Sabouraudia* **14**:181–184.

Sanyal, A. K., Das, S. K., and Banerjee, A. B., 1985, Purification and partial characterization of an exocellular proteinase from *Trichophyton rubrum*, *Sabouraudia* **23**:165–178.

Skořepová, M., and Hauck, H., 1987, Extracellular proteinases of *Trichophyton rubrum* and the clinical picture of tinea, *Mykosen* **30**:25–27.

Stahl, W. H., McQue, B., Mandels, G. R., and Siu, R. G. H., 1949, Studies on the microbiological degradation of wool. I. Sulfur metabolism, *Arch. Biochem.* **20**:422–432.

Stahl, W. H., McQue, B., Mandels, G. R., and Siu, R. G. H., 1950a, Studies on the microbiological degradation of wool. Digestion of normal and modified fibrillar proteins, *Text. Res. J.* **20**:570–579.

Stahl, W. H., McQue, B., and Siu, R. G. H., 1950b, Studies on the microbiological degradation of wool. II. Nitrogen metabolism, *Arch. Biochem.* **27**:211–220.

Staib, F., 1985, Pleural fluid as nutrient substratum for *Aspergillus fumigatus* and *A. flavus*. Submerged growth in pleural fluid and extracellular proteolysis in pleural fluid agar, *Zentralbl. Bakteriol. Mikrobiol. Hyg. [A]* **260**:566–571.

Takiuchi, I., Higuchi, D., and Negi, M., 1981, The effect of keratinase on human epidermis, especially on stratum corneum, *Jpn. J. Dermatol.* **91**:119–125.

Takiuchi, I., Higuchi, D., Sei, Y., and Koga, M., 1982, Isolation of an extracellular proteinase (keratinase) from *Microsporum canis*, *Sabouraudia* **20**:281–288.

Takiuchi, I., Higuchi, D., Sei, Y., and Koga, M., 1983, Immunological studies of an extracellular keratinase, *J. Dermatol. (Tokyo)* **10**:327–330.

Takiuchi, I., Sei, Y., Takagi, H., and Negi, M., 1984, Partial characterization of the extracellular keratinase from *Microsporum canis*, *Sabouraudia* **22**:219–224.

Volkheimer, G., Schulz, F. H., Aurich, I., Strauch, S., Beuthin, K., and Wendlandt, H., 1968, Persorption of particles, *Digestion* **1**:78–80.

Wawrzkiewicz, K., Lobarzewski, J., and Wolski, T., 1987, Intracellular keratinase of *Trichophyton gallinae*, *J. Med. Vet. Mycol.* **25**:261–268.

Weary, P. E., and Canby, C. M., 1967, Keratinolytic activity of *Trichophyton schoenleinii*, *Trichophyton rubrum* and *Trichophyton mentagrophytes*, *J. Invest. Dermatol.* **42**:240–248.

Weary, P. E., and Canby, C. M., 1969, Further observations on the keratinolytic activity of *Trichophyton schoenleinii* and *Trichophyton rubrum*, *J. Invest. Dermatol.* **53**:58–63.

Weary, P. E., Canby, C. M., and Cowley, E. P., 1965, Keratinolytic activity of *Microsporum canis* and *Microsporum gypseum*, *J. Invest. Dermatol.* **44**:300–310.

Yu, R. J., Harmon, S. R., and Blank, F., 1968, Isolation and purification of an extracellular keratinase of *Trichophyton mentagrophytes*, *J. Bacteriol.* **96**:1435–1436.

Yu, R. J., Harmon, S. R., and Blank, F., 1969a, Hair digestion by a keratinase of *Trichophyton mentagrophytes*, *J. Invest. Dermatol.* **53**:166–171.

Yu, R. J., Harmon, S. R., Wachter, P. E., and Blank, F., 1969b, Amino acid composition and specificity of a keratinase of *Trichophyton mentagrophytes*, *Arch. Biochem. Biophys.* **135**:363–370.

Yu, R. J., Harmon, S. R., Grappel, S. F., and Blank, F., 1971, Two cell-bound keratinases of *Trichophyton mentagrophytes*, *J. Invest. Dermatol.* **56**:27–32.

Yuan, L., and Cole, G. T., 1987, Isolation and characterization of an extracellular proteinase of *Coccidioides immitis*, *Infect. Immun.* **55**:1970–1978.

14

Dihydroxynaphthalene (DHN) Melanin and Its Relationship with Virulence in the Early Stages of Phaeohyphomycosis

Dennis M. Dixon, Paul J. Szaniszlo, and Annemarie Polak

1. INTRODUCTION

Pathogenicity can be defined as "giving rise to, or producing disease." Virulence is generally considered as the degree of pathogenicity. If a given microorganism possesses no degree of pathogenicity, then it would be considered both avirulent and nonpathogenic.

With the medically important fungi, pathogenic organisms are often grouped on the basis of virulence. Thus, there are primary fungal pathogens and opportunistic fungal pathogens. The former would include such agents as *Coccidioides immitis* and *Histoplasma capsulatum*, which are of such capability of virulence as to cause disease in the normal host, whereas the latter are of an accordingly lower degree of virulence, but appear infrequently in healthy hosts with intact immune systems.

The dematiaceous fungi are darkly colored molds and yeasts which are often grouped together simply because of the similar appearance of their pigmentation. Therefore, it is difficult to generalize about them in terms of biological properties and gross morphology. The medically important dematiaceous fungi are associated with at least three different disease categories: chromoblastomycosis, mycetoma, and phaeohyphomycosis. These categories have been summarized recently by McGinnis in concise and erudite reports (McGinnis, 1983; Fader and McGinnis, 1988) and will only be briefly considered here. Chromoblastomycosis is a localized cutaneous and subcutaneous disease characterized by verrucose lesions that typically become cauliflowerlike in gross pathological appearance. Important etiological agents include *Phialophora verrucosa*, *Fonsecaea pedrosoi*, and *Cladosporium carrionii* (Table I). In infected tissue, these fungi typically appear as spherical, dematiaceous cells with muriform septations (muriform cells). Mycetoma is typified by the triad of tumefaction of the infected appendage, draining sinus

Dennis M. Dixon • Laboratories for Mycology, Wadsworth Center for Laboratories and Research, New York State Department of Health, Albany, New York 12201–0509. *Paul J. Szaniszlo* • Department of Microbiology, University of Texas, Austin, Texas 78713. *Annemarie Polak* • F. Hoffmann-La Roche Ltd, CH 4002 Basel, Switzerland.

Table I. Selected Dematiaceous Fungi, the Human Diseases Caused,[a] and Demonstration of the Presence of DHN Melanin

Chromoblastomycosis	Phaeohyphomycosis
Cladosporium carrionii DHN[b]	*Alternaria alternata* DHN[b]
Fonsecaea compacta DHN[b]	*Aureobasidium pullulans* DHN[b]
Fonsecaea pedrosoi DHN[b]	*Bipolaris* spp.
Phialophora verrucosa DHN[b]	*Chaetomium* sp.
Rhinocladiella aquaspersa	*Cladosporium* spp.
	Curvularia spp.
Eumycotic mycetoma	*Dactylaria constricta* var. *gallopava*
Curvularia spp.	*Hendersonula toruloidea* DHN[b]
Exophiala jeanselmei DHN[b]	*Lecythophora* spp.
Leptosphaeria spp.	*Exserohilum* sp.
Madurella spp.	*Phaeoannellomyces werneckii* DHN[b]
Neotestudina rosatii	*Phialemonium* sp.
Pseudallescheria boydii	*Phialophora* spp. DHN[b]
Pyrenochaeta spp.	*Phoma* spp.
	Sarcinomyces sp.
	Scedosporium inflatum
	Scytalidium lignicola
	Wangiella dermatitidis DHN[b]
	Xylohypha bantiana DHN[b]

[a]Adapted from Ajello (1986), Dixon *et al.* (1986), Fader and McGinnis (1988), McGinnis and Fader (1988), Salkin *et al.* (1988), and Ahmad *et al.* (1985).
[b]DHN melanin pathway demonstrated by tricyclazole inhibition; adapted from Wheeler and Bell (1987). DHN melanin data are not available for the species where not indicated.

tracts, and the presence of grains that are essentially microcolonies of the fungus. Mycetoma can be caused by fungi (eumycotic mycetoma) or by actinomycetes (actinomycotic mycetoma). The eumycotic mycetomas are divided into black-grained and white-grained on the basis of grain color. Dematiaceous fungi, such as *Exophiala jeanselmei* and *Pyrenochaeta romeroi*, are examples of etiological agents of black-grained mycetoma (Table I). Finally, phaeohyphomycosis was conceptualized by Mariat, anglicized by Ajello, and expanded by McGinnis to refer to diseases produced by fungi that appear in tissue primarily as septate, dematiaceous hyphae (Dixon *et al.*, 1988). McGinnis has proposed four clinical categories of phaeohyphomycosis: superficial; cutaneous and corneal; subcutaneous; and systemic (McGinnis, 1983).

2. VIRULENCE OF THE DEMATIACEOUS FUNGI

In terms of virulence, the dematiaceous fungi range from one extreme to the other. For example, there is the relatively innocuous pathogen, *Phaeoannellomyces werneckii*, which causes superficial phaeohyphomycosis in the form of tinea nigra, and the polymorphic fungus, *Wangiella dermatitidis*, which not only can cause localized cutaneous phaeohyphomycosis, but also can disseminate to involve multiple organs and the brain. In this latter form of central nervous system (CNS) phaeohyphomycosis, mortality is usually 100% (Matsumoto *et al.*, 1984).

Since the primary factor unifying the dematiaceous fungi is their dark pigmentation, one logical first step to an understanding of the virulence of these fungi may involve the biochemistry of their melanin pigment systems. Wheeler and his collaborators have been especially active in this area, and have determined that in all probability the majority of medically important dematiaceous fungi are darkened because of the deposition of a particular type of melanin in their cell walls. This melanin, which has been termed DHN (dihydroxynaphthalene) melanin (Bell and Wheeler, 1986; Wheeler and Bell, 1987), is formed via pentaketide metabolism (Fig. 1) and is characteristic exclusively of Ascomycota, or of Fungi imperfecti having suspected ascomycetous affinities (Wheeler and Bell, 1987). To date, more than ten medically important fungi are known to form this unique type of melanin (Table I), which is the focus of this review. However, in addition to DHN melanin, a number of other fungal melanins are known and other biosynthetic pathways have been studied (for comprehensive reviews, see Bell and Wheeler, 1986; Wheeler

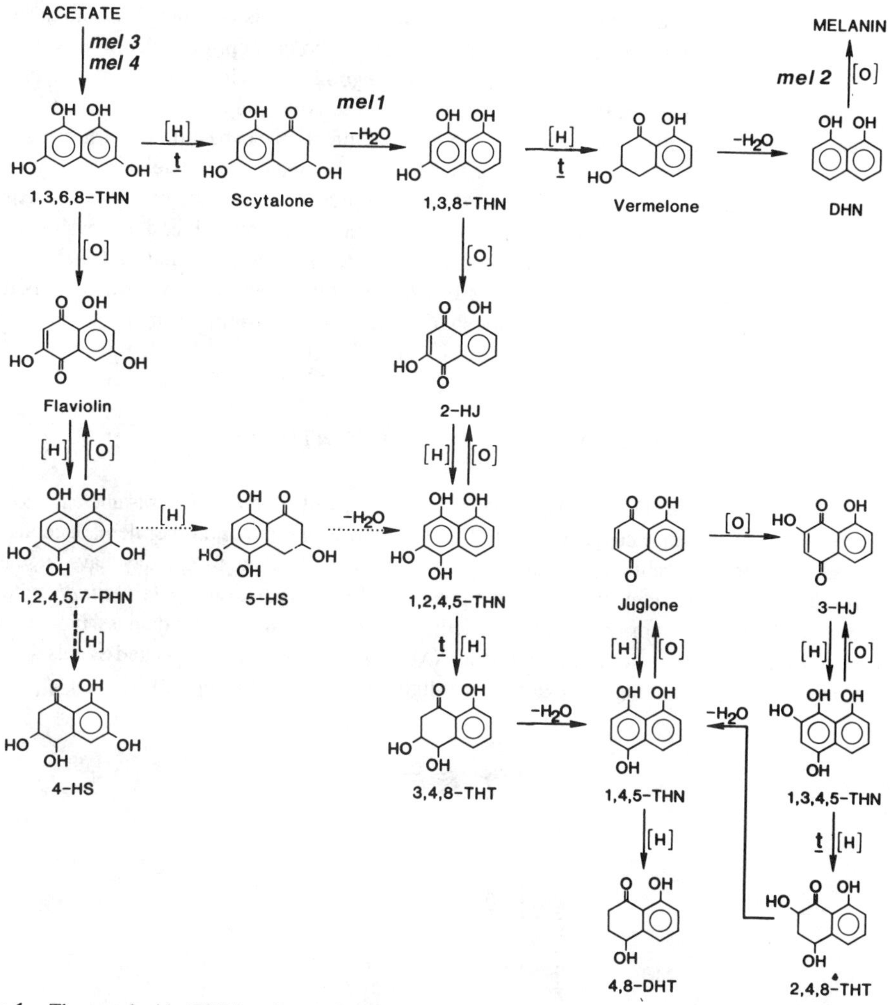

Figure 1. The pentaketide (DHN) pathway of melanin synthesis and the flaviolin and 2-HJ branch pathways in *Wangiella dermatitidis*. *Verticillium dahliae* and *Pyricularia oryzae* carry out most of the same reactions as *W. dermatitidis*, but make 4-HS as indicated by the dashed arrows. 5-HS is made and dehydrated by *W. dermatitidis* as indicated by the dotted arrows. The proposed hydroquinone intermediates have not been isolated from the three fungi since they are highly unstable in the presence of oxygen and form naphthoquinones. *mel* number = position in pathway disrupted by respective mutation; H = reduction reaction; O = oxidation reaction; −H₂O = dehydratase reaction; t = sites of tricyclazole inhibition in *W. dermatitidis*. (Reproduced and modified with permission from Wheeler and Stipanovic, 1985; pathway courtesy of M. H. Wheeler.)

and Bell, 1987). Among the medically important fungi, the best example is a type of DOPA (dihydroxyphenylalanine) melanin formed by the basidiomycete, *Filobasidiella neoformans* (anamorph = *Cryptococcus neoformans*) via the oxidation of a variety of ortho- and paradiphenols, including DOPA; the enzyme involved is a polyphenol oxidase (catecholase or laccase) (Kwon-Chung *et al.*, 1983; Polacheck *et al.*, 1982). There has been a considerable amount of interesting work attempting to determine the relationship between phenol oxidase and virulence in *Cryptococcus neoformans* (Kwon-Chung *et al.*, 1982; Kwon-Chung and Rhodes, 1986; Rhodes *et al.*, 1982). These studies of a pathogenic basidiomycete were taken as the precedent for our studies of the relationship between DHN melanin and virulence among the dematiaceous pathogens of humans.

After mutants defective in normal DHN melanin biosynthesis were derived from the human pathogen, *Wangiella dermatitidis*, we began extending the concept of melanin as a virulence factor to the Dematiaceae. This particular fungus is a good choice for experimentation because: (1) it is a zoopathogenic dematiaceous fungus that is neurotropic and quite virulent; (2) fatal infections can be established in mice; (3) CNS signs of infection can be monitored and quantified; (4) the fungus is polymorphic and has a yeastlike phase which facilitates quantification and infection studies; (5) the DHN melanin pathway is representative of many, if not all, dematiaceous pathogens of man; (6) numerous mutants of the fungus are available, or are easily derived. (For a comprehensive review of polymorphism in *W. dermatitidis*, see Geis and Jacobs, 1985.) Much of the remainder of this chapter is based upon our experiences with this well-studied organism as a model dematiaceous pathogen of man, which is extremely amenable for the study of the role of melanin in animal pathogenicity and virulence.

3. DHN MELANIN AND WANGIELLA DERMATITIDIS

The exact role that DHN melanin plays in the biology of *W. dermatitidis* is unclear. However, certain lines of evidence suggest that DHN melanin in *Wangiella* and other dematiaceous fungi enhances their survivability in nature. For example, melanized fungi may have a selective advantage over nonmelanized fungi in the presence of UV irradiation. *W. dermatitidis* wild-type strain ATCC 34100 is more resistant to the lethal effects of UV irradiation than is a biochemically and genetically uncharacterized albino strain (Mel-4) derived from it (Geis and Szaniszlo, 1984) and capable of synthesizing carotenoids in the light, but not the dark (Fig. 2). Thus, it appears that

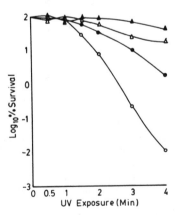

Figure 2. Survival curves of light (closed symbols)- and dark (open symbols)-incubated fungal cultures of wild-type (ATCC 34100) *W. dermatitidis* (triangles) and Mel-4 (circles) after ultraviolet irradiation (48 ergs/mm² per sec). (Reproduced with permission from Geis and Szaniszlo, 1984.)

the DHN melanin associated with *W. dermatitidis* cell walls acts as an effective protectant that guards against UV-induced mutagenesis in environments such as the upper layers of soils or the exposed surfaces of wood, where this and many other dematiaceous fungi are presumed to grow (Dixon *et al.*, 1988).

That the melanin is deposited exclusively in the cell wall of *W. dermatitidis* is based on at least three lines of evidence. First, little, if any, dark pigmentation is associated with cell homogenates of *W. dermatitidis* after removal of insoluble, darkly pigmented cell walls (Geis and Szaniszlo, unpublished data). Second, Wheeler and Stipanovic (1985) have shown that all portions of the cell walls of an albino mutant (*mel3*) derived by Geis and Szaniszlo (Geis *et al.*, 1984) appear hyaline in electron micrographs, whereas those of the parental wild type have an electron-dense outer layer and associated fibrils. Addition of scytalone, a precursor of DHN melanin, to the culture media of the albino strain allowed the mutant to become melanized and consequently reestablished in the mutant the electron opaqueness characteristic of wild-type walls (Fig. 3; courtesy of M. H. Wheeler). Finally, melanized strains of *W. dermatitidis* are much less susceptible to lysis in the presence of cell wall lytic enzymes than are melanized strains whitened by the DHN melanin biosynthesis inhibitor, tricyclazole, or are collectively albino or brown strains having lesions in their pathway leading to complete DHN melanin synthesis (Table II, Fig. 4; Jacobs and Szaniszlo, unpublished). These observations suggest that the electron-opaque outer surface seen in the electron micrographs is, in fact, melanin, which in nature is active as a resistance factor not only to the detrimental effects of irradiation, but also to microbial killing as mediated by extracellular wall lytic enzymes.

The DHN melanin deposited in cell walls of *W. dermatitidis* seems to vary quantitatively during certain periods of cell differentiation. Szaniszlo *et al.* (1983a) documented, during detailed comparisons of the cell wall compositions of the melanized wild type and a temperature-sensitive, cell-division-cycle (*cdc*) mutant, *cdc3* (Mc3), and its wild-type parent, ATCC 34100, that the melanin concentration in the cell wall of the mutant incubated at the restrictive temperature (37°C) increased to levels about five times higher than those found in the wild type treated similarly (Table III). This increase in melanin concentration correlated with the change in morphology associated with the mutant, but not the wild type when yeast cells of the mutant converted to thick-walled, unbudded, isotropically enlarged cells and thick-walled, unbudded muriform (multicellular, Mc) forms at the restrictive temperature. In contrast, when wild-type yeasts were treated identically, growth continued in the yeast morphology and only a moderate increase in cell wall melanin deposition was noted. These findings suggest that the thick-walled cells and thick-walled muriform units sometimes associated with dematiaceous pathogens *in vitro* may be more resistant to host defenses by virtue of higher concentrations of cell wall melanin, as well as possibly their altered morphologies. It is important to point out, however, that the increased cell-wall melanin deposition associated with the cell differentiation of *W. dermatitidis* converting from a yeast to isotropically enlarged, thick-walled cells or to muriform units is not required for that morphogenesis because nonmelanized Mc⁻ strains of *W. dermatitidis* convert to these alternative morphologies as efficiently as melanized strains (Geis and Jacobs, 1985).

As will be described in subsequent sections, our virulence studies have been facilitated by the availability of numerous mutant strains defective in some aspect of normal DHN melanin biosynthesis. Various lines of evidence have been used to document that at least some of these strains probably have mutations in different genes. Experimental approaches yielding these different lines of evidence have included: the use of inhibitors of DHN melanin biosynthesis, such as tricyclazole, to identify intermediates or shunt metabolites that accumulate in treated cells; cross-feeding studies employing Mel⁻ mutants that secrete intermediates able to stimulate completion of DHN melanization of other Mel⁻ strains; and cell homogenate studies to reveal the presence or absence of relevant enzymatic activities in Mel⁻ strains, compared to those of the normally melanized wild type. More recently a protoplast fusion protocol has been developed that should further clarify our virulence results by documenting that some of our mutants, which

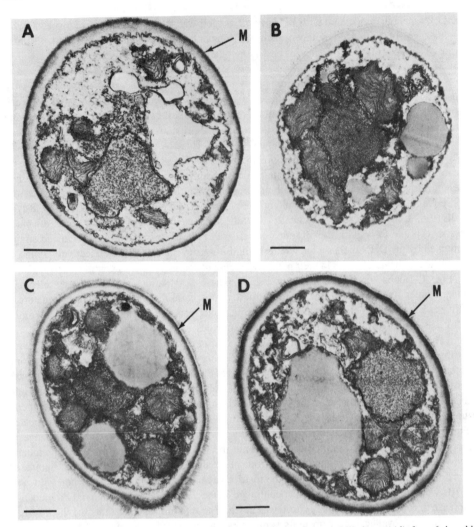

Figure 3. Electron micrographs of the wild-type isolate and albino mutant of *W. dermatitidis* from 6-day-old cultures. Fixation with OsO_4. (A) Wild-type cell of *W. dermatitidis* with melanized cell wall. (B) Albino cell of *W. dermatitidis*, *mel3*. Melanin (M) is absent. (C, D) Melanized albino cells of *W. dermatitidis*, *mel3*. Two 50-μmole additions of scytalone were made to 50-ml cultures (Czapek Dox agar supplemented with 0.1% w/v yeast extract) of the albino mutant at 24 hr (C) and 48 hr (D) after cultures were initiated. Bar = 0.5 μm (for all panels). (Reproduced with permission from Wheeler and Stipanovic, 1985.)

appear phenotypically similar, are different by virtue of their ability to complement each other and form normally melanized fusion products.

Using physiological and biochemical techniques (Geis *et al.*, 1984), three different Mel⁻ mutants were characterized to the extent that they could be genotypically designated *mel1*, *mel2*, and *mel3* (Table IV). Both *mel1* and *mel2* are strains that tend to be some shade of brown and also secrete intermediates able to cross-feed *mel3* (Fig. 5). Using extracts from solid and liquid cultures of *mel2*, sufficient amounts of DHN melanin biosynthetic intermediates were obtained to establish that *mel2* is defective in the oxidase activity required to convert dihydroxynaphthalene

Table II. Effect of Degree of DHN Pigmentation on the Resistance of W. dermatitidis to Glusulase[a]

Strain	Pigmentation	Log$_{10}$ decrease in viability[b]
Mel-F4739	Dark black	No decrease
ATCC 34100	Black	0.3
Mel-E2023	Dark brown	1.7
Mel-A722	Brown	4.0
Mel-F369	Light brown	No survivors

[a]iResistance to a 90-min treatment with a 1:10 (w/v) dilution of Glusulase (Endo Laboratories).
[b]Reflects killing due to Glusulase-induced lysis.

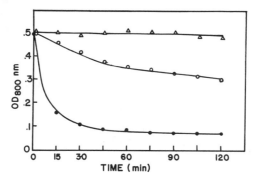

Figure 4. Lysis of tricyclazole-grown and control cells by zymolyase. Cells grown in the presence of tricyclazole (closed circles) and in its absence (open circles) were incubated in the presence of 1 mg of zymolyase 5000 (Kirin Brewery) per ml of cell suspension in 10 mM phosphate buffer, pH 7.0. The lysis of cells was monitored by measuring the decrease in absorbance at 800 nm (Kitamura and Yamamoto, 1972). Tricyclazole-grown cells incubated in the absence of zymolyase (open triangles) did not exhibit a decline in A$_{800}$. (Courtesy of C. W. Jacobs.)

Table III. Percentages of Major Cell Wall Components of Yeast (Y) and Multicellular (Mc), Sclerotic-like, Morphologies of W. dermatitidis[a,b]

Strain and treatment	Neutral sugar	Lipid	Melanin	Protein	N-Acetylglucosamine	Ash[c]	Recovery
Wt[d]Y (25°C)	72.9	3.0	4.3	8.3	2.1	3.4	94.0
cdc3[e] (25°C)	71.7	2.0	4.4	9.9	1.9	3.3	93.2
Wt Y (37°C)	73.7	2.9	5.7	8.5	3.2	3.2	97.1
cdc3 Mc (37°C)	43.0	2.2	24.8	10.8	11.8	1.7	94.3

[a]Modified from Szaniszlo et al. (1983a).
[b]Each value represents the average of two or more analyses of at least two independently derived wall samples.
[c]Ash was determined gravimetrically after ignition for 3 hr at 1000°C in a muffle furnace.
[d]Wild type, 8656.
[e]The temperature-sensitive (ts), cell-division-cycle (cdc) mutant, cdc3, was originally designated phenotypically as Mc3 because of its ability to convert from yeasts to homogeneous populations of isotropically enlarged, thick-walled, nonbudding cells and nonbudding, multicellular, muriform units at 37°C (the restrictive temperature), whereas the wild type from which it was derived continued to grow as a yeast at this same higher temperature (Roberts and Szaniszlo, 1978). Documentation that the changes in morphology associated with the mutant at the restrictive temperatures are the result of a defect in a cell-division-cycle (CDC) gene (see relevant discussion in review by Jacobs and Geis, 1985), and recent analyses of spheroplast-fusion products and segregants derived from this and other MC[-] strains, have led to the genotypic designation of Mc3 as cdc3 (Cooper, 1988).

Table IV. Characterization of Wild-Type and DHN Melanin Mutant Strains[a] of W. dermatitidis Used in Virulence Studies

Strain designation	DHN melanin mutation	Color	Enzyme deficiency	Major products accumulated	Source, origin, and reference(s)
Wild type NIH 8656, DMD 368 (ATCC 34100)	None	Black	None	None	B. H. Cooper; P. J. Szaniszlo (Geis et al., 1984; Geis and Jacobs, 1985; Oujezdsky et al., 1973)
Wild type, CM 26	None	Black	None	None	A. Polak (Dixon et al., 1987)
Mel-1 (ATCC 44502)	mel1	Red-brown	Dehydratase	Scytalone, flaviolin	Geis and Szaniszlo, NTG (Geis et al., 1984)
Mel-2 (ATCC 44503)	mel2	Dark brown	Oxidase	DHN	Geis and Szaniszlo, NTG (Geis et al., 1984)
Mel-3, DMD 369,[b] DMD 372[b] (ATCC 44504)	mel3	White	Unknown	None	Geis and Szaniszlo, spontaneous (Geis et al., 1984; Wheeler and Stipanovic, 1985)
Mel-UV10, 17, 22, 23	Unknown	White	Unknown	Unknown	Geis and Szaniszlo, UV (unpublished)

[a]Prior to biochemical or complementation analysis, all DHN melanin mutant strains were designated phenotypically as Mel−, whereas only after the relevant mutation was localized in the DHN melanin biosynthetic pathway or after the mutation was demonstrated to be distinctive by complementation analysis using protoplast fusion procedures were Mel− phenotypic strain designations changed to mel− genotypic designations.
[b]Mel-3 strains 369 and 372 are subisolates of the original mel3 mutant. DMD 369 was maintained by P.J.S. and D.M.D. whereas DMD 372 was maintained independently by ATCC, Rockville, Maryland, after deposit by Geis.

to DHN melanin (Fig. 1). Cultures of this mutant are usually dark brown and accumulate major amounts of DHN, and lesser amounts of scytalone, a precursor of DHN, and of flaviolin, 2-hydroxyjuglone (2-HJ), and 3-HJ, which accumulate by branch pathways, particularly when normal DHN synthesis is inhibited (Geis et al., 1984). In contrast to mel2, mel1 is defective in the dehydratase activity required to convert scytalone and vermelone to 1, 3, 8-trihydroxynaphthalene (1, 3, 8-THN) and DHN, respectively. Cultures of mel1 are often more reddish brown and accumulate the DHN melanin precursor scytalone and the branch metabolite flaviolin, indicating that the primary block in this mutant resides in its inability to convert scytalone to 1, 3, 8-THN (Geis et al., 1984). Finally, the mel3 strain is nonmelanized, but has all the enzymes required for the conversion of scytalone, if not 1, 3, 6, 8-THN, to DHN melanin (Wheeler and Stipanovic, 1985). Cultures of this strain do not appear to have any brown or black coloration, but may become pink if grown in the presence of light. We will refer to this type of strain as albino. The mel3 mutant does not cross-feed any other available Mel− strains of W. dermatitidis or alm or brm mutants of DHN melanin biosynthesis in Verticillium dahliae, and do not have any detectable melanin intermediates or shunt products associated with their ethyl acetate extracts (Geis et al., 1984). These particular results suggest the mutation in mel3 blocks the synthesis of 1, 3, 6, 8-THN, if not pentaketide itself.

By virtue of the absence of DHN melanin pigments in their cell walls, albino mutants of W. dermatitidis are easily isolated either as spontaneous variants of pigmented strains or after mutagenesis. Thus, numerous Mel− strains have accumulated, and are being sorted into complementation groups. Complementation analysis is facilitated using a spheroplast fusion protocol based on the fusion of albino or brown Mel− strains and the subsequent recovery of dark brown to dark black prototrophs (Fig. 6). This methodology has become particularly useful for characterizing nonmelanized strains, like mel3, which accumulate no detectable DHN melanin inter-

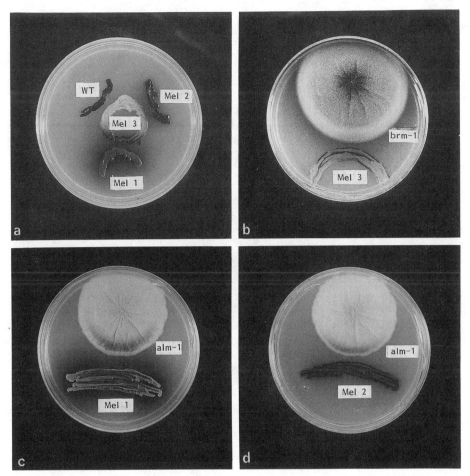

Figure 5. Cross-feeding studies with colonies of *W. dermatitidis* and *Verticillium dahliae*. The paired colonies were grown on Czapek Dox agar supplemented with 0.1% w/v yeast extract at 24°C for 6 to 10 days. (a) The *mel3* (Mel 3) mutant of *W. dermatitidis* was blackened by metabolites that diffused from the *mel1* (Mel 1) and *mel2* (Mel 2) mutants but was not affected by the wild-type (WT) strain. (b) Metabolites diffused from the *brm-1* mutant of *V. dahliae* and blackened the *mel3* mutant of *W. dermatitidis*. (c, d) Metabolites from the *mel1* and *mel2* mutants of *W. dermatitidis* blackened the *alm-1* mutant of *V. dahliae*. Sclerotia of the wild-type strain from which the mutant strains of *V. dahliae* were obtained are black. (Reproduced with permission from Geis *et al.*, 1984.)

mediates, but appear to be able to synthesize DHN from any of its known precursors. A new type of DHN melanin biosynthetic mutant, preliminarily designated as *mel4*, appears to have been discovered by this procedure (Cooper and Szaniszlo, unpublished). Precedence for such mutations among fungi that produce DHN melanin was previously established with *V. dahliae* in which mutations in at least three genes (*alm1*, *alm2*, and *alm3*) are known to give rise to nonmelanized strains. Using protoplast fusion procedures, at least three other independently derived Mel⁻ strains have been shown by either the presence or absence of complementation to be equivalent to the *mel3* strain (Cooper and Szaniszlo, unpublished). Other amino acid auxotrophic or morphological mutant strains of *W. dermatitidis* are being analyzed using melanization to select

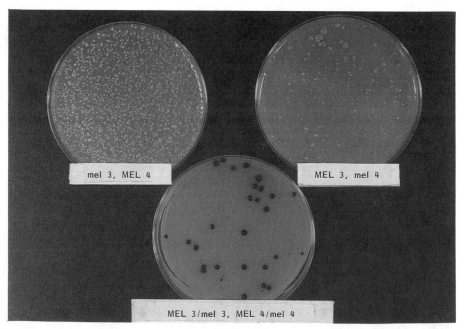

Figure 6. Genetic complementation of *mel* genes. Strains Mc2W-1 (*cdc2, met1, mel4*) and Mc3W-15 (*cdc3, ura1, mel3*) are temperature-sensitive, cell-division-cycle (*cdc*) mutants that are also nonmelanized (*mel⁻*) auxotrophs of *W. dermatitidis* ATCC 34100; *met* = methionine; *ura* = uracil. Spheroplasts of these Mel⁻ strains regenerate form albino colonies at 25°C in agarose overlaid only onto complete medium (top left, strain Mc3W-15; top right, strain Mc2W-1). When spheroplasts of these same strains are fused, they regenerate at 42°C in agarose overlaid onto minimal medium, and the pigmented fusion products are formed (bottom center). These latter conditions select for genetic complementation of the temperature-sensitive (*cdc2, cdc3*) and auxotrophic (*met1, ura1*) lesions by genes from the appropriate parental strain. The production of melanin indicates that the *mel* genes are also complementary, i.e., genes *mel3* and *mel4* are indeed different lesions in the pentaketide (DHN) pathway of melanin biosynthesis. (Courtesy of C. R. Cooper, Jr.)

fusion products from among backgrounds of white or brown parental strains (Cooper and Szaniszlo, unpublished data).

4. INITIAL STUDIES OF VIRULENCE

The strains of *W. dermatitidis* used in our studies were all derived from the same parent and represent a spectrum of infective agents that range in the extent of their characterization (Table IV). Work centered around an albino, *mel3*; we used two independently maintained cultures of *mel3*, DMD 369 and DMD 372. We were first interested in determining whether the albino mutant retained a pathogenic potential equivalent to the melanized, wild-type parent (Dixon *et al.*, 1987). Therefore, dose-response inoculation experiments were done in attempts to study both qualitative and quantitative aspects of the question—i.e., pathogenicity and virulence studies. As sometimes happens, the initial results suggested a clear-cut difference between the dematiaceous wild-type and the *mel3* strains (Fig. 7). At least in terms of the original mortality experiments, this much was found to be true.

The animal model of infection used in our experiments was that of an acute infection (Dixon, 1987). Normally, one sees in this system maximum death within 1 week following intravenous injection of high concentrations of inoculum of virulent fungi. For wild-type *W. dermatitidis*,

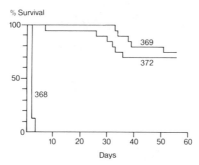

Figure 7. Comparative mortality in mice injected with *W. dermatitidis* wild-type (368) and *mel3* (369; 372) strains, 3 × 10⁷ cells per mouse, monitored for 56 days after infection. Each point represents the average from three groups of 5–10 mice each. (Reproduced with permission from Dixon *et al.*, 1987.)

100% mortality was obtained within 3 days. This demonstrated the extreme virulence of the fungus in our model, since the mortality rate was comparable to those rates seen with such fungi as *Histoplasma capsulatum* and *Cryptococcus neoformans*. Even with the latter two fungi, deaths in such tests usually are more gradual, making infection with *W. dermatitidis* an extremely acute, if not toxic type of event.

In the acute models of infection, experiments are usually terminated after 21 days (Dixon, 1987). However, at such endpoints, mice infected with the *mel3* strains of *W. dermatitidis* often began to exhibit bizarre behaviors, such as frantically running in circles in their cages, and eventually developing CNS signs of infection, such as torticollis and ataxia. Also, deaths began to occur in a linear fashion in mice infected with the *mel3* strains, especially after 30 days (compare panels a and b in Fig. 8). Thus, there was a dramatic decrease in the virulence of *mel3* relative to the parental wild type in these experiments, yet the albino mutant remained pathogenic (Fig. 8).

In attempts to elucidate the differences between the infections caused by the wild type versus the *mel3* strains, various data were collected. Time-course culture studies were conducted over the first 2 weeks of infection. The wild type and both isolates of *mel3* grew exponentially in mouse brains through the first week of infection and declined thereafter through the second week (Fig. 9). No significant quantitative differences were noted between the colony-forming units obtained from brains of mice infected with either wild type or the two *mel3* strains. Thus, the organisms devoid of melanin were able to persist and multiply to the same relative degree as were their dematiaceous, wild-type parent. Also, it is important to note, the organisms recovered from the brains of mice infected with *mel3* never developed dematiaceous pigments, and thus were not types of revertants (Dixon *et al.*, 1987).

Histopathological studies were conducted in conjunction with the quantitative culture studies (Dixon *et al.*, 1987). Mice were killed immediately after inoculation and at days 1, 2, 7, 14, and 56 of infection. Brains were fixed in formalin, sectioned, and adjacent sections stained with hematoxylin and eosin (H&E) or Grocott-Gomori methenamine silver (GMS). Individual yeastlike cells could be seen in capillaries in GMS-stained sections of brain sampled immediately after inoculation (Fig. 10A). Extravascular clusters of yeastlike cells appeared in microabscesses as discrete lesions as early as day 2 after inoculation (Fig. 10B). There were no significant quantitative differences between the lesions caused by the dematiaceous wild type and those resulting from the *mel3* strains. All the strains were capable of producing relatively the same numbers of microabscesses in brains of mice injected intravenously with comparable numbers of cells. Furthermore, the progression of the numbers of lesions relative to time was essentially the same in both groups, with an exponential rise, a plateau, and a sudden disappearance being

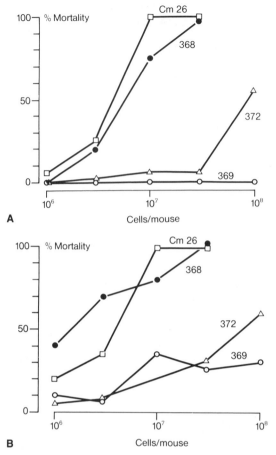

Figure 8. Mouse mortality response to dose of either wild-type (parental strain = 368; nonparental strain = Cm 26) or *mel3* (maintained independently as strains 372 and 369) *W. dermatitidis*. Mice were injected i.v. with the indicated concentrations; each data point represents an average from a minimum of five mice. Cumulative mortality calculated after either (A) 21 days or (B) 56 days.

noted. This sequence of events correlated with the trend observed by quantitative cultures of the organisms from brain (Dixon *et al.*, 1987).

Potentially significant, however, was the histopathological finding of more yeast-to-mycelial conversions associated with the lesions of mice infected with dematiaceous wild type versus those caused by *mel3* (Fig. 10C). Thus, an association between invasive hyphae and virulence must be considered. This prompted additional studies of the histopathological aspects of the infection.

5. STUDIES OF MEL3 IN ADDITIONAL MOUSE STRAINS

The initial studies utilizing the acute mouse infection model were extended to include one inbred albino mouse strain, BALB/c, and two inbred, pigmented mouse strains, DBA/2J and C57BL/6. Mice of each type were infected i.v. with graded concentrations of yeast-phase cells of

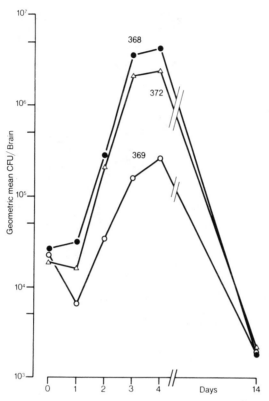

Figure 9. Comparative growth in brains following i.v. inoculation of mice with *W. dermatitidis* (1 × 10⁷ cells per mouse) wild-type (DMD 368) and *mel3* (369 and 372) strains. Closed circles = wild-type; open symbols = *mel3* strains. Each point represents the geometric mean of five determinations. (Reproduced with permission from Dixon *et al.*, 1987.)

wild type (DMD 368) and *mel3* (DMD 369). The methods used were identical to those described briefly above and in detail elsewhere (Dixon, 1987). Mortality and CNS signs of infection were recorded daily over a 20-day period. The results of these studies were summarized and compared to the results of our previous studies using Fuellinsdorf mice (Table V). The general trend demonstrated in the initial studies was repeated with the additional mouse strains. That is, the *mel3* mutant was less virulent than the dematiaceous wild type. The DBA/2J mouse was more susceptible to infection with both strains of *W. dermatitidis* tested. However, this was probably not related to mouse pigmentation since the C57BL/6 mouse was no more susceptible than the outbred albino mouse strain that was tested (BALB/c).

We extended the histopathology studies of brains of experimentally infected mice to include the three additional mouse strains. Following injection i.v. with 1 × 10⁷ cells of either wild type or *mel3*, mice of each type were sampled over a 5-day period, with two mice from each infection group subsequently being killed on each day. Brains were then fixed in formalin and stained with GMS as before (Dixon *et al.*, 1987, 1988). Again, invasive hyphae were seen in lesions of mice infected with the wild-type fungus (Table VI). However, two examples of invasive hyphae were found in brains infected with the *mel3* fungus, DMD 369, in two of three of the inbred strains of mice examined. However, one such instance was in the more susceptible mouse strain, DBA/2J, where 40% mortality occurred in groups of mice infected with this inoculum concentration.

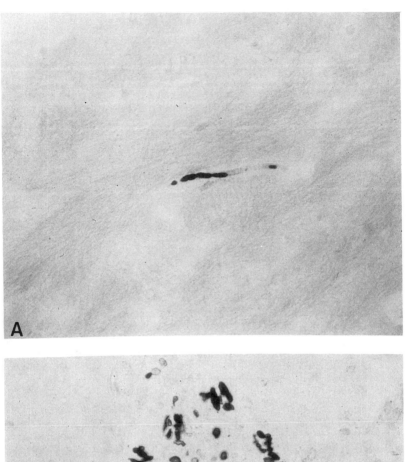

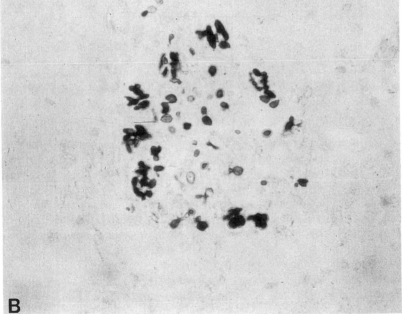

Figure 10. Representative lesions in brains of mice infected i.v. with 1×10^7 cells of *mel3* (DMD 369) *W. dermatitidis*. GMS, $\times$ 500. Yeastlike cells in a small blood vessel on the day of infection (A). Fungal morphology was predominantly yeastlike (B; focal lesion on day 4 of infection), but occasionally exhibited invasive hyphae (C; focal lesion on day 7 of infection).

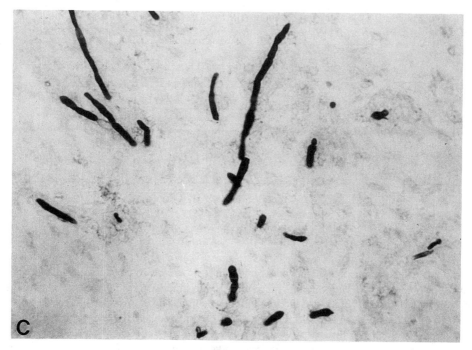

Figure 10. (*Continued*)

Thus, in three of the four instances where invasive hyphae were clearly evidenced, they were associated with virulent infections.

The additional mouse strains also were examined to compare numbers of lesions in the brains of mice infected with either the wild type or *mel3* (DMD 369). To assist with data collection, a computer morphometric program was implemented (Dixon *et al.*, 1989). This involved a camera lucida to project microscopic fields onto a digitizing pad interfaced with a

Table V. Comparison of Wild-Type (wt) and mel3 W. dermatitidis in Four Strains of Mice: Evaluation of Mortality and CNS Signs

	% mortality[a] (fraction of survivors with CNS signs)			
Fungus strain and cells/mouse	Fuellinsdorf	BALB/c	DBA/2J	C57BL/6
wt				
1×10^8	Not done	100 (—)	100 (—)	100 (—)
3×10^7	100 (—)	40 (1/6)	100 (—)	80 (0/5)
1×10^7	75 (4/4)	40 (3/7)	100 (—)	30 (1/8)
3×10^6	60 (3/80)	20 (3/10)	80 (1/3)	0 (0/10)
mel3				
1×10^8	0 (10/19)	50 (1/5)	100 (—)	50 (2/5)
3×10^7	0 (10/19)	0 (0/10)	90 (1/1)	0 (2/10)
1×10^7	0 (6/15)	0 (0/10)	40 (4/6)	0 (0/10)
3×10^6	0 (8/14)	0 (0/10)	0 (6/10)	0 (0/10)

[a]Data from Fuellinsdorf mice on day 27 from reference (Dixon *et al.*, 1987); all other data from day 20 and adapted from Dixon *et al.* (1989).

Table VI. Focal Lesions Produced in the Brains of Four Strains of Mice after Intravenous Injection of Wild-Type or Melanin-Deficient (mel3) W. dermatitidis[a]

Days post-infection	Mouse	Fungus							
		Wild type				mel3			
		ICR	C57	DBA	BALB/c	ICR	C57	DBA	BALB/c
1		7	10	7	0	3	3	22	1
2		ND[b]	23	41	10	1	15	92[c]	12
3		47[c]	32[c]	79	15	18	29[d]	105	14
4		28	72	63	45	4	44	70	17
5		7	23	ND	9	17	55[c]	61	15

[a]Adapted from Dixon *et al*. (1989).
[b]ND, no data.
[c]Invasive hyphae.
[d]Occasional hyphae.

computer-driven morphometric analytical program. Thus, lesions could be traced and quantitative morphological data collected. We were interested primarily in the numbers of lesions per representative section of brain, and the areas of brain delimited by the histopathological lesion as calculated in square micrometers using the computer program. This analysis allowed us to determine, as we had observed previously in Fuellinsdorf outbred mice, that there were no obvious differences in the numbers of lesions produced in the brains of various mouse strains infected with either strain of fungus. Also, there appeared to be no mouse strain dependence (Table VI).

Total lesion area per representative section of infected mouse brain increased in time exponentially through day 3 or 4 (Fig. 11), after which time there was a gradual decline. This

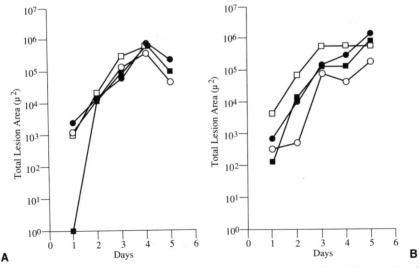

Figure 11. Total lesion area in one representative section from the brains of each of four different strains of mice infected with 1×10^7 cells of *W. dermatitidis*. Mouse strains are: ICR (open circles), C57 (closed circles), DBA/2J (open squares), BALB/c (closed squares). Fungus strains are wild type (A) and *mel3* (B). (Reproduced with permission from Dixon *et al.*, 1989.)

trend matched that of lesion number. This correlation is important since we were interested in learning if it was the total areas of the brain involved rather than the numbers of focal lesions that were important. For example, ten small microabscesses could have been misconstrued as more important than a single huge abscess involving a significant amount of the brain. Using representative sections taken from similar regions of brain and evaluating both the total numbers and total areas and perimeters of the focal lesions of fungal infection revealed that the extensiveness of fungal infection was similarly represented in both the numbers of lesions produced and the total areas of the brain infected in a given section. Thus, the addition of a computer-assisted morphometric program helped us in addressing this consideration.

6. STUDIES WITH ADDITIONAL MUTANTS IN OUTBRED MICE

The acute infection model was used to extend our virulence studies to include additional Mel⁻ strains (Table IV) derived from the wild-type parent, ATCC 34100 (DMD 368). These were *mel1* (ATCC 44502), *mel2* (ATCC 44503), UV-10, UV-17, UV-22, and UV-23. The latter mutants were derived by UV irradiation. In addition, a second wild-type strain, CM 26, was included along with DMD 368 for comparative purposes. The UV mutants were white when grown on PDA in the dark (Dixon *et al.*, 1989). The results of dose-response experiments are shown in Fig. 12. In each case, the mutants were less virulent than the same dose of either strain of wild type. At concentrations of 1×10^7 cells per mouse, the difference between the mutant and wild-type strains was particularly pronounced. At concentrations of $\geq 3 \times 10^7$ cells per mouse, Mel 1 gave mortality values of $\geq 80\%$, yet this concentration was at the plateau for the wild-type strains, and represented a huge particulate challenge. Mel 2 remained essentially avirulent at all concentrations tested. It would be tempting to equate degrees of virulence with the degrees of melanization

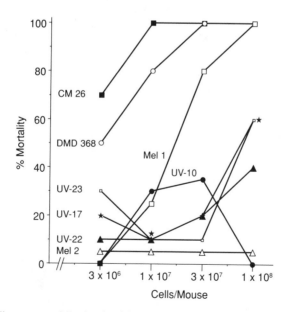

Figure 12. Mortality response following i.v. injection of *W. dermatitidis* wild type and Mel mutants in Fuellinsdorf mice. Data points represent averages at day 21 after inoculation of groups of five mice per concentration. (Reproduced with permission from Dixon *et al.*, 1989.)

seen in the mutants, but this may not be important. It may be more important to consider which intermediates may be accumulating in a given mutant and evaluate these compounds for toxicity.

7. PRELIMINARY STUDIES WITH BROTH-CULTIVATED INOCULUM

We have continued our mortality studies on the involvement of melanin in the virulence of *W. dermatitidis* by attempting to include a biochemically reconstituted albino strain in our work. Having learned that a decrease in virulence accompanied a decrease in melanization, we tested the effects of "blackening" the *mel3* mutant by growing it in the presence of scytalone, one of a number of requisite precursors of DHN melanin which it cannot synthesize (Fig. 1). Although these studies were complicated by bacterial contamination of the scytalone, they were nonetheless interesting because of what they revealed in terms of the scytalone-free (and hence contamination-free) controls. These experiments involved the liquid cultivation of *W. dermatitidis*. Inoculum of *mel3* grown in yeast nitrogen base broth (YNB; Difco) for 48 or 72 hr and then injected i.v. into Fuellinsdorf mice, produced acute mortality unlike what we had seen previously with inoculum prepared from standard Sabouraud slant cultures grown for 1 week. Thus, we began to explore the new variable of inoculum preparation as a means to investigate further the virulence mechanisms of this fungus.

Strains of *W. dermatitidis* were grown on PDA slants at 30°C for 1 week and washed from the surface with sterile saline. Inoculum was adjusted by hemocytometer to give 1×10^6 cells per ml of culture broth. The latter consisted of 20 ml YNB in 100-ml Erlenmeyer flasks. These were incubated at 30°C with shaking for 48 to 72 hr. At the end of each specified culture time, cells were collected by centrifugation and washed once in sterile physiological saline. For filtered inoculum, cell suspensions were passed through gauze.

The results, as already mentioned, were unexpected with the broth-cultured *mel3* (DMD 369) because, for the first time in our experiments, this strain resulted in 100% mortality within 3 days (Fig. 13). There was no significant difference in death rates resulting from wild type (DMD 368) versus *mel3*, nor was there any difference between the 48- and 72-hr inoculum in terms of death rates. It should be noted that there were no increases in virulence for the wild type when

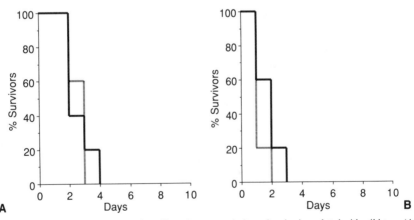

Figure 13. The effect of liquid cultivation of inoculum upon virulence in mice inoculated with wild-type (A) and *mel3* (B) *W. dermatitidis.* Solid lines represent 48-hr cultures; dotted lines represent 72-hr cultures. Cultures were centrifuged, washed, and suspended in sterile saline to give an inoculum of 1×10^8 cells administered i.v. to each mouse in groups of five.

cultivated in broth relative to agar. Thus, the increases seemed limited to the *mel3* strain which approached the constant virulence of wild type.

Microscopic examination of the broth-grown inoculum revealed a disproportionate number of microcolonies, yeast chains (pseudohyphae), and possibly moniliform hyphae (see Oujedsky *et al.*, 1983, for a description of moniliform hyphae). These results are now known to be relatively common to the broth culture of *mel3* which is somewhat more inefficient in cell separation than the wild type (Cooper, unpublished data). We considered that this morphological phase deviation could be associated with, or at least contributing to, the increased virulence and mortality seen with *mel3* as cultured for these experiments. Also, this had been theorized, based on the histopathological data presented above. Therefore, we repeated the experiments with broth-grown cultures and filtered the inoculum through gauze. Microscopic examination of the filtered inoculum revealed occasional chains of up to six yeast cells, but the hyphal filaments were removed. Interestingly, the filtration of the 48-hr-old inoculum of *mel3* did not appear to result in major differences in mortality (Fig. 13). This melanin-deficient mutant remained virulent and produced acute deaths in the experimentally infected mice. In contrast, 72-hr-old inoculum appeared to be greatly affected by the filtration step. Whereas the unfiltered inoculum resulted in 100% mortality within 3 days, the filtered inoculum did not kill any mice during this time period (Fig. 13). However, it is important to interpret these data with caution since the gauze filtration step was an inefficient means of yielding single cells for inoculation.

These data with broth-grown cultures, while obviously preliminary, could reflect our uncovering of other variables of virulence being associated with *W. dermatitidis*. For example, 1 $\times$ 10^8 cells per mouse is an enormous challenge to the host and approaches the upper limits of particulate load that can be managed by an intact host. Thus, any subtle variations in inoculum preparation or condition may be magnified *in vivo* when given at such a limiting concentration. As noted in the histopathological studies, lesions were abundant in mice injected with 1 $\times$ 10^7 cells of either the wild type or *mel3*. In those studies, we thought that the comparisons of the mortality data were most meaningful at an inoculum concentration of 3 $\times$ 10^7 cells or less per mouse. Such an approach may be prudent in interpretation of the preliminary data concerned with the broth-cultured inocula.

A fact clear from the plots of preliminary data from the broth-grown cultures is that mortality does appear to be fungus dose-dependent (Fig. 14). Also, the 48-hr-old inoculum appears to be more virulent than the 72-hr-old cultures. If these data prove to be reproducible, then it would be useful to compare the fungal strains and variables at more realistic inoculum sizes. For example, in selecting 1 $\times$ 10^7 cells per mouse, this is the concentration where histopathological lesions are clearly apparent, and thus the organism is present in large numbers in the tissue. At this concentration, the broth-grown inoculum of *mel3* appears to be less virulent than the melanized wild type when comparing the inocula grown for the same lengths of time.

With respect to the morphological form of the inoculum, it will be necessary to repeat these experiments using a more effective means of separating budding yeast cells from chains of cells or microcolonies that have not effectively completed cell separation. This was not effectively done in the preliminary studies, and may be complicating the data evaluation, although procedures based on brief sonication have been established recently to circumvent this problem (Cooper and Szaniszlo, unpublished). Also, one must consider that it is not merely the morphological form of the inoculum, but perhaps some biochemical events associated with the phase conversion that are critically important. The emergence of such events could be responsible for our data fluctuations. And, of course, one must consider the biological variations of the animal model itself. Because of this, future comparative experiments need to be conducted in tandem to allow for comparisons within the same experiment.

Critical to the evaluation of the role of melanin in the pathogenicity and virulence in the models described here will be injection of mice with Mel$^-$ strains restored in DHN melanin

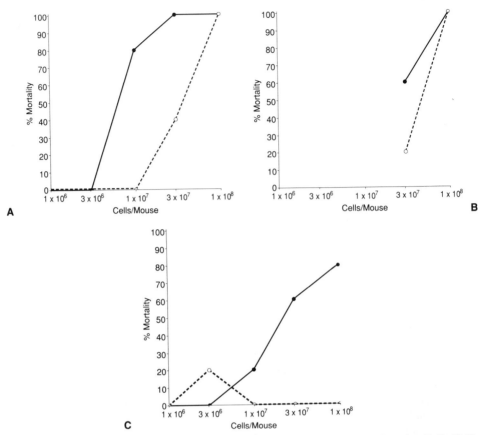

Figure 14. The effect of filtration upon virulence in mice inoculated with wild-type (A) and *mel3* (B, C) *W. dermatitidis*. Solid lines represent 48-hr cultures; dashed lines represent 72-hr cultures. Cultures were centrifuged, washed, and suspended in sterile saline to give the indicated inoculum concentrations administered i.v. to each mouse in groups of five.

Filtration (A, C) was done by passing the resuspended cells through layers of gauze; the inoculum in B was unfiltered. Mortality values were calculated after 21 days, with the last death occurring after 10 days.

deposition by use of a precursor such as scytalone, and the use of chemically demelanized strains such as those that are produced by treatment of *W. dermatitidis* with tricyclazole. This would allow a more direct evaluation of the role of melanin in the virulence and acute mortality of *W. dermatitidis*. Also, studies employing other Mel⁻ strains that accumulate larger quantities of intermediates, or shunt products, which lead to DHN melanin deposition (e.g., *mel2* and *mel3*), and of studies of the effects of these intermediates and the shunt products themselves on cell toxicity, will also be important for such future evaluations. The intermediates and shunt products of DHN melanin biosynthesis represent a variety of aromatic hydrocarbons. This suggests that wild-type strains that are less efficient in conversion of the majority of intermediates and resulting shunt products of DHN melanin biosynthesis to DHN melanin may make them more likely to produce higher levels of acute mortality than those that are more efficient in DHN melanin biosynthesis. Possibly such studies will eventually separate the acute death aspect of virulence associated with animal studies of dematiaceous fungi from those associated with the more chronic aspects relevant to the common forms of chromoblastomycosis and phaeohyphomycosis as they sometimes occur in more or less healthy humans.

8. REFERENCES

Ahmad, S., Johnson, R. J., Hillier, S., Shelton, W. R., and Rinaldi, M. G., 1985, Fungal peritonitis caused by *Lecythophora mutabilis, J. Clin. Microbiol.* 22:182–186.

Ajello, L., 1986, Hyalohyphomycosis and phaeohyphomycosis, two global disease entities of public health importance, *Eur. J. Epidemiol.* 2:243–251.

Bell, A. A., and Wheeler, M. H., 1986, Biosyntheses and functions of fungal melanins, *Annu. Rev. Phytopathol.* 24:411–451.

Cooper, C. R., 1988, Confirmation of cell-division-cycle (*CDC*) genes in *Wangiella dermatitidis* by segregant analysis of *mbc*-treated spheroplast-fusion products, *Abstr. Annu. Meet. Am. Soc. Microbiol.* 88:407.

Dixon, D. M., 1987, In vivo models: Evaluating antifungal agents, *Methods Findings Exp. Clin. Pharmacol.* 9:729–738.

Dixon, D. M., Walsh, T. J., Salkin, I. F., and Polak, A., 1986, Another dematiaceous fungus with neurotropic pathogenic potential in mammals, *Sabouraudia: J. Med. Vet. Mycol.* 25:55–58.

Dixon, D. M., Polak, A., and Szaniszlo, P. J., 1987, Pathogenicity and virulence of wild-type and melanin-deficient *Wangiella dermatitidis, J. Med. Vet. Mycol.* 25:97–106.

Dixon, D. M., Polak, A., and Walsh, T. J., 1988, Phaeohyphomycosis, in: *Proceedings of the Xth Congress of the International Society of Human and Animal Mycology* (J. M. Torres Rodriguez, ed.), J. R. Prous, Barcelona, pp. 84–92.

Dixon, D. M., Polak, A., and Conner, G. W., 1989, Mel⁻ mutants of *Wangiella dermatitidis* in mice: Evaluation of multiple mouse and fungal strains, *J. Med. Vet. Mycol.* 27:335–341.

Fader, R. C., and McGinnis, M. R., 1988, Infections caused by dematiaceous fungi: Chromoblastomycosis and phaeohyphomycosis, *Infect. Dis. Clin. N.A.* 2:925–938.

Geis, P. A., and Jacobs, C. W., 1985, Polymorphism in *Wangiella dermatitidis*, in: *Fungal Dimorphism* (P. J. Szaniszlo, ed.), Plenum Press, New York, pp. 205–223.

Geis, P. A., and Szaniszlo, P. J., 1984, Carotenoid pigments of the dematiaceous fungus *Wangiella dermatitidis*, *Mycologia* 76:268–273.

Geis, P. A., Wheeler, M. H., and Szaniszlo, P. J., 1984, Pentaketide metabolites of melanin synthesis in the dematiaceous fungus *Wangiella dermatitidis, Arch. Microbiol.* 137:324–328.

Kitamura, K., and Yamamoto, T., 1972, Purification and properties of an enzyme, zymolyase, which lyses viable yeast cells, *Arch. Biochem. Biophys. Res. Commun.* 153:403–406.

Kwon-Chung, K. J., and Rhodes, J. C., 1986, Encapsulation and melanin formation as indicators of virulence in *Cryptococcus neoformans, Infect. Immun.* 51:218–223.

Kwon-Chung, K. J., Tom, W. R., and Costa, H., 1983, Utilization of indole compounds by *Cryptococcus neoformans* to produce a melanin-like pigment, *J. Clin. Microbiol.* 18:1419–1421.

Kwon, Chung, K. J., Polacheck, I., and Popkin, T. J., 1982, Melanin-lacking mutants of *Cryptococcus neoformans* and their virulence for mice, *J. Bacteriol.* 150:1414–1421.

McGinnis, M. R., 1983, Chromoblastomycosis and phaeohyphomycosis: New concepts, diagnosis, and mycology, *Am. Acad. Dermatol.* 8:1–16.

McGinnis, M. R., and Fader, R. C., 1988, Mycetoma: A contemporary concept, *Infect. Dis. Clin. N.A.* 2:939–954.

Matsumoto, T., Padhye, A. A., Ajello, L., and Standard, P. G., 1984, Critical review of human isolates of *Wangiella dermatitidis, Mycologia* 76:232–249.

Oujezdsky, K. B., Grove, S. N., and Szaniszlo, P. J., 1973, Morphological and structural changes during the yeast-to-mold conversion of *Phialophora dermatitidis, J. Bacteriol.* 113:468–477.

Polacheck, I., Hearing, V. J., and Kwon-Chung, K. J., 1982, Biochemical studies of phenol-oxidase and utilization of catecholamines in *Cryptococcus neoformans, J. Bacteriol.* 150:1212–1220.

Rhodes, J. C., Polacheck, I., and Kwon-Chung, K. J., 1982, Phenoloxidase activity and virulence in isogenic strains of *Cryptococcus neoformans, Infect. Immun.* 36:1175–1184.

Roberts, R. L., and Szaniszlo, P. J., 1978, Temperature-sensitive multicellular mutants of *Wangiella dermatitidis, J. Bacteriol.* 135:622–632.

Salkin, I. F., McGinnis, M. R., Dykstra, M. J., and Rinaldi, M. R., 1988, *Scedosporium inflatum*, an emerging pathogen, *J. Clin. Microbiol.* 26:498–503.

Szaniszlo, P. J., Geis, P. A., Jacobs, C. W., Cooper, C. R., and Harris, J. L., 1983a, Cell wall changes associated with yeast-to-multicellular form conversion in *Wangiella dermatitidis*, in: *Microbiology—1983* (D. Schlessinger, ed.), American Society for Microbiology, Washington, D.C., pp. 239–244.

Szaniszlo, P. J., Jacobs, C. W., and Geis, P. A., 1983b, Dimorphism in pathogenic fungi, in: *Fungi Pathogenic for Humans and Animals*, Part A, *Biology* (D. H. Howard, ed.), Dekker, New York, pp. 323–436.

Wheeler, M. H., and Bell, A. A., 1987, Melanins and their importance in pathogenic fungi, in: *Current Topics in Medical Mycology*, Volume 2 (M. R. McGinnis, ed.), Springer-Verlag, Berlin, pp. 338–387.

Wheeler, M. H., and Stipanovic, R. D., 1985, Melanin biosynthesis and the metabolism of flaviolin and 2-hydroxyjuglone in *Wangiella dermatitidis*, *Arch. Microbiol.* **142**:234–241.

III

Host Response to Early Fungal Invasion

15

Invasion of Plants by Powdery Mildew Fungi, and Cellular Mechanisms of Resistance

James R. Aist and William R. Bushnell

1. INTRODUCTION

1.1. Scope and Purpose

Cereal powdery mildews are among the most intensely studied of plant diseases. Consequently, much of what is known about early host-parasite interactions has come from studies of these diseases. Most of this information has already been adequately reviewed within the past decade (Aist, 1981, 1983; Aist and Gold, 1987; Bushnell, 1982; Kunoh, 1982, 1987; Sherwood and Vance, 1982) and will not be emphasized here. We will focus, instead, on results of more recent research into early interactions between *Erysiphe graminis* and its cereal hosts, especially those interactions related to disease resistance, to provide an update on the state of our knowledge in this host-pathogen system.

1.2. Summary of Events during Early Invasion

1.2.1. Events preceding Haustorium Formation

Activity of conidia and subsequent interactions with host cells begin soon after the spore arrives at the surface of the host. The scenario presented here describes the general course of events as it occurs with *E. graminis* D.C. on compatible hosts (Aist and Israel, 1977; Bushnell and Bergquist, 1975; Fric, 1984; Kunoh, 1987; Kunoh *et al.*, 1978, 1979, 1985a).

Upon contact with a host surface, the conidium begins to secrete esterases in a fluid that obscures the surface features of the spore and spreads out over the host cuticle (Kunoh *et al.*, 1988b; Nicholson *et al.*, 1988). Surface waxes are dissolved by this fluid. The conidium also exudes a flocculent material that apparently glues it to the host surface within 20 min of contact.

James R. Aist • Department of Plant Pathology, Cornell University, Ithaca, New York 14853.   *William R. Bushnell* • USDA-ARS Cereal Rust Laboratory, University of Minnesota, St. Paul, Minnesota 55108.

At 1–2 hr after inoculation, the primary germ tube (PGT) emerges from the conidium and quickly makes contact with the host cuticle, attaching to it. The first visible host response is the development of a circular area of autofluorescence (early autofluorescence) in the host wall subjacent to the tip of the PGT, followed within 1–2 min by a subjacent cytoplasmic aggregate (CA) (Fig. 1A). Within 15 min of CA initiation, a similar, circular area of the host wall (the halo) exhibits different reactions to acidic and basic stains than those of the surrounding wall. These responses and effects occur about 2 hr after inoculation. During the next hour, a penetration peg emerges from the tip of the PGT and penetrates the host cuticle and cell wall; further visible host responses to this penetration process are not immediately apparent. However, host resistance to penetration by appressoria is temporarily and markedly enhanced at this time. Between 3 and 4 hr after inoculation, another germ tube emerges from the conidium. This germ tube will eventually produce a terminal appressorium and is therefore called the appressorial germ tube (AGT). During the fourth hour after inoculation, a papilla is deposited below the tip of the PGT (Fig. 1B).

Meanwhile, some biochemical host responses are occurring that could influence the outcome of the impending encounter between the appressorium and the host cell. During the fourth and fifth hours after inoculation, the activities of three key enzymes involved in phenolic metabolism—phenylalanine ammonia lyase (PAL), tyrosine ammonia lyase (TAL), and peroxidase—begin to increase markedly in the host cells and tissues (Green et al., 1975). Concomitantly, synthesis of cinnamic acid by the host may be enhanced (Kunoh, 1982).

Appressoria develop and mature from 6 to 10 hr after inoculation, and no new host responses have been reported to occur during this time. Subsequently, a series of events takes place at the hooked tip of the appressorium, called the primary appressorial lobe; they are much like the events that happened earlier at the tip of the PGT. First, a large CA is developed beneath the appressorial lobe, usually between 10 and 12 hr after inoculation (Fig. 1C), and produces a cell wall halo (Kunoh et al., 1985b). After another 15 to 75 min, the CA usually secretes a papilla

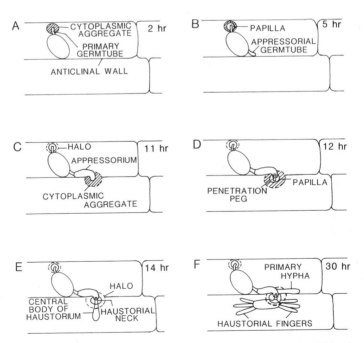

Figure 1. Primary infection structures of *Erysiphe graminis* and host responses, 2 to 30 hr after inoculation. Cross-hatched circles represent areas occupied by aggregates of host cytoplasm. Similar areas become halos in which the host wall structure and/or chemistry is altered.

below the appressorial lobe (Fig. 1D), 11 to 13 hr after inoculation. In close spatial and temporal association with papilla formation, a penetration peg grows from the appressorial lobe, through the host cuticle and wall, and into the cell lumen or into the newly forming papilla (Fig. 1D). In come cases, host resistance comes into play at this time and the peg stops growing.

1.2.2. Events following Haustorium Formation

As attack continues in susceptible cells, the penetration peg grows further, into the interior of the cell, where the tip enlarges to form the central body of a young haustorium (Fig. 1E). The CA disperses and papilla deposition is discontinued. The peg extension, which connects the appressorium to the young haustorium, is termed the haustorial neck. The haustorial central body swings sideways about 18 hr after inoculation and produces fingerlike lobes (Fig. 1F). The haustorium becomes delineated by a septum in the haustorial neck. Ultrastructurally, the haustorium resembles a mature, metabolically active hyphal cell with one nucleus and mitochondria that extend into the haustorial fingers (Hippe, 1985; Dahmen and Hobat, 1986).

The developing haustorium invaginates the host plasma membrane which persists in highly modified form and is termed the extrahaustorial membrane, forming the interface between the haustorial apparatus and host cytoplasm (Fig. 2). The haustorium continues to elongate and produce new fingers for about 5 days (Carver and Carr, 1978).

At about 24 hr after inoculation, a primary hypha (sometimes termed the elongating secondary hypha, but actually the first hypha to be formed) grows from the apex of the appressorium over the surface of the host epidermal cell (Figs. 1F, 2). The developing fungus now receives nutrients from the host, presumably through the haustorium. As growth on the host surface continues, the hypha branches, producing a hyphal colony with secondary and tertiary haustoria. The timing and number of these haustoria are influenced by length and intensity of daily light periods, probably as a consequence of host photosynthetic activity (Carver and Jones, 1988; Carver and Phillips, 1982). The colony begins to produce conidiophores and conidia after the tertiary haustorial generation.

By 4 to 6 days after inoculation, the colony sporulates heavily and becomes a highly competitive sink for nutrients that would otherwise be allocated to actively growing shoots and roots (Bushnell and Gay, 1978). Colonized tissue in susceptible hosts remains alive and undergoes a series of metabolic changes as reviewed elsewhere (Bushnell and Gay, 1978).

If the host is resistant, development of the attacking parasite can be retarded or halted completely at virtually any stage during infection and colonization of the host. We will now

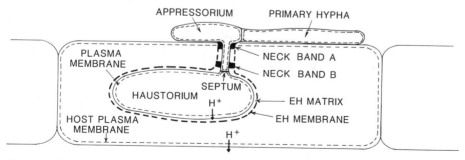

Figure 2.   A haustorium of *Erysiphe graminis* in a host cell with an attached appressorium and primary hypha on the host cell surface. Spencer-Phillips and Gay (1981) postulated that ATPase of the plasma membranes of the haustorium and host cell extrude protons (arrows), driving active transport of ions and molecules in the opposite direction, toward the haustorium. The extrahaustorial (EH) membrane lacks ATPase activity. The EH matrix is sealed by neck bands, preventing loss of matrical solutes along the haustorial neck.

discuss fungus development in relation to host responses, especially with respect to mechanisms of host resistance.

2. PRIMARY GERM TUBES

2.1. Development

The development of PGTs has been clearly summarized by Kunoh (1982). *E. graminis* is apparently the only species of *Erysiphe* that produces a PGT. Carver and Ingerson (1987) showed that PGTs will be formed even in the absence of any host structure, chemicals, or substantial conidium-surface contact. Thus, their development is not a response to deposition of the conidium onto the host. Although attachment to the host cuticle usually occurs within ca. 1 hr of PGT emergence, the PGT retains the potential for both further growth and attachment to the host for several hours (Carver and Bushnell, 1983).

2.2. Host Responses

Several responses of host cells to the attack by PGTs are now well documented. As noted above, these include the development of a circular area of host cell wall autofluorescence followed by the accumulation of a large seething mass of host cytoplasm (cytoplasmic aggregation) below the tip of the PGT (Kunoh, 1982). Subsequently, a papilla is formed at the site of the penetration pore (Kunoh, 1982). Carver and co-workers (1987) have shown that Si accumulates in these papillae, and Wolf and Fric (1981) showed that they contain protein. Early biochemical host responses—increased activity of PAL, TAL, and peroxidase (Green *et al.*, 1975) and enhanced synthesis of cinnamic acid (Kunoh, 1982) and phytoalexin (Oku and Ouchi, 1976)—are manifested apparently as autofluorescent halos and papillae (Thordal-Christensen and Smedegaard-Peterson, 1988b) and may be induced by the attempted penetration by the PGT, although such a cause-and-effect relationship has not been demonstrated. Nevertheless, this early activation of phenolic metabolism is likely to be important later, when appressoria attempt to penetrate, since autofluorescent papillae that contain phenolics have been frequently implicated in resistance to penetration (see Section 5.1).

What is certain is that the attempted penetration by the PGT does induce resistance to the subsequent penetration attempt by the appressorium (Kunoh, 1982; Thordal-Christensen and Smedegaard-Petersen, 1988a, b; Woolacott and Archer, 1984), even in compatible plant-fungus combinations. This interaction may also account for the early (6 hr after inoculation) induction of resistance by incompatible races (Oku and Ouchi, 1976), but it probably occurs too late to explain all of the induced resistance reported by Smedegaard-Petersen and co-workers (Cho and Smedegaard-Petersen, 1986; Thordal-Christensen and Smedegaard-Petersen, 1988a). In the latter studies, induced resistance was detected as early as 0.5 hr after inoculation of the inducer race, and it was well under way by 1 hr. This time frame makes it more likely that this resistance is induced by something that happens almost immediately upon deposition of the conidia, such as the early and rapid secretion of esterases by the conidia before their germination (Kunoh *et al.*, 1988b; Nicholson *et al.*, 1988).

2.3. Functions

What benefits the fungus may derive by forming a PGT has been a subject of great interest in recent years. The question has taken on new significance since the demonstrations that host cells become less susceptible to penetration from appressoria after they have been attacked by a PGT; such an effect would produce negative selection pressure that would have done away with PGTs long ago were it not for some necessary function they perform for the fungus. There is some evidence that the PGT may absorb elements, such as Ca and Si, from host cells (Kunoh, 1982; Carver et al., 1987), and absorption of Ca could be required to support the tip growth that occurs during subsequent appressorium development and penetration (Aist and Gold, 1987; Gold et al., 1986). An experimental study by Carver and Bushnell (1983) has provided strong evidence that the PGT absorbs water from the host cell, and that such water uptake is essential for the proper development and functioning of appressoria when the relative humidity is not high. Finally, Kunoh (1982) has suggested that the attachment of the PGT to the host surface helps to anchor the conidium, a function that could be very significant under windy or rainy conditions in the field.

3. CYTOPLASMIC AGGREGATION

3.1. Nature

A CA forms as more host cytoplasm streams into the region below the appressorial lobe than streams out of it. The chemical composition of the CA is important because the CA is present before and during penetration and may contain fungitoxic chemicals that are either infused into the adjacent host cell wall (halo) region or deposited as part of the papilla. Kita and co-workers (1980a) reported that the CA contains lipid, protein, and polyphenols and may be autofluorescent (Kita et al., 1981). Later, hydrolytic enzymes were localized in spherical organelles in the CA (Takahashi et al., 1985). Most recently, Russo and Bushnell (1989) found that CAs gave a positive reaction for carbohydrates, including callose, whereas they were negative for suberin, lignin, cellulose, and pectin. Of these CA components, the ones most likely to become involved in resistance to fungal penetration are phenolics, hydrolytic enzymes, and callose.

3.2. Stimuli

The nature of the stimuli that induce the CA remains elusive. Aist (1983) reviewed this topic and listed pressure from a blunt needle and needle puncture as stimuli. Recently, Russo and Bushnell (1989) have found that when a microneedle is inserted into a barley coleoptile epidermal cell, the CA forms only in the punctured cell, not in an adjacent cell. These observations suggest that the plant cell is responding to a physical, possibly tactile, stimulus, and it is reasonable to speculate that either pressure developed by an appressorium (Heintz, 1986) or the physical intrusion of a penetration peg may be sufficient to stimulate formation of a CA (Aist, 1976). However, Russo and Bushnell (1989) obtained evidence that even physical induction of a CA with a microneedle may be mediated via a chemical signal: under certain conditions, the CA would first form downstream from the puncture, then migrate upstream to the puncture site.

In the case of penetration from appressoria, it is clear that there is also a chemical signal involved. Russo and Bushnell (1989) confirmed earlier reports (Aist, 1976) that appressoria often induce a CA in both the attacked and an adjacent cell. Kunoh (1982) suggested that this signal may be related to the host wall autofluorescence that immediately precedes the initiation of the CA. In a detailed examination of this question, Kunoh and co-workers (1985a) showed that the first CA is induced long before the penetration peg is initiated in a nonhost powdery mildew system, suggesting that a chemical inducer may be released by the appressorium before the peg is extended. The chemical nature of this inducer should be investigated.

3.3. Mechanisms

The cellular processes involved in the CA response are unknown, but several probably occur, in a sequence, from the time the stimulus is received until the aggregate is well developed.

The calcium ion (Ca^{2+}) has been implicated recently as a regulator of this response. Exogenous Ca^{2+} enhanced the host cell wall autofluorescence that immediately precedes cytoplasmic aggregation (Kunoh, 1982; Kunoh *et al.*, 1985b) and also enhanced the size of the CA (Kunoh *et al.*, 1983). Chlortetracycline (CTC), a Ca^{2+} chelator and ionophore, delayed onset of the cytoplasmic aggregate and also reduced its duration in barley coleoptiles bearing an *ml-o* gene for resistance to powdery mildew (Gold *et al.*, 1986).

How Ca^{2+} can bring about the formation of a CA is uncertain, but it is known that a large influx of Ca^{2+} can stop cyclosis in plant cells (Williamson and Ashley, 1982). Such an influx, if it were localized to the penetration site, could conceivably stop cyclosis locally and cause the buildup of a mass of cytoplasm below the appressorial lobe. Thus, Ca^{2+} may be functioning as a "second messenger" that transmits the signal given by the original stimulus.

3.4. Functions

A main function of the CA is apparently to physically and chemically alter the penetration site in a permanent way. Kunoh and co-workers (1985b) reported that the CA precedes the appearance of host wall halos by at least 15 min. This result suggests to us that the CA may be responsible for halo formation, and this inference is consistent with the diameter of halos, which is typically somewhat larger than that of CAs. Another function of the CA is to deposit papillae, as summarized earlier (Aist, 1976). Recent observations by Russo and Bushnell (1989) support the deposition of wound plugs, in addition to papillae, by the CA. The localization of phenolics, hydrolytic enzymes, and callose in the CA (see Section 3.1) is consistent with this proposed function, as these components are also found in halos and papillae (see Sections 4.1 and 5.1).

4. HALO FORMATION

4.1. Nature

Halos may be defined as any locally modified region of the host cell wall around a penetration site (Aist, 1983). Because halos are usually well developed before the penetration peg is initiated (see Aist and Israel, 1977; Kunoh *et al.*, 1985b), these host wall modifications could impede growth of the penetration peg. Recent studies have shown that halos contain silicon (Carver *et al.*, 1987), proteins (Fric, 1984) including hydrolytic enzymes (Takahashi *et al.*, 1985), and phenolics (Aist *et al.*, 1988; Aist and Israel, 1986; Kunoh *et al.*, 1983, 1985b). Autofluorescence of halos (Koga *et al.*, 1980) is indicative of the phenolics in them (Aist and Israel, 1986), and Smart and co-workers (1986a) showed that raising the pH of the incubation medium caused a "negative halo" effect in which host wall autofluorescence was low or nil in the halo but enhanced elsewhere. This result suggests that the composition of phenolics in the halo is different from that in nonhalo regions of the host wall.

An interesting feature that remains unexplored is the pH of the cell wall in the halo. Conceivably, the pH of the halo could be rendered unsuitable for growth of the penetration peg by localized infusion or secretion of appropriate chemicals from the CA. While direct pH measurements have yet to be made, Kunoh and co-workers (1985b) found that halos consistently react with (positively charged) acidic dyes, whereas the nonhalo wall reacts with (negatively charged) basic dyes. Such an alteration in the binding of charged molecules by host wall components could affect the molecular communication between host and pathogen prior to initiation of the penetration peg.

4.2. Stimuli

The stimulus for halo formation seems to come from the appressorium before the penetration peg is initiated (Kunoh *et al.*, 1985a), but the nature and identity of it is unknown. The appearance of autofluorescence in the incipient halo before the CA is initiated could be related to subsequent halo induction. On the other hand, the demonstrations (Sargent and Gay, 1977; Skou, 1985) that halos are produced in response to mechanical wounds suggest that the stimulus could be physical, rather than chemical, in nature. Skou (1985) found also that prior wounding of host cells can increase the size of halos formed subsequently in response to attack by *E. graminis*.

4.3. Mechanisms

As mentioned above (Section 3.4), the halo is apparently formed by the CA. Calcium has some dramatic effects on halos: it enhances their autofluorescence (Kunoh *et al.*, 1983) and inhibits their subsequent detection by the zinc chloriodide stain (Kunoh *et al.*, 1985b). The effect on autofluorescence is probably a direct effect of Ca^{2+} on halo formation by the host CA, but the altered staining properties may be a Ca^{2+} effect on the ability of the fungus to degrade the host cell wall, since the zinc chloriodide stain is preferential for cellulose.

Investigations of host responses in barley mutants with the *ml-o* gene for resistance to powdery mildew have demonstrated genetic control of halo formation. In the mutants, halos (Skou *et al.*, 1984) are initiated earlier and attain a much larger size than those in the mother varieties (Skou, 1982b; Skou *et al.*, 1984). It was suggested that the mutation disrupts a feedback mechanism that normally limits the size to which a halo will grow (Skou, 1982a).

4.4. Functions

The presence of silicon, hydrolytic enzymes, and phenolics in halos (see Section 4.1) supports the inference that halos have the potential to prevent a penetration peg from traversing the host cell wall. And, in *ml-o* mutants of barley, the resistance to primary penetration is correlated with earliness of the halo response (Skou *et al.*, 1984). However, Aist and co-workers (1988) were able to overcome this resistance to penetration, using chlortetracycline, without affecting the halo response. Apparently, the halo is not responsible for *ml-o* resistance. Similarly, Kita and co-workers (1980b) found that the autofluorescent halo in a susceptible cultivar was present whether or not the penetration attempt was successful. The suggestion by Sargent and Gay (1977) that the halo functions to reduce water loss from the host cell at the site of penetration is an interesting possibility, at least for cereal powdery mildews.

5. PAPILLA FORMATION

5.1. Nature

A papilla is a lump of heterogeneous materials deposited by the host cytoplasm between the plasma membrane and cell wall at the site of fungal attack (Aist, 1983). Evidence continues to mount that papilla formation is a major and widespread, cellular disease resistance mechanism (Aist, 1976; Aist, 1983; Aist and Gold, 1987; Sherwood and Vance, 1982). Consequently, interest in the nature of papillae has been increasing.

Several entirely new findings deserve mention here. Takahashi and co-workers (1985) showed that hydrolytic enzymes are highly concentrated in a central layer of oversize, preformed papillae that resist penetration by *E. graminis* f. sp. *hordei* but are less concentrated in normal-size papillae that are less resistant. Ebrahim-Nesbat and co-workers (1986) localized *N*-acetylglucosamine in the outer layer of one structural type of papilla, raising the likelihood that the fungus contributes some components to the papilla. The permeability of papillae was explored by Smart and co-workers (1987), who obtained some evidence that molecules larger than 500 Da may not be able to diffuse freely through papillae. Such a restriction would block the movement of elicitors of disease resistance responses, some of which are about 1000 Da. A more complete examination of papilla permeability should be done.

More or better information is now available on several other aspects of the nature of papillae. Of course, the presence of phenolics in papillae has been increasingly emphasized. Strong correlations have been found between the presence or amount of autofluorescence (indicating phenolics) in papillae and failure of the penetration attempts (Aist *el al.*, 1988; Aist and Israel, 1986; Kita *et al.*, 1980a, b, 1981; Koga *et al.*, 1980; Russo and Bushnell, 1989; Tosa and Shishiyama, 1984b; Toyoda, 1983). Autofluorescence of papillae has now been linked to resistance to primary penetration at the gene (Aist *et al.*, 1988), cell (H. Koga; W. R. Bushnell, and R. J. Zeyer, unpublished data; Koga *et al.*, 1980), and cultivar (Koga *et al.*, 1980) levels. Although the autofluorescent materials are usually distributed evenly throughout the papilla, Koga and co-workers (1980) and Aist and Israel (1986) found that some large papillae where penetration failed had enclaves of more highly autofluorescent compounds. The chemical nature of the phenolics remains, for the most part, a mystery, but Kita and co-workers (1980a) have localized polyphenols in papillae, and Aist and co-workers (Aist *et al.*, 1988; Aist and Israel, 1986) have identified phenolic compounds, in papillae known to be resistant to penetration, as phenylpropanoids. Smart and co-workers (1986b) showed that the autofluorescent materials in oversize barley papillae are unnecessary for the papillae to be able to resist fungal penetration and inferred that there may be several, effective antifungal factors in papillae.

Further work has also been done on the elemental composition of papillae. Studies of papillae that were not chemically processed before analysis have confirmed the routine occurrence of Ca as a major element in papillae (Kunoh *et al.*, 1986; Marshall *et al.*, 1985). Kunoh and co-workers (1986) also showed that Ca and P are the major elemental components in oversize papillae that prevent fungal penetration, and they suggested that calcium phosphate accumulations in these papillae may account for this resistance.

Brief mention should be made of three other important accomplishments. Carver and co-workers (1987) conducted a time-course study of Si accumulation at penetration sites and found that Si accumulates in papillae during the penetration phase of fungal development. This means that Si is present early enough to affect the outcome of the penetration attempt. However, the same study provided evidence that Si in papillae is not necessary for them to be resistant to fungal penetration, leading to the inference that papillae may have multiple resistance factors. Smart and co-workers (1986a) provided unequivocal evidence that barley papillae contain β-1, 3-glucans, the characteristic polysaccharide of the common papilla constituent, callose. In a second paper (Smart *et al.*, 1986b), callose was shown to be unnecessary for oversize papillae in barley to resist fungal penetration. Finally, Ebrahim-Nesbat and co-workers (1986) localized papilla proteins at the ultrastructural level and found that they are concentrated in small, spherical granules within papillae and in the outermost layer of the papilla, adjacent to the host plasma membrane. The functional significance of this arrangement of proteins remains obscure.

Papillae are similar to wound plugs, which are wall appositions induced by wounding (Aist, 1976, 1983). Until recently, a careful histochemical comparison of papillae and wound plugs had not been made. Russo and Bushnell (1989) made such a comparison using epidermal cells of barley coleoptiles and found both similarities and differences. Both structures were positive for carbohydrates, callose, and protein, and negative for lignin and suberin. Only wound plugs were positive for cellulose and pectin, whereas only papillae were positive for phenolics and a basic staining material. Interestingly, the two components unique to papillae have often been implicated in resistance of barley to powdery mildew (Aist *et al.*, 1988; Aist and Israel, 1986). Plants seem to be able to tailor-make a wall apposition to either plug a wound or deter a fungal attack.

5.2. Stimuli

Because papilla formation follows a chain of host responses (i.e., early autofluorescence, cytoplasmic aggregation, halo formation), it may be anticipated that the stimuli for the prior responses are similar in some respects to those that lead to papilla formation. Thus, the stimulus apparently comes from the fungus as it penetrates the host wall, and there is reason to believe that it may be physical, chemical, or both, in nature (Aist, 1976, 1983). It has also been inferred that there may be at least two different chemical stimuli, one leading to formation of the CA, the other to papilla formation (Aist *et al.*, 1979). Recent studies have shed little light on this topic, but the histochemical results of Russo and Bushnell (1989) are pertinent. The compositional differences they found between papillae and wound plugs lead to the inference that ". . . fungal attack generates inducing factors not produced by mechanical penetration" (Russo and Bushnell, 1989).

5.3. Mechanisms and Regulation

The mechanisms thought to be involved directly in papilla formation have been reviewed extensively (Aist, 1976, 1983; Aist and Gold, 1987). One significant new finding is that the lysosomal system of the plant cell is somehow involved, because hydrolytic enzymes that are characteristic of that system are deposited during papilla formation (Takahashi *et al.*, 1985). Small granules, containing hydrolytic enzyme activity, were observed in the CA and are most likely lysosomal vesicles that fuse with the host plasma membrane, adding their contents to the growing papilla.

Aist (1983), Aist and Gold (1987), and Sherwood and Vance (1982) discussed several aspects of the regulation of papilla formation, and the reader is referred to those reviews. The major recent development has been in the demonstration that papilla formation is a Ca^{2+}-mediated secretion process (Aist *et al.*, 1988 Aist and Gold, 1987; Bayles and Aist, 1987; Gold *et al.*, 1986; Marshall *et al.*, 1985). In addition to regulating secretion *per se*, Ca^{2+} may also regulate papilla formation by its activation of β-1, 3-glucan synthetase (Bayles and Aist, 1987), since β-1, 3-glucan is a major component of callose in papillae. Aist and Gold (1987) hypothesized that a defective Ca^{2+}-regulating mechanism in *ml-o* mutants is responsible for the enhanced papilla response that is linked to powdery mildew resistance, which is consistent with the earlier suggestion by Jørgensen (1983) that ". . . the *ml-o* resistance is due to a destruction of a functional wild-type gene regulating the formation of the cell wall appositions."

Another concept that has emerged recently is that host cells may be conditioned for a more vigorous, and effective, papilla response by prior stimulation. Sahashi and Shishiyama (1986) and Thordal-Christensen and Smedegaard-Petersen (1988b) obtained strong correlative evidence that increased papilla formation is a major factor of induced resistance in barley to powdery mildew. And Kunoh and co-workers (1988a) and Thordal-Christensen and Smedegaard-Petersen (1988b) inferred that a previous attack of a host cell by *E. graminis* conditioned that cell toward effective papilla formation in response to a subsequent attack by an appressorium. Thus, papilla formation is clearly implicated as a mechanism of induced resistance in cereal powdery mildews. How this enhanced papilla response is regulated is not known.

5.4. Functions

Most of the recent work on the function of papilla formation has added to the growing and substantial evidence (Aist, 1976, 1983; Aist and Gold, 1987; Sherwood and Vance, 1982) that papilla formation is a major mechanism by which cereal hosts block primary penetration attempts by powdery mildew fungi. Correlative evidence has implicated papilla formation in resistance at several levels of specificity (Table I, Section 8). Experimental studies on the role of papilla formation in resistance to powdery mildew have been focused lately on the *ml-o* gene in barley.

Most of this evidence has demonstrated a very strong link between one or more aspects of papilla formation and resistance to penetration (Aist *et al.*, 1988; Gold *et al.*, 1986; Stolzenburg *et al.*, 1984a, b). However, the main effect of low-speed centrifugation was to inhibit papilla formation without an anticipated, corresponding increase in successful penetrations, and this anomaly remains a mystery (Stolzenburg *et al.*, 1984b).

How papillae resist a fungal penetration attempt is a question of great interest. The characteristics or components of papillae that may enable them to resist penetration were discussed in Section 5.1, and include restricted permeability, hydrolytic enzymes, phenolics, and chemical elements such as Si. Moreover, it has been inferred that papillae have several inhibitory components, any one of which can make them resistant to penetration by fungi (Aist *et al.*, 1988; Smart *et al.*, 1986b; Carver *et al.*, 1987). Finally, the early timing of papilla formation in *ml-o* barley mutants resistant to powdery mildew is a key factor in the effectiveness of the papillae in blocking penetration attempts (Gold *et al.*, 1986; Skou *et al.*, 1984). Thus, the concept has now been established experimentally that an effective papilla response requires that antifungal factors be present in the papilla during the early stages of its formation.

6. HAUSTORIA

6.1. Nutrient Uptake

If the fungus avoids papilla-associated resistance, the penetration peg enters the cell and produces a haustorium as described earlier. There, the haustorium forms with the host protoplast a unique, specialized interface which allows the fungus to compete effectively with the plant for nutrients. Radioactivity from $^{32}PO_4^{3-}$ or $^{35}SO_4^{2-}$ applied to excised leaves begins to accumulate in the fungus in large amounts when primary haustoria are partly developed, 18–22 hr after

Table I. Levels of Specificity in Relation to Development
of Erysiphe graminis and Host Response

Specificity level	Fungal development	Germ tubes and/or appressoria fail to develop properly	No haustorium	Growth of haustoria and/or hyphae stopped or retarded	Growth of haustoria and/or hyphae retarded
	Host response	None	Papilla	HR	None
		Occurrence of combination[a]			
Nonhost		+ +	±	±	±
Host inappropriate for *forma specialis*		−	+ +	+	±
Host appropriate for *forma specialis*					
Race-nonspecific					
Cell-specific		−	+ +	−	−
Organ-specific		±	+ +	+	−
Partial resistance		±	+ +	+	+ +
Race-specific		−	±	+ +	±

[a] −, combination does not occur; ±, reports of combination rare, or reports inconsistent; +, combination commonly occurs; + +, combination predominates.

inoculation, and continues to accumulate as hyphae proliferate on the host surface (Martin and Ellingboe, 1978; Stuckey and Ellingboe, 1975). Direct evidence for nutrient uptake by haustoria of *E. pisi* was obtained by Manners and Gay (1978) by labeling infected pea leaves with $^{14}CO_2$ in light, isolating haustoria *en masse* from the leaves, and measuring the amount of ^{14}C therein. Each haustorium was isolated as a complex which included the intact extrahaustorial membrane, the extrahaustorial matrix, and most of the haustorial neck (Fig. 2). The experiments were done with *E. pisi* because it produces compact haustoria which are less subject to injury during isolation than are the elongate haustoria of *E. graminis*. Labeled sucrose accumulated rapidly in haustoria even though it was not the principal sugar in the host, suggesting that sucrose was the principal carbohydrate transported from host to parasite (Gay and Manners, 1981; Manners and Gay, 1982). Rates of sucrose flux into haustoria were higher than could be explained by passive diffusion down a concentration gradient from host to parasite (Manners and Gay, 1982, 1983).

Spencer-Phillips and Gay (1981) devised a model which partially explains how the plasma membranes of the haustorium and host cell work in concert to promote flow of sugars and other substances into haustoria. They demonstrated cytochemically that the extrahaustorial membrane lacks the ATPase activity normally found on the plasma membrane of the host cell and (usually) of the haustorium. Apparently, ATPase activity is lost or never develops in the portions of the host plasma membrane that are invaginated by the developing haustorium. The transition between the membrane domains with and without ATPase activity occurs at neck band A (Fig. 2).

The absence of ATPase activity in the extrahaustorial membrane allows ATPase in the plasma membranes of the host and haustorial cells to extrude protons in the same direction (arrows, Fig. 2) without opposition by proton extrusion at the extrahaustorial membrane. The proton flow provides the driving force for active transport of ions and molecules in the opposite direction toward the haustorium. As ions and other substances move through the extrahaustorial matrix, neck bands (Fig. 2) act as a seal to keep them from leaking into apoplastic wall regions of the host cell. The impermeability of neck bands has been demonstrated with uranyl ions and an apoplastic dye (SITS) applied to isolated haustorial complexes (Gay and Manners, 1981, 1987).

Physiological evidence in support of the model was obtained in elegant experiments by Gay and co-workers (1987) using barley epidermal tissue infected with *E. graminis* f.sp. *hordei*. Fluorescein, which increases in fluorescence over the range pH 5.5–7.0, was introduced into tissues (as fluorescein diacetate) to serve as an intracellular pH indicator for host cells and haustoria. When availability of ATP was reduced by treatment of tissue with FCCP, fluorescence in both host cells and haustoria diminished sooner than in untreated controls. This conformed to the expectation that the pH of host cell and haustorium would decrease if proton extrusion diminished. Conversely, fluorescence increased when tissues were treated with fusicoccin, which specifically stimulates proton extrusion by membrane ATPase in higher plants. However, fusicoccin produced no effect when applied directly to haustorial complexes after host cells were torn open, in line with reports that fusicoccin has no effect on fungal cells and with the absence of ATPase from the extrahaustorial membrane. Gay and co-workers (1987) deduced that the increase in haustorial pH in intact tissue resulted from stimulation of ATPase at the *host* plasma membrane, in support of the model.

Mendgen and Nass (1988) used a potentiometric, cyanine dye to assess activity of haustorial mitochondria in response to sugars applied to host cells. Glucose caused a rapid decrease in fluorescence of dye-treated haustoria, apparently because increased dye uptake led to dye aggregation within mitochondria. Glucose gave a more rapid response than did other sugars. The glucose may have moved to haustoria and directly stimulated mitochondrial activity or, in line with the model of Gay and co-workers, may have stimulated host ATPase activity which, in turn, enhanced movement of sugars into haustoria. In any case, potentiometric dyes are a promising way to monitor mitochondrial activity in cells of both host and parasite (Bushnell *et al.*, 1987; Liu *et al.*, 1987).

6.2. Extrahaustorial Membrane

Other than the absence of ATPase activity, the properties of the extrahaustorial membrane are poorly understood in relation to nutrient flow toward the haustorium. The extrahaustorial membrane is thicker than the host plasma membrane (Bracker, 1968; Gay and Manners, 1981) and has been shown cytochemically to contain cellulosic polysaccharides, α-linked polysaccharides containing glucose and mannitol, and protein (Chard and Gay, 1984; Ebrahim-Nesbat *et al.*, 1982). The thickening of the membrane is distinct from papilla formation since cytoplasmic aggregates are not involved and the extrahaustorial matrix and membrane both lack callose as found in papillae.

The membrane around haustorial complexes isolated from pea was only slightly permeable to water as judged from slow swelling of the entire complex (Gil and Gay, 1977), and virtually impermeable to radioactively labeled sugars (Manners and Gay, 1980) or uranyl ions (Gay and Manners, 1987), indicating that small molecules or ions were unlikely to cross the membrane in significant amounts by passive diffusion. In contrast, the extrahaustorial membrane of haustoria from barley is highly permeable to water as indicated by rapid swelling of complexes isolated *en masse* and placed in water (Manners and Gay, 1977) or exposed individually to water by microsurgical opening of host cells (Bushnell, 1971). Shrinking of expanded complexes from barley over 10–15-min periods indicated that solutes and water diffused out of the extrahaustorial membrane at significant rates (Bushnell, 1971). If haustorial complexes isolated from pea were treated with cellulase and pectinase enzymes, they became highly permeable to water (Gil and Gay, 1977), behaving like complexes from barley. The uptake of sugars by haustoria *in situ* indicates haustoria with thickened extrahaustorial membranes in pea leaves are functional in nutrient uptake. Yet the thickening of the membranes remains puzzling and of uncertain function.

For haustoria of several rust fungi, the extrahaustorial membrane is not thickened as in powdery mildews and has been seen ultrastructurally to connect to tubular complexes which, in turn, connect to endoplasmic reticulum of host cytoplasm (Harder and Chong, 1984). The tubular complexes are thought to have a secretory role in movement of substances from host cytoplasm to the extrahaustorial matrix. Such complexes have not been reported for powdery mildews, although host vesicles and mitochondria have been reported to be in close association with the extrahaustorial membrane (Akutsu *et al.*, 1978; Kunoh and Ishizaki, 1973). Modern cryosubstitution methods which have given excellent ultrastructural images of *E. graminis* haustoria (Dahmen and Hobot, 1986; Hippe, 1985) have yet to be applied successfully for preserving host cytoplasmic structures near the extrahaustorial membrane.

6.3. Extrahaustorial Matrix

The extrahaustorial matrix, which lacks discernible ultrastructure, has been shown cytochemically to contain β-linked polysaccharides (Chard and Gay, 1984; Gil and Gay, 1977; Manners and Gay, 1983) but neither pectin compounds (Chard and Gay, 1984; Ebrahim-Nesbat *et al.*, 1982) nor callose (Gay and Manners, 1981). With *E. graminis*, the matrix is usually not much thicker than the haustorial wall (Hippe, 1985), especially near tips of haustorial fingers. However, the matrix is sometimes thick enough to be seen by light microscopy around parts of the haustorial central body. The matrix increases in volume under conditions unfavorable for growth of attached hyphae, after treatment with certain fixatives, or under hypotonic osmotic conditions (Bushnell, 1971).

So far, the extrahaustorial matrix has not been shown to have a functional role in movement of nutrients from host cell to haustorium other than as an aqueous pathway for passive diffusion. While insoluble label from $^{14}CO_2$-treated leaves is found in the matrix, it occurs only at densities similar to those in the haustorium proper (Spencer-Phillips and Gay, 1980). Gay and Manners

(1987) found that peroxidase would not enter the matrix of isolated haustorial complexes from pea, including complexes which had a broken extrahaustorial membrane. They postulated that the matrix is a gel with a pore size too small to pass the peroxidase protein molecule. Swelling and shrinking of the matrix were attributed to changes in ionic environment instead of osmotic effects. In contrast, swelling and shrinking of the matrix of *E. graminis* seem to occur osmotically without change in ionic environment (Bushnell, 1971). Nevertheless, a gel could serve to protect host or parasite from large potentially harmful molecules originating from the associated organism.

7. HYPERSENSITIVE RESPONSE

7.1. Nature

Resistance to powdery mildews is often expressed after haustorium formation by death of host cells in what is termed the hypersensitive response (HR). HR can be limited to the cell in which a primary haustorium forms, but it can occur in many cells of both the epidermis and uninvaded cells in subtending mesophyll if HR is delayed. Part or all of the epidermal cells above the dead mesophyll may remain alive and the fungus continues to grow slowly for several days (Koga *et al.*, 1978; White and Baker, 1954; Toyoda *et al.*, 1979; Hyde and Colhoun, 1975). Sometimes, HR occurs only in scattered epidermal cells where secondary haustorium formation is attempted while the primary infected cell remains alive (Carver and Williams, 1980; Tosa and Shishiyama, 1984a).

After HR, individual dead cells autofluoresce brightly under blue light excitation. First used in the study of rust diseases, autofluorescence was adopted for use in powdery mildews by Mayama and Shishiyama (1976) as a way of detecting HR in whole leaf preparations and has been widely used since. Using plasmolysis and neutral red uptake as criteria for cell viability, Koga and co-workers (1988) confirmed that autofluorescence was a reliable indicator of cell death. They postulated that vacuolar phenols are released as cells die, leading to autofluorescence. Phenols were also implicated in accumulation of silicon which occurred at HR sites only after the host cells died.

Visible events leading to HR start with disruption of long-distance patterns of cytoplasmic streaming in host cells, quickly followed by a complete halt in streaming (Bushnell, 1981). The host cell remains quiescent but turgid for 1–2 hr, then collapses. The host nucleus collapses a few minutes before the entire cell collapses (W. R. Bushnell, unpublished data). These events are similar whether they occur first in a cell with a haustorium or in a neighboring cell. The fungal haustorium collapses a few minutes before or after host cell collapse. Signs of impending cell collapse other than the halt in streaming may include localized autofluorescence of host cell walls or staining of cells with trypan blue (Wright and Heale, 1988).

7.2. Mechanisms

HR usually does not occur in powdery mildew of cereals unless at least one haustorium is present at the infection site. Is a haustorium required for induction of HR? Unfortunately, haustoria are difficult to see in collapsed epidermal cells of whole leaf mounts, especially if the haustoria are not fully developed (Asher and Thomas, 1983; Carver, 1986; Carver and Williams, 1980). Because of this limitation, reports suggesting that HR occurred in cells lacking haustoria are difficult to assess.

To ascertain how haustorium formation is related to HR with *ml–a* resistance in barley, Koga and colleagues (1989) determined the number and size of haustoria at 3-hr intervals before and during HR. The developmental sequence indicated that haustoria were present at most, if not all, HR sites; although, again, haustoria could not be seen in many cells after collapse. A similar conclusion can be reached from the detailed patterns of haustorium development described by Haywood and Ellingboe (1979) for each of several genes for resistance in wheat (assuming that deformed and collapsed haustoria and increased staining of host cells were indications of HR). We conclude, tentatively, that HR does not occur unless the penetration peg has entered the host cell and at least begun to produce a haustorium. Possible exceptions deserve further investigation, especially in nonhosts under attack by *E. graminis* (Johnson *et al.*, 1982).

Several lines of evidence indicate that host cells have an active role in the processes leading to HR. Brief heat-shock treatments before inoculation, which stopped cytoplasmic streaming for several hours, prevented HR and allowed normal development of haustoria (Hazen and Bushnell, 1983). Cytochalasin B, which disrupted normal streaming patterns, likewise prevented HR, suggesting that streaming or cytoskeletal elements in the host are necessary for HR (Hazen and Bushnell, 1983). Sodium azide, which inhibits cytochrome oxidase, also prevented HR (Hazen, 1981), indicating that respiratory activity of the host is required. Although several mRNAs and proteins are synthesized to a greater extent in resistant than in susceptible tissues (Davidson *et al.*, 1988; Manners *et al.*, 1985), we lack definitive experimental evidence that mRNA and/or protein synthesis is required for HR in powdery mildews. One of the activated genes codes for β-1, 3-glucanase (W. Jutidamrongphan, G. Mackinnon, J. M. Manners, R. S. Simpson, and K. J. Scott, personal communication) which has been found as a response product in other plant diseases (Scholtens-Toma *et al.*, Chapter 11).

Whatever the role of host cells is in generating HR, the halt in host cytoplasmic movement, disruption of the host nucleus, collapse of haustoria as well as host cell collapse in both haustorium-containing and neighboring cells suggest a toxin may be produced that moves from cell to cell. To assay for possible toxins at the cellular level, V. M. Russo and W. R. Bushnell (unpublished data) used micropipettes to remove contents from individual cells which had stopped cytoplasmic streaming due to impending HR. Either vacuolar sap or a mixture of cytoplasm and sap was transferred to individual test cells not undergoing HR. Little or no difference in toxic activity was detected between materials from cells undergoing HR and materials from control cells. If toxic factors exist, they are present at concentrations not far above the threshold required for HR.

Structural changes in the host plasma membrane in relation to HR have been monitored using plasmometric methods (Lee-Stadelmann *et al.*, 1984; O. Y. Lee-Stadelmann, W. R. Bushnell, C. M. Curran, and E. Stadelmann, unpublished data). Permeability of the membrane to the nonelectrolytes, methyl and ethyl urea, in susceptible host cells was reduced by 48 hr after inoculation, coupled with an increased tendency for the membrane to stick to host walls, resulting in concavities along the perimeter of partially plasmolysed protoplasts. These changes spread from haustorium-containing cells into neighboring cells. Similar changes occurred in resistant tissues before they underwent HR (as early as 18–20 hr after inoculation). An additional reduction in permeability occurred in haustorium-containing cells after cytoplasmic streaming stopped prior to HR, although the cells retained semipermeability until they collapsed. Lee-Stadelmann and co-workers (1984) speculated that a chemical factor moved from haustorium-containing cells to neighboring cells so that all had decreased permeability and increased membrane-wall adhesion. The altered membranes behaved as if they had increased negative charges on the surface, possibly because of an alteration in extrinsic proteins. Apparently, these changes are a part of the events that precede HR. For further ideas on how membrane alterations relate to HR, see Novacky, Chapter 17.

E. graminis can induce HR in a wide range of host and nonhost plants at all specificity levels (Table I), as will be discussed in Section 8. This suggests that the fungus has a general mechanism for eliciting HR, possibly by means of a nonspecific chemical elicitor as has been implicated in other plant diseases (see Scholtens-Toma *et al.*, Chapter 11).

7.3. Function

Does HR limit fungus development apart from other resistance factors that might be operating? When a host cell containing a primary haustorium dies before a primary hypha is produced, the consequent loss of a functional haustorium-host interface probably is a limiting factor. Fungal colonies grow only trace amounts if primary haustoria are removed micro-surgically (Bushnell *et al.*, 1967).

Correlative experimental evidence that HR of primary infected cells is a cause of reduced fungus growth is provided by the diverse treatments that inhibit HR and allow normal fungus development (heat shock, cytochalasin B, azide; Section 7.2). The evidence is not as clear when experiments are done with host-parasite combinations in which HR is normally delayed and mesophyll cells collapse. Koga and co-workers (1978) found that the amount of mesophyll collapse with different resistance genes did not correlate closely with the degree of fungal inhibition. Although thermal treatment or any of several exogenously supplied sugars delayed mesophyll death and promoted fungus growth (Toyoda *et al.*, 1978b, 1979), certain concentrations of glucose inhibited mesophyll collapse without increasing fungus development (Toyoda *et al.*, 1977). In the latter experiments, HR continued in the epidermis, suggesting that mesophyll HR was less important than epidermal HR in limiting fungus growth. Sucrose promoted growth in spite of mesophyll collapse, indicating that the factors limiting growth could be suppressed or ameliorated by sucrose.

The appressorium on the host cell surface often dies when the attached haustorium and host cell undergo HR (Shioji *et al.*, 1981; Wright and Heale, 1988) and rarely is it able to produce a new lobe for a second attempt at penetration as it often does when a papilla prevents haustorium formation without HR. Apparently, death of the haustorium and host cell is detrimental to the attached appressorium, possibly from failure of haustorial function, from physical disruption at the haustorial neck, or from toxic substances generated during HR. In the case of surviving appressoria, the septum in the haustorial neck may become sealed and protect the appressorium from harm.

In conclusion, the preponderance of available evidence for cereal powdery mildew indicates that HR of epidermal cells is a major factor in limiting fungal development. HR may: (1) limit nutrients available via haustoria; (2) physically damage appressoria or hyphae as attached haustorial necks degenerate; (3) release toxic factors from dying cells [possibly autofluorescent compounds (Shioji *et al.*, 1981; Toyoda *et al.*, 1978a)]; or (4) induce phytoalexin production by the host as demonstrated in pea mildew and implicated in barley mildew (Oku *et al.*, 1975a, b).

8. LEVELS OF SPECIFICITY

Interactions between higher plants and *E. graminis* occur at several levels of specificity, from the broad resistance of nonhosts to the race-specific resistance of appropriate hosts (Table I). A number of investigators in recent years have described development of *E. graminis* and responses of hosts and nonhosts at these various levels of specificity, showing patterns that differ from the better-known patterns at the level of race-specificity. These patterns, summarized in Table I, will be treated briefly at each level of specificity.

8.1. Nonhost

Nonhost species are defined here as those which are not hosts for any *forma specialis* of *E. graminis*. Generally in nonhosts, the fungus has difficulty producing appressorial germ tubes and appressoria, depending on the plant family. For example, germination rates for *E. graminis* f.sp. *hordei* on members of ten nonhost species outside the Graminae were about half of those on the appropriate host, barley (Table II). Rates of appressorium formation were only one-fifth those on barley leaves, and only a few germlings were able to produce penetration pegs. Consequently, few nonhost cells responded with papillae or HR. In some plant families, virtually no germination occurred as though the plant surfaces were toxic (Johnson, 1977; Johnson et al., 1982). The fungus developed further on nonhost species in the Gramineae (Table II), but appressorium development was still significantly less than on barley. Again, haustorium formation was rare and few nonhost cells responded with papillae or HR. Clearly, E. graminis is poorly adapted for germination and differentiation of appressoria on the surfaces of nonhost plants, both within and outside the Gramineae.

8.2. Host Inappropriate for Forma Specialis

Rates of germination and appressorium formation by E. graminis f.sp. hordei on hosts inappropriate for the forma specialis hordei, oats, rye, and wheat, were the same as for the appropriate host, barley (Table II). However, further development on the inappropriate hosts was usually restricted because rates of penetration peg and haustorium formation were sharply curtailed. Papillae were frequently found at sites of failed penetration. Haustorium formation and HR usually occurred at low rates (Table II) although higher rates were observed in some barley genotypes (Tosa and Shishiyama, 1984b).

8.3. Host Appropriate for Forma Specialis

8.3.1. Race-Nonspecific

8.3.1a. Cell-Specific. E. graminis is more successful in colonizing some types of epidermal cells than others, a specificity that seems to be unrelated to race-specific resistance. Most commonly reported is papilla-associated resistance (Table I) which prevents haustorium formation and tends to be highly effective in rows of epidermal cells located at a distance from stomates (Johnson et al., 1979; Koga et al., 1980; Lin and Edwards, 1974). H. Koga and colleagues (unpublished data) recently showed that this type of resistance correlated with cell length instead of cell location, as tissues with cells longer than 450 μm were resistant whereas shorter cells were susceptible regardless of distance from stomates. They speculated that the low ratio of cytoplasmic to vacuolar volume in long cells may be a factor in papilla-associated resistance. The resistance of long cells probably does not develop until leaves are fully expanded since infection rates of partially expanded leaves can be very high (McCoy and Ellingboe, 1966).

The recessive *mlo* gene in barley serves to enhance papilla-associated resistance except for stomatal subsidiary cells (Skou, 1982b). This gene confers a distinctive, general resistance that has yet to be fully overcome by E. graminis (Jørgensen, 1987). Colonies that originate in subsidiary cells develop almost normally, unlike the impeded development seen consistently with race-specific resistance.

8.3.1b. Organ-Specific. HR can occur as a race-nonspecific response that is at least partly organ-specific. HR typically occurs at a few percent of penetration sites in susceptible cereals, and can occur at more than 10% of sites in some cases, with rates higher in fifth than in first leaves (Asher and Thomas, 1983; Carver, 1986). Clearly, HR is a general resistance response to E. graminis which occurs at low levels in the absence of race-specific resistance.

There have been a few reports of reduced germination on certain leaves or parts of leaves, independent of any known genes for resistance (Douglas et al., 1984; Russell et al., 1975). In a recent study of the phenomenon on oats, Carver and Adaigbe (1990) carefully distinguished between formation of the primary and appressorial germ tubes and incubated inoculated leaves in

Table II. Development of *Erysiphe graminis* f. sp. *hordei* on Nonhosts, Hosts for Which It Is an Inappropriate Forma Specialis, and a Host for Which It Is an Appropriate Forma Specialis[a]

	Percentage of spores						
	Fungal development					Host response	
			Penetration				
Specificity level	Germination[b]	Appresorium	peg	Haustorium	Hypha	Papilla	HR
Nonhost							
Non-Gramineae (ten plant families)	35	14	8	0	0	11	2
Gramineae (big bluestem, Indian grass, maize, sorghum)	66	49	12	1	0	11	1
Host inappropriate for *forma specialis* (oats, rye, wheat)	77	71	20	3	1	40	4
Host appropriate for *forma specialis* (barley[c])	79	73	63	38	37	50	0

[a]From data of Johnson (1977) and Johnson *et al.* (1982).
[b]Germination rate based mainly on formation of appressorial germ tube.
[c]The barley cultivar was susceptible to the fungal race used.

special humidity chambers. Although differences were small, significantly more primary germ tubes formed on sixth than on first leaves and rates of appressorium formation were lower on old than on young sixth leaves. Carver and Adaigbe judged these small differences to be of doubtful biological significance. Since the fragile, short-lived conidium of E. graminis is extremely sensitive to environmental conditions, differences in surface microclimate may contribute to differences in germination. In another example of organ-specific development, Carver (1986) reported more failures of E. graminis to produce haustoria in the absence of papillae or other visible host response in fifth than in first leaves of barley.

   8.3.1c. Partial Resistance. Partial resistance (sometimes termed slow mildewing) is thought to be polygenically inherited and independent of known major race-specific genes for resistance (Jørgensen, 1987). Partial resistance is usually expressed more in adult plants than in seedlings and in these cases, could be classified under organ-specific resistance. The resistance is most often papilla-associated (Asher and Thomas, 1983; Carver and Carr, 1977; Clifford et al., 1985; Douglas et al., 1984; Parry and Carver, 1986), although failure to produce a haustorium sometimes occurs where no papilla is seen (Carver, 1986). Partial resistance has also been associated with reduced germination (Douglas et al., 1984), with HR (Asher and Thomas, 1983; Parry and Carver, 1986), and, more commonly, slowed colony development without HR (Asher and Thomas, 1983; Carver, 1987; Carver and Carr, 1978, 1980; Clifford et al., 1985). The size of haustoria can be reduced which, in turn, is suspected to limit nutrient supply to colonies (Clifford et al., 1985). Partial resistance is expressed in all the combinations of fungal development and host response listed in Table I. This is possibly a consequence of multiple resistance mechanisms which individually act at different stages of parasite development.

8.3.2. Race-Specific

   8.3.2a. Nature. Race specificity in cereal powdery mildews is controlled by gene-for-gene interactions in which corresponding genes in the host and parasite determine whether the two organisms are compatible. When any one of the many genes for resistance in the host is matched by a specific, corresponding gene for avirulence in the mildew fungus, host and parasite are incompatible (see Scholtens-Toma et al., Chapter 11). With incompatibility, fungus development is always less than with compatible host-parasite combinations. Depending on which gene is operating, the amount of development can vary from no hyphal growth to colonies with many haustoria and some sporulation. Some genes act at more than one stage of development, stopping a part of the attacking population before the primary hypha is produced, but allowing a second part to develop colonies whose development is then retarded (Hyde and Colhoun, 1975; Masri and Ellingboe, 1966; Andersen and Torp, 1986; Tosa and Shishiyama, 1984a).
   Regardless of the amount of fungal development, race-specific incompatibility is usually associated with HR, either in the primary host epidermal cell, in adjacent epidermal cells, or in subtending mesophyll cells as described in Section 7.1. Can race-specific incompatibility be expressed without HR? If a haustorium forms, its development or development of attached hyphae are only rarely reported to be affected without HR, as for example retarded and/or deformed haustorium development in barley with the Mlk gene (Bushnell and Bergquist, 1975; Masri and Ellingboe, 1966). This is in contrast to frequent reports of retarded colony development with polygenic partial resistance (Section 8.3.1c). However, several reports indicate that race-specific incompatibility inhibits fungus development before primary haustoria are formed, and without HR, usually in association with the papilla response (Johnson et al., 1979; Koga et al., 1980; Lin and Edwards, 1974; Masri and Ellingboe, 1966; McKeen and Bhattacharya, 1970). Other reports indicate that race-specific incompatibility is related to HR and not to the papilla response (Bushnell and Bergquist, 1975; White and Baker, 1984; Wright and Heale, 1988). Some

of the confusion arises from the high level of race-nonspecific, papilla-associated resistance in experimental plants (see Section 8.3.1a). Also, HR sometimes may have gone undetected, especially if autofluorescence was not used as an indicator. In reevaluating the role of HR, H. Koga and co-workers (unpublished data) concluded that papilla-associated resistance was not enhanced by *Mla* incompatibility which is, instead, associated with HR. Certainly race-specific incompatibility is not consistently expressed in the absence of HR.

8.3.2b. Mechanisms. How do gene-for-gene interactions initiate HR? Most models envision the products of a resistance gene acting as a receptor for a product of the corresponding gene for avirulence (Keen, 1982; Tepper and Anderson, 1984), although one or both of the interacting molecules may be secondary gene products. The consequence of the specific molecular interaction is to trigger events leading to host cell death. From the premise that only single genes in the host and parasite are involved, Ellingboe (1982) argued that incompatibility must be generated directly without participation of products of other genes. Others postulate that the primary interaction serves to activate a series of metabolic steps that lead to HR, possibly by activating a regulatory gene which, in turn, activates structural genes involved in HR (Tepper and Anderson, 1984).

Evidence from gene-for-gene interactions in powdery mildew of barley clearly implicates the second hypothesis. As noted earlier, several mRNAs and proteins are synthesized in greater amount in resistant than in susceptible plants, indicating that several genes are highly activated as a consequence of race-specific interaction (Davidson *et al.*, 1987, 1988; Manners *et al.*, 1985). Some of the enhanced mRNAs and proteins were activated in common by different race-specific genes, in line with the regulatory hypothesis. Furthermore, Jørgensen (1988) obtained mutants at three distinct suppressor loci that delayed HR triggered by the Mla_{12} gene. These mutants indicated that at least three loci apart from the Mla_{12} locus were responsible for metabolic processes leading to HR. Resistant reactions were also delayed by mutations at the Mla_{12} locus itself, but without loss of specificity for the corresponding avirulence gene. The Mla_{12} gene was postulated to be responsible for the initial, race-specific recognition event which, in turn, regulated expression of other genes responsible for HR. The same genes are probably turned on when HR is expressed as a general resistance response (Section 8.3.1b).

9. CONCLUDING STATEMENT

E. graminis is exquisitely adapted to develop on and infect epidermal cells of cereals. The adaptations include the differentiation of highly specialized infection structures and the ability to avoid host defenses. Our understanding of these defenses and how the fungus avoids or overcomes them is rudimentary. The preponderance of evidence indicates that the papilla response and HR are both major resistance mechanisms, although additional evidence is required for verification, especially for HR. We also need to learn more about the role of the haustorium in development of resistance responses. Do race-specific interactions take place at the interface between the haustorium and the host protoplast? Does the interruption of nutrient uptake have a role in either race-specific or race-nonspecific resistance? Also needed is an improved understanding of localized changes in host membranes in relation to papilla and halo formation, differentiation of the extrahaustorial membrane, and HR.

10. REFERENCES

Aist, J. R., 1976, Papillae and related wound plugs of plant cells, *Annu. Rev. Phytopathol.* **14**:145–163.

Aist, J. R., 1981, Development of parasitic conidial fungi in plants, in: *Biology of Conidial Fungi*, Volume 2 (G. T. Cole and B. Kendrick, eds.), Academic Press, New York, pp. 75–110.

Aist, J. R., 1983, Structural responses as resistance mechanisms, in: *The Dynamics of Host Defence* (J. A. Bailey and B. J. Deverall, eds.), Academic Press, New York, pp. 33–70.

Aist, J. R., and Gold, R. E., 1987, Prevention of fungal ingress: The role of papillae and calcium, in: *Molecular Determinants of Plant Diseases* (S. Nishimura, C. P. Vance, and N. Doke, eds.), Springer-Verlag, Berlin, pp. 47–58.

Aist, J. R., and Israel, H. W., 1977, Papilla formation: Timing and significance during penetration of barley coleoptiles by *Erysiphe graminis hordei*, *Phytopathology* **67**:455–461.

Aist, J. R., and Israel, H. W., 1986, Autofluorescent and ultraviolet absorbing components in cell walls and papillae of barley coleoptiles and their relationship to disease resistance, *Can. J. Bot.* **64**:266–272.

Aist, J. R., Kunoh, H., and Israel, H. W., 1979, Challenge appressoria of *Erysiphe graminis* fail to breach preformed papillae of a compatible barley cultivar, *Phytopathology* **69**:1245–1250.

Aist, J. R., Gold, R. E., Bayles, C. J., Morrison, G. H., Chandra, S., and Israel, H. W., 1988, Evidence that molecular components of papillae may be involved in ml-o resistance to powdery mildew, *Physiol. Mol. Plant Pathol.* **33**:17–32.

Akutsu, K., Amano, K., Doi, Y., and Yora, K., 1978, Development of cytoplasmic vesicles in and around the haustoria of barley powdery mildew (*Erysiphe graminis* f.sp. *hordei*), *Ann. Phytopathol. Soc. Jpn.* **44**: 532–538.

Andersen, J. B., and Torp, J., 1986, Quantitative analysis of the early powdery mildew infection stages on resistant barley genotypes, *J. Phytopathol.* **115**:173–186.

Asher, M. J. C., and Thomas, C. E., 1983, The expression of partial resistance to *Erysiphe graminis* in spring barley, *Plant Pathol.* **32**:79–89.

Bayles, C. J., and Aist, J. R., 1987, Apparent calcium mediation of resistance of an ml-o barley mutant to powdery mildew, *Physiol. Mol. Plant Pathol.* **30**:337–345.

Bracker, C. E., 1968, Ultrastructure of the haustorial apparatus of *Erysiphe graminis* and its relationship to the epidermal cell of barley, *Phytopathology* **58**:12–30.

Bushnell, W. R., 1971, The haustorium of *Erysiphe graminis*. An experimental study by light microscopy, in: *Morphological and Biochemical Events in Plant-Parasite Interaction* (S. Akai and S. Ouchi, eds.), Phytopathological Society of Japan, Tokyo, pp. 229–254.

Bushnell, W. R., 1981, Incompatibility conditioned by the *Mla* gene in powdery mildew of barley: The halt in cytoplasmic streaming, *Phytopathology* **71**:1062–1066.

Bushnell, W. R., 1982, Hypersensitivity in rusts and powdery mildews, in: *Plant Infection: The Physiological and Biochemical Basis* (Y. Asada, W. R. Bushnell, S. Ouchi, and C. P. Vance, eds.), Springer-Verlag, Berlin, pp. 97–116.

Bushnell, W. R., and Bergquist, S. E., 1975, Aggregation of host cytoplasm and the formation of papillae and haustoria in powdery mildew of barley, *Phytopathology* **65**:310–318.

Bushnell, W. R., and Gay, J., 1978, Accumulation of solutes in relation to the structure and function of haustoria in powdery mildews, in: *The Powdery Mildews* (D. M. Spencer, ed.), Academic Press, New York, pp. 183–235.

Bushnell, W. R., Dueck, J., and Rowell, J. B., 1967, Living haustoria and hyphae of *Erysiphe graminis* f.sp. *hordei* with intact and partly dissected host cells of *Hordeum vulgare*, *Can. J. Bot.* **45**:1719–1732.

Bushnell, W. R., Mendgen, K., and Liu, Z., 1987, Accumulation of potentiometric and other dyes in haustoria of *Erysiphe graminis* in living host cells, *Physiol. Mol. Plant Pathol.* **31**:237–250.

Carver, T. L. W., 1986, Histology of infection by *Erysiphe graminis* f.sp. *hordei* in spring barley lines with various levels of partial resistance, *Plant Pathol.* **35**:232–240.

Carver, T. L. W., 1987, Influence of host epidermal cell type, leaf age and position, on early stages of mildew colony development on susceptible and resistant oats, *Trans. Br. Mycol. Soc.* **89**:315–320.

Carver, T. L. W., and Adaigbe, M. E., 1990, Effects of host genotype, leaf age and position, and incubation humidity on germination and germling development by *Erysiphe graminis* f.sp. *avenae*, *Mycol. Res.* **94**: 18–26.

Carver, T. L. W., and Bushnell, W. R., 1983, The probable role of primary germ tubes in water uptake before infection by *Erysiphe graminis*, *Physiol. Plant Pathol.* **23**:229–240.

Carver, T. L. W., and Carr, A. J. H., 1977, Race non-specific resistance of oats to primary infection by mildew, *Ann. Appl. Biol.* **86**:29–36.

Carver, T. L. W., and Carr, A. J. H., 1978, The early stages of mildew colony development on susceptible oats, *Ann. Appl. Biol.* **89**:201–209.

Carver, T. L. W., and Carr, A. J. H., 1980, Some effects of host resistance on the development of oat mildew, *Ann. Appl. Biol.* **94**:290–293.

Carver, T. L. W., and Ingerson, S. M., 1987, Responses of *Erysiphe graminis* germlings to contact with artificial and host surfaces, *Physiol. Mol. Plant Pathol.* **30**:359–372.

Carver, T. L. W., and Jones, S. W., 1988, Colony development by *Erysiphe graminis* f.sp. *hordei* on isolated epidermis of barley coleoptile incubated under continuous light or short-day conditions, *Trans. Br. Mycol. Soc.* **90**:114–117.

Carver, T. L. W., and Phillips, M., 1982, Effects of photoperiod and level of irradiance on production of haustoria by *Erysiphe graminis* f.sp. *hordei*, *Trans. Br. Mycol. Soc.* **79**:207–211.

Carver, T. L. W., and Williams, O., 1980, The influence of photoperiod on growth patterns of *erysiphe graminis* f.sp. *hordei*, *Ann. Appl. Biol.* **94**:405–414.

Carver, T. L. W., Zeyen, R. T., and Ahlstrand, G. G., 1987, The relationship between insoluble silicon and success or failure of attempted primary penetration by powder mildew (*Erysiphe graminis*) germlings on barley, *Physiol. Mol. Plant Pathol.* **31**:133–148.

Chard, J. M., and Gay, J. L., 1984, Characterization of the parasitic interface between *Erysiphe pisi* and *Pisum sativum* using fluorescent probes, *Physiol. Plant Pathol.* **25**:259–276.

Cho, B. H., and Smedegaard-Petersen, V., 1986, Induction of resistance to *Erysiphe graminis* f.sp. *hordei* in near-isogenic barley (*Hordeum vulgare*) lines, *Phytopathology* **76**:301–305.

Clifford, B. C., Carver, T. L. W., and Roderick, H. W., 1985, The implications of general resistance for physiological investigations, in: *Genetic Basis of Biochemical Mechanisms of Plant Disease* (J. V. Groth and W. R. Bushnell, eds.), APS Press, St. Paul, pp. 43–84.

Dahmen, H., and Hobot, J. A., 1986, Ultrastructural analysis of *Erysiphe graminis* haustoria and subcuticular stroma of *Venturia inaequalis* using cryosubstitution, *Protoplasma* **131**:92–102.

Davidson, A. D., Manners, J. M., Simpson, R. S., and Scott, K. J., 1987, cDNA cloning of mRNAs induced in resistant barley during infection by *Erysiphe graminis* f.sp. *hordei*, *Plant Mol. Biol.* **8**:77–85.

Davidson, A. D., Manners, J. M., Simpson, R. S., and Scott, K. J., 1988, Altered host gene expression in near-isogenic barley conditioned by different genes for resistance during infection by *Erysiphe graminis* f.sp. *hordei*, *Physiol. Mol. Plant Pathol.* **32**:127–139.

Douglas, S. M., Sherwood, R. T., and Lukezic, F. L., 1984, Effect of adult plant resistance on primary penetration of oats by *Erysiphe graminis* f.sp. *avenae*, *Physiol. Plant Pathol.* **25**:219–228.

Ebrahim-Nesbat, F., Rohringer, R., and Heitefuss, R., 1982, Ultrastructural and histochemical studies on mildew of barley (*Erysiphe graminis* D.C. f.sp. *hordei* Marchal). II. Host cell penetration, papillae, and haustorial apparatus in the fifth leaf of susceptible plants, *Phytopathol. Z.* **105**:248–264.

Ebrahim-Nesbat, F., Heitefuss, R., and Rohringer, R., 1986, Ultrastructural and histochemical studies on mildew of barley (*Erysiphe graminis* D.C. f.sp. *hordei* Marchal). IV. Characterization of papillae in fifth leaves exhibiting adult plant resistance, *J. Phytopathol. (Berlin)* **117**:289–300.

Ellingboe, A. H., 1982, Genetical aspects of active defence, in: *Active Defense Mechanisms in Plants* (R. K. S. Wood, ed.), Plenum Press, New York, pp. 179–192.

Fric, F., 1984, Biochemical changes in barley plants in the preparasitic stage of powdery mildew-barley interaction, *Acta Phytopathol. Acad. Sci. Hung.* **19**:183–192.

Gay, J. L., and Manners, J. M., 1981, Transport of host assimilates to the pathogen, in: *Effects of Disease on the Physiology of the Growing Plant* (P. G. Ayres, ed.), Cambridge University Press, London, pp. 85–100.

Gay, J. L., and Manners, J. M., 1987, Permeability of the haustorium-host interface in powdery mildews, *Physiol. Mol. Plant Pathol.* **30**:389–399.

Gay, J. L., Salzberg, A., and Woods, A. M., 1987, Dynamic experimental evidence for the plasma membrane ATPase domain hypothesis of haustorial transport and for ionic coupling of the haustorium of *Erysiphe graminis* to the host cell (*Hordeum vulgare*), *New Phytol.* **107**:541–548.

Gil, F., and Gay, J. L., 1977, Ultrastructural and physiological properties of the host interfacial components of haustoria of *Erysiphe pisi* in vivo and in vitro, *Physiol. Plant Pathol.* **10**:1–12.

Gold, R. E., Aist, J. R., Hazen, B. E., Stolzenburg, M. C., Marshall, M. R., and Israel, H. W., 1986, Effects of calcium nitrate and chlortetracycline on papilla formation, ml-o resistance and susceptibility of barley to powdery mildew, *Physiol. Mol. Plant Pathol.* **129**:115–129.

Green, N. E., Hadwiger, L. A., and Graham, S. O., 1975, Phenylalanine ammonia-lyase, tyrosine ammonia-lyase, and lignin in wheat inoculated with *Erysiphe graminis* f.sp. *tritici*, *Phytopathology* **65**:1071–1074.

Harder, D. E., and Chong, J., 1984, Structure and physiology of haustoria, in: *The Cereal Rusts*, Volume I (W. R. Bushnell and A. P. Roelfs, eds.), Academic Press, New York, pp. 431–476.

Haywood, M. J., and Ellingboe, A. H., 1979, Genetic control of primary haustorial development of Erysiphe graminis on wheat, *Phytopathology* **69**:48–53.

Hazen, B. E., 1981, Experimental Studies on the Hypersensitive Response of Barley to Powdery Mildew. M.S. thesis, University of Minnesota.

Hazen, B. E., and Bushnell, W. R., 1983, Inhibition of the hypersensitive reaction in barley to powdery mildew by heat shock and cytochalasin B, *Physiol. Plant Pathol.* **23**:421–438.

Heintz, C., 1986, Infection mechanisms of grapevine powdery mildew (*Oidium tuckeri*): Comparative studies of the penetration process on artificial membranes and leaf epidermis, *Vitis* **25**:215–225.

Hippe, S., 1985, Ultrastructure of haustoria of *Erysiphe graminis* f.sp. *hordei* preserved by freeze-substitution, *Protoplasma* **129**:52–61.

Hyde, P. M., and Colhoun, J., 1975, Mechanisms of resistance of wheat to *Erysiphe graminis* f.sp. *tritici*, *Phytopathol. Z.* **82**:185–206.

Johnson, L. E. B., 1977, Resistance Mechanisms to Powdery Mildew Fungi (Erysiphaceae) in Nonhost and Inappropriate Host Plants, M.S. thesis, University of Minnesota.

Johnson, L. E. B., Bushnell, W. R., and Zeyen, R. J., 1979, Binary pathways for analysis of primary infection and host response in populations of powdery mildew fungi, *Can. J. Bot.* **57**:497–511.

Johnson, L. E. B., Bushnell, W. R., and Zeyen, R. J., 1982, Defense patterns in nonhost higher plant species against two powdery mildew fungi. I. Monocotyledonous species, *Can. J. Bot.* **60**:1068–1083.

Jørgensen, J. H., 1983, Experience and conclusions from the work at Risø on induced mutations for powdery mildew resistance in barley, in: *Induced Mutations for Disease Resistance in Crop Plants II*, International Atomic Energy Agency, Vienna, pp. 73–87.

Jørgensen, J. H., 1987, Three kinds of powdery mildew resistance in barley, *Barley Genet.* **5**:583–592.

Jørgensen, J. H., 1988, Genetic analysis of barley mutants with modifications of powdery mildew resistance gene *Ml-al2*, *Genome* **30**:129–132.

Keen, N. T., 1982, Specific recognition in gene-for-gene host-parasite systems, *Adv. Plant Pathol.* **1**:35–82.

Kita, N., and Toyoda, H., Shishiyama, J., 1980a, Histochemical reactions of papilla and cytoplasmic aggregate in epidermal cells of barley leaves infected by *Erysiphe graminis hordei*, *Ann. Phytopathol. Soc. Jpn.* **46**:263–265.

Kita, N., Toyoda, H., Yano, T., and Shishiyama, J., 1980b, Correlation of fluorescent appearance in papilla with unsuccessful penetration attempts in susceptible barley inoculated with *Erysiphe graminis* f.sp. *hordei*, *Ann. Phytopathol. Soc. Jpn.* **46**:594–597.

Kita, N., Toyoda, H., and Shishiyama, J., 1981, Chronological analysis of cytological responses in powdery-mildewed barley leaves, *Can. J. Bot.* **59**:1761–1768.

Koga, H., Mayama, S., and Shishiyama, J., 1978, Microscopic specification of compatible and incompatible interactions in barley leaves inoculated with *Erysiphe graminis hordei*, ,*Ann. Phytopathol. Soc. Jpn.* **44**:111–119.

Koga, H., Mayama, S., and Shishiyama, J., 1980, Correlation between the deposition of fluorescent compounds in papillae and resistance in barley against *Erysiphe graminis hordei*, *Can. J. Bot.* **58**:536–541.

Koga, H., Zeyen, R. J., Bushnell, W. R., and Ahlstrand, G. G., 1988, Hypersensitive cell death, autofluorescence, and insoluble silicon accumulation in barley leaf epidermal cells under attack by *Erysiphe graminis* f.sp. *hordei*, *Physiol. Mol. Plant Pathol.* **32**:395–409.

Kunoh, H., 1982, Primary germ tubes of *Erysiphe graminis* conidia, in: *Plant Infection: The Physiological and Biochemical Basis* (Y. Asada, W. R. Bushnell, S. Ouchi, and C. P. Vance, eds.), Springer-Verlag, Berlin, pp. 45–59.

Kunoh, H., 1987, Induced susceptibility and enhanced resistance at the cellular level in barley coleoptiles, in: *Molecular Determinants of Plant Diseases* (S. Nishimura, C. P. Vance, and N. Doke, eds.), Springer-Verlag, Berlin, pp. 59–73.

Kunoh, H., and Ishizaki, H., 1973, Incorporation of host mitochondria into the haustorial encapsulation of barley powdery mildew (II), *Ann. Phytopathol. Soc. Jpn.* **39**:42–48.

Kunoh, H., Tsuzuki, T., and Ishizaki, H., 1978, Cytological studies of early stages of powdery mildew in barley and wheat. IV. Direct ingress from superficial primary germ tubes and appressoria of *Erysiphe graminis hordei* on barley leaves, *Physiol. Plant Pathol.* **13**:327–333.

Kunoh, H., Aist, J. R., and Israel, H. W., 1979, Primary germ tubes and host cell penetrations from appressoria of *Erysiphe graminis hordei*, *Ann. Phytopathol. Soc. Jpn.* **45**:326–332.

Kunoh, H., Yamamori, K., and Ishizaki, H., 1983, Cytological studies of early stages of powdery mildew in barley and wheat. IX. Effect of various inorganic salts on the occurrence of autofluorescence at penetration sites of appressoria of *Erysiphe graminis hordei*, *Can. J. Bot.* **61**:2181–2185.

Kunoh, H., Aist, J. R., and Hayashimoto, A., 1985a, The occurrence of cytoplasmic aggregates induced by *Erysiphe pisi* in barley coleoptile cells before the host cell walls are penetrated, *Physiol. Plant Pathol.* **26**:199–208.

Kunoh, H., Kuno, K., and Ishizaki, H., 1985b, Cytological studies of the early stages of powdery mildew in barley *Hordeum vulgare* and wheat. XI. Autofluorescence and haloes at penetration sites of appressoria of *Erysiphe graminis hordei* and *Erysiphe pisi* on barley coleoptiles, *Can. J. Bot.* **63**:1535–1539.

Kunoh, H., Aist, Jr., and Israel, H. W., 1986, Elemental composition of barley coleoptile papillae in relation to their ability to prevent penetration by *Erysiphe graminis*, *Physiol. Mol. Plant Pathol.* **29**:69–78.

Kunoh, H., Katsuragawa, N., Yamaoka, N., and Hayashimoto, A., 1988a, Induced accessibility and enhanced inaccessibility at the cellular level in barley coleoptiles. III. Timing and localization of enhanced inaccessibility in a single coleoptile cell and its transfer to an adjacent cell, *Physiol. Mol. Plant Pathol.* **33**:81–93.

Kunoh, H., Yamaoka, N., Yoshioka, H., and Nicholson, R. L., 1988b, Preparation of the infection court by *Erysiphe graminis* I. Contact-mediated changes in morphology of the conidium surface, *Exp. Mycol.* **12**: 325–335.

Lee-Stadelmann, O. Y., Bushnell, W. R., and Stadelmann, E. J., 1984, Changes of plasmolysis form in epidermal cells of *Hordeum vulgare* infected by *Erysiphe graminis*: Evidence for increased membrane-wall adhesion, *Can. J. Bot.* **62**:1714–1723.

Lin, M.-R., and Edwards, H. H., 1974, Primary penetration process in powdery mildewed barley related to host cell age, cell type, and occurrence of basic staining material, *New Phytol.* **73**:131–137.

Liu, Z., Bushnell, W. R., and Brambl, R., 1987, Potentiometric cyanine dyes are sensitive probes for mitochondria in intact plant cells, *Plant Physiol.* **84**:1385–1390.

McCoy, M. S., and Ellingboe, A. H., 1966, Major genes for resistance and the formation of secondary hyphae by *Erysiphe graminis* f.sp. *hordei*, *Phytopathology* **56**:683–686.

McKeen, W. E., and Bhattacharya, P. K., 1970, Limitation of infection by *Erysiphe graminis* f.sp. *hordei* culture CR3 by the Algerian gene Mla in barley, *Can. J. Bot.* **48**:1109–1113.

Manners, J. M., and Gay, J. L., 1977, The morphology of haustorial complexes isolated from apple, barley, beet and vine infected with powdery mildews, *Physiol. Plant Pathol.* **11**:261–266.

Manners, J. M., and Gay, J. L., 1978, Uptake of ^{14}C photosynthates from *Pisum sativum* by haustoria of *Erysiphe pisi*, *Physiol. Plant Pathol.* **12**:199–209.

Manners, J. M., and Gay, J. L., 1980, Autoradiography of haustoria of Erysiphe pisi, *J. Gen. Microbiol.* **116**:529–533.

Manners, J. M., and Gay, J. L., 1982, Transport, translocation and metabolism of ^{14}C-photosynthate of the host-parasite interface of *Pisum sativum* and *Erysiphe pisi*, *New Phytol.* **91**:221–244.

Manners, J. M., and Gay, J. L., 1983, The host-parasite interface and nutrient transfer in biotrophic parasitism, in: *Biochemical Plant Pathology* (J. Callow, ed.), Wiley, New York, pp. 163–195.

Manners, J. M., Davidson, A. D., and Scott, K. J., 1985, Patterns of post-infectional protein synthesis in barley carrying different genes for resistance to the powdery mildew fungus, *Plant Mol. Biol.* **4**:275–283.

Marshall, M. R., Smart, M. G., Aist, J. R., and Israel, H. W., 1985, Chlortetracycline and barley papillae formation: Localization of calcium and alteration of the response induced by *Erysiphe graminis*, *Can. J. Bot.* **63**:876–880.

Martin, T. J., and Ellingboe, A. H., 1978, Genetic control of ^{32}P transfer from wheat to *Erysiphe graminis* f.sp. *tritici* during primary infection, *Physiol. Plant Pathol.* **13**:1–11.

Masri, S. S., and Ellingboe, A. H., 1966, Primary Infection of wheat and barley by *Erysiphe graminis*, *Phytopathology* **56**:253–378.

Mayama, S., and Shishiyama, J., 1976, Histological observation of cellular responses of barley leaves to powdery mildew infection by UV-fluorescence microscopy, *Ann. Phytopathol. Soc. Jpn.* **42**:591–596.

Mendgen, K., and Nass, P., 1988, The activity of powdery-mildew haustoria after feeding the host cells with different sugars, as measured with a potentiometric cyanine dye, *Planta* **174**:283–288.

Nicholson, R. L., Yoshioka, H., Yamaoka, N., and Kunoh, H., 1988, Preparation of the infection court by *Erysiphe graminis*. II. Release of esterase enzyme from conidia in response to a contact stimulus, *Exp. Mycol.* **12**: 336–349.

Oku, H., and Ouchi, S., 1976, Host plant accessibility to pathogens, *Rev. Plant Protection Res.* **9**:58–71.

Oku, H., Ouchi, S., Shiraishi, T., Komoto, Y., and Oki, K., 1975a, Phytoalexin activity in barley powdery mildew, *Ann. Phytopathol. Soc. Jpn.* **41**:185–191.

Oku, H., Ouchi, S., Shiraishi, T., and Baba, T., 1975b, Pisatin production in powdery mildewed pea seedlings, *Phytopathology* **65**:1263–1267.

Parry, A. L., and Carver, T. L. W., 1986, Relationship between colony development, resistance to penetration and autofluorescence in oats infected with powdery mildew, *Trans. Br. Mycol. Soc.* **86**:355–363.

Russell, G. E., Andrews, C. R., and Bishop, C. D., 1975, Germination of *Erysiphe graminis* f.sp. *hordei* conidia on barley leaves, *Ann. Appl. Biol.* **81**:161–169.

Russo, V. M., and Bushnell, W. R., 1989, Responses of barley cells to puncture by microneedles and to attempted penetration by *Erysiphe graminis* f.sp. *hordei*, *Can. J. Bot.* **67**:2912–2921.

Sahashi, N., and Shishiyama, J., 1986, Increased papilla formation, a major factor of induced resistance in the barley-*Erysiphe graminis* f.sp. *hordei* system, *Can. J. Bot.* **64**:2178–2181.

Sargent, C., and Gay, J. L., 1977, Barley epidermal apoplast structure and modification by powdery mildew contact, *Physiol. Plant Pathol.* **11**:195–205.

Sherwood, R. T., and Vance, C. P., 1982, Initial events in the epidermal layer during penetration, in: *Plant Infection: The Physiological and Biochemical Basis* (Y. Asada, W. R. Bushnell, S. Ouchi, and C. P. Vance, eds.), Springer-Verlag, Berlin, pp. 27–44.

Shioji, T., Toyoda, H., Kita, N., and Shishiyama, J., 1981, Vital behavior of parasite in fluorescing cells of powdery-mildewed barley leaves by uranine staining, *Ann. Phytopathol. Soc. Jpn.* **47**:340–345.

Skou, J. P., 1982a, Histokemiske reaktioner i bygmutanter og modersortes efter inokulering med bygmeldug, *Nord. Jordbrugsforsk.* **64**:44–48.

Skou, J. P., 1982b, Callose formation responsible for the powdery mildew resistance in barley with genes in the ml-o locus, *Phytopathol. Z.* **104**:90–95.

Skou, J. P., 1985, On the enhanced callose deposition in barley with ml-o powdery mildew resistance genes, *Phytopathol. Z.* **112**:207–216.

Skou, J. P., Jorgensen, J. H., and Lilholt, U., 1984, Comparative studies on callose formation in powdery mildew compatible and incompatible barley, *Phytopathol. Z.* **109**:147–168,

Smart, M. G., Aist, J., and Israel, H. W., 1986a, Structure and function of wall appositions. 1. General histochemistry of papillae in barley coleoptiles attacked by *Erysiphe graminis* f.sp. *hordei*, *Can. J. Bot.* **64**:793–801.

Smart, M. G., Aist, J. R., and Israel, H. W., 1986b, Structure and function of wall appositions. 2. Callose and the resistance of oversize papillae to penetration by *Erysiphe graminis* f.sp. *hordei*, *Can. J. Bot.* **64**:802–804.

Smart, M. G., Aist, J. R., and Israel, H. W., 1987, Some exploratory experiments on the permeability of papillae induced in barley coleoptiles by *Erysiphe graminis* f.sp. *hordei*, *Can. J. Bot.* **65**:745–749.

Spencer-Phillips, P. T. N., and Gay, J. L., 1980, Electron microscope autoradiography of ^{14}C photosynthate distribution at the haustorium-host interface in powdery mildew of *Pisum sativum*, *Protoplasma* **103**:131–154.

Spencer-Phillips, P. T. N., and Gay, J. L., 1981, Domains of ATPase in plasma membranes and transport through infected plant cells, *New Phytol.* **89**:393–400.

Stolzenburg, M. C., Aist, J. R., and Israel, H. W., 1984a, The role of papillae in resistance to powdery mildew conditioned by the *ml-o* gene in barley. I. Correlative evidence, *Physiol. Plant Pathol.* **25**:337–346.

Stolzenburg, M. C., Aist, J. R., and Israel, H. W., 1984b, The role of papillae in resistance to powdery mildew (*Erysiphe graminis* f.sp. *hordei*) conditioned by the ml-o gene in barley (*Hordeum vulgare*). II. Experimental evidence, *Physiol. Plant Pathol.* **25**:347–361.

Stuckey, R. E., and Ellingboe, A. H., 1975, Effect of environmental conditions on ^{35}S uptake by *Triticum aestivum* and transfer to *Erysiphe graminis* f.sp. *tritici* during primary infection, *Physiol. Plant Pathol.* **5**:19–26.

Takahashi, K., Aist, J. R., and Israel, H. W., 1985, Distribution of hydrolytic enzymes at barley powdery mildew encounter sites: Implications for resistance associated with papilla formation in a compatible system, *Physiol. Plant Pathol.* **27**:167–184.

Tepper, C. S., and Anderson, A. J., 1984, The genetic basis of plant pathogen interaction, *Phytopathology* **74**:1143–1145.

Thordal-Christensen, H., and Smedegaard-Peterson, V., 1988a, Comparison of resistance-inducing abilities of virulent and avirulent races of *Erysiphe graminis* f.sp. *hordei* and a race of *Erysiphe graminis* f.sp. *tritici* in barley, *Plant Pathol.* **37**:20–27.

Thordal-Christensen, H., and Smedegaard-Petersen, V., 1988b, Correlation between induced resistance and host fluorescence in barley inoculated with *Erysiphe graminis*, *J. Phytopathol. (Berlin)* **123**:34–46.

Tosa, Y., and Shishiyama, J., 1984a, Cytological aspects of events occurring after the formation of primary haustoria in barley leaves infected with powdery mildew, *Can. J. Bot.* **62**:795–798.

Tosa, Y., and Shishiyama, J., 1984b, Defense reactions of barley cultivars to an inappropriate *forma specialis* of the powdery mildew fungus of gramineous plants, *Can. J. Bot.* **62**:2114–2117.

Toyoda, H., 1983, Mechanisms for resistance expression in barley leaves inoculated with *Erysiphe graminis* f.sp. *hordei*, *Shokubutsu Byogai Kenkyu Kyoto* **9**:69–128.

Toyoda, H., Mayama, S., and Shishiyama, J., 1977, Inhibition of mesophyll collapse in powdery-mildewed barley leaves by glucose-supply, *Ann. Phytopathol. Soc. Jpn.* **43**:386–391.

Toyoda, H., Mayama, S., and Shishiyama, J., 1978a, Fluorescent microscopic studies on the hypersensitive necrosis in powdery-mildewed barley leaves, *Phytopathol. Z.* **92**:125–131.

Toyoda, H., Mayama, S., and Shishiyama, J., 1978b, The effects of sugar-supply on the formation of collapsed mesophyll cells and the fungal development in powdery-mildewed barley, *Phytopathol. Z.* **92**:359–367.

Toyoda, H., Yano, T., and Shishiyama, J., 1979, Delay of hypersensitive necrosis by thermal treatment and fungal development in powdery-mildewed barley leaves, *Can J. Bot.* **57**:1414–1417.

White, N. H., and Baker, E. P., 1954, Host pathogen relations in powdery mildew of barley. 1. Histology of tissue reactions, *Phytopathology* **44**:657–662.

Williamson, R. E., and Ashley, C. C., 1982, Free Ca^{2+} and cytoplasmic streaming in the alga *Chara*, *Nature* **296**:647–651.

Wolf, G., and Fric, F., 1981, A rapid staining method for *Erysiphe graminis* f.sp. *hordei* in and on whole barley leaves with a protein-specific dye, *Phytopathology* **71**:596–598.

Woolacott, B., and Archer, S. A., 1984, The influence of the primary germ tube on infection of barley by *Erysiphe graminis* f.sp. *hordei*, *Plant Pathol.* **33**:225–231.

Wright, A. J., and Heale, J. B., 1988, Host responses to fungal penetration in *Erysiphe graminis* f.sp. *hordei* infections in barley, *Plant Pathol.* **37**:131–140.

16

Induced Systemic Resistance in Plants

Nageswara Rao Madamanchi and Joseph Kuć

1. INTRODUCTION

Those investigating plant immune systems are faced with a dilemma. Disease resistance/susceptibility in plants exhibit a high degree of specificity at many levels; however, specific mechanisms for eliciting resistance and specific mechanisms for disease resistance have not been established. Based on the current literature, the presence or absence of genetic information for resistance mechanisms are not determinants of disease resistance in plants. This is true when one considers phytoalexins, hydroxyproline-rich glycoproteins, lignin, callose, peroxidases, chitinases, β-1, 3-glucanases, and pathogenesis-related proteins. The mechanisms for resistance also lack specificity in their action against specific pathogens. The chemical agents reported to elicit resistance, whether derived from pathogens or synthetic, are also notable for their lack of specificity. To add to the dilemma, pathogens can often induce systemic resistance against completely unrelated pathogens, e.g., fungi against viruses, viruses against fungi (Dean and Kuć, 1987a; Kuć, 1982, 1983, 1984, 1985a–c, 1987a; Kuć and Preisig, 1984; Tuzun and Kuć, 1989).

In the field of plant-pathogen interaction, the only well-established case for specificity at the molecular level, though limited in scope, is that reported for host-specific toxins (Durbin, 1981; Daly, 1987). A less well-established case for specificity, also limited in scope, is that reported for specific suppression of resistance (Kunoh, 1987; Oku *et al.*, 1987; Preisig and Kuć, 1987). Though specific elicitors have not yet been established, it appears that regulation of the resistance response in time and place relative to pathogen development and the effect of environment on the activity of gene products related to resistance, determine disease resistance. With the exception of the specificity of antibody production and action, this concept is equally valid in animals. Disease resistance in animals is dependent on the speed and magnitude of response and the activity of gene products as affected by the environment within or external to the animal. The ability to immunize plants and animals against disease is very strong evidence for the general existence of the potential for resistance mechanisms (Dean and Kuć, 1987a; Kuć, 1982, 1983, 1985a–c, 1987a, b; Kuć and Preisig, 1984; Kuć and Rush, 1985).

Nageswara Rao Madamanchi • Department of Plant Pathology, Physiology and Weed Science, Virginia Polytechnic Institute and State University, Blacksburg, Virginia 24061-0331.   *Joseph Kuć* • Department of Plant Pathology, University of Kentucky, Lexington, Kentucky 40546-0091.

The dilemma is further exacerbated by the observation that, unlike animals, plants can often be immunized against a broad spectrum of pathogens by inoculation with a single infectious agent or treatment with a nonspecific chemical. When infected or treated, the plant appears to activate or sensitize for activation a broad spectrum of defenses which are effective against representatives of all classes of pathogens. This lack of specificity has not precluded the survival of plants under continuous pressure from pathogens. Specificity of plant-pathogen interaction, however, is common. Pathogens of corn do not cause disease of oats, pathogens of corn may cause disease on only certain cultivars of corn, pathogens of roots may be nonpathogens of leaves, and plants at one stage of development may be resistant and at another stage of development be susceptible.

Disease resistance or susceptibility may also be a function of environmental factors such as temperature and light. The key question appears to be: How can we regulate the expression of the genetic potential already in place and maintain the activity of gene products for resistance to produce resistant plants? Plant immunization has proven successful in expressing or sensitizing for expression resistance mechanisms in plants considered genetically and economically susceptible to disease. Furthermore, immunization has proven successful in growth room, greenhouse, and field experiments without the introduction of "foreign DNA."

2. THE PHENOMENON

2.1. Plants and Pathogens with Which Induced Systemic Resistance Has Been Demonstrated

Induced systemic resistance against disease in plants following infection has been recorded but often not been documented as early as the 19th century (Chester, 1933). More recently, induced systemic resistance has been well documented against all classes of pathogens in widely diverse plants and plant tissues. Plants in which induced systemic resistance has been reported include cucumber, muskmelon, watermelon, green bean, tobacco, tomato, potato, grape, coffee, barley, pear, apple, plum, and carnation (Dean and Kuć, 1987a; Kuć, 1982, 1983, 1985a–c, 1987a, b; Kuć and Preisig, 1984; Salt and Kuć, 1985; Sequeira, 1983; Tuzun and Kuć, 1989; Ross, 1966). With two exceptions (Roberts, 1983; Wieringa-Brants and Dekker, 1987), induced resistance is not seed transmissible, but it has been demonstrated to be transmitted to regenerants via tissue culture (Kuć, 1987b; Tuzun and Kuć, 1987, 1989). The broad spectrum of pathogens and chemical agents used to induce resistance and its effectiveness against pathogens as varied as viruses, fungi, bacteria, and nematodes support the importance of both fundamental biochemical studies to reveal its molecular basis and studies to apply the technology for the control of disease in the field.

2.2. Influence of Physiological and Environmental Conditions

In cucumbers, resistance induced by infection of the first true leaf with *Colletotrichum lagenarium* or tobacco necrosis virus (TNV) persists for 3–6 weeks (Kuć and Richmond, 1977). With booster inoculation during this period, resistance lasts through flowering and fruiting. Systemic resistance cannot be induced in cucumbers, however, after the onset of fruit set (Guedes *et al.*, 1980), presumably due to a change in hormonal balance. The concentration of inducer inoculum and the number of lesions produced on the inducer leaf are directly related to the extent of protection until a saturation is attained with the inducer inoculum (Caruso and Kuć, 1977, 1979; Kuć and Richmond, 1977; Jenns and Kuć, 1980). A single lesion of *C. lagenarium* (Kuć and Richmond, 1977; Hammerschmidt, 1980), and as few as eight lesions of TNV (Jenns and Kuć, 1980) on the inducer leaf can induce systemic resistance. High concentrations of challenge inoculations reduce the effectiveness of induced resistance, particularly at low levels of inducer inoculum (Dean and Kuć, 1986a).

Removal of the inducer leaf from cucumber seedlings 72–96 hr after infection with fungi and bacteria, and 48 hr after infection with TNV does not affect induced systemic resistance in the two leaves above the inducer leaf (Caruso and Kuć, 1979; Jenns and Kuć, 1980; Kuć and Richmond, 1977). Similarly, removal of leaves above the inducer leaf after similar lag periods does not reduce protection in the excised leaves.

The signals responsible for induced systemic resistance are graft transmissible in cucumber (Jenns and Kuć, 1979) and tobacco (Tuzun and Kuć, 1985b). In cucumber, girdling the petiole of the inducer leaf prevented induction of systemic induced resistance (Guedes *et al.*, 1980). Girdling the petioles of leaves to be challenged with inducer leaf intact, prevented induction of resistance only in the girdled leaves. The source of the signal is the inducer leaf and protected but uninfected leaves do not export signal (Dean and Kuć, 1986b).

In tobacco, inducing resistance against blue mold by stem injection with *Peronospora tabacina* in plants shorter than 20 cm was found to often cause stunting due to systemic spread of the fungus, but a single inoculation of plants over 20 cm in height (Fig. 1) is effective throughout the plant's life (Cruickshank and Mandryk, 1960; Kuć and Tuzun, 1983; Tuzun *et al.*, 1984). Recently, Ye and Kuć (unpublished data) demonstrated that tobacco seedlings can be systemically protected against *P. tabacina* by limited foliar infection with *P. tabacina* followed by dipping the plants in hot water (50°C) for 30 sec to attenuate the development of the pathogen before the appearance of symptoms.

Although induced systemic resistance is stable under a broad range of environmental conditions, cucumbers can be more effectively induced in winter and tobacco in summer. Preliminary experiments by our group suggest that high nutrient levels make induction of systemic resistance in tobacco less effective.

Figure 1. Tobacco plant systemically immunized against blue mold (right) by stem injection with sporangiospores of *Peronospora tabacina* and a highly susceptible plant (left) stem-injected with water.

2.3. Efficacy

Induced systemic resistance has been found effective against a broad spectrum of fungal, bacterial, and viral diseases. It is effective under field conditions in protecting crop plants (Caruso and Kuć, 1977; Costa and Muller, 1980; Cox *et al.*, 1976; Tuzun *et al.*, 1986). Growth and yield of induced plants is either unaffected or enhanced in the absence of the pathogen (Tuzun and Kuć, 1985a; Tuzun *et al.*, 1986). Induced systemic resistance activates several natural resistance mechanisms of the host and therefore may prove very durable and stable, unlike current systemic fungicides based on a single site for action. Induced systemic resistance lasts for the life of annual plants and, since the mechanisms for resistance activated are inherent in the plants, appears as safe as resistance developed by breeding.

Induced systemic resistance is graft transmissible from root stock to scion and is transmissible via tissue culture from immunized tobacco plants (Kuć, 1987b; Tuzun and Kuć, 1987, 1989). These properties, in combination with conventional breeding for resistant plants, could have far-reaching implications for modern agriculture. Plants with multiple resistance could be rapidly developed utilizing plants resistant to a single serious disease. The ability to induce systemic resistance in even very susceptible plants may help to utilize current high-yielding, high-quality crops. The use of microorganisms as inducing agents for immunization is established. The signals for immunization may permit foliar, soil, and seed application of immunizing agents. Recent developments indicate that common, inexpensive, and nontoxic chemicals may release immunity signals and be useful for crop protection (Doubrava *et al.*, 1988; Gottstein and Kuć, 1989).

2.4. Limitations

Currently, induced systemic resistance may not be economically competitive with our present technology for disease control in modern agriculture. Technology is often lacking to administer various pathogens for immunization effectively and safely. Immunization with microorganisms, however, allows for the control of viruses and certain bacterial diseases for which chemical control is unavailable. If immunity signals and their derivatives find use in immunization, existing equipment for foliar and soil application of pesticides may be utilized. Simple compounds which release immunity signals would find wide application in the agriculture of developing nations as well as those with a highly productive agriculture (Doubrava *et al.*, 1988; Gottstein and Kuć, 1989). Such compounds convert each leaf they are applied to into a factory leaf to produce immunity signals. Unfortunately, the concept that plants can be effectively protected by immunization is generally not recognized outside the scientific community. Consequently, farmers have been unaware of the potential of induced systemic resistance in plants and thereby unable to utilize and help improve the technology.

3. MECHANISMS ACTIVATED

3.1. Penetration

Penetration of *C. lagenarium* from appressoria, but not formation of appressoria, was reduced on induced systemically resistant leaves of cucumber challenged with conidia of the fungus (Richmond *et al.*, 1979; Xuei *et al.*, 1988). When the epidermis was peeled from these leaves before challenge, resistance decreased. In light and electron microscopic studies, the epidermal cell walls appeared normal in control and immunized cucumber leaves before challenge with *C. lagenarium* (Xuei *et al.*, 1988). Penetration through the epidermal wall was evident 36 hr after challenge and was followed by hyphal proliferation and host-cell destruction in control tissues. Penetration, however, was rarely seen in immunized tissues before 72 hr, but at 24–96 hr, invaginations of the appressorial wall were observed, as well as highly electron-opaque epidermal walls beneath the appressoria. At 72 hr, thin penetration pegs were occasionally seen within the dense host wall, and at 96 hr, the peg was embedded within the underlying papilla and aggregates of callose-like material. Penetration through the epidermis was rare in immunized tissue, but when it was observed, development appeared to be similar to that in control leaves. Ultrastructural changes in immunized tissue as well as in the fungus occurred within 24 hr after challenge of the host. A correlation was evident between the comparatively few lesions observed macroscopically in immunized leaves and the sparsity of successful penetrations into host mesophyll tissue. Similar observations were made by Politis (1976) working with nonhost oats inoculated with conidia of *Colletotrichum graminicola* and Stumm and Gessler (1986) in cucumber using fluorescence microscopy. Induced resistance of tobacco to *P. tabacina*, however, appears due to restriction of the fungal development within leaves after penetration (Stolle *et al.*, 1988).

3.2. Peroxidases and Hydroxyproline-Rich Glycoproteins

A three-fold systemic increase in peroxidase activity associated with induced resistance was observed in SMR 58 cucumber after inoculation of leaf one with *C. lagenarium* (Hammerschmidt *et al.*, 1982). Since induction with several pathogens causes the increase, it is likely to be of host origin. As with induction of systemic resistance, a single lesion caused a significant increase in peroxidase activity and the increase is associated with enhanced activity of several isozymes. Although peroxidase activity increases systemically in induced plants before challenge, enzyme activity of control plants ultimately surpasses that of induced plants when symptoms are apparent.

Recently, Smith and Hammerschmidt (1988) reported at least a two-fold increase in the specific activity of soluble peroxidase extracted from intercellular wash fluids of systemically induced leaves of cucumber, muskmelon, or watermelon as compared to controls. Peroxidases induced in the three plant species appeared similar with respect to charge and molecular weight and showed a high degree of immunological cross-reactivity. Nadolny and Sequeira (1980), however, concluded that increases in peroxidase activity in tobacco, in which disease resistance has been induced by prior infiltration with heat-killed *Pseudomonas solanacearum* cells, are not involved in disease resistance.

Hydroxyproline-rich glycoproteins (HRGP) are linear molecules (Van Holst and Varner, 1984) which are rich in basic amino acids (Mazau and Esquerré-Tugayé, 1986; Mellon and Helgeson, 1982), two attributes which suggest a role in defense. They can serve to strengthen the cell wall. High levels of basic amino acids may make the cell wall a polycationic barrier which agglutinates negatively charged particles or cells like bacteria. Stermer and Hammerschmidt (1987) reported enhanced accumulation of insoluble extensin, a HRGP, and production of ethylene in cucumber seedlings following a heat-shock treatment for 40 sec at 50°C. Heat shock also induced resistance, though not systemically, in cucumber seedlings susceptible to *C. cucumerinum*. Cell walls from heat-shocked seedlings were more resistant than cell walls from unshocked controls to degradation by crude enzyme preparations from *C. cucumerinum*. Moreover, in cucumber plants with systemic disease resistance induced by a localized infection, there was a systemic increase in the insoluble extensin of the epidermis.

Higher extensin levels were associated with resistance of cucumbers to *C. cucumerinum* (Hammerschmidt *et al.*, 1984) and melons to *C. lagenarium* (Esquerré-Tugayé *et al.*, 1985). HRGP synthesis can be elicited by elicitors of both fungal and plant origin, including some which also elicit phytoalexins (Roby *et al.*, 1985). A molecular analysis of HRGP accumulation in *Phaseolus vulgaris* in response to *C. lindemuthianum* has shown that mRNAs which hybridize to a genomic clone of HRGP are increased earlier in an incompatible interaction than in a compatible interaction and that these increases are enhanced by an elicitor preparation from the pathogen (Showalter et al., 1985).

Peroxidases can cross-link cell wall polymers by catalyzing the formation of biphenyl cross-linkages (Fry, 1982; Markwalder and Neukom, 1976; Sarkanen, 1971). Thus, increase in peroxidase isozymes can increase the cross-linking of extensin molecules providing resistance to C. cucumerinum in induced systemic-resistant plants (Stermer and Hammerschmidt, 1987).

Doke (1983) suggested that activation of an O_2^--generating system in host plasma membrane may be the earliest biochemical reaction in plant cell walls associated with resistance in incompatible Phytophthora infestans and potato interactions. Buonaurio and co-workers (1987) reported that peroxidase and superoxide dismutase (SOD), a free radical scavenger, production were similar in cultivars of bean susceptible to Uromyces phaseoli, while in the hypersensitive cultivars peroxidase activity greatly exceeded that of SOD. It would be interesting to examine whether peroxidase production markedly surpasses that of SOD following induction of systemic resistance, thereby generating toxic free radicals. Besides their role in lignification, free radicals may also be involved in the metabolism of polyphenols (Lazarovits and Ward, 1982) and phytoalexins (Bostock et al., 1981, 1982).

3.3. Lignification

Hammerschmidt and Kuć (1982) demonstrated that the inhibition of penetration of C. lagenarium into systemically protected tissues and restricted development of C. cucumerinum were associated with the ability of epidermal cells to respond rapidly with the deposition of material which was histochemically defined as lignin. Using [^{14}C]t-cinnamate and [^{14}C]-L-phenylalanine, Dean and Kuć (1987b) reported more rapid lignification following challenge in induced cucumber leaves compared with noninduced leaves. Incorporation of these precursors also was approximately 50 percent greater in unchallenged protected leaf disks as compared to unprotected disks and incorporation was greatest at the periphery of disks where wounding occurred. The rate of lignification increased more rapidly in immunized as compared to control leaves after wounding by pricking the leaf surface with a pin.

3.4. Phytoalexins

Production of phytoalexins is one inducible component of a multiple-component mechanism for disease resistance in plants. Phytoalexins are low-molecular-weight compounds produced by plants in response to infection or stress. They are localized around the site of infection and are lipophilic in nature (Coxon, 1982; Kuć, 1982; Kuć and Rush, 1985; Stoessl, 1983). The synthesis and biochemical basis for the induction of phytoalexins have been reviewed (Darvill and Albersheim, 1984; Dixon et al., 1983; Ebel, 1986).

The speed and magnitude with which phytoalexins accumulate at the site of infection appear to determine disease reaction in some plant interactions with fungi and bacteria (Kuć, 1985a–c; Kuć and Rush, 1985). Recently, Sun and co-workers (1988) reported the inhibition of systemic spread of cauliflower mosaic virus (CaMV) by a photoactivatable phytoalexin, 2, 7-dihydroxycadalene (DHC), produced by cotton cultivars in response to bacterial infection. Phytoalexin accumulation was initiated rapidly and was markedly enhanced in immunized green bean seedlings challenged with compatible races of Colletotrichum lindemuthianum (Kuć, 1984). The reaction of the immunized tissue was indistinguishable from that which occurred following inoculation with an incompatible race of the fungus (Elliston et al., 1976; Kuć, 1984). Though phaseolin, kievitone, and other phytoalexins were not translocated in bean seedlings, immunization appeared to sensitize the entire seedling to rapidly accumulate the phytoalexins when challenged. This sensitization has been observed with lignification in immunized cucumber

(Hammerschmidt and Kuć, 1982; Dean and Kuć, 1987b) and with chitinases and β-1, 3-glu-canases in immunized tobacco (Tuzun et al., 1988). It was not apparent with sesquiterpenoid phytoalexin accumulation in tobacco (Stolle et al., 1988).

3.5. Chitinases and β-1, 3-glucanases

Chitinase catalyzes the hydrolysis of the β-1, 4 linkages of N-acetyl-D-glucosamine in the polymer chitin, a major component of cell walls of many fungi. The enzyme occurs in plants, although higher plants apparently do not contain a substrate for this enzyme, and the enzyme has been suggested to have a role in disease resistance against chitin-containing fungi (Boller et al., 1983). Chitinases have been reported to hydrolyze cell walls of some fungal pathogens and nonpathogens in vitro (Boller et al., 1983; Skujins et al., 1965; Schlumbaum et al., 1986; Young and Pegg, 1981).

β-1, 3-glucanase is widely present in plants (Clarke and Stone, 1962; Pegg, 1977). In contrast to chitin, β-1, 3-glucan (callose) is present endogenously in plants and is also a major component of fungal cell walls (Bartnicki-Garcia, 1968). Purified β-1, 3-glucanase from tomato has been shown to partially digest the cell walls of the pathogen Verticillium albo-atrum; the degradation of fungal cell walls is synergistically stimulated by chitinase (Young and Pegg, 1982; Skujins et al., 1965). Although extensive literature is becoming available on the accumulation and synthesis of chitinases and β-1, 3-glucanases in plants and their regulation, relatively little has been published concerning the relationship of the enzymes to induced systemic resistance.

Metraux and Boller (1986) reported the induction of chitinase activity both in the infected leaf and systemically in cucumber plants in which the first true leaf was infected by viruses, bacteria, or fungi. Boller and Metraux (1988), and Metraux and co-workers (1988) reported a 28-kDa pathogenesis-related protein (PR-protein) 7 days after inoculation in extracts of both infected and uninfected leaves of cucumber plants inoculated on the first leaf with TNV, Pseudoperonospora cubensis, C. lagenarium, or Pseudomonas lachrymans. This host-coded protein was detected in up to five leaves above the infected leaf. After purification from the intercellular fluid, the protein was identified as a chitinase. The chitinase induced by ethylene in bean plants, however, is localized in the central vacuole of the leaf cells (Boller and Vogeli, 1984). Microbially induced chitinase activity is largely intercellular (Boller and Metraux, 1988; Metraux and Boller, 1986).

The induction of systemic resistance to P. tabacina coincided with the accumulation of β-1, 3-glucanases and chitinases following stem injection of tobacco plants with the pathogen (Tuzun et al., 1989). Increased activity of β-1, 3-glucanase coincided with the development of induced resistance 15 days after stem injection. The enzyme activity increased prior to challenge, 15 and 21 days after stem injection, but the increase was not observed in water-injected plants. Enzyme activity continued to increase 2 and 6 days after challenge of induced tobacco. Very low activity was sometimes observed in control plants 2 days following challenge. The activity increased in controls 6 days after challenge, but by this time disease symptoms were apparent. The changes in the level of proteins cross-reacting with antibodies against bean β-1, 3-glucanase in Western blots corresponded to the observed changes in enzyme activity.

Constitutive levels of chitinase activity were detected in extracts of both control and stem-injected plants before challenge. Enzyme activity increased in extracts of induced but not control plants 15 and 21 days after stem injection with P. tabacina. Increased activity was evident in both induced and control plants 2 and 6 days after challenge, although higher enzyme activity was detected in induced plants 2 days after challenge. When proteins were eluted and enzyme activities assayed, chitinase activity was associated only with the regions of the gels corresponding to the major antigens observed in the Western blot. As with β-1, 3-glucanase activity, Western blot analyses agreed with determinations of enzyme activity. Since P. tabacina lacks chitin in its cell wall, it is unlikely that chitinases have a role in containing the fungus in immunized plants. The period of high β-1, 3-glucanase in immunized plants, 2 days after challenge, corresponded to the time when development of the fungus was contained in immunized plants.

3.6. Pathogenesis-Related (b) Proteins

Accumulation of soluble proteins in plants infected by viruses (Ahl and Gianinazzi, 1982; Parent and Asselin, 1984), bacteria (Gessler and Kuć, 1982), fungi (Gianinazzi *et al.*, 1980), and viroids (Granell *et al.*, 1987) has been reported. Induction of systemic resistance also tends to be accompanied by accumulation of several proteins, designated as pathogenesis-related (PR) proteins. These proteins are also produced in response to application of chemicals such as polyacrylic acid, benzoic acid derivatives (Gianinazzi and Kassanis, 1974), aspirin (White, 1979), high concentrations of various plant hormones (Van Loon, 1977, 1983), by mechanical injury and osmotic stress (Ohashi and Matsuoka, 1985), and also during flowering (Fraser, 1981). PR proteins described thus far for a number of plant species are of low molecular weight, acid-soluble, proteolysis-resistant, and have high mobility in polyacrylamide gel electrophoresis. Some of them have related structure and antigenicity (Hooft van Huijsduijnen *et al.*, 1986, 1987; Kauffmann *et al.*, 1987; Kombrink *et al.*, 1988; Legrand *et al.*, 1987; Matsuoka and Ohashi, 1984). Thirteen major proteins were detected and characterized from tobacco leaves reacting hypersensitively to tobacco mosaic virus (TMV) (Pierpoint, 1986). Although a considerable literature is accumulating on the detection of PR proteins following infection and some have been characterized as having enzyme activity, e.g., chitinases and β-1, 3-glucanases, considerably less information is available concerning PR proteins in plants with induced systemic resistance to pathogens other than viruses.

Gessler and Kuć (1982) reported the presence of a protein "E" in cucumber leaves infected with TNV and numerous bacterial and fungal pathogens. The protein had an approximate molecular weight of 16,000 and appeared rapidly in challenged leaves of immunized plants which had little or no symptoms. Metraux and co-workers (1988) reported the systemic presence of a host-coded protein in induced cucumber leaves and identified it as a chitinase. De Wit and co-workers (1986) reported the accumulation of five host-coded proteins in intercellular fluids of tomato leaves infected with *Cladosporium fulvum* (syn. *Fulvia fulva*). These proteins accumulate 2–4 days earlier in incompatible interactions than in compatible ones.

Tuzun and co-workers (1988, 1989) reported increases in levels of several PR proteins, with estimated molecular weights ranging from 14,500 to 65,000, in pH 2.8 buffer extracts of immunized tobacco. Some PR proteins increased further 2 and 6 days after challenge. Increases in PR proteins were detected in control plants only 6 days after challenge. Two proteins were apparent only in immunized plants. One of these was present prior to and after challenge and the other appeared only 6 days after challenge. Western blot analyses indicated that several of the proteins were chitinases and β-1, 3-glucanases. The patterns of PR proteins, chitinase, and β-1, 3-glucanases for tobacco systemically immunized against blue mold and TMV by foliar inoculation with TMV are very similar to those for tobacco immunized by stem injection with *P. tabacina* (Ye and Kuć, 1988a, b).

The role of PR proteins, including chitinases and β-1, 3-glucanases, in the defense of plants against microorganisms is still unclear. Fraser (1981, 1982) and Fraser and Clay (1983) observed that a temporal or quantitative correlation between the concentration of PR-1a (the commonest PR protein) and the amount of induced resistance in tobacco infected with TMV was not apparent. Resistance was observed in leaves before detectable accumulation of PR protein. Abscisic acid sprayed on plants induced resistance without inducing PR proteins. Low doses of methyl benzimidazole-2yl-carbamate caused accumulation of PR proteins but not resistance. *Nicotiana glutinosa* plants accumulated large amounts of PR protein after infection, but were more susceptible to a second inoculation. Some positive correlations have been observed, however, between the accumulation of PR proteins and induced systemic resistance (Metraux *et al.*, 1988; Tuzun *et al.*, 1988; Ye and Kuć, 1988a, b). The localization of these proteins relative to pathogens, the time course of their appearance in intact plant-pathogen systems, and their relation to development of pathogens in the plant require further elucidation. The identification of the biological activity of PR proteins, in addition to chitinases and β-1, 3-glucanases, may reveal that they are important in induced resistance with all classes of pathogens at either the regulatory level or as components of disease resistance mechanisms.

4. GENE EXPRESSION AS IT RELATES TO INDUCED SYSTEMIC RESISTANCE—APPLICATIONS OF MOLECULAR BIOLOGY

Disease resistance in plants appears to be determined by the speed and magnitude with which genetic information for resistance mechanisms is expressed and by the activity of gene products, rather than by the presence or absence of genetic information for such mechanisms (Dean and Kuć, 1987a; Kuć, 1982, 1983, 1984, 1985a, b, 1987a, b; Kuć and Preisig, 1984; Tuzun and Kuć, 1989). A genetically incompatible interaction is marked by the early recognition of the invading pathogen and activation of the defense genes resulting in the inhibition of the spread of the pathogen. In a compatible interaction the pathogen masks recognition, recognition is lacking, or the pathogen suppresses the defense response. The plant often responds to the pathogen by the widespread accumulation of defense gene products when the pathogen has already established itself in the plant. The delayed plant response may be due to endogenous plant signals or elicitors released from damaged plant and/or microbial cells in the plant. If the spread of the pathogen could be prevented until the plant can respond to the presence of the pathogen, theoretically, a compatible interaction would be converted into an incompatible interaction. This is achieved by inoculating the plant with an avirulent pathogen, or a virulent pathogen whose spread is restricted by controlled inoculations. Induced systemic resistance, as it exists in nature or practiced in plant research laboratories, is based on this principle. The ability of plants to respond to pathogens away from the site of induction presupposes the presence of a translocatable factor or signal. The genes involved in the molecular recognition of the pathogen, the chemical nature of the signal, and the regulation of signal and gene product formation are not established. However, several of the genes encoding the enzymes involved in the accumulation of gene products associated with resistance have been cloned and characterized in the last few years. The induction of several of these genes, each by several stimuli, confirmed the previous observations that disease resistance is multifaceted and regulated by more than one mechanism (Dean and Kuć, 1987a; Kuć, 1982, 1983, 1984, 1985a–c, 1987a, b; Kuć and Preisig, 1984; Tuzun and Kuć, 1989).

Corbin and co-workers (1987) characterized three different transcripts of HRGP in bean induced by a fungal elicitor, wounding, or infection with *C. lindemuthianum*. These transcripts are encoded by separate genes and HRGP genes are differentially regulated in terms of both the kinetics of transcript accumulation and the overall pattern of activation in wounded hypocotyls and hypocotyls infected with an incompatible or a compatible race of the fungus *C. lindemuthianum*. The HRGP genes are activated at a distance from the site of infection.

The operation of several distinct signal systems would allow the plant to activate similar defense responses under very different biological circumstances, related to the expression of different forms of disease resistance including localized hypersensitivity, induced immunity, and lesion limitation as well as prevention of infection after mechanical damage (Corbin *et al.*, 1987; Dean and Kuć, 1987a).

To determine whether activation of specific genes relates to the expression of corresponding defense responses, Lawton and Lamb (1987) analyzed the transcripts completed *in vitro* in isolated nuclei from suspension-cultured bean cells treated with an elicitor. Induction of the synthesis and accumulation of phenylalanine-ammonia lyase (PAL), chalcone synthetase (CHS), and HRGP mRNAs in elicited cells and wounded or infected hypocotyls reflected marked stimulation of the transcription of the corresponding genes. Transcriptional activation occurred not only in directly infected tissue but also in distant, hitherto uninfected tissue. Lagramini and co-workers (1987) reported the cloning of cDNA for tobacco peroxidase. Using these clones as probes should help reveal the role of peroxidase in infection and development. Davidson and co-workers (1988), working with near-isogenic cultivars of resistant and susceptible barley, demonstrated that genes for resistance function as regulators for the synthesis of common host mRNAs during the early stages of infection. Ecker and Davis (1987) are among many who have suggested that ethylene is a signal for plants to activate defense genes, based on their observation of large increases of PAL, 4-coumarate CoA-ligase, CHS, and HRGP mRNAs in carrot roots treated with ethylene.

Another area where much progress has occurred due to molecular biology technology is in defining the role of PR proteins. Carr and co-workers (1985) observed that mRNAs encoding the PR1 family of proteins are present only in tobacco plants that are synthesizing these proteins, indicating that their synthesis is regulated at the transcriptional level. Somssich and co-workers (1986) reported that regulation of PR protein synthesis occurs at the transcriptional level in parsley cell suspension cultures treated with a cell wall preparation from *Phytophthora megasperma* f.sp. *glycinea*. Several PR proteins were characterized as chitinases and glucanases (Legrand *et al.*, 1987; Kauffmann *et al.*, 1987; Kombrink *et al.*, 1988). Pfitzner and Goodman (1987) reported the presence of multiple PR-protein genes in the genomes of tobacco and tomato. Hooft van Huijsduijnen and co-workers (1986, 1987) have shown that steady-state levels of acidic and basic chitinases are increased 100-fold in tobacco after TMV infection, and synthesis of acidic and basic chitinases is controlled at the transcriptional level.

Induction of systemic resistance against viral diseases using molecular biology technology has been reported. Powell Abel and co-workers (1986) observed that transgenic tobacco plants expressing a cDNA coding for the coat protein gene of the U1 strain of TMV were resistant to disease caused by the same virus. The plants expressing the coat protein also had delayed symptom development after inoculation with a severe strain of TMV, PV230, which is immunologically related to the U1 strain. The presence of coat protein on the challenge virus was necessary for maximum protection (Nelson *et al.*, 1987). Similar observations were made by Tumer and co-workers (1987) in tomato and tobacco plants expressing a chimeric gene coding for alfalfa mosaic virus (AlMV) coat protein.

5. CONCEPTS AND APPLICATIONS FOR THE FUTURE

In the future, isolation and characterization of the *cis*-acting sequences and *trans*-acting regulatory proteins responsive to different stimuli will provide a direct approach for analysis of the molecular mechanisms underlying intercellular transduction signals. Expression in transgenic cells of chimeric genes encoding reporter proteins regulated by different stress-responsive gene promoters will provide bioassays for factors involved in the signal pathways (Corbin *et al.*, 1987). Genetic transformation experiments may reveal whether the basic endogenous signaling mechanisms for induction of host gene expression are the same in unrelated plant species. Possible consensus sequences regulating stress-related genes in plants can be derived from comparative studies of the putative regulatory regions of specifically induced members of multigene families and genes encoding metabolically unrelated enzymes (Dixon *et al.*, 1987).

Because of the lag time involved in the expression of disease resistance in induced systemic resistance studies, this system may be ideal for tracking the signal involved in the activation of defense genes removed from the site of initial inoculation. Use of techniques like organomercurial affinity chromatography in this system may permit the determination of the spatial and temporal prevalence of mRNAs involved in induced resistance.

It seems possible that plants can be modified such that the genes involved in defense response are activated by developmental signals. Such modifications may impose a burden on the energy budget of plants, and this can conceivably be circumvented by engineering the expression of these genes in specific tissues and organs (Lamb *et al.*, 1986; Rosahl *et al.*, 1987). It may be possible to enhance induced resistance by mobilizing the secondary messengers involved in signal transduction. Progress in identification of signals involved in disease response may ultimately lead to the use of inexpensive and harmless chemicals that can activate the disease response genes or sensitize the plant to respond only when and where infected, thus conserving energy. Recent evidence is available to support this suggestion (Doubrava *et al.*, 1988; Gottstein and Kuć, 1989). Induced systemic resistance has been proven effective against a broad spectrum of pathogens in the greenhouse and field. Continuous and coordinated efforts by biochemists, physiologists, plant pathologists, agronomists, extension specialists, and farmers will be required to understand the molecular basis of the phenomenon and to exploit it for maximum benefit to all of us.

6. REFERENCES

Ahl, P., and Gianinazzi, S., 1982, b-protein as a constitutive component in highly (TMV) resistant interspecific hybrids of *N. glutinosa* × *N. debneyi*, *Plant Sci. Lett.* **26**:173–181.

Bartnicki-Garcia, S., 1968, Cell wall chemistry, morphogenesis, and taxonomy of fungi, *Annu. Rev. Microbiol.* **22**:87–108.

Boller, T., and Metraux, J. P., 1988, Extracellular localization of chitinase in cucumber, *Physiol. Mol. Plant Pathol.* **33**:11–16.

Boller, T., and Vogeli, U., 1984, Vacuolar localization of ethylene-induced chitinase in bean leaves. *Plant Physiol.* **74**:442–444.

Boller, T., Gehri, A., Mauch, F., and Vogeli, U., 1983, Chitinase in bean leaves: Induction by ethylene, purification, properties, and possible function, *Planta* **157**:22–31.

Bostock, R., Kuć, J., and Laine, R., 1981, Eicosapentaenoic and arachidonic acids from *Phytophthora infestans* elicit fungitoxic sesquiterpenes in potato, *Science* **212**:67–69.

Bostock, R., Laine, R., and Kuć, J., 1982, Factors affecting the elicitation of sesquiterpenoid phytoalexin accumulation by eicosapentaenoic and arachidonic acids in potato, *Plant Physiol.* **70**:1417–1424.

Buonaurio, R., Torre, G. D., and Montalbini, P., 1987, Soluble superoxide dismutase (SOD) in susceptible and resistant host-parasite complexes of *Phaseolus vulgaris* and *Uromyces phaseoli*, *Physiol. Mol. Plant Pathol.* **31**:173–184.

Carr, J. P., Dixon, D. C., and Klessig, D. F., 1985, Synthesis of pathogenesis-related proteins in tobacco is regulated at the level of mRNA accumulation and occurs on membrane-bound polysomes, *Proc. Natl. Acad. Sci. USA* **82**:7999–8003.

Caruso, F., and Kuć, J., 1977, Field protection of cucumber, watermelon and muskmelon against *Colletotrichum lagenarium* by *Colletotrichum lagenarium*, *Phytopathology* **67**:1290–1292.

Caruso, F., and Kuć, J., 1979, Induced resistance of cucumber to anthracnose and angular leaf spot by *Pseudomonas lachrymans* and *Colletotrichum lagenarium*, *Physiol. Plant Pathol.* **14**:191–201.

Chester, K. S., 1933, The problem of acquired physiological immunity in plants, *Q. Rev. Biol.* **8**:129–154, 275–324.

Clark, A. E., and Stone, B. A., 1962, β-1, 3-glucan hydrolases from the grape vine (*Vilis vinifera*) and other plants, *Phytochemistry* **1**:175–188.

Corbin, D. R., Sauer, N., and Lamb, C. J., 1987, Differential regulation of a hydroxyproline-rich glycoprotein gene family in wounded and infected plants, *Mol. Cell. Biol.* **7**:4337–4344.

Costa, A. S., and Muller, G. W., 1980, Tristeza control by cross protection: A U.S.-Brazil cooperative success, *Plant Dis.* **64**:538–541.

Cox, J. E., Fraser, L. R., and Broadbent, P., 1976, Stem pitting of grape fruit. Field protection by the use of mild strains: An evaluation of trials in two climatic districts, in: *Proc. 7th Conf. Int. Org. Citrus Virol.* (E. C. Calavan, ed.), University of Florida Press, Gainesville, pp. 68–70.

Coxon, D., 1982, Phytoalexins from other plants, in: *Phytoalexins* (J. Bailey and J. Mansfield, eds.), Blackie, Glasgow, pp. 106–132.

Cruickshank, I. A. M., and Mandryk, M., 1960, The effect of stem infestation of tobacco with *Peronospora tabacina* Adam on foliage reaction to blue mold, *J. Aust. Inst. Agric. Sci.* **26**:369–372.

Daly, J. M., 1987, Toxins as determinants of plant diseases, in: *Molecular Determinants of Plant Diseases* (S. Nishimura, C. P. Vance, and N. Doke, eds.), Japan Scientific Societies Press/Springer-Verlag, Tokyo/Berlin, pp. 119–126.

Darvill, A. G., and Albersheim, P., 1984, Phytoalexins and their elicitors—A defense against microbial infection in plants, *Annu. Rev. Plant Physiol.* **35**:243–275.

Davidson, A. D., Manners, J. M., Simpson, R. S., and Scott, K. J., 1988, Altered host gene expression in near-isogenic barley conditioned by different genes for resistance during infection by *Erysiphe graminis* f.sp. *hordei, Physiol. Mol. Plant Pathol.* **32**:127–139.

Dean, R. A., and Kuć, J., 1986a, Induced systemic protection in cucumber: The effect of inoculum density on symptom development caused by *Colletotrichum lagenarium* in previously infected and uninfected plants, *Phytopathology* **76**:186–189.

Dean, R. A., and Kuć, J., 1986b, Induced systemic resistance in cucumber: The source of the "signal," *Physiol. Plant Pathol.* **28**:227–233.

Dean, R. A., and Kuć, J., 1987a, Immunization against disease: The plants fight back, in: *Fungal Infection of Plants* (G. F. Pegg and P. G. Ayers, eds.), Cambridge University Press, London, pp. 383–410.

Dean, R. A., and Kuć, J., 1987b, Rapid lignification in response to wounding and infection as a mechanism for induced systemic protection in cucumber, *Physiol. Mol. Plant Pathol.* **31**:69–81.

De Wit, P. J. G. M., Buurlage, M., and Hammond, K. E., 1986, The occurrence of host-pathogen-and infection-specific proteins in the apoplasts of *Cladosporium fulvum* (syn. *Fulvia fulva*) infected tomato, *Physiol. Mol. Plant Pathol.* **29**:159–172.

Dixon, R. A., Gerrish, C., Lamb, C. J., and Robbins, M. P., 1983, Elicitor mediated induction of chalcone isomerase in *Phaseolus vulgaris* cell suspension cultures, *Planta* **159**:561–569.

Dixon, R. A., Bolwell, G. P., Hamdan, M. A. M. S., and Robbins, M. P., 1987, Molecular biology of induced resistance, in: *Genetics and Plant Pathogenesis* (P. R. Day and G. J. Jellis, eds.), Blackwell, Oxford, pp. 245–259.

Doke, N., 1983, Generation of superoxide anion by potato tuber protoplasts during the hypersensitive response to hyphal wall components of *Phytophthora infestans* and specific inhibition of the reaction by supressors of hypersensitivity, *Physiol. Plant Pathol.* **23**:359–367.

Doubrava, N. S., Dean, R. A., and Kuć, J., 1988, Induction of systemic resistance to anthracnose caused by *Colletotrichum lagenarium* in cucumber by oxalate and extracts from spinach and rhubarb leaves, *Physiol. Mol. Plant Pathol.* **33**:69–80.

Durbin, R., 1981, *Toxins in Plant Disease*, Academic Press, New York.

Ebel, J., 1986, Phytoalexin synthesis: The biochemical analysis of the induction process, *Annu. Rev. Phytopathol.* **24**:235–264.

Ecker, J. R., and Davis, R. W., 1987, Plant defense genes are regulated by ethylene, *Proc. Natl. Acad. Sci. USA* **84**:5202–5206.

Elliston, J., Kuć, J., and Williams, E., 1976, Protection of bean against anthracnose by *Colletotrichum* species nonpathogenic on bean, *Phytopathol. Z.* **86**:117–126.

Esquerré-Tugayé, M. T., Mazau, D., Pelissier, D., Roby, D., Rumeau, D., and Toppan, A., 1985, Induction by elicitors and ethylene of proteins associated to the defense of plants, in: *Cellular and Molecular Biology of Plant Stress* (J. L. Key and T. Kosuge, eds.), Liss, New York, pp. 459–473.

Fraser, R. S. S., 1981, Evidence for the occurrence of the 'pathogenesis-related' proteins in leaves of healthy tobacco plants during flowering, *Physiol. Plant Pathol.* **19**:69–76.

Fraser, R. S. S., 1982, Are 'pathogenesis-related' proteins involved in acquired systemic resistance of tobacco plants to tobacco mosaic virus? *J. Gen. Virol.* **58**:305–313.

Fraser, R. S. S., and Clay, C. M., 1983, Pathogenesis-related proteins and acquired systemic resistance: Causal relationship or separate effects? *Neth. J. Plant Pathol.* **89**:283–292.

Fry, S. C., 1982, Isodityrosine, a new cross-linking amino acid from plant cell wall glycoprotein, *Biochem. J.* **204**:449–455.

Gessler, C., and Kuć, J., 1982, Appearance of a host protein in cucumber plants infected with viruses, bacteria and fungi, *J. Exp. Bot.* **33**:58–66.

Gianinazzi, S., and Kassanis, B., 1974, Virus resistance induced in plants by polyacrylic acid, *J. Gen. Virol.* **23**:1–9.

Gianinazzi, S., Ahl, P., Cornu, A., Scalla, R., and Cassini, R., 1980, First report of host b-protein appearance in response to fungal infection in tobacco, *Physiol. Plant Pathol.* **16**:337–342.

Gottstein, H. D., and Kuć, J., 1989, The induction of systemic resistance to anthracnose in cucumber by phosphates, *Phytopathology* **79**:176–179.

Granell, A., Belles, J. M., and Conejerp, V., 1987, Induction of pathogenesis-related proteins in tomato by citrus exocortis, viroid, silver ion and ethephon, *Physiol. Mol. Plant Pathol.* **31**:83–90.

Guedes, M. E., Richmond, S., and Kuć, J., 1980, Induced systemic resistance to anthracnose in cucumber as influenced by the location of the inducer inoculation with *Colletotrichum lagenarium* and onset of flowering and fruiting, *Physiol. Plant Pathol.* **17**:229–233.

Hammerschmidt, R., 1980, Lignification and related phenonolic metabolism in the induced systemic resistance of cucumber to *Colletotrichum lagenarium* and *Cladosporium lagenarium*, Ph.D. thesis, University of Kentucky.

Hammerschmidt, R., and Kuć, J., 1982, Lignification as a mechanism for induced systemic resistance in cucumber, *Physiol. Plant Pathol.* **20**:61–71.

Hammerschmidt, R., Nuckles, E., and Kuć, J., 1982, Association of peroxidase activity with induced systemic resistance in cucumber to *Colletotrichum lagenarium*, *Physiol. Plant Pathol.* **20**:73–82.

Hammerschmidt, R., Lamport, D. T. A., and Muldoon, E. P., 1984, Cell wall hydroxy-proline enhancement and lignin deposition as an early event in the resistance of cucumber to *Cladosporium cucumerinum*, *Physiol. Plant Pathol.* **24**:43–47.

Hooft van Huijsduijnen, R. A. M., Van Loon, L. C., and Bol, J. F., 1986, cDNA cloning of six mRNAs induced by TMV infection of tobacco and a characterization of their translation products, *EMBO J.* **5**:2057–2061.

Hooft van Huijsduijnen, R. A. M., Kauffmann, S., Brederode, F. T., Cornelissen, B. J. C., Legrand, M., Fritig, B., and Bol, J. F., 1987, Homology between chitinases that are induced by TMV infection of tobacco, *Plant Mol. Biol.* **9**:411–420.

Jenns, A., and Kuć, J., 1979, Graft transmission of systemic resistance of cucumber to anthracnose induced by *Colletotrichum lagenarium* and tobacco necrosis virus, *Phytopathology* **69**:753–756.

Jenns, A., and Kuć, J., 1980, Characteristics of anthracnose resistance induced by localized infection with tobacco necrosis virus, *Physiol. Plant Pathol.* **17**:81–91.

Kauffmann, S., Legrand, M., Geoffroy, P., and Fritig, B., 1987, Biological function of 'pathogenesis-related' proteins: Four PR proteins of tobacco have 1, 3-β-glucanase activity, *EMBO J.* **6**:3209–3212.

Kombrink, E., Schroder, M., and Hahlbrock, K., 1988, Several "pathogenesis-related" proteins in potato are 1, 3-β-glucanases and chitinases, *Pro. Natl. Acad. Sci. USA* **85**:782–786.

Kuć, J., 1982, Plant immunization-mechanisms and practical implications, in: *Active Defense Mechanism in Plants* (R. K. S. Wood, ed.), Plenum Press, New York, pp. 157–178.

Kuć, J., 1983, Induced systemic resistance in plants to diseases caused by fungi and bacteria, in: *The Dynamics of Host Defense* (J. A. Bailey and B. J. Deverall, eds.), Academic Press, New York, pp. 191–221.

Kuć, J., 1984, Phytoalexins and disease resistance mechanisms from a perspective of evolution and adaptation, in: *Origin and Development of Adaptation*, Pitman, London, pp. 100–118.

Kuć, J., 1985a, Increasing crop productivity and value by increasing disease resistance through non-genetic techniques, in: *Forest Potentials: Productivity and Value* (R. Ballard, P. Sarnum, G. A. Ritchie, and J. K. Winjum, eds.), Weyerhaeuser Company, Centralia, WA, pp. 147–190.

Kuć, J., 1985b, Induced systemic resistance to plant disease and phytointerferons—Are they compatible? *Fitopatol. Bras.* **10**:17–40.

Kuć, J., 1985c, Expression of latent genetic information for disease resistance in plants, in: *Cellular and Molecular Biology of Plant Stress* (J. L. Key and T. Kosuge, eds.), Liss, New York, pp. 303–318.

Kuć, J., 1987a, Plant immunization and its applicability for disease control, in: *Innovative Approaches to Plant Disease Control* (J. Chet, ed.), Wiley, New York, pp. 255–274.

Kuć, J., 1987b, Translocated signals for plant immunization, *Ann. N.Y. Acad. Sci.* **494**:221–223.

Kuć, J., and Preisig, C., 1984, Fungal regulation of disease resistance mechanisms in plants, *Mycologia* **76(5)**:767–784.

Kuć, J., and Richmond, S., 1977, Aspects of the protection of cucumber against *Colletotrichum lagenarium* by *Colletotrichum lagenarium*, *Phytopathology* **67**:533–536.

Kuć, J., and Rush, J. S., 1985, Phytoalexins, *Arch. Biochem. Biophys.* **236**:455–472.

Kuć, J., and Tuzun, S., 1983, Immunization for disease resistance in tobacco, *Rec. Adv. Tobacco Sci.* **9**:179–213.

Kunoh, H., 1987, Induced susceptibility and enhanced resistance at the cellular level in barley coleoptiles, in: *Molecular Determinants of Plant Diseases* (S. Nishimura, C. P. Vance, and N. Doke, eds.), Japan Scientific Societies Press/Springer-Verlag, Tokyo/Berlin, pp. 59–71.

Lagramini, L. M., Burkhart, W., Moyer, M., and Rothstein, S., 1987, Molecular cloning of complementary DNA encoding the lignin-forming peroxidase from tobacco: Molecular analysis and tissue-specific expression, *Proc. Natl. Acad. Sci. USA* **84**:7542–7546.

Lamb, C. J., Bell, J. N., Cramer, C. C., Dildine, S. L., Grand, C., Hedrick, S. A., Ryder, T. B., and Showalter, A. M., 1986, Molecular response of plants to infection, in: *Biotechnology for Solving Agricultural Problems* (P. C. Augustine, H. D. Danforth, and M. R. Bakst, eds.), Nijhoff, The Hague, pp. 237–251.

Lawton, M. A., and Lamb, C. J., 1987, Transcriptional activation of plant defense genes by fungal elicitor, wounding, and infection, *Mol. Cell. Biol.* **7**:335–341.

Lazarovits, G., and Ward, E. W. B., 1982, Polyphenol oxidase activity in soybean hypocotyls at sites inoculated with *Phytophthora megasperma* f.sp. glycinea, *Physiol. Plant Pathol.* **21**:227–236.

Legrand, M., Kauffmann, S., Geoffroy, P., and Fritig, B., 1987, Biological function of pathogenesis-related proteins: Four tobacco pathogenesis-related proteins are chitinases, *Proc. Natl. Acad. Sci. USA* **84**:6750–6754.

Markwalder, H. U., and Neukom, H., 1976, Diferulic acid as a possible cross link in hemicelluloses from wheat germ, *Phytochemistry* **15**:836–837.

Matsuoka, M., and Ohashi, Y., 1984, Biochemical and serological studies of pathogenesis-related proteins of *Nicotiana* spp., *J. Gen. Virol.* **64**:2209–2215.

Mazau, D., and Esquerré-Tugayé, M. T., 1986, Hydroxyproline rich glycoprotein accumulation in the cell walls of plants infected by various pathogens, *Physiol. Mol. Plant Pathol.* **29**:147–157.

Mellon, J. E., and Helgeson, J. P., 1982, Interaction of a hydroxyproline-rich glycoprotein from tobacco callus with potential pathogens, *Plant Physiol.* **70**:401–405.

Metraux, J. P., and Boller, T., 1986, Local and systemic induction of chitinase in cucumber plants in response to viral, bacterial and fungal infections, *Physiol. Mol. Plant Pathol.* **28**:161–169.

Metraux, J. P., Streit, L., and Staub, T., 1988, A pathogenesis-related protein in cucumber is a chitinase, *Physiol. Mol. Plant Pathol.* **33**:1–9.

Nadolny, L., and Sequeira, L., 1980, Increases in peroxidases are not directly involved in induced resistance in tobacco, *Physiol. Plant Pathol.* **16**:1–8.

Nelson, R. S., Powell Abel, P., and Beachy, R. N., 1987, Lesions and virus accumulation in inoculated transgenic tobacco plants expressing the coat protein gene of tobacco mosaic virus, *Virology* **158**:126–132.

Ohashi, Y., and Matsuoka, M., 1985, Synthesis of stress proteins in tobacco leaves, *Plant Cell Physiol.* **26**:473–480.

Oku, H., Shiraishi, T., and Ouchi, S., 1987, Role of specific suppressors in pathogenicity of *Mycosphaerella* species, in: *Molecular Determinants of Plant Diseases* (S. Nishimura, C. P. Vance, and N. Doke, eds.), Japan Scientific Societies Press/Springer-Verlag, Tokyo/Berlin, pp. 145–156.

Parent, J. G., and Asselin, A., 1984, Detection of pathogenesis-related proteins (PR or b) and of other proteins in the intercellular fluid of hypersensitive plants infected with tobacco mosaic virus, *Can. J. Bot.* **62**:564–569.

Pegg, G. F., 1977, Glucanohydrolases of higher plants: A possible defence mechanism against parasitic fungi, in: *Cell Wall Biochemistry Related to Specificity in Host-Pathogen Interaction* (B. Solheim and J. Raa, eds.), Universitaesforlaget, Tromso, Norway, pp. 305–342.

Pfitzner, U. M., and Goodman, H. M., 1987, Isolation and characterization of cDNA clones encoding pathogenesis-related proteins from tobacco mosaic virus infected plants, *Nucleic Acid Res.* **15**:4449–4465.

Pierpoint, W. S., 1983, The major proteins in extracts of tobacco leaves that are responding hypersensitively to virus-infection, *Phytochemistry* **22**:2691–2697.

Pierpoint, W. S., 1986, The pathogenesis-related proteins of tobacco leaves, *Phytochemistry* **25**:1595–1601.

Politis, D. J., 1976, Ultrastructural study of penetration of maize leaves by *Colletotrichum graminicola*, *Physiol. Plant Pathol.* **3**:465–471.

Powell Abel, P., Nelson, R. S., De, B., Hoffmann, N., Rogers, S. G., Fraley, R. T., and Beachy, R. N., 1986, Delay of disease development in transgenic plants that express the tobacco mosaic virus coat protein gene, *Science* **232**:738–743.

Preisig, C. L., and Kuć, J., 1987, Phytoalexins, elicitors, enhancers, suppressors, and other considerations in the regulation of R-gene resistance to *Phytophthora infestans* in potato, in: *Biochemical and Molecular Determinants of Plant Disease* (S. Nishimura, C. P. Vance, and N. Doke, eds.), Japan Scientific Societies Press/Springer-Verlag, Tokyo/Berlin, pp. 203–221.

Richmond, S., Kuć, J., and Elliston, J., 1979, Penetration of cucumber leaves by *Colletotrichum lagenarium* is reduced in plants systemically protected by previous infection with the pathogen, *Physiol. Plant Pathol.* **14**:329–338.

Roberts, D. A., 1983, Acquired resistance to tobacco mosaic virus transmitted to the progeny of hypersensitive tobacco, *Virology* **124**:161–163.

Roby, D., Toppan, A., and Esquerré-Tugayé, M. T., 1985, Cell-surfaces in plant microorganism interactions. V. Elicitors of fungal and plant origin trigger the synthesis of ethylene and of cell-wall hydroxyproline rich glycoprotein in plants, *Plant Physiol.* **77**:700–704.

Rosahl, S., Schell, J., and Willmitzer, L., 1987, Expression of a tube-specific storage protein in transgenic tobacco plants: Demonstration of an esterase activity, *EMBO J.* **6**:1155–1159.

Ross, A. F., 1966, Systemic effects of local lesion formation, in: *Viruses of Plants* (A. B. R. Beemster and J. Dijkstra, eds.), North-Holland, Amsterdam, pp. 127–150.

Salt, S. D., and Kuć, J., 1985, Elicitation of disease resistance in plants by the expression of latent genetic information, in: *Bioregulators for Pest Control* (P. A. Hedin, ed.), American Chemical Society, Washington, D.C., pp. 47–68.

Sarkanen, K. V., 1971, Precursors and their polymerization, in: *Lignins: Occurrence, Formation, Structure and Reactions* (K. V. Sarkanen and C. H. Ludwig, eds.), Wiley, New York, pp. 99–163.

Schlumbaum, A., Mauch, F., Vogeli, U., and Boller T., 1986, Plant chitinases are potent inhibitors of fungal growth, *Nature* **324**:365–367.

Sequeira, L., 1983, Mechanisms of induced resistance in plants, *Annu. Rev. Microbiol.* **37**:51–79.

Showalter, A. M., Bell, J. N., Cramer, C. L., Bailey, J. A., Varner, J. E., and Lamb, C. J., 1985, Accumulation of hydroxyproline-rich glycoprotein mRNAs in response to fungal elicitor and infection, *Proc. Natl. Acad. Sci. USA* **82**:6551–6555.

Skujins, J. J., Potgeiter, H. J., and Alexander, M., 1965, Dissolution of fungal cell walls by a streptomycete chitinase and β-(1, 3)-glucanase, *Arch. Biochem. Biophys.* **111**:358–364.

Smith, J., and Hammerschmidt, R., 1988, Comparative study of acidic peroxidases associated with induced resistance in cucumber, muskmelon and watermelon, *Physiol. Mol. Plant Pathol.* **33**:255–261.

Somssich, I. E., Schmelzer, E., Bollmann, J., and Hahlbrock, K., 1986, Rapid activation by fungal elicitor of genes encoding "pathogenesis-related" proteins in cultured parsley cells, *Proc. Natl. Acad. Sci. USA* **83**:2427–2430.

Stermer, B. A., and Hammerschmidt, R., 1987, Association of heat shock induced resistance to disease with increased accumulation of insoluble extensin and ethylene synthesis, *Physiol. Mol. Plant Pathol.* **31**:453–461.

Stoessl, A., 1983, Secondary plant metabolites in preinfectional and postinfectional resistance, in: *The Dynamics of Host Defense* (J. Bailey and B. Deverall, eds.), Academic Press, New York, pp. 71–122.

Stolle, K., Zook, M., Shain, L., Hebard, F., and Kuć, J., 1988, Restricted colonization by *Peronospora tabacina* and phytoalexin accumulation in immunized tobacco leaves, *Phytopathology* **78**:1193–1197.

Stumm, D., and Gessler, C., 1986, Role of papillae in the induced systemic resistance of cucumbers against *Colletotrichum lagenarium*, *Physiol. Plant Pathol.* **29**:405–410.

Sun, T. J., Melcher, U., and Essenberg, M., 1988, Inactivation of cauliflower mosaic virus by a photoactivatable cotton phytoalexin, *Physiol. Mol. Plant Pathol.* **33**:115–126.

Tumer, N. E., O'Connel, K. M., Nelson, R. S., Sanders, P. R., Beachy, R. N., Fraley, R. T., and Shah, D. M., 1987, Expression of alfalfa mosaic virus coat protein gene confers cross-protection in transgenic tobacco and tomato plants, *EMBO J.* **6**:1181–1188.

Tuzun, S., and Kuć, J., 1985a, A modified technique for inducing systemic resistance to blue mold and increasing growth of tobacco, *Phytopathology* **75**:1127–1129.

Tuzun, S., and Kuć, J., 1985b, Movement of a factor in tobacco infected with *Peronospora tabacina* Adam which systemically protects against blue mold, *Physiol. Plant Pathol.* **26**:321–330.

Tuzun, S., and Kuć, 1987, Persistence of induced systemic resistance to blue mold in tobacco plants derived via tissue culture, *Phytopathology* **77**:1032–1035.

Tuzun, S., and Kuć, J., 1989, Induced systemic resistance to blue mold, in: *Blue Mold of Tobacco* (W. E. McKeen, ed.), American Phytopathological Society Press, St. Paul, pp. 177–200.

Tuzun, S., Nesmith, W., and Kuć, J., 1984, The effect of stem injections with *Peronospora tabacina* and metalaxyl treatment on growth of tobacco and protection against blue mold in the field, *Phytopathology* **74**:804.

Tuzun, S., Nesmith, W., Ferriss, R. S., and Kuć, J., 1986, Effects of stem injections with *Peronospora tabacina* on growth of tobacco and protection against blue mold in the field, *Phytopathology* **76**:938–941.

Tuzun, S., Rao, M. N., Vogeli, U., Schardl, C. L., and Kuć, J., 1988, Early accumulation of β-1, 3-glucanases, chitinases and other b-proteins in tobacco immunized against blue mold, (abstract), *Phytopathology* **78**:1556.

Tuzun, S., Rao, M. N., Vogeli, U., Schardl, C. L., and Kuć, J., 1989, Induced systemic resistance to blue mold: Early induction and accumulation of β-1, 3-glucanases, chitinases, and other pathogenesis-related proteins (b-proteins) in immunized tobacco, *Phytopathology* **79**:979–983.

Van Holst, G., and Varner, J. E., 1984, Reinforced polypropylene II conformation in a hydroxyproline-rich cell wall glycoprotein from carrot root, *Plant Physiol.* **74**:247–251.

Van Loon, L. C., 1977, Induction by 2-chloroethylphosphonic acid of viral like lesions, associated proteins, and systemic resistance in tobacco, *Virology* **80**:417–420.

Van Loon, L. C., 1983, The induction of pathogenesis-related proteins by pathogens and specific chemicals, *Neth. J. Plant Pathol.* **89**:265–273.

White, R. F., 1979, Acetylsalicylic acid (aspirin) induces resistance to tobacco mosaic virus in tobacco, *Virology* **99**:410–412.

Wieringa-Brants, D. H., and Dekker, W. C., 1987, Induced resistance in hypersensitive tobacco against tobacco mosaic virus by injection of intercellular fluid from tobacco plants with systemic acquired resistance, *J. Phytopathol.* **118**:165–170.

Xuei, X. L., Jarlfors, U., and Kuć, J., 1988, Ultrastructural changes associated with induced systemic resistance of cucumber to disease: Host response and development of *Colletotrichum lagenarium* in systemically protected leaves, *Can. J. Bot.* **66**:1028–1038.

Ye, X. S., and Kuć, J., 1988a, Systemic resistance to *Peronospora tabacina* induced by tobacco mosaic virus on tobacco cv. Ky 14, *Phytopathology* **78**:1547 (abstr.).

Ye, X. S., and Kuć, J., 1988b, Induced resistance to blue mold by *Peronospora tabacina* and tobacco mosaic virus elicits a similar pattern of b-proteins, *Phytopathology* **78**:1527 (abstr.).

Young, D. H., and Pegg, G. F., 1981, Purification and characterization of 1, 3-β-glucanase hydrolases from healthy and *Verticillium albo-atrum* infected tomato plants, *Physiol. Plant Pathol.* **19**:391–417.

Young, D. H., and Pegg, G. F., 1982, The action of tomato and *Verticillium albo-atrum* glycosidases on the hyphal wall of *V. albo-atrum*, *Physiol. Plant Pathol.* **21**:411–423.

17

The Plant Membrane and Its Response to Disease

Anton Novacky

1. INTRODUCTION

Increased electrolyte leakage in numerous host-pathogen combinations has led to the concept that "changes in permeability appear to be universally characteristic of diseased plant tissue regardless of disease type or the nature of the pathogenic agent" (Wheeler, 1978). A change in membrane permeability is one of the first detectable symptoms. In some studies with toxic pathogen-produced metabolites (phytotoxins) that reproduce disease symptoms of pathogens, e.g., *Cochliobolus victoriae* toxin (victorin), permeability alterations were found shortly after contact with the host cells was established. It was hypothesized that all subsequent pathological changes may be triggered by this primary event (Wheeler and Luke, 1963). Because the plasma membrane is the first site of interaction between invading microorganisms and the plant protoplast, it became obvious that the plasma membrane plays an important role in disease development. This role was suggested to be even more dramatic in the case of host-selective toxins, toxins injurious only to hosts susceptible to the parental fungal species, because the plasma membrane participates in the recognition process (Daly, 1984). It is becoming more obvious, however, that massive electrolyte leakage may result from general cell injury rather than from the alterations of specific sites at the plasma membrane (e.g., damage caused by oxygen free radical-induced membrane lipid peroxidation) (Thompson *et al.*, 1987).

It is important to bear in mind the fact that, in addition to the necrotrophic pathogens which produce toxins and induce massive electrolyte leakage, there are numerous biotrophic pathogens, e.g., rusts and mildews, that affect permeability in a slightly different way. Biotrophic pathogens affect passive permeability and cause changes in water transport and the transport of nonelectrolytes that depend on the lipophily of individual transported species as was documented in the classical work of Thatcher (1942). Thatcher analyzed cell plasmolysis and deplasmolysis under different nonelectrolyte plasmolytica (for critical review see Wheeler and Hanchey, 1968). The recent plasmometric study of Lee-Stadelmann and co-workers (personal communication) of *Hordeum vulgare* infected with a compatible race of the powdery mildew fungus, *Erysiphe*

Anton Novacky • Department of Plant Pathology, University of Missouri, Columbia, Missouri 65211.

graminis f.sp. *hordei*, demonstrates changes in the permeability to nonelectrolytes. Plasmometric measurements indicated that the infection process did not affect permeability to water and urea, but reduced passive permeability to more lipophilic nonelectrolytes. This study illustrates that membrane alterations in biotrophic pathogenesis are less dramatic than in necrotrophic pathogenesis. The subtlety of these changes is probably the reason for the minimal attention the phenomenon has received from plant pathologists.

In this chapter I will discuss some case histories with ample experimental evidence for the interaction between toxic products of pathogens and host plasma membranes. The focus will be on fungal pathogenesis (compatible combinations) as well as interactions between membranes and fungal elicitors of the resistance response (incompatible combinations).

2. MEMBRANE STRUCTURE AND FUNCTION

The present concept of membrane structure has evolved from the model of Singer and Nicholson (1972). The fluid phospholipid bilayer (lipid matrix) serves as a skeletal structure of the membrane in which several types of proteins are embedded: the H^+-ATPases, carrier proteins, single ion channels, receptors, and coupling proteins (Fig. 1). Any of these in addition to the phospholipids, could pose as targets (binding sites) or be ports of entry for the fungal toxic metabolites. The plasma membrane ATPase, the major protein component ("a pump") found in eukaryotic cells, transports Na^+ in animal cells and H^+ in plant and prokaryotic cells and H^+ and Na^+ in eubacterial cells. Several membrane transport studies during the last two decades have demonstrated that the H^+-ATPase is the center of the energy-dependent ("active") transport across the plant plasma membrane contributing to the generation of electrical transmembrane potentials up to -300 mV. For a description of membrane potential measurements and their interpretation, see Spanswick (1981) and Novacky and co-workers (1989). Present understanding of this transporting mechanism in plant tissues is derived from Mitchell's chemiosmotic principle (Slayman, 1985). Mitchell's hypothesis was developed for the qualitatively different eubacterial ATPase, the F_0F_1 ATP synthetase, that is located in membranes of mitochondria, chloroplasts, or prokaryotic cells. However, this concept may be applied also to the function of the plasma membrane H^+-ATPase which transfers H^+ during ATP hydrolysis in a manner similar to H^+ translocation by the F_0F_1-ATPase but in the energetically uphill direction. The translocation of proton by the H^+-ATPase generates a proton gradient (Mitchell's "proton motive force") which is responsible for the reentry of H^+ into the cell. Reentry of H^+ is coupled to solute transport mechanisms (secondary active transport), such as cotransport of ions, amino acids, and sugars or export of various ions from the cell via antiport (Poole, 1978; Reinhold and Kaplan, 1984). In addition to the plasma membrane H^+-ATPase, plant cells have an H^+-ATPase located at the tonoplast, endoplasmic reticulum, and Golgi apparatus. These ATPases differ fundamentally from the plasma membrane H^+-ATPase in structure and function and seem to be closely related to the eubacterial F_0F_1-type ATPase by an ancestral gene common with archaebacteria (Nelson and Taiz, 1989). Their physiological function, however, is similar: transfer of H^+ from the cytoplasm into the vacuole (Sze, 1984; Rea and Sanders, 1987).

3. INTERACTION WITH THE PLASMA MEMBRANE H⁺-ATPASE

3.1. Fusicoccin

The nonselective toxic product of *Fusarium amygdali*, fusicoccin (FC), a diterpene glucoside, attracted the attention of plant physiologists after reports of its dramatic stimulation of cell

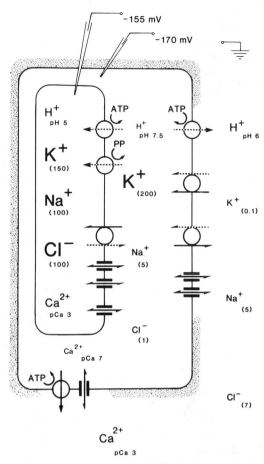

Figure 1. Model of plant cell ion compartmentation in relation to membrane potential measurements with glass micropipettes (microelectrodes). First micropipette shown is crossing both the plasma membrane and the tonoplast with its tip in the vacuole and the second one crossing only the plasma membrane with a tip in the cytosol. Ion concentrations within the cell (millimolar concentrations in parentheses) result from the primary active, energy-dependent, transport of the electrogenic pump (H^+-ATPase), the secondary active transport shown as an H^+/K^+ cotransport, H^+/K^+ countertransport (antiport), and a passive movement of ions through ion channels. In addition to H^+-ATPases (regulating an ion-determined pH 5.0 in the vacuole and pH 7.5 in the cytoplasm) a Ca^{2+}-ATPase is present (maintaining pCa 3.0 in the vacuole and pCa 7.0 in the cytoplasm). (Reproduced from Blatt, 1989, with permission of Academic Press.)

enlargement in common bioassays for auxin activity (Marrè *et al.*, 1971). Since this discovery, FC has become the best known fungal toxin. All activities of FC have been used in numerous plant physiological studies (Marrè, 1979) and have been linked to the primary stimulation of the plasma membrane H^+-ATPase which results in membrane hyperpolarization (i.e., more negative membrane potential), extracellular acidification, influx of potassium and other cations, uptake of chloride and other anions, and uptake of glucose, sucrose, and amino acids (Marré, 1979). Direct interference with the H^+-ATPase was obviously indicated. In parallel to a search for the auxin-binding protein (Libbenga *et al.*, 1986), an intensive attempt to identify the postulated FC-binding site began. Dohrmann and co-workers (1977) reported the existence of an FC-binding protein from maize coleoptiles, and attempts to isolate it continued until 1989 when de Boer and co-workers (1989) successfully isolated and purified the FC-binding proteins from the plasma membrane of oat roots. The 29- and 31-kDa proteins were directly correlated to the [^{3}H]-FC binding activity. The stoichiometry (1:2) of the two protein bands suggested to the authors that the native form of the FC-binding protein is a 92-kDa heterotrimer. These results are complemented by the report of Meyer and co-workers (1989) on the photoaffinity labeling of a 34-kDa plasma membrane polypeptide of *Arabidopsis thaliana* with the tritiated azido analogue of FC. Both studies clearly established the existence of an FC-binding protein located at the plasma membrane. The identity of the FC-binding protein(s) and the relation of binding to FC stimulation of the H$^+$-ATPase were reported by Aducci and Ballio (1989).

Two possibilities for the binding *vis-à-vis* stimulation were discussed by de Boer and co-workers (1989). The activated FC-binding protein may stimulate the H$^+$-ATPase by an induced protein kinase-mediated phosphorylation similar to that described for the bacterial toxin, syringomycin, by Bidwai and Takemoto (1987), or the FC-binding protein might be an outwardly rectifying K$^+$ channel as suggested by the electrophysiological experiments of Blatt (1988) and supported by the flux studies of Clint and MacRobbie (1984). FC binding would result in channel closing and thus in the observed K$^+$ efflux reduction. But the "plugging" of the potassium channel alone cannot be a reason for the H$^+$-ATPase stimulation. It is now well established that plasma membrane single ion channels are voltage-dependent (Hedrich and Schroeder, 1989). Thus, in the case of FC the channel closing may be a consequence of the FC-induced membrane hyperpolarization. Similarly, the FC-stimulated phosphorylation of the 33-kDa polypeptide found by Tognoli and Basso (1987) could be a result of the FC-induced K$^+$-dependent pH increase.

3.2. HS Toxin (Helminthosporoside)

HS toxin is a host-selective toxin produced by *Helminthosporium sacchari*, the causal agent of eyespot disease of sugarcane. The toxin consists of a central bicarbocyclic sesquiterpene aglycone linked to two residues of 5-*O*-(β- galactofuranosyl)-β-galactofuranoside (Macko *et al.*, 1981). Strobel (1973) reported isolation of the HS-toxin receptor protein (48 kDa) from susceptible lines of sugarcane and a similar but slightly defective protein from resistant lines. He postulated a direct interaction with the plasma membrane binding site. This report was questioned by several authors concerning the purity of the toxin preparations and the highly variable bioassays (see review of Yoder, 1980; Daly, 1981). Hence, the existence of an HS-toxin receptor and its role in toxin action must await further investigation. Membrane potential studies, however, suggest the possibility that the HS-toxin binding protein might indeed be present at the plasma membrane of susceptible sugarcane (Schröter *et al.*, 1985). HS toxin depolarized the membranes of susceptible sugarcane cells. The energy-dependent component of membrane potential was suppressed by the toxin. Under anoxic conditions in the dark, when the plasma membrane H$^+$-ATPase was not operating and membranes exhibited only the diffusion potential, the toxin treatments did not affect the membrane potential. Only when oxygen was available, i.e., the H$^+$-ATPase functioned again, was the toxin-induced depolarization observed. The interaction with the H$^+$-ATPase is not yet clear: under light conditions when more energy is supplied to the membrane or in the presence of fusicoccin, the toxin-induced depolarization is much less pronounced (Schröter *et al.*, 1985).

It has been reported that the toxin may be converted to a nontoxic lower homologue by removal of one or more galactose units from the toxin molecule (Livingston and Scheffer, 1983; Macko *et al.*, 1982; Macko, 1983). A nontoxic low homologue minus one galactose unit did not affect the membrane potential. However, it did affect the activity of the complete HS toxin. Membrane depolarization was prevented when HS toxin was combined with the nontoxic lower homologue in a 1:30 ratio (Fig. 2). This observation is consistent with reports that sugarcane

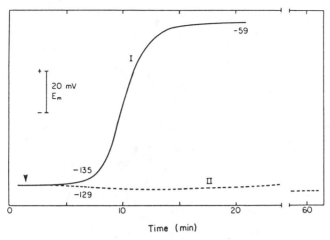

Figure 2. Effect of a nontoxic low homologue of HS toxin on the action of HS toxin on the membrane potential of sugarcane leaf cell: 1 μM HS toxin + 30 μM low homologue of HS toxin (dashed line), 1 μM HS toxin (solid line). (Reproduced from Schröter *et al.*, 1985, with permission of Academic Press.)

leaves are protected against HS toxin by nontoxic lower homologues (Livingston and Scheffer, 1981; Duvick *et al.*, 1984), and may support the existence of the toxin binding site at the plasma membrane: a nontoxic low homologue may occupy a site and thus prevent the binding of the HS toxin molecule.

Membrane potential experiments at different temperatures also suggest a possible interference of HS toxin with the plasma membrane. At high temperatures (above 30°C), sugarcane was less sensitive to HS toxin (Byther and Steiner, 1975; Strobel, 1979), and membranes did not depolarize. Membrane repolarization or depolarization occurred after shifting the temperature upwards and downwards around 30°C (Schröter *et al.*, 1985).

4. INTERACTION WITH A SINGLE ION CHANNEL

Zinniol

Interference of a phytotoxin with single ion channel activity was demonstrated by Thuleau and co-workers (1988) in a study of the binding of zinniol to protoplasts and membranes from sensitive plant host lines. Zinniol [1, 2-bis(hydroxymethyl)-3-methoxy-4-methyl-5-(3-methyl-3-butenyloxy) benzene] is produced by fungi from the *Alternaria* group and causes symptoms similar to those induced by *Alternaria* spp. (Barash *et al.*, 1981). [3H]-Zinniol binds to carrot protoplasts and to microsomes (Fig. 3). The H^+-ATPase in protoplasts is not affected by binding; furthermore, the toxin does not affect Ca^{2+}-ATPase activity in isolated microsomes. A marked stimulation of Ca^{2+} influx by zinniol, however, was observed when the Ca^{2+} flux was monitored in protoplasts (Fig. 4). Toxin not only stimulated calcium entry, but also inhibited (by 50%) the action of a calcium channel blocker (e.g., verapamil) in zinniol-sensitive plants. The authors consider zinniol to be a calcium channel agonist. But the target of the toxin and calcium channel blockers are not exactly the same: the data suggest the existence of at least two distinct sites, one of which is common to both toxin and the blockers, and the other specific to the blockers. Zinniol is an example of a toxin that binds to a specific plasmalemma protein and interferes with its function. The control of various cellular processes, including cellular organization, is thought to

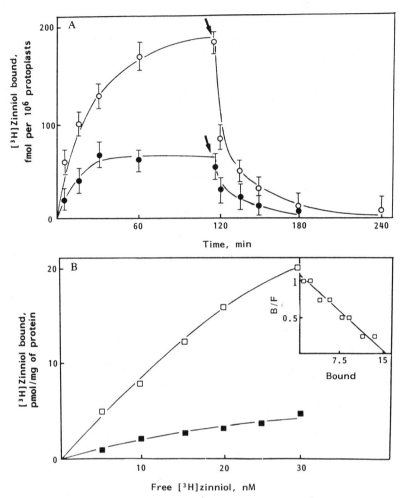

Figure 3.   Binding of [³H] zinniol to carrot protoplasts and microsomes. (A) Kinetics of [³H] zinniol binding to sensitive (●) and resistant (○) lines of carrot protoplasts. Addition of unlabeled zinniol is indicated by arrows (B) Equilibrium binding of [³H] zinniol to microsomes from sensitive lines in the presence (■) or absence (□) of unlabeled zinniol. (Insert) Scatchard plot of the specific binding component: bound pmoles per mg. of protein; bound/free (B/F), ml/mg. (Reproduced from Thuleau *et al.*, 1988, with permission of the authors.)

be directed by calcium (Kauss, 1987). A sudden change in the cytosolic calcium concentration may have far-reaching consequences for the cell. In the case of zinniol, toxin-binding and opening of calcium channels are perhaps the only steps necessary for the pathogen to kill the cell.

5.   THE PLASMA MEMBRANE AS A POSSIBLE TOXIN TARGET

5.1.   Tentoxin

Tentoxin (cyclo-*N*-methyl-L-alanyl-L-leucyl-*N*-methyl-trans-dehydrophenylalanyl-glycyl) is an example of a nonselective toxin produced by a fungus (*Alternaria alternata*) which has a

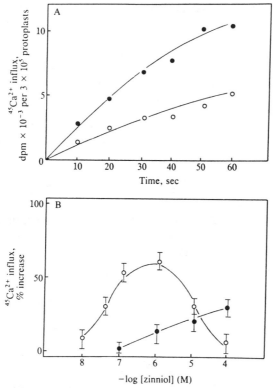

Figure 4. (A) $^{45}Ca^{2+}$ influx into carrot protoplasts in the absence (○) or presence (●) of 1μM zinniol. (B) Effect of zinniol on Ca^{2+} influx in sensitive (○) and resistant (●) cell lines. (Reproduced from Thuleau *et al.*, 1988, with permission of the authors.)

highly specific host range. It causes chlorosis in seedlings of many plant species, much more than are hosts of the fungus, indicating that the pathogen might be able to extend its host range by acquiring the ability to synthesize and export toxins (Durbin, 1988). The observation of seedling chlorosis and toxin action on stomatal movement (Durbin and Uchytil, 1976) prompted research into toxin interference with the photosynthetic apparatus. Discovery that the chloroplast-coupling factor (CF_1) is inhibited by the binding of tentoxin to a single site on the (α or β) ATPase subunits (Steele *et al.*, 1976) made this toxin the first phytotoxin with a known "site of action."

Binding to the catalytic β subunit inhibited or stimulated the ATPase activity depending on toxin concentration while photophosphorylation was inhibited by binding to the nucleotide binding subunit of CF_1 and by interference of the toxin with photosynthetic electron flow (Klotz, 1988). In agreement with this idea, the membrane depolarization by tentoxin found in *Lemna gibba* and *Egeria densa* (Böcher and Novacky, 1981; Dahse *et al.*, 1988) was thought to result from an inhibition of energy transfer to the plasma membrane following the primary toxin action on photophosphorylation. However, membrane depolarization by tentoxin in some instances, e.g., in the dark, is obviously unrelated to photophosphorylation as is the counteraction of the fusicoccin-induced opening of stomata in the dark (Durbin and Uchytil, 19736. In subsequent studies, several additional membrane effects of tentoxin were detected that could not be explained as an inhibition of photophosphorylation (Klotz, 1988). Perhaps the most important of them is the effect of tentoxin in picomolar concentrations on ion transport in wheat roots, because concentrations in the picomolar range are two orders of magnitude lower than active concentrations of plant hormones. Tentoxin in picomolar concentrations not only decreased the potassium influx into roots but also increased the translocation of potassium to the shoots (Erdei and Klotz, 1988). The tentoxin effect was noticed only in plants of an intermediate potassium status in which K^+ uptake is (secondary) active and obeys a negative feedback regulation. Furthermore, the secondary active efflux, and not the passive influx of calcium across the plasma membrane, was affected by tentoxin. Tentoxin was active at a concentration of 1 pM *in vivo* as well as in the *in vitro* enzyme assays indicating that the toxin is reaching its targets in the cell. The effective concentration range of tentoxin is rather wide: from micromolar in membrane potential experiments to picomolar in K^+ ion transport experiments.

An important property of tentoxin was found in experiments with lipid bilayer membranes and with unilamellar liposomes. Tentoxin increased the K^+ conductivity of lipid bilayer membranes. This and the pattern of its partition between immiscible solvents were used to propose the interaction of tentoxin with the lipid matrix as a possible new mechanism of facilitated ion transport (Klotz *et al.*, 1987). Experiments in which tentoxin caused a dramatic leakage of K^+ from K^+-preloaded unilamellar liposomes (Hartung *et al.*, 1987) were compatible with these findings (Fig. 5). This property might be an important prerequisite for rapid penetration across membranes by the toxin (Klotz, 1988).

5.2. Victorin

The action of the *Cochliobolus victoriae* toxin, victorin, at the plasma membrane of susceptible oat cultivars was postulated in the early years after the toxin discovery by Meehan and Murphy (1947) when a rapid (5 min) induction of electrolyte leakage followed toxin treatment (Wheeler, 1978). All other toxin-induced alterations were observed much later. Electrophysiological measurements (Novacky and Hanchey, 1974; Gardner *et al.*, 1974) revealed a rapid (2–4 min) irreversible membrane depolarization in susceptible cells. These observations correlated well with the compartmental analysis of changes in permeability in victorin-treated tissues (Keck and Hodges, 1973) and supported the hypothesis that the plasma membrane is the site

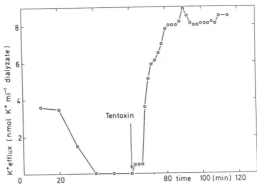

Figure 5. Effect of tentoxin (75 μM) on K^+ leakage from unilamellar liposomes preloaded with K^+. (Reproduced from Hartung *et al.*, 1987, with permission of Elsevier.)

of victorin action. However, evaluation of the site of action was difficult because only crude toxin preparations in concentrated form were used.

Victorin was purified in 1985 (Wolpert *et al.*, 1985; Macko *et al.*, 1985). Victorin C, a cyclic peptide, is recognized as the major host-selective toxin produced by the fungus. Cell membranes of susceptible oats depolarized after addition of the purified toxin (6–125 nM) following a lag period of 10–15 min (Novacky *et al.*, 1986). The lag period was independent of the pH but shortened to 6.5 min by an increased toxin concentration (1.25 μM). The victorin-induced membrane depolarization caused a loss of the energy-dependent component of the membrane potential. After a few minutes the membrane potential began to decline to a level below the diffusion potential indicating the onset of electrolyte leakage. Both membrane potential and membrane resistance were measured in subsequent experiments with root cap cells. Parallel to membrane depolarization, specific membrane resistance decreased indicating an increase in electrical conductance (Novacky *et al.*, 1989). With a new patch clamp technique (Hedrich and Schroeder, 1989), single ion channel activities of susceptible oat protoplasts were studied. Voltage-dependent channels with conductances ranging between 23 and 150 pS were identified. No victorin-induced alterations in the activity of single ion channels, however, were detected. Channels inactive at the time of toxin application were not reopened. Hence, it was concluded that victorin does not form new or activate present membrane channels (Mayer and Novacky, 1989).

Victorin may interact with the recently identified 100-kDa victorin-binding protein (Wolpert and Macko, 1989). This event would lead to alterations (e.g., in intracellular pH) detected then as a membrane depolarization. The location of the binding protein has not been established and, therefore, the possibility of a binding site at the plasma membrane remains unknown.

5.3. Cercosporin

Cercosporin, a nonspecific toxin, is produced by members of the genus *Cercospora* which cause leaf spot diseases on several hosts. It was first isolated in 1957 by Kuyama and Tamura from *Cercospora kikuchii* and later detected in other *Cercospora* species (see Daub, 1982a). The toxin was characterized independently by two groups (Lousberg *et al.*, 1971; Yamazaki and Ogawa, 1972). Cercosporin produces symptoms on a wide range of host plants. Structurally it is related to several photosensitive compounds that absorb light to form an electron-excited state. This energy is subsequently transferred to oxygen generating one of two highly reactive products, singlet oxygen ($^1O_2$) or superoxide (O_2^-), that produce the damage. Photosensitizing properties of cercosporin are expressed only after irradiation. In the dark, cercosporin is nontoxic even at high concentrations.

Electrolyte leakage from several plant species was detected after a 30-min exposure to cercosporin (Macri and Vianello, 1979) suggesting that this effect could be a secondary one. Subsequent studies, however, showed unequivocally that membrane damage was the primary toxic effect. Lipid peroxidation caused by cercosporin was demonstrated *in vitro* by Covallini and co-workers (1979) and *in vivo* by Daub (1982b). Later the generation of singlet oxygen or superoxide by cercosporin was shown by Daub and Hangarter (1983). In an electron spin resonance analysis of tobacco cell membranes treated with cercosporin, Daub and Briggs (1983) found that cercosporin caused a marked increase in the ratio of saturated to unsaturated fatty acids, an increase in the phase transition temperature, and, hence, a decrease in membrane fluidity causing electrolyte leakage. Quenchers of singlet oxygen delayed cell death caused by cercosporin (Daub, 1982a) and the superoxide generated by cercosporin was scavenged by superoxide dismutase (Daub and Hangarter, 1983). Thus, the mode of action of cercosporin through singlet oxygen and superoxide was confirmed.

The cercosporin effect on plant host cell membranes may be a rather unique mechanism of toxin action, but the attack of activated oxygen species on cell membranes is not (see Section 6.2).

6. RECOGNITION OF AN INCOMPATIBLE PATHOGEN

6.1. Alterations of Membrane Potential

Plant recognition of certain structures produced by incompatible pathogenic fungi ("elicitors"), leading to the induction of a cascade of biochemical events which culminates in the expression of resistance, is a very likely concept of the incompatible combination (Yoshikawa, 1983). The existence and location of putative recognition sites ("receptors") in the plant are unknown. Nevertheless, rapid membrane depolarization after treatment of potato cells with the elicitors from the cell wall of *Phytophthora infestans* (Kota and Stelzig, 1977), or tobacco cells with the elicitors of *P. parasitica* var. *nicotianae* and cantaloupe cells with the elicitors of *Colletotrichum lagenarium* (Pelissier *et al.*, 1986), and agglutination of potato protoplasts (Peters *et al.*, 1978) have been interpreted as evidence of the elicitor action at the cell surface.

Mayer and Ziegler (1988) found, similarly to previous reports, immediate depolarization after application of a highly purified glucan fraction (average molecular mass of 4–8 kDa) isolated from cell walls of *P. megasperma* f.sp. *glycinea* (Ziegler and Pontzen, 1982). Membranes depolarized only transiently and after a repolarization they hyperpolarized (Fig. 6). A comparison of the concentration dependence for the elicitation of phytoalexin synthesis with that for the elicitation of membrane depolarization revealed that, while an elicitor concentration of 0.1 μg/ml induced nearly maximum phytoalexin synthesis, it did not depolarize membranes. The pH dependence of phytoalexin synthesis and membrane depolarization were also different. Although phytoalexin synthesis has an optimum at pH 6.0, membranes depolarized similarly at all pH values tested. Mayer and Ziegler (1988) rightly doubt that the membrane depolarization has a role in the induction process. They also rule out the possibility of elicitor uptake via the H^+ co-transport system since an increase in the elicitor concentration did not increase the amplitude of membrane depolarization. The authors did find, however, a correlation between values of membrane hyperpolarization and phytoalexin synthesis: the concentration range of elicitor that induced maximum phytoalexin synthesis also hyperpolarized membranes. The membrane hyperpolarization may be interpreted as a signal transmission that induces the phytoalexin synthesis. However, as the authors point out, numerous metabolic processes (e.g., alterations of the cytosolic pH) may cause a change in membrane potential values; hence, these interpretations should be considered with caution.

Low and Heinstein (1986) chose another approach to investigate the effect of a phytoalexin-elicitor stimulation. The authors used fluorescent probes to monitor the effect of an elicitor extracted from a wilt-causing pathogen, *Verticillium dahliae*, on the membrane potential and the

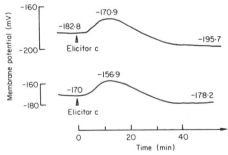

Figure 6. Effect of an elicitor from the cell wall of *Phytophthora megasperma* f.sp. *glycinea* on the membrane potential of soybean cotyledonary cells. Responses of two cells are shown. (Reproduced from Mayer and Ziegler, 1988, with permission of Academic Press.)

cytosolic pH in suspension cell cultures of cotton (*Gossypium arboreum*). They monitored, in parallel, changes in fluorescence intensity of the membrane potential-sensitive dyes, oxonol and carbocyanine, and the pH-sensitive dye, pyranine, after application of the elicitor. Within a few min after the contact with the elicitor extract (10 μl), a dramatic decline in fluorescence intensity occurred (Fig. 7). The lag period between elicitor addition and the onset of the fluorescence decrease was 4–5 min. The abrupt change in fluorescence indicates a highly cooperative response experienced by most cells in the fluorescence cuvette. A gradual increase in elicitor concentration up to 100 μl shortened the lag period to less than 30 sec. Since similar responses were obtained with two membrane potential probes and an ion-selective probe, the authors speculate that the elicitation involves the opening and/or closing of a single ion channel(s) in the plasma membrane. Although this might be true, it is difficult to evaluate data obtained with a heterogeneous elicitor preparation. However, there is at least one other report on elicitor-induced change of membrane permeability (Köhle *et al.*, 1985).

The involvement of calcium during the elicitation process was suggested by Young and Kauss (1983). They found a rapid release of $^{45}Ca^{2+}$ from soybean suspension culture cells after treatment with the elicitor chitosan. This release occurred more quickly than the chitosan-induced leakage of intracellular electrolytes indicating that the calcium was released primarily from the cell wall and/or plasma membrane. The authors suggested that chitosan release by the fungus may cause a change in calcium concentration of plant cell wall and plasma membrane. But they did not rule out a release of calcium from an intracellular site since it was reported that chitosan can penetrate plant cells (Hadwiger *et al.*, 1981).

6.2. Membranes as Targets of Activated Oxygen

Species of activated oxygen—superoxide (O_2^-); the perhydroxy radical ($HO_2^{\cdot}$), a protonated form of superoxide; hydrogen peroxide (H_2O_2); the hydroxy radical ($OH^{\cdot}$); and a nonradical excited state of oxygen, singlet oxygen ($^1O_2$)—are generated in animal and plant cells as a by-product of numerous enzymatic reactions (Elstner, 1982). Because they are highly reactive and thus capable of causing serious damage, activated oxygen species within the cell are under strict control of two protective enzymes, superoxide dismutase and catalase, and several antioxidants such as glutathione, ascorbate, tocopherol, and polyamines (Elstner, 1982; Freeman and Crapo, 1982; Gregory and Fridovich, 1973). It is increasingly evident that oxygen free radicals are involved in processes of oxidative deterioration during plant senescence and wounding and, not surprisingly, also in plant pathogenesis. Mainly the unsaturated fatty acids of membrane lipids are the primary targets of oxygen free radicals although numerous macromolecules can be damaged. Once such an attack occurs, it triggers a self-perpetuating wave of free radical production (Thompson *et al.*, 1987).

An increased level of oxygen free radicals during fungal pathogenesis or hypersensitive reaction may result from the increased production of free radicals or the inhibition of protective enzymes or antioxidants. For example, the previously mentioned membrane hyperpolarization of soybean cells after elicitor stimulation (Mayer and Ziegler, 1988) may well reflect elicitor-induced alterations that either disturb the system of free radical control or induce free radical generation. Involvement of the superoxide anion in the fungal hypersensitive reaction was convincingly demonstrated in potato tuber tissue infected with an incompatible race of *P. infestans* (Doke, 1985, and therein). Superoxide production was detected as the reduction of extracellular cytochrome *c* by aged potato disks. An increase in reduction activity was detected 1 hr after inoculation. It increased for 3–4 hr and then declined. Very little reduction activity was detected in tissues infected with a compatible race of the fungus (Doke, 1983a). A similar increase in cytochrome *c* and/or nitroblue tetrazolium reduction was observed after treatments of potato disks with hyphal cell wall components (an HR elicitor). A significant increased reduction of

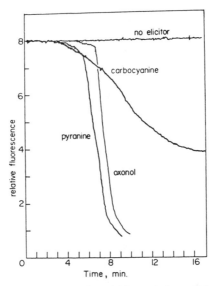

Figure 7. Effect of an elicitor extract from *Verticillium dahliae* on the intracellular pH and membrane potential of suspension culture cells of cotton (*Gossypium arboreum*). Cells were incubated with pH-sensitive (pyranine) or membrane potential-sensitive (carbocyanine and oxonol VI) dyes. An addition of elicitor caused a sharp decrease of dye fluorescence of all three dyes used. (Reproduced from Low and Heinstein, 1986, with permission of Academic Press.)

cytochrome *c* was found already 2 min after addition of hyphal wall components when the same experimental protocol was used with isolated protoplasts. This indicates that the superoxide generating system is activated immediately after contact with an elicitor (Doke, 1983b). The dependence of cytochrome *c* reducing activity on NADPH was found later in membrane fractions isolated from wounded potato tuber tissues and it was concluded that superoxide generation is dependent on NADPH (Doke, 1985). The author suggested that an NADPH oxidase system may be activated by the hypersensitive situation. In a subsequent study, superoxide generation in response to *P. infestans* infection was found also in leaves (Chai and Doke, 1987). In this tissue superoxide generation occurred in two stages, before and after fungal penetration. The first increase was a nonspecific one that developed in response to either the compatible or incompatible race or to the zoospore germination fluid. The second stage occurred only after inoculation with an incompatible race and was related to the hypersensitive response. Leaves of nonhost plants (bean, soybean, cowpea, and tobacco) inoculated with the fungus or treated with the germination fluid also exhibited superoxide generation. However, in these leaves the second stage of superoxide generation was absent. Doke's fungal experiment, together with the reports on superoxide generation in plants undergoing hypersensitive response against viruses (Doke and Ohashi, 1988) and bacteria (Keppler and Novacky, 1987), strongly suggest that the generation of superoxide represents a general defense system against incompatible pathogens.

7. CONCLUSION

From several case histories presented in this chapter it is obvious that situations where the primary site of pathogen interference with the membrane is unequivocally established are rather limited. The case of fusicoccin might be a unique exception with the well-established existence of fusicoccin-binding sites at the plasma membrane and demonstrated relationship between fusicoccin binding and the activity of H^+-ATPase. In several other situations, e.g., HS toxin or elicitors of phytoalexins, there is only indirect evidence for an interaction with the plasma membrane.

Another mechanism by which fungal pathogens can affect host cell membranes is the triggering of oxygen free radicals. Membrane damage by activated oxygen so far described indicates that this could be a more general phenomenon. The strong leakage of electrolytes found in many diseased tissues may be a consequence of membrane lipid peroxidation induced by oxygen free radicals. This assumption might be supported by the recent finding that victorin treatment of susceptible oat leaves increased the level of superoxide and lipid peroxidation (Klotz *et al.*, 1989).

Very important parallels of oxygen free radical production in animal defense systems are reported in cells such as neutrophils and macrophages (Badwey and Karnovsky, 1980; Light *et al.*, 1981; Chapter 20).

ACKNOWLEDGMENT. Research in the author's laboratory has been supported by grants from the National Science Foundation.

8. REFERENCES

Aducci, P., and Ballio, A., 1989, Mode of action of fusicoccin: The role of specific receptors, in: *Phytotoxins and Plant Pathogenesis* (A. Graniti, R. D. Durbin, and A. Ballio, eds.), Springer-Verlag, Berlin, pp. 143–150.

Badwey, J. A., and Karnovsky, M. L., 1980, Active oxygen species and the functions of phagocytic leukocytes, *Annu. Rev. Biochem.* **49**:695–726.

Barash, I., Mor, H., Netzler, D., and Kashman, Y., 1981, Production of zinniol by *Alternaria dauci* and its phytotoxic effect on carrot, *Physiol. Plant Pathol.* **19**:7–16.

Bidwai, A. P., and Takemoto, J. Y., 1987, Bacterial phytotoxin, syringomycin, induces a protein kinase-mediated phosphorylation of red beet plasma membrane polypeptides, *Proc. Nat. Acad. Sci. USA* **84**:6755–6759.

Blatt, M. R., 1990, A primer in plant electrophysiological methods, in: *Methods in Plant Biochemistry*, Volume 19 (K. Hostettmann, ed.), Academic Press, New York (in press).

Böcher, M., and Novacky, A., 1981, Effect of tentoxin on the membrane potential of *Lemna gibba* G1, *Plant Sci. Lett.* **23**:269–276.

Byther, R. S., and Steiner, G. W., 1975, Heat induced resistance of sugarcane to *Helminthosporium sacchari* and helminthosporoside, *Plant Physiol.* **56**:425–419.

Chai, H. B., and Doke, N., 1987, Superoxide anion generation: A response of potato leaves to infection with *Phytophthora infestans*, *Phytopathology* **77**:645–649.

Clint, G. M., and MacRobbie, E. A. C., 1984, Effects of fusicoccin in "isolated" guard cells of *Commelina communis* L., *J. Exp. Bot.* **35**:180–192.

Covallini, A., Bindoli, A., Macri, F., and Vianello, A., 1979, Lipid peroxidation induced by cercosporin as a possible determinant of its toxicity, *Chem. Biol. Interact.* **28**:139–146.

Dahse, I., Müller, E., Liebermann, B., and Eichhorn, M., 1988, The membrane potentials as indicator for transport and energetic processes of leaf cells of the aquatic plant *Egeria densa*. III. The effect of specific energy transfer inhibitor tentoxin and its derivative dihydrotentoxin, *Biochem. Physiol. Pflanz.* **183**:59–66.

Daly, J. M., 1981, Mechanism of action, in: *Toxins in Plant Disease* (R. D. Durbin, ed.), Academic Press, New York, pp. 331–394.

Daly, J. M., 1984, The recognition in plant disease, *Annu. Rev. Phytopathol.* **22**:273–307.

Daub, M. E., 1982a, Cercosporin, a photosensitizing toxin from *Cercospora* species, *Phytopathology* **72**:370–374.

Daub, M. E., 1982b, Peroxidation of tobacco membrane lipids by the photosensitizing toxin, cercosporin, *Plant Physiol.* **69**:1361–1364.

Daub, M. E., and Briggs, S. P., 1983, Changes in tobacco cell membrane composition and structure caused by cercosporin, *Plant Physiol.* **71**:763–766.

Daub, M. E., and Hangarter, R. P., 1983, Light-induced production of singlet oxygen and superoxide by the fungal toxin, cercosporin, *Plant Physiol.* **73**:855–857.

de Boer, A. H., Watson, B. A., and Cleland, R. E., 1989, Purification and identification of the fusicoccin binding protein from oat root plasma membrane, *Plant Physiol.* **89**:250–259.

Dohrmann, U., Hertel, U., Pesci, P., Cocucci, S. M., and Marrè, E., 1977, Localization of "in vitro" binding of the fungal toxin fusicoccin to plasma-membrane-rich fractions from corn coleoptiles, *Plant Sci. Lett.* **9**:291–299.

Doke, N., 1983a, Involvement of superoxide anion generation in the hypersensitive response of potato tuber tissues to infection with an incompatible race of *Phytophthora infestans* and to the hyphal wall components, *Physiol. Plant Pathol.* **23**:345–357.

Doke, N., 1983b, Generation of superoxide anion by potato tuber protoplasts during the hypersensitive response to hyphal wall components of *Phytophthora infestans* and specific inhibition of the reaction by suppressor of hypersensitivity, *Physiol. Plant Pathol.* **23**:359–367.

Doke, N., 1985, NADPH-dependent O_2^- generation in membrane fractions isolated from wounded potato tubers inoculated with *Phytophthora infestans, Physiol. Plant Pathol.* **27**:311–322.

Doke, N., and Ohashi, Y., 1988, Involvement of an O_2^- generating system in the induction of necrotic lesions on tobacco leaves infected with tobacco mosaic virus, *Physiol. Mol. Plant Pathol.* **32**:163–175.

Durbin, R. D., 1988, The role of microbial toxins in plant-pathogen specificity, in: *Physiology and Biochemistry of Plant-Microbial Interactions* (N. T. Keen, T. Kosuge, and L. L. Walling, eds.), American Society of Plant Physiologists, pp. 96–102.

Durbin, R. D., and Uchytil, T. F., 1976, The effect of tentoxin on fusicoccin-induced stomatal opening, *Phytopathriedit* **15**:62–63.

Duvick, J. P., Daly, J. M., Kratky, Z., Macko, V., Acklin, W., and Arigoni, D., 1984, Biological activity of the isomeric forms of *Helminthosporium sacchari* toxin and of homologues produced in culture, *Plant Physiol.* **74**:117–122.

Elstner, E. F., 1982, Oxygen activation and oxygen toxicity, *Annu. Rev. Plant Physiol.* **33**:73–96.

Erdei, L., and Klotz, M. G., 1988, Growth and internal ion concentrations in seedlings of winter wheat are affected by 1pM tentoxin, *Physiol. Plant* **73**:295–298.

Freeman, B. A., and Crapo, J. D., 1982, Biology of disease, free radicals and tissue injury, *Lab. Invest.* **47**:412–426.

Gardner, J. M., Scheffer, R. P., and Higinbotham, N., 1974, Effects of host-specific toxins on electropotentials of plant cells, *Plant Physiol.* **54**:246–249.

Gregory, E. M., and Fridovich, I., 1973, Oxygen toxicity and the superoxide dismutase, *J. Bacteriol.* **114**:1193–1197.

Hadwiger, L. A., Beckman, J. M., and Adams, M. J., 1981, Localization of fungal components in the pea-*Fusarium* interaction detected immunochemically with antichitosan and anti-fungal cell wall antisera, *Plant Physiol.* **67**:170–175.

Hartung, W., Ullrich-Eberius, C. I., and Novacky, A., 1987, Effect of phytotoxins tentoxin, HmT-toxin and HV-toxin on K^+ efflux from unilamellar liposomes, *Plant Sci.* **49**:9–13.

Hedrich, R., and Schroeder, J. I., 1989, The physiology of ion channels and electrogenic pumps in higher plants, *Annu. Rev. Plant Physiol.* **40**:539–569.

Kauss, H., 1987, Some aspects of calcium-dependent regulation in plant metabolism, *Annu. Rev. Plant Physiol.* **38**:47–72.

Keck, R. W., and Hodges, T. K., 1973, Membrane permeability in plants: Changes induced by host-specific pathotoxins, *Phytopathology* **63**:226–230.

Keppler, L. D., and Novacky, A., 1987, The initiation of membrane lipid peroxidation during bacteria-induced hypersensitive reaction, *Physiol. Mol. Plant Pathol.* **30**:233–245.

Klotz, M. G., 1988, The action of tentoxin on membrane processes in plants, *Physiol. Plant* **74**:575–582.

Klotz, M. G., Müller, E., and Liebermann, B., 1987, Potassium transport through lipid bilayer membranes facilitated by tentoxin dimers. A new mechanism of ion carrier transport? *Biophys. Chem.* **27**:183–189.

Klotz, M. G., Hoffmann, R., and Novacky, A., 1989, The critical role of the hydroxyl radical in microbial infection of plants, in: *Plant Membrane Transport: The Current Position* (J. Dainty, M. I. DeMichelis, E. Marre, and F. Rasi-Caldogno, eds.), Elsevier Amsterdam, New York, Oxford, pp. 657–662.

Köhle, H., Jeblick, W., Poten, F., Blashek, W., and Kauss, H., 1985, Chitosan-elicited callose synthesis in soybean cells as a Ca^{2+}-dependent process, *Plant Physiol.* **77**:544–551.

Kota, D. A., and Stelzig, D. A., 1977, Electrophysiology as a means of studying the role of elicitors in plant disease resistance, *Proc. Am. Phytopathol. Soc.* **4**:216–217 (Abstr.).

Kuyama, S., and Tamura, T., 1957, Cercosporin. A pigment of *Cercosporina kikuchiana* Matsumoto et Tomoyasu. I. Cultivation of fungus, isolation and purification of pigment, *J. Am. Chem. Soc.* **79**:5727–5736.

Libbenga, K. R., Maan, A. C., Linde, P. C. G., and Menners, A. M., 1986, Auxin receptors, in: *Hormones, Receptors and Cellular Interactions in Plants* (C. M. Chadwick and D. R. Garrod, eds.), Cambridge University Press, London, pp. 1–68.

Light, D. R., Walsh, C., O'Callangham, M. O., Goetzl, E. J., and Tauber, A. I., 1981, Characteristics of the cofactor requirements for the superoxide-generating NADPH oxidase of human polymorphonuclear leucocytes, *Biochemistry* **20**:1468–1476.

Livingston, R. S., and Scheffer, R. P., 1981, Isolation and characterization of host-selective toxin from *Helminthosporium sacchari*, *J. Biol. Chem.* **256**:1705–1710.

Livingston, R. S., and Scheffer, R. P., 1983, Conversion of *Helminthosporium sacchari* toxin to toxoids by β-galactofuranoside from *Helminthosporium, Plant Physiol.* **72**:530–534.

Lousberg, R. J. J. C., Wiss, U., Salemink, C. A., Arnone, A., Merlini, L., and Nasini, G., 1971, The structure of cercosporin, a naturally occurring quinone, *Chem. Commun.* **1971**:1463–1464.

Low, P. S., and Heinstein, P. F., 1986, Elicitor stimulation of the defense response in cultured plant cells monitored by fluorescent dyes, *Arch. Biochem. Biophys.* **249**:472–479.

Macko, V., 1983, Structural aspects of toxins, in: *Toxins and Plant Pathogenesis* (J. M. Daly and B. D. Deverall, eds.), Academic Press, New York, pp. 41–80.

Macko, V., Wolpert, T. J., Acklin, W., Jaun, B., Seibl, J., Meili, J., and Arigoni, D., 1985, Characterization of victorin C, the major host-selective toxin from Cochliobolus victoriae: structure of degradation products, *Experientia* **41**:1366–1370.

Macko, V., Goodfriend, K., Wachs, T., Renwick, J. A. A., Acklin, W., and Arigoni, D., 1981, Characterization of the host-specific toxins produced by *Helminthosporium sacchari*, the causal organism of eyespot disease of sugarcane, *Experientia* **37**:923–924.

Macko, V., Grinnalds, C., Golay, J., Arigoni, D., Acklin, W., Weibel, F., and Hildebrand, C., 1982, Characterization of lower homologues of host-specific toxins from *Helminthosporium sacchari, Phytopathology* **72**:942 (Abstr.).

Macko, V., Wolpert, T. J., Acklin, W., Jaun, B., Seible, J., Meili, J., and Arigoni, D., 1985, Characterization in victorin C, the major host-selective toxin from *Cochliobolus victoriae*: Structure and degradation products, *Experientia* **41**:1366–1370.

Macri, F., and Vianello, A., 1979, Photodynamic activity of cercosporin on plant tissues, *Plant Cell Environ.* **2**:267–271.

Marrè, E., 1979, Fusicoccin: A tool in plant physiology, *Annu. Rev. Plant Physiol.* **30**:273–288.

Marrè, E., Lado, P., Rasi Caldogno, F., and Colombo, R., 1971, Fusicoccin as a tool for the analysis of auxin action, *Rend. Accad. Naz. Lincei* **50**:45–49.

Mayer, M. G., and Novacky, A., 1989, Do host-specific toxins affect single ion channels? in: *Phytotoxins and Plant Pathogenesis* (A. Graniti, R. D. Durbin, and A. Ballio, eds.), Springer-Verlag, Berlin, pp. 433–435.

Mayer, M. G., and Ziegler, E., 1988, An elicitor from *Phytophthora megasperma* f.sp. *glycinea* influences the membrane potential of soybean cotyledonary cells, *Physiol. Mol. Plant Pathol.* **33**:397–407.

Meehan, F., and Murphy, H. C., 1947, Differential phytotoxicity of metabolic by-products of *Helminthosporium victoriae, Science* **106**:270–271.

Meyer, C., Feyererabend, M., and Weiler, E. W., 1989, Fusicoccin-binding proteins in *Arabidopsis thaliana* (L.) Heynh. Characterization, solubilization and photoaffinity labeling, *Plant Physiol.* **89**:692–699.

Nelson, N., and Taiz, L., 1989, The evolution of H+-ATPases, *Trends Bioch. Sci.* **14**:113–116.

Novacky, A., and Hanchey, P., 1974, Depolarization of membrane potentials in oat roots treated with victorin, *Physiol. Plant Pathol.* **4**:161–165.

Novacky, A., Ullrich-Eberius, C. I., Wolpert, T., and Macko, V., 1986, Effect of victorin C on oat cell membranes: Electrophysiology, *Plant Physiol. Suppl.* **80**:64.

Novacky, A., Ullrich-Eberius, C. I., and Ball, E., 1989, Interactions of phytotoxins with plant cell membranes: Electrophysiology and ion flux-induced pH change, in: *Phytotoxins and Plant Pathogenesis* (A. Graniti, R. D. Durbin, and A. Ballio, eds.), Springer-Verlag, Berlin, pp. 151–166.

Pelissier, B., Thibaud, J. B., Grignon, C., and Esquerre-Tugaye, M. T., 1986, Cell surface in plant-microorganism interactions. VII. Elicitor preparations from two fungal pathogens depolarize plant membranes, *Plant Sci.* **46**:103–109.

Peters, B. M., Cribbs, D. H., and Stelzig, D. A., 1978, Agglutination of plant protoplasts by fungal cell wall glucans, *Science* **201**:364–365.

Poole, R. J., 1978, Energy coupling for membrane transport, *Annu. Rev. Plant Physiol.* **29**:437–460.

Rea, P. A., and Sanders, D., 1987, Tonoplast energization: Two H+ pumps, one membrane, *Physiol. Plant* **71**:131–141.

Reinhold, L., and Kaplan, A., 1984, Membrane transport of sugars and amino acids, *Annu. Rev. Plant Physiol.* **35**:45–83.

Schröter, H., Novacky, A., and Macko, V., 1985, Effect of *Helminthosporium sacchari*-toxin on cell membrane potential of susceptible sugarcane, *Physiol. Plant Pathol.* **26**:165–174.

Singer, S. J., and Nicholson, G. L., 1972, The fluid mosaic model of the structure of cell membranes, *Science* **175**:720–731.

Slayman, C. L., 1985, Plasma membrane proton pumps in plants and fungi, *BioScience* **35**:34–37.

Spanswick, R. M., 1981, Electrogenic ion pumps, *Annu. Rev. Plant Physiol.* **32**:267–289.

Steele, J. A., Uchytil, T. F., Durbin, R. D., Bhatnagar, P. K., and Rich, D. H., 1976, Chloroplast coupling factor 1: A species-specific receptor for tentoxin, *Proc. Natl. Acad. Sci. USA* **73**:2245–2248.

Strobel, G. A., 1973, The helminthosporoside-binding protein of sugarcane, its properties and relationship to susceptibility to eyespot disease, *J. Biol. chem.* **248**:1321–1328.

Strobel, G. A., 1979, The relationship between membrane ATPase activity in sugarcane and heat-induced resistance to helminthosporoside, *Biochem. Biophys. Acta* **554**:460–468.

Sze, H., 1984, H^+-translocating ATPases of the plasma membrane and tonoplast of plant cells, *Physiol. Plant* **61**:683–691.

Thatcher, F. S., 1939, Osmotic and permeability relations in the nutrition of fungus parasites, *Am. J. Bot.* **26**:449–458.

Thatcher, F. S., 1942, Further studies of osmotic and permeability relations in parasitism, *Can. J. Res. Sect. C* **20**:283–311.

Thompson, J. E., Legge, R. L., and Barber, R. F., 1987, The role of free radicals in senescence and wounding, *New Phytol.* **105**:317–344.

Thuleau, P., Graziana, A., Rossignol, M., Kauss, H., Auriol, P., and Ranjeva, R., 1988, Binding of the phytotoxin zinniol stimulates the entry of calcium into plant protoplasts, *Proc. Natl. Acad. Sci. USA* **85**:5932–5935.

Tognoli, L., and Basso, B., 1987, The fusicoccin-stimulated phosphorylation of a 33 kDa polypeptide in cells of *Acer pseudoplatanus* as influenced by extracellular and intracellular pH, *Plant Cell Environ.* **10**:233–239.

Wheeler, H., 1978, Disease alterations in permeability and membranes, in: *Plant Disease*, Volume III (J. G. Horsfall and E. B. Cowling, eds.), Academic Press, New York, pp. 327–347.

Wheeler, H., and Hanchey, P., 1968, Permeability phenomena in plant disease, *Annu. Rev. Phytopathol.* **6**:331–350.

Wheeler, H., and Luke, H. H., 1963, Microbial toxins in plant disease, *Annu. Rev. Microbiol.* **17**:223–242.

Wolpert, T. J., and Macko, V., 1989, Specific binding of victorin to a 100 Kd protein from oats, *Proc. Natl. Acad. Sci. USA* **86**:4092–4096.

Yamazaki, S., and Ogawa, T., 1972, The chemistry and stereochemistry of cercosporin, *Agric. Biol. Chem.* **36**:1707–1718.

Yoder, O. C., 1980, Toxins in pathogenesis, *Annu. Rev. Phytopathol.* **18**:103–129.

Yoshikawa, M., 1983, Macromolecules, recognition, and the triggering of resistance, in: *Biochemical Plant Pathology* (J. A. Callow, ed.), Wiley, New York, pp. 267–298.

Young, D. H., and Kauss, H., 1983, Release of calcium from suspension-cultured *Glycine max* cells by chitosan, other polycations, and polyamines in relation to effects on membrane permeability, *Plant Physiol.* **73**:698–702.

Ziegler, E., and Pontzen, R., 1982, Specific inhibition of glucan-elicited glyceollin accumulation in soybeans by an extracellular mannan-glycoprotein of *Phytophthora megasperma* f.sp. *glycinea*, *Physiol. Plant Pathol.* **20**:321–331.

18

The Fungal Spore

Reservoir of Allergens

J. P. Latgé and S. Paris

1. INTRODUCTION

The presence of fungal spores in the air has been known for a long time. In the mid-19th century, Pasteur and numerous other botanists used various spore traps and sampling techniques to investigate "floating matter of the air" (Ainsworth, 1986). Since then, surveys of fungal populations in the atmosphere have been performed continuously in many parts of the world and have provided a very long list of about 200,000 identified mold species (Lacey, 1981; Koivikko, 1983; Wilken-Jensen and Gravesen, 1984).

A remarkable feature of the air spora is the ubiquitous occurrence of the same fungal species. However, the concentration and type of spores may vary transiently with the environmental and climatic conditions. The average number of spores in the air is usually about 10^4–10^5 spores/m^3 with peak concentrations of up to 10^6 spores/m^3 during windy periods following wet weather (Lacey, 1981). Most airborne fungi are saprophytic conidial fungi with pigmented spores; the most frequently detected are dematiaceous *Cladosporium* and *Alternaria*. Ascomycetes and Basidiomycetes are also found in high numbers but are often underestimated in most spore-trapping techniques because of their inability to grow on most media. Spores are present indoors and outdoors. The spore content of the indoor air is usually lower than that of the outdoor air, but similar fungal species are found in both environments (Lacey, 1981).

Spores present in the air are regularly and continuously inhaled by humans. Consequently, it is understandable that most fungi present in high concentrations in the air have been suspected to be the causative agents of respiratory diseases (rhinitis, asthma) in humans. Once inhaled into the upper respiratory tract, most of the spores are directed toward the stomach through ciliary beating and destroyed in the digestive tract. However, some spores remain in the nasal mucosa or in the bronchial system where they are phagocytosed and lysed by macrophages and polymorpho-nuclear cells. The fungal components thus released in the body fluids induce the synthesis of specific immunoglobulins (IgG, IgM, IgE, IgA) by B lymphocytes. These different antibody types can recognize similar or distinct antigens (Mackenzie, 1983; Kurup *et al.*, 1988; O'Neil *et*

J. P. Latgé and S. Paris • Mycology Unit, Pasteur Institute, 75015 Paris, France.

al., 1988; Longbottom, 1988). Indeed, serological tests demonstrate that most individuals have significant amounts of IgG and IgM directed toward airborne fungi (Latgé *et al.*, unpublished; Notermans *et al.*, 1988). However, only a few individuals, genetically predisposed or transiently immunosuppressed, respond to repeated exposure to fungi by the production of high levels of IgE antibodies (Roitt *et al.*, 1985). IgE was recognized in the early 1960s as the immunoglobulin responsible for the immediate hypersensitivity response (Type I of Gells and Combs classification), commonly called allergy. The immune mechanisms involved in an allergy crisis are quite well established (Fig. 1). IgEs produced after the first exposure to specific antigens (allergens) bind to specific receptors on mast cells or basophils (Fig. 2) which are the main target cells of IgE-mediated allergic reactions (Schleimer *et al.*, 1985). With a second exposure, the allergens bind to the IgEs fixed on mast cells. The bridges formed between one allergen molecule and two IgE molecules induce the degranulation of mast cells (Fig. 2). These granules release numerous pharmacologically active mediators, causing the clinical symptoms of anaphylaxis such as asthma or rhinitis.

Most of these immune mechanisms have been elucidated with inhaled allergens originating from pollen or house dust mite. Conversely, although about 100 fungal species have been associated with different forms of allergy (Table I), the role of airborne spores in triggering allergy is poorly understood (Hoffman, 1984). This is essentially due to (1) the necessity of using a methodology specifically adapted to study fungal allergens and (2) the poor biochemical and immunological characterization of fungal allergens (Latgé and Paris, 1988).

This chapter will selectively review the methods available to investigate fungal allergens and give an update of our knowledge on the characterization of spore allergens. The reader should remember that the isolation of one pure fungal allergen which has yet to be achieved remains the essential goal of all fungal immunoallergologists in order to elucidate the immunological mechanisms associated with fungal allergy.

2. METHODOLOGY AVAILABLE TO STUDY FUNGAL SPORE ALLERGY

2.1. Culture

Mass production of spores is a prerequisite to the preparation of spore extracts to be evaluated for allergenic activity. The absence of any easy way to produce large amounts of ascospores and basidiospores *in vitro* is probably responsible for the lack of studies on these groups of fungi. Conversely, deuteromycetous conidia are easily produced aerially on agar media. Minimal culture media such as V8 or malt extract have been successfully used for the production of conidia of *Alternaria* and *Cladosporium*. Conidia should be collected under vacuum or with a paint brush. Floating the plates with water should be avoided since medium components and mycelial metabolites released during growth would be recovered and may interfere with spore allergens (van der Heide *et al.*, 1985; Gumowski *et al.*, 1985). Age and viability of the spore should also be taken into account since aging of the spore may lead to the appearance or disappearance of allergens. A high percentage of airborne spores are dead (Lacey, 1981), suggesting that young viable spores may not be the best source of spore allergens. The strain and media should also be characterized as variations have been detected in allergens produced by different strains of *Cladosporium* and *Alternaria* cultured in different media (Salvaggio and Aukrust, 1981; Sward-Nordmo *et al.*, 1984; Helm *et al.*, 1987; Steringer *et al.*, 1987; Latgé and Paris, unpublished).

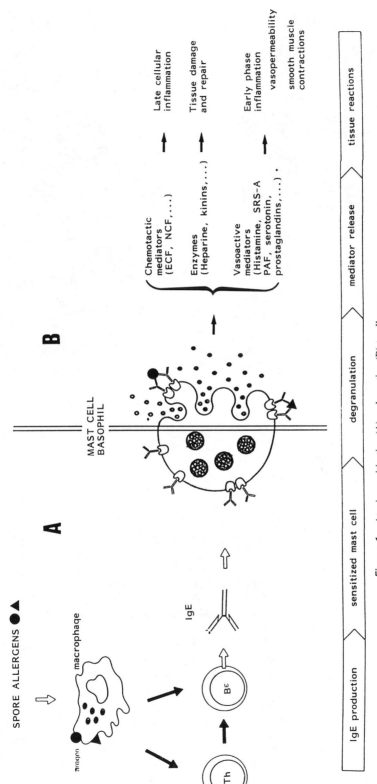

Figure 1. Atopic sensitization (A) and reaction (B) to allergens.

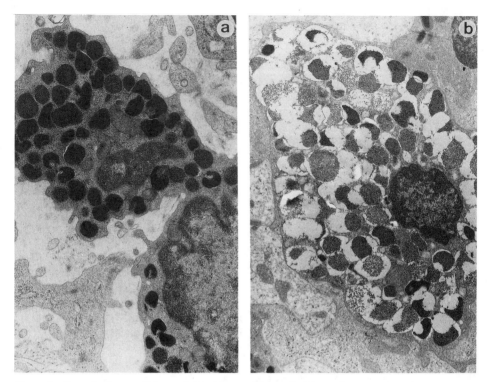

Figure 2.   Example of mast cell degranulation. (a) normal mast cell of the nasal mucosa of untreated guinea pig; (b) degranulated mast cell of the nasal mucosa of sensitized guinea pig challenged with extracts of *Cladosporium*. (From Bouziane *et al.*, 1988).

2.2. Spore Extracts

Several methods, summarized in Table II, have been used to extract spore allergens. In spite of their heterogeneity, the methods used can be separated into two groups: passive methods in which nondisrupted spores are incubated in various solutions, and active methods in which spores are broken with glass beads in a cell homogenizer using different buffers.

The passive methods are derived from techniques used to extract pollen or mite allergens but have not been optimized for spore allergen extraction. Recent studies with *C. cladosporioides* conidia have demonstrated that the amount of products released during the extraction procedure depends on the composition of the solution used and on the concentration of conidia (Bouziane *et al.*, 1989). Maximal release was obtained with a low concentration of conidia (0.1 mg spore dry wt/ml) and 0.1 N NaOH, Coca's solution, 50 mM Tris, or 0.9% NaCl solution or after a short ultrasonication (Bouziane *et al.*, 1989). Kinetic studies have demonstrated that maximal sugar and protein release is attained 3 to 4 hr after the beginning of incubation (Fig. 3). Longer incubation did not modify the concentration of sugar released but provoked a decrease in the amount of protein recovered due to the activity of extracellular serine-proteases. After 24 hr of incubation, only 10 to 25% of the maximal protein concentration remains. Proteolysis can be blocked by phenylmethylsulfonyl fluoride (PMSF) or by using a 5–50 mM Tris buffer pH 9.0. Such proteolytic degradation may interfere with the allergenic potency of the extracts and could explain, for example, why the concentration of the major allergen of *Cladosporium* (Ag 32) can vary from 1000% to 5% of a standard reference depending on the extraction procedure (Salvaggio and Aukrust, 1981). On the other hand, proteases may play a role in allergy. Fungal components

Table I. Fungal Genera Reported to Be Associated with Allergy[a]

Absidia	Cryptostroma	Inonotus	Polystictus
Agaricus	Curvularia	Leptosphaeria	Psilocybe
Agrocybe	Dacrymyces	Leptosphaerulina	Puccinia
Alternaria	Daldinia	Lycogala	Rhizopus
Amanita	Debaryomyces	Lycoperdon	Rhodotorula
Ankistrodesmus	Dicoccum	Macrosporium	Saccharomyces
Armillaria	Didymella	Malassezia	Scleroderma
Arthrinium	Drechslera	Microsphaera	Scopulariopsis
Aspergillus	Epicoccum	Mucor	Serpula
Aureobasidium	Epidermophyton	Mycogone	Sphaerotheca
Boletinellus	Erysiphe	Naematolona	Spondylocladium
Boletus	Eurotium	Neochloris	Sporobolomyces
Botrytis	Fomes	Neurospora	Sporotrichum
Bracteacoccus	Fugus	Nigrospora	Stemonitis
Calvatia	Fuligo	Paecilomyces	Stemphylium
Candida	Fusarium	Papularia	Stereum
Cantharellus	Ganoderma	Penicillium	Tilletiopsis
Cephalosporium	Geastrum	Phoma	Trichoderma
Chaetomium	Geotrichum	Phytophthora	Trichophyton
Chlorophyllum	Gibberella	Piptoporus	Trichothecium
Cladosporium	Gliocladium	Pisolithus	Ustilago
Claviceps	Gnomonia	Pleurotus	Verticillium
Coniosporium	Graphium	Podaxis	Xylaria
Coprinus	Hormidium	Polyporus	Xylobolus
Cryptococcus	Hypholoma		

[a]Allergens have been demonstrated in the spores of all genera underlined.

partially hydrolyzed by proteases may become powerful allergens; proteases themselves are also known for their immunoreactivity (Yuan and Cole, 1987). Other fungi will react differently from *Cladosporium*. For example, in the case of *Alternaria*, no proteolytic degradation of the exudate has been detected (Vijay *et al.*, 1987; Paris *et al.*, 1990a). Maximal release of proteins and sugars occurred with a high concentration of conidia (75 mg spore dry wt/ml) after 24 hr of incubation at 4°C (Paris *et al.*, unpublished).

Several authors have recently reported that these passive methods can be advantageously replaced by cell disruption methods, which have been previously used to extract fungal antigens (Kauffman *et al.*, 1984; Liengswangwong *et al.*, 1987; Paris *et al.*, 1990a; Bouziane and Latgé, unpublished). As with exudates, the extraction buffer influences the protein yield. In the case of *C. cladosporioides*, 50 mM $NaHCO_3$, PBS, 50 mM Tris, and 50 mM Tris + 1 mM EDTA + 1% polyvinlypyrrolidone (PVP) have been used for extraction of proteins yielding 0.8, 0.9, 1.3, and 4.0% of total spore dry weight, respectively. For all fungal species tested in our laboratory, the amount of protein extracted is higher after conidial disruption than with passive methods (Paris *et al.*, 1990a; Bouziane and Latgé, unpublished). The extraction buffer has to be chosen carefully because spore disruption induces the release of endogenous proteases which will slowly degrade the extract during storage (Fig. 4). In the case of *C. cladosporioides*, the use of 5–50 mM Tris pH 9.0 is sufficient to inhibit the proteolysis. Proteolytic inhibitors such as PMSF or *N-p-*tosyl-L-lysine chloromethyl ketone (TLCK) or ion chelators such as EDTA can be advantageously added to the cell disruption buffer to block proteolysis. With dematiaceous fungi, the use of a phenolic complexing agent such as PVP is also recommended because it decreases the protein precipitation induced by melanin-related compounds without interfering with the allergenic potency of the extract (Latgé and Paris, 1988; Latgé *et al.*, 1988; Paris *et al.*, 1990a).

Table II. Methods of Extraction Used to Isolate Spore Allergens

Fungus	Passive methods[a]	Active methods[b]
Deuteromycetes		
Alternaria	Water[c]	NaHCO$_3$ 0.3% + NaCl 0.9%[e]
	PBS + 0.1% phenol + 0.2% Tween 80, 48 hr at 4°C[d]	NaHCO$_3$ 50 mM + NaCl 0.15 M + EDTA 5 mM + PMSF 1 mM PVP
	NAHCO$_3$ 0.3% + NaCl 0.9% + phenol 0.5%, 24 hr at 4°C[e]	NaCl 0.15 M 1%[e]
	PBS, 48 hr at room temperature[f]	Water[f]
Aspergillus		
A. fumigatus	PBS, 1 hr at 4°C[g]	PBS[h]
Xerophilic Aspergillus		NaHCO$_3$ 50 mM[h]
Cladosporium	0.125 NH$_4$HCO$_3$ pH 7.9, 18 hr at 6°C[i]	PBS
	Water	NaHCO$_3$ 50 mM pH 8.0
	NaHCO$_3$ 50 mM pH 8.0	Tris 5–50 mM pH 9.0
	Tris 50 mM pH 9.0	Tris 50 mM pH 9.0 + EDTA 1 mM + PVP 1% } [k]
	NaOH 0.1 N	
	NaHCO$_3$ 0.3% + NaCl 0.9% + phenol 0.5%	
	Detergents (NP40, SDS, alkyl cetrimide) 1% in NaHCO$_3$ 50 mM pH 8.0 } [j]	
Epicoccum	PBS or water, ultrasonication (30 min 100 W)	
Penicillium	Water, 48 hr at 4°C[l]	
	PBS + 0.1% phenol + 0.2% Tween 80, 48 hr at 4°C[d]	
Ascomycetes		
Basidiomycetes	Ethylether-treated or untreated spores in 0.125 M NH$_4$HCO$_3$ pH 7.3, 16–24 hr at 4°C[n]	Water[m]
		Ethylether-treated or untreated spores in 0.125 M NH$_4$HCO$_3$ pH 8.1 ± aprotamine
		Tris-HCl 20 mM + 1% Triton X-100 } [o]

[a]Passive methods: incubation of the spores in solution without disruption of the spores and recovery of the liquid extract by filtration.
[b]Active methods: breakage of the spores with a glass bead cell homogenizer and recovery of the centrifugation supernatant.
[c]Solomon et al. (1980).
[d]Licorish et al. (1985).
[e]Paris et al. (1990a).
[f]Kauffman et al. (1984).
[g]Rijckaert (1981).
[h]Latgé (unpublished).
[i]Swärd-Nordmo et al. (1984).
[j]Bouziane et al. (1989).
[k]Bouziane and Latgé (unpublished).
[l]Portnoy et al. (1987).
[m]Burge et al. (1985).
[n]Santilli et al. (1985), Lehrer et al. (1986).
[o]Liengswangwong et al. (1987).

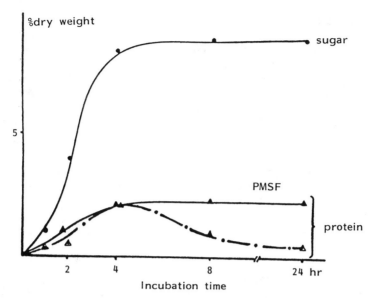

Figure 3. Sugars and proteins released from conidia of *C. cladosporioides* incubated at 20°C in a mixture of 0.9% NaCl, 0.3% NaHCO₃, and 0.5% phenol + 1 mM PMSF. (From Bouziane *et al.*, 1989.)

2.3. Testing for Allergenicity

2.3.1. Clinical Tests

Clinical tests are performed on patients with a history of aggravation of respiratory allergic symptoms in the peak mold season.

Skin prick test (SPT) is the usual method employed in allergy diagnosis (Dreborg *et al.*,

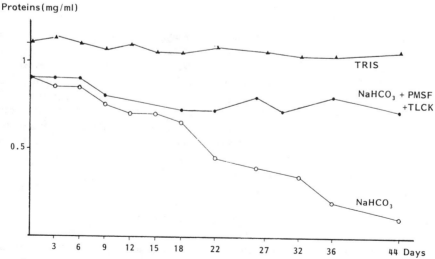

Figure 4. Protein concentration of *C. cladosporioides* extracts (obtained by glass bead, conidial disruption in different buffers) during storage at 4°C.

1986; Malling, 1985a,b, 1987). Concentration of *Alternaria* and *Cladosporium* extracts employed successfully in these tests vary from 100 to 500 μg protein/ml (Paris *et al.*, 1990a; Gumowski *et al.*, unpublished). These concentrations are much higher than the doses usually found in commercial extracts which never exceed 50 μg/ml (Paris and Latgé, unpublished).

Inhalation challenge (nasal or bronchial provocation) can also be performed either with liquid extracts or with conidia previously killed by irradiation (Licorish *et al.*, 1985; Fadel *et al.*, 1986; Gumowski *et al.*, 1987). Ranges of 10^5–4×10^5 conidia of *Alternaria*, 10^5–5×10^7 conidia of *Penicillium*, and 10^4–10^6 conidia of *Cladosporium* induced immediate allergic response in sensitized patients (Licorish *et al.*, 1985; Gumowski *et al.*, unpublished). Provocative doses of spores are within natural exposure ranges.

Because of irritability problems due to intact spores during inhalation challenge, or non-specific skin reactions due to nonallergenic fungal metabolites and the variability inherent in each individual, dose–response curves should be performed on all patients. Anergic individuals or patients allergic to inhalants other than fungi should also be tested.

2.3.2. In Vitro Tests

The dosage and characterization of spore allergens require the use of *in vitro* tests. The basic method to quantify specific IgE for a given allergen is the radioallergosorbent test (RAST). The principle of this assay is shown in Table III. Use of 5 and 500 μg/ml of protein extracts of *C. cladosporioides* and *A. alternata*, respectively, is necessary to saturate CNBr-activated disks and to measure the specific IgE in a 10^{-1} serum dilution (Bouziane *et al.*, unpublished; Paris, unpublished). The main disadvantage of this method is the use of CNBr-activated disks which are only able to covalently bind NH_2-allergens, thus excluding most polysaccharidic and lipidic antigens.

The enzyme-linked immunosorbent assay (ELISA) can be used as an alternative method to the RAST assay. The former has the advantage of taking into account polysaccharidic as well as protein antigens (Sepulveda *et al.*, 1979) (Table III). In this case, 10 ng/ml to 10 μg/ml of extract can be used to coat wells and anti-IgE monoclonal antibody conjugates have been advantageously used to measure specific IgE directed against several fungi. The sensitivity obtained with ELISA is comparable to that with RAST (Table III). Modifications of the ELISA involve the substrate of the enzyme conjugate: the use of a fluorescent β-galactosidase substrate such as 4-methylumbel-liferyl β-D-galactoside (Labrousse *et al.*, 1982; Labrousse and Avrameas, 1987) has greatly increased the sensitivity of IgE detection using a galactosidase-labeled anti-IgE. This technique has been advantageously used to detect *Cladosporium*- and *Alternaria*-specific IgE (Latgé and Paris, unpublished). Other modifications are related to the sandwich or double antibody techniques. The BALISA method developed by Kurup (1986) is one of the most promising since specific IgE in 10^{-2}–10^{-3} serum dilutions can be detected. However, the efficiency of this method and the double antibody method of Sepulveda *et al.* (1979) has only been demonstrated with allergic bronchopulmonary aspergillosis (ABPA) sera characterized by a high concentration of IgE but not with sera from patients allergic to *Aspergillus* (Kurup, 1986). Similarly, sandwich methods using rabbit anti-*Cladosporium* and anti-*Alternaria* IgG to coat the wells or the BALISA method did not give better results than the single antibody technique using an anti-IgE monoclo-nal conjugate (Latgé and Paris, unpublished).

The identification of individual allergens from inside a crude spore extract can be made using various electrophoretic techniques described elsewhere (Dreborg *et al.*, 1986). Two techniques are most widely used: cross-radioimmunoelectrophoresis (CRIE) and Western blot-ting. CRIE has the advantage of being highly sensitive and quantitative. For example, 3 μg of protein from *Alternaria* is sufficient to visualize specific allergens. The main disadvantage of this method is the revelation of only IgEs that give IgG in the rabbit system. Also, due to the use of different rabbit antisera and the lack of internal standards, comparison of results obtained in

Table III. Different In Vitro Methods Used to Measure Specific IgEs against Fungal Allergens

Method	Antigen concentration (μg protein/ml)	Patient serum dilution	System used to detect bound IgE
RAST	500 (Alternaria)[a]	10^{-1}	^{125}I–polyclonal anti-IgE
	5 (Cladosporium)[b]	10^{-1}	^{125}I–polyclonal anti-IgE
	25 (Alt, Clad, Asp)[c]	5×10^{-1}	^{125}I–polyclonal anti-IgE
	10,000 (Aspergillus)[d]	3×10^{-1}	β-Galactosidase–polyclonal anti-IgE + ONPG
ELISA	1 (Cladosporium)[e]	10^{-1}	β-Galactosidase–polyclonal anti-IgE + ONPG
	0.01 (Cladosporium)[e]	10^{-1}	β-Galactosidase–monoclonal anti-IgE + ONPG[f]
	10 (Alternaria)[e]	10^{-1}	β-Galactosidase–monoclonal anti-IgE + ONPG[f]
	1 (Cladosporium)[e]	10^{-2}	β-Galactosidase–monoclonal anti-IgE + umbelliferyl galactoside
	1 (Cladosporium)[e]	10^{-1}	Peroxidase–monoclonal anti-IgE + OPD[g]
	10 (Epicoccum)[h]	3×10^{-2}	Alkaline phosphatase–monoclonal anti-IgE + pNPP
	100 (Aspergillus)[j]	10^{-1}–10^{-2}	Rabbit anti-IgE alkaline phosphatase–anti-rabbit IgG + pNPP
	1 (Aspergillus)[j]	10^{-3} (ABPA)	Goat anti-human IgE, biotinylated rabbit anti-goat, avidin–biotinylated peroxidase + OPD
	1 (Cladosporium)[k] (Alternaria)	10^{-1}	Rabbit anti-Cladosporium IgG or rabbit anti-Alternaria IgG, β-galactosidase–polyclonal anti-IgE + ONPG

[a]Paris (unpublished)
[b]Bouziane et al. (1989a)
[c]Tee et al. (1987)
[d]Kauffman et al. (1984)
[e,k]Latgé (unpublished).
[f]Biosys, France.
[g]Stallergènes/Biomérieux, France
[h]Portnoy et al. (1987).
[i]Sepulveda et al. (1979).
[j]Kurup (1986), Kurup et al. (1988).

different laboratories is very difficult. CRIE has gradually been replaced by immunostaining of nitrocellulose blots following SDS–PAGE or isoelectrofocusing where the allergen molecular weight or isoelectric point can be determined (Bengtsson *et al.*, 1986; Tovey and Baldo, 1987).

These techniques which have been widely used to study inhalant allergens do not give results which always correlate well. This discrepancy between results of different tests appears even more important during the course of fungal allergy studies (Latgé and Paris, 1988). Consequently, the diagnosis of fungal allergy can only be assessed after performing several *in vivo* and *in vitro* tests.

3. CHARACTERIZATION OF SPORE ALLERGENS

Fungal allergy studies remain limited to very few fungal models (Salvaggio and Aukrust, 1981; Bush and Yunginger, 1987). Moreover, the fungal material used in these studies has very often an undefined commercial origin, or, when specified, is composed of a mixture of spores and mycelia. Consequently, our knowledge of spore allergens remains very limited and is focused almost exclusively on conidial allergens of *Cladosporium* and *Alternaria*.

3.1. Alternaria

Most of the studies on *Alternaria* allergens have been performed on the mycelium of *A. alternata* (= *A. tenuis*) (Vijay *et al.*, 1979, 1985; Nyholm and Lowenstein, 1982; Budd *et al.*, 1983a,b; Bush *et al.*, 1983; Steringer *et al.*, 1987). Only three studies deal with conidia (Solomon *et al.*, 1980; Hoffman *et al.*, 1981; Licroish *et al.*, 1985). Conidial and mycelial extracts gave similar values in SPT in most patients. However, some patients react more with the spore extract than with the mycelial extract (Solomon *et al.*, 1980; Hoffman *et al.*, 1981; Fadel *et al.*, 1986). RAST inhibition studies indicated that spore and mycelial extracts have common allergens and are present in higher amounts in the mycelial extracts (Fig. 5). On the other hand, the spore extract contains specific allergens which are not found in the mycelial extract (Fig. 5) (Paris and Fitting, unpublished). Moreover, histamine release experiments have shown that spores are more reactive than mycelia (Fadel *et al.*, 1986). All these observations justify studying allergens from conidia.

In our laboratory, among the 17 strains tested belonging to 7 species of *Alternaria* (*A. tenuissima*, *A. chartarum*, *A. dentritica*, *A. oleracea*, *A. brassicicola*, *A. alternata*, *Alternaria* sp.) only one strain of *A. alternata* and one strain of *A. brassicicola* had both high sporulation capacities and high allergen content in their conidia (Paris, unpublished). Extracts used for testing allergenicity were obtained after disruption of the conidia in 50 mM $NaHCO_3$ buffer or by incubation in Coca's solution. Both extracts can be used to diagnose *Alternaria* allergy. Glass bead cell disruption gave the highest protein yield. However, SPT, RAST, and CRIE demonstrated that passive extracts contain the highest specific allergen activity, i.e., amount of allergen per milligram of protein extract. This result may be partly due to the presence in the disrupted conidial extract of proteolytic enzymes which can degrade the allergens. Two to three allergens only were identified by CRIE and three by immunoblotting in extracts obtained by extraction in the absence of a proteolytic inhibitor. In contrast, proteases are absent in the exudates. Similarly, the disruption of conidia in the presence of proteolytic inhibitors (PMSF, EDTA) and PVP resulted in the detection of the highest number of allergens: eight in CRIE and eight in Western blots using a human serum pool (Fig. 6). Such extracts have the highest allergenic potency (Paris *et al.*, 1990). At least 13 allergens which react with more than 25% of the patient sera could be identified in these extracts separated by SDS–PAGE and immunoblotted with individual patient sera (Fig. 7). Most spore allergens are present in the mycelium (Fig. 7). However, five allergens (95, 82, 56, 48, 36 kDa) detected in the conidial extract can be considered as spore-specific.

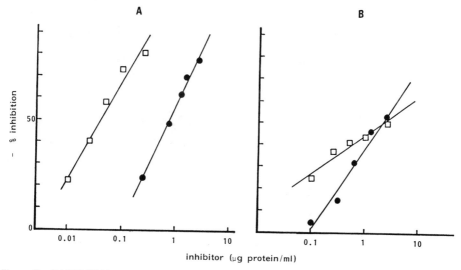

Figure 5. RAST inhibition studies of passive extracts of conidia and mycelia of *A. alternata* (IP 1563) obtained after 24 hr of incubation at 4°C. (A) Mycelial extract bound to CNBr disks; (B) conidial extract bound to CNBr disks. Serum inhibition with conidial extract (●) or with mycelial extract (□). *r* values are 0.9.

Among these spore specific allergens, the 56 kDa is the only major allergen (defined as a fraction which binds IgE from at least 50% of the sera tested). Since the spore is the only inhaled propagule and is responsible for the allergy crisis, it is surprising that out of the four major allergens only one is spore-specific (56 kDa), two are found only in the mycelium (85, 42 kDa), and one (31 kDa) is present in both extracts (Paris *et al.*, 1990b). Two hypotheses can be proposed to explain the presence of more major allergens in the mycelium than in the spores. First, allergens are more easily extracted from the mycelium than from spores because the high melanin content of spores would induce precipitation and/or inactivation of allergens. The fact that use of PVP significantly increased the yield of allergenic proteins suggested the deleterious effect of melanin pigments on the recovery of spore allergens during extraction. Second, our *in vitro* culture conditions may not permit the expression of the most potent allergens. Two studies have suggested that conidia found in nature are much more allergenic than conidia produced in laboratory conditions. Inhalation challenges undertaken with *in vitro*-produced conidia could not be compared to responses obtained with similar doses of conidia present in the natural environment (Licorish *et al.*, 1985). Agarwal *et al.* (1983) found a much higher content of the major allergen Alt I in "wild" conidia than expected from calculations made with "laboratory" conidia. These differences could explain the low concentration of Alt I in the conidia produced *in vitro*. Nevertheless, this allergen is the most reactive allergen in *Alternaria in vitro* extracts: 84% of patients' sera recognized this molecule (Yunginger *et al.*, 1980).

Alt I has been characterized by CRIE in which it migrates toward the anode and gave the largest precipitin peak recognized by IgE (Bush *et al.*, 1983; Nyholm *et al.*, 1983; Aukrust *et al.*, 1985). It is similar to the 31-kDa allergen we isolated recently (Paris *et al.*, 1990b). Yunginger *et al.* (1980) have isolated Alt I by gel exclusion chromatography in a peak with an estimated molecular mass between 25 and 50 kDa. This result is in agreement with our SDS–PAGE molecular mass estimation of Alt I. Alt I is a glycoprotein composed of at least five isoelectric variants in the range of 4–4.3 (Yunginger *et al.*, 1980). Under reducing conditions it appears as a dimeric molecule with two subunits of approximately 19 and 20 kDa (Paris et al., 1990b). Besides

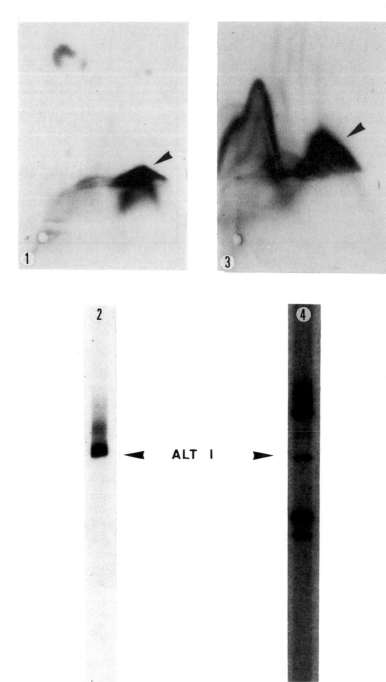

Figure 6.   CRIE (1, 3) and immunoblots (2, 4) of *A. alternata* extracts obtained by incubation of spores in Coca's solution (1, 2), or by disruption in carbonate buffer with 1% PVP, 1 mM PMSF, and 5 mM EDTA (3, 4). Arrowheads correspond to Alt I.

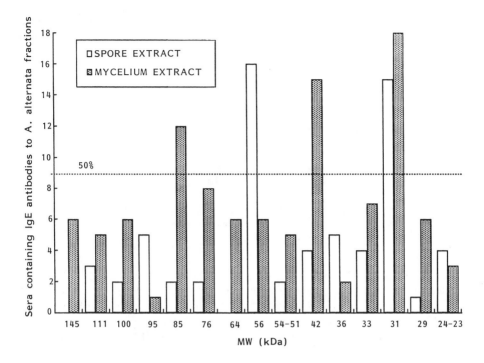

Figure 7. Histogram showing the frequency of IgE binding of 19 individual sera to spore and mycelium extracts of *A. alternata*.

being the major allergen, Alt I has another advantage over other major conidial allergens such as the 56 kDa: it can be extracted from the mycelium at higher concentrations than from the conidia (Aukrust *et al.*, 1985; Paris *et al.*, 1990a). This property will facilitate its purification since it is much easier to mass-produce mycelium in a fermenter than to recover high amounts of conidia from petri cultures.

Purification of Alt I has already been undertaken by fractionation of mycelial extracts using anion-exchange chromatography (Yunginger *et al.*, 1980; Paris *et al.*, unpublished). On a Mono-Q HPLC column the fraction containing the 31 kDa is eluted with 0.4 M NaCl at a pH between 4 and 5 (Fig. 8) (Paris and Debeaupuis, unpublished). This preparation is currently being used to produce monoclonal antibodies with the aim of characterizing this and other *Alternaria* allergenic preparations.

3.2. Cladosporium

Previous studies with *Cladosporium* have been limited mainly to *C. herbarum* mycelium allergens (Bush and Yunginger, 1987). In order to identify conidial allergens of this genus, research has recently been undertaken in our laboratory on the conidia of *C. cladosporioides*, which is the most prevalent airborne species of *Cladosporium* (Lacey, 1981).

For skin testing, conidia were either disrupted in a 5-50 mM Tris buffer or incubated in Coca's solution for 24 hr at 4°C. Intracellular extracts of disrupted conidia were more allergenic

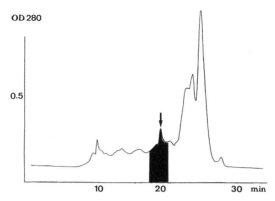

Figure 8. Elution profile of *A. alternata* mycelial extract on a Mono-Q, anionic exchange column. Alt I was eluted (arrow) with 0.4 M NaCl from an extract in 25 mM Tris (pH 7.5) + 0.1% CHAPS.

than passive extracts. However, at 0.5 mg protein/ml of extract, most of the patients studied reacted to both extracts (Gumowski *et al.*, unpublished). Studies using extracts from *C. cladosporioides* and *C. herbarum* demonstrated that 80% of patients reacted to both species. RAST inhibition studies confirmed that allergenic potencies of *C. cladosporioides* and *C. herbarum* are only slightly different (Bouziane *et al.*, unpublished). A good correlation between RAST and other *in vivo* clinical tests was obtained from preadolescent patients who showed highest reactions to *Cladosporium* extracts (Koivikko, 1984; Nüsslein *et al.*, 1987; Gumowski *et al.*, unpublished). In contrast, more than 50% of the adult patients with positive skin reactions had negative RAST and negative nasal or bronchial provocations to these same extracts (Gumowski *et al.*, unpublished). Several hypotheses can be formulated to explain this discrepancy between *in vivo* and RAST results. At the concentration studied (250–500 μg protein/ml), the fungal extracts could induce a nonspecific degranulation of mast cells in the absence of specific IgEs. However, *in vivo* studies with animal models have demonstrated that extracts of *C. cladosporioides* do not contain metabolites capable of inducing nonspecific degranulation of mast cells in nonsensitized animals (Bouziane *et al.*, 1988). Several allergens do not bind to CNBr-activated disks. Landmark and Aukrust (1985), using fused rocket radioimmunoelectrophoresis on fractions obtained by gel permeation HPLC, have demonstrated that high-molecular-weight allergens do not bind to cellulose RAST disks. Recently, we obtained positive ELISA results with spore extracts previously absorbed to a CNBr-Sepharose column (Latgé, unpublished). Moreover, polysaccharides of *Cladosporium* which have high IgE binding affinities (Swärd-Nordmo *et al.*, 1988b) would not bind to CNBr-activated disks. IgEs located at the skin level and circulating IgEs may have different specificities. Positive skin tests can also precede the appearance of circulating IgEs (Malling *et al.*, 1985, 1986). IgG and IgE recognizing the same fungal epitopes could interfere (Lynch *et al.*, 1975). This seems highly improbable in *Cladosporium* where similar RAST values have been obtained with patient sera and purified IgE from the same patients isolated on an anti-IgE monoclonal affinity column (Latgé, unpublished).

All these clinical studies indicate that the diagnosis of a positive allergy to *Cladosporium* must be done with care. Similar difficulties in the diagnosis of respiratory allergy to *Cladosporium* using commercial extracts have been pointed out by several authors (Aas *et al.*, 1980; Malling, 1985a,b; Duc *et al.*, 1986). Inhalation challenge assay and RAST should be associated with skin test to ascertain the diagnosis.

The localization of allergens in conidia have been performed using physical and chemical techniques to microdissect the conidia (Bouziane *et al.*, 1989; Latgé *et al.*, 1988). The conidia of *C. cladosporioides* are characterized by a bilayered cell wall covered by a fibrous network of interwoven rodlets (10 × 150–350 nm) (Fig. 9). The components of this rodlet layer could be of primary importance in the case of fungal respiratory allergy since they are the first components in contact with the nasal mucosa and may trigger any allergy crisis. The rodlet layer can be removed by ultrasonication (80 W, 30 min) or incubation in various chemical solutions (e.g., 1% SDS, Coca's solution, 0.1 N NaOH) at low spore concentration (0.1 mg/ml) (Fig. 9). The removal of the rodlet layer releases surface wall components into the extract. Other extraction procedures also release conidial components without disrupting the rodlet configuration. These extracts, obtained with spore concentrations higher than 0.4 mg/ml, contain exclusively internal components originating either from inner layers of the wall or from the cytoplasm (Latgé *et al.*, unpublished; Bouziane *et al.*, 1989). Extracts isolated by conidial disruption would not be expected to contain exclusively intracytoplasmic components. In fact, immunocytochemical study of this fraction using the GLAD method of Larsson (1984) demonstrated that this extract contained not only cytoplasmic molecules but also wall components (Latgé and Paris, 1988). All these extracts obtained passively or actively under different conditions contained high amounts of allergens [detected either *in vitro* by the RAST or *in vivo* by the passive cutaneous anaphylaxis test (Bouziane *et al.*, 1988, 1989)]. These results indicate that allergens are present on the surface of the conidia as well as inner compartments. Immunolabeling experiments have confirmed that allergens are present in the outer and inner layer of the wall and in the cytoplasm of the conidia (Latgé *et al.*, unpublished) (Fig. 10).

The characterization of *Cladosporium* allergens has focused on a highly allergenic fraction, obtained by conidial disruption in a mixture containing 50 mM Tris, 1 mM EDTA, and 1% PVP (Latgé *et al.*, unpublished). Immunoblots using this extract and sera from allergic patients demonstrated that more than 30 allergenic proteins can be detected in this extract with a range of molecular mass between 18 and 150 kDa (Fig. 11). Most of the allergens have a molecular mass between 40 and 150 kDa, which also corresponds to the zone of the immunoblot recognized by

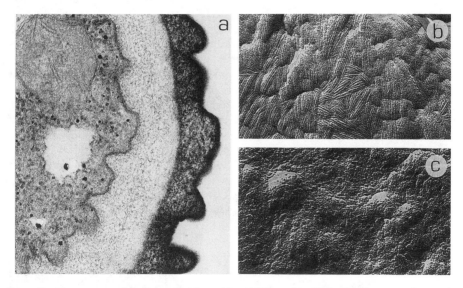

Figure 9. Ultrastructure of *C. cladosporioides* conidia. (a) Bilayered conidial wall with an outer electron-dense layer. Presence (b) or absence (c) of rodlet fascicles on the surface of conidia of *C. cladosporioides*. Conidia were either untreated (b), or incubated with 0.1 N NaOH (c). (Adapted from Bouziane *et al.*, 1989, and Latgé *et al.*, 1988.)

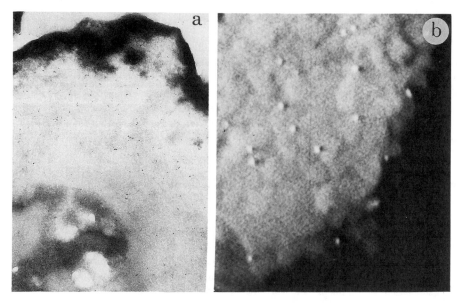

Figure 10. IgE-binding sites localized on thin sections (a) and on the surface (b) of *C. cladosporioides* conidia. Allergens are localized by sequential incubation with patient serum, and anti-human IgE rabbit IgG, and anti-rabbit IgG conjugated to 5-nm (a) and 40-nm (b) gold.

IgG. A high degree of heterogeneity was demonstrated between each individual serum, which makes it difficult to select major allergens based on immunoblot analyses (Fig. 11). In spite of using reducing electrophoretic conditions, no allergen was demonstrated in the region of 13 kDa. An allergen of this molecular mass (Ag 32) was identified by Aukrust and Borch (1979) as the major allergen of *Cladosporium*. The reason for not detecting this allergen in our conidial extracts is unknown. Ag 32 could be specific to the mycelium of *C. herbarum*. This hypothesis seems highly improbable because similar SPT results have been obtained with extracts from *C. herbarum* mycelium and *C. cladosporioides* conidia. Ag 32 has been isolated from extracts obtained after 16–24 hr of incubation of conidia in ammonium bicarbonate (Aukrust and Borch, 1979). These conditions are known to induce the release of high concentrations of proteases (Bouziane *et al.*, 1989). Consequently, another explanation of the presence of a 13-kDa allergen in the extracts would be that this molecule is the product of partial proteolytic degradation of allergens having higher molecular masses.

An intermediate allergen of 18kDa recognized by about 25% of the patients' sera has also been identified by immunoblots of SDS–PAGE gels (Fig. 11). This molecule may correspond to Ag 54 which is the only *Cladosporium* allergen purified and chemically characterized to date (Swärd-Nordmo *et al.*, 1985, 1988a,b). Ag 54 is easily isolated by gel filtration on Ultrogel Aca 54. It has an estimated molecular mass of 19 kDa and a pI of 5. It contains 20% protein and 80% carbohydrate. The latter is composed of mannose, galactose, and glucose in the proportion 1 : 0.6 : 1.3. Sugar and protein are O-glycosidically linked between threonine and mannose. The carbohydrate moiety is essentially composed of a highly branched galactomannan with a core of mannose (1,2 and 1,6 linked) and glucose (1,4 and 1,6 linked). In addition, 1,6-linked galacto-furanose side chains have been detected bound to C-2 of the 1,6-linked mannose and C-3 of the 1,6 and 1,2-linked mannose units. Removal of the main part of the carbohydrate moiety resulted in a sharp decline of the allergenic activity (Swärd-Nordmo *et al.*, 1985, 1988a,b).

The heterogeneity between each patient serum indicates that purified allergens would not be

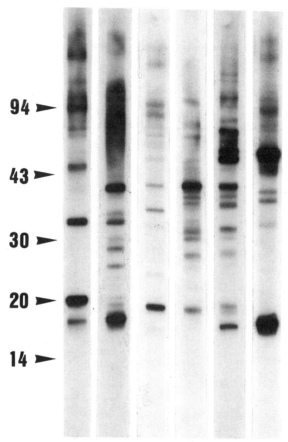

94 ►

43 ►

30 ►

20 ►

14 ►

Figure 11. Nitrocellulose immunoblots of conidial extracts of *C. cladosporioides* [obtained by cell disruption in 50 mM Tris (pH 9.0) + 1% PVP + 1 mM EDTA] showing the heterogeneous distribution of the allergens recognized by different patient sera.

of any use for allergy diagnosis. The goal of the purification experiments using *Cladosporium* is different from that involving *Alternaria*. Isolation of a single allergen from *Cladosporium* is not advantageous but enrichment of the extracted allergens is desirable in order to eliminate nonspecific positive skin tests obtained with our extracts. Size exclusion chromatography was not appropriate since all peaks (e.g., four in the case of Sepharose CL6B which included the exclusion peak with an estimated mass of 4×10^6 Da) contained allergenic activity. Similar results were obtained by Landmark and Aukrust (1985) who demonstrated that *C. herbarum* extracts contained allergens mainly with a range between 10 and 300 kDa. Immunoblot analysis and chromatography results obtained with *Cladosporium* are in disagreement with the general concept (Dreborg *et al.*, 1986) that allergens must be small enough (10 to 70 kDa) to allow them to cross membranes. In contrast to size exclusion chromatography, ion-exchange columns are very useful for isolating *C. cladosporioides* allergens. Using a Mono-Q Sepharose column, two peaks eluted at 0.2 and 0.8 M NaCl and contained allergenic activity, whereas the neutral fraction without any activity could be discarded (Fig. 12). These two peaks can be further separated by chromatofocusing on a Mono-P column. The 0.8 M fraction contained one allergen peak which eluted between pH 3.8 and 4.5, while the 0.2 M fraction had two peaks which eluted at pH 5.5 and 5.1 (Fig. 12).

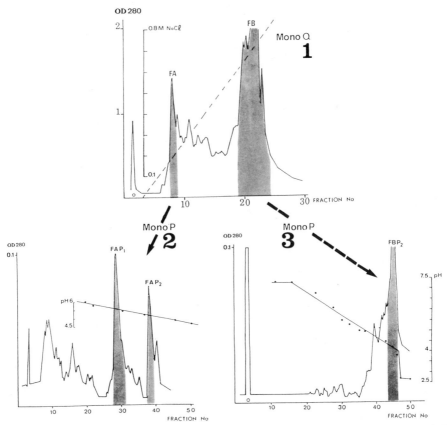

Figure 12. Elution profiles of *C. cladosporioides* extracts (obtained by conidial disruption in a solution containing 50 mM Tris + 1 mM EDTA + 1% PVP) using anion-exchange chromatography (1), followed by chromatofocusing (2, 3). Elution buffers were either 25 mM Tris (pH 7.5) + NaCl (1), or polybuffer (2, 3). Shaded areas contained allergens. (Latgé and Schrével, unpublished.)

4. OTHER FUNGI

The major mycelial allergens of *A. fumigatus* have been fairly well characterized (Wallenbeck *et al.*, 1984; Bengtsson *et al.*, 1986; Longbottom, 1986, 1988; Longbottom and Austwick, 1986). Among the 15 allergens detected by CRIE and immunoblot, a heat-labile non-Con A-binding protein with a mass between 18 and 24 kDa exhibited the strongest IgE binding. Although mycelial allergens of *A. fumigatus* have been extensively studied, only two papers have investigated the conidial allergens of this species (Stevens *et al.*, 1970; Kauffman *et al.*, 1984). The relevant allergens of *Aspergillus* conidia remain to be isolated. This need for further investigation is reinforced by recent studies suggesting that the role played by *Aspergillus* spore allergens is probably underestimated. The recent multiplication of humidifiers and air-conditionng apparatuses in most modern buildings has led to the appearance of a new population of spores in these indoor environments. Xerophilic *Aspergillus* often become dominant in these air-conditioned buildings (Molina, 1986) and seem responsible for some occupational respiratory diseases. The allergens of these *Asperigillus* species are totally unknown. Moreover, xerophilic

Aspergillus such as *A. glaucus*, *A. restrictus*, *A. versicolor*, *A. penicillioides*, and *A. repens* are present in house dust and in association with mites (van de Lustgraaf, 1978; Latgé and Samson, unpublished). Specific antibodies directed toward different xerophilic *Aspergillus* species react positively with house dust and mite allergenic extracts (Latgé, unpublished). Similarly, some patients allergenic to house dust or mites have positive RAST with xerophilic fungi suggesting that they have been sensitized after repeated exposure to these fungi (Latgé and Fitting, unpublished). Allergens are quickly released within 20 min of incubation of xerophilic fungi in PBS (Rijckaert and Broers, 1980). If the ecological relationship between xerophilic fungi and house dust mites is well established, further investigation is needed to understand the allergenic association between these two inhabitants of dry environments.

Very few other studies have been undertaken with saprophytic deuteromycetes. *Epicoccum* and *Penicillium* have been shown to contain allergens (Licorish *et al.*, 1985; Portnoy *et al.*, 1987). It should be pointed out that, although *Penicillium* is frequently found in the atmosphere, no studies have examined the characteristics of the mycelial or conidial allergens of this genus.

Studies been performed with basidiomycetes in localized areas where basidiospore storms are regularly observed followed by an increased seasonal occurrence of acute asthma (Santilli *et al.*, 1985; Lehrer *et al.*, 1986; Sprenger *et al.*, 1986). The different species investigated are gilled mushrooms (e.g. *Armillara, Amanita, Agaricus, Pleurotus, Coprinus, Psilocybe*), polypore mushrooms (e.g., *Ganoderma, Inonotus*), and puffballs (e.g., *Scleroderma, lycoperdon, Geastrum*). Basidiospore extracts usually give stronger skin test reactions than mycelial extracts (Lehrer *et al.*, 1986). However, in *Coprinus quadrifidus* and *Pleurotus ostreatus*, cap and stalk extracts contained allergens similar to those present in spore extracts (Weissman *et al.*, 1987; Davis *et al.*, 1988). The allergenic potency of basidiospore extracts depended on the origin of the spores and method of extraction, and differed from species to species (Lopez *et al.*, 1985; Liengswangwong *et al.*, 1987). Asthmatic patients have a higher incidence of immediate skin reaction to basidiospore extracts than patients with allergic rhinitis (Lehrer *et al.*, 1986; Santilli *et al.*, 1985). Skin tests and RAST correlate in most but not all species (Butcher *et al.*, 1987). Defatting of spores with ethyl ether followed by homogenization in 0.125 M NH_4CO_3 resulted in the highest allergen yield (Liengswangwong *et al.*, 1987). Basidiospores originating from nature gave extracts with a higher potency than basidiospores produced *in vitro* (Lehrer *et al.*, 1986). Puffball (particularly *Scleroderma*) extracts gave the highest reactivity. However, RAST inhibition studies indicated the presence of allergenic determinants shared by most species, the most potent inhibitor being *Psilocybe cubensis* (O'Neil *et al.*, 1988). Cross-reactivity between Basidiomycetes and Deuteromycetes has not been investigated, although the work of Lehrer *et al.* (1986) would suggest possible association between allergens of Basidiomycetes and dematiaceous conidial fungi. The degree of variation of allergens present in Basidiomycetes has not been appreciated until now.

Allergenicity of airborne ascospores has been examined in a single report by Burge *et al.* (1985). Positive skin reactions were obtained with *Leptosphaeria, Leptosphaerulina, Gnomonia, Gibberella*, and *Chaetomium* using distilled water extracts of sporulating mycelia of these different genera. Important cross-reactivity was observed between *Leptosphaeria* and *Alternaria*.

5. PERSPECTIVES FOR FUTURE RESEARCH

The number of allergenic fungi studied to date remains far too limited. Research should be extended to fungi other than the dematiaceous, saprophytic deuteromycetes, such as *Cladosporium* and *Alternaria*, which are the only species researched extensively at the moment. Even in the new fungal models studied, the allergen molecules are poorly characterized. Purification and isolation of allergens should be pursued. Their characterization will permit standardization of the fungal extracts which has not yet been realized, although attempts have been initiated with *Alternaria* (Helm *et al.*, 1987). Such standardization is essential for clinical practice, and particularly for specific diagnosis and immunotherapy. Such goals can be attained with mono-specific or monoclonal antibodies directed toward major fungal allergens. An understanding of cellular immune reactions involved in allergy, recognition of immunodominant epitopes, especially those involving the carbohydrate moiety, attachment of the allergens to IgE on their target cells, the competition between IgE and other immunoglobulins or plasmatic factors with the allergens also requires the availability of pure, allergenic molecules.

6. REFERENCES

Aas, K., Leegaard, J., Aukrust, L., and Grimmer, O., 1980, Immediate type hypersensitivity to common molds, comparison of different diagnostic materials, *Allergy*, 35:443–451.

Agarwal, M. K., and Swanson, M. C., Reed, C. E., and Yunginger, J. W., 1983, Immunochemical quantitation of airborne short ragweed, *Alternaria*, antigen E, and Alt-I allergens: A two year prospective study, *J. Allergy Clin. Immunol.* 72:40–45.

Ainsworth, G. C., 1986, *Introduction to the History of Medical and Veterinary Mycology*, Cambridge University Press, London.

Aukrust, L., and Borch, S. M., 1979, Partial purification and characterization of two *Cladosporium herbarum* allergens, *Int. Arch. Allergy Appl. Immunol.* 60:68–79.

Aukrust, L., Borch, S. M., and Einarsson, R., 1985, Mould allergy—Spores and mycelium as allergen source, *Allergy* 40:43–48.

Bengtsson, A., Karlsson, A., Rolfsen, W., and Einarsson, R., 1986, Detection of allergens in mould and mite preparations by a nitrocellulose electroblotting technique, *Int. Arch. Allergy Appl. Immunol.* 80:383–390.

Bouziane, H., Latgé, J. P., Prévost, M. C., Chevance, L. G., and Paris, S., 1988, Nasal allergy in guinea pigs, *Mycopathologia* 101:181–186.

Bouziane, H., Latgé, J. P., Mécheri, S., Fitting, C., and Prévost, M. C., 1989, Release of allergens from *Cladosporium* conidia, *Int. Arch. Allergy Appl. Immunol.* 88:261–266.

Budd, T. W., Kuo, C. Y., Yoo, T. J., McKenna, W. R., and Cazin, J., 1983a. Antigens of *Alternaria*. I. Isolation and partial characterization of a basic peptide allergen, *J. Allergy Clin. Immunol.* 71:277–282.

Budd, T. W., Kuo, C. Y., Cazin, J., and Yoo, T. J., 1983b, Allergens of *Alternaria*: Further characterization of a basic allergen fraction, *Int. Arch. Allergy Appl. Immunol* 71:83–87.

Burge, H., Gold, M. D., Muilenberg, M., and Solomon, W., 1985, Allergenicity of airborne ascospores, *J. Allergy Clin. Immunol.* 75:118.

Bush, R. K., and Yunginger, J. W., 1987, Standardization of fungal allergens, *Clin. Rev. Allergy* 5:3–21.

Bush, R. K., Voss, M. J., and Bashiran, S., 1983, Detection of *Alternaria* allergens by crossed radioimmuno-electrophoresis, *J. Allergy Clin. Immunol.* 71:239–244.

Butcher, B. T., O'Neil, C. E., Reed, M. A., Altman, L. C., Lopez, M., and Lehrer, S. B., 1987, Basidiomycete allergy: Measurement of spore specific IgE antibodies, *J. Allergy Clin. Immunol.* 80:803–809.

Davis, W. E., Horner, W. E., Salvaggio, J. E., and Lehrer, S. B., 1988, Basidiospore allergens: Analysis of *Coprinus quadrifidus* spore, cap and stalk extracts, *Clin. Allergy* 18:261–267.

Dreborg, S., Einarsson, R., and Longbottom, J. L., 1986, The chemistry and standardization of allergens, in: *Handbook of Experimental Immunology*, Volume 1, 4th ed. (D. M. Weir and L. A. Herzenberg, eds.), Blackwell, Oxford, pp. 10.1–10.28.

Duc, J., Kolly, M., and Pécoud, A., 1986, Fréquence des allergènes respiratoires impliqués dans la rhinite et l'asthme bronchique de l'adulte. Etude prospective, *Schweiz. Med. Wochenschr.* 116:1205–1210.

Fadel, R., Paris, S., Fitting, C., Rassemont, R., and David, B., 1986, A comparison of extracts from *Alternaria* spores and mycelium, *J. Allergy Clin. Immunol.* 77:242.

Gumowski, P., Bernardini, D., Grange, F., and Girard, J. P., 1985, Synthetic non proteinic and anergic culture medium for the growth of *Candida albicans*, in: *Annu. Meet. Eur. Acad. Allergy Clin. Immunol* Stockholm, June 2–5, Abstracts, p. 180.

Gumowski, P. I., Grange, F., and Girard, J. P., 1987, Asthmes intrinsèques et réactivité spécifique aux moisissures, *Med. Hyg.* 45:153–157.

Helm, R. M., Squillace, D. L., Aukrust, L., Borch, S. M., Baer, H., Bush, R. K., Lowenstein, H., Znairowski, R., Nitchuk, W., and Yunginger, J. W., 1987, Production of an international reference standard *Alternaria* extract. I. Testing of candidate extracts, *Int. Arch. Allergy Appl. Immunol.* **82**:178–189.

Hoffman, D. R., 1984, Mould allergens, in: *Mould Allergy* (Y. Al-Doory and J. F. Domson, eds.), Lea & Febiger, Philadelphia, pp. 104–116.

Hoffman, D. R., Kozak, P. P., Gillman, S. A., Cummins, L. H., and Gallup, J., 1981, Isolation of spore specific allergens from *Alternaria*, *Ann. Allergy* **46**:310–316.

Kauffman, H. F., Heide, S., van der, Beaumont, F., Monchy, J. G. R., and de Vries, K., 1984, The allergenic and antigenic properties of spore extracts of *Aspergillus fumigatus*: A comparative study of spore extracts with the mycelium and culture filtrate extracts, *J. Allergy Clin. Immunol.* **73**:567–573.

Koivikko, A., 1983, Patients exposure to moulds, in: *Mould Allergy Workshop* (T. Foucard and S. Dreborg, eds.), Pharmacia, Uppsala, pp. 35–41.

Koivikko, A., 1984, Mould allergy in Finland, in: *Atlas of Moulds in Europe Causing Respiratory Allergy* (K. Wilken-Jensen and S. Gravesen, eds.) ASK Publ., Copenhagen, pp. 90–91.

Kurup, V. P., 1986, Enzyme-linked immunosorbent assay in the detection of specific antibodies against *Aspergillus* in patient sera, *Zantralbl. Bakteriol. Hyg. A* **261**:509–516.

Kurup, V. P., Ramasamy, M., Greenberger, P. A. and Fink, J. N., 1988, Isolation and characterization of a relevant *Aspergillus fumigatus* antigen with IgG and IgE binding activity, *Int. Arch. Allergy Appl. Immunol.* **86**:176–182.

Labrousse, H., and Avrameas, S., 1987, A method of quantification of a colored or fluorescent signal in enzyme immunoassay by photodensitometry, *J. Immunol. Methods* **103**:9–14.

Labrousse, H., Guesdon, J. L., Ragimbeau, J., and Avrameas, S., 1982, Miniaturization of β-galactosidase immunoassays using chromogenic and fluorogenic substrates, *J. Immunol. Methods* **48**:132–147.

Lacey, J., 1981, The aerobiology of conidial fungi, in: *Biology of Conidial Fungi*, Volume 1 (G. T. Cole and B. Kendrick, eds.), Academic Press, New York, pp. 373–416.

Landmark, E., and Aukrust, L., 1985, High-performance liquid chromatography of *Cladosporium herbarum*. Identification of allergens with immunological techniques, *Int. Arch. Allergy Appl. Immunol.* **78**:71–76.

Larsson, L. I., 1984, Labelled antigen detection methods, in: *Immunolabelling for Electron Microscopy* (J. M. Polak and I. M. Varndell, eds.), Elsevier, Amsterdam, pp. 123–128.

Latgé, J. P., and Paris, S., 1988, Allergens of *Alternaria* and *Cladosporium*, in: *Fungal Antigens* (E. Drouhet, G. T. Cole, L. De Repentigny, J. P., Latgé, and B. Dupont, eds.), Plenum Press, New York pp. 237–258.

Latgé, J. P., Bouziane, H., and Diaquin, M., 1988, Ultrastructure and composition of the conidial wall of *Cladosporium cladosporioides*, *Can. J. Microbiol.* **34**:1325–1329.

Lehrer, S. B., Lopez, M., Butcher, B. T., Olson, J., Reed, M., and Salvaggio, J. E., 1986, Basidiomycete mycelia and spore allergen extracts: skin test reactivity in adults with symptoms of respiratory allergy, *J. Allergy Clin. Immunol.* **78**:478–485.

Licorish, K., Novey, H. S., Kozak, P., Fairshter, R. D., and Wilson, A. F., 1985, Role of *Alternaria* and *Penicillium* spores in the pathogenesis of asthma, *J. Allergy Clin. Immunol.* **76**:819–825.

Liengswangwong, V., Salvaggio, J. E., Lyon, F. L., and Lehrer, S. B., 1987, Basidiospore allergens: Determination of optimal extraction methods, *Clin. Allergy* **17**:191–198.

Longbottom, J. L., 1986, Antigens and allergens of *Aspergillus fumigatus*. II. Their further identification of partial characterization of a major allergen (Ag3), *J. Allergy Clin. Immunol.* **78**:18.

Longbottom, J. L., 1988, Allergens of *Aspergillus fumigatus*, in: *Fungal Antigens* (E. Drouhet, G. T. Cole, L. De Repentigny, J. P. Latgé, and B. Dupont, eds.), Plenum Press, New York pp. 223–236.

Longbottom, J. L., and Austwick, P. K. C., 1986, Antigens and allergens of *Aspergillus fumigatus*. I. Characterization by quantitative immunoelectrophoretic methods, *J. Allergy Clin. Immunol.* **78**:9–18.

Lopez, M., Butcher, B. T., Salvaggio, J. E., Olson, J. S., Reed, M. A., McCants, M. L., and Lehrer, S. B., 1985, Basidiomycete allergy: What is the best source of antigen? *Int. Arch. Allergy Appl. Immunol.* **77**:169–170.

Lynch, N. R., Dunand, P., Newcomb, R. W., Chai, H., and Bigley, J., 1975, Influence of IgG antibody and glycopeptide allergens on the correlation between the radioallergosorbent test (RAST) and skin testing or bronchial challenge with *Alternaria*, *Clin. Exp. Immunol.* **22**:35–46.

Mackenzie, D. W. R., 1983, Serodiagnosis, in: *Fungi pathogenic for Humans and Animals*. B1. *Pathogenicity and Detection* (D. H. Howard, ed.), Dekker, New York, pp. 121–218.

Malling, H. J., 1985a, Diagnosis and immunotherapy of mould allergy. II. Reproducibility and relationship between skin sensitivity estimated by endpoint titration and histamine equivalent reaction using skin prick test and intradermal test, *Allergy* **40**:354–362.

Malling, H. J., 1985b, Reproducibility of skin sensitivity using a quantitative skin prick test, *Allergy* **40**:400–404.

Malling, H. J., 1987, Quantitative skin prick testing: Dose-response of histamine- and allergen-induced wheal reactions, *Allergy* **42**:196–204.

Malling, H. J., Agrell, B., Croner, S., Dreborg, S., Foucard, T., Kjellman, M., Koivikko, A., Roth, A., and Weeke, B., 1985, Diagnosis and immunotherapy of mould allergy. I. Screening for mould allergy, *Allergy* **40**:108–114.

Malling, H. J., Dreborg, S., and Weeke, B., 1986, Diagnosis and immunotherapy of mould allergy. III. Diagnosis of *Cladosporium* allergy by means of symptom score, bronchial provocation test, skin prick test, RAST, CRIE and histamine release, *Allergy* **41**:57–67.

Molina, C., 1986, *Humidifiers and air-conditioners diseases*, INSERM, Paris.

Notermans, S., Dufrenne, J., Wijnands, L. M., and Engel, H. W. B., 1988, Human serum antibodies to extra cellular polysaccharides (EPS) of moulds, *J. Med. Vet. Mycol.* **26**:41–48.

Nüsslein, H. G., Zimmerman, T., Baum, M., Fuchs, C., Kölble, K., and Kalden, J. R., 1987, Improved *in vitro* diagnosis of allergy to *Alternaria tenuis* and *Cladosporium herbarum*, *Allergy* **42**:414–422.

Nyholm, L., and Lowenstein, H., 1982, Identification and characterization of allergens from *Alternaria alternata*, *Allergy* **37** (Suppl. 1):40.

Nyholm, L., Lowenstein, H., and Yunginger, J. W., 1983, Immunochemical partial identity between two independently identified and isolated major allergens from *Alternaria alternata* (Alt-1 and Ag 1), *J. Allergy Clin. Immunol.* **71**:461–467.

O'Neil, C. E., Hughes, J. M., Butcher, B. T., Salvaggio, J. E., and Lehrer, S B., 1988, Basidiospore extracts: Evidence for common antigenic/allergenic determinants, *Int. Arch. Allergy Appl. Immunol.* **85**:161–166.

Paris, S., Fitting, C., Ramirez, E., Latgé, J. P., and David, B., 1990a, Comparison of different extraction methods of *Alternaria* allergens, *J. Allergy Clin. Immunol.* **85**:941–948.

Paris, S., Fitting, C., Latgé, J. P., Guinnepain, M. T., Herman, D., and David, B., 1990b, Comparison of conidial and mycelial allergens of *Alternaria alternata*, *Int. Arch. Allergy Appl. Immunol.* (in press).

Portnoy, J., Chapman, J., Burge, H., Muilenberg, M., and Solomon, W., 1987, *Epicoccum* allergy: Skin reaction patterns and spore/mycelium disparities recognized by IgG and IgE ELISA inhibition, *Ann. Allergy* **59**:39–43.

Rijckaert, G., 1981, Fast releasing allergens from some organisms living in house dust, Ph.D. thesis, Nijmegen University, The Netherlands.

Rijckaert, G., and Broers, J. L. V., 1980, Time dependent release of allergens from some xerophilic fungi, *Allergy* **35**:679–682.

Roitt, I. M., Brostoff, J., and Male, D. K., 1985, *Immunology*, Gower Medical, London.

Salvaggio, J., and Aukrust, L., 1981, Mold-induced asthma, *J. Allergy Clin. Immunol.* **68**:327–346.

Santilli, J., Rockwell, W. J., and Collins, R. P., 1985, The significance of the spores of the Basidiomycetes (mushrooms and their allies) in bronchial asthma and allergic rhinitis, *Ann. Allergy* **55**:469–471.

Schleimer, R. P., Fox, C. C., Naclerio, R. M., Plant, M., Creticos, P. S., Togias, A. G., Warner, J. A., Kagey-Sobotka, A., and Lichtenstein, L. M., 1985, Role of human basophils and mast cells in the pathogenesis of allergic diseases, *J. Allergy Clin. Immunol.* **76**:369–374.

Sepulveda, R., Longbottom, J. L., and Pepys, J., 1979, Enzyme linked immunosorbest assay (ELISA) for IgG and IgE antibodies to protein and polysaccharide antigens of *Aspergillus fumigatus*, *Clin. Allergy* **9**:359–371.

Solomon, W. R., Burge, H. A., and Muilenberg, L., 1980, Allergenic properties of *Alternaria* spore, mycelium, and "metabolic" extracts, *J. Allergy Clin. Immunol.* **65**:229.

Sprenger, J. D., Altman, T. C., O'Neil, C. E., Lehrer, S. B., and Ayars, G. H., 1986, Skin test reactivity to basidiospores in adults in Seattle with respiratory allergies, *J. Allergy Clin. Immunol.* **77**:200.

Steringer, I., Aukrust, L., and Einarsson, R., 1987, Variability of antigenicity/allergenicity in different strains of *Alternaria alternata*, *Int. Arch. Allergy Appl. Immunol.* **84**:190–197.

Stevens, E. A. M., Hilvering, S., and Orie, N. G. M., 1970, Inhalation experiments with extracts of *Aspergillus fumigatus* on patients with allergic aspergillosis and aspergilloma, *Thorax* **25**:11.

Swärd-Nordmo, M., Almeland, T. L., and Aukrust, L., 1984, Variability in different strains of *Cladosporium herbarum* with special attention to carbohydrates and contents of two important allergens (Ag-32 and Ag-54), *Allergy* **39**:387–394.

Swärd-Nordmo, M., Wold, J. K., Smestad Paulsen, B., and Aukrust, L., 1985, Purification and partial characterization of the allergen Ag-54 from *Cladosporium herbarum*, *Int. Arch. Allergy Appl. Immunol.* **78**:249–255.

Swärd-Nordmo, M., Smestad Paulsen, B., and Wold, J. K., 1988a, The glycoprotein allergen Ag-54 (cla hII) from *Cladosporium herbarum*. Structural studies of the carbohydrate moiety, *Int. Arch. Allergy Appl. Immunol.* **85**:288–294.

Swärd-Nordmo, M., Smestad Paulsen, B., and Wold, J. K., 1988b, The glycoprotein allergen Ag-54 (cla hII) from *Cladosporium herbarum*. Further biochemical characterization, *Int. Arch. Allergy Appl. Immunol.* **85**:295–301.

Tee, R. D., Gordon, D. J., and Newman Taylor, A. J., 1987, Cross-reactivity between antigens of fungal extracts: Studies by RAST-inhibition and immunoblot technique, *J. Allergy Clin. Immunol.* **79**:627–633.

Tovey, E. R., and Baldo, B. A., 1987, Characterization of allergens by protein blotting, *electrophoresis* **8**:452–463.

van der Heide, S., Kauffman, H. F., and de Vries, K., 1985, Cultivation of fungi in synthetic and semi-synthetic liquid medium. II. Immunochemical properties of the antigenic and allergenic extracts, *Allergy* **40**:592–598.

van de Lustgraaf, B., 1978, Ecological relationships between xerophilic fungi and house dust mites (Acarida: Pyroglyphidae), *Oecologia (Berlin)* **33**:315–359.

Vijay, H. M., Huang, H., Young, N. M., and Bernstein, I. L., 1979, Studies on *Alternaria* allergens. I. Isolation of allergens from *Alternaria tenuis* and *Alternaria solani*, *Int. Arch. Allergy Appl. Immunol.* **60**:229–239.

Vijay, H. M., Young, N. M., Jackson, G. E. D., White, G. P., and Bernstein, I. L., 1985, Studies on *Alternaria* allergens. V. Comparative biochemical and immunological studies of three isolates of *Alternaria tenuis* cultured on synthetic media, *Int. Arch. Allergy Appl. Immunol.* **78**:37–42.

Vijay, H. M., Young, N. M., and Bernstein, I. L., 1987, Studies on *Alternaria* allergens. VI. Stability of the allergen components of *Alternaria tenuis* extracts under a variety of storage conditions, *Int. Arch. Allergy Appl. Immunol.* **83**:325–328.

Wallenbeck, I., Aukrust, L., and Einarsson, R., 1984, Antigenic variability of different strains of *Aspergillus fumigatus*, *Int. Arch. Allergy Appl. Immunol.* **73**:166–172.

Weissman, D. N., Halmepuro, L., Salvaggio, J. E., and Lehrer, S. B., 1987, Antigenic/allergenic analysis of basiodiomycete cap, mycelium and spore extracts, *Int. Arch. Allergy Appl. Immunol.* **84**:56–61.

Wilken-Jensen, K., and Gravesen, S., 1984, *Atlas of Moulds in Europe Causing Respiratory Allergy*, ASK Publ., Copenhagen.

Yuan, L., and Cole, G. T., 1987, Isolation and characterization of an extracellular proteinase of *Coccidioides immitis*, *Infect. Immun.* **55**:1970–1978.

Yunginger, J. W., Jones, R. T., Nesheim, M. E., and Geller, M., 1980, Studies on *Alternaria* allergens. III. Isolation of a major allergenic fraction (ALT-I), *J. Allergy Clin. Immunol.* **66**:138–147.

19

Conidia of Coccidioides immitis

Their Significance in Disease Initiation

Garry T. Cole and Theo N. Kirkland

1. INTRODUCTION

Airborne fungal spores (sporangiospores, ascospores, basidiospores, and conidia) are ubiquitous (Gregory, 1973; Cole and Samson, 1984) and a source of constant aggravation to individuals suffering from inhalant allergy (Hoffman, 1984). Mold components of the air spora also represent a threat to immunocompromised patients, particularly those undergoing chemotherapy for cancer or individuals presenting with advanced stages of acquired immunodeficiency syndrome (AIDS; Bronnimann *et al.*, 1987; Musial *et al.*, 1988). Fungi which cause respiratory infections in humans are typically opportunistic pathogens (Cox, 1983). *Coccidioides immitis*, however, is somewhat of an anomaly among the systemic fungal pathogens. The saprobic phase of this imperfect fungus inhabits arid soil in regions of the southwestern United States, Mexico, Central and South America (Pappagianis, 1988). Its tiny, air-dispersed conidia, if inhaled, can initiate a human respiratory disease known as valley fever (coccidioidomycosis), which is characterized by a unique parasitic cycle (Cole and Sun, 1985; Zimmer and Pappagianis, 1986a). Results of clinical studies of coccidioidomycosis have suggested that immunologically compromising illnesses, although present in some patients who have contracted this mycosis (Deresinski and Stevens, 1975; Roberts, 1984; Bronnimann *et al.*, 1987), are not prerequisite to infection (Pappagianis, 1975; Rutala and Smith, 1978; Cox, 1983). *Coccidioides*, therefore, is potentially a primary pathogen. It is also an appropriate experimental model for examining both the nature of host interaction with the fungal propagule and the mechanisms by which the pathogen is able to survive in this new, hostile environment. The focus of this chapter is on the infectious conidium of *C. immitis*, and in particular the conidial wall. Products derived from the cell wall of *C. immitis* have been demonstrated to elicit host immune responses which are directed against the pathogen, while other wall-associated products have been suggested to play pivotal roles in survival of the fungus *in vivo* (Cole and Sun, 1985; Cole *et al.*, 1987; Yuan and Cole, 1987). For example, we have shown that components of a water-soluble, conidial wall fraction are reactive with patient

Garry T. Cole • Department of Botany, University of Texas, Austin, Texas 78713.   *Theo N. Kirkland* • Veterans Administration Medical Center, San Diego, California 92161.

sera both in immunodiffusion assays (Cole *et al.*, 1987) and in enzyme-linked immunosorbent assays (ELISAs; Cole *et al.*, 1988a,b,c, 1989). We have also demonstrated that this soluble conidial wall fraction is a potent elicitor of proliferative response of sensitized murine T lymphocytes *in vitro*. The hydrophobic, outer layer of the conidial wall, on the other hand, has been suggested to serve as a passive barrier to destructive enzymes released by the host and may, thereby, contribute to the survival of the pathogen *in vivo* (Drutz and Huppert, 1983). Components of this same outer conidial wall layer may also be capable of modulating the host's cell-mediated immune defense system. We present evidence in this chapter that a solubilized fraction of the isolated outer conidial wall material contains macromolecular components which can apparently suppress T-cell proliferation and superoxide production by alveolar macrophage. Coccidioidomycosis is one of the most difficult of the mycoses to treat and cure (Catanzaro, 1985). This may in part be due to the formidable nature of the infectious conidia.

2. SAPROBIC AND PARASITIC CYCLES OF COCCIDIOIDES IMMITIS

Both the saprobic and parastic cycles of this fungal pathogen have been reproduced *in vitro* and are illustrated in Fig. 1.

2.1. Conidiogenesis

Conidia of *C. immitis* form simply by septation and disarticulation of fertile hyphae (Kwon-Chung, 1969; Sun *et al.*, 1979; Cole and Sun, 1985). Diagnostic features of this developmental process are the endogenous nature of conidial differentiation (i.e., within the preexisting, fertile hyphal wall) and autolysis of adjacent hyphal compartments prior to disarticulation of the chain of cells (Fig. 1). This mechanism of conidiogenesis, referred to as enteroarthric development (Cole and Samson, 1979), is recognized in other, possibly related conidial fungi such as *Malbranchea* (Currah, 1985; Sigler and Carmichael, 1976). The significance of the details of enteroarthric development relative to this discussion is that certain autolysins, which most likely play important roles in the breakdown of alternate cells of the chain and in eventual rupture of the outer, fertile hyphal wall, appear to be trapped between the outer and inner conidial wall layers (Fig. 2). Circumstantial evidence for this is based on the observation that removal of the outer wall by mechanical shearing of the cells results in release of multiple, wall-associated proteinases (Cole *et al.*, 1988c). Some of these same proteolytic enzymes may also participate in disease initiation, as discussed later (Section 6.1).

2.2. Conidial Wall Ultrastructure

Conidia of *C. immitis* are difficult to chemically fix and embed for thin-sectioning and electron microscopy, presumably because of the thickness and hydrophobicity of the cell envelope. The outer, electron-dense wall layer (OCW; Fig. 2A), which was derived from the fertile hyphal wall (Fig. 1), forms a cylindrical sleeve which partially encompasses the inner wall layer of the conidium (ICW). Amorphous material is visible between the two wall layers (IMS; Fig. 2A), perhaps binding the thin cylindrical OCW to the thicker ICW. The latter also forms the two end walls or septa of the conidium, which originally separated the arthroconidium from the adjacent autolytic cells (Fig. 1). Remnants of partially digested wall material of the degenerate cells are visible on the surface of the septal walls (SW) in Fig. 2A.

Conidia which are inoculated onto agar medium containing basal salts with dextrose and ammonium acetate as the sole sources of organic carbon and nitrogen (Levine, 1961; Levine *et al.*, 1960; Brosbe, 1967), and incubated at 39°C in the presence of 20% CO_2/80% air (Lones and

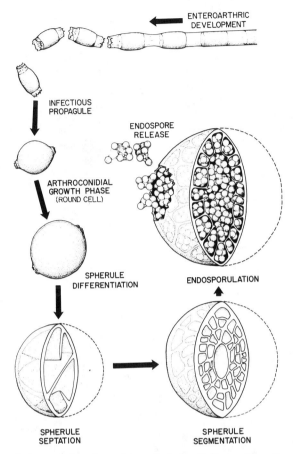

Figure 1. Diagrammatic interpretation of morphogenesis during first-generation arthroconidium–spherule–endospore transformation in *Coccidioides immitis*. (After Cole and Sun, 1985.)

Peacock, 1960), begin to swell within 18–24 hr (Cole and Sun, 1985). The apparently inelastic, electron-dense, outer wall layer of these round cells ruptures while the thin inner layer continues to grow by intussusception of new wall material (Fig. 2B). We suggest that this same developmental step occurs *in vivo* (Sun *et al.*, 1986). This growth phase is the first visible morphogenetic event leading to differentiation of the spherule, a coenocyte which at maturity may be as much as 60 μm in diameter (Fig. 1). Rupture of the hydrophobic outer wall layer during round cell formation (Fig. 1) results in the natural release of soluble macromolecules trapped between the inner and outer conidial walls, including certain proteinases mentioned above. This event apparently also leads to partial hydration of the inner wall layer (Cole *et al.*, 1983) and probably results in an increase in ion transport and molecular migration through the conidial envelope.

2.3. Host Infection

Conidia of *C. immitis* are small enough (3–6 × 2–4 μm) to be carried in the airstream deep into the respiratory tract, even to the alveoli (Cole and Samson, 1984). BALB/c mice (males, 23–27 g) inoculated intranasally with different numbers of viable arthroconidia have been shown to

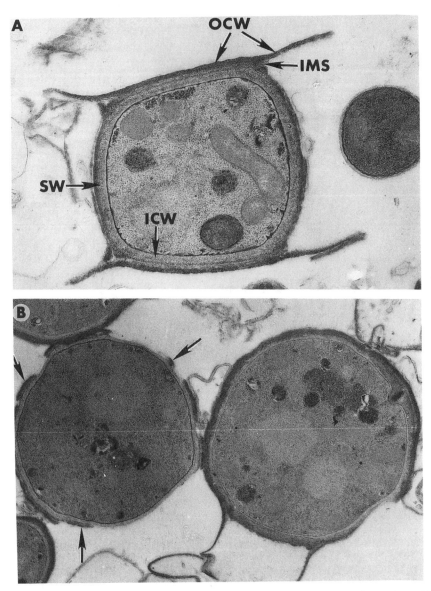

Figure 2. Thin sections of arthroconidium (A) and round cells (B). The latter were formed from arthroconidia in modified Converse medium (Levine, 1961) after incubation at 39°C in the presence of 20% CO_2/80% air for 18–24 hr. Arrows in (B) represent fragments of the naturally ruptured, outer conidial wall layer. ICW, inner conidial wall; IMS, intermural space; OCW, outer conidial wall; SW, septal wall. A, × 21,280; B, × 11,500.

develop pulmonary lesions, typical of coccidioidomycosis (Huppert *et al.*, 1976; Sun *et al.*, 1986). Death of these animals occurred when the inoculum contained as few as 100 cells (Fig. 3). The virulence of *C. immitis* conidia is further underscored by the fact that this pathogen is one of ten etiologic agents most frequently transmitted to laboratory workers when accidentally exposed to the contents of petri plate cultures of the saprobic phase (Johnson *et al.*, 1964; Wedum *et al.*, 1969; Drutz and Huppert, 1983).

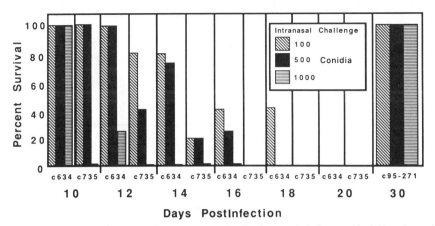

Figure 3. Survival of BALB/c mice (males, 23–27 g each) after intranasal challenge with viable arthroconidia (100, 500, or 1000 cells as indicated) from two strains isolated from patients with disseminated coccidioidomycosis (c634 and c735), and a strain identified as a temperature-sensitive, auxotrophic mutant (c95-271).

Several strains of *C. immitis* were tested for pathogenicity using intranasally challenged BALB/c mice. Our results suggest that different levels of virulence exist among the isolates (Fig. 3). As expected, a temperature-sensitive, auxotrophic mutant [c95-271 (Walch and Walch, 1967)], which was unable to convert *in vitro* from the saprobic to the parasitic phase when grown at 39°C, was avirulent even when the intranasal challenge dose was increased to 1000 cells (Fig. 3). Of significance in the case of this mutant, however, is that conidia do convert to round cells *in vitro* at 39°C but fail to develop further at this temperature and die. As also demonstrated by virulent strains, the outer conidial wall of the mutant ruptures during round cell formation, releasing macromolecules which were trapped beneath the hydrophobic layer. The cell envelope of strain c95-271 was shown to contain immunoreactive components also found in the virulent strains (see Section 3.5). Advantage was taken of this fact by using strain c95-271 to sensitize BALB/c mice against *Coccidioides* conidial wall-associated antigens for cellular immunoassays, as described in Section 3.4.

Inhalation of *C. immitis* is the most common route of infection of humans, although cases of primary cutaneous inoculation (Jacobs, 1980), maternal–fetal transmission (Bernstein *et al.*, 1981), and development of localized infection at various anatomical sites of trauma (Pappagianis, 1985) have also been reported. Symptomatic coccidioidomycosis resulting from dissemination of the pathogen from these latter foci may occur long after the injury, which suggests "persistence of viable *C. immitis* organisms even in the absence of overt illness . . ." (Pappagianis, 1985). Approximately 25,000 to 100,000 new cases of coccidioidomycosis are diagnosed each year on the basis of skin test reactivity (Drutz and Huppert, 1983). Most infections are subclinical or produce subacute respiratory symptoms (Werner *et al.*, 1972; Lundergan *et al.*, 1985; Bronnimann *et al.*, 1987). A survey of coccidioidomycosis cases among university students in an endemic region of Arizona indicated that infected individuals developed symptomatic illness that necessitated medical attention at a rate of 0.4% per year (Kerrick *et al.*, 1985). Calhoun and co-workers (1985) reported that renal and cardiac transplant recipients had coccidioidal infections detected at a frequency of 3 and 6% per year, respectively. On the other hand, a recent report of an unpredicted high annual frequency of symptomatic coccidioidoymcosis (27%) among human immunodeficiency virus (HIV)-infected patients in the Tucson area (Bronnimann *et al.*, 1987) has raised an intriguing question: Are many cases of coccidioidal illness in these and other immunocompromised patients who have been previously exposed to *C. immitis* caused by reactivation of latent infection, or are they the result of primary infection (Seltzer *et al.*, 1986; Pappagianis, 1987; Bronnimann *et al.*, 1987)? Perhaps this same question should also be asked of other fungal diseases frequently encountered in AIDS patients, particularly cryptococcosis and candidosis (Musial *et al.*, 1988).

2.4. The Parasitic Cycle

Details of morphogenesis of the pathogen *in vivo* closely resemble developmental aspects reported from *in vitro* studies (Donnally and Yunis, 1974; Miyaji and Nishimura, 1984; Sun *et al.*, 1986). Conidia undergo an extended growth phase, first becoming round cells, and then large, multinucleate spherules (Figs. 1 and 2). This increase in cell volume is accompanied by rapid and apparently synchronous nuclear divisions (Sun and Huppert, 1976; Cole and Sun, 1985). The arrest of karyokinesis is followed by segmentation of the cytoplasm, which occurs by centripetal growth of the innermost wall layer of the spherule envelope (Fig. 1; Sun *et al.*, 1979; Cole and Sun, 1985). Segmentation of the spherule continues until the protoplasm has been compartmentalized into uninucleate endospores. The outer wall layer of the spherule stretches as endospore differentiation proceeds until the envelope finally ruptures, allowing clusters of endospores to escape (Fig. 1). Individual endospores (2–3 μm in diameter) function in dissemination of the pathogen *in vivo*. The mycelial phase does not normally occur within tissues of the host (Puckett, 1954).

3. TECHNIQUES OF ISOLATION AND CHARACTERIZATION OF CONIDIAL WALL FRACTIONS

We first developed a technique of mechanical fractionation of the conidial wall and release of water-soluble and -insoluble components (Cole *et al.*, 1982, 1983). This procedure avoids use of harsh solvents or enzymatic treatment during the fractionation procedure and the antigenic fractions liberated from the wall may, therefore, be maintained in their native form. The insoluble material released during cell fractionation is composed of the outer, hydrophobic wall layer and the inner, primarily polysaccharide-containing conidial wall (Fig. 2A). In order to further separate and test the biological activity of these insoluble fractions, detergent or alkali extraction procedures were subsequently employed. The water-soluble components, on the other hand, were easily fractionated using various chromatographic and preparative gel electrophoresis procedures. Both the water-soluble and solubilized outer and inner wall fractions have been tested for reactivity in humoral and cellular immunoassays. Purified antigens with proteolytic activity have been isolated from the soluble wall fractions and characterized by appropriate enzyme assays (Cole *et al.*, 1989). Confirmation that the isolated antigens are, in fact, derived from the conidial wall was conducted by raising antibody in rabbits against the purified fractions, and then using the antisera as immunoprobes for localization of specific macromolecules on thin sections of *C. immitis* by immunoelectron microscopy (Cole *et al.*, 1987; Yuan *et al.*, 1988). These procedures are briefly described below.

3.1. Wall Fractionation

A method originally developed for isolation of conidial wall layers of *Aspergillus niger* (Cole *et al.*, 1979) was later used for our studies of *C. immitis* (Cole *et al.*, 1982, 1983). Conidia were vacuum-harvested from plate cultures of the saprobic phase which was grown at 30° for 30–60 days. The outer sleeve of wall material was removed from cells by a shearing process as the conidia were forced under pressure through the needle valve of a Ribi cell fractionator (model RF-1, refrigerated; Sorvall, Inc., Norwalk, Conn.; Fig. 4). The valve of the fractionator is adjustable which permits regulation of the pressure and the shearing force applied to the cells. A jet of chilled nitrogen gas directed at the needle valve allows the temperature to be regulated as the cells are sheared. The fractionation procedure was conducted at 5–10°C. As the outer conidial wall layer was stripped from the cells, water-soluble macromolecules were released from the envelope of the sheared arthroconidia. The viability of conidia subjected to this mechanical fractionation was shown by dilution plating to be the same as that of untreated conidia (85–90%; Cole *et al.*, 1983). Thus, the cells were not ruptured or killed by the shearing process, so there was no exposure of the isolated fractions to proteolytic enzymes from the cytoplasm. In order to reduce the effects of wall-associated proteinases released during fractionation on other components, conidia were suspended in sterile distilled water containing 10^{-3} M phenylmethylsulfonyl fluoride (PMSF). If our intent was to isolate active, wall-associated proteinases, PMSF was not used since it is an irreversible inhibitor of certain proteolytic enzymes (Turini *et al.*, 1969). The sheared cells were later subjected to glass bead homogenization (Braun homogenizer, model MSK; Bronwill Scientific Inc., Rochester, N.Y.) which yielded both an inner conidial wall fraction (ICWF) and a cytosol fraction (Cole *et al.*, 1983; Fig. 4). Some contamination of the inner wall fraction with remnants of outer wall material (Fig. 4), and mixing of cell wall macromolecules with the cytosol fraction during this step were unavoidable. The soluble conidial wall fraction (SCWF) and outer conidial wall fraction (OCWF) were separated by differential centrifugation (Cole *et al.*, 1983). The inner wall and cytosol fractions obtained by homogenization of sheared conidia were isolated in the same manner.

The OCWF and ICWF were further fractionated by nonionic detergent and sodium hydroxide extraction, respectively. In the case of OCWF, the lyophilized wall material was suspended in 0.05 M sodium phosphate buffer containing 1.8% (w/v) *N*-octyl-β-D-glucopyranoside (Calbiochem, La Jolla, Calif.). The mixture was stirred continuously while incubated at 4°C for 96 h. Sodium azide (0.02%, w/v) was added to prevent microbial growth. The pellet and supernatant were collected after centrifugation. The pellet was washed with buffer and frozen, while the supernatant was dialyzed against distilled water and the retentate was isolated and lyophilized. Alkali extraction of the ICWF was conducted at room temperature using 1 N NaOH (1 mg sample/ ml) with continuous agitation for 3–12 hr (Cole *et al.*, 1985a; Cole and Sun, 1985). After centrifugation (27,000g), the supernatant was dialyzed against distilled water and the retentate was isolated and lyophilized. The biological activities of the ICWF and OCWF extracts are discussed in Sections 4.2–4.5.

3.2. Chromatographic and Electrophoretic Isolation Procedures

Components of the SCWF were separated by molecular size differences on a Sephacryl S-300 (Pharmacia) chromatographic column (318-ml bed volume), yielding products with enhanced immunoreactivity when tested in humoral and cellular immunoassays (Cole *et al.*, 1987). Further separation of the SCWF has been performed by high-pressure liquid chromatography (HPLC) using Superose 12 HR 10/30 preparative gel filtration (Pharmacia) and Protein Pak DEAE 5PW (Waters) ion-exchange columns (Cole *et al.*, 1988c). Purified components have been used to raise polyclonal antibody in rabbits which were then employed for further isolation of specific antigens from the SCWF and other crude preparations of *C. immitis* by immunoaffinity chromatography (Cole *et al.*, 1989; Cox and Britt, 1986b). Lectin-affinity chromatography, using concanavalin A (Con A) covalently linked to Sepharose 4B (Cox and Britt, 1986a), has also been used for successful fractionation of conidial wall preparations (Kruse and Cole, 1990).

We have recently used preparative gel electrophoresis followed by HPLC gel filtration for purification of additional components of the conidial wall preparations (Cole *et al.*, 1989). Initial

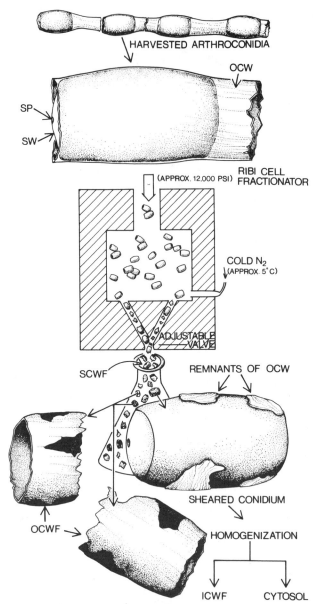

Figure 4. Illustration of steps in the isolation of arthroconidial wall and cytosol fractions of *C. immitis*. Arthroconidia harvested from plate cultures are first subjected to cell shearing in a Ribi fractionator, followed by isolation of the water-soluble and outer, hydrophobic conidial wall fractions (SCWF and OCWF, respectively) from the sheared, arthroconidial suspension, and then isolation of the inner conidial wall fraction (ICWF) and cytosol fraction from the sheared and homogenized arthroconidia. OCW, outer conidial wall; SP, septal pore; SW, septal wall.

separation was conducted by SDS–PAGE under reducing or nonreducing conditions. For example, the alkali extract of the ICWF was first fractionated by Con A-affinity chromatography and the bound components were then separated by SDS–PAGE (7.5% gel) under reducing conditions. The lane containing the standards and the adjacent lane to which the sample had been added were cut from the gel and stained by Coomassie and/or silver staining techniques (Merril *et al.*, 1984). The gel bands of interest were then located in the unstained gel by their careful realignment next to the stained lanes. The band(s) of interest were excised from the former, minced, and subjected to electroelution (Hunkapiller *et al.*, 1983; Cole *et al.*, 1989) in an electro-separation system (Elutrap; Schleicher and Schuell Inc., Keene, N.H.). The electroeluted sample was either removed from the Elutrap and dialyzed against distilled water, or electrodialyzed in the Elutrap against 10 mM ammonium acetate (pH 8.9; adjusted with ammonium hydroxide) for 5 hr at 100 V. After lyophilization, the electroeluted, ICWF-derived sample was subjected to HPLC gel filtration which yielded a purified, conidial wall antigen reactive with patient IgM antibody. The steps of this isolation procedure are summarized in Fig. 5.

Simultaneous detection of proteolytic activity and the relative molecular mass (M_r) of crude and purified fractions has been performed by substrate gel electrophoresis (Resnick *et al.*, 1987; Cole *et al.*, 1989). Briefly, polyacrylamide is copolymerized with 1% gelatin. The substrate used is chosen on the basis of results of spectrophotometric assays of enzyme–substrate interaction, and the ability of the substrate to copolymerize with polyacrylamide. The solubilized sample in the presence of bromophenol blue (0.025%), SDS (0.4%) and glycerol (20%) was subjected to electrophoresis. The gel was washed with Triton X-100 (2.5%) to remove the SDS and transferred to incubation buffer, which was adjusted to the optimal pH for the enzyme(s) of interest. The gel was incubated at 37°C for 12 hr and stained by Coomassie blue R-250. The clear region of the gel represented the site where enzyme digestion of the gelatin had taken place (Fig. 6).

3.3. Humoral Immunoassays

The antigen content of the SCWF and solubilized outer and inner conidial wall fractions of *C. immitis* has been analyzed using techniques of two-dimensional immunoelectrophoresis (2D-IEP) (Huppert *et al.*, 1978, 1979; Cole *et al.*, 1983, 1985a, 1986, 1988c; Reiss *et al.*, 1985). Huppert and co-workers (1978, 1979) first developed a 2D-IEP reference system for *C. immitis* antigens. Coccidioidin (CDN) was used as the reference antigen, which is a skin test-active substance employed for many years to survey human populations for exposure to this pathogen (Smith *et al.*, 1948). CDN was prepared as two separate products (Huppert, 1983): a pooled, broth culture filtrate of several strains of *Coccidioides* mycelia (F fraction), and a toluene lysate of the washed, cellular retentate (L fraction). Anti-CDN (F+L) was derived from a hyperimmunized burro (Huppert *et al.*, 1978, 1979) and used in 2D-IEP plates to develop our own CDN–anti-CDN reference system (Cole *et al.*, 1988c). The precipitin peaks are identified in Fig. 7A,B according to the previously established systems (Huppert *et al.*, 1978, 1979; Cole *et al.*, 1983). The antigen–antibody reference system has proved most useful for monitoring successive steps of physico-chemical fractionation of soluble, or solubilized wall material. For example, tandem 2D-IEP has been used to identify antigens comprising chromatographically- or electrophoretically-purified fractions on the basis of their fusion with known precipitin peaks in the CDN–anti-CDN reference system (Fig. 7C,D). Alternatively, the specificity of adsorption of a precipitin peak in the reference system by rabbit antiserum raised against a purified fraction can be examined by intermediate gel immunoelectrophoresis (Fig. 7E,F). CDN is an appropriate reference antigen for studies of wall antigens since, on the basis of immunoelectron microscopy (Section 3.5), most of the antigenic components of CDN are located in the cell envelope (Cole *et al.*, 1986, 1987). The reference antigen has also been shown to elicit most clinically significant immunological responses in coccidioidomycosis (Huppert, 1983). CDN contains the majority of antigens found

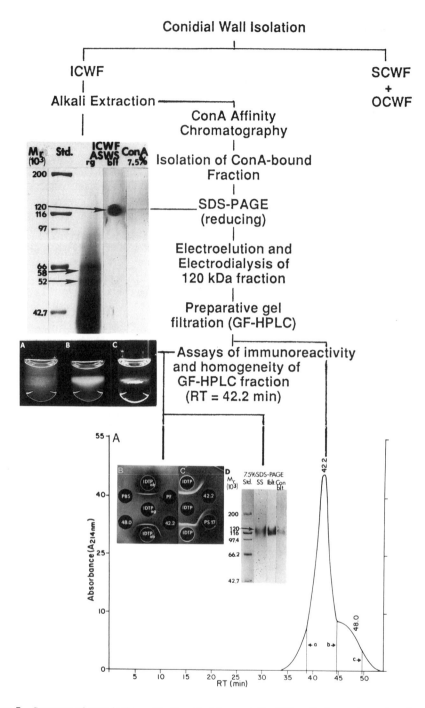

Figure 5. Summary of steps in the purification of a tube precipitin (TP) antibody-reactive antigen from the alkali-soluble, water-soluble, inner conidial wall fraction (ICWF-ASWS) of *C. immitis*. Crude ICWF-ASWS was examined by SDS–PAGE under reducing conditions (7.5% gel) and blotted using Con A conjugated to peroxidase. Con A affinity bound fraction of ICWF-ASWS was examined in the same gel. The electroeluted, 120-kDa fraction was subjected to GF-HPLC (A). The HPLC fraction (major peak; RT = 42.2 min) was examined for immunoreactivity in the IDTP assay (B, C) and tube precipitin (TP) test (A, control; B and C, HPLC fraction after incubation with IDTP-positive serum at 37°C for 2 and 24 hr, respectively). The HPLC fraction was examined for homogeneity by SDS-PAGE (D) (7.5% reducing gel) using techniques of silver staining (SS), immunoblotting (Iblt.; IDTP-Positive serum from patient), and Con A blotting (Conblt.) for identification of the 120-kDa band.

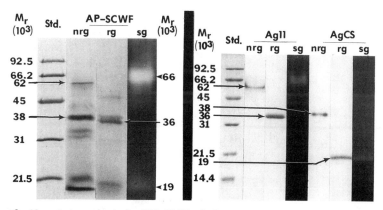

Figure 6. Nonreducing and reducing SDS–PAGE gels (lanes nrg and rg, respectively) and substrate gels (lanes sg) with Coomassie blue stain of the acetone precipitate of SCWF (AP-SCWF) and purified fractions of SCWF identified as Ag11 and AgCS. Estimated M_r of samples and standards (lane Std.) are shown. (From Cole *et al.*, 1989.)

in the conidial wall fractions and is, therefore, a valid immunologic reference complex for our 2D-IEP studies.

Serological investigations of patients suffering from coccidioidomycosis have revealed a consistent and predictable immunological pattern (Sawaki *et al.*, 1966). Delayed-type cutaneous hypersensitivity is demonstrated within 2 to 5 weeks after primary infection. This is followed by production of tube precipitin (TP) antibody (Smith *et al.*, 1950, 1956), which is of the IgM class (Pappagianis *et al.*, 1965; Sawaki *et al.*, 1966). This early IgM antibody response is short-lived, apparently disappearing within 6 months (Cox *et al.*, 1984). Production of complement-fixing (CF) antibody (Smith *et al.*, 1950, 1956), which is of the IgG class (Pappagianis *et al.*, 1965; Sawaki *et al.*, 1966), is indicative of disseminated coccidioidomycosis and a measure of the severity of this fungal disease. Serologic assays, therefore, are important diagnostic and prognostic aids in coccidioidomycosis (Sawaki *et al.*, 1966). Huppert and Bailey (1963, 1965a,b) identified the crude antigenic fractions of *C. immitis* responsible for the CF- and TP-antibody reactions and, using a simple immunodiffusion (ID) technique, developed a diagnostic method for detection of these separate antibodies in patient sera. The IDCF and IDTP tests are routine clinical procedures in the diagnosis of coccidioidomycosis and have been used as valuable reference tests during steps of purification of CF and TP antigens (Cox *et al.*, 1987; Cox and Britt, 1987; Zimmer and Pappagianis, 1988; Kaufman *et al.*, 1985; Kruse and Cole, 1990; Cole *et al.*, 1990).

The major limitation of the ID test is its relatively low sensitivity, at least in comparison to the ELISA (Ausubel *et al.*, 1989). In concert, however, these two humoral immunoassays have provided important information on reactivity of patient sera with antigens derived from cell walls of *C. immitis* (Cole *et al.*, 1988a, 1990). For example, we were able to show that IDTP-positive patient sera have high levels of IgM antibody which binds to a purified TP antigen (TP-Ag) when tested in the indirect ELISA using an anti-IgM-specific, secondary antibody conjugate (Cole *et al.*, 1990). The IDTP-negative, but IDCF-positive sera, on the other hand, showed low IgM affinity, or in some cases, no binding of IgM antibodies to this same purified TP antigen. The reactivity of IgG antibody from IDTP-positive patient sera to the TP-Ag was generally low when measured by the indirect ELISA using an anti-IgG-specific secondary antibody conjugate. Control sera (IDTP- and IDCF-negative) showed neither IgM nor IgG binding to the TP-Ag antigen. Thus, the correlation of serological responses in the ID assay and ELISA has provided valuable information on antibody binding to the isolated TP-Ag (See Sections 4.4, 4.5).

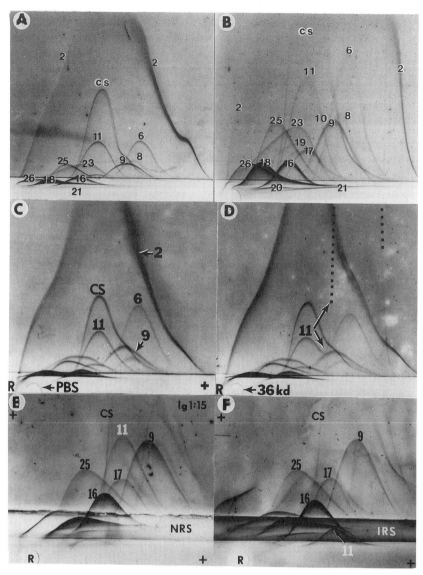

Figure 7. Two-dimensional immunoelectrophoresis (2D-IEP) gels (A, B), tandem 2D-IEP gels (C, D), and intermediate gel 2D-IEP plates (E, F) of *C. immitis* fractions. (A, B) 2D-IEP of coccidioidin (CDN) in cathodal well against hyperimmunized burro antiserum in upper gel diluted 1:10 and 1:15 with electrophoresis buffer, respectively. Precipitin peaks of reference system identified by established number/letter code. (From Cole *et al.*, 1988c.) (C, D) tandem 2D-IEP gels of reference antigen (R = CDN) in cathodal well and purified Ag11 (36 kd) isolated by immunoaffinity chromatography and added to anodal well. The anodal well of the control gel contained PBS. (From Cole *et al.*, 1989.) (E, F) Intermediate gel contained either normal rabbit serum (NRS) or immune rabbit serum (IRS) raised against the purified Ag11. (From Yuan *et al.*, 1988.)

The competitive inhibition ELISA procedure has also been employed in determinations of relative affinity of patient sera to a purified antigen compared to a reference antigen complex (Cole *et al.*, 1990). This technique involves preadsorption of a test serum with a selected reagent prior to reacting the serum sample with the test antigen bound to the wells of the microtiter plate. We have recently used this procedure to identify a major epitope of the TP-Ag which is discussed in Section 4.5.

3.4. Cellular Immunoassays

Our investigations of the ability of cell wall fractions of *C. immitis* to stimulate T-lymphocyte responses have involved both immune lymph node (LN) cells and T-cell lines in antigen-specific proliferation assays (Cole *et al.*, 1987, 1988a). In brief, (BALB/c × DBA/2) F1 (CD2F1) female mice were immunized subcutaneously with 10^6 live, attenuated round cells (Fig. 1) of the mutant strain (c95-271) described above (Section 2.3). The mice were boosted 1 month later with 10^6 formalin-fixed cells of the same strain in complete Freund adjuvant inoculated into the footpads and base of the tail. Eight days later, the draining inguinal and para-aortic lymph nodes were removed, cell suspensions prepared, and immune LN cells cultured (Corradin and Chiller, 1977). The cells were maintained in 96-well plates (Costar) with or without the isolated cell wall fraction. In addition, the same wall fractions were tested in proliferation assays against LN cells from mice which had been immunized with complete Freund adjuvant alone as a control for non-specific, mitogenic activity (Cole *et al.*, 1988a). After incubation for 72 hr (37°C, 5% CO_2), [^{3}H] thymidine (1 μCi in 10 μl) was added to each well and the cells were cultured for an additional 18 hr. The contents of the wells were then harvested onto paper filters using an automatic harvester, the filters were suspended in scintillation fluid, and the cells were counted in a liquid scintillation counter. The data are expressed as the mean change in counts per minute (Δcpm = mean cpm in presence of wall fraction − mean cpm in absence of wall fraction).

The immune lymphocytes were passed over nylon wool to enrich for T cells (Julius *et al.*, 1974). The cell suspensions contained less than 5% B lymphocytes, as judged by anti-immuno-globulin fluorescence staining and flow microfluorimetry (Cole *et al.*, 1988a). Nevertheless, we preferred to compare the reactivity of various wall fractions with T cells alone, since these lymphocytes appear to participate in limiting the spread of coccidioidal infection and preventing reinfection (Beaman *et al.*, 1977, 1979). We recently developed an SCWF, T-cell line derived from mice immunized as above. The LN cells were alternately cultured in the presence of the SCWF and without the SCWF for 1-week periods in order to identify and isolate the reactive clone (Kimoto and Fathman, 1980). The cell line has been shown to be stable for more than 1 year. The cells all express the L3T4 and Thy1 antigens, indicating that they are helper T cells. As expected, they are *H-2* restricted and respond to products of *C. immitis* only in the context of *Ia^d*.

3.5. Immunoelectron Microscopy

Rabbit antisera were raised separately against CDN and SCWF and the sera were fractionated on a DEAE-Affigel Blue column to obtain IgG preparations which were free of detectable proteolytic activity (Cole *et al.*, 1987). Thin sections of resin-embedded conidia of *C. immitis*, mounted on nickel grids, were first etched with 3% H_2O_2, washed with buffer, immersed in drops of filtered 10% chicken egg albumin, and then reacted with one of the IgG fractions (Cole *et al.*, 1987). After several buffer washes, the sections were reacted with a solution of colloidal gold conjugated with goat anti-rabbit IgG. Following additional washes and poststaining, the sections were examined in the transmission electron microscope. Sections of conidia which were reacted with anti-CDN and anti-SCWF showed very similar patterns of immunolabel (Cole *et al.*, 1986, 1987). Most of the gold particles were located on the ICW and intermural space between the ICW and OCW (cf. Figs. 2A and 8A). Certain regions of the cytoplasm showed clusters of immunolabel. Essentially no label was associated with the outermost layer of the conidial wall. These comparable patterns of immunolabel using either anti-CDN or anti-SCWF further attest to the validity of CDN as a reference antigen for our identification of cell wall macromolecules. The anti-SCWF immunoprobe also densely labeled the wall of parasitic cells of *C. immitis*, including young spherules (Fig. 8B) and endospores released from mature spherules (Fig. 9). These results suggested that at least some immunogenic components of the SCWF are common to cell walls of the saprobic and parasitic cycles. The SCWF, therefore, is an appropriate fraction to evaluate the significance of cell wall-associated, immunoreactive macromolecules on the course of coccidioidal infection of the host.

Immunoelectron microscopy (IEM) has also helped to validate our use of the mutant strain c95-271 for cellular immunoassays of conidial wall fractions. The antigenic content of the SCWF obtained from a virulent strain (c735, a recent isolate from a patient with disseminated coccidioidomycosis) and the antigenic content of the same conidial wall fraction of strain c95-271 appeared to be comparable on the basis of their examination in our 2D-IEP, CDN–anti-CDN reference system. The pattern of immunolabel on thin sections of conidia of strain c95-271 using antibody raised against SCWF of the virulent strain was essentially identical to the distribution of label described above (cf. Figs. 8A and 10A). When conidia of the mutant strain were grown under conditions which permitted cells of the wild-type strain to convert to the spherule–endospore phase [i.e., glucose–salts medium (Levine, 1961), 39°C, 20% CO_2–80% air], round cells of c95-271 were formed but no segmentation or endosporulation occurred (Fig. 10B,C). As in the case of virulent conidia, however, the outer, inelastic conidial wall layer of strain c95-271 ruptured during this growth phase (Fig. 10C) which permitted release of water-soluble products (SCWF) from the intermural space and inner wall layer. It follows, therefore, that inoculation of BALB/c mice with conidia of the attenuated mutant (c95-271) results in sensitization of the murine immune system to the majority of immunogenic components of the SCWF.

4. OUTER AND INNER CONIDIAL WALL FRACTIONS

The composition of immunoreactivity of these two components of the conidial envelope have been examined. The wall fractions were isolated during successive steps of cell shearing and homogenization (Fig. 4) which yielded the hydrophobic, outer sleeve of wall material (OCWF), and thick, inner wall layer (ICWF). Each component of the conidial envelope has been obtained in large quantities which permitted experimentation using various solubilization methods. The solubilized subfractions have been tested for reactivity in the cellular and humoral immunoassays described above.

4.1. Composition and Ultrastructure

Each wall fraction was subjected to compositional analysis (Cole *et al.*, 1983) and the data are summarized in Tables I and II. The major components of the OCWF are protein and lipid. Because of the insolubility of the outer wall fraction, results of standard protein assays [e.g., Lowry (Lowry *et al.*, 1951), Bradford (Bradford, 1976)] were unreliable even after treatment of the wall preparation with 1 N NaOH at 100°C (Cole *et al.*, 1983). Protein estimates, therefore, were based on micro-Kjeldahl analyses. Total nitrogen calculations took into account the presence of amino sugars, and the percent protein was computed using the method of Kurkela *et al.* (1980). The chemical composition of the ICWF was clearly distinguished from the OCWF (Table I). The percentage of total neutral carbohydrate of the ICWF was almost three times that of

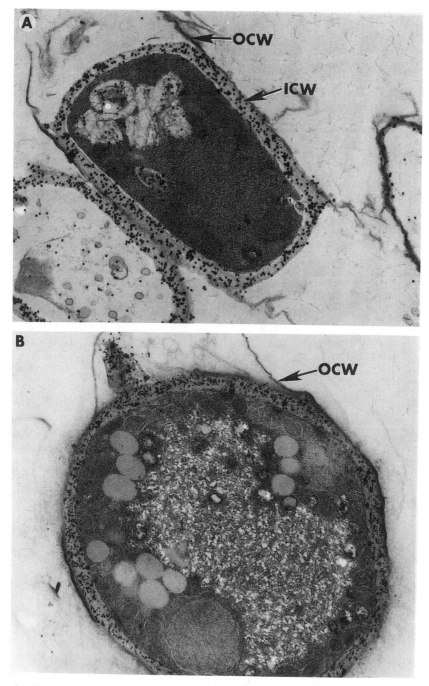

Figure 8. Thin sections of arthroconidium (A) and round cell (B; cf. Fig. 1) reacted with rabbit anti-SCWF serum followed by anti-rabbit IgG conjugated with 20-nm colloidal gold particles. SCWF derived from strain C634. ICW, inner conidial wall; OCW, outer conidial wall. A, × 23,400; B, × 12,000.

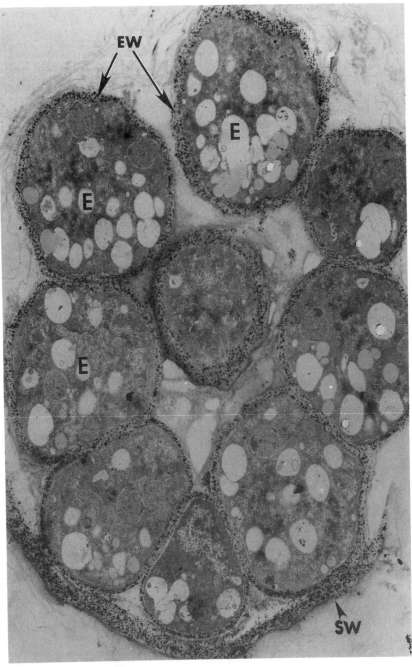

Figure 9. Thin section of endosporulating spherule reacted with rabbit anti-SCWF serum as in Fig. 8. E, endospores; EW, endospore wall, SW, ruptured spherule envelope. × 8000.

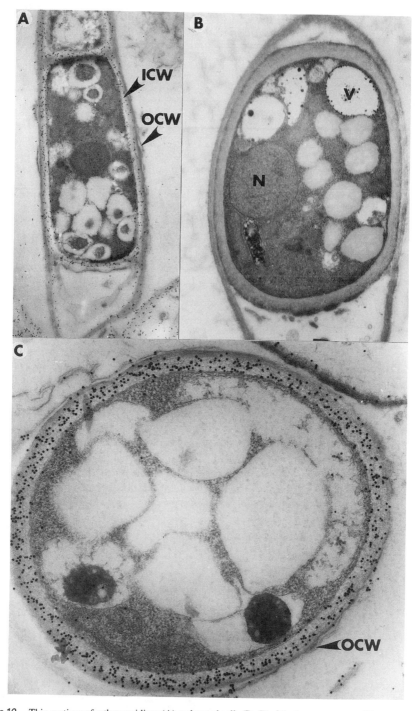

Figure 10. Thin sections of arthroconidium (A) and round cells (B, C) of the temperature sensitive, auxotrophic mutant strain (c95-271). (A, C) Reacted with rabbit antiserum raised against the SCWF derived from virulent strain 735, followed by anti-rabbit IgG conjugated with 15-nm colloidal gold particles. (B) Reacted with preimmune rabbit serum followed by secondary antibody/gold conjugate. ICW, inner conidial wall; N, nucleus; OCW, outer conidial wall; V, vacuole. A, × 13,160; B, × 18,000; C, ×37,960.

Table I. Summary of the Chemical Composition
of Arthroconidial Wall Fractions

	Conidial wall fractions[a]		
	OCWF (%)	ICWF (%)	SCWF (%)
Total neutral carbohydrate, by GLC[b]	12.0[b]	32.3	31.6
Total hexosamine	1.7	21.9	0.6
Peptides	49.8[c]	27.4	28.5
Lipids			
Readily extractable	15.4	4.5	7.4
Bound	9.7	12.9	5.2
Other	NM[d]	NM	20.0[e]
Ash	0.1	0.02	NM
Recovery	88.7	99.0	93.3

[a]Values expressed as percentages of total dry weight of each wall fraction. OCWF, outer conidial wall fraction; ICWF, inner conidial wall fraction. SCWF, soluble conidial wall fraction (see Fig. 4). From Cole *et al.* (1983).
[b]Determined by summation of major monosaccharides identified by gas/liquid chromatographic (GLC) analysis.
[c]Determined by micro-Kjeldahl method (Kabat and Mayer, 1961; Campbell *et al.*, 1970).
[d]NM, not measurable.
[e]Lyophilized pigment component obtained from methanol–H_2O layer during isolation of readily extractable lipids from SCWF.

the OCWF. As expected, most of the hexosamine was detected in the ICWF. Results of chitinase digestion of the OCWF and ICWF suggested that chitin was concentrated in the latter fraction (Cole *et al.*, 1983). Comparison of the monosaccharide content (Table II) revealed that the two wall fractions contained mannose, glucose, and galactose in order of abundance. In addition, 3-*O*-methylmannose (3-*O*-MM) was detected as a prominent monosaccharide of the ICWF, but occurred only in trace amounts in the OCWF. Interest in this particular sugar stems from reports that 3-*O*-MM may be unique to *C. immitis* among the systemic fungal pathogens (Wheat *et al.*, 1983; Cole *et al.*, 1985a) and appears to be reactive with sera from coccidioidomycosis patients (Cole *et al.*, 1990), as described below (Section 4.5).

Ultrastructural examination of the OCWF has revealed "rodlet-fascicles" which appear as clusters of microfibrils in carbon/platinum replicas of the wall fraction. Rodlets have been

Table II. Monosaccharide Composition
of Arthroconidial Wall Fractions

	Conidial wall fractions[a]		
Monosaccharide	OCWF (%)	ICWF (%)	SCWF (%)
Mannose	64.6	45.8	30.3
Glucose	22.5	35.1	58.7
Galactose	11.5	7.1	7.9
3-*O*-Methylmannose	1.4	12.0	3.1

[a]Values expressed as percentages of total neutral carbohydrate determined by gas/liquid chromatography. OCWF, outer conidial wall fraction; ICWF, inner conidial wall fraction; SCWF, soluble conidial wall fraction. From Cole *et al.* (1983).

observed on the surface of dry, air-dispersed conidia of numerous imperfect fungi (Cole, 1973; Cole and Pope, 1981; Beever *et al.*, 1979; Cole *et al.*, 1983; Cole and Sun, 1985). These individual wall microfibrils have been reported to be composed mainly of protein (Beever *et al.*, 1979), but the rodlet layer may actually be a complex of glycoproteins (Hashimoto *et al.*, 1976) or lipoproteins (Cole and Pope, 1981). Rodlets seem to contribute significantly to hydrophobicity of the conidial wall and thus to dissemination of these air-dispersed propagules (Beever and Dempsey, 1981). The striking ultrastructural feature of the ICWF was the presence of a well-defined complex of intertwined fibrils which are much larger than rodlets and are digested by incubation of the wall fraction with chitinase (Cole *et al.*, 1983). These microfibrils are presumed to form the chitin framework of the conidial envelope. No rodlets were observed in electron microscopic preparations of the ICWF. However, thin and elongated fibrils persisted after chitinase digestion and were suggested to represent β-glucan components of the ICWF. The presence of β-1,3-glucan in the inner spherule wall has been reported (Hector and Pappagianis, 1982).

4.2. Suppression of Phagocytosis and Lymphocyte Proliferation

Drutz and Huppert (1983) proposed that the hydrophobic, outer conidial wall layer of *C. immitis* has antiphagocytic properties and the authors demonstrated that its removal by sonication prompts phagocytosis ($> 25\%$) by human polymorphonuclear neutrophils (PMNs). However, in spite of this increased level of phagocytosis there was no significant improvement in the intracellular killing of sheared conidia compared to intact cells (Drutz and Huppert, 1983). The dry and apparently resilient outer conidial wall layer may represent an important passive and/or active barrier of the pathogen against attack by the host cellular defense system.

We used the LN proliferation assay to originally evaluate cellular immunoreactivity of soluble walls and cytosol preparations from both arthroconidia and parasitic cell types (Cole *et al.*, 1987). The SCWF was shown to elicit the highest LN-cell response of the fractions tested (Table III), which was even greater than when immune LN cells were exposed to the spherules (strain c95-271) used for immunization of the BALB/c mice (Table IV; see Section 3.4). We were surprised to find that when intact, formalin-killed conidia were exposed to sensitized LN cells by

Table III. Immune LN Cell Proliferation in Presence of Selected Fractions of C. immitis[a]

Ag[b]	Mean Δ cpm ± SEM at Ag concn[c] (μg/ml):			
	100	50	25	12.5
SCWF	75,965 ± 16,049	90,899 ± 4,766	101,291 ± 1,640	129,054 ± 9,062
Arthroconidial cytosol	33,560 ± 3,927	87,536 ± 2,632	102,563 ± 8,925	86,533 ± 6,519
Spherule cytosol	56,570 ± 1,253	46,093 ± 5,693	50,944 ± 4,737	38,575 ± 6,574
Endospore cytosol	55,364 ± 6,380	62,093 ± 4,736	35,406 ± 2,196	35,919 ± 3,208
Spherule culture supernatant	102,024 ± 10,736	80,697 ± 6,710	89,996 ± 4,944	59,753 ± 3,651
Glucose–yeast extract	ND[d]	0	0	0

[a]LN, lymph node. From Cole *et al.* (1987).
[b]Antigens (Ag) tested. SCWF, soluble conidial wall fraction. Cytosol fractions obtained by glass bead homogenization of intact cells followed by centrifugation (81,500g) and lyophilization of supernatant. Spherules and endospores obtained by inoculating Converse liquid medium (Levine, 1961) with arthroconidia and incubating the cultures at 39°C in the presence of 20% CO_2–80% air. Supernatant of the 48-hr spherule culture was obtained by centrifugation (1020g). The 1% glucose–0.5% yeast extract medium was that used in culturing arthroconidia.
[c]Counts per minute of [^{3}H]thymidine-labeled LN cells exposed to Ag minus counts per minute without Ag. The data are mean values for triplicate determinations. Lyophilized Ag preparations were added to LN cell culture media. SEM, standard errors of the mean.
[d]ND, not done.

Table IV. Immune LN Cell Proliferation in Presence
of Intact and Sheared Arthroconidia[a]

	Mean Δcpm at cell concn[c] (No./ml):		
Cell type (strain)	10^6	10^5	10^4
Spherules (c95–271)[b]	73,399	35,189	34,767
Arthroconidia (c634)	0	33,818	14,170
Sheared arthroconidia (c634)	62,506	27,915	14,786

[a]LN, lymph node. From Cole *et al.* (1987).
[b]Spherules from strain used to immunize BALB/c mice.
[c]The data are mean values for triplicate determinations. The mean without Ag was 2811
cpm. Standard errors were less than 10% of the mean. Formalin-killed fungal cells were
suspended in LN cell culture medium.

simply suspending the arthroconidia in LN cell culture media at a concentration of 10^6 cells/ml, no stimulation of LN cell proliferation was noted (Table IV). The sheared conidia, on the other hand, were almost as stimulatory as spherules at 10^6 cells/ml. This discrepancy in response to intact and sheared conidia suggested that the outer conidial wall blocks access of LN cells to underlying wall-associated antigens, or perhaps the outer wall layer actively inhibits LN cell proliferation. To test this second hypothesis, we added increasing concentrations of the isolated OCWF to the LN culture media containing sensitized LN cells, plus either formalin-killed spherules (c95-271) or the SCWF (Table V). The OCWF was able to effectively block cell proliferation in both cases at concentrations of 50–100 μg/ml. These data suggest that the OCWF actively suppresses T-cell response to the infectious conidium rather than just acts as a passive barrier to the host immune system.

Individuals with self-limited coccidioidomycosis probably manifest strong cell-mediated immune responses to the pathogen, while patients with the chronic or progressive form of this disease suffer from depressed cell-mediated immunity (Cox, 1986, 1988; Cox and Pope, 1987). Cox and co-workers have shown that the high susceptibility of BALB/c mice to pulmonary coccidioidal infection is due to acquired suppression of cell-mediated immune reactivity (Cox *et al.*, 1988; Cox and Kennel, 1988). They proposed that the acquired anergy is associated with, and

Table V. Effect of Outer Conidial Wall Fraction (OCWF) on Immune
LN Proliferative Response to C. immitis Antigens[a]

	Mean Δcpm at OCWF concn[c] (μg/ml):			
Antigens[b]	0	10	50	100
Spherules (c95–271),[d] 10^6 cells/ml	25,391	12,997	572	519
SCWF,[d] 50 μg/ml	27,291	22,759	1,540	189

[a]Immune lymph node (LN) cells from (BALB/c × DBA/2) F1 mice immunized 3× with formalin-fixed
spherules (strain c95–271). LN cells (4 × 10^5) cultured for 72 hr and then pulsed for 18 hr with 1 μCi
[3H]thymidine.
[b]Antigens tested.
[c]Counts per minute (cpm) of [3H]thymidine-labeled LN cells without OCWF minus counts per minute when
exposed to OCWF. Lyophilized OCWF preparations were added to LN cell culture media. The mean value
without antigen or OCWF was 1015 cpm. This value was subtracted from all values derived from experiments.
The data are mean values for triplicate experiments and the standard errors were < 10% of the mean.
[d]Spherules produced by strain c95–271 added to LN cell culture media at a concentration of 10^6 cells/ml. SCWF
(soluble conidial wall fraction obtained from strain c634) prepared as described (Cole *et al.*, 1987) and added to
LN cell culture media at a concentration of 50 μg/ml.

perhaps attributable to, the activation of a splenic cell population(s) by circulating *C. immitis* antigens (Cox and Kennel, 1988). Antigen-specific suppression of T-lymphocyte reactivity, which had also been demonstrated by earlier investigators (Ibrahim and Pappagianis, 1973), was induced in mice injected intravenously with an alkali extract of the isolated mycelial wall. It has been suggested that the suppression-inducing, soluble wall fraction contains two antigens previously isolated and identified by 2D-IEP as Ag2 (Fig. 7) and an incomplete precipitating antigen (IPA) (Cox, 1988; Cox and Britt, 1986b). A marked increase in serum levels of these antigens has been shown in BALB/c mice after intranasal inoculation with arthroconidia (Cox, 1988). The circulating *Coccidioides* cell wall antigens were detected by ELISA in sera of infected mice and could be passively transferred to normal, syngeneic mice by intravenous injection with sera from the infected donors. The result was suppression of the delayed-type hypersensitivity response of recipients to immunization with CDN (Cox, 1988; Cox and Kennel, 1988). These data suggest an immunosuppressive role for Ag2 and/or IPA (Cox, 1988). Ag2 has been identified as a component of the OCWF (Cole *et al.*, 1985b). We have also isolated both Ag2 and what appears to be the equivalent of IPA from the ICWF (Kruse and Cole, 1990). It would seem that coccidioidomycosis patients are exposed to these two antigens early in the course of infection. Additional investigations are necessary to determine whether the two purified antigens, either together or separately, are responsible for the observed anergy in experimental animals.

4.3. Influence on Superoxide Anion Release by Host Phagocytes

Investigations were conducted in collaboration with M. D. Robertson (See Chapter 21) to determine whether the OCWF and components of a detergent extract of this conidial wall fraction could influence spontaneous release of superoxide anion by phagocytes isolated from rat lungs. Crucial for effectiveness of the host cellular immune defense system are the microbicidal products of the oxygen-dependent reactions (respiratory burst) of phagocytic cells. These products are commonly released in the presence of a potential pathogen and include superoxide anion (O_2^-), singlet oxygen, hydrogen peroxide, and hydroxyl radicals (Robertson *et al.*, 1987; see also Chapter 21). In adition to fungicidal effects, release of O_2^- further triggers chain reactions of the respiratory burst leading to production of hydrogen peroxide (Klebanoff, 1980). Robertson and co-workers (1987) have shown that conidia of *Aspergillus fumigatus* in the presence of phagocytic cells from lungs of mice and rats failed to trigger an increase in O_2^- release, while zymosan or control conidia of the saprophyte *Penicillium ochrochloron* stimulated release of a substantial increase of this oxygen intermediate. A diffusate obtained from *A. fumigatus* conidia, released from cells when simply suspended in Hanks' balanced salt solution (HBSS) at 37°C for 6 hr, demonstrated a similar inhibitory effect. We compared spontaneous release of O_2^- from rat alveolar macrophages in the presence of the OCWF of *C. immitis*, as well as an *N*-octyl-β-D-glucopyranoside (OG)-soluble extract of the OCWF (OG-Sol) and corresponding detergent-insoluble fraction of the OCWF (OG-Insol) (Table VI). Alveolar macrophages were obtained from lungs of PVG rats by bronchoalveolar lavage. The cell population comprised > 98% macrophages. Superoxide anion was measured according to the method of Johnston (1981). Production of O_2^- can be detected by its ability to reduce cytochrome *c*, a reaction accompanied by an increase in spectrophotometric absorbance of the reaction mixture measured at a wavelength of 550 nm. However, reduction of cytochrome *c* under conditions of our experiments was not specific for O_2^-. The required specificity was achieved by using superoxide dismutase, an enzyme for which O_2^- is the only known substrate. Each assay was run with or without superoxide dismutase and only the portion of the reduced cytochrome *c* which was inhibited by superoxide dismutase was considered to reflect the nanomoles of superoxide released (Table VI). We concluded that the detergent-soluble fraction of OCWF can significantly inhibit the spontaneous release of O_2^- from rat alveolar macrophages. The protein composition of both the crude OCWF

Table VI. Effect of Outer Conidial Wall Fractions
of C. immitis on Spontaneous Release of Superoxide
Anion by Rat Alveolar Macrophages

Treatment	Superoxide anion (nmoles; mean)
HBSS[a]	16.60
OCWF[b]	
5 μg	14.35
10 μg	13.60
50 μg	11.05
OG-Sol/OCWF[c]	
5 μg	11.45
10 μg	11.05
50 μg	6.40
OG-Insol/OCWF[d]	
5 μg	15.80
10 μg	15.75
50 μg	13.20

[a]Exposure of phagocytes to Hanks' balanced salt solution (HBSS) gave a measure of the
spontaneous release of superoxide anion.
[b]Outer conidial wall fraction (OCWF) added to reaction mixture at concentrations of 5,
10, and 50 μg in 1.5 ml HBSS.
[c]Octylglucopyranoside-soluble fraction of OCWF (5–50 μg in 1.5 ml HBSS).
[d]Octylglucopyranoside-insoluble fraction of OCWF (5–50 μg in 1.5 ml HBSS).

and detergent-soluble fraction of OCWF was examined by SDS–PAGE (Fig. 11). The major components of the OG-soluble fraction were estimated to be in the range of 23–60 kDa. Attempts are underway to further fractionate this wall preparation and test its components on inhibition of spontaneous release of O_2^- from zymosan-stimulated alveolar macrophages obtained from both rats and humans.

4.4. Patient Antibody Reactivity to Heat-Stable Macromolecules

As indicated earlier in this chapter, patients presenting with primary coccidioidal infection typically have high titers of anti-*Coccidioides* IgM precipitin antibodies (Huppert, 1983). Early detection of Igm response to coccidioidal infection has been performed by reacting patient sera with CDN in the classical TP test (Pappagianis *et al.*, 1961b; Smith *et al.*, 1948, 1956, 1950). This serologic test has been replaced by an IDTP assay (Huppert and Bailey, 1965) in which precipitin antibodies have been detected to antigens derived from *C. immitis* fractions, including an alkali-soluble, water-soluble (ASWS) fraction of the inner conidial wall (Kruse *et al.*, 1990). Detection of IgM precipitin antibody to *C. immitis* antigen is a valuable aid to diagnosis of early coccidioidal infection. However, our knowledge of the specific nature of the TP antigen(s) is still incomplete. Only recently have investigators been successful in at least partial purification and characterization of antigens responsible for the early IgM patient antibody response (Cox and Britt, 1986b; Galgiani *et al.*, 1988). Our focus has again been on the conidial wall as a source of antigen which elicits humoral reactivity, perhaps in the earliest stages of pulmonary infection.

Immunoreactive components of the ICWF were obtained by solubilization of the wall preparation with 1 N NaOH using a modification of the method of Cox *et al.* (1984). Two separate precipitin bands were formed by immunodiffusion when the ICWF-ASWS was reacted with the patient reference serum (Fig. 12). Reactivity of this crude fraction was retained in the IDTP assay

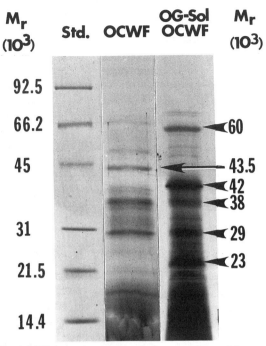

Figure 11. SDS–PAGE gel (14%) separation, under reducing conditions, of the outer conidial wall fraction (OCWF) and octyl-β-D-glucopyranoside-soluble fraction (OG-Sol) of the OCWF. Molecular mass (M_r) standards (Std.) are also shown.

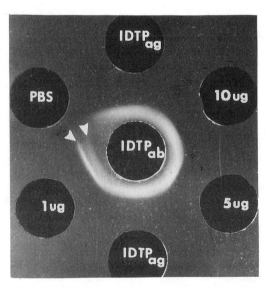

Figure 12. Immunoreactivity of the alkali extract of the inner conidial wall fraction (ICWF-ASWS). Wells of the ID plate contain human reference antibody (ab) and reference TP antigen (ag). The concentrations of the ICWF-ASWS added to the test wells were 10, 5, and 1 μg in 30 μl PBS. Arrowheads indicate two precipitin bands.

even after heating to 60°C for 30 min. The procedure used for isolation of the 120-kDa, precipitin antibody-reactive component of the ICWF-ASWS was previously outlined in Fig. 5. As pointed out in Section 3.3, the TP-Ag showed high affinity for IgM antibody from coccidioidomycosis patients who had been previously determined to be IDTP-positive, but relatively low affinity for IgG, as defined by the ELISA. Pronase digestion and heating (100°C, 5 min) had no apparent effect on immunoreactivity of the TP-Ag, while periodate oxidation resulted in total loss of its IDTP activity (Kruse and Cole, 1990). Further discussion of the carbohydrate composition of the TP-Ag and the possible role of a specific sugar component of this antigen in serologic reactivity of the heat-stable macromolecule is presented below (Section 4.5).

A second, heat-stable antigen of the ICWF-ASWS and mycelial culture filtrate fractions is Ag2 (Cole *et al.*, 1985a), as revealed by tandem 2D-IEP in our CDN–anti-CDN reference system (Fig. 13A). The Ag2 component of the ICWF-ASWS was distinguished from the 120 kDa TP-Ag by its low affinity for Con A (Fig. 13B). The technique of advancing-line IEP was used to establish

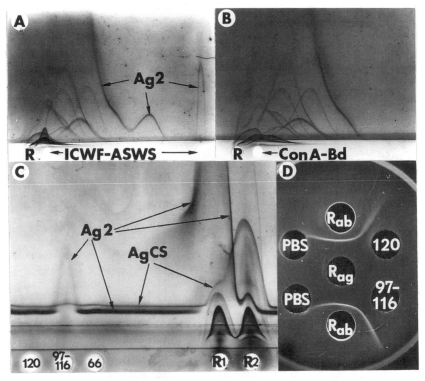

Figure 13. Antigenic composition of subfractions of ICWF-ASWS and mycelial culture filtrate which showed reactivity with patient tube precipitin antibody. Tandem 2D-IEP gels (A, B) revealed that the original ICWF-ASWS contained Ag2 (A), while the Con A-bound fraction of ICWF-ASWS (ConA-Bd) lacked Ag2 (B). The reference antigen (R) added to the cathodal wells was CDN. The burro, anti-CDN immunoglobulin in the upper gel was diluted 1:10 with electrophoresis buffer. The advancing-line IEP gel (C) of different molecular weight fractions electroeluted from reducing gels of the mycelial culture filtrate revealed that the 97- to 116-kDa fraction contained Ag2, while the 120-kDa fraction (i.e., TP-Ag described by Kruse and Cole, 1990; Cole *et al.*, 1990), and 66-kDa fraction (Cole *et al.*, 1989) lacked Ag2. The advancing lines were identified as Ag2 and AgCS on the basis of their fusion with precipitin peaks of the reference antigens (R1, CDN; R2, SCWF). Immunoreactivity of the 120-kDa and 97- to 116-kDa fractions with patient serum in the IDTP assay is shown in D. R$_{ab}$, IDTP reference antibody; R$_{ag}$, IDTP reference antigen.

that a 97- to 116-kDa fraction of the mycelial culture filtrate, isolated from SDS–PAGE gels (7.5%) by electroelution, corresponds to Ag2 (Fig. 13C). The electroeluted 97- to 116-kDa band was also shown to be reactive with sera from coccidioidomycosis patients in theIDTP assay (Fig. 13D). We subsequently isolated a 110-kDa fraction from 7.5% SDS–PAGE gel separations of the mycelial culture filtrate and ICWF-ASWS, showed that it corresponded to Ag2, and that it was reactive with sera from coccidioidomycosis patients in the IDTP assay (Kruse and Cole, 1990; Cole *et al.*, 1990). It appears, therefore, that at least two heat-stable macromolecules of ICWF-ASWS and mycelial culture filtrate, identified as 120- and 110 kDa fractions, are recognized by sera from patients in early stages of coccidioidal infection. These data contradict an earlier report by Cox and Britt (1986b) who suggested that Ag2 is not reactive with IDTP antibody. They provided evidence that the incomplete precipitating antigen (IPA), identified in 2D-IEP gels of their ASWS cell wall extracts, was the component reactive with IgM precipitin antibody. In a subsequent study (Dolan *et al.*, 1989), the IPA was used in ELISAs to identify reactive hybridomas derived from spleen cells of BALB/c mice immunized with toluene-induced lysates of spherule-phase cells (spherulin; SPH). A monoclonal antibody (MAb) detected a diffusely staining band with a range of 130–330 kDa in immunoblots of SDS–PAGE reducing gel separations of IPA, SPH, and CDN. However, this same MAb also reacted with a 110-kDa band, which the authors suggested is an epitope common to their IPA and Ag2. The latter antigen was previously demonstrated to be a component of both SPH and CDN (Cox *et al.*, 1984; Cox and Britt, 1987). It is possible that the IPA identified by Cox and Britt (1986b) as the IDTP antigen is a heterogeneous fraction containing both the precipitin antibody-reactive Ag2 and 120-kDa glycoprotein described above.

Zimmer and Pappagianis (1989) reported isolation of a fraction that was responsible for production of the antigen–antibody precipitate in the classical TP test. They first used pronase digestion of an antigen–antibody precipitate produced by the reaction of CDN with human sera from coccidioidomycosis patients to obtain the crude antigen. The washed "button" was incubated wth 400 μg of pronase per ml of buffer (pH 7.4) at 37°C for 72 hr and then heated at 60°C for 30 min to terminate the reaction. The digest was then subjected to gel filtration and ion-exchange chromatography to isolate the TP-Ag. The fractions with IDTP reactivity detected after DEAE chromatography had molecular sizes of 225 and 140 kDa, based on gel filtration estimates. A potential problem in this isolation procedure is that the exhaustive digestion of the crude fraction with pronase may have resulted in some alteration in composition of the 120-kDa TP-Ag which we have isolated, and total or partial loss of the 110-kDa TP-Ag (Kruse and Cole, 1990).

Examination of 14% SDS–PAGE gel separations of the Con A-bound, alkali extract of the ICWF revealed three additional components estimated as 31, 66, and 19 kDa (Kruse and Cole, 1990). All three fractions are immunoreactive with patient antibody in the ELISA after heating (60°C, 30 min) and are, therefore, identified as heat-stable antigens. The major 31-kDa fraction showed high affinity for Con A, reactivity with IgG in the ELISA, but no detectable reactivity with precipitin (IgM) antibody in the IDTP assay. Both the isolated 66- and 19-kDa components showed reactivity with patient IgG and IgM antibody in the ELISA (Cole *et al.*, 1989), but were not reactive in the IDTP assay. Patient IgM precipitin antibody, which was isolated from IDTP-positive sera eluted over a solid-phase immunosorbent column containing the purified 120-kDa TP-Ag (Cole *et al.*, 1990), did not bind to either the 66- or 19-kDa fraction in the ELISA. We conclude that the alkali-solubilized extract of the ICWF contains several patient IgM and IgG antibody-reactive components.

4.5. Patient IgM Antibody Reactivity to 3-O-Methyl-D-Mannose

It is not surprising that complex glycoproteins were released from the ICWF by alkali solubilization (Kruse and Cole, 1990), considering the abundance of protein and carbohydrate in the ICWF (Table I). On the basis of an earlier suggestion of the presence of a mannan–protein complex interspersed throughout the wall of parasitic cells of *C. immitis* (Hector and Pappagianis, 1982), we expected to detect such complexes in the ICWF-ASWS. Gas chromatography/mass spectroscopy (GC/MS) examination of the monosaccharide composition of the 120-kDa, TP-Ag has revealed 3-*O*-MM (Cole *et al.*, 1990). As pointed out earlier (Section 4.1), this sugar has been suggested to be unique to *C. immitis* among the systemic fungal pathogens (Cole *et al.*, 1985a; Wheat *et al.*, 1983), and known to be associated with biologically active fractions of CDN (Hassid *et al.*, 1943; Pappagianis *et al.*, 1961a; Scheer *et al.*, 1970; Anderson *et al.*, 1971; Wheat *et al.*, 1978). This latter association led us to speculate that 3-*O*-MM may be partly responsible for the serologic reactivity of the TP-Ag. To test this hypothesis, synthesized 3-*O*-MM was used in competitive inhibition ELISAs by preadsorption of human serum samples with the monosaccharide prior to antibody binding to the TP-Ag (Cole *et al.*, 1990). Results of this assay, shown in Fig. 14A, demonstrate that almost total blocking of antibody adsorption occurred at a concentration of 0.5 μg sugar/μl serum. Purity of the 3-*O*-MM was confirmed by GC/MS analysis. Preadsorption of patient sera with xylose, mannose, glucosamine, and 3-*O*-methylglucose standards had virtually no effect on binding of serum samples to the TP-Ag. Antigen-specific patient antibody obtained from a solid-phase immunosorbent column containing the purified 120-kDa glycoprotein was also reacted with 3-*O*-MM. The patient precipitin antibody reactivity with the TP-Ag was blocked by preadsorption with 3-*O*-MM. The TP-Ag-reactive antibody fraction obtained by SPIA was characterized as IgM in the ELISA (Kruse and Cole, 1990; Cole *et al.*, 1990). Reactivity of this human antibody fraction with the inner conidial wall and spherule was confirmed by IEM (Fig. 14B,C). This same antibody fraction has also been tested for reactivity in the ELISA with synthesized 3-*O*-MM conjugated with bovine serum albumin (BSA) and bound to wells of microtitration plates which were provided by M. B. Goren (National Jewish Center for Immunology and Respiratory Medicine, Denver, Colorado). The TP-Ag-specific patient antibody was reactive with the synthesized sugar-BSA conjugate in the ELISA.

The potential application of 3-*O*-MM suggested from these data is to use the synthesized sugar conjugated with an appropriate protein (e.g., BSA) as an antigen in a diagnostic ELISA for detection of anti-*Coccidioides* IgM precipitin antibody. A possible problem is that 3-*O*-methylated heteromannans have been reported in certain prokaryotes (Nimmich, 1970), including the Mycobacteriales (Maitra and Ballou, 1974). Patient serologic cross-reactivity to CDN and tuberculin have been reported (Salvin, 1959). To further explore the possibility of common carbohydrate epitopes in CDN and tuberculin, we are presently examining the reactivity of tuberculosis patient sera with the TP-Ag and 3-*O*-MM.

5. WATER-SOLUBLE CONIDIAL WALL FRACTION

The principal goals of our investigations of *C. immitis* have been to purify, locate, and ultimately characterize antigens of this pathogen which elicit patient humoral and cellular immune responses in hopes of improving diagnostic reagents, better understanding host–parasite relationships, and contributing to the possible development of a vaccine against coccidioidomycosis. Ideally, one would like to analyze water-soluble antigens which had not been degraded by the organism's proteolytic enzymes, or chemically or enzymatically altered during the extraction process. The antigens must be available in reasonable quantity and not mixed with a large excess of nonantigenic material. The soluble conidial wall fraction (SCWF), released after the hydrophobic outer conidial wall is removed in the Ribi cell fractionator (Fig. 4), satisfies these criteria. This soluble fraction is equivalent to the outer conidial wall fraction II (OCW II) described in our original reports (Cole *et al.*, 1982, 1983, 1985a,b; Cole and Sun, 1985).

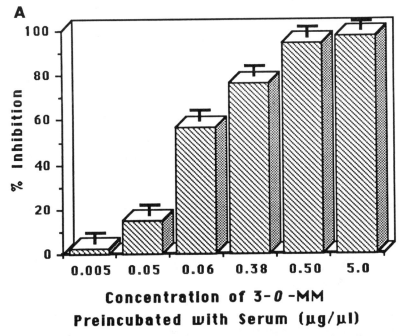

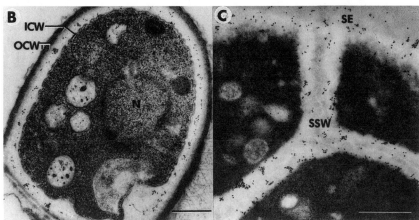

Figure 14. Reactivity of coccidioidomycosis patient tube precipitin antibody (IgM) with synthesized 3-*O*-methyl-*D*-mannose (A) and with thin-sectioned arthroconidia (B) and spherules (C). Panel A shows results of competitive-inhibition ELISA of IDTP-positive, patient serum preincubated with different concentrations of 3-*O*-MM prior to assay of IgM antibody binding to the TP-Ag (120 kDa)-coated wells of the microtitration plate. ICW, inner conidial wall; OCW outer conidial wall; SE, spherule envelope; SSW, segmented spherule wall. B, × 23,180; C, × 30,000 (from Cole *et al.*, 1990.)

5.1. Composition

Results of examination of the gross chemical composition of the SCWF were presented in Tables I and II. An approximate 1:1 ratio of protein:carbohydrate content was determined, with glucose as the major monosaccharide detected by GLC (Cole *et al.*, 1983). Another major component of the SCWF was an amber-colored pigment which was isolated in the methanol–

water layer during lipid extraction. The chemical nature of this pigment is still unknown. The protein content of the SCWF, analyzed by SDS–PAGE, is characterized as a complexity of polypeptides with a range of sizes from < 14 kDa (pigmented fraction) to > 120 kDa. The major polypeptides in reducing gels (12.5%) were identified as 48-, 36-, and 19-kDa bands (Cole et al., 1987). A 21-kDa component of the SCWF reported in our original analysis (Cole et al., 1987) was later shown to correspond to the 19-kDa proteinase isolated from SCWF and described in an earlier publication (Cole et al., 1989).

5.2. Identification and Localization of Antigenic Components

Antigenic components of the SCWF have been identified by tandem- and advancing-line IEP using our CDN–anti-CDN reference system (Cole et al., 1983, 1985a, 1987, 1988c; Cole and Sun, 1985; see Section 3.3). The major antigens were identified in IEP gels using the standardized designation system (Huppert et al., 1978, 1979; Cole et al., 1988c) and the corresponding molecular size of each isolated fraction is presented in Table VII. Comparison of the antigenic content of the solubilized outer and inner conidial wall layers (OCWF and ICWF, respectively), the conidial cytosol, and the SCWF by immunoelectrophoresis in the CDN–anti-CDN reference system has clearly revealed that the SCWF has the greatest diversity of components (Cole et al., 1985a; Cole, 1986). As previously pointed out (Section 3.5), antigens of the SCWF were located by IEM and are primarily associated with the wall of both saprobic and parasitic cell types (Figs. 8, 9). It is not surprising that the wall contains an abundance of water-soluble, antigenic components since earlier investigations have indicated that the most significant biological

Table VII. Antigenic Components of the Soluble
Conidial Wall Fraction Identified in the CDN–
Anti-CDN Reference System and by SDS–Page[a]

Reference antigen no.	Molecular mass in reducing gel[b] (kDa)	Reference
2	110	Kruse and Cole (1990)
3	—	
6	—	
7	—	
8	—	
9	—	
11	36	Yuan and Cole (1987)
12 (= CS)[c]	19	Cole et al. (1989)
16	—	
18	48	Pappagianis and Zimmer (1988)
23	—	
TP (= IPA)[d]	120	Kruse and Cole (1990)

[a]Antigens of the SCWF identified by established numbering/lettering system of coccidiodin (CDN)–anti-CDN immunoelectrophoresis reference system (Huppert et al., 1978, 1979; Cole et al., 1988c). Isolated antigens also characterized by SDS–PAGE.

[b]Molecular mass estimates based on analysis of SDS–PAGE gels under reducing conditions.

[c]Antigen CS (Cole et al., 1989b) is equivalent to Ag12 in the CDN–anti-CDN reference system (Huppert et al., 1978, 1979).

[d]An incomplete precipitin arc (IPA) is formed by reaction of burro anti-CDN with the isolated 120-kDa tube precipitin (TP) antigen in the CDN–anti-CDN reference system (unpublished data). This precipitin arc was not previously recognized in the burro reference system.

activities associated with coccidioidomycosis in both patients and experimental animals are elicited by antigens of cell wall origin (Cox *et al.*, 1984; Cox, 1983; Huppert, 1983). Our gel filtration and preparative SDS–PAGE gel separations of the SCWF have yielded antigens suspected to play significant roles in the disease process (e.g., potent elicitors of sensitized T-cell response; proteinases which may be involved in tissue invasion), as well as macromolecules which may be of diagnostic potential (e.g., TP-Ag and AgCS; Cole *et al.*, 1985a, 1989, 1990).

5.3. T-Cell Response

As discussed previously in reference to Table III, draining LN cells from mice immunized with *C. immitis* spherules respond very well to the SCWF. Fractionation of the SCWF by gel filtration yielded products with enhanced immunoreactivity in the LN proliferation assay (Fig. 15) and, in most cases, less complex antigenic composition (Cole *et al.*, 1987). It appears, however, that the fractions containing the highest and lowest molecular weight components (i.e., FR1a and FR6, respectively) have low levels of immunoreactivity. The most reactive fractions (FRs 2, 3a, 3b) were examined by SDS–PAGE (Fig. 16A) and shown to contain different, prominent polypeptides as revealed by Coomassie blue stain. Some of the components of these GF fractions

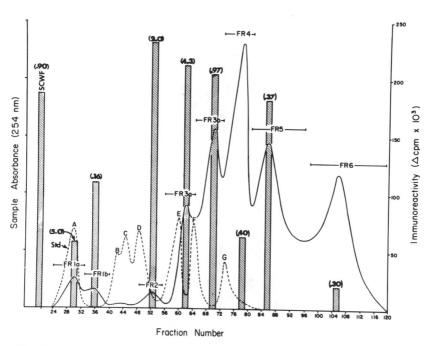

Figure 15. Chromatographic elution profile of SCWF (solid line) from Sephacryl S-300 column. Sample fractions were monitored at 254 nm. Elution profile of standard (std) mixture (dotted line) monitored at 280 nm is superimposed. Bars represent immunoreactivity of SCWF and each pooled column fraction in immune LN proliferation assay at a sample concentration of 10 μg/ml cell culture medium (see Table III). Numbers in parentheses above each bar represent protein/carbohydrate ratios. Components of the standard mixture are blue dextran (2000 kDa), β-amylase from sweet potato (200 kDa), alcohol dehydrogenase from yeast (150 kDa), bovine serum albumin (66 kDa), carbonic anhydrase from bovine erythrocytes (29 kDa), cytochrome *c* from horse heart (12.4 kDa), and aprotinin from bovine lung (6.5 kDa) (A through G, respectively). (From Cole *et al.*, 1987.)

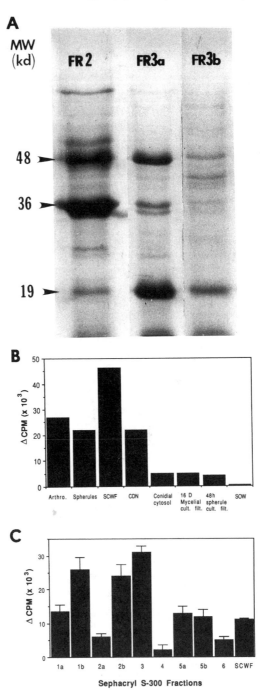

Figure 16.   SDS–PAGE gel separation of Sephacryl S-300 fractions (FRs 2, 3a, 3b) of the SCWF (A) (cf. Fig. 15) and T-cell reactivity of the crude (B), and Sephacryl S-300 fractions (C) examined in the T-cell proliferation assay using the SCWF T-cell line. Derivation of T-cell line described in text. Mycelial and spherule culture filtrates obtained from liquid cultures grown at 30 and 39°C, respectively, as described (Yuan *et al.*, 1988). Δcpm, counts per minute of [³H]thymidine-labeled T cells exposed to antigen minus counts per minute without antigen. Standard errors in B were less than 10% of the mean. The data are mean values of triplicate determinations. CDN, coccidioidin; SCWF, soluble conidial wall fraction; SOW, spherule outer wall fraction.

have been identified as antigens of the CDN–anti-CDN reference system (Table VII). In order to further characterize the cellular immunoreactivity of the soluble conidial wall material, an SCWF T-cell line was derived from BALB/c × DBA/2 (F1) mice immunized with *C. immitis* spherules as previously described (Section 3.4). The relative reactivity of this cell line with various *C. immitis* preparations is summarized in Fig. 16B. In all experiments the antigens were tested at a number of concentrations (5–50 μg/ml) and the maximal responses are shown. The standard deviations were less than 10% of the mean. Several points about these data should be emphasized. The SCWF cell line retains the ability to respond to sheared arthroconidia and, to a lesser extent, spherules. However, the T-cell line is very selective in its response to soluble *C. immitis* wall antigens. The SCWF is consistently the most reactive antigen complex, and is more reactive than CDN, the conidial cytosol, mycelial, or spherule culture filtrates, or the spherule outer wall (SOW) fraction (Cole *et al.*, 1988b). The clear difference in reactivity of the T-cell line to the SCWF and SOW suggests that T-cell stimulatory antigens in the conidial phase are not shared by the spherule phase. To examine which components of the SCWF elicit T-lymphocyte proliferative response, gel filtration subfractions of the SCWF obtained from the Sephacryl S-300 column were tested for reactivity in the T-cell line (Fig. 16C). The relative antigenicity correlates well with previous data obtained from proliferation assays using LN cells and the same gel filtration fractions (Fig. 15). Current investigations are focused on isolation and characterization of specific T-cell-stimulating antigenic components of FRs 2, 3a, and 3b.

5.4. Patient Antibody Response to Specific Antigens

Differences in humoral reactivity to three antigens purified from the SCWF (Ag11, 66 kDa-Ag, AgCS) have been demonstrated using serum from 21 coccidioidomycosis patients (Fig. 17). All patients tested were positive in the IDCF assay (Cole *et al.*, 1989). Two antigens (Ag11 and 66 kDa-Ag) were isolated from the SCWF by immunoaffinity chromatography using the respec-

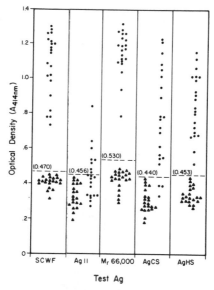

Figure 17.   Results of ELISA with control serum samples (▲) and coccidioidomycosis patient serum samples (●) adsorbed to the indicated crude or purified test antigens (Ag) (10 μg/ml) which were bound to wells of microdilution plates. Goat anti-human IgG (H+L) conjugated to peroxidase was used for the detection of adsorbed antibody. The value for the mean O.D. of the control serum plus twice the standard deviation of the mean for each antigen tested is shown above the dashed line.

tive, monospecific antiserum (Yuan and Cole, 1987; Yuan *et al.*, 1988; Cole *et al.*, 1989). Rabbit antiserum was raised against Ag11 which was purified by gel filtration (Yuan and Cole, 1987; Yuan *et al.*, 1988), and against the 66 kDa-Ag isolated by electrophoretic separation and electroelution from SDS–PAGE gels (Cole *et al.*, 1989). Isolation of antigens directly from agarose gels (e.g., 2D-IEP gels) was avoided because agarose induces the formation of antibodies against a 68-kDa protein which is contained in commercial agarose (Aoues and Huault, 1988). Ag11 corresponds to the 36-kDa component of the SCWF (Table VII). The 66 kDa-Ag was not identified as an antigen of the CDN–anti-CDN reference system (Cole *et al.*, 1989). AgCS was isolated by electroelution of a 38-kDa band from SDS–PAGE gel separations (nondenaturing) of a 75% acetone precipitation of the SCWF (Cole *et al.*, 1988c, 1989). AgCS was identified as a 19 kDa fraction by SDS–PAGE under reducing conditions (Table VII). A fourth antigen which we tested in the ELISA (Fig. 17) was identified as a *C. immitis*-specific, heat-stable (HS) antigen isolated from the mycelial culture filtrate. This antigenic fraction was provided by L. Kaufman (Centers for Disease Control, Atlanta, Ga.) and described in earlier reports by this same investigator and co-workers (Kaufman and Standard, 1987; Kaufman *et al.*, 1985). The serum samples screened by the ELISA showed a clear difference in the range of O.D. values for Ag11 and the 66 kDa-Ag. As noted in the following discussion (Section 6.1), these two antigens have proteolytic activity and Ag11 has been shown to be capable of cleavage of human immunoglobulins (Yuan and Cole, 1987; Cole *et al.*, 1987, 1989). One possible contributing factor to the low levels of patient antibody binding to Ag11 was that cleavage of immunoglobulins occurred during incubation of serum samples with the proteinase. However, preincubation of Ag11 with PMSF, which irreversibly binds to the proteinase (Yuan and Cole, 1987), had no effect on the antigen–antibody binding capacity as determined by the ELISA. An explanation for the poor antibody response to Ag11 may simply be the paucity of antigen presented to the host during the parasitic cycle. Although Ag11 is a major component of the SCWF, we have shown by IEM that it is localized within the inner wall complex of endosporulating spherules (see Section 6.1) and that, on the basis of ELISA studies, relatively little antigen was released into the growth media of the parasitic phase (Yuan *et al.*, 1988). In support of this latter observation. Zimmer and Pappagianis (1986b) did not detect a 36-kDa band (reducing gels) in their immunoblot analysis of extracellular proteins released during growth of the spherule–endospore phase. In the same study, however, the authors did report a reactive band of approximately 66-kDa in immunoblots of the 28 hr spherule–endospore culture filtrate using both CF- and TP-positive patient serum samples. We have shown that the 66 kDa proteinase is a component of the membranous, spherule outer wall (SOW) fraction and was reactive with test sera on the basis of immunoblot analyses (Cole *et al.*, 1988a). In contrast to Ag11, we suggest that the uniformly high levels of IgG binding to the 66 kDa-Ag in the ELISA (Fig. 17) are due to presentation of this antigen to the host during early infection and throughout the course of the disease. Additional studies are necessary to evaluate the diagnostic and prognostic potential of the 66 kDa-Ag.

AgCS also occurs in both the SCWF of the infectious phase and the SOW fraction of the proliferating phase of the disease (Cole *et al.*, 1988b). We have shown that AgCS is a heat-stable antigen, probably specific for *C. immitis*, and the major component of the SDS–PAGE gel separation of the HS antigen which was reported by Kaufman and co-workers (1985, 1987) and used for immunoidentification of *C. immitis* (Cole *et al.*, 1985a, 1989). These latter investigations detected a single precipitin line in immunodiffusion plates when anti-HS antibody raised in rabbits was reacted with the *C. immitis* culture filtrate or exoantigen. They demonstrated that this IDHS assay was immunodiagnostic of *C. immitis* cultures. We have recently established that AgCS reacts with anti-HS in ID plates to form a precipitin line which fuses with the IDHS preciptin. Similar O.D. values were obtained in the ELISA for individual serum samples tested with AgCS and the IDHS antigen (i.e., AgHS in Fig. 17). Binding of patient antibody to the IDHS antigen was inhibited in the ELISA by preincubation of serum samples with AgCS (Cole *et al.*, 1989). On the basis of results of their ID tests, Kaufman and co-workers (1985) reported that antibodies to the IDHS antigen are infrequently found in human serum samples from patients with coccidioidomycosis and that this reagent is, thus, of little serodiagnostic value. Our ELISA data, on the contrary, suggest that the IDHS antigen and purified AgCS warrant further investigation to evaluate their immunodiagnostic potential.

6. POSSIBLE BIOLOGICAL FUNCTIONS OF ISOLATED CONIDIAL WALL COMPONENTS

Mackenzie (1983) has pointed out that few studies of isolated fungal antigens have focused on the biological nature or function of these macromolecules. The author has further emphasized; "In view of the gains that can be anticipated in acquiring an understanding of the mechanisms of pathogenicity, and being able to ascribe greater significance to the results of serological tests, this failure to link studies on the chemical nature of antigens with analyses of e.g., their enzymatic nature, is a surprising one."

6.1. Wall-Associated Proteinases

Several recent reports have focused on characterization of proteinases of the saprobic and parasitic phases of *C. immitis* (Cole *et al.*, 1989; Nziramasanga and Lupan, 1985; Lupan and Nziramasanga, 1986; Yuan and Cole, 1987, 1989; Resnick *et al.*, 1987), and the possible roles of these products in both morphogenesis of the pathogen and host tissue invasion (Yuan *et al.*, 1988). Certain proteinases have been shown to be capable of collagenolytic and elastinolytic activity *in vitro* (Nziramasanga and Lupan, 1985; Lupan and Nziranmasanga, 1986; Yuan and Cole, 1987; Resnick *et al.*, 1987). It has also been speculated that one or more of the proteinases expressed during the parasitic cycle may participate in endospore release from mature spherules (Fig. 1), a developmental process which is pivotal for dissemination of the pathogen in host tissue (Yuan *et al.*, 1988; Yuan and Cole, 1989; Resnick *et al.*, 1987).

Three proteinases of *C. immitis* have been isolated from fractions of the saprobic phase and identified by substrate gel electrophoresis (Fig. 6). A summary of the molecular weight data obtained from SDS–PAGE and gel filtration examinations of these purified proteinases is presented in Table VIII. The serologic reactivity of these same proteolytic antigens was previously discussed and summarized in Fig. 17. Both the 19-kDa and 66-kDa proteinases, identified as AgCS and the 66 kDa-Ag, respectively, were isolated from cell wall fractions (e.g., SCWF and

Table VIII. Summary of Data on Molecular Mass and Substrate Activity of Isolated Proteinases of C. immitis[a]

	Molecular Mass (kDa) by:				Proteolytic activity[b]		
Proteolytic fraction	SDS–PAGE (nonreducing)	SDS–PAGE (reducing)	Substrate gel	GF[c]	Gelatin	IgG	sIgA
Ag11	62	36	66	60	+	+	+
AgCS	38	19	19 (38)[e]	39,19[f]	+	+	+
66 kDa-Ag[d]	58	66	60	56	+	−	−

[a]From Cole *et al.* (1989).
[b]Ability or inability of proteinases to digest gelatin and human immunoglobulins is indicated by + and −, respectively.
[c]Gel filtration (GF) chromatography.
[d]66 kDa-Ag, immunoaffinity-purified from 5-day mycelial culture filtrate fraction.
[e]Major proteolytic fraction appeared to be 19-kDa fraction; digestion band at 38 kDa also observed.
[f]Two absorbance peaks were recorded by gel filtration for AgCS with estimated sizes of 39 and 19 kDa.

ICWF-ASWS; Cole *et al.*, 1989; Kruse and Cole, 1990) and shown to be released into the culture media during growth of both the saprobic and parasitic phases of the pathogen. This latter observation suggests that these proteolytic antigens are presented to the host during growth and endosporulation of spherules and the host, therefore, may respond with production of specific antibody. The data in Fig. 17 suggest that the host does respond to the presence of these antigens *in vivo*. Further investigations of the immunological and physicochemical features of the 19- and 66-kDa proteinases are in progress. We have recently obtained in amino-terminal sequence for the 19-kDa proteinase which is as follows: Ala-Ser-Thr-Ala-Asp-Leu-Ser-Tyr-Asp-Thr-His-Tyr-Asp-Asp-Pro-Ser-Leu-Pro-Leu-His-Gly-Gly-Val-Thr-His-Asp-Gly-Lys-Asn-Gly-Met-Ile-Thr-Lys-Gly-Tyr-Asn. The first nine amino acids of this sequence match part of the amino-terminal sequence for a 21-kDa serine proteinase of *C. immitis* described by Resnick and co-workers (1990). It seems that these two products are the same proteinase.

The 36-kDa proteinase (Ag11) was originally isolated from the SCWF and is characterized by a broad substrate specificity, optimal activity at 35–40°C (pH 8.0), an isoelectric point of pH 4.5, and inhibition by organofluorides, *N*-tosyl-L-phenylalanine chloromethylketone, chymostatin, and α-1-antitrypsin (Yuan and Cole, 1987). The most intriguing feature of this chymotrypsin-like serine proteinase is its localization in the segmentation wall apparatus of developing spherules (Fig. 1; Yuan *et al.*, 1988). We raised monospecific antiserum in rabbits against the purified proteinase for use in IEM. Immunolabel was most concentrated on the wall ingrowths of the spherule envelope (i.e., early stage of spherule septation; Fig. 1) and the resulting segmentation apparatus which subdivides the mature spherule into endospore-containing compartments. This association of Ag11 with the endogenous wall material of the parasitic cells may account for the low levels of the antigen detected in the spherule culture media (Yuan *et al.*, 1988), as well as low levels of the patients' antibody response to Ag11 (Fig. 17). The total wall fraction of endosporulating spherules was isolated, treated with PMSF to inactivate any residual 36-kDa proteinase in the wall isolates, and used as a test substrate for the purified chymotrypsin-like proteinase isolated from the SCWF. The same test was conducted using mycelial wall and presegmented spherule wall fractions. Only the endosporulating spherule wall was significantly digested. Active, 36-kDa proteinase was isolated from intact and viable, endosporulating spherules by brief extraction of cells with 1% octyl-β-D-thioglucoside, a nonionic detergent. We suggested that the chymotypsin-like proteinase may be partly responsible for autolysis of the segmentation apparatus of mature spherules which results in endospore release (Yuan *et al.*, 1988).

The immunoprobe used for electron microscopy detected Ag11 in the spherule wall at all stages of cell development, as well as in the wall of liberated endospores (i.e., second-generation spherule initials). We proposed that temporal regulation of the activity of this proteinase may involve a wall-associated inhibitor which should be present in the spherule envelope, at least prior to endospore release. Results of our search for the proteinase inhibitor are discussed below.

6.2. Wall-Associated Proteinase Inhibitor

Both the cytosol of the mycelial phase and wall of segmented spherules have been used as sources of a 5-kDa peptide which has been shown to efficiently block activity of the 36-kDa proteinase (Yuan and Cole, 1989). The inhibitor was purified to apparent homogeneity by a three-step process which included trichloroacetic acid (TCA) precipitation, gel filtration, and reverse-phase HPLC. The peptide is heat-stable and tolerant to low pH. The inhibitor and 36-kDa proteinase were shown to react in a 1:1 stoichiometry and the dissociation constant (K_i) of the enzyme–inhibitor complex is 2.3×10^{-8}M. Results of kinetic studies of the inhibition of the *C. immitis* proteinase indicated that enzyme inactivation occurs by competitive inhibition. A possible function of the low-molecular-weight inhibitor, as suggested for certain plant inhibitors (Wilson, 1986), is the regulation of proteolysis during specific stages of cellular development.

The inhibitor was suggested to be involved in regulation of the 36-kDa proteinase activity in the segmentation apparatus of endosporulating spherules (Yuan and Cole, 1989).

Monospecific antibody has been raised in rabbits against the 5-kDa inhibitor. The anti-36-kDa and anti-5-kDa sera have been used for localization of the respective antigens by conjugation with immunogold probes of different sizes (Fig. 18A,B). Both the proteinase and inhibitor were colocalized in the spherule wall and most concentrated in the segmentation apparatus. TCA

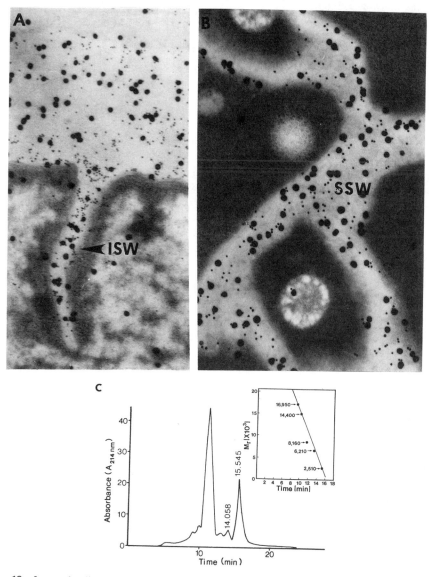

Figure 18. Immunolocalization (A, B) and chromatographic isolation (C) of an inhibitor of the 36-kDa wall-associated proteinase. (A, B) Double-labeled spherule walls in which the larger particle (40 nm) identified the proteinase inhibitor, and the smaller particle (15 nm) identified the proteinase. (C) Gel filtration–HPLC fractionation of a trichloroacetic acid (TCA) extract of the isolated wall of segmented spherules (Yuan and Cole, 1989). Insert is a plot of molecular weights of standards separated using the same conditions as employed for the sample. SSW, segmented spherule wall; ISA, initial ingrowth of the segmentation apparatus (Yuan *et al.*, 1988). A, × 34,450; B, × 40,000.

precipitates of the endosporulating spherule wall obtained using the method previously reported (Yuan and Cole, 1989) were further fractionated by gel filtration rather than by reverse-phase HPLC (Fig. 18C). The fractions were tested for inhibition of the 36-kDa proteinase activity in the presence of a synthetic peptide substrate [*N*-benzoyl-L-tyrosine ethyl ester (BTEE); Yuan and Cole, 1989]. As previously demonstrated, proteinase inhibitor activity was associated with a fraction of approximately 5 kDa (RT = 14.1 min; Fig. 18C). On the other hand, activity of the proteinase was significantly enhanced in the presence of a gel filtration fraction identified by a retention time of 15.6 min (approx. 2.5 kDa). We have previously shown that the primary and secondary sites of cleavage of angiotensin I by the 36-kDa proteinase are Phe-His and Tyr-Ile, respectively, typical of a chymotrypsin-like proteinase (Yuan and Cole, 1989). We also determined that 0.15 μg of inhibitor was required to block digestion of a fixed amount of angiotensin I by 1.0 μg of proteinase, which is consistent with the 1:1 stoichiometry of inhibition. Preliminary investigations of the amino acid composition of the 2.5-kDa gel filtration fraction (Fig. 18C) indicated the presence of large amounts of phenylalanine. It is possible, therefore, that this spherule wall-associated peptide could serve as a natural substrate of the 36-kDa proteinase, is incorporated into the segmentation apparatus of mature spherules just prior to endospore release, and effectively competes with the 5-kDa inhibitor for binding to the active site of the enzyme. Kinetic studies of the proteinase–inhibitor–2.5-kDa peptide interaction are in progress.

The ultimate goal of these investigations is to determine the mechanism(s) of regulation of endosporulation in *C. immitis*. Since endospore release is an important virulence factor for *C. immitis*, cell products which regulate this process represent potential targets for anti-*Coccidioides* drugs. Continuation of our investigations of endosporulation currently involves techniques of molecular biology related to recombinant DNA studies of *C. immitis*. Both the construction of a cDNA expression library for *C. immitis* and isolation of a cDNA clone which may code for the 36-kDa proteinase have been reported (S. Zhu, A. Eakin, Y. Yuan, J. H. McKerrow, and G. T. Cole, *Abstr. Annu. Meet. Am. Soc. Microbiol.* 1989, F-76, p. 470) and provide the basis for these future investigations.

7. SUMMARY

Human exposure to airborne, infectious arthroconidia of *C. immitis* can lead to a range of clinical forms of coccidioidomycosis from a benign, self-limited infection to a severe, progressive, and often fatal mycosis involving pulmonary and extrapulmonary tissues. Immunologically compromising illnesses, although present in some patients, are not prerequisite to infection by *C. immitis*. Inhaled arthroconidia are small enough to reach the alveoli. Conidia appear to be resistant to host defenses and capable of suppressing cellular immune responses which may partly account for the severity of this disease. We have focused our attention on the conidial wall as a source of biologically active macromolecules which participate in various aspects of pathogen–host interactions. It has been possible to isolate soluble, antigenic complexes from the conidial envelope, characterize their reactivity in humoral and cellular immunoassays, and obtain purified components from these wall fractions using various separation techniques. Some of these purified components have been characterized on the basis of their chemical composition, antigenic identity in standardized immunoelectrophoresis and immunodiffusion tests as well as in immunoassays which measure relative antibody adsorption and T-cell response to the homogeneous fractions. Certain purified components have also demonstrated proteolytic activity and evidence has been presented that at least one of these proteinases may participate in morphogenesis of the parasitic phase and/or in host invasion. Our recent development of a cDNA expression library and the availability of a genomic library for *C. immitis* have permitted our initial application of recombinant DNA technology to further characterization of these purified, conidial wall-associated fractions. This approach will ultimately lead to an understanding of gene regulation of antigen/proteinase expression during arthroconidium–spherule–endospore development *in vitro*. In turn, the results of these investigations will contribute to our comprehension of the pattern of host immunological response to conidial products and possibly to the identification of factors which contribute to disease initiation.

ACKNOWLEDGMENTS.   The authors are grateful to S. H. Sun and K. R. Seshan for contributions of data on virulence of *C. immitis* strains and the immunoelectron micrographs used in this chapter.

These investigations were supported by Public Health Service Grant AI19149 from the National Institute of Allergy and Infectious Diseases.

8. REFERENCES

Anderson, K. L., Wheat, R. W., and Conant, N. F., 1971, Fractionation and composition studies of skin-test active components of sensitins from *Coccidioides immitis*, *Appl. Microbiol.* **22**:294–299.

Aoues, A., and Huault, C., 1988, Production of monospecific antisera by immunization with precipitin line leads to misinterpretation of immunoblot. *Electrophoresis* **9**:103–104.

Ausubel, F. M., Brent, R., Kingston, R. E., Moore, D. D., Seidman, J. G., Smith, J. A., and Struhl, K. (eds.), 1989, *Current Protocols in Molecular Biology*, Volumes 1 and 2, Wiley, New York.

Beaman, L., Pappagianis, D., and Benjamin, E., 1977, Significance of T cells in resistance to experimental murine coccidioidomycosis, *Infect. Immun.* **17**:580–585.

Beaman, L., Pappagianis, D., and Benjamin, E. 1979, Mechanisms of resistance to infection with *Coccidioides immitis* in mice, *Infect. Immun.* **23**:681–685.

Beever, R. E., and Dempsey, G. P., 1978, Function of rodlets on the surface of fungal spores, *Nature* **272**:608–610.

Beever, R. E., Redgwell, R. J., and Dempsey, G. P., 1979, Purification and chemical characterization of the rodlet layer of *Neurospora crassa* conidia, *J. Bacteriol.* **140**:1063–1070.

Bernstein, D. I., Tipton, J. R., Shoot, S. F., and Cherry, J. D., 1981, Coccidioidomycosis in a neonate; maternal–infant transmission, *J. Pediatr.* **99**:752–754.

Bradford, M. M., 1976, A rapid and sensitive method for the quantitation of microgram quantities of protein utilizing the principle of protein dye binding. *Anal. Biochem.* **72**:248–254.

Bronnimann, D. A., Adam, R. D., Galgiani, J. N., Habib, M., Peterson, E. A., Porter, B., and Bloom, J. W., 1987, Coccidioidomycosis in the acquired immunodeficiency syndrome, *Ann. Intern. Med.* **106**:372–379.

Brosbe, E. A., 1967, Use of refined agar for the *in vitro* propagation of the spherule phase of *Coccidioides immitis*, *J. Bacteriol.* **93**:497–498.

Calhoun, D. L., Galgiani, J. N., Zukoski, C., and Copeland, J. G., 1985, Coccidioidomycosis in recent renal or cardiac transplant recipients, in: *Coccidioidomycosis: Proceedings of the 4th International Conference on Coccidioidomycosis* (H. E. Einstein and A. Catanzaro, eds.), National Foundation for Infectious Diseases, Washington, D.C., pp. 312–318.

Campbell, D. H., Garvey, J. S., Cremer, N. E., and Sussdorf, D. H., 1970, *Methods in Immunology*, Benjamin, New York.

Catanzaro, A., 1985, Treatment of pulmonary coccidioidomycosis, in: *Coccidioidomycosis: Proceedings of the 4th International Conference on Coccidioidomycosis* (H. E. Einstein and A. Catanzaro, eds.), National Foundation for Infectious Diseases, Washington, D.C., pp. 451–457.

Cole, G. T., 1973, A correlation between rodlet orientation and conidiogenesis in Hyphomycetes, *Can. J. Bot.* **51**:2413–2422.

Cole, G. T., 1986, Models of cell differentiation in conidial fungi, *Microbiol. Rev.* **50**:95–132.

Cole, G. T., and Pope, L. M., 1981, Surface components in *Aspergillus niger* conidia, in: *The Fungal Spore: Morphogenetic Controls* (G. Turian and H. R. Hohl, eds.), Academic Press, New York, pp. 195–215.

Cole, G. T., and Samson, R. A., 1984, The conidia, in: *Mould Allergy* (Y. Al-Doory and J. F. Domson, eds.), Lea & Febiger, Philadelphia, pp. 66–103.

Cole, G. T., and Sun, S. H., 1985, Arthroconidium–spherule–endospore transformation in *Coccidioides immitis*, in: *Fungal Dimorphism: With Emphasis on Fungi Pathogenic for Humans* (P. J. Szaniszlo and J. L. Harris, eds.), Plenum Press, New York, pp. 281–333.

Cole, G. T., Sekiya, T., Kasai, R., Yokohama, T., and Nozawa, Y., 1979, Surface ultrastructure and chemical composition of the cell walls of conidial fungi, *Exp. Mycol.* **3**:132–156.

Cole, G. T., Sun, S. H., and Huppert, M., 1982, Isolation and ultrastructural examination of conidial wall components of *Coccidioides* and *Aspergillus*, *Scanning Electron Microsc.* **1982**(IV):1677–1685.

Cole, G. T., Pope, L. M., Huppert, M., Sun, S. H., and Starr, P., 1983, Ultrastructure and composition of conidial wall fractions of *Coccidioides immitis*, *Exp. Mycol.* **7**:297–318.

Cole, G. T., Chinn, J W., Pope, L. M., and Starr, P., 1985a, Characterization and distribution of 3-*O*-methylmannose in *Coccidioides immitis*, in: *Coccidioidomycosis: Proceedings of the 4th International Conference on Coccidioidomycosis* (H. E. Einstein and A. Catanzaro, eds.), National Foundation for Infectious Diseases, Washington, D.C., pp. 130–145.

Cole, G. T., Pope, L. M., Huppert, M., Sun, S. H., and Starr, P., 1985b, Wall composition of different cell types of *Coccidioides immitis*, in: *Coccidioidomycosis: Proceedings of the 4th International Conference on Coccidioidomycosis* (H. Einstein and A. Catanzaro, eds.), National Foundation for Infectious Diseases, Washington, D.C., pp. 112–129.

Cole, G. T., Starr, M. E., Sun, S. H., and Kirkland, T. N., 1986, Antigen identification in *Coccidioides immitis*, in: *Microbiology–1986* (L. Leive, ed.), American Society for Microbiology, Washington, D.C., pp. 159–164.

Cole, G. T., Kirkland, T. N., and Sun, S. H., 1987, An immunoreactive, water-soluble conidial wall fraction of *Coccidioides immitis*, *Infect. Immun.* **55**:657–667.

Cole, G. T., Kirkland, T. N., Franco, M., Zhu, S., Yuan, L., Sun, S. H., and Hearn, V. N., 1988a, Immunoreactivity of a surface wall fraction produced by spherules of *Coccidioides immitis*, *Infect. Immun.* **56**:2695–2701.

Cole, G. T., Seshan, K. R., Franco, M., Bukownik, E., Sun, S. H., and Hearn, V. N., 1988b, Isolation and morphology of an immunoreactive outer wall fraction produced by spherules of *Coccidioides immitis*, *Infect. Immun.* **56**:2686–2694.

Cole, G. T., Sun, S. H., Dominguez, J., Yuan, L., Franco, M., and Kirkland, T. N., 1988c, Wall-associated antigens of *Coccidioides immitis*, in: *Fungal Antigens: Isolation, Purification and Detection* (E. Drouhet, G. T., Cole, L. de Repentigny, J.-P. Latgé, and B. Dupont, eds.), Plenum Press, New York, pp. 395–415.

Cole, G. T., Zhu, S., Pan, S., Yuan, L., Kruse, D., and Sun, S. H., 1989, Isolation of antigens with proteolytic activity from *Coccidioides immitis*, *Infect. Immun.* **57**:1524–1534.

Cole, G. T., Kruse, D., Zhu, S., Seshan, K. R., and Wheat, R. W., 1990, Composition, serologic reactivity, and immunolocalization of a 120 kilodalton tube precipitin antigen of *Coccidioides immitis*, *Infect. Immun.* **58**:179–188.

Corradin, G., and Chiller, H. M., 1977, Lymphocyte specificity to protein antigens. I. Characterization of the antigen-induced *in vitro* T-cell dependent proliferative response with lymph node cells from primed mice, *J. Immunol.* **119**:1048–1053.

Cox, R. A., 1983, Cell-mediated immunity, in: *Fungi Pathogenic for Humans and Animals*, Part B, *Pathogenicity and Detection* (D. Howard, ed.), Marcel Dekker, New York, pp. 61–98.

Cox, R. A., 1986, Acquired resistance to coccidioidomycosis and modulation of host response, in: *Microbiology– 1986* (L. Leive, ed.), American Society for Microbiology, Washington, D.C., pp. 172–176.

Cox, R. A., 1988, Immunosuppression by cell wall antigen of *Coccidioides immitis*, *Rev. Infect Dis.* **10**(Suppl. 2):5415–5418.

Cox, R. A., and Britt, L. A., 1986a, Isolation and identification of an exoantigen specific for *Coccidioides immitis*, *Infect. Immun.* **52**:138–143.

Cox, R. A., and Britt, L. A., 1986b, Isolation of a coccidioidin component that reacts with immunoglobulin M precipitin antibody, *Infect. Immun.* **53**:449–453.

Cox, R. A., and Britt, L. A., 1987, Antigenic identity of biologically active antigens in coccidioidin and spherulin, *Infect. Immun.* **55**:2590–2596.

Cox, R. A., and Kennell, W., 1988, Suppression of T-lymphocyte response by *Coccidioides immitis* antigen, *Infect. Immun.* **56**:1424–1429.

Cox, R. A., and Pope, R. M., 1987, Serum-mediated suppression of lymphocyte tranformation responses in coccidioidomycosis, *Infect. Immun.* **55**:1058–1062.

Cox, R. A., Huppert, M., Starr, P., and Britt, L. A., 1984. Reactivity of an alkali-soluble, water-soluble antigen of *Coccidioides immitis* with anti-*Coccidioides* immunoglobulin M precipitin antibody, *Infect. Immun.* **43**:502–507.

Cox, R. A., Britt, L. A., and Michael, R. A., 1987, Isolation of *Coccidioides immitis* F antigen by immunoaffinity chromatography with monospecific antiserum, *Infect. Immun.* **55**:227–232.

Cox, R. A., Kennell, W., Boncyk, L., and Murphy, J. W., 1988, Induction and expression of cell-mediated immune responses in inbred mice infected with *Coccidioides immitis*, *Infect. Immun.* **56**:13–17.

Currah, R.S., 1985, Taxonomy of the Onygenales: Arthrodermataceae, Gymnoascaceae, Myxotrichaceae, and Onygenaceae, *Mycotaxon* **24**:1–216.

Deresinski, S. C., and Stevens, D. A., 1975, Coccidioidomycosis in compromised hosts: Experience at Stanford University Hospital, *Medicine (Baltimore)* **54**:377–395.

Dolan, M. J., Cox, R. A., Williams, V., and Woolley, S., 1989, Development and characterization of a monoclonal antibody against the tube precipitin antigen of *Coccidioides immitis*, *Infect. Immun.* **57**:1035–1039.

Donnally, W. H., and Yunis, E. J., 1974, The ultrastructure of *Coccidioides immitis*. Study of a human infection, *Arch. Pathol.* **94**:227–232.

Drutz, D. J., and Huppert, M., 1983, Coccidioidomycosis: Factors affecting the host–parasite interaction, *J. Infect. Dis.* **147**:372–390.

Galgiani, J. N., Dugger, K. D., Ampel, N. M., Sun, S. H., and Law, J. H., 1988, Extraction of serologic and delayed hypersensitivity antigens from spherules of *Coccidioides immitis*, *Diagn. Microbiol. Infect. Dis.* **11**:65–80.

Gregory, P. H. 1973, *The Microbiology of the Atmosphere*, 2nd ed., Leonard Hill, Aylesbury, England.

Hashimoto, T., Wu-Yuan, C. D., and Blumenthal, H. J., 1976, Isolation and characterization of the rodlet layer of *Trichophyton mentagrophytes* microconidial wall, *J. Bacteriol.* **127**:1543–1549.

Hassid, W. Z., Baker, E. E., and McCready, R. M., 1943. An immunologically active polysaccharide produced by *Coccidioides immitis* Rixford and Gilchrist, *J. Biol. Chem.* **149**:303–311.

Hector, R., and Pappagianis, D., 1982, Enzymatic degradation of the walls of spherules of *Coccidioides immitis*, *Exp. Mycol.* **6**:136–152.

Hoffman, D. R., 1984, Mould allergens, in: *Mould Allergy* (Y. Al-Doory and J. F. Domson, eds.), Lea & Febiger, Philadelphia, pp. 104–116.

Hunkapiller, M. W., Lujan, E., Ostrander, F., and Hood, L. E., 1983, Isolation of microgram quantities of proteins from polyacrylamide gels for amino acid sequence analysis, *Methods Enzymol.* **91**:227–236.

Huppert, M., 1983, Antigens used for measuring immunological reactivity, in: *Fungi Pathogenic for Humans and Animals*, Part B, *Pathogenicity and Detection* (D. Howard, ed.), Marcel Dekker, New York, pp. 219–302.

Huppert, M., and Bailey, J. W., 1963, Immunodiffusion as a screening test for coccidioidomycosis serology, *Sabouraudia* **2**:284–291.

Huppert, M., and Bailey, J. W., 1965a, The use of immunodiffusion tests in coccidioidomycosis. I. The accuracy and reproducibility of the immunodiffusion test which correlates with complement fixation, *Am. J. Clin. Pathol.* **44**:364–368.

Huppert, M., and Bailey, J W., 1965b, The use of immunodiffusion tests in coccidioidomycosis. II. An immunodiffusion test as a substitute for the tube precipitin test, *Am. J. Clin. Pathol.* **44**:369–373.

Huppert, M., Sun, S. H., Gleason-Jordan, I., and Vukovich, K. R., 1976, Lung weight parallels disease severity in experimental coccidioidomycosis, *Infect. Immun.* **14**:1356–1368.

Huppert, M., Spratt, N. S., Vukovich, K. R., Sun, S. H., and Rice, E. H., 1978, Antigenic analysis of coccidioidin and spherulin determined by two-dimensional immunoelectrophoresis, *Infect. Immun.* **20**:541–551.

Huppert, M., Adler, J. P., Rice, E. H., and Sun, S. H., 1979, Common antigens among systemic fungi analyzed by two-dimensional immunoelectrophoresis, *Infect. Immun.* **23**:479–485.

Ibrahim, A. B., and Pappagianis, D., 1973, Experimental induction of anergy to coccidioidin by antigens of *Coccidioides immitis*, *Infect. Immun.* **7**:786–794.

Jacobs, P. H., 1980, Cutaneous coccidioidomycosis, in: *Coccodioidomycosis: A Text* (D. A. Stevens, ed.), Plenum Medical, New York, pp. 213–224.

Johnson, J. E., Perry, J. E., Fekety, F. R., Kadull, P. J., and Cluff, L. E., 1964, Laboratory-acquired coccidioidomycosis: A report of 210 cases, *Ann. Intern. Med.* **60**:941–956.

Johnston, R. B., 1981, Secretion of superoxide anion, in: *Methods for Studying Mononuclear Phagocytes* (D. O. Adams, P. J. Edelson, and H. Koren, eds.), Academic Press, New York, pp. 489–497.

Julius, M. H., Simpson, E., and Hertzenberg, L. A., 1974, A rapid method for the isolation of functional thymus-derived murine lymphocytes, *Eur. J. Immunol.* **3**:645–649.

Kabat, E. A., and Mayer, M. M., 1961, *Experimental Immunochemistry*, Thomas, Springfield, Ill.

Kaufman, L., and Standard, P. G., 1987, Specific and rapid identification of medically important fungi by exoantigen detection, *Annu. Rev. Microbiol.* **41**:209–225.

Kaufman, L., Standard, P. G., Huppert, M., and Pappagianis, D., 1985, Comparison and diagnostic value of the coccidioidin heat-stable (HS and tube precipitin) antigens in immunodiffusion, *J. Clin. Microbiol.* **22**:515–518.

Kerrick, S. S., Lundergan, L. L., and Galgiani, J. N., 1985, Coccidioidomycosis at a university health service, *Am. Rev. Respir. Dis.* **131**:100–102.

Kimoto, M., and Fathman, C. G., 1980, Antigen-reactive T-cell clones. 1. Transcomplementing hybrid I-A gene products function effectively in antigen presentation, *J. Exp. Med.* **152**:759–770.

Klebanoff, S. J., 1980, Oxygen intermediates and the microbicidal event, in: *Mononuclear Phagocytes* (R. van Furth, ed.), Nijhoff, The Hague, pp. 1105–1137.

Kruse, D., and Cole, G. T., 1990, Isolation of tube precipitin antibody-reactive fractions of *Coccidioides immitis*, *Infect. Immun.* **58**:169–178.

Kurkela, R., Koivurinta, J., and Kuusinen, R., 1980, Non-protein nitrogen compounds in the higher fungi–A review, *Food Chem.* **5**:109–130.

Kwon-Chung, K. J., 1969, *Coccidioides immitis*: Cytological study on the formation of the arthrospores, *Can. J. Genet. Cytol.* **11**:43–53.

Levine, H. B., 1961, Purification of the spherule–endospore phase of *Coccidioides immitis*, *Sabouraudia* **1**:112–115.

Levine, H. B., Cobb, J. M., and Smith, C. E., 1960, Immunity to coccidioidomycosis induced in mice by purified spherule, arthrospore, and mycelial vaccines, *Trans. N. Y. Acad. Sci.* **22**:436–449.

Lones, G. W., and Peacock, C. L., 1960, Role of carbon dioxide in the dimorphism of *Coccidioides immitis*, *J. Bacteriol.* **79**:308–309.

Lowry, O. H., Rosebrough, N. J., Farr, A. J., and Randall, R. J., 1951, Protein measurement with Folin phenol reagent, *J. Biol. Chem.* **193**:265–275.

Lundergan, L. L., Kerrick, S. S., and Galgiani, J. N., 1985, Coccidioidomycosis at a university outpatient clinic: A clinical description, in: *Coccidioidomycosis: Proceedings of the 4th International Conference on Coccidioidomycosis* (H. E. Einstein and A. Catanzaro, eds.), National Foundation for Infectious Diseases, Washington, D.C., pp. 47–54.

Lupan, D. M., and Nziramasanga, P., 1986, Collagenolytic activity of *Coccidioides immitis*, *Infect. Immun.* **51**:360–361.

Mackenzie, D. W. R., 1983, Serodiagnosis, in: *Fungi Pathogenic for Humans and Animals*, Part B, *Pathogenicity and Detection* (D. Howard, ed.), Marcel Dekker, New York, pp. 121–218.

Maitra, S. K., and Ballou, C. E., 1974, Multiple forms of the methylmannose polysaccharide (MMP) from mycobacteria, *Fed. Proc. Am. Soc. Exp. Biol.* **33**:1452.

Merril, G. R., Goldman, D., and Van Keuren, M. L., 1984, Gel protein stains: Silver stain, *Methods Enzymol.* **104**:441–447.

Miyaji, M., and Nishimura, K., 1984, The parasitic cycle of *Coccidioides immitis* in murine coccidioidomycosis, *Jpn. J. Med. Mycol.* **25**:76–83.

Musial, C. E., Cockerill, F. R., and Roberts, G. D., 1988, Fungal infections of the immunocompromised host: Clinical and laboratory aspects, *Clin. Microbiol. Rev.* **1**:349–364.

Nimmich, W., 1970, Occurrence of 3-*O*-methylmannose in lipopolysaccharides of *Klebsiella* and *Escherichia coli*, *Biochem. Biophys. Acta* **215**:189–191.

Nziramasanga, P., and Lupan, D. M., 1985, Elastase activity of *Coccidioides immitis*, *J. Med. Microbiol.* **19**:109–114.

Pappagianis, D., 1975, Opportunism in coccidioidomycosis, in: *Proceedings on Opportunistic Fungal Infections* (E. W. Chick, A. Balows, and M. L. Furcolow, eds.), Charles C. Thomas, Springfield, Ill., pp. 221–229.

Pappagianis, D., 1985, The phenomenon of locus minoris resistentiae in coccidioidomycosis, in: *Coccidioidomycosis: Proceedings of the 4th International Conference on Coccidioidomycosis* (H. E. Einstein and A. Catanzaro, eds.), National Foundation for Infectious Diseases, Washington, D.C., pp. 310–329.

Pappagianis, D., 1987, Coccidioidomycosis reactivated in "cured" immunosuppressed patients, *Mycol. Observer* **7**:5.

Pappagianis, D., 1988, Epidemiology of coccidioidomycosis, *Curr. Top. Med. Mycol.* **2**:199–228.

Pappagianis, D., Putman, E. W., and Kobayashi, G. S., 1961a, Polysaccharides of *Coccidioides immitis*, *J. Bacteriol.* **82**:714–723.

Pappagianis, D., Smith, C. E., Kobayashi, G. S., and Saito, M. J., 1961b, Studies of antigens from young mycelia of *Coccidioides immitis*, *J. Infect. Dis.* **108**:35–44.

Pappagianis, D., Lindsey, N. J., Smith, C. E., and Saito, M. J., 1965, Antibodies in human coccidioidomycosis: Immunoelectrophoretic properties, *Proc. Soc. Exp. Biol. Med.* **118**:118–122.

Puckett, T. F., 1954, Hyphae of *Coccidioides immitis* in tissues of the human host, *Am. Rev. Tuberc.* **70**:320–327.

Reiss, E., Huppert, M., and Cherniak, R., 1985, Characterization of protein and mannan polysaccharide antigens of yeasts, moulds and Actinomycetes, *Curr. Top. Med. Mycol.* **1**:172–207.

Resnick, S., Pappagianis, D., and McKerrow, J. H., 1987, Proteinase production by the parasitic cycle of the pathogenic fungus *Coccidioides immitis*, *Infect. Immun.* **55**:2807–2815.

Resnick, S., Zimmer, B., Pappagianis, D., Eakin, A., and McKerrow, J., 1990, Purification and amino-terminal sequence analysis of the complement-fixing and precipitin antigens from *Coccidioides immitis*, *J. Clin. Microbiol.* **28**:385–388.

Roberts, C. J., 1984, Coccidioidomycosis in acquired immune deficiency syndrome. Depressed humoral as well as cellular immunity, *Am. J. Med.* **76**:734-736.

Robertson, M. D., Seaton, A., Milne, L. J. R., and Raeburn, J. A., 1987, Suppression of host defenses by *Aspergillus fumigatus, Thorax* **42**:19–25.

Rutala, P. J., and Smith, J. W., 1978, Coccidioidomycosis in potentially compromised hosts: The effect of immunosuppressive therapy on dissemination, *Am. J. Med. Sci.* **275**:283–295.

Salvin, S., 1959, Current concepts of diagnosis and skin hypersensitivity in the mycoses, *Am. J. Med.* **27**:97–114.

Sawaki, Y., Huppert, M., Bailey, J. W., and Yagi, V., 1966, Patterns of human antibody reactions in coccidioidomycosis, *J. Bacteriol,* **91**:422–427.

Scheer, E., Terai, T., Kulkarni, S., Conant, N. F., Wheat, R., and Lowe, E. P., 1970, An unusual reducing sugar from *Coccidioides immitis, J. Bacteriol.* **103**:525–526.

Seltzer, J., Broaddus, V. C., Jacobs, R., and Golden, J. A., 1986, Reactivation of *Coccidioides* infection, *West J. Med.* **145**:96–98.

Sigler, L., and Carmichael, J. W., 1976, Taxonomy of *Malbranchea* and some other hyphomycetes with arthroconidia, *Mycotaxon* **4**:349–488.

Smith, C. E., Whiting, E. G., Baker, E. E., Rosenberger, H. G., Beard, R. R., and Saito, M. T., 1948, The use of coccidioidin, *Am. Rev. Tuberc.* **57**:330–360.

Smith, C. E., Saito, M. J., Beard, R. R., Kepp, R. M., Clark, R. W., and Eddie, B. U., 1950, Serological tests in the diagnosis and prognosis of coccidioidomycosis, *Am. J. Hyg.* **52**:1–21.

Smith, C. E., Saito, M. J., and Simons, S. A., 1956, Pattern of 39,500 serologic tests in coccidioidomycosis, *J. Am. Med. Assoc.* **160**:546–552.

Sun, S. H., and Huppert, M., 1976, A cytological study of morphogenesis in *Coccidioides immitis, Sabouraudia* **14**:185–198.

Sun, S. H., Sekhon, S. S., and Huppert, M., 1979, Electron microscopic studies of saprobic and parasitic forms of *Coccidioides immitis, Sabouraudia* **17**:265–273.

Sun, S. H., Cole, G. T., Drutz, D. J., and Harrison, J. L., 1986, Electron-microscopic observations of the *Coccidioides immitis* parasitic cycle in vivo, *J. Med. Vet. Mycol.* **24**:183–192.

Turini, P., Kurooka, S., Steer, M., Corbascio, A. N., and Singer, T. R., 1969, The action of phenylmethylsulfonyl fluoride on human acetylcholinesterase, chymotrypsin, and trypsin, *J. Pharmacol. Exp. Ther.* **167**:98–104.

Walch, H. A., and Walch, R. K., 1967, Studies with induced mutants of *Coccidioides immitis,* in: *Coccidioidomycosis: Proceedings of 2nd Coccidioidomycosis Symposium,* University of Arizona Press, Phoenix, pp. 339–347.

Wedum, A. G., and Kruse, R. H., 1969, Assessment of risk of human infection in the microbiological laboratory, in: Publication No. 30, 2nd ed., U.S. Department of the Army, Ft. Detrick, Md., pp. 1–83.

Werner, S. B., Pappagianis, D., Heindl, I., and Mickel, A., 1972, An epidemic of coccidioidomycosis among archeology students in northern California, *N. Engl. J. Med.* **286**:507–512.

Wheat, R. W., Su Chung, K. S., Omellas, E. P., and Scheer, E. R., 1978, Extraction of skin test activity from *Coccodioides immitis* mycelia by water, perchloric acid, and aqueous phenol extraction, *Infect. Immun.* **19**:152–159.

Wheat, R. W., Woodruff, W. W., and Haltiwanger, R. S. 1983, Occurrence of antigenic (species-specific?) partially 3-*O*-methylated heteromannans in cell wall and soluble (nonwall) components of *Coccidioides immitis* mycelia, *Infect. Immun.* **41**:728–734.

Wilson, K. A., 1986, Role of proteolytic enzymes in the mobilization of protein reserves in the germinating dicot seed, in: *Plant Proteolytic Enzymes* (M. J. Dalling, ed.), CRC Press, Boca Raton, pp. 19–47.

Yuan, L., and Cole, G. T., 1987, Isolation and characterization of a secretory proteinase of *Coccidioides immitis, Infect. Immun.* **55**:1970–1978.

Yuan, L., and Cole, G. T., 1989, Characterization of a proteinase inhibitor isolated from the fungal pathogen *Coccidioides immitis, Biochem. J.* **257**:729–736.

Yuan, L., Cole, G. T., and Sun, S. H., 1988, Possible role of a proteinase in the endosporulation of *Coccidioides immitis, Infect. Immun.* **56**:1551–1559.

Zimmer, B. L., and Pappagianis, D., 1986a, Taxonomic and physiologic characteristics of *Coccidioides immitis,* in: *Microbiology–1986* (L. Leive, ed.), American Society for Microbiology, Washington, D.C., pp. 165–168.

Zimmer, B. L., and Pappagianis, D., 1986b, Comparison of immunoblot analyses of spherule–endospore-phase extracellular protein and mycelial-phase antigen of *Coccidioides immitis, Infect. Immun.* **53**:64–70.

Zimmer, B. L., and Pappagianis, D., 1988, Characterization of a soluble protein of *Coccidioides immitis* with activity as an immunodiffusion complement fixation antigen, *J. Clin. Microbiol.* **26**:2250–2256.

Zimmer, B. L., and Pappagianis, D., 1989, Immunoaffinity isolation and partial characterization of the *Coccidioides immitis* antigen detected by the tube precipitin and immunodiffusion–tube precipitin tests, *J. Clin. Microbiol.* **27**:1759–1766.

20

Cell-Mediated Host Response to Fungal Aggression

Alayn R. Waldorf

1. INTRODUCTION

Mononuclear phagocytes and granulocytes readily kill many potentially infectious fungi, even in the nonimmune individual. Phagocytic cells, macrophages, monocytes, or polymorphonuclear leukocytes (PMN) interact with fungi in several ways. If the fungal particle is phagocytized, it may either be killed, or in some instances remain viable and replicate or grow within or even grow out of the phagocyte. However, some fungi are too large to be ingested by a single phagocyte and so must be handled by extracellular defense mechanisms.

The function of the phagocytic system, as a part of the innate defense system, is difficult to separate from the phagocytosis that is mediated by acquired immune mechanisms. Since products of lymphocytes (i.e., lymphokines, antibodies) can influence any or all of the fixed and circulating phagocytes, an assessment of the innate ability of a host to phagocytize and kill an organism is difficult. The situation with the opportunistic fungi is particularly difficult since even "normal" humans and animals are commonly exposed to these organisms and their antigens. It is the rare individual that has had no previous contact with the ubiquitous conidia and spores of these organisms. Similarly, exposure to *Candida* occurs by casual exposure or because of the commensal relationship of *Candida*, or other yeasts expressing similar antigens, which colonize the gastrointestinal tract. The wide variation and disagreement in the results of studies investigating innate host responses may be partially explained by variations in previous exposure of individuals to the organisms and, therefore, an unequal contribution to the immune status at the time of the study.

There are several unique features of the fungi which make their interaction with phagocytic cells particularly interesting for study. Many fungi are opportunistic, they are saprophytic or commensal organisms, and usually infect only patients predisposed by some underlying disease or treatment. Usually, distinct defense mechanisms are defective and it is thus possible to utilize this to determine which aspects of the host's defense are of importance in controlling the infection. Second, although there are limited numbers of fungi which have been shown to cause

Alayn R. Waldorf • Department of Biological Sciences, California State University, Hayward, California 94542.

disease in man and animals, the host must contend with several forms of each of these organisms. In the dimorphic fungi this appears obvious. Yet, even with the nondimorphic fungi such as *Aspergillus* or *Rhizopus*, the host must contend with several forms of the organism to successfully eliminate them: a resting, metabolically inactive conidium or spore, which swells and germinates, and finally the hyphal organism which invades tissue. Each form displays different surface features and elicits different host responses, and is usually handled differently by the phagocytic cells.

The ultimate outcome of the host–parasite interaction is influenced by a variety of factors. The route and magnitude of the exposure to the host, quantity and metabolic state of the host's phagocytic cells, presence of nonspecific factors such as complement, presence of specific opsonic antibodies and lymphokines, and presence of specific virulence factors produced by the fungi all contribute to the interaction of fungus and host. This chapter will review fungicidal mechanisms of phagocytes and summarize recent literature describing interactions between fungi and phagocytes.

2. FUNGICIDAL MECHANISMS

2.1. Oxygen-Dependent Fungicidal Mechanisms

There are two general types of antimicrobial mechanisms available to leukocytes: oxygen-dependent and oxygen-independent. Phagocytosis of microorganisms by leukocytes is accompanied by a dramatic increase in oxygen consumption, which is linked to the enzyme system NAD(P)H oxidase (for a review see Beaman and Beaman, 1984; Klebanoff, 1980). This reaction, called the respiratory burst, leads to the production of several potentially fungicidal products including superoxide anion (O_2^-), singlet oxygen, hydrogen peroxide (H_2O_2), and hydroxyl radicals (OH^-). Ferrous ion can react with H_2O_2 to generate OH^-, which then reacts with iodide to form the toxic products of the ferrous ion system (Klebanoff, 1982).

When the phagosome containing the fungi fuses with lysosomes or granules, myeloperoxidase (MPO) is released into the vacuole. Hydrogen peroxide in the presence of halide ions and MPO produces microbicidal reactions due to the halogenation of proteins, leading to aldehyde formation and peptide cleavage and, perhaps, the production of hypochlorous acid (HOCl) (Beaman and Beaman, 1984). Interaction of HOCl with certain substances present in the organism, leukocytes, or surrounding milieu can result in the production of other compounds such as chloramines which may also be toxic for fungi. MPO is present in the granules of PMN, bone marrow promonocytes, amd most blood monocytes. However, this enzyme disappears during the maturation of the latter cells to macrophages. Thus, tissue macrophages lack MPO and its resulting antifungal reaction products.

Phagocytes respond to ingested particles by this respiratory burst, and the toxic oxygen metabolites are released into phagocytic vacuoles or to the exterior of the cell, under certain conditions. This degranulation to the exterior of the phagocyte is of crucial importance with organisms that are too large to be completely phagocytized such as hyphae, *Blastomyces* yeasts, and *Coccidioides* spherules.

The congenital disorder chronic granulomatous disease (CGD) has provided an important opportunity to examine the role of oxidative metabolism in phagocyte antimicrobial studies. PMN and monocytes of affected individuals are usually totally deficient in the ability to generate O_2^- and H_2O_2. Thus, their phagocytes have selectively impaired microbicidal abilities (Klebanoff, 1982). In addition to severe bacterial infections, these individuals have frequent occurrences of severe *Aspergillus* and *Candida* infections.

Hereditary MPO deficiency, an autosomal recessive trait, wherein affected individuals have

PMN and monocytes totally devoid of MPO activity, has indicated the importance of the MPO system as an antimicrobial component of these cells.

2.2. Oxygen-Independent Fungicidal Mechanisms

The fusion of lysosomes with the plasma membrane of phagosomes, called degranulation, results in the release of lysosomal contents. The ability of leukocytes to kill some fungi in the absence of oxidative metabolism may reflect the presence in certain phagocytes, of proteins and peptides with intrinsic antifungal activity. PMN granules or lysosomes contain microbicidal and hydrolytic digestive enzymes including: acid hydrolases, lysozyme, neutral proteases, cationic proteins, phospholipase, and lactoferrin (for review see Beaman and Beaman, 1984). Macrophages have many analogous antifungal components. The role of the granule contents in the killing of fungi is not as clearly defined as the microbicidal activity of the oxygen metabolites.

3. ROLE OF THE MONONUCLEAR PHAGOCYTE

In vitro studies on the capacity of macrophages to kill fungi, or to inhibit growth or germination, have produced conflicting results. This may be due in part to differences in the antifungal activity of macrophages from different anatomical sites, and from different animals, as well as the use of different assays for measuring inhibition of germination or killing. In addition, mononuclear cells are often elicited or cultured before their use in experimental assays. In contrast, the technique for obtaining alveolar macrophages usually does not involve an eliciting agent, and these macrophages are rarely cultured. For these reasons, and because many of the fungi begin as primary pulmonary infections, alveolar macrophage antifungal activities are examined separately from those of peritoneal macrophages and peripheral blood monocytes.

3.1. Alveolar Macrophage Defense

3.1.1. Alveolar Macrophage Defense against Conidia, Spores, and Blastoconidia of Opportunistic Fungi

In most instances, the opportunistic fungal infections which cause systemic disease are initiated by inhalation of airborne conidia and spores. The first phagocytic cells to encounter these inhaled or aspirated fungal particles are the alveolar macrophages and their importance in host defense has long been known (Merkow *et al.*, 1968, 1971). Following intranasal inoculation of *Aspergillus* conidia, alveolar macrophages form an efficient early defense system against resting conidia, by rapid killing of *Aspergillus* conidia by normal alveolar macrophages both *in vitro* and *in vivo* (Waldorf *et al.*, 1984a,b; Schaffner *et al.*, 1982, 1983; Kurup, 1984). In contrast, cortisone treatment of macrophages significantly impairs the ability of macrophages to prevent germination of *Aspergillus* conidia (Merkow *et al.*, 1971; Schaffner *et al.* 1982, 1983; Waldorf *et al.*, 1984b). Alveolar macrophages from cortisone-treated mice phagocytize conidia, but unlike normal macrophages, lysosomal granules do not fuse with the phagosomes containing the conidia, conidia germinate, and the fungus readily pierces the phagocyte. An additional study of the suppression of the macrophage's antifungal activity by glucocorticoids indicates that glucocorticoid effects are due to a reduction in the neutral proteases in the macrophage lysosomes (Schaffner, 1985).

Alveolar macrophages have the capacity to kill *Aspergillus* conidia by both oxidative and nonoxidative mechanisms, since they are killed at equal rates under aerobic and anaerobic conditions (Schaffner *et al.*, 1983). However, the conidia must become metabolically active before killing of *Aspergillus* conidia can occur, perhaps by increasing the permeability of the conidial wall and allowing better activity of the macrophage products (Levitz *et al.*, 1986). The antimicrobial system composed of ferrous ion, H_2O_2, and iodide has potent antifungal activity against *Aspergillus* conidia and *Rhizopus* spores (Levitz and Diamond, 1984). In cell-free conditions, the ferrous ion system has comparable activity to the MPO system of PMN against these two fungal particles.

Diabetes mellitus, especially with ketoacidosis, has been the most commonly recognized underlying disease associated with mucormycosis (Meyer and Armstrong, 1973). Animal models of mucormycosis in diabetic animals have been used to determine the influence of diabetes in predisposition to mucormycosis, as well as determine the relative importance of different cell populations in host defense against the causative agents (for review see Waldorf, 1986; Waldorf and Diamond, 1984; Waldorf *et al.*, 1983, 1984a; Smith, 1976; Lundborg and Holma, 1972). In these animal studies, it has been found that normal alveolar macrophages form an efficient defense against spores by inhibiting spore germination (Lundborg and Holma, 1972; Waldorf *et al.*, 1984b). In contrast, tissues of diabetic or cortisone-treated mice following intranasal inoculation, *Rhizopus* spores readily germinate. Although the *Rhizopus* spores are prevented from germinating in normal animals, they are not killed, evidenced by the fact that *Rhizopus* spores remain viable in lung tissue of normal mice for as long as 10 days after intranasal inoculation, without germinating. This is in contrast to *Aspergillus* conidia which are removed within 2 days of inoculation. Thus, alveolar macrophages from normal, diabetic, and cortisone-treated mice are unable to kill *Rhizopus* spores whether *in vitro* or *in vivo* (Waldorf *et al.*, 1984b).

The mechanism by which alveolar macrophages inhibit *Rhizopus* spore germination remains unknown. Hydrogen peroxide, in cell-free assays, significantly inhibits *Rhizopus* spore germination and would appear to be an important oxidative mechanism by which alveolar macrophages mediate the inhibition of spore germination (Waldorf, unpublished observation). Studies in our laboratory have shown that diabetic alveolar macrophages are unable to inhibit spore germination and produce significantly less H_2O_2 in response to opsonized *Rhizopus* spores. These results suggest that the decreased production of H_2O_2 in diabetic alveolar macrophages would constitute an important defect in host defense against mucormycosis (Waldorf, unpublished observations).

Candida infections in the lung are uncommon except as terminal complications even though *C. albicans* is frequently recovered from sputum cultures. The difference in the relative frequency of candidiasis in the gastrointestinal tract and lower respiratory tract in immunocompromised patients may reflect differences in innate tissue resistances, in the recruitment of the systemic host defense factors to the involved sites, or in quantitative characteristics of fungal exposure. Alveolar macrophages have relatively weak oxidative candidacidal activity. However, murine lung lavage fluid has been shown to contain a protein with anti-*Candida* activity (Nugent and Flick, 1987). This soluble factor appears to contribute to lung defenses by reducing fungal viability and by reducing adherence of *C. albicans* to tissue surfaces.

Cationic proteins from rabbit alveolar macrophages also have direct candidacidal activity (Patterson-Delafield *et al.*, 1981). The most active leukocyte peptides killed 99% of *C. albicans* within 20 min. These cationic peptides inhibited oxygen consumption and increased *Candida* cell permeability. Whether or not these macrophage peptides are released and function *in vivo* is unknown.

Cryptococcus neoformans typically initiates infection after inhalation of its yeast cells or basidiospores into the upper or lower respiratory tract. In tissue, organisms have a polysaccharide capsule that plays a major role in virulence of the organism by inhibiting phagocytosis (Kozel and Gotschlich, 1982; Kozel and Mastroianni, 1976), activating complement (Kozel and Pfrommer, 1986), and binding antibody (Kozel *et al.*, 1984). Mouse and pig alveolar macrophages appear to be unable to kill *Cryptococcus* yeasts following phagocytosis except under certain conditions, such as a high macrophage-to-yeast ratio, the presence of fresh serum, and the addition of endotoxin, where fungistasis is then seen (Granger and Perfect, 1986; Kitz *et al.*, 1984).

3.1.2. Alveolar Macrophage Defense against Hyphae

The tissue form of many of the opportunistic fungi is characterized by hyphal forms that are too large to be ingested completely by phagocytic cells. Therefore, any antifungal activity of macrophages must be carried out extracellularly. Because of the difficulty in obtaining alveolar macrophages, few studies have been performed with these tissue macrophages, although they are the first phagocytic cell to encounter the fungi which begin as pulmonary infections. Normal alveolar macrophages attach to and damage *Rhizopus* hyphae in the absence of complete ingestion (Waldorf *et al.*, 1984a). Diabetic alveolar macrophages also induce a similar range of macrophage-mediated damage.

3.1.3. Alveolar Macrophage Defense against Dimorphic Forms

The systemic pathogenic fungi are typically acquired via the respiratory tract from where they may disseminate. Blastomycosis is initiated by the inhalation of conidia that evolve into yeasts within the lung. The mechanisms of resistance to *Blastomyces* appear to be complex. Alveolar macrophages from normal mice show only modest killing of *Blastomyces* yeast, and in contrast to what is found with immune peritoneal macrophages, immune alveolar macrophages also show only modest activity against *Blastomyces* (Sugar *et al.*, 1986). Studies with human alveolar macrophages have surprisingly different results. Immune alveolar macrophages had both an increased phagocytic ability and an increased inhibition of yeast intracellular growth, compared to nonimmune alveolar macrophages (Bradsher *et al.*, 1987). The disparity between these two studies is most likely due to technical differences in the assays as well as species differences in the source of the macrophages.

Coccidioides immitis, which is the causative agent of coccidioidomycosis, initiates an infection after inhalation of a single arthroconidium. Following inhalation, the arthroconidia convert into the *in vivo* spherule and endosphere phases. Alveolar macrophages from rhesus macaques phagocytize arthroconidia and endospores of *C. immitis*, but are unable to reduce their viability or stop their progression into spherules, apparently because the organism is able to inhibit phagosome/lysosome fusion (Beaman and Holmberg, 1980). However, if murine alveolar macrophages are incubated with recombinant gamma interferon during their interaction with *C. immitis*, there is enhanced antifungal action of macrophages, due to increased fusion of phagosomes with lysosomes (Beaman, 1987). The increased killing action of the treated alveolar macrophages is similar to that seen when macrophages are incubated with immunized lymphocytes.

Macrophages are the primary host cells for *Histoplasma capsulatum* The yeast form of the organism not only survives within macrophages but also multiplies intracellularly. Thus, the macrophage, which usually eliminates organisms from the primary site of infection, becomes a site of continued fungal growth and aids in dissemination of the yeast to other tissues. Intracellular killing of *H. capsulatum* by nonimmune alveolar macrophages does not seem to occur although a lysosome-rich fraction of normal rabbit alveolar macrophages inhibits protein synthesis and decreases viability of the yeast (Calderone and Peterson, 1979). If extrapolation from results using peritoneal macrophages is valid, *H. capsulatum* survives in macrophages by failing to trigger the respiratory burst of macrophages. Thus, rather than inhibiting superoxide generation or inactivating the anion, the yeasts cell appears to avoid the toxic effects of superoxide by failing to trigger its release (Eissenberg and Goldman, 1987).

Paracoccidioidomycosis also is initiated following the inhalation of conidia and/or hyphal fragments, which then convert to a yeast. Alveolar macrophages from normal mice have a low but significant ability to kill *Paracoccidioides* yeast (Brummer *et al.*, 1988). Exposure of alveolar macrophages to lymphokines or gamma interferon significantly enhanced the fungicidal action of the macrophages to *Paracoccidioides*, thus suggesting a crucial role of lymphokine or interferon-mediated activation of alveolar macrophages in host defense against these pathogens. The study of Brummer *et al.*, (1988) indicates that the enhanced killing by alveolar macrophages is via an oxygen-independent mechanism.

3.2. Peritoneal Macrophage and Peripheral Blood Monocyte

3.2.1. Peritoneal Macrophage and Peripheral Blood Monocyte Defense against Conidia, Spores, and Blastoconidia of Opportunistic Fungi

The mechanism(s) by which monocytes kill fungi, inhibit intracellular growth, or inhibit germination of conidia and spores is not clear. The combination of MPO, H_2O_2, and a halide appears to play an important role in the anticonidial activity of human blood monocytes to *Aspergillus* conidia and *Rhizopus* spores (Levitz and Diamond, 1984). In cell-free systems, H_2O_2, HOCl, and OH^- have all been shown to kill *Aspergillus* conidia (Levitz and Diamond, 1985a) However, H_2O_2 appears to play the central role in defense against *A. fumigatus* conidia by human peripheral blood monocytes. The MPO–H_2O_2–halide system and an MPO-independent pathway, the ferrous ion–H_2O_2–halide system, both contribute significantly to the oxidative killing of *Aspergillus* conidia (Washburn *et al.*, 1987).

Peripheral blood monocytes can also kill *Aspergillus* conidia by mechanisms other than oxidative systems. Monocytes from patients with CGD, which are unable to generate comparable amounts of reactive oxygen intermediates, are presumably able to kill *Aspergillus* conidia via neutral proteases (Schaffner, 1985).

Peritoneal macrophages are capable of phagocytizing and, to a limited degree, killing ingested *Candida*. However, candidacidal activities of phagocytes show considerable variability. The differences in these results, which measure candidacidal activity of host effector cells, are in large part methodological. For example, we have found that staining methods used to estimate killing of *Candida* usually underestimate values when compared to direct plating methods. In addition, staining techniques have led to discrepancies in the intracellular killing and the role of inhibition of phagosome–lysosome fusion as a means by which *Candida* yeast evade macrophage killing (Mor and Goren, 1987).

Activated macrophages appear to play a role in resistance against disseminated candidiasis. Resident peritoneal macrophages do not readily kill *Candida albicans*, yet activated macrophages kill two to three times more *C. albicans* blastoconidia (Kagaya and Fukazawa, 1981; Sasada and Johnson, 1980; Kolotila *et al.*, 1987; Bistoni *et al.*, 1986) The enhanced candidacidal effect is reflected by an enhanced capacity to release O_2^- in activated macrophages. The role of oxidative metabolism in the killing of *Candida* blastoconidia by macrophages was confirmed using scavengers of toxic oxygen metabolites. Candidacidal activity of resident or activated macrophages is inhibited effectively by superoxide dismutase, which removes O_2^-, and less effectively but significantly by catalase, which removes H_2O_2 (Lehrer, 1970; Sasada and Johnson, 1980).

Cryptococcus can be killed intracellularly by macrophages; however, killing depends on serum opsonins, both immunoglobulin and complement components, and activation of the macrophages. Encapsulated organisms require anticapsular antibodies, as normal serum IgG is not adequate for opsonization and subsequent engulfment by macrophages (Kozel *et al.*, 1984; Ikeda *et al.*, 1984). Moreover, only if encapsulated cryptococci are opsonized with anticapsular IgG or complement can peritoneal macrophages kill these yeasts (Levitz and DiBenedetto, 1988).

If the encapsulated yeasts are not opsonized with specific IgG, the macrophages must be activated with recombinant gamma interferon for significant killing of the yeasts. Moreover, an additional nonadherent cell, probably a lymphocyte, is necessary for killing in this system.

In contrast to the results of peritoneal macrophages, cerebrospinal fluid macrophages are unable to inhibit growth of *Cryptococcus neoformans* although they are activated to produce increased amounts of H_2O_2 (Perfect *et al.*, 1988). Results of several studies suggest that the respiratory burst does not play a primary role in killing of phagocytized *C. neoformans* (Levitz and DiBenedetto, 1988). However, oxidative mechanisms may still play a role in some situations by damaging organisms and thus making them more susceptible to nonoxidative damage.

3.2.2. Peritoneal Macrophage and Peripheral Blood Monocyte Defense against Hyphae

Normal human peripheral blood monocytes can damage and apparently kill *Rhizopus* and *Aspergillus* hyphae by oxidative microbicidal systems, even though the hyphae are too large to be ingested completely (Diamond *et al.*, 1982,1983) Neither complement nor immunoglobulin is required for damage to the hyphae. By the use of oxygen metabolite inhibitors, it would appear that the $MPO-H_2O_2$–halide system is of primary importance in damaging both *Aspergillus* and *Rhizopus* hyphae. Monocytes from patients with CGD are unable to damage *Rhizopus* hyphae, although they can damage *Aspergillus* hyphae.

Hyphal damage by monocytes from patients with complete hereditary MPO deficiency presumably occurs because of alternative oxidative or nonoxidative antihyphal mechanisms (Diamond *et al.*, 1982). These differences in ability to damage *Rhizopus*, but not *Aspergillus*, hyphae also may explain, in part, the predisposition of these patients for aspergillosis but not mucormycosis. *In vivo*, alternative mechanisms to the MPO system inducing hyphal damage are particularly relevant because granule-associated MPO is lost during differentiation of monocytes into macrophages. This suggests that oxidative microbicidal products independent of the MPO system may be important (Lehrer, 1978).

That *Aspergillus* hyphae are damaged by monocytes from CGD patients indicates that, at least under some conditions, there exists the potential for activity of nonoxidative antihyphal mechanisms (Diamond *et al.*, 1983) Oxygen-independent anti-*Aspergillus* hyphal activity of CGD monocytes is inhibited by polyanions but not by polycations of similar size and charge density. Whole cell lysates and separated granule fractions from CGD monocytes cause hyphal damage which also is inhibitable by polyanions. In contrast, comparable fractions from normal monocytes damage hyphae only in the presence of additional halide and an H_2O_2-generating system. Therefore, potentially fungicidal cationic proteins may be present in monocytes from CGD patients. Such cationic proteins might have microbicidal effects by themselves. Even if present in suboptimal levels for fungicidal activity, cationic proteins can interact synergistically with oxidative mechanisms.

In contrast to *Aspergillus*, nonoxidative mechanisms, including granule-associated cationic proteins, do not appear to be involved in damage of *Rhizopus* hyphae by monocytes (Diamond *et al.*, 1982). Separated monocyte lysosomal granules also do not effect *Rhizopus* hyphal viability unless an H_2O_2 source and halides are added to supplement granule-associated MPO.

Normal human blood monocytes also attach to and damage *Candida albicans* hyphae, inducing a reduction in radiolabeled cytosine or uracil uptake by hyphae, as evidenced by light and electron microscopy (Diamond *et al.*, 1982; Diamond and Haudenschild, 1981). Again, it would appear that the $MPO-H_2O_2$–halide system is most important, as monocytes from patients with CGD are unable to damage *C. albicans* hyphae. In contrast, monocytes from patients with hereditary MPO deficiency can damage *Candida* hyphae (Diamond and Haudenschild, 1981). Therefore, MPO-independent mechanisms may play a role in damage to *Candida* hyphae by monocytes in addition to oxidative mechanisms. It has been speculated that the loss in oxidative antifungal mechanisms may be compensated for by an increase in other mechanisms including certain enzymes, such as alkaline phosphatase (Larrocha *et al.*, 1982).

3.2.3. Peritoneal Macrophage and Peripheral Blood Monocyte Defense against Dimorphic Forms

Monocytes and macrophages have been shown to be important cells in host defense against blastomycosis. Peripheral blood monocytes and monocyte-derived macrophages can kill *Blastomyces* conidia (Drutz and Frey, 1985), while peripheral blood monocyte monolayers and cultured macrophages have been shown to inhibit *B. dermatitidis* yeast replication (Brummer and Stevens, 1982). Mononuclear cells appear to kill the yeast phase of *B. dermatitidis* by the oxygen-dependent $MPO-H_2O_2$–halide and iron–H_2O_2–halide systems (Sugar *et al.*, 1983, 1984).

Since cell-mediated immunity plays a dominant role in the immune reponse to *B. dermatitidis*, and macrophages play an important role in cell-mediated immunity, host defense studies of blastomycosis often use activated or immune macrophages. Peripheral blood monocytes from blastomycosis patients have increased phagocytosis and decreased growth of intracellular yeasts in contrast to cells of normal volunteers (Bradsher *et al.*, 1985, 1987). Similar results are observed with sensitized and nonsensitized macrophages from murine models of blastomycosis (McDaniel and Cozad, 1983). These studies suggest that lymphocytes secrete lymphokines that activate the macrophages to greater microbicidal function. Confirming this hypothesis, peritoneal macrophages activated by recombinant gamma interferon have been shown to kill extracellular *Blastomyces* yeast by as yet undescribed oxygen-independent mechanisms (Brummer and Stevens, 1987; Brummer *et al.*, 1985a)

Like *Blastomyces*, acquired immunity to infection by the intracellular fungus *Histoplasma capsulatum* is also mediated by cells. Mononuclear phagocytes from immunized or nonimmunized animals restrict the intracellular growth of the yeast only when the phagocytes are activated by immunized lymphocytes, by lymphokines from immune lymphocytes, or by recombinant gamma interferon (Wu-Hsieh and Howard, 1984, 1987). Although the activated macrophages can restrict intracellular growth of the yeast, the yeast are not killed.

The mechanism of immunity in paracoccidioidomycosis has yet to be elucidated (for a review see Franco, 1987). Leukocytes from patients infected by *Paracoccidioides brasiliensis* showed phagocytosis and killing of these organisms. Serum from these patients increased the killing of the yeast and was specific for *P. brasiliensis* (Restrepo and Velez, 1975). Murine natural killer-like cells can limit the growth of *P. brasiliensis* and also appear to be increased in number during infection with this organism (Jimenez and Murphy, 1984; Mota *et al.*, 1988)

4. ROLE OF THE POLYMORPHONUCLEAR LEUKOCYTE

4.1. Polymorphonuclear Leukocyte Defense against Conidia, Spores, and Blastoconidia of Opportunistic Fungi

The cause of the significant increase in opportunistic fungal infections observed in the last decade is unknown, but is believed to be the result of an increase in aggressive cytotoxic therapy and its resulting neutropenia (Stahel *et al.*, 1982). In addition to the association of invasive aspergillosis and pulmonary mucormycosis with neutropenia and CGD, data from experimental animal models suggest an important role for the PMN in host defense against these organisms. Normal mice expose to the inhalation of *Aspergillus* conidia or *Rhizopus* spores are resistant to lethal infection (Waldorf *et al.*, 1984a,b; Schaffner *et al.*, 1982; Waldorf and Diamond, 1984). However, an increased susceptibility to fatal pulmonary aspergillosis or mucormycosis develops when animals are suppressed by subcutaneous inoculations of cortisone, lymphoid leukemia, X-irradiation, cytotoxic drugs, fed deficient diets, or diabetes is induced by injection of alloxan or streptozotocin.

Although myelosuppression is the major risk factor for invasive aspergillosis in humans, it causes only a minor deleterious effect on the natural resistance of laboratory animals. The short-term myelosuppression achievable in laboratory animals does not involve the pool of tissue macrophages that has a turnover rate of several months (van Furth, 1970) and thus does not affect the function of the reticuloendothelial system. The roles of myelosuppression and neutropenia become evident only after *Aspergillus* conidia escape from the reticuloendothelial system and start mycelial growth (Schaffner *et al.*, 1982). Similarly, if alveolar macrophages are altered by the induction of diabetes, they can no longer prevent *Rhizopus* spore germination, and the importance of PMN in defense becomes apparent (Waldorf *et al.*, 1984a,b). Neutrophils have a limited ability in preventing spore and conidium germination or killing of these fungal particles. Human PMN *in vitro* do not kill phagocytized resting (metabolically inactive) *Aspergillus* conidia (Schaffner *et al.*, 1982; Levitz *et al.*, 1986; Levitz and Diamond, 1985b). However, when the conidia are incubated to induce swelling prior to germination (and become metabolically active), approximately 30% of the metabolically active conidia are killed by neutrophils (Levitz and Diamond, 1985a). Although in cell-free systems the MPO–H_2O_2–halide system effectively kills *Aspergillus* conidia, this system is ineffective against these conidia in intact PMN. It appears that resting *Aspergillus* conidia fail to trigger the respiratory burst in PMN, in comparison to swollen *Aspergillus* conidia (Levitz and Diamond, 1985a). Unlike the situation with *Coccidioides*, the reduced generation of oxidative products is not due to a defect in phagosome–lysosome fusion.

In addition to not stimulating the respiratory burst, resting *Aspergillus* conidia are more resistant to killing by singlet oxygen, H_2O_2, and HOCl, compared to preincubated conidia (Levitz and Diamond, 1985a). Resting *Aspergillus* conidia and *Rhizopus* spores are also resistant to neutrophil cationic proteins which are capable of damaging swollen conidia and spores (Levitz *et al.*, 1986).

In animal models, *Candida albicans* is efficiently eliminated from the lower respiratory tract, presumably by the influx of neutrophils into lung tissue (Nugent and Onofrio, 1983). *In vitro* data appear to confirm this, although studies aimed at quantitative determination of the candidacidal activities of phagocytes show considerable variations. Neutrophils are able to ingest and kill *Candida* blastoconidia (Richardson and Smith, 1983; Cockayne and Odds, 1984; Antley and Hazen, 1988; for review see Waldorf, 1986). Moreover, the PMN is the most efficient of the leukocyte populations in their candidacidal activity.

Studies of the biochemical killing mechanisms of phagocytes have indicated that several candidacidal substances occur in PMN. The MPO–H_2O_2–halide system, in particular the toxic product HOCl, is lethal for *C. albicans* blastoconidia (Klebanoff, 1980; Wagner *et al.*, 1986). PMN from MPO-deficient patients phagocytize *Candida* but their ability to kill the ingested *Candida* is reduced (Lehrer, 1970). Neutrophils from CGD patients are unable to damage *Candida* blastoconidia.

Alternative candidacidal mechanisms also exist in PMN. Human MPO-deficient PMN kill *C. parapsilosis* and *C. pseudotropicalis* by a mechanism completely independent of MPO, iodination, and H_2O_2 (Lehrer, 1972). It has been shown that *C. albicans* blastoconidia are susceptible to several cationic peptides purified from rabbit granulocytes (Selsted *et al.*, 1985). It is possible that these cationic peptides are an important component of the antifungal defense mechanism of neutrophils.

Like the other opportunistic fungi, a critical element of defense against *Cryptococcus neoformans* appears to be the PMN. After experimental respiratory infection of animals, organisms are rapidly destroyed by an influx of neutrophils, with surviving yeasts generating large capsules. Phagocytosis by PMN of encapsulated cryptococci is dependent on opsonization of the yeast cell by opsonic ligands of the complement cascade (Kozel *et al.*, 1984) The inactive decay product of the C3 component of complement binds to the perimeter of the *Cryptococcus* capsule and is necessary for attachment of PMN prior to their phagocytosis (Kozel *et al.*, 1988) Cryptococci that are phagocytized by human PMN are killed by mechanisms dependent on the generation of H_2O_2 (Diamond *et al.*, 1972) and by nonoxidative mechanisms (Miller and Kohn, 1983).

4.2. Polymorphonuclear Leukocyte Defense against Hyphae

As mentioned above (Section 4.1), the importance of the PMN in the host defense against *Aspergillus* and *Rhizopus*, only becomes obvious after *Aspergillus* conidia and *Rhizopus* spores escape control by the macrophage and monocyte cell lines and start mycelial growth. Once these cells are induced to swell prior to initiation of germination, an impressive ability of PMN to damage these organisms is evident. Human PMN can damage and kill hyphae of *Aspergillus* and *Rhizopus* (Diamond and Clark, 1982; Schaffner *et al.*, 1986). Of importance in damaging hyphae by human PMN are the oxidative microbicidal mechanisms.

In addition to oxidative mechanisms, PMN can also kill hyphae by the action of microbicidal granule substances (Levitz *et al.*, 1986; Selsted *et al.*, 1985). However, antihyphal mechanisms of intact leokocytes do not seem to be identical for *Aspergillus* and *Rhizopus* hyphae. Interactions of several potential oxidative and nonoxidative antihyphal mechanisms, *in vivo*, may define the host's ability to limit fungal infections. These differences may be related to the availability of alternative, nonoxidative mechanisms active against *Rhizopus* and not *Aspergillus* in host PMN. Differing susceptibility of *Aspergillus* and *Rhizopus* hyphae to granule–associated cationic proteins, acting alone or together with other constituents (such as elastase to lysozyme), might explain such a process. When concentrations of oxidative or nonoxidative substances are limiting or suboptimal, interactions of mechanisms may have important implications in activity of defense mechanisms against opportunistic mycoses in the intact host.

Following germ tube formation of blastoconidia, *Candida albicans* characteristically form pseudohyphae which are too large to be ingested by phagocytic cells. Human PMN can damage and kill *C. albicans* (Cockayne and Odds, 1984; Diamond and Clark, 1982; Diamond *et al.*, 1980; for review see Waldorf, 1986). Even in the absence of serum, neutrophils attach to and spread over the surfaces of hyphae, which are too large to be completely phagocytized, and induce morphologic changes and metabolic impairments in the hyphae. Moreover, where *C. albicans* germ tubes and hyphae are completely phagocytized, significantly more hyphal forms are killed than blastoconidia (Cockayne and Odds, 1984). Although both opsonized and unopsonized *Candida* are damaged, there is a delay in intracellular activation of PMN to unopsonized *Candida* hyphae (Levitz *et al.*, 1987). Unopsonized *Candida* hyphae are recognized via ligands which are capable of activating receptors on the PMN. Although unopsonized hyphae do not elicit membrane depolarization, they do elicit a delayed time course of actin polymerization, delayed rises in cystolic calcium and O_2^- release, but no delayed degranulation (Lyman *et al.*, 1987; Kolotila and Diamond, 1988). Of central importance in damaging *Candida* hyphae by PMN are the oxidative mechanisms, including ammonium chloride and sodium hypochlorite or HOCl (Wagner *et al.*, 1986; Levitz *et al.*, 1987; Lyman *et al.*, 1987) in addition to granule–associated cationic proteins (Diamond and Clark, 1982).

4.3. Polymorphonuclear Leukocyte Defense against Dimorphic Forms

Blastomyces dermatitidis typically causes suppurative and granulomatous lesions that usually contain significant numbers of PMN. Interactions between *Blastomyces* and PMN have been investigated in several laboratories. *Blastomyces* conidia are effectively phagocytized and killed by PMN, as are the hyphal forms of this organism (Drutz and Frey, 1985). However, as pointed out by Drutz and Frey (1985), hyphae of this organism are seldom present under conditions encountered in an actual infection. Ingestion of conidia by PMN is augmented by the presence of complement, and killing occurs by predominantly oxidative mechanisms.

In contrast to conidia, the yeast phase of *B. dermatitidis* is generally too large to be ingested by PMN, and is considerably more difficult for PMN to kill. Moreover, virulent strains of *Blastomyces* yeasts are resistant to PMN killing (Drutz and Frey, 1985). The yeasts, when killed, are done so predominantly by extracellular attachment and degranulation, whereas conidia are first ingested, then killed. Attachment to the yeast forms is enhanced by serum and complement.

In a system analogous to that of activated macrophages, *Blastomyces* antigen-elicited and activated PMN have an increased fungicidal activity against *Blastomyces* yeast as compared to nonspecifically elicited PMN (Brummer *et al.*, 1985b; Brummer and Stevens, 1984). In addition, the activated PMN also has an increased respiratory burst to *Blastomyces* yeast. The increased respiratory burst and killing of yeast by antigen-induced PMN appears to be due to H_2O_2 and OH^-, or its metabolites (Brummer *et al.*, 1985b). A similar enhanced killing of *B. dermatitidis* yeast by nonspecifically elicited PMN was seen when the PMN were activated by recombinant gamma interferon (Morrison *et al.*, 1987). These authors suggest that gamma interferon plays a role in the communication between T lymphocytes and PMN with respect to antifungal mechanisms.

Unlike their inactivity against conidia of the opportunistic fungi, human PMN are able to inhibit the growth of arthroconidia of *Coccidioides immitis*, presumably by the action of H_2O_2 on the arthroconidia (Galgiani *et al.*, 1982, 1984, 1988). Confirming these results, PMN from a patient with CGD were unable to inhibit arthroconidia growth.

Circulating PMN from patients with paracoccidioidomycosis have normal levels of phagocytic activity, yet they are unable to kill the ingested *Paracoccidioides brasiliensis* yeast (Goihman-Yahr *et al.*, 1980). Moreover, the more severe the clinical form of the disease, the greater is the failure of intracellular killing of the yeast. Paracoccidioidomycosis patients also have a high frequency of certain histocompatibility antigens, suggesting that genetic factors influence susceptibility and the type of disease seen. Whether this is related to defects in innate defense mechanisms is unknown at this time.

4.4. Chemotaxis of Neutrophils in Response to Fungi

PMN play a major role in host defense against hyphal forms of opportunistic fungi and against arthroconidia of *Coccidioides* and conidia of *Blastomyces*. The mechanisms by which spores and conidia induce mobilization and migration of PMN early in the infectious processes are, therefore, important. *Aspergillus* conidia and *Rhizopus* spores, as stimulators, in the absence of serum, are ineffective in inducing PMN migration *in vitro* (Waldorf and Diamond, 1985). Fresh *A. fumigatus* conidia or *R. oryzae* spores, or hyphae from these organisms, have a reduced serum–activating activity. In contrast, preincubated, swollen conidia or spores do generate serum–derived chemotactic activity for PMN. In addition, *Rhizopus* hyphae can generate chemotactic factors for neutrophils directly (Chinn and Diamond, 1982; Marx *et al.*, 1982). Thus, the normal host apparently responds with a neutrophilic inflammatory response, stimulated by complement activation, only after the initiation of the conidial or spore germination process. As the germination process continues and hyphae are formed, the stimulation of the host's inflammatory response, by the activation of complement, returns to levels comparable to those induced by fresh conidia. However, the hyphae then directly generate a chemotactic factor for PMN.

Blastomyces typically causes suppurative and granulomatous lesion that usually contain significant numbers of PMN. Broth culture filtrates of *B. dermatitidis* stimulated granulocyte chemotactic activity significantly better than did filtrates from *Histoplasma capsulatum* and *Cryptococcus neoformans*, two fungi that usually do not elicit PMN in infected tissue (Sixbey *et al.*, 1979). Yeast cells of *Blastomyces* generate both complement-mediated and serum-independent chemotaxins which induce PMN and monocyte migration (Thurmond and Mitchell, 1984).

Although *P. brasiliensis* activates the complement system through the alternative pathway, these complement components do not appear to play a role in the migration of cells to lesions (Calich *et al.*, 1985). Neutrophils chemotax in response to a soluble macrophage factor released by the interaction of the fungus and peritoneal macrophages. Thus, it is suggested that *P. brasiliensis* initially reacts with alveolar macrophages inducing the release of peptides which attract PMN to the focus, and thereby increasing the local defense (Boscardin *et al.*, 1985).

5. CONCLUSION

When fungi succeed in penetrating the epithelium and establishing themselves deep within the interstitium of the tissues, the phagocytes assume an important defensive role in the mycoses, as is the case for most infectious diseases. Attempts to define the mechanisms of host defense against fungal pathogens have been the focus of intense investigation over the past several years. Although numerous factors may contribute to the early host response to fungi, their fate at the hands of phagocytic leukocytes figures importantly in host resistance.

In summary, several distinct lines of defense against the fungi are known in the normal host. Macrophages kill *Aspergillus* conidia and prevent germination of *Rhizopus* spores. PMN kill *Candida* blastoconidia, *Coccidioides* arthroconidia, and *Blastomyces* conidia. PMN also damage the hyphal forms of *Aspergillus*, *Rhizopus*, and *Candida*, and when activated the yeast form of *Blastomyces*. Monocytes and macrophages, especially when activated, damage, kill, or inhibit growth of *Cryptococcus* and the tissue forms of *Coccidioides*, *Histoplasma*, *Blasotmyces*, and *Paracoccidioides*. Thus, PMN and monocytes or macrophages provide distinctly different components of host defenses against the different forms of the same organism and also against the opportunistic and pathogenic fungi. Moreover, the phagocytic cells appear to damage the different fungal particles by different mechanisms.

Because of the ubiquity of some of the fungi, particularly the opportunistic fungal conidia and spores, clearance of conidia may be mechanical or, in some situations perhaps, analogous to the clearance of damaged host cells, neither inflammatory infiltrates nor immunological responses being essential. Clearance of these spores and conidia probably takes place frequently, and it would be important for the recognition process involved to be subtle and non–specific, since it would be undesirable to use mechanisms that entail unnecessary inflammatory reactions or specific immune reponses. However, when germination of the opportunistic fungi occurs, or if conidia from the pathogenic fungi are inhaled, a more serious threat to the host is seen and the host responds via the PMN and its more potent antimicrobial arsenal. When the tissue forms of the pathogenic fungi emerge, the host again fortifies its antimicrobial arsenal by immunologically activating macrophages and PMN via lymphokines. Thus, the interactions of several potential oxidative and nonoxidative antifungal mechanisms and of several phagocytic cell lines define the ability of the host to limit fungal infections.

6. REFERENCES

Antley, P. P., and Hazen, K. C., 1988, Role of yeast cell growth temperature on *Candida albicans* virulence in mice, *Infect. Immun.* **56**:2884–2890.

Beaman, L., 1987, Fungicidal activation of murine macrophages by recombinant gamma interferon, *Infec. Immun.* **55**:2951–2955.

Beaman, L., and Beaman, B. L., 1984, The role of oxygen and its derivatives in microbial pathogenesis and host defense, *Annu. Rev. Microbiol.* **38**:27–48.

Beaman, L., and Holmberg, C. A., 1980, In vitro response of alveolar macrophages to infection with *Coccidioides immitis*, *Infect. Immun.* **28**:594–600.

Bistoni, F., Vecchiarelli, A., Cenci, E., Puccetti, P., Marconi, P., and Cassone, A., 1986, Evidence for macro-phage–mediated protection against lethal *Candida albicans* infection, *Infect. Immun.* **51**:668–674.

Boscardin, R. N., Brandao, H., and Balla, A., 1985, Bronchoalveolar lavage findings in pulmonary paracoccidi-oidomycosis, *Sabouraudia* **23**:143–146.

Bradsher, R. W., Ulmer, W. C., Marmer, D. J., Townsend, J. W., and Jacobs, R. F., 1985, Intracellular growth and phagocytosis of *Blastomyces dermatitidis* by human immune and nonimmune monocytes–derived macro-phages, *J. Infect. Dis.* **151**:57–64.

Bradsher, R. W., Balk, R. A., and Jacobs, R. F., 1987, Growth inhibition of *Blastomyces dermatitidis* in alveolar and peripheral macrophages from patients with blastomycosis, *Am. Rev. Respir. Dis.* **135**:412–417.

Brummer, E., and Stevens, D. A., 1982, Opposite effects of human monocytes, macrophages, and poly-morphonuclear neutrophils on replication of *Blastomyces dermatitidis* in vitro, *Infect. Immun.* **36**:297–303.

Brummer, E., and Stevens, D. A., 1984, Activation of murine polymorphonuclear neutrophils for fungicidal activity with supernatants from antigen–stimulated immune spleen cell cultures, *Infect. Immun.* **45**: 447–452.

Brummer, E., and Stevens, D. A., 1987, Fungicidal mechanisms of activated macrophages: Evidence for nonoxida-tive mechanisms for killing of *Blastomyces dermatitidis*, *Infect. Immun.* **55**:3221–3224.

Brummer, E., Morrison, C. J., and Stevens, D. A., 1985a, Recombinant and natural gamma interferon activation of macrophages in vitro: Different dose requirements for induction of killing activity against phagocytizable and nonphagocytizable fungi, *Infect. Immun.* **49**:724–730.

Brummer, E., Sugar, A. M., and Stevens, D. A., 1985b, Enhanced oxidative burst in immunologically activated but not elicited polymorphonuclear leukocytes. Correlates with fungicidal activity, *Infect. Immun.* **49**:369–401.

Brummer, E., Hanson, L. H., Restrepo, A., and Stevens, D. A., 1988, In vivo and in vitro activation of pulmonary macrophages by IFN-gamma for enhanced killing of *Paracoccidioides brasiliensis* or *Blastomyces dermati-tidis*, *J. Immunol.* **140**:2786–2789.

Calderone, R. A., and Peterson, R. L., 1979, Inhibition of amino acid uptake and incorporation into *Histoplasma capsulatum*, *J. Reticuloendothel. Soc.* **26**:11–19.

Calich, V. L. G., Vaz, C. A. C., and Burger, E., 1985, PMN chemotactic factor produced by glass-adherent cells in the acute inflammation caused by *Paracoccidioides brasiliensis*, *Br. J. Exp. Pathol.* **66**:57-65.

Chinn, R. Y. W., and Diamond, R. D., 1982, Generation of chemotactic factors by *Rhizopus oryzae* in the presence and absence of serum: Relationship to hyphal damage mediated by human neutrophils and effects of hyperglycemia and ketoacidosis, *Infect. Immun.* **38**:1123–1129.

Cockayne, A., and Odds, F. C., 1984, Interactions of *Candida albicans* yeast cells, germ tubes, and hyphae with human polymorphonuclear leukocytes in vitro, *J. Gen. Microbiol.* **130**:465–471.

Diamond, R. D., and Clark, R. A., 1982, Damage to *Aspergillus fumigatus* and *Rhizopus oryzae* hyphae by oxidative and nonoxidative microbicidal products of human neutrophils in vitro, *Infect. Immun.* **38**:487–495.

Diamond, R. D., and Haudenschild, C. C., 1981, Monocyte mediated serum-independent damage to hyphal and pseudohyphal forms of *Candida albicans* in vitro, *J. Clin. Invest.* **67**:173–182.

Diamond, R. D., Root, R. K., and Bennett, J. E., 1972, Factors influencing killing of *Cryptococcus neoformans* by human leukocytes in vitro, *J. Infect. Dis.* **125**:367–376.

Diamond, R. D., Clark, R. A., and Haudenschild, C. C., 1980, Damage to *Candida albicans* hyphae and pseudohyphae by the myeloperoxidase system and oxidative products of neutrophil metabolism in vitro, *J. Clin. Invest.* **66**:908–917.

Diamond, R. D., Haudenschild, C. C., and Erickson, N. F. III, 1982, Monocyte-mediated damage to *Rhizopus oryzae* in vitro, *Infect. Immun.* **38**:292–297.

Diamond, R. D., Huber, E., and Haudenschild, C. C., 1983, Mechanisms of destruction of *Aspergillus fumigatus* hyphae mediated by human monocytes, *J. Infect. Dis.* **147**:474–483.

Drutz, D. J., and Frey, C. L., 1985, Intracellular and extracellular defenses of human phagocytes against *Blastomyces dermatitidis* conidia and yeasts, *J. Lab. Clin. Med.* **105**:737–750.

Eissenberg, L. G., and Goldman, W. E., 1987, *Histoplasma capsulatum* fails to trigger release of superoxide from macrophages, *Infect. Immun.* **55**:29–34.

Franco, M., 1987, Host–parasite relationships in paracoccidioidomycosis, *J. Med. Vet. Mycol.* **25**:5–18.

Galgiani, J. N., Hayden, R., and Payne, C. M., 1982, Leukocyte effects on the dimorphism of *Coccidioides immitis*, *J. Infect. Dis.* **146**:56–63.

Galgiani, J. N., Payne, C. M., and Jones J. F., 1984 Human polymorphonuclear leukocyte inhibition of incorpora-tion of chitin precursors into mycelia of *Coccidioides immitis*, *J. Infect. Dis.* **149**:404–412.

Galgiani, J. N., Hewlett, E. L., and Friedman, R. L., 1988, Effects of adenylate cyclase toxin from *Bordetella pertusis* on human neutrophil interactions with *Coccidioides immitis* and *Staphylococcus aureus*, *Infect. Immun.* **56**:751–755.

Goihman-Yahr, M., Essenfeld–Yahr, E., Albornoz, M. C., Yarzabal, L., Gimenez, M. H., San-Kanski, B., and Ocanto, A., 1980, Defect of in vitro digestive ability of polymorphonuclear leukocytes in paracoccidioidomycosis, *Infect. Immun* **28**:557–598.

Granger, D. L., and Perfect, J. R., 1986, Macrophage-mediated fungistasis in vitro: Requirements for intracellular and extracellular cytotoxicity, *J. Immunol.* **136**:672–680.

Ikeda, R., Shinoda, T., Kagaya, K., and Fukazawa, Y., 1984, Role of serum factors in the phagocytosis of weakly or heavily encapsulated *Cryptococcus neoformans* strains by guinea pig peripheral blood leukocytes, *Microbiol. Immunol.* **28**:51–61.

Jimenez, B. E., and Murphy, J. W., 1984, In vitro effects of natural killer cells against *Paracoccidioides brasiliensis* yeast, *Infect. Immun.* **46**:552–558.

Kagaya, K., and Fukazawa, Y., 1981, Murine defense mechanism against *Candida albicans* infection. II. Opsonization, phagocytosis and intracellular killing of *C. albicans*, *Microbiol. Immunol.* **25**:807–818.

Kitz, D. J., Johnson, C. R., Kobayashi, G. S., Medoff, C., and Little, J. R., 1984, Growth inhibition of *Cryptococcus neoformans* by cloned cultured murine macrophages, *Cell. Immunol.* **88**:489–500.

Klebanoff, S. J., 1980, Oxygen metabolism and the toxic properties of phagocytes, *Ann. Intern. Med.* **93**:480–485.

Klebanoff, S. J., 1982, The iron–hydrogen peroxide–iodide cytotoxic system, *J. Exp. Med.* **156**:1262–1269.

Kolotila, M. P., and Diamond, R. D., 1988, Stimulation of neutrophil actin polymerization and degranulation by opsonized and unopsonized *Candida albicans* hyphae and zymosan, *Infect. Immun.*, **56**:2016–2022.

Kolotila, M. P., Smith, W. C., Rogers, A. L., 1987, Candidacidal activity of macrophages from three mouse strains as demonstrated by a new method: Neutral red staining, *J. Med. Vet. Mycol.* **25**:283–290.

Kozel, T. R., and Gotschlich, E. C., 1982, The capsule of *Cryptococcus neoformans* passively inhibits phagocytosis of the yeast by macrophages, *J. Immunol.* **129**:1675–1680.

Kozel, T. R., and Mastroianni, R. P., 1976, Inhibition of phagocytosis by cryptococcal polysaccharide: Dissociation of the attachment and ingestion phases of phagocytosis, *Infect. Immun.* **14**:62–67.

Kozel, T. R., and Pfrommer, G. S. T., 1986, Activation of the complement system by *Cryptococcus neoformans* leads to binding of iC3b to the yeast, *Infect. Immun.* **52**:1–5.

Kozel, T. R., Highison, B., and Stratton, C. J., 1984, Localization on encapsulated *Cryptococcus neoformans* of serum components opsonic for phagocytosis by macrophages and neutrophils, *Infect. Immun.* **43**:574–579.

Kozel, T. R., Pfrommer, G. S., Guerlain, A. S., Highison, B. A., and Highison, G. J., 1988, Strain variation on phagocytosis of *Cryptococcus neoformans*: Dissociation of susceptibility to phagocytosis from activation and binding of opsonic fragments of C3, *Infect. Immun.* **56**:2794–2800.

Kurup, V. P., 1984, Interaction of *Aspergillus fumigatus* spores with pulmonary alveolar macrophages of rabbits, *Immunobiology* **166**:53–61.

Larrocha, C., Fernandez De Castro, M., Fontan, G., Viloria, A., Fernandez–Chacon, J., and Jimenez, C., 1982, Hereditary myeloperoxidase deficiency: Study of 12 cases, *Scand. J. Haematol.* **29**:389–397.

Lehrer, R. I., 1970, Measurement of candidacidal activity of specific leukocyte types in mixed cell populations. I. Normal, myeloperoxidase deficient, and chronic granulomatous disease neutrophils, *Infect. Immun.* **2**:41–47.

Lehrer, R. I., 1972, Functional aspects of a second mechanism of candidacidal myeloperoxidase-linked and myeloperoxidase–independent candidacidal mechanisms, *J. Clin. Invest.* **55**:338–346.

Lehrer, R. I., 1978, Metabolism and microbicidal function, *Ann. Intern. Med.* **88**:79–91.

Levitz, S. M., and Diamond, R. D., 1984, Killing of *Aspergillus fumigatus* spores and *Candida albicans* yeast phase by the iron–hydrogen peroxide–iodide cytotoxic system: Comparison with the myeloperoxidase–hydrogen peroxide–halide system, *Infect. Immun.* **43**:1100–1102.

Levitz, S. M., and Diamond, R. D., 1985a, Mechanisms of resistance of *Aspergillus fumigatus* conidia by neutrophils in vitro, *J. Infect. Dis.* **152**:33–42.

Levitz, S. M. and Diamond, R. D., 1985b, A rapid colorimetric assay of fungal viability with the tetrazolium salt MTT, *J. Infect. Dis.* **152**:938–945.

Levitz, S. M., and DiBenedetto, D. J., 1988, Differential stimulation of murine resident peritoneal cells by selectively opsonized encapsulated and acapsular *Cryptococcus neoformans*, *Infect. Immun.* **56**:2544–2551.

Levitz, S. M., Selsted, M. E., Ganz, T., Lehrer, R. I., Diamond, R. D., 1986, In vitro killing of spores and hyphae of *Aspergillus fumigatus* and *Rhizopus oryzae* by rabbit neutrophil cationic peptides and bronchoalveolar macrophages, *J. Infect. Dis.* **154**:483–489.

Levitz, S. M., Lyman, C. A., Murata, T., Sullivan, J. A., Mandell, G. L., and Diamond, R. D., 1987, Cytosolic calcium changes in individual neutrophils stimulated by opsonized and unopsonized *Candida albicans* hyphae, *Infect. Immun.* **55**:2783–2788.

Lundborg, M., and Holma, B., 1972, In vitro phagocytosis of fungal spores by rabbit lung macrophages, *Sabouraudia* **10**:152–156.

Lyman, C. A., Simons, E. R., Melnick, D. A., and Diamond, R. D., 1987, Unopsonized *Candida albicans* hyphae stimulate a neutrophil respiratory burst and a cystosolic calcium flux without membrane depolarization, *J. Infect. Dis.* **156**:770–776.

McDaniel, L. S., and Cozad, G. C., 1983, Immunomodulation by *Blastomyces dermatitidis*: Functional activity of murine peritoneal macrophages, *Infect. Immun.* **40**:733–740.

Marx, R. D., Forsyth, K. R., and Hentz, S. K., 1982, Mucorales species activation of a serum leukotactic factor, *Infect. Immun.* **38**:1217–1222.

Merkow, L. L., Prado, M., Epstein, S. M., Verney, E., and Sidransky, H., 1968, Lysosomal stability during phagocytosis of *Aspergillus flavus* spores by alveolar macrophages of cortisone-treated mice, *Science* **160**:79–81.

Merkow, L. P., Epstein, S. M., Sidransky, H., Verney, E., and Prado, M., 1971, The pathogenesis of experimental pulmonary aspergillosis. An ultrastructural study of alveolar macrophages after phagocytosis of *Aspergillis flavus* spores in vivo, *Am. J. Pathol.* **62**:57–74.

Meyer, R. D., and Armstrong, D., 1973, Mucormycosis-changing status, *CRC Crit. Rev. Clin. Lab. Sci.* **4**:421–451.

Miller, G. P. G., and Kohn, S., 1983, Antibody dependent leukocyte killing of *Cryptococcus neoformans*, *J. Immunol.* **131**:1455–1459.

Mor, N., and Goren, M.B., 1987, Discrepancy in assessment of phagosome–lysosome fusion with two lysosomal markers in murine macrophages infected with *Candida albicans*, *Infect. Immun.* **55**:1663–1667.

Morrison, C. J., Brummer, E., Isenberg, R. A., and Stevens, D. A., 1987, Activation of murine polymorphonuclear neutrophils for fungicidal activity by recombinant gamma interferon, *J. Leukocyte Biol.* **41**:434–440.

Mota, N. G. S., Peracoli, M. T. S., Mendes, R. P., Gattass, C. R., Marques, S. A., Soares, A. M. V. C., Izatto, I. C., and Rezkallah–Iwasso, M. T., 1988, Mononuclear cell subsets in patients with different clinical forms of paracoccidioidomycosis, *J. Med. Vet. Mycol.* **26**:105–111.

Nugent, K. M., and Flick, R. B., Jr., 1987, Candidacidal factors in murine bronchoalveolar lavage fluid, *Infect. Immun.* **55**:541–546.

Nugent, K. M., and Onofrio, J. M., 1983, Pulmonary tissue resistance to *Candida albicans* in normal and in immunosuppressed mice, *Am. Rev. Respir. Dis.* **128**:909–914.

Patterson-Delafield, J., Martinez, R. J., and Lehrer, R. I., 1981, Microbiocidal cationic proteins in rabbit alveolar macrophages: Amino acid composition and functional attributes, *Infect. Immun.* **31**:723–731.

Perfect, J. R., Hobbs, M. M., Granger, D. L., and Durack, D. T., 1988, Cerebrospinal fluid macrophage response to experimental cryptococcal meningitis: Relationship between in vivo and in vitro measurements of cytotoxicity, *Infect. Immun.* **56**:849–854.

Restrepo, M. A., and Velez, H., 1975, Effctos de la fagocitosis in vitro sobre el *Paracoccidioides brasiliensis*, *Sabouraudia* **13**:10–21.

Richardson, M. D., and Smith, H., 1983, Ultrastructural features of phagocytosis and intracellular killing of *Candida albicans* by mouse polymorphonuclear phagocyte monolayers, *Mycopathologia* **83**:97–102.

Sasada, M.,and Johnson, R. B., 1980, Macrophage microbicidal activity. Correlation between phagocytosis-associated oxidative metabolism and the killing of *Candida* by macrophages, *J. Exp. Med.* **152**:85–98.

Schaffner, A., 1985, Therapeutic concentrations of glucocorticoids suppress the antimicrobial activity of human macrophages without impairing their responsiveness to gamma interferon, *J. Clin. Invest.* **76**:1755–1764.

Schaffner, A., Douglas, H., and Braude, A. I., 1982, Selective protection against conidia by mononuclear and against mycelia by polymorphonuclear phagocytes in resistance to *Aspergillus*. Observations on these two lines of defense in vivo and in vitro with human and mouse phagocytes, *J. Clin. Invest.* **69**:617–631.

Schaffner, A., Douglas, H., Braude, A. I., and Davis, C. E., 1983, Killing of *Aspergillus* spores depends on the anatomical source of the macrophage, *Infect. Immun.* **42**:1109–1115.

Schaffner, A. Davis, C. E., Schaffner, T., Market, M., Douglas, H., and Braude, A. I., 1986, In vitro susceptibility of fungi to killing by neutrophil granulocytes discriminates between primary pathogenicity and opportunism, *J. Clin. Invest.* **78**:511–524.

Selsted, M. F., Szklarek, D., Ganz, T., and Lehrer, R. I., 1985, Activity of rabbit leukocyte peptides against *Candida albicans*, *Infect. Immun.* **49**:202–206.

Sixbey, J. W., Fields, B. T., Sun, C. N., Clark, R. A., and Nolan, C. M., 1979, Interactions between human granulocytes and *Blastomyces dermatitidis*, *Infect. Immun.* **23**:41–44.

Smith, J. M., 1976, In vivo development of spores of *Absidia ramosa*, *Sabouraudia* **14**:11–15.

Stahel, R. A., Vogt, P., Schuler, G., Ruttner, J. R., Frick, P., and Oelz, O., 1982, Systemic fungal infections in hematological malignancies; a growing problem, *J. Infect.* **5**:269–275.

Sugar, A. M., Chahal, R. S., Brummer, E., and Stevens, D. A., 1983, Susceptibility of *Blastomyces dermatitidis* to products of oxidative metabolism, *Infect. Immun.* **41**:908–912.

Sugar, A. M., Chahal, R. S., Brummer, E., and Stevens, D. A., 1984, The iron–hydrogen peroxide–iodide system is fungicidal: Activity against the yeast phase of *Blastomyces dermatitidis*, *J. Leukocyte Biol.* **36**:545–548.

Sugar, A. M., Brummer, E., and Stevens, D. A., 1986, Fungicidal activity of murine broncho-alveolar macrophages against *Blastomyces dermatitidis*, *J. Med. Microbiol.* **21**:7–11.

Thurmond, L. M., and Mitchell, T. G., 1984, *Blastomyces dermatitidis* chemotactic factor: Kinetics of production and biological characterization evaluated by a modified neutrophil chemotaxis assay, *Infect. Immun.* **46**:87–93.

van Furth, R., 1970, The origin and turnover of promonocytes, monocytes and macrophages in normal mice, in *Mononuclear Phagocytes* (R. van Furth, ed.), Blackwell, Oxford, pp. 151–190.

Wagner, D.K., Collins-Lech, C., and Sohnle, P. G., 1986, Inhibition of neutrophil killing of *Candida albicans* pseudohyphae by substances which quench hypochlorous acid and chloramines, *Infect. Immun.* **51**:731–735.

Waldorf, A. R., 1986, Host–parasite relationship in opportunistic mycoses, *CRC Crit. Rev. Microbiol.* **13**:133–172.

Waldorf, A. R., and Diamond, R. D.,1984, Cerebral mucormycosis in diabetic mice after intrasinus challenge, *Infect. Immun.* **44**:194–195.

Waldorf, A. R., and Diamond, R. D., 1985, Neutrophil chemotactic responses induced by *Rhizopus* spores and *Aspergillus* conidia, *Infect. Immun.* **48**:458–463.

Waldorf, A. R., Halde, C., and Vedros, N. A., 1983, Passive immunization in murine mucormycosis, *Mycopathologia* **83**:149–155.

Waldorf, A. R., Ruderman, N., and Diamond, R. D., 1984a, Specific susceptibility to mucormycosis in murine diabetes and bronchoalveolar macrophage defense against *Rhizopus*, *J. Clin. Invest.* **74**:150–160.

Waldorf, A. R., Levitz, S. M., and Diamond, R. D., 1984b, In vivo bronchoalveolar macrophage defense against *Rhizopus oryzae* and *Aspergillus fumigatus*, *J. Infect. Dis.* **150**:752–760.

Washburn, R. G., Gallin, J. I., and Bennett, J. E., 1987, Oxidative killing of *Aspergillus fumigatus* proceeds by parallel myeloperoxidase–dependent and independent pathways, *Infect. Immun.* **55**:2088–2092.

Wu-Hsieh, B., and Howard, D. H., 1984, Inhibition of growth of *Histoplasma capsulatum* by lymphokine-stimulated macrophages, *J. Immun.* **132**:2593–2597.

Wu-Hsieh, B., and Howard, D. H., 1987, Inhibition of the intracellular growth of *Histoplasma capsulatum* by recombinant murine gamma interferon, *Infect. Immun.* **55**:1014–1016.

21

Suppression of Phagocytic Cell Responses by Conidia and Conidial Products of Aspergillus fumigatus

Maura D. Robertson

1. INTRODUCTION

Out of the 115 recognized species and varieties of aspergilli, described by Raper and Fennell (1965), the one most commonly associated with disease in both man and animals is *Aspergillus fumigatus*. Our understanding of why *A. fumigatus* is a problem to certain individuals is very limited although it has been suggested that conidia of *A. fumigatus* may be particularly resistant to host defense network.

In this chapter I intend to focus on the interaction of conidia and conidial products of *A. fumigatus* with phagocytic cells. I will try to highlight possible areas where mechanisms of resistance by the conidia, to this component of the early host defense system, may be operating.

Conidia of A. fumigatus—Growth and Dispersal

Although the natural habitat of *A. fumigatus* is the soil, using decaying organic matter as a source of food, its ability to grow over a temperature range from below 20°C up to 50°C enables it to adapt to many different situations (Raper and Fennell,1965). Conidia of *A. fumigatus* are generally 2.5–3 μm in diameter and are chiefly liberated by bending and breaking of conidium-covered stems and leaves and during handling or eating of hay or straw.

There is a seasonal variation in the incidence of conidia of *A. fumigatus* in the atmosphere which tends to correlate with the presence of decaying organic matter. Generally, conidium counts of *A. fumigatus* are low, accounting for approximately 6% of the total fungi present in the air spora, with increased levels occurring during the winter months (Lacey, 1981; Mullins *et al.*, 1976; Vernon and Allan, 1980; Noble and Clayton, 1963; Solomon and Burge, 1975)

Maura D. Robertson • Institute of Occupational Medicine, Edinburgh EH8 9SU, Scotland.

2. PATHOGENICITY OF A. FUMIGATUS

Aspergillus illustrates particularly well the importance of the two factors in pathogenicity,the inherent properties of the organisms and the defenses of the individual host.

2.1. Pathogenicity in Animals

Conidia of *A. fumigatus* act as primary pathogens in many animals, particularly those that feed on hay, straw, and grain. A number of years ago, it was shown that high levels of airborne conidia cause acute respiratory distress in various species of birds including newly hatched chicks (Bardana, 1980; Austwick, 1965: McDiarmid, 1955) and penguins (Appleby, 1962). Austwick (1963) showed that high concentrations of *A. fumigatus* in moldy straw may cause the death of adult poultry within 24 hr. Respiratory disease caused by *Aspergillus* has also been found in young lambs (Austwick *et al.*, 1960) and piglets, (Austwick, 1965) and can cause mycotic abortion in cattle (Austwick and Venn, 1962). A survey of the frequency of lesions containing *A. fumigatus* in the lungs of 49 healthy dairy cows at slaughter found that 36 of them showed evidence of *A. fumigatus*-related lesions (Austwick, 1962).

2.2. Pathogenicity in Man

Conidia of *A. fumigatus* rarely act as primary pathogens in man. Instead, they have the unusual ability to manifest themselves in a variety of ways, especially in people with underlying disease (Pepys, 1981). In common with many other fungi, conidia of *A. fumigatus* can be allergenic (Lacey, 1981) provoking exacerbations of asthma especially during high seasonal concentrations of mold sporulation (Radin *et al.*, 1983; Beaumont *et al.*, 1984). Some of these asthmatic patients go on to develop a hypersensitivity reaction, allergic bronchopulmonary aspergillosis (Hinson *et al.*, 1952). Patients with cystic fibrosis are also at risk of developing bronchial colonization by *A. fumigatus*; some of these patients show evidence of allergic bronchopulmonary aspergillosis (Mearns *et al.*, 1967; Laufer *et al.*, 1984). Inhalation of large numbers of airborne conidia by both atopic and nonatopic subjects can result in the development of hypersensitivity pneumonitis (allergic alveolitis) (Pepys, 1969; Seaton and Morgan, 1984). Conidia may also grow saprophytically within old pulmonary cavities to form an aspergilloma (Orie *et al.*, 1960).

Finally, the conidium may act as a true opportunistic pathogen taking advantage of the presence of local lung disease, such as infarct, tumor, pneumonia, or of a generalized impairment of the host defense network to become either localized invasive or disseminated disease. *A. fumigatus* is one of the commonest opportunistic pathogens found in the lungs of immunosuppressed patients. Patients on high-dose corticosteroids and those with either defective granulocyte function or neutropenia are particularly at risk from invasive aspergillosis (Warren and Warnock, 1982; Young *et al.*, 1972).

In common with other fungi, *A. fumigatus* produces a wide range of substances including some with mycotoxic and antibiotic properties (Wilson, 1971; Trivedi and Rao, 1979; McCowen *et al.*, 1951).

3. ASPERGILLUS FUMIGATUS IN THE LUNG

The normal portal of entry of *A. fumigatus* into the body is the lung (Marsh *et al.*, 1979). Conidia of *A. fumigatus* can, because of their small size (< 5 μm) and aerodynamic properties, bypass the upper respiratory tract defenses and may reach distal regions of the lung (Lacey, 1981;

Morgan, 1984). In this region the host defenses rely on phagocytic cells to remove the conidia efficiently (Gee and Khandwala, 1977; DuBois, 1985). If, however, the phagocytic cells are unable to clear the conidia quickly, germination and fungal colonization may occur (Domer and Carrow, 1983). A possible relationship between isolation of A. fumigatus and lung disease was shown in 1959 by Pepys et al., who cultured conidia of A. fumigatus more frequently from the sputa of asthmatic patients than from the sputa of people with other lung diseases. Mullins and Seaton (1978), who were able to show that conidia of A. fumigatus were recovered from postmortem lung specimens more frequently than would have been anticipated from their prevalence in the air spora, suggested that this fungus may have a specific ability to resist the natural defenses of the lung.

Experimental Animal Studies

Experimental animal studies have also shown that conidia of A. fumigatus appear to be fairly resistant to the normal host defenses of the lung (Thurston et al., 1979; Smith, 1972; Williams et al., 1981; Lehman and White, 1976). Treatment of animals with corticosteroids prior to challenge with conidium of A. fumigatus, substantially impairs their host defenses and renders them highly susceptible to disseminated infection (White, 1977). Sandhu and co-workers (1970) exposed mice, by inhalation, to seven different species of Aspergillus (A. fumigatus, A. flavus, A. tamarii, A. nidulans, A. nidulans var. dentatus, A. niger, A. terreus) and found that corticosteroid treatment lowered the host resistance to all seven species. However, A. fumigatus was the most pathogenic causing the highest mortality in the corticosteroid group (13 out of 15) as well as in the untreated group (4 out of 15).

A study on the pathogenesis of experimental pulmonary aspergillosis in normal- and cortisone-treated rats by Turner and co-workers (1976) showed that rats were resistant to infection following intratracheally administered conidia of A. fumigatus even though they developed a subacute interstitial pneumonia. Although the lesions were more severe in animals whose inflammation was suppressed with corticosteroids, there was no evidence of hyphal growth in the pulmonary tissue. They postulated that phagocytosis followed by digestion of conidium, was of paramount importance in the control of fungal infections (Turner et al., 1976).

4. PHAGOCYTIC CELLS VERSUS ASPERGILLUS FUMIGATUS

The importance of the phagocytic cell in the eradication of conidia of A. fumigatus was highlighted in 1897 by Rénon who injected conidia of A. fumigatus into the peritoneal cavities of guinea pigs and rabbits; for comparison he used conidia of A. niger which he considered to be nonpathogenic. After 3 hr. he examined the peritoneal exudates and found that the majority of conidia of A. niger were then within the leukocytes. However, the majority of conidia of A. fumigatus were still lying free within the exudate (Fig. 1). Rénon also demonstrated that conidia of A. fumigatus were still present though not germinating in the lymphatic sac of frogs 35 hr. after injection. Rénon suggested that the "poor response" of the leukocytes was the reason for the pathogenicity of A. fumigatus.

In the 90 years since the pioneering work of Rénon, our knowledge of the role that phagocytic cells play in the eradication of A. fumigatus has only recently begun to progress. The published studies on the ability of phagocytic cells to kill conidia of A. fumigatus have produced divergent results. Kurup (1981) found that rabbit alveolar macrophages killed 80% of conidia of A. flavus, 60% of A. fumigatus, and only 30% of of A. niger following a 2 hr. incubation in vitro. Schaffner and co-workers (1983), who compared the ability of phagocytic cells, from different anatomical sites, to kill the three fungi, showed that alveolar macrophages from mice and rabbits

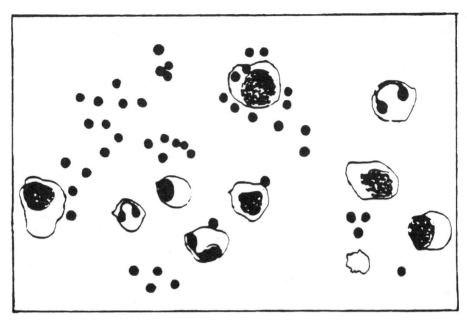

Figure 1.   A copy of the original illustration made by Rénon in 1897 of the interaction between conidia and cells, found in the peritoneal exudate which was obtained from guinea pigs 3 hr after receiving an intraperitoneal injection of conidia of *A. fumigatus*. The majority of conidia are lying free in the exudate.

were more efficient than peritoneal macrophages in the prevention of germination of all three aspergilli. Schaffner and co-workers also found that the alveolar macrophages from rabbits killed 82–90% of conidia of *A. fumigatus* within 30 hr. while, in contrast, the ability of peritoneal macrophages to inhibit germination of the ingested conidia was substantially less. The results of another study comparing natural immunity to *A. fumigatus in vivo* with the action of phagocytes against the organism *in vitro*, led Schaffner and co-workers (1982) to suggest that monocytes may be concerned with killing conidia while PMN may be responsible for the eradication of hyphae, apparently in the absence of a specific immune response. Diamond and co-workers (1978, 1983) have also shown that PMN as well as monocytes can damage hyphae. Lehrer and Jan (1970) found that even after incubation for 3 hr with human monocytes and PMN, the phagocytosed opsonized conidia of *A. fumigatus* remained viable. Experiments *in vitro* by Kurup (1984) showed that pulmonary alveolar macrophages from normal rabbits were relatively ineffective at killing conidia of *A. fumigatus* in 4 hr when compared with the high percentage of killed conidia of *A. flavus* and *A. niger* (18.9% compared with 90 and 82%, respectively). Waldorf and co-workers (1984) showed that alveolar macrophages of mice challenged *in vivo* continued to kill conidia of *A. fumigatus* when cultured *in vitro*. The results of two case reports have shown that phagocytic cells obtained from patients with recurrent *Aspergillus* infections were unable to kill conidia of *A. fumigatus* efficiently (Pagani *et al.*, 1981; Fietta *et al.*, 1984). This led Fietta and co-workers (1984) to suggest that oxygen-independent mechanisms could play a basic role in *Aspergillus* killing.

In order to give us a clearer understanding of what might be happening between conidia of *A. fumigatus* and phagocytic cells, it is perhaps important to go back and dissect (in a fairly simplified manner) the stages which occur during the phagocytic process. The process of phagocytosis can for ease of description be arbitrarily divided into four phases:

- Recognition and Chemotaxis
- Attachment
- Ingestion
- Killing

4.1. Recognition and Chemotaxis

Phagocytic cells have the capacity to discriminate between foreign and self materials as well as alterations to self. When phagocytic cells, first encounter foreign microorganisms, whether by chance or by specific attraction to the cell surface of the microorganism, they are able to secrete or initiate the production of mediators which can mobilize other phagocytic cells to the site of inflammation (Wilkinson, 1976). One of the most potent chemotactic factors generated during phagocyte activation is the complement component C5a (Whaley, 1985). Many microorganisms (including *A. fumigatus*) have been shown to directly stimulate the production of chemotactic factors from the complement system as well as the fibrinolytic and kinin generating systems (Marx and Flaherty, 1976; Edwards, 1976; Gallin, 1976).

4.2. Attachment

The attachment process involves the apposition of the phagocytic cell membrane and the surface of the microorganism. Attachment can occur in the absence of cellular metabolism and is dependent upon the interplay of surface forces (surface tension, surface free energy), cell surface receptors, structural properties of the plasma membrane, as well as the molecular nature of the microorganisms (Stendahl *et al.*, 1981; Absolom, 1986). *A. fumigatus* which has been found to activate complement directly via the alternate pathway may become coated with C3b which could perhaps result in the attachment of the conidia to the phagocytic cell in the absence of specific antibody (Marx and Flaherty,1976; Edwards, 1976).

The ability of phagocytes to adhere to various surfaces in the absence of antibody or complement is a well-known phenomenon (Karnovsky *et al.*, 1975; Weir and Ögmundsdöttir, 1977). Lectinlike receptors which can recognize carbohydrates in the surfaces of microorganisms are thought to be part of a more primitive nonspecific system of recognition (Weir *et al.*, 1979) and it has recently been shown that conidia of *A. fumigatus* can bind to phagocytic cells via these particular receptors (Kan and Bennett, 1988).

Robertson and co-workers (1987c) looked at the ability of monocytes and PMN obtained from six control human subjects, to become cell-associated with conidia of *A. fumigatus* which had been opsonized in 5% autologous serum. They found that after incubation *in vitro* for 1 hr at 37°C, more than 80% of conidia of *A. fumigatus* became cell-associated with both human monocytes and PMN. It was not possible, using light microscopy, to distinguish between attached and ingested conidia.

Reactive Oxygen Intermediates

During attachment of the conidia to the surface of the cell, perturbation of the plasma membrane may induce a phagocytic cell to undergo a respiratory burst with the production of substantial quantities of superoxide anion and hydrogen peroxide. These products may interact to form a variety of short-lived oxygen species (e.g., hydroxyl radical, singlet oxygen) and longer-lived oxidants (e.g., hypochlorous acid, monochloramine). These reactive oxygen intermediates constitute key components of the oxygen-dependent antimicrobial and cytocidal mechanisms of phagocytic cells (Fantone and Ward, 1982; Karnovsky and Badwey, 1986).

The potential importance of reactive oxygen intermediates and of hydrogen peroxide in particular for the eradication of conidia and hyphae has been demonstrated with cell-free systems of myeloperoxidase–hydrogen peroxide–iodide (Lehrer, 1969; Diamond and Clark, 1982) and ferrous iron–hydrogen peroxide–iodide (Levitz and Diamond, 1984) both of which are capable of killing *A. fumigatus*. Although Lehrer and Jan (1970) found that the myeloperoxide–hydrogen peroxide–iodide system was efficient at killing conidia of *A. fumigatus in vitro*, they also showed that conidia of *A. fumigatus* were much more resistant to the myeloperoxidase–hydrogen peroxide–chloride system. As chloride is much more readily available to phagocytic cells than iodide, they suggested that the relative ineffectiveness of chloride in the system could underlie the apparent inability of human neutrophils to kill ingested conidia of *A. fumigatus*.

Washburn and co-workers (1987) have looked at the relative importance of reactive oxygen intermediates, in the killing of conidia of *A. fumigatus*, by monocytes from control subjects and from patients with either chronic granulomatous disease or myeloperoxidase deficiency. By selectively adding to their test system substances which scavenge reactive oxygen intermediates (catalase, taurine, mannitol) and also a hydrogen peroxide generating system (glucose–glucose oxidase), they found that the myeloperoxidase–hydrogen peroxide–halide system and a myeloperoxidase-independent pathway, probably the ferrous ion–hydrogen peroxide–halide system, each appear to contribute significantly to the oxidative killing of conidia of *A. fumigatus*.

In an attempt to identify the mechanisms used by this fungus to resist killing by phagocytic cells, Robertson and co-workers (1987b) have examined the effect of conidia of *A. fumigatus* on the ability of the cell to produce reactive oxygen intermediates. The chemiluminescent responses of human PMN to opsonized conidia of *A. fumigatus* were initially looked at (Robertson *et al.*, 1988) and compared with the chemiluminescent responses of PMN to a positive control zymosan and conidia of the nonpathogenic fungus *Penicillium ochrochloron* which are of a similar size to the conidia of *A. fumigatus* (Fig. 2). It was found that conidia of *A. fumigatus* elicited significantly lower chemiluminescent response by human PMN when compared with either zymosan or *P. ochrochloron* ($p < 0.001$). Robertson and co-workers (1987b) then looked at the release by

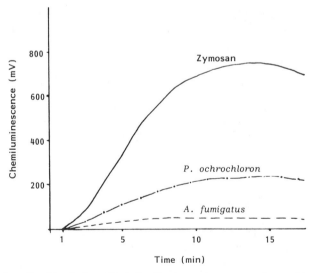

Figure 2. Typical profile of luminol-amplified chemiluminescent responses of normal human PMN to the following triggers, opsonized in human AB sera: zymosan (5 mg), conidia of *A. fumigatus*, and conidia of *P. ochrochloron* at a conidial/cell ratio of 10:1; continuously measured (in mV) over a 20 min period.

phagocytic cells from mice and rats of the specific reactive oxygen intermediates superoxide anion and hydrogen peroxide in response to challenge by conidia of *A. fumigatus* and *P. ochrochloron*. *Corynebacterium parvum*-stimulated mouse peritoneal exudate cells, which are known to spontaneously release high levels of reactive oxygen intermediates, released significantly less superoxide anion ($p < 0.001$) in response to conidia of *A. fumigatus* than in response to the zymosan control or conidia of *P. ochrochloron* (see Fig. 3). Both zymosan and *P. ochrochloron* conidia were associated with a significant increase in release of superoxide anion that was significantly greater than the spontaneous release by cells ($p < 0.001$). Both zymosan and *P. ochrochloron* slightly reduced the spontaneous release of hydrogen peroxide by *C. parvum*-stimulated mouse peritoneal exudate cells (see Fig. 4). Conidia of *A. fumigatus* produced a reduction of hydrogen peroxide release with increasing conidium/cell ratios. When the results from the individual conidium/cell ratios were combined, the release of hydrogen peroxide was significantly lower in response to *A. fumigatus* ($p < 0.001$) than to zymosan or *P. ochrochloron*. A similar trend has also been reported by Levitz and Diamond (1985) who found that conidia of *A. fumigatus* induced human PMN to release significantly less superoxide anion, hydrogen peroxide, and hypochlorous acid that did *Candida albicans* or zymosan. However, the amount of superoxide anion generated by PMN in response to conidia of *A. fumigatus* in the study by Levitz and Diamond (1985) was still significantly more than the spontaneous release by PMN alone.

4.3. Ingestion

Attachment of microorganisms via specific receptors can trigger the cell to undergo the process of ingestion which is highly complex and involves a coordinated interaction of the plasma membrane with contractile elements in the cytoplasm (Zuckerman and Douglas, 1979; Yin and Stossel, 1982). During ingestion of microorganisms, phagocytic cells release many antimicrobial substances including reactive oxygen intermediates and lysosomal enzymes (Fantone *et al.*, 1987; Ögmundsdóttir and Weir, 1980; Silverstein *et al.*, 1977).

Ingestion is often an important step in the killing of most microorganisms and it has often been assumed that conidia of *A. fumigatus* are (ingested) phagocytosed prior to being killed (Kurup, 1981; Schaffner, *et al.*, 1982). However, Robertson and co-workers (1987c) have shown that conidia of *A. fumigatus* are relatively resistant to ingestion by human and mouse phagocytes. While initial studies using light microscopy did not allow differentiation of ingested surface-bound conidium, further work using the Nomarski optics revealed that a substantial number of

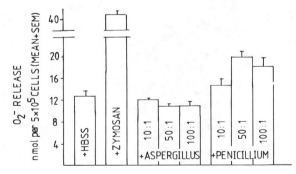

Figure 3. The effect of adding opsonized fungal conidia of *A. fumigatus* and *P. ochrochloron* at increasing conidium/cell ratios on the release of superoxide anion by *C. parvum*-induced mouse peritoneal exudate cells (5×10^5 cells/2 hr). (From Robertson *et al.*, 1987b.)

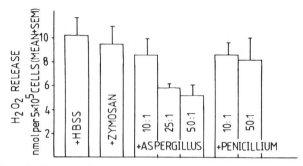

Figure 4. The effect of adding opsonized fungal conidia of *A. fumigatus* and *P. ochrochloron* at increasing conidium/cell ratios on the release of hydrogen peroxide by *C. parvum*-induced mouse peritoneal exudate cells (5 × 10^5 cells/2 hr). (From Robertson *et al.*, 1987b.)

conidia of *A. fumigatus* appeared to remain on the surface of the cell (see Fig. 5). These observations proved difficult to quantitate but were supported by further observations using scanning electron microscopy, which showed that a high proportion of conidia of *A. fumigatus* were bound to the surface of the cells *in vivo* (see Fig. 6), while conidia of the control fungus *P. ochrochloron* were becoming ingested (Fig. 7).

The difficulties involved in discriminating between attached and ingested particles have been discussed by van Furth and Diesselhoff-den Dulk, (1980) who stated that:

> even the application of electron microscopy within sections cannot solve this problem, because particles trapped between folds of the cell membrane can seem in tangential sections to lie inside the cell.

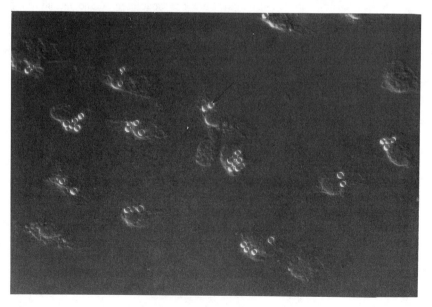

Figure 5. Photomicrographs, using Nomarski differential interference contrast optics, of the cell association of conidia of *A. fumigatus* opsonized in autologous serum with mouse peritoneal exudate cells following incubation at 37°C for 3 hr. Conidia are indicated with arrows; original magnification ×400.

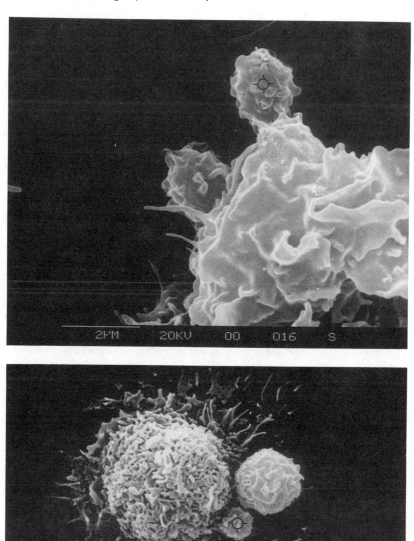

Figure 6. Scanning electron micrographs of the cell association of fungal conidia following the interaction *in vivo* for 1.5 hr of mouse peritoneal exudate cells and conidia of *A. fumigatus*. ○, attached conidia. (From Robertson *et al.*, 1987c.)

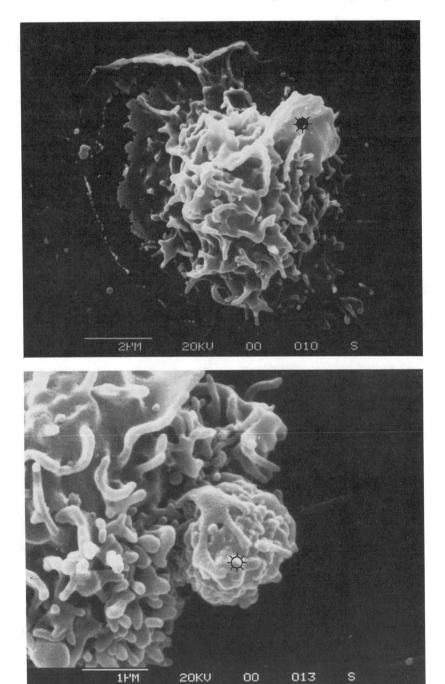

Figure 7. Scanning electron micrographs of the cell-association of fungal conidia following the interaction *in vivo* for 1.5 h of mouse peritoneal exudate cells and conidia of *P. ochrochloron*. ○, conidia partially ingested; ●, fully ingested. (From Robertson *et al.*, 1987c.)

Levitz and Diamond (1985), using transmission electron microscopy of peroxidase-stained sections of conidia of *A. fumigatus* which had been incubated with PMN at a conidium/cell ratio of 10:1 in PBS for 30 min, apparently found that phagosome–lysosome fusion had occurred and that greater than 95% of the cell-associated conidia of *A. fumigatus* were intracellular; it is difficult to assess the validity of these findings since no examples of this work were actually given in the paper. In contrast, the suggestion that conidia of *A. fumigatus* may be resistant to ingestion (Robertson *et al.*, 1985) is supported by the work of Kurup (1984) who, using rat alveolar macrophages *in vitro*, could show no evidence of phagosome–lysosome fusion following phagocytosis of *A. fumigatus*. This may also be the explanation of the finding of Lehrer and Jan (1970) who showed that apparently phagocytosed conidia (as judged by light microscopy) were nevertheless resistant to killing. Prior opsonization of the conidia in normal rabbit serum, rabbit anti-*A. fumigatus* sera, complement, or lung lavage fluid had no real effect on the phagocytosis or killing of the conidia (Kurup, 1984).

4.4. Killing

The assay used by Robertson (1988) to measure killing of conidia of *A. fumigatus* involves incubating the conidia with phagocytic cells for 3 hr at 37°C. The conidia are then disrupted from the cells and the colony-forming units measured (see Fig. 8). The results given in Table I show that conidia of *A. fumigatus* opsonized in autologous sera were significantly more resistant to killing *in vitro* by rat alveolar macrophages, than were similarly opsonized conidia of the control *P.*

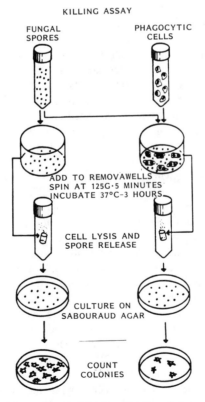

Figure 8. Illustration of the basic procedure used for the estimation of phagocytic killing of fungal conidia.

Table I. Killing of Fungal Conidia,
Opsonized in Autologous Serum,
by Rat Alveolar Macrophages
Incubated at 37°C for 3 hr

	% killing (mean, SD)	
Experiment	A. fumigatus	P. ochrochloron
1	11.5 (0.60)*	16.5 (1.5)
2	25.1 (1.52)*	33.5 (5.5)
3	15.9 (0.78)*	34.2 (0.41)

*Percentage of A. fumigatus killed significantly less than P. ochrochloron ($p < 0.01$).

ochrochloron ($p < 0.01$) even though the number of conidia becoming cell-associated was similar; percentage mean of cell-associated conidia ($\pm$ SEM) : A. fumigatus 86.3 ($\pm$ 3.5), P. ochrochloron 83.6 ($\pm$ 2.9).

Although there are difficulties in comparing the results of one group with those of others because of differences in experimental procedure, such as conidium/cell ratios, incubation periods, and methods of estimating germination, a reasonable amount of agreement can be found. The percentage killing of A. fumigatus found using rat alveolar macrophages is similar to that found by Kurup (1984) who used rabbit alveolar macrophages and by Waldorf and co-workers (1984) who used mice alveolar macrophages. Robertson and co-workers (1987a) also looked at the ability of human alveolar macrophages to kill conidia of A. fumigatus. They found that conidia of A. fumigatus opsonized in 5% pooled normal sera were significantly more resistant to killing by human pulmonary macrophages that were similarly opsonized conidia of P. ochrochloron ($p < 0.02$). Approximately 46% more conidia of P. ochrochloron were killed by the macrophages that were conidia of A. fumigatus. Although Robertson (1988a) found that human alveolar macrophages appeared to be more efficient at killing conidia of A. fumigatus than were rat alveolar macrophages, this may have been related to differences in their state of functional activation. The alveolar macrophages from rats used in the study were obtained from specific-pathogen-free animals and were usually in a resting state whereas the human pulmonary macrophages were obtained from lungs of people (mainly city dwellers) who would have been constantly exposed to normal environmental agents (and would perhaps have been more activated). Previous studies have shown PVG rat alveolar macrophages, which had been activated by quartz inhalation, to be more efficient at killing conidia of A. fumigatus than were alveolar macrophages from the nonstimulated animals (Donaldson et al., 1987).

When Robertson and co-workers (1988b) measured the ability of human monocytes and PMN to kill conidia of A. fumigatus, they found that conidia of A. fumigatus, opsonized in autologous serum, were significantly more resistant to killing by monocytes ($p < 0.025$) and PMN ($p < 0.001$) than were similarly opsonized conidia of the control P. ochrochloron (see Fig. 9). Human monocytes were significantly more efficient than PMN at killing conidia of A. fumigatus ($p < 0.001$) and P. ochrochloron ($p < 0.002$). The results of the study by Lehrer and Jan (1970), who showed that a mixture of monocytes and neutrophils was incapable of killing conidia of A. fumigatus in vitro after a 3-hr incubation, are in contrast to the findings of a study by Robertson and co-workers (1988) where both human PMN and monocytes were found to kill conidia of A. fumigatus. However, they found that PMN were not nearly as efficient as monocytes at killing conidia of A. fumigatus, a finding which is in keeping with the results of Schaffner and co-workers (1982), who previously showed that mononuclear phagocytes were important in the

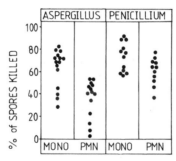

Figure 9.   The percentage of conidia of *A. fumigatus* and *P. ochrochloron*, opsonized in 5% autologous serum, which were killed by human monocytes (MONO) and polymorphonuclear leukocytes (PMN) following incubation at 37°C for 3 hr.

killing of conidia while PMN were more important in the eradication of hyphae. Moreover, it is clear from clinical evidence that PMN play an important part in preventing invasive aspergillosis, in that the disease occurs particularly in patients with neutropenia and chronic granulomatous disease (Fraser *et al.*, 1979; Chusid *et al.* 1981).

When Robertson and co-workers (1987a) opsonized conidia of *A. fumigatis* in serum which had been heat-treated to remove complement (56°C, 30 min), the ability of human lung macrophages to kill conidia of *A. fumigatus* was substantially increased (approximately 74%) when compared with non-heat-treated sera ($p < 0.001$). No increased killing of similarly opsonized conidia of *P. ochrochloron* was found. Thus, instead of promoting killing, the presence of heat-labile serum components actually enhanced the resistance of *A. fumigatus* to attack by phagocytic cells. The other serum component which is destroyed by heating at 56°C for 30 min is IgE. However, as increased killing of *A. fumigatus* was also found using heat-treated serum which did not contain high levels of IgE, it seemed unlikely that IgE was causing this effect. Like Kurup (1984), Robertson (1988) found that specific antibody to *A. fumigatus* in the sera had no effect on the killing of *A. fumigatus* by cells from mice specifically sensitized (to *A. fumigatus*) or by cells from naive mice.

5. EFFECT OF CONIDIAL PRODUCTS ON PHAGOCYTIC CELL FUNCTION

To learn why *A. fumigatus* appeared to be resistant to phagocytosis and also failed to produce a significant respiratory burst by phagocytic cells, Robertson and co-workers (1987b,d, 1988) looked at the possibility that *A. fumigatus* might produce a substance which could have a direct effect on phagocytosis and cellular production of reactive oxygen intermediates. Studies carried out to test this hypothesis showed that diffusates from conidia of *A. fumigatus* significantly inhibited ($p < 0.0025$) the phagocytosis of antibody-coated ^{51}Cr-labeled sheep red blood cells by *C. parvum*-stimulated mouse peritoneal exudate cells (Robertson *et al.*, 1987c). Further studies also showed that diffusates from *A. fumigatus* greatly reduced ($p < 0.001$) the spontaneous release of superoxide anion and hydrogen peroxide (Table II) by *C. parvum*-stimulated mouse peritoneal exudate cells (Robertson *et al.*, 1987b). Using bronchoalveolar lavage cells obtained from rats whose lungs had been treated with *C. parvum*, this inhibitory effect was found to be strikingly dependent on the concentration of diffusate (Fig. 10). No corresponding effect was found with diffusates from conidia of *P. ochrochloron*. The inhibitory component of the conidium diffusate of *A. fumigatus* was released immediately the conidia were put into suspen-

Table II. The Effect of Adding Conidium
Diffusates at a 1:4 Dilution on the Spontaneous
Release of Superoxide Anion and Hydrogen
Peroxide by C. parvum-Stimulated Mouse
Peritoneal Exudate Cells

Treatment	Superoxide anion	Hydrogen peroxide
	(nmoles/5 $\times$ 10⁵ cells/2 hr)	
HBSS[a]	14.38 (0.26)[b]	18.82 (1.63)
A. fumigatus	6.73 (0.43)*	4.16 (0.52)*
P. ochrochloron	14.40 (0.67)	19.43 (2.25)

[a]The addition of Hanks's balanced salt solution (HBSS) provides a measure of
the spontaneous release.
[b]Mean (SEM) of three separate experiments.
*Response to A. fumigatus significantly lower than HBSS or P. ochrochloron (p
< 0.001).

sion and continued to be released after the conidia had been washed. The conidial diffusates were shown to be affecting the production of reactive oxygen intermediates and not simply scavenging them once they had been produced. In addition, Robertson (1988) confirmed that the diffusates were not cytotoxic.

These observations,which supported the earlier findings using conidia of A. fumigatus. were consistent with a report by Müllbacher and co-workers (1985) who found that gliotoxin isolated from a 3-day culture supernatants of A. fumigatus had an inhibitory effect on phagocytosis of carbon particles by mouse peritoneal exudates. The same group in another study has also shown that gliotoxin inhibited the basal rate of hydrogen peroxide production by human PMN (Eichner et al., 1986). However, unlike the results reported by Robertson and co-workers (1987b, c) using conidium diffusates of A. fumigatus, they found that gliotoxin could not inhibit phorbol myristate acetate-triggered production of hydrogen peroxide. The difference between gliotoxin and the conidium diffusate described by Robertson and co-workers is that gliotoxin could not be isolated until the conidia had been in culture for at least 3 days, at which time mycelial growth would be expected to be abundant. In contrast, the diffusate described by Robertson and co-workers

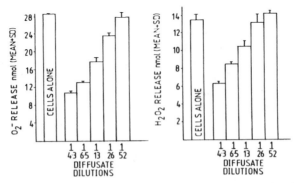

Figure 10. The effect of increasing dilution of conidial diffusates of A. fumigatus diluted in Hank's balanced salt solution (HBSS) on the spontaneous release of superoxide anion and hydrogen peroxide by phorbol myristate acetate-triggered C. parvum-induced rat bronchoalveolar lavage cells (2.5 $\times$ 10⁵ cells/2 hr) compared with cells alone (HBSS added). (From Robertson et al., 1987b.)

diffused from the respirable-size conidia as soon as the conidia were put into suspension and its quick release suggests that it is a constituent present on or close to the surface of the conidium rather than a metabolite which is produced as the result of germination.

In an attempt to pinpoint the effector mechanism of the inhibitory component of the conidium diffusate of A. fumigatus, its effect on cell movement, a fundamental aspect of phagocytic cell activation, was examined (Robertson et al., 1987d). In order for a phagocytic cell to become activated thereby enhancing its phagocytic ability, a rearrangement of the cell membrane must occur (Zuckerman and Douglas, 1979; Griffin, 1984). This involves a complex process which requires a coordinated interaction of the plasma membrane with contractile elements in the cytoplasm (Stossel, 1976). The changes are reflected in the enhanced ability of the cell to spread over a surface to migrate in response to chemical stimuli. Robertson and co-workers (1987d) have shown that conidium diffusates of A. fumigatus significantly decreased ($p < 0.001$) the capacity of C. parvum-stimulated mouse peritoneal exudate cells to spread on glass as compared to control HBSS (see Fig. 11). In addition, Robertson and co-workers (1987d) showed that conidium diffusates of A. fumigatus could reduce the number of PMN migrating toward a know chemoattractant by approximately 50%. In contrast, conidium diffusates of P. ochrochloron had no effect.

Thus, it appears that A. fumigatus conidia release a diffusable substance that has several effects of considerable biological importance in ensuring their survival in the lung and therefore enhancing their pathogenic potential. Possibly the diffusate is the first mechanism by which the conidium reduces the efficiency of the phagocyte, while gliotoxin, which is produced once germination has taken place, may be a second line of antiphagocytic defense used by A. fumigatus to establish itself and remain within the lung.

The actions of the A. fumigatus conidial diffusate would appear to be similar to those of a group of fungal metabolites called cytochalasins (Greek cytos a cell, chalasis relaxation) (Carter, 1967). Cytochalasin B, the most extensively studied of the metabolites, has a molecular weight of less than 500 and has been shown to inhibit phagocytosis, cell spreading, and chemotaxis (Allison et al., 1971; Malawista et al., 1971) and like the conidium diffusate, is not directly toxic to the cells. Cytochalasin B has been shown to act by affecting the orientated microfilaments of the substratum-associated regions of the plasma membrane and the reordering of the lipid membrane, causing depolarization of the plasma membrane (Diaz et al., 1984). The region of the plasma membrane in which the conidium diffusate of A. fumigatus exerts its influence has still to be established. Washburn and co-workers (1986) obtained water-soluble supernatants from seven day old Sabouraud agar cultures of A. fumigatus by washing the cultures with Hank's balanced salt solution and using a wooden applicator. They also prepared other supernatants by washing the solid cultures of A. fumigatus with distilled water and shaking with glass beads. In each case it seems likely that substances from both the mycelial mat as well as the conidia may have been eluted. They found that these A. fumigatus supernatants had antiphagocytic properties and could interfere with complement activation.

Another study by Chaparas and co-workers (1986) found that 3-day-old liquid culture supernatants of A. fumigatus could inhibit mitogen-induced lymphocytic transformation and therefore appear to possess immunosuppressive activity.

It is perhaps important to remember at this point that the pathogenicity of A. fumigatus should be regarded as incidental to its normal saprophytic life cycle in nature (San-Blas, 1982). These biological properties of the fungus may have evolved as a defense mechanism in its natural habitat, the soil, in order to protect itself against phagocytosis by its natural enemies, the amoebae. Support for this hypothesis comes from the work of Old and Darbyshire (1978), who conducted feeding trials to assess the ability of the giant vampyrellid amoeba (Arachnula impatiens Cienk), to attach and lyse fungal conidia, and found that conidia of a few fungal species were resistant to ingestion; among those was A. niger.

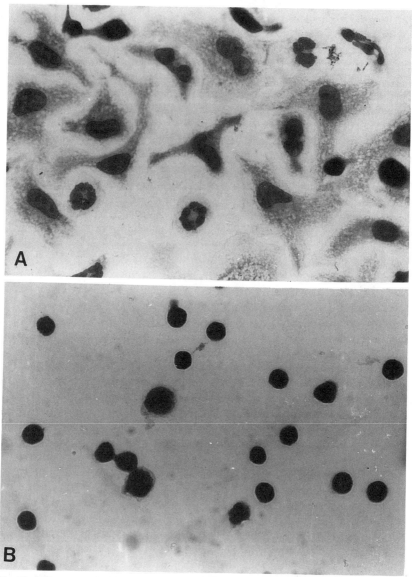

Figure 11. Photomicrographs of the spreading of *C. parvum*-stimulated mouse peritoneal exudate cells following incubation at 37°C for 1 hr in RPMI-10% fetal calf serum containing (A) HBSS or (B) *A. fumigatus* diffusate (1:2
dilution). Original magnification ×600. (From Robertson *et al.*, 1987d.)

6. SUMMARY

The results of the research to date have helped us in our understanding of *A. fumigatus* and
disease in that they have shown that conidia of *A. fumigatus* do use mechanisms to resist the host
defense network. They have been found to be relatively resistant to both phagocytosis and killing
by both human and animal phagocytic cells. This property may in part be due to the ability of
conidia to produce substances including a low-molecular-weight diffusate that inhibits phago-

cytosis, production of reactive oxygen intermediates, phagocyte migration, and macrophage spreading. It is speculated that this property of the fungus may have evolved as a mechanism of defenses against its natural enemies in the soil, the amoebae. In addition, it has been shown that *A. fumigatus* does activate complement via the alternative pathway and that the presence of heat-labile serum components (possibly complement) is utilized by the fungus to resist killing by phagocytic cells. It, therefore, seems possible that *A. fumigatus* may be using complement opsonization to its advantage in that binding via complement receptors may enable the diffusate to have more intimate contact with the cell membrane, thereby enhancing its inhibitory potential. If, however, exogenous heat-labile complement components are removed, the binding to phagocytic cells will be via other receptors, e.g., lectinlike receptors. It may be that the position of binding within the cell membrane as well as the avidity of the conidia to those other receptors does not permit the conidium diffusates to exert such an inhibitory effect on phagocyte function. An example of another microorganism using complement components to its advantage is the protozoan *Babesia rodhaini*, which utilizes C3b to gain access, via C3b receptors, to the erythrocyte cytoplasm where it multiplies (Jack and Ward, 1980). It, therefore, seems possible that complement or other heat-labile serum component, may paradoxically enhance the ability of *A. fumigatus* conidia to survive and, therefore, germinate.

How do these observations fit with the clinical manifestations associated with *A. fumigatus*? The antiphagocytic properties which have been described may help to explain why *A. fumigatus* can overcome the already defective host defenses of the immunocompromised individual and also why it is the predominant fungus associated with the development of mycetomas. Also, *A. fumigatus* has a predilection for the airways of subjects with asthma and with cystic fibrosis, two conditions in which the airway is likely to contain an exudate rich in serum components (including complement), perhaps providing ideal conditions for the survival of inhaled *Aspergillus* conidia.

Further research is now required to look into these mechanisms of resistance; the work to date has shown there is a definite effect; future work has to be directed at finding the cause!

7. REFERENCES

Absolom, D. R., 1986, Measurement of surface properties of phagocytes, bacteria and other particles, *Methods Enzymol.* **132**:16–95.

Allison, A. C., Davies, P., and de Petris, S., 1971, Role of contractile microfilaments in macrophage movement and endocytosis, *Nature New Biol.* **232**:153–155.

Appleby, E. C., 1962, Mycosis of the respiratory tract in penguins, *Proc. Zool. Soc.* **139**:495–501.

Austwick, P. K. C., 1962, The presence of *Aspergillus fumigatus* in the lungs of dairy cows, *Lab. Invest.* **II**:1065–1072.

Austwick, P. K. C., 1963, Ecology of *Aspergillus fumigatus* and the pathogenic phycomycetes, *Recent Prog. Microbiol.* **8**:644–651.

Austwick, P. K. C., 1965, Pathogenicity, in *The Genus Aspergillus* (K. B. Raper and D. I. Fennell, eds.), Williams & Wilkins, Baltimore, pp. 82–126.

Austwick, P. K. C., and Venn, J. A. J., 1962, Mycotic abortion in England and Wales 1954–1960, *Proc. 4th Int. Congr. Anim. Reprod.* **3**:562–568.

Austwick, P. K. C., Gitter, M., and Watkins, C. V., 1960, Pulmonary aspergillosis in lambs, *Vet. Rec.* **72**:19–21.

Bardana, J. E. J., 1980, The clinical spectrum of aspergillosis—Part 1. Epidemiology, pathogenicity, infection in animals and immunology of *Aspergillus*, *CRC Crit. Rev. Clin. Lab. Sci.* **13(1)**:21–83.

Beaumont, F., Kauffman, H. F., Sluiter, H. J., and de Vries, K., 1984, Environmental aerobiological studies in allergic bronchopulmonary aspergillosis, *Allergy* **39**:183–193.

Carter, S. B., 1967, Effect of cytochalasins on mammalian cells, *Nature* **21**:261–264.

Chaparas, S. D., Morgan, P. A., Holobaugh, P., and Kim, S. J., 1986, Inhibition of cellular immunity by products of *Aspergillus fumigatus*, *J. Med. Vet. Mycol.* **24**:67–76.

Chusid, M. J., Sohnle, P., Fink, J. N., and Shea, M. L., 1981, A genetic defect of granulocyte oxidative metabolism in a man with disseminated aspergillosis, *J. Lab. Clin. Med.* **97**:730–738.

Diamond, R. D., and Clark, R. A., 1982, Damage to *Aspergillus fumigatus* and *Rhizopus oryzae* hyphae by oxidative and non-oxidative microbicidal products of human neutrophils *in vitro*, *Infect. Immun.* **38**:487–495.

Diamond, R. D., Krzesicki, R., Epstein, B., and Jao, W., 1978, Damage to hyphal forms of fungi by human leukocytes *in vitro*, *Am. J. Pathol.* **91**:313–323.

Diamond, R. D., Huber, E., and Haudenschild, C. C., 1983, Mechanisms of destruction of *Aspergillus fumigatus* hyphae mediated by human monocytes, *J. Infect. Dis.* **147**:474–483.

Diaz, B., Niubo, E., Companioni, M., Ancheta, O., and Kouri, J., 1984, Effects of cytochalasin B and of deoxyglucose on phagocytosis-related changes in membrane potential in rat peritoneal macrophages, *Exp. Cell Res.* **150**:494–498.

Domer, J. A., and Carrow, E. W., 1983, Immunity to fungal infections, in: *Host Defenses to Intracellular Pathogens* (T. K. Eisenstein, P. Actor, and H. Friedman, eds.), Plenum Press, New York, pp. 383–408.

Donaldson, K., Brown, G. M., Robertson, M. D., Slight, J., and Seaton, A., 1987, Effector functions of bronchoalveolar leukocytes from rats exposed to coalmine dust by inhalation, *Thorax* **42**:748 (abstr.).

DuBois, R. M., 1985, The alveolar macrophage, *Thorax* **40**:321–327.

Edwards, J. H., 1976, A quantitative study on the activation of the alternative pathway of complement by mouldy hay dust and thermophilic actinomycetes, *Clin. Allergy* **6**:155–164.

Eichner, R. D., Salami, M., Wood, P. R., and Mullbacher, A., 1986, The effect of gliotoxin upon macrophage function, *Int. J. Immunopharmacol.* **8**:789–797.

Fantone, J. C., and Ward, P. A., 1982, Role of oxygen-derived free radicals and metabolites in leukocyte-dependent inflammatory reactions. Review article, *Am. J. Pathol.* **107**:397–418.

Fantone, J. C., Feltner, D. E., Brieland, J. K., and Ward, P. A., 1987, Phagocytic cell-derived inflammatory mediators and lung disease, *Chest* **91**:428–435.

Fietta, A., Sacchi, F., Mangiarotti, P., Manara, G., and Grassi, G., 1984, Defective phagocyte *Aspergillus* killing associated with recurrent *Aspergillus* infections, *Infection* **12**:10–13.

Fraser, D. W., Ward, J. I., Ajello, L., and Phikaytis, B. D., 1979, Aspergillosis and other systemic mycoses. The growing problem, *J. Am. Med. Assoc.* **242**:1631–1635.

Gallin, J. I., 1976, The role of chemotaxis in the inflammatory–immune response of the lung, in: *Immunologic and Infectious Reactions in the Lung* (C. H. Kirkpatrick and H. Y. Reynolds, eds.), Dekker, New York, pp. 161–178.

Gee, J. B. L., and Khandwala, A. S., 1977, Mortality, transport and endocytosis in lung defense cells, in: *Respiratory Defense Mechanisms*, Part II (J. D. Brain, D. F. Proctor, and L. M. Reid, eds.), Dekker, New York, pp. 927–983.

Griffin, F. M., Jr., 1984, Activation of macrophage complement receptors for phagocytosis, in: *Macrophage Activation* (D. O. Adams and M. G. Hanna, Jr., eds.), Plenum Press, New York, pp. 57–70.

Hinson, K. F. W., Moon, A. J., and Plummer, N. S., 1952, Bronchopulmonary aspergillosis, *Thorax* **7**:317–333.

Jack, R. M., and Ward, P. A., 1980, *Babesia rodhaini* interactions with complement: Relationship to parasitic entry into red cells, *J. Immunol.* **124**:1566–1573.

Kan, V. L., and Bennett, J. E., 1988, Lectin-like attachment sites of murine pulmonary alveolar macrophages bind *Aspergillus fumigatus* spores, *J. Infect. Dis.* **158**:407.

Karnovsky, M. L., and Badwey, J. A., 1986, Respiratory burst during phagocytosis: An overview, *Methods Enzymol.* **132**:353–354.

Karnovsky, M. L., Lazdins, J., and Simmons, S. R., 1975, Metabolism of activated mononuclear phagocytes at rest and during phagocytosis, in: *Mononuclear Phagocytes in Immunity, Infection, and Pathology* (R. van Furth., ed.), Blackwell, Oxford, pp. 423–438.

Kurup, V. P., 1981, *In vitro* infection of rabbit alveolar macrophages with *Aspergillus* spores, *Abstr. Ann. Meet. Am. Soc. Microbiol.* **81**:317.

Kurup, V. P., 1984, Interactions of *Aspergillus fumigatus* spores and pulmonary alveolar macrophages of rabbits, *Immunobiology* **166**:53–61.

Lacey, J., 1981, The aerobiology of conidial fungi, in: *Biology of Conidial Fungi* (G. T. Cole and B. Kendrick, eds.), Academic Press, New York, pp. 373–416.

Laufer, P., Fink, J. N., Bruns, W. T., Unger, G. F., Kalbfleish, J. H., Greenberger. P. A., and Patterson, R., 1984, Allergic bronchopulmonary aspergillosis in cystic fibrosis, *J. Allergy Clin. Immunol.* **73**:44–48.

Lehman, P. F., and White, L. O., 1976, Acquired immunity to *Aspergillus fumigatus*, *Infect. Immun.* **13**:1296–1298.

Lehrer, R. I., 1969, Antifungal effects of peroxidase systems, *J. Bacteriol.* **99**:361–365.

Lehrer, R. I., and Jan, R. G., 1970, Interaction of *Aspergillus fumigatus* spores with human leukocytes and serum, *Infect. Immun.* **1**:345–350.

Levitz, S. M., and Diamond, R. D., 1984, Killing of *Aspergillus fumigatus* spores and *Candida albicans* yeast phase by the iron–hydrogen peroxide–iodide cytotoxic system: Comparison with the myeloperoxidase–hydrogen peroxide–halide system, *Infect. Immun.* **43**:1100–1102.

Levitz, S. M. and Diamond, R. D., 1985, Mechanisms of resistance of *Aspergillus fumigatus* conidia to killing by neutrophils *in vitro*, *J. Infect. Dis.* **152**:33–42.

McCowen, M. C., Callender, M. E., and Lawlis, J. F., Jr., 1951, Fumagillin (H-3), a new antibiotic with amebicidal properties, *Science* **113**:202–203.

McDiarmid, A., 1955, Aspergillosis in free living wild birds, *J. Comp. Pathol. Ther.* **65**:246–249.

Malawista, S. E., Gee, J. B. L., and Bensch, K. G., 1971, Cytochalasin B reversibly inhibits phagocytosis: Functional metabolic and ultrastructural effects in human blood leukocytes and rabbit alveolar macrophages, *J. Biol. Med.* **44**:286–300.

Marsh, P. B., Millner, P. D., and Kla, J. M., 1979, A guide to the recent literature on aspergillosis as caused by *Aspergillus fumigatus*, a fungus frequently found in self-heating organic matter, *Mycopathologia* **69**:67–81.

Marx, J. J., and Flaherty, D. K., 1976, Activation of the complement sequence by extracts of bacteria and fungi associated with hypersensitivity pneumonitis, *J. Allergy Clin. Immunol.* **57**:328–334.

Mearns, M., Longbottom, J. L., and Batten, J. C., 1967, Precipitating antibodies to *Aspergillus fumigatus* in cystic fibrosis, *Lancet* **1**:538–539.

Morgan, W. K. C., 1984, The deposition and clearance of dust from the lungs, in: *Occupational Lung Diseases*, 2nd ed. (W. K. C. Morgan and A. Seaton, eds.), Saunders, Philadelphia, pp. 77–96.

Müllbacher, A., Waring, P., and Eichner, R. D., 1985, Identification of an agent in cultures of *Aspergillus fumigatus* displaying anti-phagocytic and immunomodulatory activity *in vitro*, *J. Gen. Microbiol.* **131**:1251–1258.

Mullins, J., and Seaton, A., 1978, Fungal spores in lung and sputum, *Clin. Allergy* **8**:525–533.

Mullins, J., Harvey, R., and Seaton, A., 1976, Sources and incidence of airborne *Aspergillus fumigatus* (Fres.), *Clin. Allergy* **6**:209.

Noble, W. C., and Clayton, Y. M., 1963, Fungi in the air of hospital wards, *J. Gen. Microbiol.* **32**:397–402.

Ögmundsdöttir, H. M. and Wier, D. M., 1980, Mechanisms of macrophage activation: Review, *Clin. Exp. Immunol.* **40**:223–234.

Old, K. M., and Darbyshire, J. F., 1978, Soil fungi as food for giant amoebae, *Soil Biol. Biochem.* **10**:93–100.

Orie, N. G. M., De Vries, G. A., and Kikstra, A., 1960, Growth of *Aspergillus* in the human lung; aspergilloma and aspergillosis, *Am. Rev. Respir. Dis.* **82**:649–662.

Pagani, A., Spalla, R., Ferrari, F. A., Duse, M., Lenzi, L., Bretz, U., Baggiolini, M., and Siccardi, A. G., 1981, Defective *Aspergillus* killing by neutrophil leukocytes in a case of systemic aspergillosis, *Clin. Exp. Immunol.* **43**:201–207.

Pepys, J., 1969, Hypersensitivity diseases of the lung due to fungi and other organic dusts, *Monogr. Allergy* **4**:1–199.

Pepys, J., 1981, Fungi in pulmonary allergic diseases, in: *Immunology of Human Infection*, Part 1 (A. J. Nahmias and R. J. O'Reilly, eds.), Plenum Press, New York, pp. 561–584.

Pepys, J., Riddell, R. W., Citronk, M., Clayton, Y. M., and Short, E. I., 1959, Clinical and immunologic significance of *Aspergillus fumigatus* in the sputum, *Am. Re. Respir. Dis.* **80**:167–180.

Radin, R. C., Greenberger, P. A., Patterson, R., and Ghory, A., 1983, Mould counts and exacerbations of allergic bronchopulmonary aspergillosis, *Clin. Allergy* **13**:271–275.

Raper, K. B., and Fennell, D. I., (eds.), 1965, *The Genus Aspergillus*, Williams & Wilkins, Baltimore.

Rénon, L., 1897, Etude sur l'aspergillose chez les animaux et chez l'homme, Masson, Paris.

Robertson, M. D., 1988, Host defences against *Aspergillus fumigatus*, Ph.D. Thesis, University of Edinburgh.

Robertson, M. D., Raeburn, J. A., Gormley, I. P., and Seaton, A., 1985, Do phagocytic cells ingest spores of *Aspergillus fumigatus*? *Thorax* **40**:237. (abstr.).

Robertson, M. D., Kerr, K. M., and Seaton, A., 1987a, The effect of complement on the killing of *Aspergillus fumigatus* by lung macrophages, *Thorax* **42**:213.

Robertson, M. D., Seaton, A., Milne, L. J. R., and Raeburn, J. A., 1987b, Suppression of host defences by *Aspergillus fumigatus*, *Thorax* **42**:19–25.

Robertson, M. D., Seaton, A., Milne, L. J. R., and Raeburn, J. A., 1987c, Resistance of spores of *Aspergillus fumigatus* to ingestion by phagocytic cells, *Thorax* **42**:466–472.

Robertson, M. D., Seaton, A., Raeburn, J. A., and Milne, L. J. R., 1987d, Inhibition of phagocytic migration and spreading by spore diffusates of *Aspergillus fumigatus*, *J. Med. Vet. Mycol.* **25**:389–396.

Robertson, M. D., Brown, D. M., MacLaren, W. M., and Seaton, A., 1988, Fungal handling by phagocytic cells from asthmatic patients sensitized and non-sensitized to *Aspergillus fumigatus*, *Thorax* **43**:224.

San-Blas, G. 1982, The cell wall of fungal human pathogens: Its possible role in host–parasite relationships: A review, *Mycopathologia* **79**:159–184.

Sandhu, D. K., Sandhu, R. S., Damodaran, V. N., and Randhawa, H. S., 1970, The effect of cortisone on bronchopulmonary aspergillosis in mice exposed to spores of various *Aspergillus* species, *Sabouraudia* **8**:32–38.

Schaffner, A., Douglas, H., and Braude, A. I., 1982, Selective protection against conidia by mononuclear and against mycelia by polymorphonuclear phagocytes in resistance to *Aspergillus*, *J. Clin. Invest.* **69**:617–631.

Schaffner, A., Douglas, H., Braude, A. I., and Davis, C. E., 1983, Killing of *Aspergillus* spores depends on the anatomical source of the macrophage, *Infect. Immun.* **42**:1109–1115.

Seaton, A., and Morgan, W. K. C., 1984, Hypersensitivity pneumonitis, in: *Occupational Lung Diseases*, 2nd ed. (W.K. C. Morgan and A. Seaton, eds.) Saunders, Philadelphia, pp. 564–608.

Silverstein, S. C., Steinman, R. M., and Cohn, Z. A., 1977, Endocytosis, *Ann. Rev. Biochem.* **46**:669–722.

Smith, G. R., 1972, Experimental aspergillosis in mice: Aspects of resistance, *J. Hyg.* **70**:741–754.

Solomon, W. R., and Burge, H. P., 1975, *Aspergillus fumigatus* levels in and out of doors in urban air, *J. Allergy Clin. Immunol.* **55**:90–91.

Stendahl, O., Dahlgren, C., Edebo, M., and Öhman, L., 1981, Recognition mechanisms in mammalian phagocytosis, *Monogr. Allergy* **17**:12–27.

Stossel, T. P., 1976, The mechanism of phagocytosis, *J. Reticuloendothel. Soc.* **19**:237–245.

Thurston, J. R., Cysewski, S. J.,and Richard, J. L., 1979, Exposure of rabbits to spores of *Aspergillus fumigatus* or *Penicillium* sp. Survival of fungi and microscopic changes in the respiratory and gastrointestinal tract, *Am. J. Vet. Res.* **40**:1443–1449.

Trivedi, L. S., and Rao, K. K., 1979, Production of cellulolytic enzymes by *Aspergillus fumigatus*, *Indian J. Exp. Biol.* **17**:671–674.

Turner, K. J., Hackshaw, R., Papadimitriou, J., and Perrott, J., 1976, The pathogenesis of experimental pulmonary aspergillosis in normal and cortisone-treated rats, *J. Pathol.* **118**:65–73.

van Furth, R., and Diesselhoff-den Dulk, M. M. C., 1980, Method to prove ingestion of particles by macrophages with light microscopy, *Scand. J. Immunol.* **12**:265–269.

Vernon, D. R. H., and Allan, F., 1980, Environmental factors in allergic bronchopulmonary aspergillosis, *Clin. Allergy* **10**:217–227.

Waldorf, A. R., Levitz, S. M., and Diamond, R. D., 1984, *In vivo* bronchoalveolar macrophage defence against *Rhizopus oryzae* and *Aspergillus fumigatus*, *J. Infect. Dis.* **150**:752–760.

Warren, R. E., and Warnock, D. W., 1982, Clinical manifestations and management of aspergillosis in the compromised patient, in: *Fungal Infection in the Compromised Patient* (D. W. Warnock and M. D. Richardson, eds.), Wiley, New York, pp. 119–153.

Washburn, R. G., Hammer, C. H., and Bennett, J. E., 1986, Inhibition of complement by culture supernatants of *Aspergillus fumigatus*, *J. Infect. Dis.* **154**:944–951.

Washburn, R. G., Gallin, J. I., and Bennett, J. E., 1987, Oxidative killing of *Aspergillus fumigatus* proceeds by parallel myeloperoxidase-dependent and independent pathways, *Infect. Immun.* **55**:2088–2092.

Weir, D. M., 1980, Surface carbohydrates and lectins in cellular recognition, *Immunol. Today* **1**:45–51.

Weir, D. M., and Ögmundsdöttir, H. M., 1977, Non-specific recognition mechanisms by mononuclear phagocytes, *Clin. Exp. Immunol.* **30**:323–329.

Weir, D. M., Graham, L. M., and Ögmundsdöttir, H. M., 1979, Binding of mouse peritoneal macrophages to tumour cells by a 'lectin-like' receptor, *J. Clin. Lab. Immunol.* **2**:51–54.

Whaley, K., 1985, An introduction to the complement system, in: *Methods in Complement for Clinical Immunologists* (K. Whaley, ed.), Churchill Livingstone, Edinburgh, pp. 1–20.

White, L. O., 1977, Germination of *Aspergillus fumigatus* conidia in the lungs of normal and cortisone-treated mice, *Sabouraudia* **15**:37–41.

Wilkinson, P. C., 1976, Recognition and response in mononuclear and granular phagocytes, *Clin. Exp. Immunol.* **25**:355–366.

Williams, D. M., Weiner, M. H., and Drutz, D. J., 1981, Immunologic studies of disseminated infection with *Aspergillus fumigatus* in the nude mouse, *J. Infect. Dis.* **143**:726–733.

Wilson, B. J., 1971, Miscellaneous *Aspergillus* toxins, in: *Microbial Toxins*, Volume VI (A. Ciegler, S. Kadis, and S. J. Ajl, eds.), Academic Press, New York, pp. 207–295.

Yin, H. L. and Stossel, T. P., 1982, The mechanism of phagocytosis, in: *Phagocytosis—Past and Future* (M. L. Karnovsky and L. Bolis, eds.), Academic Press, New York, pp. 13–27.

Young, R. C., Jennings, A., and Bennett, J.E., 1972, Species identification of invasive aspergillosis in man, *Am. J. Clin. Pathol.* **58**:554–557.

Zuckerman, S. H., and Douglas, S. D., 1979, Dynamics of the macrophage plasma membrane, *Annu. Rev. Microbiol.* **33**:267–307.

IV

Molecular Aspects of Disease Initiation

22

Molecular Approaches to the Analysis of Pathogenicity Genes from Fungi Causing Plant Disease

Robert C. Garber

1. INTRODUCTION

Fungi are relatively simple eukaryotes; most reproduce rapidly and are easy to grow. Fungi in general have small genomes; during vegetative growth many are haploid and some are uninucleate. In numerous cases their life cycle can be completed in the laboratory and the products of meiosis analyzed individually, facilitating classical genetics. The fungi which are responsible for plant diseases are interesting biologically and important economically. In sum, plant pathogenic fungi would appear to be excellent subjects for genetic and biochemical analysis, and attractive organisms for the application of molecular approaches to dissecting the phenomenon of pathogenesis. However, there is a disappointingly short list of examples for which we understand, even in part, the molecular basis for fungal–plant interactions. One reason is the great diversity of fungi and the many routes by which they have achieved their successes as pathogens. As a result of this diversity, no fungus has emerged as an overall model for the analysis of pathogenicity toward plants, as the yeasts *Saccharomyces cerevisiae* and *Schizosaccharomyces pombe* or the filamentous fungi *Neurospora crassa* and *Aspergillus nidulans* have for other areas of cell and molecular biology. However, methods in recombinant DNA technology are making significant contributions to the study of plant diseases caused by fungi, and progress in several host–pathogen systems has been rapid in recent years. The goal of this chapter is to consider recent progress and prospects for future developments in understanding the genes which fungi employ in inciting plant disease.

The scope of this chapter will be confined to the fungal partner in the plant–pathogen relationship. Furthermore, it will emphasize fungal genes responsible for *pathogenicity factors*: compounds produced by the fungus which influence its ability to cause disease. Numerous molecules have been proposed as candidates for pathogenicity factors, though in few cases have

Robert C. Garber • Biotechnology Center and Department of Plant Pathology, Ohio State University, Columbus, Ohio 43210. Present address: Springer-Verlag, New York, New York 10010.

their roles been established conclusively. When such a "candidate molecule" is a polypeptide, then the gene which encodes it may be regarded as a putative pathogenicity gene. When—as is often the case—the pathogenicity factor is not a polypeptide, then genetic data must be established which tie a specific locus to the production of a proposed pathogenicity factor. Such a connection establishes the locus as a putative pathogenicity gene, particularly if the absence or alteration of the gene affects the ability of the pathogen to cause disease. It should be noted that for the purposes of this chapter, a distinction will not be made between the qualitative and quantitative contributions to disease made by pathogenicity factors and virulence factors, respectively (Yoder, 1980): if a gene product or a metabolite can be rigorously connected to the ability of the fungus to incite disease, I will treat it as a pathogenicity factor.

Plant pathogenic fungi cause disease in numerous ways. Furthermore, the characteristics of a disease may be defined differently at the molecular level, the cellular level, the whole plant/ whole pathogen level, and the population level. Thus, the processes and gene products which are involved in the ability of a fungus to cause disease are, not surprisingly, many and varied. Some are relatively well known, such as the ability to enzymatically breach the plant's cuticle, the production of toxic compounds which damage plant cells, and the capacity of a fungus to tolerate antimicrobial compounds synthesized in the plant in response to infection. Candidate pathogenicity factors and their associated genes have been identified for each of these phenomena. We can also recognize genes whose products contribute to processes required for disease, but whose role in pathogenesis may be less obvious: the adhesion of a spore to the surface of a plant; the ability to assess the topography of the host surface; or the formation of complex structures involved in penetration (e.g., appressoria) or in reproduction (e.g., asexual or sexual spores).

Perhaps the most important conceptual advance which molecular biology has brought to the analysis of fungal plant pathogens is the capacity to rigorously determine whether the product or process resulting from expression of a putative pathogenicity gene is in fact required for pathogenesis. Genes can now be isolated, altered *in vitro*, and replaced in the fungus from which they came. If a gene can be altered or deleted without having an effect on the phenomenon under investigation, a possible role for the gene in the process is eliminated. Alternatively, if a change in the gene results in a new "disease phenotype," then the gene's role in disease is powerfully demonstrated, since under careful experimental conditions other factors in the interaction are held nearly constant. A tool of such precision is without precedent in the study of plant disease, and will change forever our ideas about what constitutes an acceptable standard of evidence regarding the role of a molecule in disease. The remainder of this chapter will consider some of the "molecular methods" which are now practical in the fungi, and will introduce by example some of the approaches which have been taken in the isolation of fungal pathogenicity genes.

2. TECHNOLOGY FOR ISOLATION OF FUNGAL GENES

Contemporary experimental biology advances primarily by the vigorous development of a small number of model systems: organisms usually characterized by small size, rapid reproduction, and tractable genetics. Among the fungi, the budding yeast *Saccharomyces cerevisiae* and the fission yeast *Schizosaccharomyces pombe* have been the dominant models (Botstein and Fink, 1988). In contrast to these two yeasts, most plant pathogenic fungi are filamentous: they grow in plants as multicellular hyphal tubes, not as single cells. Furthermore, many yeast genes and the sequences which regulate them do not function in plant pathogenic fungi. As a result, much of the sophisticated technology developed for *Saccharomyces cerevisiae* has been useful as a paradigm for orienting technical progress in other fungi, but has not been directly transferable. The two filamentous fungi that have served as models for the development of gene cloning and gene transfer methods for plant pathogenic species have been the Ascomycetes *Neurospora crassa* and

Aspergillus nidulans. Both fungi have well-established genetic maps and mutant collections (Fungal Genetics Stock Center, 1988). While there are no plant pathogenic fungi for which a comparable background exists, many are amenable to genetic development (Sidhu, 1988). The following descriptions of gene cloning methods will outline some of the important contemporary techniques, and will point out their use in fungi where appropriate.

2.1. Cloning by Transformation

Transformation refers to the uptake and expression of exogenous DNA by a living cell. The yeast *Saccharomyces cerevisiae* was the first fungus to be reliably transformed (Hinnen *et al.*, 1978). Transformation of yeast has been routine for a decade, and has revolutionized the study of yeast biochemistry and genetics. Contrary to initial expectations, several important aspects of yeast transformation are apparently not applicable to most other fungi. These include the ability to maintain transforming DNA by autonomous replication of a plasmid or minichromosome in fungal cells [reported only for *Saccharomyces cerevisiae* (Stinchcomb *et al.*, 1980), *Schizosaccharomyces pombe* (Beach and Nurse, 1981), and recently, *Ustilago maydis* (Tsukuda *et al.*, 1988)] and the use of yeast promoters to express genes in heterologous systems. Yeast promoters generally do not function in filamentous fungi. Efforts by several laboratories, however, have resulted in successful integrative transformation in a number of filamentous fungi. As with yeast, the first to be transformed were those with the most highly developed genetic systems: *Neurospora crassa* (Case *et al.*, 1979) and *Aspergillus nidulans* (Ballance *et al.*, 1983; Tilburn *et al.*, 1983; Yelton *et al.*, 1984). The vectors and procedures developed for *Neurospora* and *Aspergillus* have led to the transformation of many more filamentous fungi (reviewed in Rambosek and Leach, 1987). In contrast to the situation with yeast promoters, one of the findings of transformation studies in filamentous fungi has been that promoters which function in one filamentous fungus can often be used to confer expression of a marker gene in other genera and species (Turgeon *et al.*, 1987; Rodriguez and Yoder, 1987; Panaccione *et al.*, 1988; Parsons *et al.*, 1987).

One of the most powerful applications of transformation is the ability to isolate a gene directly from a DNA library by complementation of a strain that lacks functional copy of the gene of interest. This procedure is generally referred to as "cloning by transformation" (Fig. 1). It requires transformation efficiencies high enough to generate thousands of independent transformants, because even with the small, haploid genomes typical of fungi, a minimum of 2000 independent transformants is required to adequately represent an entire genome when cosmid clones are used for transformation (Yoder *et al.*, 1986). In yeast, high-frequency transformation ($> 10^3/\mu g$ DNA) is achieved through the use of autonomously replicating plasmids. With most filamentous fungi, transformation occurs by integration of transforming DNA into chromosomes. The efficiency of integrative transformation among filamentous fungi varies from fewer than $1/\mu g$ DNA to greater than $5 \times 10^4/\mu g$ DNA, which means that cloning by transformation ranges from very laborious to a minor experimental component. Recovery of a gene from the transformed strain for further analysis is straightforward in those few systems where autonomous replication of transforming DNA can be achieved: it is a simple matter to transform *E. coli* directly with DNA isolated from the transformed fungal cell. Alternatively, when the transforming DNA containing the gene of interest has integrated into the fungal cell's chromosomes, recovery can be achieved by constructing a DNA library from the transformed strain and using adjacent vector sequences as a hybridization probe, since vector DNA will usually not hybridize to DNA of the untransformed fungal cell.

In principle, any DNA sequence which can be identified genetically can be isolated using transformation, as long as its presence in the transformed cell can be selected for directly, or detected by screening. The great power of cloning by transformation lies in the fact that it presupposes only a defined locus whose DNA sequence is short enough to fit on a single clone in

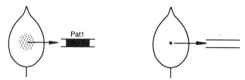

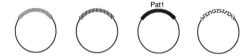

Fungal strain with pathogenicity gene (Pat1) causes a lesion on leaf;
another strain lacking a functional copy of Pat1 is non-pathogenic.

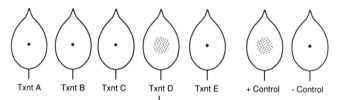

DNA library is prepared from the fungal strain containing Pat1.

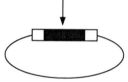

Txnt A   Txnt B   Txnt C   Txnt D   Txnt E   + Control   - Control

The strain lacking a functional copy of Pat1 is transformed with the library;
transformants (Txnt A-E) are assayed for pathogenicity.

Transforming DNA is recovered from the pathogenic transformant (TxntD)
and is analyzed for the presence of Pat1.

Figure 1. Gene cloning by transformation.

the library. In plant–pathogen interactions, where the biochemical bases for many of the interesting components of disease (e.g., specificity, virulence, hypersensitivity) are not usually known, the ability to isolate genes in the absence of knowledge about gene products is invaluable. Gene cloning by transformation is routine in yeast (Rose, 1987) and has been achieved in both *N. crassa* (Akins and Lambowitz, 1985; Paietta *et al.*, 1987) and *A. nidulans* (Yelton *et al.*, 1985). Among plant pathogenic fungi, at least two genes associated with pathogenicity have been cloned by transformation: the pisatin demethylase gene of the pea pathogen *Nectria haematococca* (Weltring *et al.*, 1988; see Section 3.2) and *b* mating type alleles of the corn smut fungus *Ustilago maydis* (Kronstad and Leong, 1989; see Section 4.2). In addition, the *Trp1* gene of the southern corn leaf blight pathogen *Cochliobolus heterostrophus* was cloned by transformation of *E. coli* (Turgeon *et al.*, 1986), and a mating type allele of *C. heterostrophus* has been cloned by transformation of *C. heterostrophus* (Turgeon *et al.*, 1988).

2.2. Cloning by Heterologous Hybridization

A frequently exploited approach to gene cloning involves starting with a clone which is already available from another organism. If the species under investigation and the species from which the clone is available are closely related, it is likely that sufficient conservation of DNA sequence will exist to permit isolation of the gene by screening a DNA library, using the heterologous sequence as a hybridization probe (Fig. 2). In fact, some genes are so strongly conserved that startling taxonomic distances my be leapt. For example, a gene in the heat shock gene family of *Drosophila* (*hsp*70) was used to isolate an *hsp*70-related sequence from the fungus *U. maydis* (Holden *et al.*, 1985) whose regulatory signals in turn conferred expression of a prokaryotic hygromycin B phosphotransferase gene in *U. maydis* (Wang *et al.*, 1988). The method of cloning by heterologous hybridization is frequently used to compare the sequence of a given gene from related organisms for phylogenetic analysis, but it also finds a role in the development of a transformation system. A cloned gene from one organism may not function in

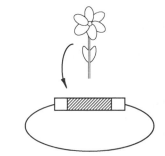

Cloned gene of interest is available from another organism.

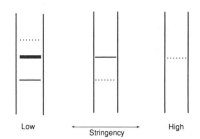

Low   Stringency   High

The "heterologous" clone is used as a hybridization probe to demonstrate the presence of related sequence(s) in DNA from the organism of interest; this may require relaxed hybridization or washing conditions.

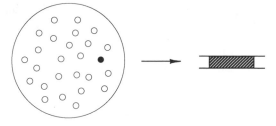

The heterologous gene serves as a probe to identify hybridizing clones in a DNA library made from the organism of interest.

The "homologous" gene from the organism of interest is available for analysis, or as a selectable marker in transformation experiments.

Figure 2. Gene cloning by heterologous hybridization.

another because of differences in the regulatory sequences that govern the gene's expression. If, however, the DNA sequence of the gene is sufficiently conserved to allow for its identification by DNA–DNA hybridization, a clone may be isolated which should function when reintroduced into the organism of interest by transformation. For example, a gene which complements an auxotrophic mutant may serve as the basis for transformation (Faugeron et al., 1989). In work with the maize pathogen *Colletotrichum graminicola*, an ascomycete, a mutant β-tubulin gene was cloned by heterologous hybridization, using the β-tubulin gene of *N. crassa* as a probe (Panaccione et al., 1988). In this instance, the two genes were found to transform *C. graminicola* to benomyl resistance with comparable efficiency.

2.3. Cloning with Oligonucleotide Probes

With the development of protein sequencing technology now capable of determining the amino acid sequence of extremely small quantities of protein, it has become increasingly practical to determine the partial amino acid sequence of purified proteins, which in turn are used to predict the DNA sequences encoding the protein. Oligonucleotides as short as 15–20 bases (5–7 amino acids) may be used successfully as radioactive probes to screen DNA libraries by hybridization (Kinnaird et al., 1982), and longer oligonucleotides may often be obtained (decreasing the likelihood of spurious hybridization). Because the genetic code is degenerate (more than one triplet specifies a given amino acid), mixtures of oligonucleotides are generally prepared to increase the probability that at least a portion of the probe will match the DNA sequence of the gene encoding the protein under investigation (Wallace and Miyada, 1987; Wood, 1987) (Fig. 3). The use of oligonucleotides to clone genes in fungi has received relatively little application. One reason is that protein which is sufficiently pure to be sequenced is also sufficiently pure for use in generating antibodies which can in turn be used to screen cDNA expression libraries (see Section 2.4).

2.4. cDNA Cloning by Differential Hybridization or Antibody Screening

In cases where there is evidence that an mRNA of interest is transcribed at a particular stage of development or in response to a specific induction signal, it is possible to take advantage of differential gene expression to isolate a cDNA clone. A cDNA library constructed from induced cells will be enriched for the message. If such a library is screened with probes made of total cDNA from both the induced and noninduced states, clones representing the differentially expressed gene should hybridize more strongly to the probe made of cDNA from the induced state (Fig. 4). Because other genes may be coinduced with the gene of interest, additional evidence proving the identity of the differentially expressed clone is required. For example, the sequence of the cDNA clone itself may provide sufficient evidence for the gene's identity. Alternatively, the cDNA clone may be used as a probe to identify a specific mRNA which in turn can be translated *in vitro* into a protein which will immunoprecipitate with antibodies specific to the gene product of interest.

An increasingly popular method of isolating cDNA clones corresponding to a particular gene is by antibody screening. The primary prerequisite is a protein of sufficient purity to generate antibodies. A cDNA library is constructed in a vector such as λgt11 (Huynh et al., 1985), which is a bacteriophage expression vector: clones in λgt11 are expressed in *E. coli* as fusions between the *E. coli* β-galactosidase protein and the protein encoded by the insert. The antibody to the protein of interest is then used to identify fusion polypeptides with antigenic determinants (epitopes) which are recognized by the antibody. Detection of the positive clones is achieved most often by incubating a filter containing the fusion protein–antibody complex with an anti-IgG ("second antibody") conjugated with alkaline phosphatase or with peroxidase (Mierendorf et al.,

MetAspHisGlyLys

Partial amino acid sequence of polypeptide is determined.

Met Asp His Gly Lys

| ATG GAT CAT GGT AAA |
| ATG GAC CAC GGC AAG |
| ATG GAT CAC GGG AAA |
| ATG GAC CAT GGA AAG |
| etc |

Mixture of oligonucleotides is synthesized based on deduced DNA sequence.

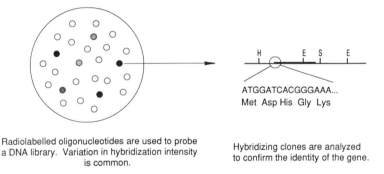

ATGGATCACGGGAAA...
Met Asp His Gly Lys

Radiolabelled oligonucleotides are used to probe
a DNA library. Variation in hybridization intensity
is common.

Hybridizing clones are analyzed
to confirm the identity of the gene.

Figure 3.   Gene cloning with oligonucleotide probes.

1987). Incubation in the presence of appropriate substrates which respond to alkaline phosphatase or peroxidase by producing a colored product completes the detection process (Fig. 5). The procedure of antibody screening, also referred to as immunoscreening, has been used with plant pathogenic fungi to isolate cDNAs encoding the pathogenicity determinants cutinase and pectate lyase (see Section 3.1).

2.5. Cloning by RFLP Mapping and Chromosome Walking

Restriction fragment length polymorphisms (RFLPs) are differences found among individual organisms in the length of a particular DNA fragment generated by digestion with a restriction enzyme. The difference may be due to the insertion or deletion of a block of DNA within the fragment, or to an alteration in one or more bases which results in the gain or loss of a restriction enzyme recognition site. RFLPs have experienced rapid development as genetic markers for the

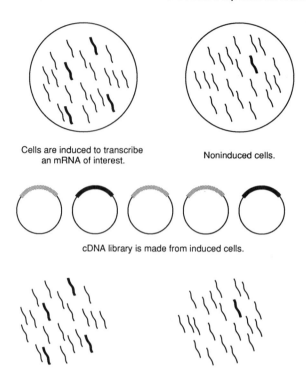

Cells are induced to transcribe
an mRNA of interest.   Noninduced cells.

cDNA library is made from induced cells.

Radiolabelled hybridization probes are made from RNAs of induced and noninduced cells.

Hybridized with   Hybridized with
"Induced" Probe   "Noninduced" Probe

Probes are used independently to screen the cDNA library made from induced cells;
clones specific to the induced probe (boxed) potentially represent the message of interest.

Figure 4.   cDNA cloning by differential hybridization.

construction of genetic maps (Botstein *et al.*, 1980), and as starting points for chromosome walks leading to the isolation of genes which are otherwise difficult to reach (Young *et al.*, 1988) (Fig. 6). Among the fungi, RFLP mapping has been pursued vigorously for only a handful of species. The *N. crassa* RFLP map began with a relatively small number of 5S ribosomal RNA gene markers (Metzenberg *et al.*, 1985), and is growing steadily (Metzenberg and Grotelueschen, 1987–1988). The oomycete *Bremia lactucae* has been the subject of the most extensive RFLP map development among plant pathogenic fungi (see Section 4.3).

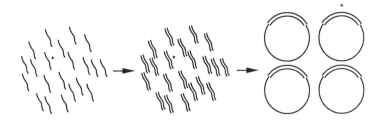

cDNA library is constructed in an expression vector, from cells transcribing an mRNA of interest (*).

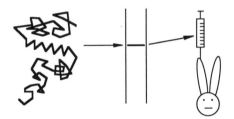

Purified protein of interest is used to generate antibody

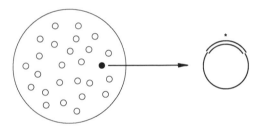

cDNA library is induced to express fusion proteins in E. coli; transferred to filter; screened with antibody. Hybridizing clones are analyzed to confirm identity of the gene.

Figure 5. cDNA cloning by antibody screening.

3. ISOLATION OF PATHOGENICITY GENES WITH KNOWN PRODUCTS

To date, the most productive approach to the cloning of fungal pathogenicity genes has been to work with genes whose products have already been identified biochemically as playing a role in disease, particularly genes whose products are enzymes. Genes whose products are known polypeptides are often more amenable to cloning than genes encoding unknown products. This is especially true if the polypeptide is available in relatively purified form, or if the fungus lacks an efficient transformation system to facilitate cloning. As described in Sections 2.3 and 2.4, the availability of purified protein frequently means that such genes may be isolated by antibody screening of a cDNA expression library or by using a partial amino acid sequence to generate an oligonucleotide probe for screening a genomic or cDNA library. The use of these approaches is well documented in other biological systems, but only a handful of examples have been reported with plant pathogenic fungi.

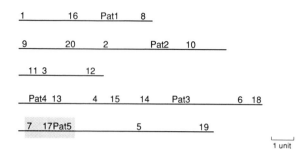

RFLP clones (1-20) and uncloned genes of interest (Pat1-Pat5) are mapped onto chromosomes.

RFLP marker (17) which maps close to a gene of interest (Pat5) may be used as the starting point for a chromosome walk using overlapping phage or cosmid clones.

Walk is complete when the gene of interest is found on a clone; identification of the clone may be based on transformation of a strain lacking Pat5.

Figure 6.   Cloning by RFLP mapping and chromosome walking.

3.1. Cell Wall-Degrading Enzymes

The ability of fungi to penetrate the complex external barrier of plants during pathogenesis attracted considerable attention to the mechanisms by which fungi may degrade (Kolattukudy and Crawford, 1987) or otherwise breach (Aist and Williams, 1971) cell walls. A growing literature describes the diversity of fungal enzymes which are capable of decomposing components of plant cell walls. The study of at least four such enzymes has advanced to the point of gene cloning: cutinases, pectate lyases, cellulases, and ligninases.

The gene encoding a cutinase from the pea pathogen *Fusarium solani* f.sp. *pisi* was cloned by taking advantage of the inducibility of cutinase production in this fungus (Soliday *et al.*, 1984). A cDNA library from induced mycelium of *F. solani pisi* was screened with radiolabeled cDNA prepared from either induced or noninduced mycelium. cDNA clones hybridizing to induced but not to noninduced probes were determined to be authentic cutinase clones based upon their ability to hybridize to *F. solani pisi* mRNA which, upon cell-free translation, reacted to anti-cutinase

IgG. In addition, the amino acid sequence predicted from the putative cutinase cDNA matched a partial amino acid sequence determined from purified *F. solani pisi* cutinase. Subsequent to the cloning of the *F. solani pisi* cutinase gene, homologous genes from the pepper pathogen *Colletotrichum capsici* and the papaya pathogen *C. gloeosporoides* were cloned in the same laboratory (Ettinger *et al.*, 1987). The *C. capsici* gene was cloned by screening a λgt11 cDNA library made from a culture induced for cutinase production with anticutinase antibodies. The cutinase gene from *C. gloeosporoides* was isolated from a genomic DNA library by heterologous hybridization to the *C. capsici* gene. The approaches taken to isolate these cutinase genes demonstrate several of the methods outlined in Section 2 of this chapter. More recently, work on cutinase gene expression has focused on its induction in *F. solani pisi* by cutin monomers (Woloshuk and Kolattukudy, 1986), apparently through transcriptional activation (Podila *et al.*, 1988).

Research on pectin-degrading enzymes has been extensive in plant pathogenic bacteria (reviewed in Collmer and Keen, 1986), and numerous genes encoding these enzymes have been cloned from bacteria. In fungi, a pectate lyase has been purified from *F. solani pisi* (Crawford and Kolattukudy, 1987) and from *A. nidulans* (Dean and Timberlake, 1989a). This enzyme has been implicated in the penetration of pea by *F. solani pisi* (Köller *et al.*, 1982). The pectate lyase gene from *A. nidulans* was isolated by immunoscreening a cDNA library with antibody raised from purified protein. This cDNA clone was then used as a probe to isolate the genomic copy (Dean and Timberlake, 1989b).

Cellulases and ligninases are produced by several fungi which are intermediate between saprophytes and parasites in their behavior. *Trichoderma* produces enzymes which are capable of degrading cellulose and hemicellulose, but not lignin (Knowles *et al.*, 1987). *Trichoderma* is not generally considered to be a severe plant pathogen, but is responsible for degradation of cellulose-containing products, particularly in tropical environments. *T. reesei* is known to make both exocellulases and endoglucanases. At least four cellulase genes have been cloned from *T. reesei*: two exocellulases or cellobiohydrolases, *cbh1* (Shoemaker *et al.*, 1983; Teeri *et al.*, 1983) and *cbh2* (Teeri *et al.*, 1987; Chen *et al.*, 1987); and two endoglucanases, *egl1* (Penttilä *et al.*, 1986; van Arsdell *et al.*, 1987) and *egl3* (cited in Knowles *et al.*, 1987). Because these enzymes are inducible, their cloning was performed in a manner similar to that used for the cutinase gene of *F. solani pisi*: a cDNA library made from mRNA of induced cells was screened differentially with cDNA probes made from induced or from noninduced cultures. The identity of clones which hybridized to the induced probe was confirmed by their ability to specifically hybridize to mRNA which could direct the translation *in vitro* of anticellulase immunoprecipitable material (van Arsdell *et al.*, 1987).

Lignin is a phenylpropanoid-based polymer which is exceptionally resistant to chemical degradation and serves to strengthen plant cell walls and protect cellulose from microbial attack (Tien, 1987). Enzymatic degradation of lignin in nature is achieved by several fungi, mostly basidiomycetes (Agrios, 1988). Among these are the so-called white-rot fungi, which cause significant economic loss in timber-producing trees and their products. The white-rot fungus *Phanerochaete chrysosporium* has been the focus of research on the enzyme ligninase, which depolymerizes lignin (Tien, 1987). Recent progress in the purification of ligninase from *P. chrysosporium* has permitted the cloning of several ligninase genes (Tien and Tu, 1987; de Boer *et al.*, 1987; Walther *et al.*, 1988). The genes were cloned by screening expression libraries with antibodies raised against purified ligninase isoenzymes.

All of the fungal cell wall-degrading enzymes whose genes have been cloned are believed to play a significant role in the ability of the fungus to penetrate or colonize its host. In no case, however, has this been conclusively established by an experiment which proves or refutes the role of the enzyme in disease. For example, if a gene encoding a wall-degrading enzyme is specifically deleted or otherwise mutated in a pathogen, is the fungus now unable to colonize its normal host? Alternatively, if the gene is transferred to a fungus which is not otherwise pathogenic on the natural host of the fungus from which the gene was cloned, is the transformant now able to act as a pathogen? Both of these questions can be approached experimentally, the latter by transformation and the former by using gene replacement methods. Gene replacement takes advantage of the property of homologous recombination by targeting transforming DNA to its homologous copy on a chromosome. Certain patterns of crossover events during integration of homologous DNA can result in replacement or disruption of the resident copy by the gene on the transformation vector (Fig. 7). A gene disruption experiment may be complicated by the presence of two or more related genes in the same organism encoding the same enzyme, such as cutinases (Ettinger *et al.*, 1987) or ligninases (Tien and Tu, 1987). Both copies of the gene would need to be disrupted, or an alternative organism identified which has only one functional copy of the gene. Using gene replacement methods to determine whether the pathogenicity of a fungus is altered by the introduction or removal of a single gene is a powerful and elegant experimental approach which can provide strong support or can clearly refute the role of a gene product in disease.

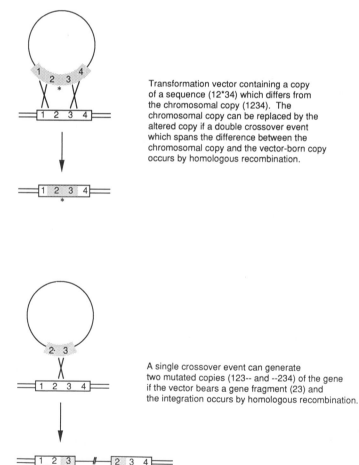

Transformation vector containing a copy of a sequence (12*34) which differs from the chromosomal copy (1234). The chromosomal copy can be replaced by the altered copy if a double crossover event which spans the difference between the chromosomal copy and the vector-born copy occurs by homologous recombination.

A single crossover event can generate two mutated copies (123-- and --234) of the gene if the vector bears a gene fragment (23) and the integration occurs by homologous recombination.

Figure 7.   Models for gene replacement and gene disruption.

3.2. Phytoalexin-Detoxifying Enzymes

A long-standing hypothesis in the field of plant pathology states that certain low-molecular-weight antimicrobial compounds, generically referred to as phytoalexins, are synthesized by plants after exposure to microorganisms as an active defense response to infection (VanEtten and Pueppke, 1976). There are numerous examples in the scientific literature of correlations between levels of phytoalexin present in host tissue and the plant's resistance to fungal infection. Similarly, fungi pathogenic on a particular host plant may show less sensitivity *in vitro* to the host's phytoalexin than a nonpathogen of that plant. The controlled demonstration that phytoalexin synthesis is a defense mechanism, or that phytoalexin tolerance by a fungus is required for pathogenicity, has been an elusive goal.

Research on the role of the phytoalexin pisatin in the interaction between pea and the pathogen *Nectria haematococca* has taken a biochemical, genetic, and recently a molecular genetic approach. *N. haematococca* mating population VI synthesizes a cytochrome P-450 monooxygenase which demethylates pisatin (Matthews and VanEtten, 1983) and converts it into a compound which is less inhibitory to *N. haematococca* growth *in vitro* (VanEtten *et al.*, 1975). Genetic analysis has demonstrated several genes, collectively called PDA (pisatin-demethylating activity) genes, which are responsible for the ability of *N. haematococca* to detoxify pisatin (Kistler and VanEtten, 1984; Tegtmeier and VanEtten, 1982). The inheritance of levels of pathogenicity in *N. haematococca* correlates strongly with the inheritance of pisatin-demethylating activity (summarized in VanEtten and Kistler, 1988). Recently, a PDA gene from *N. haematococca* was cloned by transformation of *Aspergillus nidulans* (Weltring *et al.*, 1988). The absence of an efficient transformation system for *N. haematococca* and the absence of pisatin-demethylating activity in the efficiently transformed *A. nidulans* were factors in its choice. A cosmid library containing DNA from a pisatin-demethylating strain of *N. haematococca* was used to transform *A. nidulans*. One transformant out of 1250 tested showed the ability to demethylate ^{14}C-labeled pisatin. This approach to cloning a putative fungal pathogenicity gene contrasts with that used to isolate the cutinase gene from the same fungus, and was selected in part because antibodies to the gene product were not available for the product of the PDA gene.

4. ISOLATION OF PATHOGENICITY GENES WITH UNKNOWN PRODUCTS BUT GENETICALLY ESTABLISHED ROLES IN DISEASE

Other classes of pathogenicity genes include those where the candidate molecule is not a polypeptide and hence not the direct product of the pathogenicity gene, as well as those where the pathogenicity determinant is completely unknown. Such genes have been identified genetically, and include important determinants of specificity and virulence/avirulence (Ellingboe, 1976). Vigorous gene-cloning efforts are under way with several "unknown product" pathogenicity genes.

4.1. Host-Specific Toxins

Fungal toxins represent a classic group of "candidate molecules" whose role in disease has been debated for decades. Many toxins produced by fungi appear to have a greater effect on hosts than on nonhosts, a specificity which naturally suggests that the toxin molecule is important in pathogenesis. Among the most intensively studied are the toxins produced by fungi in the genus *Cochliobolus*. This genus of cereal pathogens includes species which produce chemically diverse host-specific toxins, including linear polyketols (HMT toxin, made by *C. heterostrophus*) and cyclic peptides (HC toxin, made by *C. carbonum*). Toxin production in *Cochliobolus* is known to be controlled by single loci in *C. heterostrophus*, *C. carbonum*, and *C. victoriae* from genetic data (Lim and Hooker, 1971; Scheffer et al., 1967; Yoder and Gracen, 1975). In no case is the gene product of a *Cochliobolus* toxin locus known with certainty. In *C. carbonum*, however, the toxin locus cosegregates with the activity of two cyclic peptide synthetases believed to be responsible for HC-toxin biosynthesis (Walton, 1987). Two approaches to the isolation of *Cochliobolus* toxin loci are currently being taken. On the assumption that the HC-toxin synthetases may be the products of the *C. carbonum* toxin locus, partial amino acid sequences have been used to generate oligonucleotide probes for screening *C. carbonum* DNA libraries (J. Walton, personal communication). With *C. heterostrophus*, no such gene product information is available. As a result, an attempt to clone the *Tox1* locus of *C. heterostrophus* is being made by transformation: a cosmid library of DNA from a strain containing the *TOX1* allele (race T) is being transferred into a near-isogenic strain of *C. heterostrophus* containing the *tox1* allele (race O) (G. Turgeon, personal communication). Transformants are assayed individually for the ability to synthesize HMT. This approach makes no assumption about the nature of the *Tox1* locus except that it will be expressed in the *tox1* strain and that the functional gene product comes from *TOX1* and not *tox1*.

In other fungal plant pathogens which produce host-specific toxins, the appropriate approach to gene cloning will depend upon (1) the availability of genetic data implicating a single locus in toxin production and (2) the availability of biochemical information which might identify a gene product that could in turn lead to isolation of the locus, as is being pursued for the HC-toxin gene.

4.2. Mating Type Loci

Mating type in fungi refers to the genetic factors which control a broad spectrum of activities that are associated with the passage of cells between different stages of the life cycle (Herskowitz, 1988). For example, the mating type of a cell may determine its ability to enter meiosis, to fuse with another cell and undergo karyogamy, to undergo hyphal growth, and so forth. In heterothallic Ascomycetes and Basidiomycetes, mating type is traditionally considered to be a primary determinant of sexual reproduction: strains of appropriate mating type are able to mate and give rise to the cells in which meiosis occurs.

Characterization of the genes which regulate these phenomena using classical genetics has identified "mating type loci." The diversity of events which are influenced by mating type genes indicates that they function as master regulatory loci (Day and Garber, 1988; Herskowitz, 1987). In yeast, the molecular biology of the mating type locus is known in extraordinary detail (reviewed by Herskowitz, 1987), including the nature of the gene products encoded by the mating type locus and many of the genes which themselves regulate expression of the mating type locus.

For the purposes of this chapter, the mating type locus is of interest because it acts as a determinant of pathogenicity in some fungi, and because the cloning of mating type genes is an active research front which illustrates the diversity of techniques that can be brought to bear on the problem of cloning a fungal locus whose product is completely unknown.

Mating type loci are most clearly pathogenicity determinants in the smut fungi, where they determine the ability of two haploid, nonpathogenic cells to fuse, form a dikaryon, and grow as pathogenic hyphae. This complex of behaviors may be controlled by one locus, as in the smut *Ustilago violacea* (Day and Garber, 1988), or two loci, as in the smut *U. maydis* (Banuett and Herskowitz, 1988).

Mating type loci have recently been cloned from the ascomycetes *Neurospora crassa* (Glass et al., 1988) and *Cochliobolus heterostrophus* (Turgeon et al., 1988), and the basidiomycetes *Schizophyllum commune* (Giasson et al., 1989) and *Ustilago maydis* (Kronstad and Leong, 1989). The genes have been cloned by several approaches. The *N. crassa* mating type gene was isolated based on its proximity to *un-3*, a temperature-sensitive mutation which could be cloned by complementation. The mating type locus was so tightly linked that it occurred on the same cosmid which complemented the *un-3* mutation. The *S. commune* mating type allele *Aα 4* was cloned by cosmid clone walking from the tightly linked gene *PAB 1*. The identities of the cloned mating type alleles from both *N. crassa* and *S. commune* were verified by transformation: mating type activity was altered in most or all of the isolates transformed with clones containing putative mating type alleles.

Mating type alleles of *C. heterostrophus* and *U. maydis* were cloned directly from genomic libraries by transformation. In the case of *C. heterostrophus*, a library of cosmid clones made from a strain containing the *MAT1* allele was used to transform a strain containing the *MAT2* allele. Each transformant was assayed for changes in mating behavior, and a transformant was identified which was no longer heterothallic: it mated with both *MAT1* and *MAT2* testers. An allele (*b1*) of the *U. maydis* mating type locus was selected from a cosmid library by transforming a diploid of mating type *b2/b2* and screening for a transformant which showed mycelial growth. This phenotype was known from previous work with triploids to be a characteristic of *b1/b2/b2* strains.

The molecular mechanisms by which mating type loci regulate mating behavior are still unknown in fungi, with the exception of *S. cerevisiae*. It is not yet clear whether the sequences of the mating type genes themselves will suggest mechanisms, or will reveal common features of these genes when compared among different fungi.

4.3. Avirulence Genes

The controversy regarding the role of phytoalexins in disease resistance cited in Section 3.2 is modest in comparison to the literature debating the basis of host specificity in plant disease. At the heart of the specificity question is the gene-for-gene hypothesis, which postulates corresponding genes in the host and pathogen regulating the overall outcome of the interaction (Flor, 1971). Although the details of the hypothesis are beyond the scope of this chapter, there is good evidence that such genes [called avirulence genes (*Avr*) in the pathogen, and resistance genes in the host] may be identified through classical genetics (Gabriel *et al.*, 1988). These genes are important, since many significant pathogens are able to overcome the action of plant resistance genes by evolving populations containing appropriate *Avr* alleles that enable the pathogen to cause disease on previously resistant hosts.

Ironically, the diseases for which fungal *Avr* genes and host resistance genes have been well characterized are those in which the fungus is recalcitrant to exploitation by molecular biology. In many cases, the fungus is an obligate pathogen (rust, powdery mildew, downy mildew) for which a transformation system has not been developed. The *Avr* genes of the downy mildew *Bremia lactucae* have been extensively investigated at the level of conventional genetics (Michelmore *et al.*, 1984; Norwood *et al.*, 1983), and a detailed RFLP map of *B. lactucae* is being constructed with the aim of facilitating the isolation of *Avr* genes that are tightly linked to RFLP clones (Hulbert and Michelmore, 1988). Linkage which was estimated to represent a distance of approximately 150kb between an *Avr* locus and a cloned RFLP marker has been reported (Hulbert *et al.*, 1988), a distance that may permit chromosome walking by overlapping phage or cosmid clones (Spoerel and Kafatos, 1987). Identification of a clone containing the *Avr* gene using this method will presumably require a transformation system for *B. lactucae*.

5. PROSPECTS FOR PROGRESS

There is a paradox concerning the analysis of pathogenicity genes from fungi, alluded to at the beginning of this chapter. On the one hand, plant pathogenic fungi present an embarrassment of riches to the investigator: they produce a dazzling array of secondary metabolites, sophisticated penetration and colonization structures, reproductive patterns, and so on. The genes underlying many of these compounds and behaviors are legitimately considered to be pathogenicity genes. However, the very diversity of strategies that fungi have developed in order to succeed as pathogens may preclude the emergence of an overall conceptual model for how pathogenicity genes work in fungi, or an experimental model to focus research efforts on one system. For example, the insights gained from a detailed understanding of the significance of toxin genes in disease may not shed much light on the regulation of genes encoding phytoalexin-detoxifying enzymes. This is not necessarily an unpleasant state of affairs; it simply means that genes will need to be isolated and their role in the disease process evaluated independently, rather than expecting them to fit into a preexisting pattern.

What approaches will be most useful in orienting future research on fungal pathogenicity genes? At least three, not mutually exclusive, are emerging. The first, and most developed to date, is to study the regulation of a gene in the organism from which it was isolated. This approach is exemplified by the work on the cutinase gene of *Fusarium solani* f.sp. *pisi*. Eventually, it will involve dissection of the entire suite of pathogenicity genes of an organism. This approach will probably confirm the role of a number of gene products which have long been proposed as pathogenicity determinants. It is the regulation of their expression that remains largely unknown at present and is likely to provide sharper insights into the role of these genes in disease.

A second approach is to transfer genes into nonpathogens or into pathogens with different hosts. This addresses the question of how many genes are required to convert a nonpathogen into a pathogen, or whether a pathogenicity gene can also be thought of as a host- or organ-specificity gene. For example, would transferring the genes that are required for HMT production in *Cochliobolus heterostrophus* to the maize pathogen *C. carbonum* enable the latter to be as virulent on Texas male sterile cytoplasm maize as the toxin-producing race T of *C. heterostrophus*? Would an *Avr* gene from the potato pathogen *Phytophthora infestans* allow the soybean pathogen *P. megasperma* f.sp. *glycinea* to infect susceptible potatoes? This approach will certainly require careful planning and probably approval by regulatory agencies.

A third and extremely underexploited route to the problem of isolating fungal pathogenicity genes is through mutant analysis. Compared to yeast, *Aspergillus*, or *Neurospora*, very little progress has been made in characterizing interesting mutants in plant pathogenic fungi. Active work taking this approach includes the isolation of spore morphology mutants in the rice pathogen *Magnaporthe grisea* (J. Hamer, personal communication), appressorium-deficient mutants of *Colletotrichum lindemuthianum* (R. Staples, personal communication), nonpathogenic mutants of *Fulvia fulva* (Oliver, *et al.*, 1988), and mating type mutants of *Ustilago maydis* (Banuett and Herskowitz, 1988). As transformation systems become more widespread among plant pathogenic fungi, the ability to clone genes which complement mutations will grow. The great value of mutant analysis in model fungal systems should be a spur to encourage ingenuity in designing assays which identify mutants altered in significant components of pathogenicity.

The isolation of pathogenicity genes from plant pathogenic fungi is in a state of rapid change, as indicated by the large number of unpublished results cited in this chapter. The spread of transformation technology is driving progress in a large number of species, and we may expect the isolation of pathogenicity genes to extend to more difficult systems such as obligate pathogens or fungi without sexual stages, and to more difficult genes such as those responsible for host specificity.

DEDICATION.   To Laurel Elizabeth Garber.

6. REFERENCES

Agrios, G. N., 1988, *Plant Pathology*, 3rd ed., Academic Press, New York, pp. 71–72.

Aist, J. R., and Williams P. H., 1971, The cytology and kinetics of cabbage root hair penetration by *Plasmodiophora brassicae*, *Can. J. Bot.*, **49**:2023–2034.

Akins, R.A., and Lambowitz, A.M., 1985, General method for cloning *Neurospora crassa* nuclear genes by complementation, *Mol. Cell. Biol.* **5**:2272–2278.

Ballance, D. J., Buxton, F. P., and Turner, G., 1983, Transformation of *Aspergillus nidulans* by the orotidine-5′-phosphate decarboxylase gene of *Neurospora crassa*, *Biochem. Biophys. Res. Commun.* **112**:284–289.

Banuett, F., and Herskowitz, I., 1988, *Ustilago maydis*, smut of maize, in: *Genetics of Plant Pathogenic Fungi* (G. S. Sidhu, ed.), Academic Press, New York, pp. 427–455.

Beach, D., and Nurse, P., 1981, High-frequency transformation of the fission yeast *Schizosaccharomyces pombe*, *Nature* **290**:140–142.

Botstein, D., and Fink, G. R., 1988, Yeast: An experimental organism for modern biology, *Science* **240**:1439–1443.

Botstein, D., White, R. L., Skolnick, M., and Davis, R. W., 1980, Construction of a genetic linkage map in man using restriction fragment length polymorphisms, *Am. J. Hum. Genet.* **32**:314–331.

Case, M. E., Schweizer, M., Kushner, S. R., and Giles, N. H., 1979, Efficient transformation of *Neurospora crassa* by utilizing hybrid plasmid DNA, *Proc. Natl. Acad. Sci. USA* **76**:5259–5263.

Chen, C. M., Gritzali, M., and Stafford, D. W., 1987, Nucleotide sequence and deduced primary structure of cellobiohydrolase II of *Trichoderma reesei*, *Bio/Technology* **5**:274–278.

Collmer, A., and Keen, N. T., 1986, The role of pectic enzymes in plant pathogenesis, *Annu. Rev. Phytopathol.* **24**:383–409.

Crawford, M. S., and Kolattukudy, P. E., 1987, Pectate lyase from *Fusarium solani* f.sp.*pisi*: Purification, characterization, *in vitro* translation of the mRNA, and involvement in pathogenicity, *Arch. Biochem. Biophys.* **258**:196–205.

Day, A. W., and Garber, E. D., 1988, *Ustilago violacea*, anther smut of the Caryophyllaceae, in: *Genetics of Plant Pathogenic Fungi* (G. S. Sidhu, ed.), Academic Press, New York, pp. 457–482.

Dean, R. A., and Timberlake, W. E., 1989a, Production of cell wall-degrading enzymes by *Aspergillus nidulans*: A model system for fungal pathogenesis of plants, *Plant Cell* **1**:265–273.

Dean, R. A., and Timberlake, W. E., 1989b, Regulation of the *Aspergillus nidulans* pectate lyase gene (*pel*A), *Plant Cell* **1**:275–284.

de Boer, H. A., Zhang, Y. Z., Collins, C., and Reddy, C. A., 1987, Analysis of nucleotide sequences of two ligninase cDNAs from a white-rot filamentous fungus, *Phanerochaete chrysosporium*, *Gene* **60**:93–102.

Ellingboe, A. H., 1976, Genetics of host–parasite interactions, in: *Physiological Plant Pathology* (R. Heitefuss and P. H. Williams, eds.), Springer-Verlag, Berlin, pp. 761–778.

Ettinger, W. F., Thukral, S. K., and Kolattukudy, P. E., 1987, Structure of cutinase gene, cDNA, and the derived amino acid sequence from phytopathogenic fungi, *Biochemistry* **26**:7883–7892.

Faugeron, G., Goyon, C., and Grégoire, A., 1989, Stable allele replacement and unstable non-homologous integration events during transformation of *Ascobolus immersus*, *Gene* **76**:109–119.

Flor, H. H., 1971, Current status of the gene-for-gene concept, *Annu. Rev. Phytopathol* **9**:275–296.

Fungal Genetics Stock Center, 1988, Catalogue of strains, *Fungal Genetics Newsletter Suppl.* 35.

Gabriel, D. W., Loschke, D. C., and Rolfe, B. G., 1988, Gene-for-gene recognition: The ion channel defense model, in: *Molecular Genetics of Plant– Microbe Interactions* (R. Palacios and D. P. S. Verma, eds.), APS Press, St. Paul, pp. 3–14.

Giasson, L., Specht, C. A., Milgrim, C., Novotny, C. P., and Ullrich, R. C., 1989, Cloning and comparison of Aα mating-type alleles of the Basidiomycete *Schizophyllum commune*, *Mol. Gen. Genet.* **218**:72–77.

Glass, N. L., Vollmer, S. J., Staben, C., Grotelueschen, J., Metzenberg, R. L., and Yanofsky, C., 1988, DNAs of the two mating-type alleles of *Neurospora crassa* are highly dissimilar, *Science* **241**:570–573.

Herskowitz, I., 1987, A master regulatory locus that determines cell specialization in yeast, *Harvey Lect.* **81**:67–92.

Herskowitz, I., 1988, The life cycle of the budding yeast, *Saccharomyces cerevisiae*, *Microbiol. Rev.* **52**:536–553.

Hinnen, A., Hicks, J. B., and Fink, G. R., 1978, Transformation of yeast, *Proc. Natl. Acad. Sci. USA* **75**:1929–1933.

Holden, D., Wang, J., and Leong, S. A., 1985, Analysis of heat shock genes from *Ustilago maydis*, Abstract, First International Congress on Plant Molecular Biology, Savannah, Georgia.

Hulbert, S. H., and Michelmore, R. W., 1988, DNA restriction fragment length polymorphism and somatic variation in the lettuce downy mildew fungus, *Bremia lactucae*, *Mol. Plant–Microbe Interact* **1**:17–24.

Hulbert, S. H., Ilott, T. W., Legg, E. J., Lincoln, S. E., Lander, E. S., and Michelmore, R. W., 1988, Genetic analysis of the fungus, *Bremia lactucae*, using restriction fragment length polymorphisms, *Genetics* **120**:947–958.

Huynh, T. V., Young, R. A., and Davis, R. W., 1985, Constructing and screening cDNA libraries in λgt10 and λgt11, in: *DNA Cloning*, Volume 1 (D. M. Glover, ed.), IRL Press, Washington, D.C., pp. 49–78.

Kinnaird, J. H., Keighren, M. A., Kinsey, J. A., Eaton, M., and Fincham, J. R. S., 1982, Cloning of the *am* (glutamate dehydrogenase) gene of *Neurospora crassa* through the use of a synthetic DNA probe, *Gene* **20**:387–396.

Kistler, H. C., and VanEtten, H. D., 1984, Three non-allelic genes for pisatin demethylation in the fungus *Nectria haematococca*, *J. Gen. Microbiol.* **130**:2595–2603.

Knowles, J., Lehtovaara, P., Penttilä, M., Teeri, T., Harkki, A., and Salovuori, I., 1987, The cellulase genes of Trichoderma, *Antonie van Leeuwenhoek J. Microbiol. Serol.* **53**:335–341.

Kolattukudy, P. E., and Crawford, M. S., 1987, The role of polymer degrading enzymes in fungal pathogenesis, in: *Molecular Determinants of Plant Diseases* (S. Nishimura, C. P. Vance, and N. Doke, eds.), Springer-Verlag, Berlin, pp. 75–96.

Köller, W., Allan, C. R., and Kolattukudy, P. E., 1982, Role of cutinase and cell wall degrading enzymes in infection of *Pisum sativum* by *Fusarium solani* f.sp. *pisi*, *Physiol. Plant Pathol.* **20**:47–60.

Kronstad, J. W., and Leong, S. A., 1989, Isolation of two alleles of the *b* locus of *Ustilago maydis*, *Proc. Natl. Acad. Sci. USA* **86**:978–982.

Lim, S. M., and Hooker, A. L., 1971, Southern corn leaf blight: Genetic control of pathogenicity and toxin production in race T and race O of *Cochliobolus heterostrophus*, *Genetics* **69**:115–117.

Matthews, D. E., and VanEtten, H. D., 1983, Detoxification of the phytoalexin pisatin by a fungal cytochrome P-450, *Arch. Biochem. Biophys.* **224**:494–505.

Metzenberg, R. L., and Grotelueschen, J., 1987, A restriction polymorphism map of *Neurospora crassa*: More data, *Fungal Genet. Newsl.* **34**:39–44.

Metzenberg, R. L., and Grotelueschen, J., 1988, Restriction polymorphism maps of *Neurospora crassa*: Updates, *Fungal Genet. Newsl.* **35**:30–35.

Metzenberg, R. L., Stevens, J. N., Selker, E. U., and Morzycka-Wroblewska, E., 1985, Identification and chromosomal distribution of 5S rRNA genes in *Neurospora crassa*, *Proc. Natl. Acad. Sci. USA* **82**:2067–2071.

Michelmore, R. W., Norwood, J. M., Ingram, D. S., Crute, I. R., and Nicholson, P., 1984, The inheritance of virulence in *Bremia lactucae* to match resistance factors 3, 4, 5, 6, 8, 9, 10, and 11 in lettuce (*Lactuca sativa*), *Plant Pathol.* **33**:301–315.

Mierendorf, R. C., Percy, C., and Young, R. A., 1987, Gene isolation by screening λgt11 libraries with antibodies, *Methods Enzymol.* **152**:458–469.

Norwood, J. M., Michelmore, R. W., Crute, I. R., and Ingram, D. S., 1983, The inheritance of specific virulence in *Bremia lactucae* (downy mildew) to match resistance factors 1, 2, 4, 6, and 11 in *Lactuca sativa* (lettuce), *Plant Pathol.* **32**:177–186.

Oliver, R. P., Roberts, I. N., McHale, M., Coddington, A., Talbot, N., Lewis, B., Kenyon, L., Turner, J., and Hamouri, B. E., 1988, Molecular approaches for the study of the pathogenicity of *Fulvia fulva*, in: *Molecular Genetics of Plant–Microbe Interactions* (R. Palacios and D. P. S. Verma, eds.), APS Press, St. Paul, pp. 263–264.

Paietta, J. V., Akins, R. A., Lambowitz, A. M., and Marzluf, G. A., 1987, Molecular cloning and characterization of the cys-3 regulatory gene of *Neurospora crassa*, *Mol. Cell. Biol.* **8**:2506–2511.

Panaccione, D. G., McKiernan, M., and Hanau, R. M., 1988, *Colletotrichum graminicola* transformed with homologous and heterologous benomyl-resistance genes retains expected pathogenicity to corn, *Mol. Plant–Microbe Interact.* **1**:113–120.

Parsons, K. A., Chumley, F. G., and Valent, B., 1987, Genetic transformation of the fungal pathogen responsible for rice blast disease, *Proc. Natl. Acad. Sci. USA* **84**:4161–4165.

Penttilä, M., Lehtovaara, P., Nevalainen, H., Bhikhabhai, R., and Knowles, J., 1986, Homology between cellulase genes of *Trichoderma reesei*: Complete nucleotide sequence of the endoglucanase I gene, *Gene* **45**:253–263.

Podila, G. K., Dickman, M. B., and Kolattukudy, P. E., 1988, Transcriptional activation of a cutinase gene in isolated fungal nuclei by plant cutin monomers, *Science* **242**:922–925.

Rambosek, J., and Leach, J., 1987, Recombinant DNA in filamentous fungi: Progress and prospects, *CRC Crit. Rev. Biotechnol.* **6**:357–393.

Rodriguez, R. J., and Yoder, O. C., 1987, Selectable markers for transformation of the fungal plant pathogen *Glomerella cingulata* f.sp. *phaseoli* (*Colletotrichum lindemuthianum*), *Gene* **54**:73–81.

Rose, M. D., 1987, Isolation of genes by complementation in yeast, *Methods Enzymol.* **152**:481–504.

Scheffer, R. P., Nelson, R. R., and Ullstrup, A. J., 1967, Inheritance of toxin production and pathogenicity in *Cochliobolus carbonum* and *Cochliobolus victoriae*, *Phytopathology* **57**:1288–1291.

Shoemaker, S., Schweikart, V., Ladner, M., Gelfand, D., Kwok, S., Myambo, K., and Innis, M., 1983, Molecular cloning of exocellobiohydrolase I derived from *Trichoderma reesei* strain L27, *Bio/Technology* **1**:691.

Sidhu, G. S., (ed.), 1988, *Genetics of Plant Pathogenic Fungi*, Academic Press, New York.

Soliday, C. L., Flurkey, W. H., Okita, T. W., and Kolattukudy, P. E., 1984, Cloning and structure determination of cDNA for cutinase, an enzyme involved in fungal penetration of plants, *Proc. Natl. Acad. Sci. USA* **81**:3939–3943.

Spoerel, N. A., and Kafatos, F. C., 1987, Isolation of full-length genes: Walking the chromosome, *Methods Enzymol.* **152**:598–603.

Stinchcomb, D. T., Thomas, M., Kelly, J., Selker, E., and Davis, R. W., 1980, Eukaryotic DNA segments capable of autonomous replication in yeast, *Proc. Natl. Acad. Sci. USA* **77**:4559–4563.

Teeri, T., Salovuori, I., and Knowles, J., 1983, The molecular cloning of the major cellulase gene from *Trichoderma reesei*, *Bio/Technology* **1**:696.

Teeri, T., Lehtovaara, P., Kauppinen, S., Salovuori, I., and Knowles, J., 1987, Homologous domains in *Trichoderma reesei* cellulolytic enzymes: Gene sequence and expression of cellobiohydrolase II, *Gene* **51**: 43–52.

Tegtmeier, K. J., and VanEtten, H. D., 1982, The role of pisatin tolerance and degradation in the virulence of *Nectria haematococca* on peas: A genetic analysis, *Phytopathology* **72**:608–612.

Tien, M., 1987, Properties of ligninase from *Phanerochaete chrysosporium* and their possible applications, *CRC Crit. Rev. Microbiol.* **15**:141–168.

Tien, M., and Tu, C.-P. D., 1987, Cloning and sequencing of a cDNA for a ligninase from *Phanerochaete chrysosporium*, *Nature* **326**:520–523.

Tilburn, J., Scazzocchio, C., Taylor, G. C., Zabicky-Zissman, J. H., Lockington, R. A., and Davies, R. W., 1983, Transformation by integration in *Aspergillus nidulans*, *Gene* **26**:205–221.

Tsukuda, T., Carleton, S., Fotheringham, S., and Holloman, W. K., 1988, Isolation and characterization of an autonomously replicating sequence from *Ustilago maydis*, *Mol. Cell. Biol.* **8**:3703–3709.

Turgeon, B. G., MacRae, W. D., Garber, R. C., Fink, G. R., and Yoder, O. C., 1986, A cloned tryptophan-synthesis gene from the ascomycete *Cochliobolus heterostrophus* functions in *Escherichia coli*, yeast and *Aspergillus nidulans*, *Gene* **42**:79–88.

Turgeon, B. G., Garber, R. C., and Yoder, O. C., 1987, Development of a fungal transformation system based on selection of sequences with promoter activity, *Mol. Cell. Biol.* **7**:3297–3305.

Turgeon, B. G., Ciuffetti, L., Schäfer, W., and Yoder, O. C., 1988, Isolation of the mating type locus of *Cochliobolus heterostrophus*, in: *Molecular Genetics of Plant–Microbe Interactions* (R. Palacios and D. P. S. Verma eds.), APS Press, St. Paul, pp. 265–266.

van Arsdell, J. N., Kwok, S., Schweikart, V. L., Ladner, M. B., Gelfand, D. H., and Innis, M., 1987, Cloning, characterization, and expression in *Saccharomyces cerevisiae* of endoglucanase I from *Trichoderma reesei*, *Bio/Technology* **5**:60–64.

VanEtten, H. D., and Kistler, H. C., 1988, *Nectria haematococca*, in: *Genetics of Plant Pathogenic Fungi* (G. S. Sidhu, ed.), Academic Press, New York, pp. 189–206.

VanEtten, H. D., and Pueppke, S. G., 1976, Isoflavonoid phytoalexins, in: *Biochemical Aspects of Plant–Parasitic Relationships* (J. Friend and D. R. Threlfall, eds.), Academic Press, New York, pp. 239–289.

VanEtten, H. D., and Pueppke, S. G., and Kelsey, T. C., 1975, 3,6a-Dihydroxy-8,9-methylenedioxypterocarpan as a metabolite of pisatin produced by *Fusarium solani* f.sp. *pisi*, *Phytochemistry* **14**:1103–1105.

Wallace, R. B., and Miyada, C. G., 1987, Oligonucleotide probes for the screening of recombinant DNA libraries, *Methods Enzymol.* **152**:432–442.

Walther, I., Kalin, M., Reiser, J., Suter, F., Fritsche, B., Saloheimo, M., Leisola, M., Teeri, T., Knowles, J. K. C., and Fietcher, A., 1988, Molecular analysis of a *Phanerochaete chrysosporium* lignin peroxidase gene, *Gene* **70**:127–137.

Walton, J. D., 1987, Two enzymes involved in biosynthesis of the host-selective phytotoxin HC-toxin, *Proc. Natl. Acad. Sci. USA* **84**:8444–8447.

Wang, J., Holden, D. W., and Leong, S., 1988, Gene transfer system for *Ustilago maydis* based on resistance to hygromycin B, *Proc. Natl. Acad. Sci. USA* **85**:865–869.

Weltring, K.-M., Turgeon, B. G., Yoder, O. C., and VanEtten, H. D., 1988, Isolation of a phytoalexin-detoxification gene from the plant pathogenic fungus *Nectria haematococca* by detecting its expression in *Aspergillus nidulans*, *Gene* **68**:335–344.

Woloshuk, C. P., and Kolattukudy, P. E., 1986, Mechanism by which contact with plant cuticle triggers cutinase gene expression in the spores of *Fusarium solani* f.sp. *pisi*, *Proc. Natl. Acad. Sci. USA* **83**:1704–1708.

Wood, W. I., 1987, Gene cloning based on long oligonucleotide probes, *Methods Enzymol.* **152**:443–447.

Yelton, M. M., Hamer, J. E., and Timberlake, W. E., 1984, Transformation of *Aspergillus nidulans* by using a trpC plasmid, *Proc. Natl. Acad. Sci. USA* **81**:1470–1474.

Yelton, M. M., Timberlake, W. E., and van den Hondel, C. A. M. J. J., 1985, A cosmid for selecting genes by complementation in *Aspergillus nidulans*: Selection of the developmentally regulated yA locus, *Proc. Natl. Acad. Sci. USA* **82**:834–838.

Yoder, O. C., 1980, Toxins in pathogenesis, *Annu. Rev. Phytopathol.* **18**:103–129.

Yoder, O. C., and Gracen, V. E., 1975, Segregation of pathogenicity types and host-specific toxin production in progenies of crosses between races T and O of *Helminthosporium maydis* (*Cochliobolus heterostrophus*), *Phytopathology* **65**:273–276.

Yoder, O. C., Weltring, K., Turgeon, B. G., Garber, R. C., and VanEtten, H. D., 1986, Technology for molecular cloning of fungal virulence genes, in: *Biology and Molecular Biology of Plant–Pathogen Interactions* (J. Bailey, ed.), Springer-Verlag, Berlin, pp. 371–384.

Young, N. D., Zamir, D., Ganal M. W., and Tanksley, S. D., 1988, Use of isogenic lines and simultaneous probing to identify DNA markers tightly linked to the Tm-2a gene in tomato, *Genetics* **120**:579–585.

23

Current Status of the Molecular Basis of Candida Pathogenicity

David R. Soll

1. INTRODUCTION

All organisms have evolved to succeed in a particular niche, and fungi pathogenic to man are no exception. Many colonize healthy hosts at low levels as commensals. When the host becomes physiologically or immunologically compromised, or in some cases, even when the host exhibits no obvious predisposing conditions, these fungi multiply and penetrate tissue, causing infection and, in an ever-increasing number of cases, death. Our primary interest is the capacity of these fungi to cause disease, and our ultimate goals are to control their growth and invasiveness, and to irradicate them in infected hosts. The application of new technologies of molecular biology will, without doubt, play a crucial role in this endeavor since it will be through them that we ultimately elucidate the genes, regulatory pathways, and gene products which combine to make these fungi such successful pathogens. In this review, the discussion will focus first upon the basic biology of one of the more prominent of these fungi, *Candida albicans*. The progress which has been made as well as the potential which now exists for applying molecular genetic techniques to the pathogenesis and epidemiology of this opportunistic pathogen will be reviewed.

2. THE BASIC BIOLOGY OF CANDIDA DIMORPHISM

Before considering the application of molecular genetic techniques to *C. albicans*, it would be of benefit to review the basic biology of the organism and highlight those characteristics which may be basic to its pathogenesis. *C. albicans* is a dimorphic yeast which multiplies either in a budding form (Fig. 1A) indistinguishable from diploid *Saccharomyces cerevisiae* (Bedell *et al.*, 1980; Gow *et al.*, 1986; Herman and Soll, 1984) or in an elongate hyphal form (Fig. 1B; Herman and Soll, 1984; Soll *et al.*, 1978). Most strains can grow in a simple medium containing biotin, salts, a carbon source such as glucose, and a nitrogen source other than nitrate (Odds, 1988). In budding growth, an outpocketing of the plasma membrane and cell wall expands into a round to

---

David R. Soll • Department of Biology, University of Iowa, Iowa City, Iowa 52242.

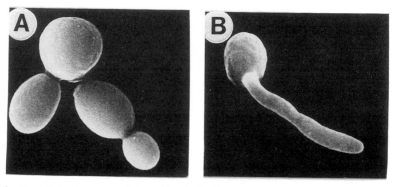

Figure 1. The cellular phenotypes of dimorphism in *C. albicans*. Scanning electron micrographs of a budding cell (A) and a hypha-forming cell (B).

ellipsoidal daughter cell. At the time of initial evagination, a filament ring forms under the plasma membrane at the mother–daughter cell junction, then disappears (Soll and Mitchell, 1983) as the chitin-containing septal ring grows inwardly to form the complete septum separating mother and daughter cell (Mitchell and Soll, 1979b). Interestingly, the final septum contains a central pore (Odds, 1984). During bud growth, there appear to be two mechanisms of wall expansion, a general system which is evenly distributed throughout the wall and a localized apical expansion zone at the proximal end of the expanding bud (Staebell and Soll, 1985; Soll *et al.*, 1985). These two expansion mechanisms are regulated by a refined temporal program (see Fig. 2) in which the apical zone shuts down when the daughter cell reaches two-thirds its final size, and the general zone shuts down when the daughter cell reaches its final size (Staebell and Soll, 1985; Soll, 1986). Changes in F-actin distribution during bud growth include the localization of F-actin granules

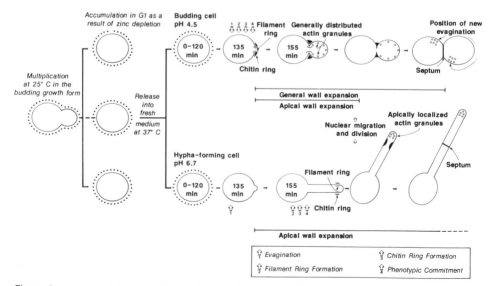

Figure 2. A temporal diagram of landmark events during bud and hypha formation under the regime of pH-regulated dimorphism. Note that the time of evagination is the same in the two populations, but the timing of filament ring formation, chitin ring formation, and phenotypic commitment differ. See Soll (1986).

throughout the cytoplasmic cortex of the daughter cell followed by relocalization at the next site of evagination (Anderson and Soll, 1986). During daughter cell growth, the nucleus migrates to the mother–daughter cell neck, then divides, one daughter nucleus reentering the mother cell and one entering the daughter cell (Bedell *et al.*, 1980). Microtubules span the axis of the mother–daughter cell during nuclear migration and division (Soll, 1988). In most respects, this scenario, which is diagrammed in Fig. 2, is similar to that of bud formation in *Saccharomyces cerevisiae* (Byers, 1981).

The capacity to form a hypha is without doubt one major pathogenic attribute of infectious yeast like *C. albicans*. Variants incapable of hypha formation exhibit decreased virulence (Sobel *et al.*, 1984). In stained sections of infected tissue, hyphae can be seen weaving through tissue, generating branches, and producing pockets of budding cells through lateral budding (e.g., Luna and Tortoledo, 1985). Although hyphae can be found in the absence of any obvious pathogenesis (e.g., Arendorf and Walker, 1979), several studies have demonstrated increased hypha formation associated with disease (e.g., Budtz-Jorgensen *et al.*, 1975). There seems to be little doubt that the hypha has evolved as a mechanism for tissue penetration (Odds, 1988) and therefore, understanding the regulation of hypha formation may provide us with new strategies of therapy.

Hypha formation can be induced in budding cells by a number of diverse methods including (1) dilution into serum-containing medium (Barlow and Alersley, 1974), (2) dilution into starvation medium containing *N*-acetylglucosamine (Mattia *et al.*, 1982; Simonetti *et al.*, 1974; Shepherd *et al.*, 1980b) as well as other related compounds (Dabrowa *et al.*, 1976), and (3) dilution of stationary phase or starved cells into fresh medium at high pH under the regime of pH-regulated dimorphism (Buffo *et al.*, 1984; Mattia and Cassone, 1979; Soll and Herman, 1983 see Fig. 2). For most strains, the conditions of high temperature (above 32°C) and high pH (above 6.2) are conducive to hypha formation (Buffo *et al.*, 1984), but many strains exist which form hyphae at low temperature and low pH (e.g., variant M10; Bedell and Soll, 1979). Using temperature and pH to regulate phenotype under the regime of "pH-regulated dimorphism" (Buffo *et al.*, 1984), it has been possible to compare the temporal programs of bud and hypha formation because of the unusual degree of temporal parallelism in diverging cell populations obtained by this method (Soll, 1984, 1986). Under the regime of pH-regulated dimorphism, a population of cells which has entered stationary phase at 25°C under zinc limiting conditions (Bedell and Soll, 1979; Soll, 1985) is separated into two subpopulations (Fig. 2). One subpopulation is diluted into fresh medium at 37°C, pH 4.5 (low pH) and the other into fresh medium with the same nutrient and salt composition, the same temperature, but at pH 6.7 (high pH). At low pH, the unbudded singlets, which are blocked in G1, semisynchronously form buds *en masse* after an average lag period of 125 to 135 min., and at high pH, they form hyphae with the same synchrony after roughly the same average lag period (Buffo *et al.*, 1984). Using this method, it has been demonstrated that during hypha formation at high pH, the dynamics of actin localization at the site of the new evagination (Anderson and Soll, 1986) as well as the morphology of the membrane and cell wall are superficially similar to the localization and timing observed during bud formation, at low pH. However, in contrast to budding cells, the filament ring under the plasma membrane (Soll and Mitchell, 1983), and associated chitin ring on the plasma membrane (Mitchell and Soll, 1979b) are rarely observed at the mother–daughter cell junction immediately after evagination (Soll, 1986). Instead, as the hypha continues to elongate, the filament ring forms 20 to 30 min after evagination under the plasma membrane and the chitin ring over the plasma membrane, on average roughly 2 μm from the mother–daughter cell junction (Soll and Mitchell, 1983). It has been suggested that the filament ring forms at the apical growth zone of an emerging daughter cell, and the temporal delay during hypha formation leads to the change in filament ring position which in turn results in the change in septum position (Soll and Mitchell, 1983; Soll, 1986; Fig. 2). In addition, during hyphal growth, actin granules are localized at the hyphal tip rather than distributed evenly in the cortex, as is the case in an expanding bud (Anderson and Soll,

1986), and the level of general wall expansion is dramatically reduced, resulting in continuous apical expansion as long as the cell continues to grow in the hyphal form (Staebell and Soll, 1985). The temporal and spatial differences in the developmental cytology of bud and hypha which are outlined in Fig. 2 are accompanied by differences in the proportions of wall components (Chattaway *et al.*, 1968), as well as by a number of other qualitative differences (e.g., Shepherd *et al.*, 1980a; Niimi *et al.*, 1980; Braun and Calerone, 1978; Stewart *et al.*, 1988; Kaur *et al.*, 1988), but few of these differences have been demonstrated to be absolutely requisite for the genesis of the alternative growth forms (Anderson and Soll, 1986). In several cases, the experimental procedures employed for generating the alternative phenotypes for comparison have been too disparate to conclude that the physiological differences which were measured were involved in phenotypic determination (Soll, 1984, 1985).

To date, we have no firm handle on the regulatory processes involved in the choice between the genesis of a bud and the genesis of a hypha. The rigorous comparisons of the developing bud and developing hypha have provided us with no more than lists of phenotypic characteristics of the alternative growth forms, and we are still ignorant of the molecular switch which sends a cell down one track for bud formation, or another for hypha formation (see later section on the molecular biology of dimorphism). Recently, it was demonstrated that an increase in intracellular pH accompanies hypha formation, but the two studies to date on this subject (Stewart *et al.*, 1988; Kaur *et al.*, 1988) differ in the measured dynamics of this increase, and the change in pH has not been fully explored to distinguish between cause and effect in relation to phenotypic differentiation.

The temporal parallelism in diverging cell populations achieved by pH-regulated dimorphism (Fig. 2) also provides a method for determining the point of phenotypic commitment during bud and hypha formation (Mitchell and Soll, 1979a). Cells were shifted from low to high pH to pinpoint the time of commitment to the bud phenotype, and alternatively, cells were shifted from high to low pH to pinpoint the time of commitment to the hypha phenotype. The commitment times differed between the diverging populations, and, interestingly, coincided with the formation of the filament rings and incipient chitin-containing ring (Soll and Mitchell, 1983; Soll, 1986). These temporal correlations are diagrammed in Fig. 2. The timing of filament ring formation may therefore be fundamental to the difference in daughter cell shape, and this hypothesis is supported by the phenotype of *S. cerevisiae* mutants with disrupted filament rings. At nonpermissive temperature, a strain of *S. cerevisiae* A364A carrying a temperature-sensitive mutation in cdc10 forms a daughter cell which is superficially similar to a pseudohypha. A scanning electron micrograph of this hyphalike daughter cell taken by A. Gosney and J. Forshyke was recently published in a review article by A.E. Wheals on the yeast cell cycle (Wheals, 1987); this photo could easily substitute for a pseudohypha-forming and perhaps even a hypha-forming cell under the regime of pH-regulated dimorphism. The cdc10 gene codes for a component of the transient filament ring, and cdc10 mutants do not form this ring at the mother–daughter cell junction in *S. cerevisiae*. However, it is not immediately obvious why the absence of the ring would lead to an elongate rather than round daughter cell since one would not expect the absence of the ring to lead to changes in the zones of wall deposition (Staebell and Soll, 1985), which seem the most likely candidates for explaining the difference in cell shape (Soll, 1986). The phenotype of this cdc10 mutant points to the possible usefulness of employing cell division cycle mutations in elucidating the regulatory pathways basic to dimorphism in organisms like *C. albicans*. Fortunately, the cloned cdc10 gene hybridizes to a single *Eco*RI fragment on a Southern blot, and is now being cloned from *C. albicans* (B. Morrow, J. Anderson, J. Pringle, and D. R. Soll, unpublished results). Antiserum specific to the *S. cerevisiae* gene product selectively stains the neck region of budding *C. albicans* and a band at the position of the ring in a hypha, 2 μm on average from the junction of mother cell and tube (J. Anderson, J. Pringle, and D.R. Soll, unpublished observations). It therefore seems reasonable to suggest that studies of the genes involved in bud development and the cell division cycle in *Saccharomyces* will provide us with tools for unraveling the complex developmental pathways involved in the decision to make a bud or hypha in *C. albicans*. It also seems apparent that the evolution of the hypha in pathogenic yeast incorporated changes in the basic processes involved in bud formation.

3. THE BASIC BIOLOGY OF HIGH-FREQUENCY SWITCHING IN C. ALBICANS

It has become increasingly apparent in recent years that the bud and hypha phenotypes do not represent the entirety of the phenotypic repertoire of *C. albicans* and related species. For the past 50 years, researchers have reported phenotypic variability and instability in *C. albicans* which was distinguished primarily by colony phenotype. A review of the literature prior to 1968 can be found in an article by Brown-Thomsen (1968). Although continued reports of phenotypic instability of colony morphology continued after this time (e.g., Ireland and Sarachek, 1968; Saltarelli, 1973; Mackinnon, 1969), it was not until careful analyses were performed on the heritability of this phenomenon that a picture of phenotypic switching at both the colony and cellular levels emerged (Slutsky *et al.*, 1985, 1987; Soll and Kraft, 1988; Soll, 1988). The first strain characterized in detail was *C. albicans* strain 3153A, a common laboratory strain employed by a number of laboratories worldwide. Cells of wild type 3153A form smooth white colonies (Fig. 3A1) on the amino acid-rich agar used in the original studies (Slutsky *et al.*, 1985). However, when cells from an original smooth white colony are plated, variant colonies (Fig. 3A) are formed at a frequency of roughly 10^{-4}. If 3153A cells are treated with a low dose of ultraviolet (UV) irradiation, killing roughly 10% of the cell population, the frequency of variants increases by roughly two orders of magnitude (Slutsky *et al.*, 1985). This increase cannot be due to enrichment of a preexisting subpopulation of resistant cells in light of the low percent kill. Surprisingly, variants emanating spontaneously or after UV irradiation continue to generate variants with phenotypes other than their own at frequencies close to 10^{-2} (Fig. 3B). The variants included in the 3153A switching system are: original smooth, star, ring, irregular wrinkle, stipple, hat, fuzzy, and revertant smooth (Fig. 3A). Each variant can switch to most of the other variants at relatively high frequency (Fig. 3B). Original smooth and revertant smooth differ in switching frequency. The former exhibits a frequency of 10^{-4} and the latter 10^{-2}. Revertants from a spontaneous irregular wrinkle colony have been obtained which exhibit both the original smooth phenotype and reduced switching frequency, suggesting that they represent true revertants (Slutsky *et al.*, 1985). In the original study of the 3153A system, it was demonstrated that the variant colony morphologies resulted from differences in cellular phenotypes in the colony domes. In general, the more wrinkled a colony phenotype was, the greater the proportion of hyphae or pseudohyphae in the colony dome (Soll *et al.*, 1987b). In the fuzzy phenotype (Fig. 3A7), hyphal bundles emerge from the colony dome as aerial mycelia.

At roughly the same time that the 3153A switching system was characterized (Slutsky *et al.*, 1985), Pomes and associates (1985) reported a UV-induced smooth–rough transition which they interpreted to be the result of mitotic recombination. Interestingly, this same hypothesis for the appearance of the star phenotype (originally M10 phenotype) in strain 3153A led to the discovery of reversible high-frequency switching in this strain (see discussion in Soll, 1990a). Subsequent to their original report, Pomes and associates demonstrated spontaneous high-frequency switching in two putative benomyl-induced variants (Pomes *et al.*, 1987; Nombela *et al.*, 1987). The phenotypes they described included rough and smooth, but it must be kept in mind that the composition of the agar influences the colony phenotypes of variants, and the studies by Slutsky and associates (1985) and by Nombela and associates (1987) employed different agars, so the phenotypes in the different strains examined cannot be compared. What has emerged from these original studies as well as studies of vaginal and systemic infections (Soll *et al.*, 1987a, 1988) is

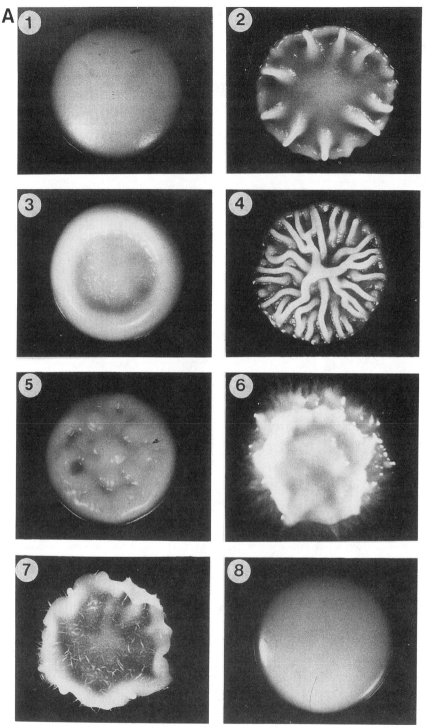

Figure 3. The switching system of *C. albicans* strain 3153A. (A) Colony morphologies of the switch pheno-types: 1, original smooth; 2, star; 3, ring; 4, irregular wrinkle; 5, stipple; 6, hat; 7, fuzzy; 8, revertant smooth. (B) Frequencies of variant colony formation by each of the switch phenotypes. For details, see Slutsky *et al.* (1985).

B

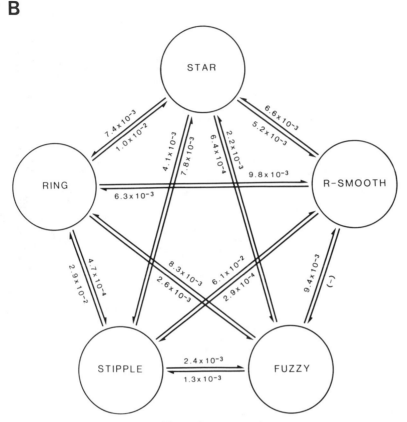

Figure 3.   *(Continued)*

that the majority of *C. albicans* and *C. tropicalis* strains are either in a high-frequency mode of switching, or can be stimulated to enter this mode.

Perhaps the most interesting switching system so far identified in *C. albicans* is the white–opaque transition in WO-1, a strain isolated from the blood and lungs of a bone marrow transplant patient at the University of Iowa Hospitals and Clinics in 1985 (Slutsky *et al.*, 1987; Anderson and Soll, 1987; Anderson *et al.*, 1989; Rikkerink *et al.*, 1988). When cells from a white colony of strain WO-1 are plated, a majority will form white colonies and a minority gray, or "opaque," colonies, and when cells from an opaque colony are plated, a majority will form opaque colonies and a minority white colonies. The difference between white and opaque colonies can be accentuated by the addition of any of a number of vital stains to the agar which differentially stain white and opaque cells: phloxine B differentially stains opaque colonies red (Anderson and Soll, 1987; Fig. 4A), fast green differentially stains opaque colonies gray (Fig. 4B), methylene blue differentially stains opaque cells blue-gray and white cells light blue (Fig. 4C), and bismuth differentially stains opaque colonies black (Rikkerink *et al.*, 1988). Both opaque and white colonies exhibit sectors of alternative phenotype (e.g., opaque sectors in a white colony in Fig. 4D are differentially stained by methylene blue). When white colonies are aged in parafilm-wrapped plates, opaque sectors form perfusely around the edge of the colony (Fig. 4E,F). The frequency of opaque colony-forming cells in white colonies was measured to be between 10^{-4} and 10^{-2}, and the frequency of white colony-forming cells in opaque colonies between 10^{-2} and 10^{-1}

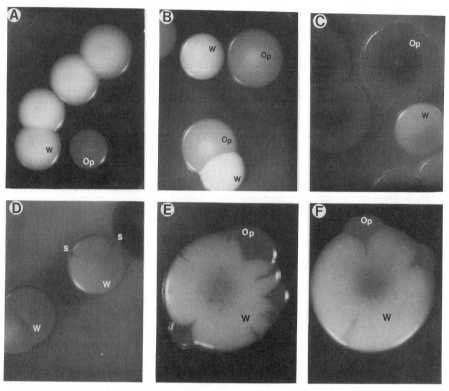

Figure 4. The "white-opaque transition" in *C. albicans* strain WO-1. (A) Switch from white (W) to opaque (Op) (agar contained phloxine B which preferentially stains opaque cells deep red). (B) Switch from opaque (Op) to white (agar contained fast green which preferentially stains opaque cells gray). (C) Switch from opaque (Op) to white (W) (agar contained methylene blue which preferentially stains white cells light blue and opaque cells blue-gray). (D) Opaque sectors (S) in a white (W) colony (methylene blue). (E) High-frequency sectoring (Op) at the periphery of a white (W) colony which was aged under airtight conditions (phloxine B). (F) Opaque sectors (Op) at the periphery of a white (W) colony (phloxine B).

in the original study (Slutsky *et al.*, 1987). By applying the Luria-Debruck fluctuation formula, Rikkerink and associates (1988) measured the spontaneous rate of switching in liquid microcultures to be 5×10^{-4} in the white-to-opaque direction and less than 10^{-4} in the opaque-to-white direction. They also found that by incubating opaque cells at high temperature (34°C, for periods in excess of 13 hr. in nutrient medium), they could obtain mass conversion to white (Rikkerink *et al.*, 1988). Incubation at high temperature in water did not result in induced switching, suggesting that the cells may have to grow and divide in order to switch (Rikkerink *et al.*, 1988). In addition, it has been demonstrated that UV irradiation stimulates switching in both the white-to-opaque and opaque-to-white directions, as well as subsequent sectoring in both directions (Morrow *et al.*, 1989).

The white-opaque transition has a dramatic effect on the phenotype of budding cells (Slutsky *et al.*, 1987; Anderson and Soll, 1987). Cells growing in the budding form in the white phase are similar to cells in the budding form in most other strains of *C. albicans* (Fig. 5A). However, cells in the budding form in the opaque phase exhibit a unique phenotype. Rather than the round shape of white cells, opaque cells exhibit a bean shape (Fig. 5C). In addition, they are twice as large, twice as heavy, and can exhibit a different budding pattern (Slutsky *et al.*, 1987). When viewed in thin sections by transmission electron microscopy, they exhibit one or more large

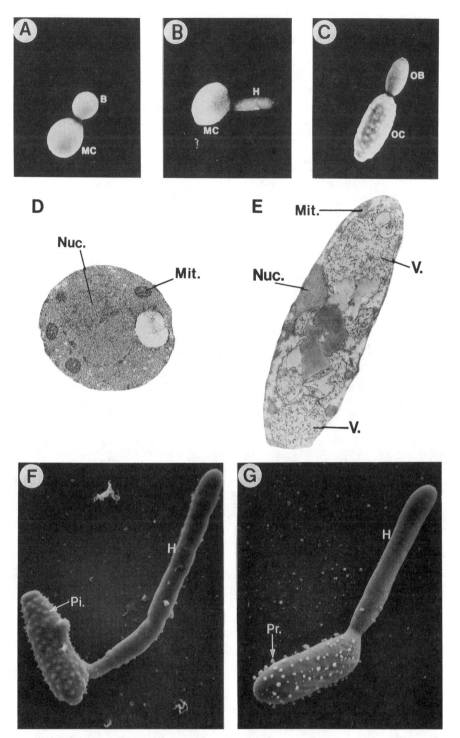

Figure 5. Cellular phenotypes in the white-opaque transition. (A) Scanning electron micrograph (SEM) of a white cell (MC) forming a bud (B). (B) SEM of a white cell (MC) forming a hypha (H). (C) SEM of an opaque cell (OC) forming a bud (OB). (D) Transmission electron micrograph (TEM) of a white cell in the budding phase (Nuc., nucleus; Mit., mitochondrion). (E) TEM of an opaque cell (V., vacuoles with spaghetti-like material inside). (F, G) SEMs of opaque cells forming hyphae (H); note the pimples (Pi.) in F and the protrusions (Pr.) in G on the mother cell's surface.

membrane-bound vacuoles containing spaghetti-like material of unknown origin (Fig. 5E). These large vacuoles press mitochondria and nucleus into the cytoplasmic cortex (Fig. 5E), a cytoplasmic morphology quite distinct from white cells (Fig. 5D). When viewed by scanning electron microscopy, opaque cells exhibit a unique surface morphology which includes pimples, small protrusions, or pits covering the entire surface(Fig. 5C,F,G). White budding and hypha-forming cells exhibit smooth surfaces (Fig. 5A,B). The protrusions emanating from the pimples at times appear surprisingly similar to virus-like particles (Fig. 6A–C), but because discrete channels can sometimes be seen penetrating the wall of the pimple (Fig. 6A, unfilled arrow), it is more likely that the protrusions may represent cytoplasmic protrusions penetrating the channel. Indications of double membranes bounding some of these protrusions support this suggestion (Anderson and Soll, 1990).

Opaque buds do not exhibit pimples, protrusions, or pits until they have reached roughly half their final volume (Anderson and Soll, 1987). An opaque-specific antiserum (generated against opaque cells and absorbed with white cells) stains the surface of opaque cells in a punctate fashion

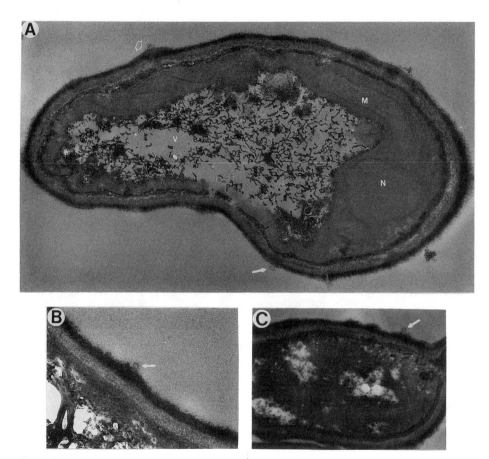

Figure 6. Transmission electron micrographs of viruslike particles embedded in the wall of the opaque cell. (A) Profile of an entire opaque cell in which a pimple with a visible channel is noted by the unfilled arrow and a viruslike particle on the cell surface is noted by the filled arrow (N, nucleus; M, mitochondrion; V, vacuole). (B, C) Viruslike particles (noted by arrow) on the surface of opaque cells. Note that white cells lack these particles.

reminiscent of the pimple pattern observed by scanning electron microscopy (Anderson and Soll, 1987). In spite of the increased size and unique morphology of opaque cells, their DNA content is approximately the same as white cells, and ethidium bromide-stained gels of EcoRI fragments and Southern blot hybridization patterns with the midrepeat sequence Ca3 (see later section for more detailed discussion) are identical for white and opaque cell DNA (Hicks et al., in preparation). OFAGE patterns of white and opaque cell chromosomes are also identical (Rikkerink et al., 1988; Soll, Hicks, Morrow, and Donelson, unpublished observations).

White and opaque cells differ not only in morphology, but also in their basic physiology and environmental constraints in the bud-to-hypha transition. C. albicans strain 3153A and strain WO-1 in the white phase will assimilate glucose, 2-keto-D-gluconate, xylose, adonitol, xylitol, galactose, sorbitol, methyl-D-glucoside, N-acetyl-D-glucosamine, maltose, sucrose, and trehalose (Soll, 1990a). WO-1 in the opaque phase loses the capacity to assimilate adonitol, xylitol, methyl-D-glucoside, and trehalose (Soll, 1990a). When opaque cells switch back to white, they regain the capacity to assimilate these latter sugars.

When white cells which have entered stationary phase at 25°C in the budding form are diluted into fresh medium at 37°C, pH 6.7, they form hyphae like most other strains of C. albicans (Slutsky et al., 1987; Anderson et al., 1989). In contrast, when opaque cells are treated in the same manner, they continue to bud. However, if opaque cells are allowed to adhere to the glass wall of a perfusion chamber and are perfused with fresh medium at 37°C, pH 6.7, they form hyphae (Fig. 5F,G; Anderson et al., 1989). Depending upon the clone, the proportion of cells forming hyphae varies from 1 to 74%. In addition, opaque cells will form extremely long hyphae when distributed onto a sheet of cultured skin cells (Anderson et al., 1989). Therefore, opaque cells are capable of forming hyphae, but the normal environmental constraints on the bud–hypha transition observed for white cells have been altered for opaque cells.

Since opaque budding cells exhibit surface pimples and punctuate surface staining with opaque-specific antiserum, hyphae formed by opaque cells were examined for the same unique characteristics. Surprisingly, opaque hyphae are smooth (Fig. 5F,G), and devoid of surface staining by the opaque-specific antiserum (Anderson et al., 1989). This has led to the suggestion that switching from white to opaque has a dramatic effect on the expression of genes involved in the budding phenotype, but no obvious effect on the expression of genes specific to hypha formation. Although the hyphae formed by opaque cells are morphologically identical to hyphae formed by white cells, they appear to remain genetically opaque since a reduction in pH and temperature leads to the formation by hyphae of opaque buds (Anderson et al., 1989).

It therefore seems reasonable to conclude that C. albicans has at least two levels of phenotypic variability built into its developmental repertoire for pathogenesis (Soll, 1990a). At the basic level of dimorphism, the cell can change phenotype for tissue penetration. At the more general level of high-frequency switching, the cell can heritably but reversibly change a number of fundamental physiological and cellular characteristics, modify the constraints on the dimorphic transition, and in some cases change its budding phenotype. It is as if the variability due to switching is superimposed upon the basic developmental capacity to differentiate back and forth between the bud and hypha phenotype (Soll, 1990a). Switching occurs at the site of infection (Soll et al., 1987a) and may provide a colonizing yeast population with variants with an adaptive advantage if the physiology of the host changes. It may also afford the population with multiple phenotypes for penetrating different body locations, or the variability necessary to escape drug treatment or the immune system. Switching has been demonstrated to dramatically affect adhesion (Kennedy et al., 1988), antigenicity (Anderson and Soll, 1987), drug resistance (Slutsky, 1986; Soll et al., 1988, in preparation), susceptibility to neutrophil killing in vitro (M. Kolotila and R. Diamond, 1990), acid protease secretion (T. Ray, C. Payne, and D. R. Soll, in preparation), virulence in mouse models (T. Ray, C. Payne, B. Morrow, and D. R. Soll, in preparation), and penetration of keratinized tissue (T. Ray et al., in preparation).

4. BASIC BIOLOGICAL CHARACTERISTICS OTHER THAN DIMORPHISM AND SWITCHING WHICH MAY PLAY ROLES IN PATHOGENESIS

Dimorphism and switching together still do not provide the entire explanation for the pathogenic success of *C. albicans* and other dimorphic yeast. To begin with, the capacity of *C. albicans* to colonize the healthy individual as a commensal at low levels represents an important pathogenic characteristic. The commensal is no doubt in competition with other microorganisms of the microflora, basically bacteria, and this may explain in part the increase in *Candida* infections after treatment with antibacterial drugs such as tetracycline (Caruso, 1964). Because of the importance of anchoring to mucosa before penetration of tissue as well as the possible role of adhesion to plastic (McCourtie and Douglas, 1981; Rotrosen *et al.*, 1986) in nosocomial infections, attention must also be paid to the adhesive properties of *C. albicans* (see Chapter 7). In addition, excretion of acid protease (MacDonald and Odds, 1980; Ruchel, 1981) and phospholipase (Pugh and Cawson, 1977) has been implicated in tissue penetration, and variants which have lost the capacity to excrete acid protease exhibit decreased virulence (Kwon-Chung *et al.*, 1985). Budding cells have been demonstrated to invade the gastrointestinal mucosa through progressive digestion of the mucus barrier and microvillus layer, then penetration of the epithelial cells by extracellular digestion rather than hyphal penetration (Cole *et al.*, 1988). Hormone receptors have recently been identified in *C. albicans* (Loose *et al.*, 1981; Powell *et al.*, 1984) and it is not outrageous to suggest that they may play a role in responsiveness to hormone fluctuations in the host, although cycling levels of yeast during the menstrual cycle have not been convincingly demonstrated (see discussion in Odds, 1988). Recently, it was demonstrated that the *C. albicans* surface contains a complement receptor (Heidenreich and Dierich, 1985; Edwards *et al.*, 1986) which is a glycoprotein (Linehan *et al.*, 1988). The role of this receptor is not understood, but it is likely that it plays a pathogenic role since nonpathogenic species are devoid of the receptor (Heidenreich and Dierich, 1985). *C. albicans* has been demonstrated to suppress the immune system (Domer *et al.*, 1988) and undergo rapid antigenic changes (Anderson and Soll, 1987; Poulain *et al.*, 1983) which may represent a strategy for escaping the immune system similar to that in trypanosomes (Borst, 1983) and *Salmonella* (Silverman and Simon, 1983). Different strains of *C. albicans* exhibit variable sensitivity to a number of antifungal agents (e.g., Defever *et al.*, 1982; Dick *et al.*, 1980; Horsburgh *et al.*, 1982; Warnock *et al.*, 1983), and the organism has the potential for rapid changes in sensitivity through switching (Soll *et al.*, 1989; Slutsky, 1986). *C. albicans* also appears to be more resistant to general environmental insults than related nonpathogenic yeasts like *S. cerevisiae*, and this is most evident in the increased difficulty in removing its wall by digestive enzymes. This list of potential pathogenic characteristics, in addition to dimorphism and switching, has been briefly reviewed in order to remind the reader of the types of pathogenic traits which can now be addressed at the molecular level.

5. THE MOLECULAR GENETICS OF C. ALBICANS

To date, there has been no indication that *C. albicans* has a sexual cycle (Whelan, 1987). The majority of strains so far analyzed appear to be diploid (Olaiya and Sogin, 1979; Riggsby *et al.*, 1982) although it has been demonstrated that both aneuploidy (Hilton *et al.*, 1985) and tetraploidy (Suzuki *et al.*, 1986) can be achieved. Balanced lethals have been demonstrated when tested for (Whelan and Soll, 1982), suggesting that haploidization would have been highly unlikely in the recent history of the strains examined. Because of the absence of a sexual cycle, standard genetic methods have not been possible. Linkage studies have been successful employing spheroblast fusion techniques (Poulter *et al.*, 1981; Sarachek *et al.*, 1981; Kakar *et al.*, 1983), mitotic segregation studies (Whelan and Magee, 1981; Whelan and Soll, 1982), and hybridization to electrophoretically separated chromosomes (Magee *et al.*, 1988).

5.1.   The C. albicans Genome

The DNA content of *C. albicans* is approximately 37 fg (Riggsby *et al.*, 1982; Whelan and Magee, 1981; Soll *et al.*, 1981), very close to that of diploid *S. cerevisiae* (Lauer *et al.*, 1977). Monitoring DNA reassociation kinetics, Riggsby and associates (1982) found that 13% of cellular DNA reassociated rapidly, and that 3% represented foldback DNA. To further characterize the repetitive sequences in whole cell DNA, Wills and associates (1984) purified repetitive sequences (Cot 1.59) and performed a minicot analysis on this fraction. They obtained a two-component curve consisting of a major slow reassociating component and a minor fast reassociating component with reiteration frequencies of roughly 17 and 280, respectively (Wills *et al.*, 1984). Mitochondrial DNA sequences appear to represent a significant portion of the major component. The mitochondrial DNA is circular, is composed of 41-kb pairs, and contains a large inverted duplication (Wills *et al.*, 1985). The proportion of the nuclear genome composed of repetitive sequences has been estimated to be between 1 and 2% (Laurer *et al.*, 1977). Several genomic midrepeat sequences have been cloned. Wills and colleagues (1984) identified three *Eco*RI fragments containing repeat sequences which were likely tandemly repeated ribosomal cistrons. Scherer and Stevens (1988) isolated a species-specific moderately repetitive sequence, clone 27A, which is repeated roughly 10 times in the genome of *C. albicans* and hybridizes to all but one of the OFAGE-separable chromosomes. Southern blot hybridization patterns with this probe are stable within a strain, but vary between strains, making it a valuable probe for assessing strain relatedness, a point which will be returned to later in this review. Hicks and associates (1990) isolated two species-specific midrepeat sequences, Ca3 and Ca7. Ca3 is repeated roughly 15 times in the genome (Hicks *et al.*, 1988), and exhibits stability within a strain and variability between strains similar to that of clone 27A (Scherer and Stevens, 1988). Ca7 is repeated roughly 15 times in the genome, is present on all OFAGE-separable chromosomes, and is telomeric (Hicks *et al.*, 1990). In contrast to 27A and Ca3, the Southern blot hybridization pattern with Ca7 is not stable within most strains of *C. albicans* after only a limited number of generations (Hicks *et al.*, 1990; Soll *et al.*, 1987a, 1988).

Magee and collaborators recently employed three methods for chromosome separation (Magee *et al.*, 1988): orthogonal field gel electrophoresis (OFAGE), field inversion gel electrophoresis (FIGE), and contour-clamped homogeneous field gel electrophoresis (CHEF). They have been able to resolve ten chromosomal bands in a number of strains, and have concluded that *C. albicans* contains seven chromosome pairs with polymorphism not only between strains, but between homologous chromosomes in the same strain (Magee *et al.*, 1988). Other studies have suggested eight pairs of chromosomes (J. Hicks, personal communication). By hybridizing cloned genes to blots of separated chromosomes, Magee and associates (1988) have been able to assign them to particular chromosomes: TUB1, rDNA, and GAL1 to chromosome 1; SOR9 and MGL1 to chromosome 1 or 2; ADE2, SOR2, and URA3 to chromosome 3; LYS2 to chromosome 5; and ADE1, BEN[R], MTX[R], and TRP1 to chromosome 6. As more genes are cloned, they will no doubt continue to be assigned to linkage groups by hybridization to blots of electrophoretically separated chromosomes, and linkage groups will expand when genes within linkage groups defined by parasexual methods are cloned and assigned to a chromosome by hybridization. Hopefully, a single nomenclature will evolve quickly for the chromosomes of *C. albicans*.

5.2.   Integrative Transformation in C. albicans

Recently, a transformation system was developed which was designed after the standard system in *Saccharomyces*. Kurtz and co-workers (1986) generated a genomic library of Sau3 fragments (5–20 kb) in YEp13, an *S. cerevisiae* shuttle vector which carries the LEU2 selectable marker. They used this library to transform *S. cerevisiae* strains which were ade2⁻ and leu⁻. They selected leucine prototrophs and selected from these variants which were adenine prototrophs. To ensure that the adenine prototrophs were due to complementation and not spontaneous reversion, they demonstrated plasmid loss. Three *C. albicans* insertion fragments (pMK3, pMC1, and pMC2) responsible for the transformation of *S. cerevisiae* were inserted into the *Cla*I site of the plasmid pBR322, which contains an amphacilin-resistant gene. These ADE2$^+$-containing plasmids were then used to transform ade2$^-$ strains of *C. albicans* by either a *Neurospora crassa* (Case *et al.*, 1980) or *S. cerevisiae* (Beggs, 1978) transformation protocol. Since ade2$^-$ strains form pink to red colonies and ADE2$^+$ strains form white colonies on indicator agar, screening of transformants consisted of isolating white colonies. Two major white colony phenotypes were observed, petite and large. The large white colonies represented stable ADE2$^+$ strains. The transformation frequency was found to be between 0.5 and 5.0 transformants/μg DNA, and the frequency per viable spheroplast between 3 and 6 × 10^6 spheroblasts, roughly ten times the spontaneous reversion rate. Transformation number increased linearly with increasing DNA, and vector DNA was identified in transformant DNA. The plasmid DNA was found to integrate at the ADE2 gene site in one of the two homologous chromosomes carrying ade2$^-$. Since the initial use of ADE2 as a selectable marker for transformation, URA3 and LEU2 have also been used successfully (Kurtz *et al.*, 1987, 1988). In a review of the molecular genetics of *C. albicans*, Kurtz and colleagues (1988) point out that the low frequency of transformants is only a problem if a highly revertable mutant is employed.

By screening for ADE2$^+$ transformants which were unstable in the absence of selection and which contained free plasmids, Kurtz and co-workers (1987) have isolated an autonomously replicating sequence from the *C. albicans* genome. CARS (*Candida* autonomously replicating sequence) increases the rate of transformation and generates transforming sequences which are oligomers containing head-to-tail tandem repeats. Occasionally, these transforming sequences were observed to integrate into chromosomes. CARS will prove to be extremely useful in assessing the effect of overexpression of particular genes.

5.3. Using the Transformation System to Investigate *C. albicans* Pathogenesis

Perhaps the most important use of the *C. albicans* transformation system will be to identify genes involved in pathogenic processes, and to disrupt putative pathogenic genes by site-specific mutagenesis in order to assess phenotype. In the first approach, stable mutants with altered pathogenic traits (e.g. loss of hypha formation, loss of adhesion to mucosa, absence of acid protease secretion) can be rescued by transformation with vectors containing a selectable marker in addition to genomic DNA fragments. The sequence responsible for the rescue can then be recloned. In this approach, if the wild-type gene is trans-dominant, then integration or site-specific recombination is not necessary. In the second approach, cloned genes can be altered *in vitro* by deletion, insertion, or point mutation, and gene replacement or insertional mutagenesis carried out. Kelly and associates have demonstrated one-step gene disruption of the URA3 gene, leading to ura3 mutants. Their method was that of Rothstein (1983), which was developed for *S. cerevisiae*. They inserted the cloned ADE2 gene into the cloned URA3 gene and linearized the DNA, which facilitates gene replacement. They then transformed ade2 mutants with disrupted DNA, and selected ADE$^+$ transformants. Since *C. albicans* is diploid, replacement would be on only one of the chromosome homologues. This was demonstrated by Southern blot hybridization, and the heterozygote was converted to a homozygote by mitotic recombination stimulated by UV irradiation (Kelly *et al.*, 1987).

5.4. Cloning Genes Which May Play a Role in Pathogenesis

As previously noted, one way to clone genes involved in pathogenic processes is to transform stable mutants defective in putative pathogenic characteristics with vectors containing genomic DNA fragments and a selectable marker, then screen for wild-type transformants. One method which has been useful in rescue experiments in other organisms has been the use of drug-resistance markers (e.g. DeLozanne and Spudich, 1987). Transforming DNA can contain a drug-resistant marker which is expressed in both eukaryotes and prokaryotes, which facilitates screening of transformants and cloning of the fragment in bacteria. Unfortunately, drug-resistant genes have not been useful for this purpose in *C. albicans*, and to date most of the *Candida* genes have been cloned as a result of their capacity to be expressed in *S. cerevisiae* or by cross-hybridization to homologous sequences in *S. cerevisiae* (Kurtz *et al.*, 1988). Seven genes (URA3, HIS3, TRP1, ADE2, GAL1, LEU2, and CHS1) have been cloned by complementation of mutant *S. cerevisiae* or *E. coli* genes (Kurtz *et al.*, 1988; Jenkinson *et al.*, 1988); four genes (DFR1, SOR2, SOR9, AND STA1) have been cloned as a result of unusual, selectable phenotypes which were generated when they were introduced into *Saccharomyces* (Kurtz *et al.*, 1988); and three genes (ACT1, TUB2, and the λ-G protein gene) have been cloned by cross-hybridization with homologous *S. cerevisiae* genes (Hicks, personal communication). Many other *C. albicans* genes which exhibit homology with *S. cerevisiae* genes (e.g. cdc4, 10, 11, and 12; J. Morrow, J. Anderson, J. Pringle, and D. R. Soll, unpublished observations) were in the process of being cloned at the time this chapter was being written, and were therefore not included in this discussion. However, it is clear that the list of genes cloned as a result of homology with *S. cerevisiae* genes will continue to expand rapidly in the next few years with the increased attention now being directed at the molecular biology of *C. albicans* and the influx of molecular biologists now interested in *C. albicans* and trained in *S. cerevisiae* molecular biology. It should be noted that to date the majority of genes which have been cloned are involved in basic cellular metabolism or cell architecture and are not immediate candidates for genes with selective roles in pathogenesis.

6. THE MOLECULAR BIOLOGY OF DIMORPHISM

Although the technologies have been available for investigating the molecular biology of dimorphism for several years, we still know virtually nothing about the involvement of differential gene expression in the phenotypic decision between the alternative growth forms. There have been numerous studies documenting antigenic differences between budding cells and hypha-forming cells (Syverson *et al.*, 1975; Ho *et al.*, 1979; Hopwood *et al.*, 1986; Brawner and Cutler, 1986; Smail and Jones, 1984; Sundstrom *et al.*, 1987), but few clear instances in which budding and hypha-forming cells contain different gene products (Soll, 1986, 1990a). The molecular identity of most of the antigens differentially expressed in budding and hypha-forming cells is either not known or demonstrated to involve changes in carbohydrates. In no case has it been demonstrated that the antigenic difference represents a case of differential gene transcription or translation.

There have been several studies comparing the proteins synthesized during bud and hypha formation, using one- (1D-PAGE) and two-dimensional polyacrylamide gel electrophoresis (2D-PAGE). Using pH-regulated dimorphism (Fig. 2) to obtain temporal parallelism in the diverging populations, it was demonstrated that protein synthesis begins roughly 50 min after stationary-phase cells are diluted into fresh medium for both budding and hypha-forming populations evaginating semisychronously on average after 135 min. (Brummel and Soll, 1982; Finney *et al.*, 1985). After the initiation of protein synthesis, the majority of the 374 polypeptides monitored by 2D-PAGE continued to be synthesized after evagination. However, 17 polypeptides were synthesized only prior to evagination, and 60 polypeptides were synthesized only after evagination, demonstrating that a very detailed program of gene expression accompanies bud and hypha formation (Finney *et al.*, 1985). However, the programs were similar in the diverging populations. Of the 374 polypeptides monitored, only 2 appeared to be specific for hypha-forming cells and only 2 appeared to be specific for budding cells. Employing a mutant which made buds at both pH's, it was demonstrated that there was only one putative bud-specific polypeptide and only one putative hypha-specific polypeptide. However, there was no indication whether even this difference was due to post-translational modification or differential degradation or excretion. Using temperature to regulate protein synthesis, Brown and Chaffin (1981) found 5 polypeptides out of 230 monitored by 2D-PAGE which were specific for budding cells and no polypeptides specific for hypha-forming cells. However, this study did not take into account the changes associated with the program of evagination, so the differences observed could have been due to temporal differences between the diverging populations. In a study again using temperature to regulate phenotype, Ahrens and co-workers (1983) demonstrated that there were 4 polypeptides specific for high-temperature cultures (which induced hypha formation) and 3 polypeptides specific for low-temperature cultures (which remained budding cells), but it was not demonstrated that the differences in polypeptides were due to temperature or phenotype. Finally, Manning and Mitchell (1980a) used prolonged labeling regimes with [^{35}S]sulfate and identified 10 polypeptides which were putatively hypha-specific, but immunoabsorption studies and mutant analysis indicated that these differences may have been due to posttranslational modifications. In this study, they did identify several putative bud-specific polypeptides. In a subsequent immunological study, they identified 83 putative hypha-specific polypeptides, but all were present in a strain forming buds at high temperature (Manning and Mitchell, 1980b). They also found that 11 of the putative bud-specific antigens were not present in the strain forming buds at high temperature, suggesting either dramatic differences in gene expression between strains or differences in posttranslational modification of groups of polypeptides.

Perhaps the most crucial lesson which can be learned from the immunological and PAGE studies so far performed is that the two methods are inadequate for studying differential gene transcription during dimorphism. In addition, it is clear that temperature, pH, sampling time in synchronized populations, and strain can affect the pattern of protein synthesis, and proper controls including the use of mutants must be carefully applied when concluding that a gene product is in fact phenotype-specific. Unfortunately, no reports have been published to date of cloned genes differentially expressed in the alternative bud and hypha phenotypes, although concerted effort is now under way in a number of laboratories to this end. There are a number of methods for cloning such genes including subtraction hybridization (Timberlake, 1980) or direct "plus–minus" screening (St. John and Davis, 1981). In either case, if genes do exist which are differentially expressed they can be cloned and tested for expression by Northern blot hybridization using different conditions for phenotypic regulation, morphological mutants, and the regime of pH-regulated dimorphism to assess the temporal relationship of gene expression and bud or hypha development.

Recently, Russell and colleagues (1987) reported that the DNA of budding cells contained twice the amount of 5-methyldeoxycytidine as hypha-forming cells. The authors suggested that this difference may reflect greater gene activity in hypha-forming cells, but direct studies on protein synthesis do not support this conclusion (Finney *et al.*, 1985). Unfortunately, the methods used to grow cells in the budding and hyphal forms in this DNA methylation were not well controlled, and the phase of growth (log versus stationary) not defined. It should be realized that the molecular biology of dimorphism is not limited to the regulation of genes expressed solely in one phenotype or the other. The argument has previously been developed (Soll, 1986; Odds, 1988; Chapter 5) that major aspects of hypha formation involve changes in the timing or proportion of events common to bud formation. Therefore, there may very well be regulatory processes involved in hypha formation which selectively modify the timing of transcription or translation of a gene also expressed in budding cells, a form of developmental heterochromy (Soll, 1990b). The difference in the timing of filament ring formation between a budding and hypha-forming cell (Soll and Mitchell, 1983) represents an excellent candidate for a gene expressed in both phenotypes, but with altered timing (Soll, 1986).

There may also be posttranslational levels of phenotypic regulation. For instance, general expansion of the budding wall represents roughly 33% of total wall expansion during initial bud growth, but roughly 5% during initial hypha growth under the regime of pH-regulated dimorphism (Staebell and Soll, 1985). However, one can increase the rate of the general expansion mechanism by decreasing the pH, which leads to an increase in hypha diameter. This suggests that the machinery is in place but depressed. Therefore, it is important that we do not lose sight of other mechanisms which may be involved in the regulation of dimorphism in our quest for genes differentially expressed in dimorphism.

7. THE MOLECULAR BIOLOGY OF SWITCHING

Unfortunately, we know as little about the molecular biology of switching as we do about the molecular biology of dimorphism. Besides the question of phenotypic regulation, which includes the differential expression of genes in different switch phenotypes, we have the added problem of determining the molecular basis of a reversible switch (Soll, 1988). There are a number of examples of switching systems in both prokaryotes and eukaryotes which have been elucidated at the molecular level (Shapiro, 1983; Berg and Howe, 1988). In *Salmonella*, inversion of a DNA sequence of roughly 1000 bp containing a promoter for an operon including a gene for H2 antigen and a gene for a repressor for an H1 antigen gene uncouples the promoter and operon (Silverman and Simon, 1983). This results in the cessation of H2 antigen and H1 repressor synthesis, and synthesis of H1 antigen. Subsequent inversion results in H2 antigen synthesis and H1 repression. The inversion is exact, leading to a reversible phase transition between the two antigenic states. In *S. cerevisiae*, two unexpressed loci (HML and HMR) possess the alternative mating type genes a and α, and an expressed locus MAT possesses either the a or α gene. The MAT locus can switch from a to α, or α to a, by the transposition of mating-type information from the alternative silent locus (Oshima and Takano, 1971; Hicks *et al.*, 1977; Hicks and Herskowitz, 1976). This precise conversion is not reciprocal; therefore, neither of the silent loci loses its original information and the cell retains the capacity to switch once again. Besides a number of precise, reversible switching systems, there are also a number of transposition systems in prokaryotes, lower eukaryotes, and higher eukaryotes which, in some cases, involve transposition to preferred sites (e.g., Toussaint and Resibois, 1983; Kleckner, 1983; Roeder and Fink, 1983; Rubin, 1983; Federoff, 1983). All of these elucidated systems serve as possible models for the mechanism basic to *C. albicans* switching, and have provided a framework for developing experimental strategies, although mechanisms other than transposition such as reversible DNA methylation (Konieczny and Emerson, 1984), reversible gene duplication, aneuploidy, metastable physiological switches which do not involve DNA rearrangement, and "cortical inheritance"-type changes (Sonneborn, 1963) have not been ruled out.

Only a few observations suggest that the switching mechanism involves an event involving DNA. First, UV irradiation stimulates switching in strain 3153A (Slutsky *et al.*, 1985) and the white-opaque transition (Morrow *et al.*, 1989), in the latter case in both directions. Second, temperature-stimulated switching from opaque to white will not occur when cells are suspended in water, in the absence of cell multiplication (Rikkerink *et al.*, 1988). Temperature-stimulated switching from opaque to white is also depressed dramatically in growth medium in the presence of 10 mM hydroxyurea (M. Bergen and D. R. Soll, unpublished observations), which inhibits DNA replication but not initial cell growth (Bedell *et al.*, 1980). This probably represents the extent of information on the possible DNA basis of switching. Switch phenotypes contain roughly the same amounts of DNA (Slutsky *et al.*, 1987), identical ethidium bromide-stained patterns of *Eco*RI-digested DNA separated on agarose gels, and identical OFAGE (Rikkerink *et al.*, 1988) and TAFE (B. Morrow and D. R. Soll, unpublished observations)-separated chromosome patterns (examples of the TAFE patterns of white and opaque cell chromosomes are presented in

Fig. 7, columns 1–4). Therefore, no drastic changes in the basic genome resulting from switching are discernible.

Based upon the hypothesis that switching may be due to transposition and may involve midrepeat sequences, Hicks and associates developed a simple method for cloning repetitive DNA sequences (Hicks *et al.*, 1990). A library of *Eco*RI-digested DNA from *C. albicans* strain 3153A was constructed in λ-gt10. The library was screened with ^{32}P-labeled nick-translated unfractionated DNA of strain 3153A. Plaques exhibiting strong signals were presumed to contain repetitive DNA, and those exhibiting weak signals unique sequence DNA. Two moderately repetitive DNA sequences, Ca3 and Ca7, were demonstrated to be distinct from ribosomal and mitochondrial sequences. The Southern blot hybridization pattern of the first sequence, Ca3, was found to be stable within a strain and similar in switch phenotypes of a single strain (Hicks *et al.*, 1990; Soll *et al.*, 1987a; Soll, 1990a). In contrast, the second moderately repetitive sequence, Ca7, was found to be highly mobile in strain WO-1 and all pathogenic strains tested, but the Southern blot hybridization patterns did not correlate with switch phenotypes. The pattern changes after a limited number of cell divisions, but in strain 3153A, the pattern is stable, even when switch phenotypes are compared, indicating that Ca7 transposition has nothing to do with high-frequency switching.

Since nonswitching variants are obtainable in strain WO-1 (Soll and Shmid, unpublished results), rescue experiments should provide us with genes involved in high-frequency switching. In addition, subtraction hybridization or plus–minus screening should allow identification and

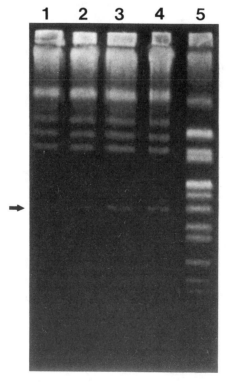

Figure 7.   TAFE separation of chromosomes from *C. albicans* strain WO-1. Lanes: 1, white WO-1; 2, opaque WO-1; 3, white WO-1; 4, opaque WO-1; 5, *S. cerevisiae*. Arrow points to low-molecular-weight "chromosome." Note the constancy of the chromosomal pattern in the switch phenotypes, which were isolated in sequence.

cloning of genes differentially expressed in switch phenotypes. The demonstration of an opaque-specific antigen (Anderson and Soll, 1987) suggests that differential gene expression may indeed be involved in the white-opaque transition.

8. APPLYING MOLECULAR TECHNOLOGIES TO THE BASIC QUESTIONS OF CANDIDA EPIDEMIOLOGY AND DISEASE

To date, molecular genetic techniques have not afforded us with any insights into the pathogenic traits of *C. albicans*, but they have provided us with tools for answering a number of basic questions related to the actual disease. It is surprising that some of the most fundamental questions related to the epidemiology of candidiasis remain unanswered. For instance, are commensal strains responsible for subsequent infections? Do commensals from different body locations represent the same strain? Do commensal strains change with time, or aging? Or, does an individual retain the same commensal strain throughout life? In recurrent infections, is the individual reinfected by the same strain or by a sequence of different strains? Do strains cycle through switch phenotypes during prolonged colonization, after drug treatment, or at the onset of infection? Is there strain-relatedness for infections of the same body location?

Although there have been numerous attempts to answer some of these questions, researchers have been stymied until recently by the lack of a rapid and precise method for assessing whether strains are genetically related. Previous methods which were used to assess strain relatedness included (1) separation of strains into serotype groups A and B (Hasenclever and Mitchell, 1961; Hasenclever *et al.*, 1961); (2) resistance to chemicals (McCreight and Warnock, 1982; Odds and Abbott, 1980); (3) sensitivity to yeast killer factor (Polonelli *et al.*, 1983); (4) assimilation patterns for sugars and nitrogen sources (Odds and Abbott, 1980; Williamson *et al.*, 1986); and (5) colony or streak morphology (Brown-Thomsen, 1968; Hunter *et al.*, 1989). All of these tests assessed the relatedness of either the physiology or development of separate strains, not genetic relatedness. Therefore, they all ran the risk of cogrouping genetically unrelated strains with similar physiologies, and separating switch phenotypes of the same strain exhibiting quite disparate physiologies.

In the past few years, a number of methods have begun to be developed for assessing genetic relatedness using relatively straightforward DNA technologies. To begin with, several laboratories simultaneously described the use of restriction fragment patterns of total cell DNA for assessing relatedness (Scherer and Stevens, 1987; Soll *et al.*, 1987b; Cutler *et al.*, 1988). In this straightforward method, total cell DNA is isolated and digested with one of a number of endonucleases, including *Msp*I (Cutler *et al.*, 1988) or *Eco*RI (Scherer and Stevens, 1987; Soll *et al.*, 1987b). Fragments are then separated by agarose gel electrophoresis and the gel either photographed directly with 300-nm transillumination or stained with ethidium bromide and photographed. In Fig. 8, the *Eco*RI digest patterns are presented for strain 3153A (lanes 2 and 10); strain HMI(S)12 (lane 3), a strain isolated from a lesion of the oral mucosa; strain G62 (lane 4); strain WO5 (lane 5), a mouth isolate which exhibits the white-opaque transition; strain WO-1 (lane 6), an isolate from the blood and lungs of a bone marrow transplant patient which exhibits the white-opaque transition; and HMH9, HMH5, and HMH1 (lanes 7, 8, and 9, respectively), commensal isolates from healthy mouths. It is immediately obvious that the patterns of this random set of isolates are dissimilar except for HMH9 and HMH1. Interestingly, HMH9 and HMH1 were isolated from healthy individuals working in the same building.

In the most critical test of the usefulness of restriction fragment patterns, Scherer and Stevens (1987) demonstrated that (1) the major bands visualized by 300-nm transillumination through an orange filter were ribosomal, (2) the pattern was stable within a strain over many generations, (3) patterns were roughly species-specific, and (4) subgroups could be distinguished

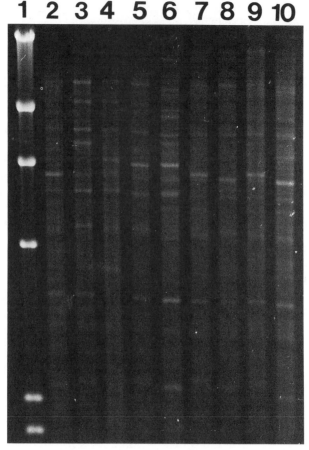

Figure 8. Distinguishing strains of *C. albicans* by comparing the ethidium bromide-stained patterns of total cell DNA digested with *Eco*RI and separated on agarose gels. Lanes: 1, molecular weight markers; 2, strain 3153A (common laboratory strain); 3, strain HMI(S)12, isolated from an oral lesion; 4, strain G62, isolated from a systemic infection; 5, strain WO5, isolated from a healthy mouth; 6, strain WO-1, isolated from a systemic infection; 7, strain HMH9, isolated from a healthy mouth; 8, strain HMH5, isolated from a healthy mouth; 9, strain HMH1, isolated from a healthy mouth; 10, strain 3153A.

within a species. By comparing Southern blot hybridization patterns of *Eco*RI-digested DNA probed with cloned *S. cerevisiae* ribosomal DNA, Magee and colleagues (1987) could distinguish 6 different polymorphic groups in 12 analyzed *C. albicans* isolates. These patterns were also distinguishable in ethidium bromide-stained gels.

Olivo and colleagues have demonstrated that polymorphisms exist in mitochondrial DNA, and *Eco*RI fragments of the mitochondrial genome (Wills *et al.*, 1985) can be combined to generate a combination probe to assess strain relatedness. They found that four patterns were distinguishable in 26 *C. albicans* isolates for which *Hae*III-digested DNA was probed with the mitochondrial probe. Although the stability of the mitochondrial patterns was not directly tested in these studies, it was demonstrated that surveillance cultures from the site of infection resulted in a constant mitochondrial restriction pattern (Olivo *et al.*, 1987).

Strain relatedness has also been assessed by polymorphisms in the patterns of OFAGE-separated chromosomes. Magee and Magee (1987) compared four strains and found four different patterns. Besides indicating the high degree of polymorphism in chromosome size including polymorphisms in chromosomal homologues, the studies to date (Magee and Magee, 1987; Snell *et al.*, 1987; Rikkerink *et al.*, 1987; Hicks *et al.*, 1990) demonstrate the usefulness of the technique and the need for a far more detailed assessment of its application to questions of strain relatedness.

To date, the method most heavily applied to studies of strain relatedness in *C. albicans* has been Southern blot hybridization of *Eco*RI-digested whole cell DNA with cloned moderately repetitive sequences (Soll *et al.*, 1987a, 1988, 1989; Scherer and Stevens, 1988; Hicks *et al.*, 1990). In this method, total cell DNA is isolated (e.g., Cryer *et al.*, 1975; Scherer and Stevens, 1987) and digested with a restriction enzyme (e.g., *Eco*RI). The resulting mixture of DNA fragments is then separated by agarose gel electrophoresis and transferred to nitrocellulose or another hybridization membrane according to the general methods developed by Southern (1975). The hybridization probe (e.g., a hybrid lambda bacteriophage or plasmid containing a moderately repetitive sequence) is then radiolabeled with ^{32}P by nick translation (Rigby *et al.*, 1977). The membrane, nick-translated probe, and excess denatured thymus or salmon DNA are added together to a hybridization bag and incubated at roughly 65°C. The membrane is then washed free of unhybridized probe and fluorographed.

The two moderately repetitive *C. albicans* sequences which have been used for assessing strain relatedness are 27A (Scherer and Stevens, 1988) and Ca3 (originally referred to as JH3; Soll *et al.*, 1987a, 1989; Hicks *et al.*, 1988). Both are species-specific, exhibiting either very weak or undetectable hybridization to most other species. Probe 27A hybridizes to six of seven OFAGE-separable chromosomes and between 10 and 20 *Eco*RI restriction fragments by Southern blot hybridization (Scherer and Stevens, 1988). When four *C. albicans* strains were allowed to multiply through roughly 300 generations, and six clones of each strain analyzed, one of the 24 clones showed a change in the Southern blot pattern with 27A. This demonstrates that within a strain, the 27A pattern is relatively stable. However, Scherer and Stevens (1988) also found that one of ten spontaneous 5-fluorocytosine-resistant variants exhibited a change. Although relatively stable within a strain, the 27A pattern differs between most strains, making it an excellent tool for determining relatedness. Of 20 independently isolated strains from hospital patients, Scherer and Stevens (1988) found that the majority differed in their 27A pattern and that the differences in hybridization patterns were much greater than the differences in ethidium bromide staining patterns, indicating that Southern blot hybridization patterns with moderately repetitive sequences are a more powerful tool than ethidium bromide-stained patterns for assessing strain unrelatedness.

The characteristics of probe Ca3 are quite similar to those of 27A, and to date no one has definitely demonstrated that these probes are distinct, although the *Eco*RI sites in restriction maps for Ca3 and 27A do not completely correspond. Using 4 *Eco*RI subfragments of Ca3 as probes for Southern blot hybridization with *C. albicans* restriction digests, Hicks and co-workers (1990) identified the left junctional sequence of Ca3 by the fact that it hybridizes with equal intensity to 10 to 12 genomic fragments, a characteristic of a dispersed, single copy repeat sequence. One subfragment, B, hybridized to half of the fragments also identified by the left junctional sequence, and to several of higher copy number, some larger than the actual B fragment, suggesting that there is polymorphism of Ca3 within a strain. The subfragment C hybridizes to a very large band besides itself and one other *Eco*RI restriction fragment. However, there is extreme variability when hybridized to Hind III digests, suggesting that a subregion of fragment C containing the Hind III site undergoes rearrangement without affecting the *Eco*RI sites. Finally, subfragment A is the most conserved part of Ca3, exhibiting only two polymorphic forms.

As in the case of 27A, the Southern blot hybridization pattern with Ca3 is relatively stable within a strain, varies between most strains, and is species-specific (Hicks *et al.*, 1990; Soll *et al.*,

1989). Hicks and co-workers (1990) demonstrated stability of the Ca3 pattern in subclones separated by more than 100 combined generations in strain 3153A and WO-1, and in a more recent analysis (Schmid and Soll, unpublished observations), it was found to be stable in strain WO-1 after 300 generations. In addition, stability was demonstrated for subclones of single strains exhibiting different switch phenotypes generated in the laboratory (Hicks *et al.*, 1990) or at the same site of infection (Soll *et al.*, 1987a). On the other hand, patterns vary between most strains. In Fig. 9A, Southern blot hybridization patterns of *Eco*RI-digested DNA of ten random commensal strains isolated from the mouths of different healthy individuals are presented for comparison.

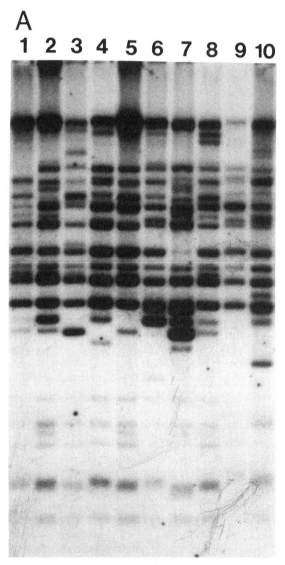

Figure 9. An analysis of relatedness of (A) ten independently isolated strains from the mouths of healthy individuals (lanes 1–10) and (B) ten independently isolated strains from immunocompromised patients with systemic infections (lanes 1–10; lane 11, laboratory strain 3153A).

All of the patterns differed in at least one band position or band intensity. In an analysis of ten strains isolated from immunocompromised patients with systemic infections, all Southern blot hybridization patterns (Fig. 9B, lanes 1–10) of *Eco*RI-digested DNA probed with Ca3 differed. To date, the number of strains exhibiting different *Eco*RI patterns is 24, and the number of *Hind*III patterns even greater (Schmid and Soll, unpublished). If we assume that these strains are equally distributed in nature, then the probability of isolating any single *Eco*RI pattern is 1 in 24 or 0.04, and the probability of isolating two strains in sequence or simultaneously which exhibit the same *Eco*RI pattern is 0.0016 (0.04 × 0.04), roughly 2 in 1000. Therefore, if isolates

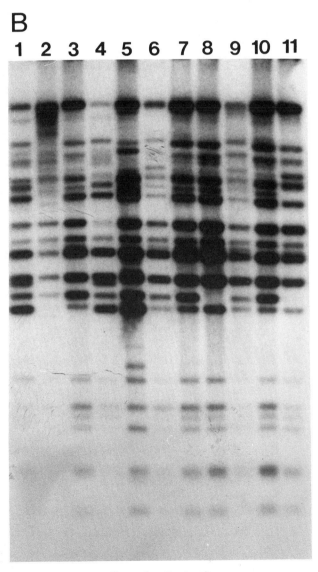

Figure 9. (*Continued*)

from the mouth and vagina of a patient obtained at the same time, or isolates from the vaginal canal during successive episodes of vaginal candidiasis, exhibit identical Ca3 Southern blot hybridization patterns, one can conclude that the two isolates very likely represent offsprings of a single infecting strain. If two isolates exhibit different patterns, one can be relatively secure that they represent different strains. These decisions on relatedness will provide us with answers to the fundamental questions proposed at the outset of this section.

A method is now in progress to assess quantitatively how related strains really are by analyzing both the pattern and intensity of bands (Schmid *et al.*, 1990). In this method, the sizes of hybridized fragments (bands on Southerns) are compared to those of laboratory strain 3153A, which has been carefully characterized as a standard. The strengths of bands are divided into four classes, 1 to 4. Class 1 represents the weakest and class 4 the strongest band intensity (Fig. 10). Only bands larger than 1.85×10^3 bp are included in the analysis since smaller bands are too weak to provide reproducible markers. The similarity of two patterns is then assessed by the following equation:

$$S_{AB} = \frac{\sum_{i=1}^{K} (a_1 + b_i|a_i - b_i|)}{\sum_{i=1}^{K} (a_i + b_i)}$$

where S_{AB} is the similarity value, a_i and b_i are the class strengths of band i for strain a and b (no band is counted as class O), and K is the number of bands. Similarity values can be used to develop dendrograms according to the method of Li (1981). In Fig. 11A, ten strains isolated from the mouths of healthy individuals and laboratory strain 3153A are grouped in a dendrogram according to similarity values. In Fig. 11B, strain relatedness is presented in a histogram. Using the 3153A pattern as reference, all strains assessed in this manner are now being entered into a computer memory bank which can be accessed in the development of dendrograms and relatedness histograms as the data base grows (Schmid *et al.*, 1990).

Although these computer assisted methods are still being developed and have not yet been applied to pathogenic and epidemiological problems, the basic method of comparing Southern blot hybridization patterns with Ca3 to assess strain relatedness has been applied with some success. A few of the more interesting studies are outlined below.

8.1. Phenotypic Variability and Genetic Homogeneity in Vaginal Infections

Candida samples were cloned from the vaginal canals of 11 patients with acute *C. albicans* vaginitis (Soll *et al.*, 1987a). Four of the eleven samples contained multiple colony phenotypes. By comparing Ca3 Southern blot hybridization patterns, it was demonstrated that the multiple phenotypes in each of the four samples represented a genetically homogeneous strain in each case. It was further demonstrated that in the 4 cases in which multiple phenotypes were present at the site of infection, cells from one phenotype could switch to the alternative phenotype at high frequency. Of the 11 samples, 9 were in a high-frequency mode of switching.

The usefulness of the Ca3 probe in this study was to demonstrate that multiple phenotypes at the site of infection were in fact switch phenotypes of the same strain. Without genetic proof of relatedness, one could not have excluded the possibility that the multiple phenotypes represented multiple strains with similar switching repertoires. Indeed, in this study (Soll *et al.*, 1987a) it was found that only one genetic strain was prevalent at the site of infection, and that all strains were *C. albicans*. Most surprising was that 2 of the 11 strains exhibited no switching distinguishable by colony morphology. If 2 strains which were not actively switching caused acute vaginitis with symptoms similar to those caused by 9 strains actively switching, one may ask what role, if any, is

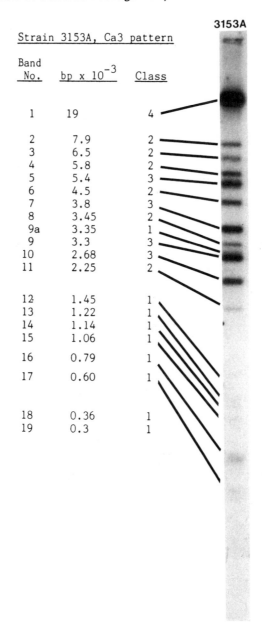

Strain 3153A, Ca3 pattern

3153A

Band No.	bp x 10^{-3}	Class
1	19	4
2	7.9	2
3	6.5	2
4	5.8	2
5	5.4	3
6	4.5	2
7	3.8	3
8	3.45	2
9a	3.35	1
9	3.3	3
10	2.68	3
11	2.25	2
12	1.45	1
13	1.22	1
14	1.14	1
15	1.06	1
16	0.79	1
17	0.60	1
18	0.36	1
19	0.3	1

Figure 10. The reference system for quantitatively assessing relatedness by band position and intensity. The reference system is based on the Ca3 Southern hybridization pattern with *Eco*RI-digested DNA of strain 3153A. Band intensities are scored 1 through 4, with 4 being the most intense. The lower-molecular-weight class 1 bands are not visible in the photo.

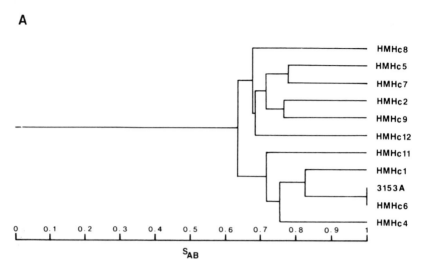

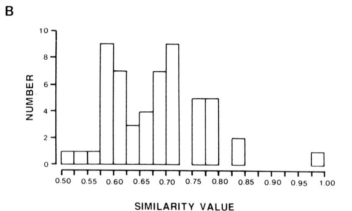

Figure 11. An analysis of relatedness for 10 strains isolated from the mouths of healthy individuals (HMH, human mouth healthy) and strain 3153A. (A) Dendogram based on similarity values (S_{AB}). (B) A histogram of the similarity values between the 11 strains. See text for calculation of similarity value. These plots were computer generated.

played by switching in vaginal candidiasis. Perhaps the answer may be in the evolution, potential severity, and recurrence of the disease. Far more detailed studies will be necessary for an answer to this question.

8.2. Phenotypic Variability and Genetic Invariability during Successive Episodes of Vaginal Candidiasis

An unusually high proportion of yeast vaginitis patients have more than a single episode within a year's time (Sobel, 1984, 1985), and it has been assumed that the source of the reinfecting strain is the flora of the gastrointestinal tract (Miles *et al.*, 1977). In a single case study of a recurrence patient, samples were obtained from roughly 15 body locations at the time of three

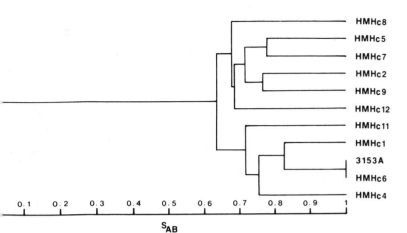

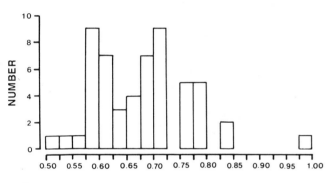

An analysis of relatedness for 10 strains isolated from the mouths of healthy individuals (HMH, h healthy) and strain 3153A. (A) Dendogram based on similarity values (S_{AB}). (B) A histogram of the lues between the 11 strains. See text for calculation of similarity value. These plots were computer

witching in vaginal candidiasis. Perhaps the answer may be in the evolution, potential d recurrence of the disease. Far more detailed studies will be necessary for an answer stion.

enotypic Variability and Genetic Invariability during Successive isodes of Vaginal Candidiasis

iusually high proportion of yeast vaginitis patients have more than a single episode ar's time (Sobel, 1984, 1985), and it has been assumed that the source of the reinfecting ie flora of the gastrointestinal tract (Miles *et al.*, 1977). In a single case study of a patient, samples were obtained from roughly 15 body locations at the time of three

All of the patterns differed in at least one band position or band intensity. In an analysis of ten strains isolated from immunocompromised patients with systemic infections, all Southern blot hybridization patterns (Fig. 9B, lanes 1–10) of *Eco*RI-digested DNA probed with Ca3 differed. To date, the number of strains exhibiting different *Eco*RI patterns is 24, and the number of *Hind*III patterns even greater (Schmid and Soll, unpublished). If we assume that these strains are equally distributed in nature, then the probability of isolating any single *Eco*RI pattern is 1 in 24 or 0.04, and the probability of isolating two strains in sequence or simultaneously which exhibit the same *Eco*RI pattern is 0.0016 (0.04 × 0.04), roughly 2 in 1000. Therefore, if isolates

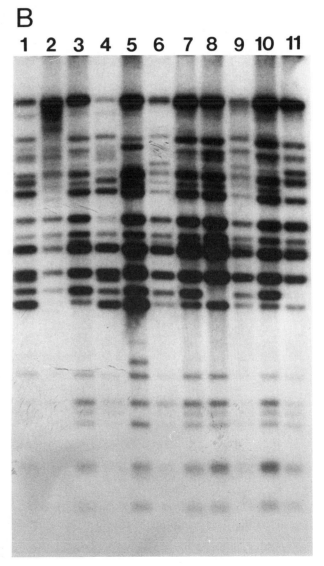

Figure 9. (Continued)

from the mouth and vagina of a patient obtained at the same time, or isolates from the vaginal canal during successive episodes of vaginal candidiasis, exhibit identical Ca3 Southern blot hybridization patterns, one can conclude that the two isolates very likely represent offsprings of a single infecting strain. If two isolates exhibit different patterns, one can be relatively secure that they represent different strains. These decisions on relatedness will provide us with answers to the fundamental questions proposed at the outset of this section.

A method is now in progress to assess quantitatively how related strains really are by analyzing both the pattern and intensity of bands (Schmid et al., 1990). In this method, the sizes of hybridized fragments (bands on Southerns) are compared to those of laboratory strain 3153A, which has been carefully characterized as a standard. The strengths of bands are divided into four classes, 1 to 4. Class 1 represents the weakest and class 4 the strongest band intensity (Fig. 10). Only bands larger than 1.85×10^3 bp are included in the analysis since smaller bands are too weak to provide reproducible markers. The similarity of two patterns is then assessed by the following equation:

$$S_{AB} = \frac{\sum_{i=1}^{K} (a_1 + b_i|a_i - b_i|)}{\sum_{i=1}^{K} (a_i + b_i)}$$

where S_{AB} is the similarity value, a_i and b_i are the class strengths of band i for strain a and b (no band is counted as class O), and K is the number of bands. Similarity values can be used to develop dendrograms according to the method of Li (1981). In Fig. 11A, ten strains isolated from the mouths of healthy individuals and laboratory strain 3153A are grouped in a dendrogram according to similarity values. In Fig. 11B, strain relatedness is presented in a histogram. Using the 3153A pattern as reference, all strains assessed in this manner are now being entered into a computer memory bank which can be accessed in the development of dendrograms and relatedness histograms as the data base grows (Schmid et al., 1990).

Although these computer assisted methods are still being developed and have not yet been applied to pathogenic and epidemiological problems, the basic method of comparing Southern blot hybridization patterns with Ca3 to assess strain relatedness has been applied with some success. A few of the more interesting studies are outlined below.

8.1. Phenotypic Variability and Genetic Homogeneity in Vaginal Infections

Candida samples were cloned from the vaginal canals of 11 patients with acute *C. albicans* vaginitis (Soll et al., 1987a). Four of the eleven samples contained multiple colony phenotypes. By comparing Ca3 Southern blot hybridization patterns, it was demonstrated that the multiple phenotypes in each of the four samples represented a genetically homogeneous strain in each case. It was further demonstrated that in the 4 cases in which multiple phenotypes were present at the site of infection, cells from one phenotype could switch to the alternative phenotype at high frequency. Of the 11 samples, 9 were in a high-frequency mode of switching.

The usefulness of the Ca3 probe in this study was to demonstrate that multiple phenotypes at the site of infection were in fact switch phenotypes of the same strain. Without genetic proof of relatedness, one could not have excluded the possibility that the multiple phenotypes represented multiple strains with similar switching repertoires. Indeed, in this study (Soll et al., 1987a) it was found that only one genetic strain was prevalent at the site of infection, and that all strains were *C. albicans*. Most surprising was that 2 of the 11 strains exhibited no switching distinguishable by colony morphology. If 2 strains which were not actively switching caused acute vaginitis with symptoms similar to those caused by 9 strains actively switching, one may ask what role, if any, is

Strain 3153A, Ca3 pattern

Band No.	bp x 10^{-3}	Class
1	19	4
2	7.9	2
3	6.5	2
4	5.8	2
5	5.4	3
6	4.5	2
7	3.8	3
8	3.45	2
9a	3.35	1
9	3.3	3
10	2.68	3
11	2.25	2
12	1.45	1
13	1.22	1
14	1.14	1
15	1.06	1
16	0.79	1
17	0.60	1
18	0.36	1
19	0.3	1

Figure 10. The reference system for quantitatively assessing relatednes reference system is based on the Ca3 Southern hybridization pattern with Band intensities are scored 1 through 4, with 4 being the most intense. The lo not visible in the photo.

Figure
human r
similarit
generate

played
severity
to this

8.2.

Ar
within a
strain is
recurre

successive infections as well as interphase times immediately following drug therapy (Soll *et al.*, 1989). The dynamics of colonization are presented in Table I. At the time of the first infection, extremely high colonization was evident in the vulva-vaginal, anal-rectal regions and at the back of the tongue. After successful clotrimazole treatment (following infection #1), the mouth sites exhibited the only significant colonization. Eighty-seven days after the first infection and eighty days after the termination of the first drug treatment, infection recurred (Table I). This time the patient was successfully treated with butaconazole. Thirty-two days after the second infection and twenty-five days after termination of drug treatment, the patient presented with a third infection (Table I). Several questions were posed in this single study. First, did the oral, vulva-vaginal, and anal-rectal isolates represent three strains, or a single strain? Second, were the three successive infections due to one strain or three strains? Third, did phenotypic switching play a role in recurrence? First, it was demonstrated by Ca3 Southern blot patterns that there were three different strains of *C. albicans* colonizing the patient at the time of the first vaginal infection: one strain in the mouth, a second under the breast, and a third in the vulva-vaginal and anal-rectal regions (Fig. 12). Second, the vulva-vaginal strains isolated at the time of the three infections represented a single recurring strain. Third, the single infecting strain changed colony morphology in successive vaginal infections: medium-sized smooth white (infection #1) → small irregular edged (infection #2) → medium-sized smooth white (infection #3). It was quite obvious during the study that without a method for assessing genetic relatedness, the mouth and genital strains could have been construed as representing the same strain, and the isolates from the first and second vaginal infection different strains. Surprisingly, *in vitro* testing of drug sensitivity did not support the prediction that the changes in switch phenotype between the first and second, and second and third infections were the result of selection for drug-resistant switch phenotypes (Soll *et al.*, 1988).

This single case study demonstrates the power of combining (1) full body mycoflora analyses, (2) assessments of original colony phenotypes and switching repertoires, (3) drug sensitivity tests, and (4) assessments of strain relatedness by comparisons of Southern blot hybridization patterns with a species-specific moderately repetitive sequence probe. However, any generalizations about the relatedness of strains in different body locations and in successive episodes of vaginitis will require a far greater sample size than 1.

8.3. Multiple Strains during the Course of a Single Systemic Infection

The probes 27A and Ca3 are species-specific, and therefore can only be used for assessing strain relatedness within the species *C. albicans*. However, species-specific moderately repetitive sequences have also been isolated from *C. krusei* and *C. tropicalis* (Hicks *et al.*, 1990). In a study of a bone marrow transplant patient with a systemic *Candida* infection, the *C. tropicalis* probe Ct13.8 was employed (Soll *et al.*, 1988). The history of isolation, colony morphology, and drug regimens are presented in Fig. 13. A strain with a smooth white colony phenotype was isolated from the urine on the first day of the study. Two days later, a strain with an irregular edge colony was isolated from the blood. One day later, a strain with a fuzzy colony was isolated from skin blisters. Two days later, and at the time amphotericin B was increased to a therapeutic level, the irregular edge and fuzzy phenotypes were cloned from the blood. Although the fuzzy phenotype was apparently removed from the blood by amphotericin B after one day of treatment, the irregular edge phenotype remained. Five days after initiation of amphotericin B treatment, 5-fluorocytosine treatment was added. This treatment removed the irregular edge phenotype from the blood. The patient, who was self-administering clotrimazole orally, stopped of her own volition. Soon afterwards, irregular edge, fuzzy, and a new bumpy halo phenotype were isolated from her throat. Finally, a smooth white phenotype was isolated from the throat. Using the two *C. albicans*-specific probes Ca3 and Ca7, it was demonstrated that the original smooth white strain

Table I. History of Colonization during Recurrent Vaginal Infections of One Patient[a]

Body location	Clotrimazole		Butaconazole		
	A. 0 days Vag. Inf. #1	B. 29 days Interphase	C. 87 days Vag. Inf. #2	D. 101 days Interphase	E. 119 days Vag. Inf. #3
Cheek	0	1 (Mo11)	5 (Mo15)	0	16 (Mo25)
Saliva	1	4 (Mo10)	1 (Mo17)	—	9 (Mo28)
Back of tongue	93 (Mo5)	6 (Mo12)	12 (Mo18)	2 (Mo22)	11 (Mo24)
Nasal canal	—	0	0	0	—
Ear canal	—	0	0	0	—
Armpit	0	0	0	0	—
Under breast	3 (Mo1)	0	0	0	—
Nipple	—	0	0	0	—
Navel	—	—	0	0	0
Groin	13 (Mo6)	1 (Mo13)	1 (Mo16)	—	0
Vulvar region	678 (Mo4)	0	474 (Mo21)	0	2000 (Mo26)
Vaginal wall	2870 (Mo3)	0	786 (Mo19)	0	3000 (Mo29)
Vaginal pool	7460 (Mo2)	0	376 (Mo20)	0	—
Anus	46 (Mo8)	0	62 (Mo14)	0	13 (Mo23)
Deep rectum (stool)	16 (Mo7)	0	0	0	2 (Mo27)
Back of knee	—	0	0	0	—
Between toes	1 (Mo9)	—	0	0	—
Vaginal wet mount	pos.	neg.	pos.	neg.	pos.

[a]The number of colonies obtained by plating original samples on five agar plates is presented for each body location. A dash indicates that no sample was taken from that body location. The label of a clone of each original sample which was used for further experimentation is given to the right of the colony number. Each drug treatment (indicated over arrows between A and B, and C and D) was initiated on the date noted below A and C and was administered vaginally for 7 days. Observation of yeast in the vaginal canal in wet mounts obtained from the vaginal pool at the time of examination is noted by "pos." at the bottom of each column; observation of no yeast is noted by "neg."

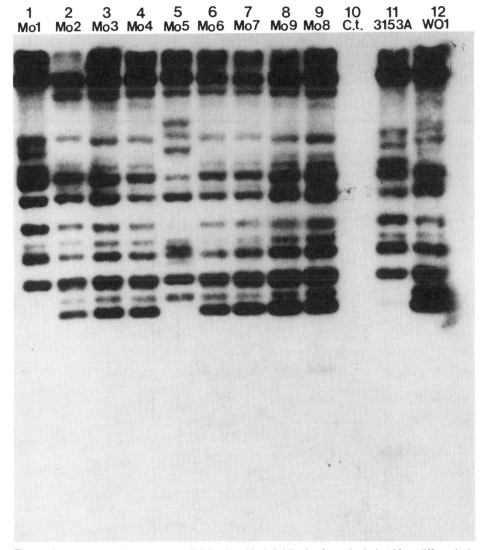

Figure 12.   An analysis of relatedness by Ca3 Southern blot hybridization for strains isolated from different body locations at the time of the first vaginal infection in a series of recurrent infections. Lanes: 1, Mo1 (under breast); 2, Mo2 (vaginal pool); 3, Mo3 (vaginal wall); 4, Mo4 (vulvar area); 5, Mo5 (back of tongue); 6, Mo6 (groin); 7, Mo7 (deep rectal-stool); 8, Mo9 (between toes); 9, Mo8 (anus); 10, *C. tropicalis* ATCC 34139; 11, *C. albicans* 3153A; 12, *C. albicans* WO-1. Note that all isolates exhibit complex patterns with Ca3 demonstrating that they are of the species *C. albicans*. Three patterns are distinguishable in the isolates and can be grouped: pattern 1, Mo1; pattern 2, Mo2, Mo3, Mo4, Mo6, Mo7, Mo9, and Mo8; pattern 3, Mo5. For details see Soll *et al.*, (1989).

isolated from urine and the final smooth white isolated from the throat represented two distinct strains of *C. albicans*. All other strains were not of the species *C. albicans*. Using the *C. tropicalis*-specific probe Ct13.8 (Fig. 14), it was demonstrated that all of the isolates but the two smooth white strains (lanes 4 and 11, respectively), which were already typed as *C. albicans*, were *C. tropicalis*. The irregular edge isolates (lanes 5, 8, and 9) and the fuzzy isolates (lanes 6 and 7) represented two distinct strains. The two bumpy halo isolates differed by one band and

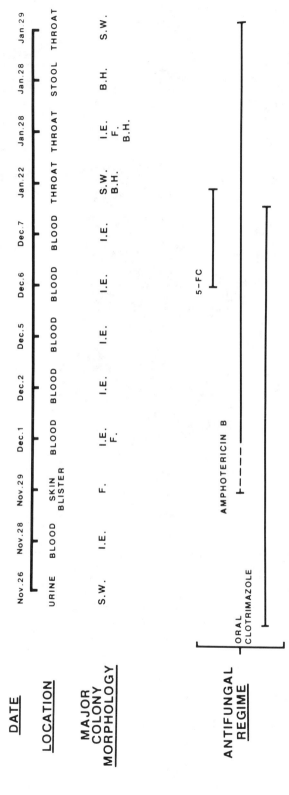

Figure 13. A temporal diagram of the dates of isolation of *Candida*, major colony phenotypes, and drug regimens during the course of a single systemic infection. Dates are presented when *Candida* colonization was detected. The dashed line represents time between test dose and therapeutic dose of amphotericin B. 5-FC, fluorocytosine; S. W., smooth white; F., fuzzy; B. H., bumpy halo. For details, see Soll *et al.* (1988).

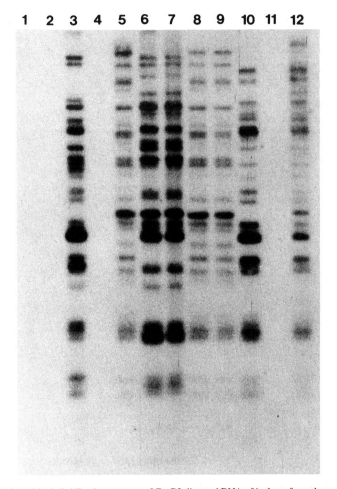

Figure 14. Southern blot hybridization patterns of *Eco*RI-digested DNA of isolates from the systemic infection in Fig. 13 with the probe Ct13.8, a moderately repetitive sequence cloned from *C. tropicalis*. Lanes: 1, *C. albicans* 3153A; 2, *C. albicans* WO-1; 3, *C. tropicalis* ATCC 34139; 4, Ur-1 (smooth white isolated Nov. 26 from urine); 5, B-6 (irregular edge isolated Nov. 28 from blood); 6, Sk-1 (fuzzy isolated Nov. 29 from skin blister); 7, B-1 (fuzzy isolated Dec. 1 from blood); 8, B-4 (irregular edge isolated Dec. 1 from blood); 9, B-13 (irregular edge isolated Dec. 7 from blood); 10, Th-14 (bumpy halo isolated Jan. 22 from throat); 11, Th-8 (smooth white isolated Jan. 29 from throat); 12, St-1 (bumpy halo isolated Jan. 28 from stools). Strains Ur-1 and Th-8 were of the species *C. albicans* and therefore exhibited no hybridization to Ct13.8. At least three strains were clearly distinguishable in the other isolates: pattern 1 for B-6, B-4, and B-13, all irregular edge; pattern 2 for Sk-1 and B-1, both fuzzy; and pattern 3 for Th-14 and St-1, both bumpy halo (although one band difference was evident).

presumably represented a single strain which had undergone a rearrangement or transposition of Ct13.8 which did not affect colony morphology. All of the *C. tropicalis* strains were in a high-frequency mode of switching and exhibited the degree of Ct13.8 polymorphism observed for Ca3 in *C. albicans*. This study demonstrated that during the course of this systemic infection, five different strains of *Candida* involving two different species colonized this individual. It also appeared that differences in sensitivity to amphotericin B accounted for the selective removal of the fuzzy but not irregular edge from the blood. *In vitro* analysis of the drug sensitivity of the two strains supported this conclusion (Soll *et al.*, 1988). Again, it should be pointed out that no generalizations can be made from a single case study. However, this study does demonstrate the power of combining methods for assessing colony phenotype, switching repertoires, genetic relatedness of strains, and drug sensitivity in order to understand strain dynamics during the course of an individual fungal infection.

9. CONCLUDING REMARKS

It is clear from this discussion that the technologies of molecular genetics are being applied to *C. albicans* (Kurtz *et al.*, 1988) and have already begun to impact on the way we are assessing the role of the organism in disease (e.g., Soll *et al.*, 1987a, 1988, 1989; Scherer and Stevens, 1988). Methods have been developed for integrative transformation (Kurtz *et al.*, 1986), gene cloning (Kurtz *et al.*, 1988), and gene disruption (Kelly *et al.*, 1987). Although these methods have not yet been applied directly to genes involved in the pathogenesis of *C. albicans*, they are now available, and will no doubt be used to explore the mechanisms involved in switching and dimorphism. They will also be used to clone and explore the regulation of genes encoding putative pathogenic factors. The application of molecular probes to assess strain relatedness will allow us to obtain answers to several of the most basic questions of *Candida* epidemiology and pathogenesis. It is therefore possible that the impasse which seems to have set in regarding our capacity to reverse the frequency of candidiasis, the level of recurrence, the increase in systemic infections, and the increasing level of certified deaths caused by the disease, estimated to exceed 300 in 1986 in the United States alone (Odds, 1988), may be at an end.

Although I have focused upon one system in order to provide a unified review of the pertinent biology, molecular genetics and epidemiology, it should be realized that most other infectious fungi must be, and in many cases are being, attacked with the same molecular genetic tools reviewed in this chapter. The application of molecular genetic technologies to infectious fungi heralds a new period in medical mycology and requires all of us in the field to become, if we are not already, literate in the area of molecular biology.

On the other hand, there must be words of caution. The application of molecular genetic techniques means nothing without a very firm biological base. In our exuberance to apply these new techniques to an infectious organism, we may fail to realize that a very large body of information has amassed regarding the biology of dimorphism and switching, drug sensitivity, adhesion, physiology, virulence, epidemiology, and pathogenesis. Much of this information was obtained in a scientific fashion and can only provide the reductionists with the necessary context to apply their tools. In a sense, these earlier studies will provide the molecular biologists with the direction they need to know where to look. To the more biologically oriented researcher in this field, two things must be considered. First, we have in no way done more than scratch the surface of this organism's phenotypic repertoire, and it would be foolish to abandon this absolutely necessary pursuit for the more stylish technologies of molecular biology. On the other hand, these technologies are not difficult, and in the best of all worlds, an adaptive approach is warranted in which all levels of analysis are applied. In the end, our role is to understand the beast and cure the disease, and the technologies represent no more than the tools for this end.

ACKNOWLEDGMENTS.   The author is indebted to N. Schroeder, D. Kruse, H. Vawter, and M. Lohman for assistance in preparing the manuscript, and to J. Hicks, P. Magee, J. Anderson, B. Morrow, J. Schmid, and M. Bergen for sharing unpublished results. The author is especially indebted to Dr. R. Galask for his collaboration in the clinical studies involving healthy individuals and patients with acute yeast vaginitis. The research from the author's laboratory was supported by Public Health Science Grant AI 23922 from the National Institutes of Health and by a grant from the Iowa High Technology Council.

10. REFERENCES

Ahrens, J. C., Daneo-Moore, L., and Buckley, H. R., 1983, Differential protein synthesis in *Candida albicans* during blastospore formation at 24.5°C and during germ tube formation at 37°C, *J. Gen. Microbiol.* **129**:1133–1139.

Anderson, J. M., and Soll, D. R., 1986, Differences in actin localization during bud and hypha formation in the yeast *Candida albicans*, *J. Gen. Microbiol.* **132**:2035–2047.

Anderson, J. M., and Soll, D. R., 1987, Unique phenotype of opaque cells in the white-opaque transition of *Candida albicans*, *J. Bacteriol.* **169**:5579–5588.

Anderson, J., Mihalik, R., and Soll, D. R., 1990, Ultrastructure and antigenicity of the unique cell wall pimple of the *Candida* opaque phenotype, *J. Bacteriol.* **172**:224–235.

Anderson, J., Cundiff, L., Schnars, B., Gao, M., Mackenzie, I., and Soll, D. R., 1989, Hypha formation in the white-opaque transition of *Candida albicans*, *Infect. Immun.* **57**:458–467.

Arendorf, T. M., and Walker, D. M., 1979, Oral candidal populations in health and disease, *Br. Dent. J.* **147**:267–272.

Barlow, A. J. E., and Aldersley, T., 1974, Factors present in serum and seminal plasma which promote germ-tube formation and mycelial growth of *Candida albicans*, *J. Gen. Microbiol.* **82**:261–272.

Bedell, G. W., and Soll, D. R., 1979, Effects of low concentrations of zinc on the growth and dimorphism of *Candida albicans*: Evidence for zinc-resistant and -sensitive pathways for mycelium formation, *Infect. Immun.* **26**:348–354.

Bedell, G. W., Werth, A., and Soll, D. R., 1980, The regulation of nuclear migration and division during synchronous bud formation in released stationary phase cultures of the yeast *Candida albicans*, *Exp. Cell. Res.* **127**:103–113.

Beggs, J. D., 1978, Transformation of yeast by a replicating hybrid plasmid, *Nature* **275**:104–109.

Berg, D., and Howe, M. (eds.), 1988, *Mobile DNA*, ASM Press, Washington, D.C.

Borst, P., 1983, Antigenic variation in trypanosomes, in: *Mobile Genetic Elements* (J. Shapiro, ed.), Academic Press, New York, pp. 622–659.

Braun, P. C., and Calderone, R. A., 1978, Chitin synthesis in *Candida albicans*: Comparison of yeast and hyphal forms, *J. Bacteriol.* **135**:1472–1477.

Brawner, D. L., and Cutler, J. E., 1986, Variability in expression of cell surface antigens of *Candida albicans* during morphogenesis, *Infect. Immun.* **51**:337–343.

Brown, L. A., and Chaffin, W. L., 1981, Differential expression of cytoplasmic proteins during yeast bud and germ tube formation in *Candida albicans*, *Can. J. Microbiol.* **27**:580–585.

Brown-Thomsen, J., 1968, Variability in *Candida albicans* (Robin) Berkhout, *Hereditas* **60**:355–398.

Brummel, M., and Soll, D. R., 1982, The temporal regulation of protein synthesis during synchronous bud or mycelium formation in the dimorphic yeast *Candida albicans*, *Dev. Biol.* **89**:211–224.

Budtz-Jorgensen, E., Stenderup, A., and Grabowski, M., 1975, An epidemiologic study of yeasts in elderly denture wearers, *Comm. Dent. Oral Epidemiol.* **3**:115–119.

Buffo, J., Herman, M. A., and Soll, D. R., 1984, A characterization of pH-regulated dimorphism in *Candida albicans*, *Mycopathologia* **85**:21–30.

Byers, B., 1981, Cytology of the yeast life cycle, in: *The Molecular Biology of the Yeast Saccharomyces: Life Cycle and Inheritance* (J. N. Strathern, E. W. Jones, and J. R. Broach, eds.), Cold Spring Harbor Laboratory, Cold Spring Harbor, N.Y., pp. 59–96.

Caruso, L. J., 1964, Vaginal moniliasis after tetracycline therapy, *Am. J. Obstet. Gynecol.* **90**:374–378.

Case, M. E., Schweiger, M., Kushner, S. R., and Giles, N. H., 1980, Efficient transformation of *Neurospora crassa* utilizing hybrid plasmid DNA, *Proc. Natl. Acad. Sci. USA* **77**:5259–5263.

Chattaway, F. W., Homes, M. R., and Barlow, A. J. E., 1968, Cell wall composition of the mycelial and blastospore forms of *Candida albicans*, *J. Gen. Microbiol.* **51**:367–376.

Cole, G. T., Seshan, K. R., Pope, L. M., and Yancey, R. J., 1988, Morphological aspects of gastrointestinal tract invasion by *Candida albicans* in the infant mouse, *J. Vet. Med. Mycol.* **26**:173–185.

Cryer, D. R., Eccleshall, R., and Marmur, J., 1975, Isolation of yeast DNA, *Methods Cell Biol.* **12**:39–44.

Cutler, J., Glee, P. M., and Horn, H., 1988, *Candida albicans* and *Candida stellatoidea*-specific DNA fragments, *J. Clin. Microbiol.* **26**:1720–1724.

Dabrowa, N., Taxer, S. S., and Howard, D. H., 1976, Germination of *Candida albicans* induced by proline, *Infect. Immun.* **13**:830–835.

Defever, K. S., Whelan, W. L., Rogers, A. L., Beneke, E. S., Veselenak, J. M., and Soll, D. R., 1982, *Candida albicans* resistance to 5-fluorocytosine: Frequency of partially resistant strains among clinical isolates, *Antimicrob. Agents Chemother.* **22**:810–815.

DeLozanne, A., and Spudich, J. A., 1987, Disruption of the *Dictyostelium* myosin heavy chain gene by homologous recombination, *Science* **236**:1086–1091.

Dick, J. D., Merz, W. G., and Saral, R., 1980, Incidence of polyene-resistant yeasts recovered from clinical specimens, *Antimicrob. Agents Chemother.* **18**:158–163.

Domer, J., Elkins, K., Ennist, D., and Baker, P., 1988, Modulation of immune responses by surface polysaccharides of *Candida albicans*, *Rev. Infect. Dis.* **10**:419–422.

Edwards, J. E., Gaither, T. A., O'Shea, J., Rotrosen, D., Lawley, T. L., Wright, S. A., Frank, M. M., and Green, I., 1986, Expression of specific binding sites on *Candida* with functional and antigenic characteristics of human complement receptors, *J. Immunol.* **137**:3577–3583.

Federoff, N., 1983, Controlling elements in maize, in: *Mobile Genetic Elements* (J. Shapiro, ed.), Academic Press, New York, pp. 1–63.

Finney, R., Langtimm, C. J., and Soll, D. R., 1985, The programs of protein synthesis accompanying the establishment of alternative phenotypes in *Candida albicans*, *Mycopathologia* **91**:3–15.

Gow, N. A. R., Henderson, G., and Gooday, G. V., 1986, Cytological interrelationships between the cell cycle and duplication cycle of *Candida albicans*, Microbios **47**:97–105.

Hasenclever, H. F., and Mitchell, W. O., 1961, Antigenic studies of *Candida*. I. Observation of two antigenic groups in *Candida albicans*, *J. Bacteriol.* **82**:547–577.

Hasenclever, H. F., Mitchell, W. O., and Loewe, J., 1961, Antigenic studies of *Candida*. II. Antigenic relation of *Candida albicans* group A and group B to *Candida stellatoidea* and *Candida tropicalis*, *J. Bacteriol.* **82**:574–577.

Heidenreich, F., and Dierich, M. P., 1985, *Candida albicans* and *Candida stellatoidea*, in contrast to other *Candida* species, bind iC3b and C3d but not C3b, *Infect. Immun.* **50**:598–600.

Herman, M. A., and Soll, D. R., 1984, A comparison of volume growth during bud and mycelium formation in *Candida albicans*: A single cell analysis, *J. Gen. Microbiol.* **130**:2219–2228.

Hicks, J., and Herskowitz, I., 1976, Interconversion of yeast mating types. I. Direct observations of the action of homothallism (HO) gene, *Genetics* **83**:245–258.

Hicks, J. B., Strathern, J. N., and Herskowitz, I., 1977, The cassette model of mating-type interconversion, in: *DNA Insertion Elements, Plasmids, and Episomes* (A. J. Bukkari, J. A. Shapiro, and S. L. Adhya, eds.), Cold Spring Harbor Laboratory, Cold Spring Harbor, N.Y., pp. 457–462.

Hicks, J. B., McEachern, M. J., Rutchenko-Bulgac, E., Schmid, J., and Soll, D. R., 1988, DNA repeat sequences in *Candida* yeasts and their use in strain identification, (submitted).

Hilton, C., Markie, D., Corner, B., Rikkerink, E., and Poulton, R., 1985, Heat shock induces chromosome loss in the yeast *Candida albicans*, *Mol. Gen. Genet.* **200**:162–168.

Ho, Y. M., Mun, H. N., and Huang, C. T., 1979, Antibodies to germinating and yeast cells of *Candida albicans* in human and rabbit sera, *J. Clin. Pathol.* **32**:399–405.

Hopwood, V., Poulain, D., Fortier, B., Evans, G., and Vernes, A., 1986, A monoclonal antibody to a cell wall component of *Candida albicans*, *Infect. Immun.* **54**:222–227.

Horsburgh, C. R., Kirkpatrick, C. H., and Teutsch, C. B., 1982, Ketoconazole and the liver, *Lancet* **1**:860.

Hunter, P. R., Fraser, C. A. M., and MacKenzie, D. W. R., 1989, Morphotype markers of virulence in human candidal infections, *J. Med. Microbiol.* **28**:85–91.

Ireland, R., and Sarachek, A., 1968, A unique minute-rough colonial variant of *Candida albicans*, *Mycopathol. Mycol. Appl.* **35**:346–360.

Jenkinson, H. F., Shep, G. P., and Shepherd, M. G., 1988, Cloning and expression of the 3-isopropylonalate dehydrogenase gene from *Candida albicans*, *FEMS Microbiol. Lett.* **49**:285–288.

Kakar, S. N., Partridge, R., and Magee, P. T., 1983, A genetic analysis of a wide variety of auxotrophs and demonstration of linkage and complementation, *Genetics* **104**:241–251.

Kaur, S., Mishra, P., and Prasad, R., 1988, Dimorphism-associated changes in intracellular pH of *Candida albicans*, *Biochem. Biophys. Acta* **972**:277–282.

Kelly, R., Miller, S. M., Kurtz, M. B., and Kirsch, D. R., 1987, Directed mutagenesis in *Candida albicans*: One-step gene disruption to isolate ura3 mutants, *Mol. Cell. Biol.* **7**:199–207.

Kennedy, M. J., Rogers, A. L., Hanselman, L. R., Soll, D. R., and Yancey, R. J., 1988, Variation in adhesion and cell surface hydrophobicity in *Candida albicans* white and opaque phenotypes, *Mycopathologia* **102**:149–156.

Kleckner, N., 1983, Transposon Tn 10, in: *Mobil Genetic Elements* (J. Shapiro, ed.), Academic Press, New York, pp. 261–298.

Kolotila, M. P., and Diamond, R. D., 1990, Effects of neutrophils and in vitro oxidants on survival and phenotypic switching of *Candida albicans* WO-1, *Infect. Immun.* **58**:1174–1179.

Konieczny, S. F., and Emerson, C. P., 1984, 5-Azacytidine induction of stable mesodermal stem cell lineages from 10T1/2 cells: Evidence for regulatory genes controlling determination, *Cell* **38**:791–800.

Kurtz, M. B., Cortelyou, M. W., and Kirsch, D. R., 1986, Integrative transformation of *Candida albicans*, using a cloned *Candida* ADE2 gene, *Mol. Cell. Biol.* **6**:142–149.

Kurtz, M. B., Cortelyou, M. W., Miller, S. M., Lai, M., and Kirsch, D. R., 1987, Development of autonomously replicating plasmids for *Candida albicans*, *Mol. Cell. Biol.* **7**:209–217.

Kurtz, M. B., Kirsch, D. R., and Kelly, R., 1988, The molecular genetics of *Candida albicans*, *Microbiol. Sci.* **5**:58–63.

Kwon-Chung, K. J., Lehman, D., Good, D., and Magee, P. T., 1985, Genetic evidence of the role of extracellular proteinase in the virulence of *Candida albicans*, *Infect. Immun.* **49**:571–575.

Lauer, G. O., Roberts, T. M., and Klotz, L. C., 1977, Determination of the nuclear DNA content of *Saccharomyces cerevisiae* and implications of the organization of DNA in yeast chromosomes, *J. Mol. Biol.* **14**:507–526.

Li, W. H., 1981, Simple method for constructing phylogenetic trees from distance matrices, *Proc. Natl. Acad. Sci. USA* **78**:1085–1089.

Linehan, L., Wadsworth, E., and Calderone, R., 1988, *Candida albicans* C3d receptor, isolated by using a monclonal antibody, *Infect. Immun.* **56**:1981–1986.

Loose, D. S., Schurman, D., and Feldman, D., 1981, A corticosteroid binding protein and endogenous ligand in *C. albicans* indicating a possible steroid-receptor system, *Nature* **293**:477–479.

Luna, M. A., and Tortoledo, M. E., 1985, Histologic identification and pathological patterns of disease due to *Candida*, in: *Candidiasis* (G. P. Bodey and V. Fainstein, eds.), Raven Press, New York, pp. 13–51.

McCourtie, J., and Douglas, L. J., 1981, Relationship between cell surface composition of *Candida albicans* and adherence to acrylic after growth on different carbon sources, *Infect. Immun.* **32**:1234–1241.

McCreight, M. C., and Warnock, D. W., 1982, Enhanced differentiation of isolates of *Candida albicans* using a modified resistogram method, *Mykosen* **25**:589–598.

MacDonald, F., and Odds, F. C., 1980, Inducible proteinase of *Candida albicans* in diagnostic serology and in the pathogenesis of systemic candidiasis, *J. Med. Microbiol.* **13**:423–436.

Mackinnon, J. E., 1969, Dissociation in *Candida albicans*, *J. Infect. Dis.* **66**:59–77.

Magee, B. B., and Magee, P. T., 1987, Electrophoretic karyotypes and chromosome numbers in *Candida* species, *J. Gen. Microbiol.* **133**:425–430.

Magee, B. B., D'Souza, T. M., and Magee, P. T., 1987, Strain and species identification by restriction fragment polymorphisms in the ribosomal DNA repeat of *Candida* species, *J. Bacteriol.* **169**:1639–1643.

Magee, B. B., Koltin, Y., Gorman, J. A., and Magee, P. T., 1988, Assignment of cloned genes to the seven electrophoretically separated *Candida albicans* chromosomes, *Mol. Cell. Biol.* **8**:4721–4726.

Manning, M., and Mitchell, T. G., 1980a, Analysis of cytoplasmic antigens of the yeast and mycelial phases of *Candida albicans* by two-dimensional electrophoresis, *Infect. Immun.* **30**:484–495.

Manning, M., and Mitchell, T. G., 1980b, Morphogenesis of *Candida albicans* and cytoplasmic proteins associated with differences in morphology, strain or temperature, *J. Bacteriol.* **144**:258–273.

Mattia, E., and Cassone, A., 1979, Inducibility of germ-tube formation in *Candida albicans* at different phases of yeast growth, *J. Gen. Microbiol.* **113**:439–442.

Mattia, E., Carruba, G., Angiolella, L., and Cassone, A., 1982, Induction of germ tube formation by N-acetyl-D-glucosamine in *Candida albicans*: Uptake of inducer and germinative response, *J. Bacteriol.* **152**:555–562.

Miles, O. R., Olsen, L., and Roger, A., 1977, Recurrent vaginal candidiasis. Importance of an intestinal reservoir, *J. Am. Med. Assoc.* **238**:1836–1837.

Mitchell, L. H., and Soll, D. R., 1979a, Commitment to germ tube or bud formation during release from stationary phase in *Candida albicans*, *Exp. Cell Res.* **120**:167–179.

Mitchell, L. H., and Soll, D. R., 1979b, Temporal and spatial differences in septation during synchronous mycelium and bud formation by *Candida albicans*, *Exp. Mycol.* **3**:298–309.

Morrow, B., Anderson, J., Wilson, E., and Soll, D. R., 1989, Bidirectional stimulation of the white-opaque transition of *Candida albicans* by ultraviolet irradiation, *J. Gen. Microbiol.* **135**:1201–1208.

Niimi, M., Niimi, K., Tokunaga, J., and Nakayama, H., 1980, Changes in cyclic nucleotide levels and dimorphic transition in *Candida albicans*, *J. Bacteriol.* **142**:1010–1014.

Nombela, C., Pomes, R., and Gil, C., 1987, Protoplasts fusion hybrids from *Candida albicans* morphological mutants, *CRC Crit. Rev. Microbiol.* **15**:79–85.

Odds, F. C., 1984, Demonstration of a septal pore in budding *Candida albicans* yeast cells, *Sabouraudia* **22**:505–507.

Odds, F. C., 1988, *Candida and Candidosis: A Review and Bibliography*, Baillière Tindale, London.

Odds, F. C., and Abbott, A. B., 1980, A simple system for the presumptive identification of *Candida albicans* and differentiation of strains within the species, *Sabouraudia* **18**:301–318.

Olaiya, A. F., and Sogin, S. J., 1979, Ploidy determination of *Candida albicans*, *J. Bacteriol.* **140**:1043–1049.

Olivo, P. D., McManus, E. J., Riggsby, W. S., and Jones, J. M., 1987, Mitochondrial DNA polymorphism in *Candida albicans*, *J. Infect. Dis.* **156**:214–215.

Oshima, Y., and Takano, I., 1971, Mating types in *Saccharomyces*: Their convertability and homothallism, *Genetics* **67**:327–335.

Polonelli, L., Archibusacci, C., Sesito, M., and Morace, G., 1983, Killer system: A simple method for differentiating *Candida albicans* strains, *J. Clin. Microbiol.* **17**:774–780.

Pomes, R., Gil, C., and Nombela, C., 1985, Genetic analysis of *Candida albicans* morphological mutants, *J. Gen. Microbiol.* **131**:2107–2113.

Pomes, R., Gil, C., Cabetas, M. D., and Nombela, C., 1987, Variability of colonial morphology in benomyl-induced morphological mutants from *Candida albicans*, *FEMS Microbiol. Lett.* **48**:225–259.

Poulain, D., Tronchin, G., Vernes, A., Popeye, R., and Biguet, J., 1983, Antigenic variation of *Candida albicans in vivo* and *in vitro*—Relationships between antigens and serotypes, *Sabouraudia* **21**:99–112.

Poulter, R., Jeffery, K., Hubbard, M. J., Shepherd, M. G., and Sullivan, P. A., 1981, Parasexual genetic analysis of *Candida albicans* by spheroplast fusion, *J. Bacteriol.* **146**:833–840.

Powell, B. L., Frey, C. L., and Drutz, D. J., 1984, Identification of a 17-estradiol binding protein in *Candida albicans* and *Candida* (Torulopsis) *glabrata*, *Exp. Mycol.* **8**:304–313.

Pugh, D., and Cawson, R. A., 1977, The cytochemical localization of phospholipase in *Candida albicans* infecting the chick chorio-allantoic membrane, *Sabouraudia* **15**:29–35.

Rigby, P. W. J., Deckman, M., Rhodes, C., and Berg, P., 1977, Labeling deoxynucleic acid to high specificity activity *in vitro* by nick translation with DNA polymerase I, *J. Mol. Biol.* **113**:237–251.

Riggsby, W. S., Torres-Bauza, L. J., Wills, J. W., and Townes, T. M., 1982, DNA content, kinetic complexity and the ploidy question in *Candida albicans*, *Mol. Cell. Biol.* **2**:853–862.

Rikkerink, E. H. A., Magee, B. B., and Magee, P. T., 1988, Opaque-white phenotype transition: A programmed morphological transition in *Candida albicans*, *J. Bacteriol.* **170**:895–899.

Roeder, G. S., and Fink, J., 1983, Transposable elements in yeast, in: *Mobile Genetic Elements* (J. Shapiro, ed.), Academic Press, New York, pp. 300–328.

Rothstein, R. S., 1983, One-step gene disruption in yeast, *Methods Enzymol. Part C* **101**:202–211.

Rotrosen, D., Gibson, T. R., and Edwards, J. E., 1986, Adherence of *Candida* species to intravenous catheters, *J. Infect. Dis.* **147**:594.

Rubin, G. M., 1983, Dispersed repetitive DNA's in *Drosophila*, in: *Mobile Genetic Elements* (J. Shapiro, ed.), Academic Press, New York, pp. 329–361.

Ruchel, R., 1981, Properties of purified proteinase from the yeast *Candida albicans*, *Biochem. Biophys. Acta* **659**:99–113.

Russell, P. J., Welsch, J. A., Rachlin, E. M., and McCloskey, J. A., 1987, Different levels of DNA methylation in yeast and mycelial forms of *Candida albicans*, *J. Bacteriol.* **169**:4393–4395.

St. John, T. P., and Davis, R. W., 1981, The organization and transcription of the galactose gene cluster of *Saccharomyces*, *J. Mol. Biol.* **152**:285–315.

Saltarelli, C. G., 1973, Growth stimulation and inhibition of *Candida albicans* by metabolic by-products, *Mycopathol. Mycol. Appl.* **51**:53–63.

Sarachek, A., Rhoads, D. D., and Schwarzhoff, R. H., 1981, Hybridization of *Candida albicans* through fusion of protoplasts, *Arch. Microbiol.* **129**:1–8.

Scherer, S., and Stevens, D. A., 1987, Application of DNA typing methods to epidemiology and taxonomy of *Candida* species, *J. Clin. Microbiol.* **25**:675–679.

Scherer, S., and Stevens, D. A., 1988, A *Candida albicans* dispersed, repeated gene family and its epidemiologic applications, *Proc. Natl. Acad. Sci. USA* **85**:1452–1456.

Schmid, J., Voss, E., and Soll, D. R., 1990, Computer-assisted methods for assessing strain relatedness in *Candida albicans* by fingerprinting with the moderately repetitive sequence Ca3, *J. Clin Microbiol.* **28**:1236–1243.

Shapiro, J. A. (ed.), 1983, *Mobile Genetic Elements*, Academic Press, New York.

Shepherd, M. G., Ghazali, H. M., and Sullivan, P. A., 1980a, N-Acetyl-D-glucosamine kinase and germ-tube formation in *Candida albicans*, *Esp. Mycol.* **4**:147–159.

Shepherd, M. G., Yin, C. Y., Ram, S. P., and Sullivan, P. A., 1980b, Germ tube induction in *Candida albicans*, *Can. J. Microbiol.* **26**:21–26.

Silverman, M., and Simon, M., 1983, Phase variation and related systems, in: *Mobile Genetic Elements* (J. Shapiro, ed.), Academic Press, New York, pp. 537–557.

Simonetti, N., Strippoli, V., and Cassone, A., 1974, Yeast-mycelial conversion induced by N-acetyl-D-glucosamine in *Candida albicans*, *Nature* **250**:344–346.

Slutsky, B., 1986, A characterization of two high frequency switching systems in the dimorphic yeast *Candida albicans*, Ph.D. thesis, University of Iowa, Iowa City.

Slutsky, B., Buffo, J., and Soll, D. R., 1985, High frequency switching of colony morphology in *Candida albicans*, *Science* **230**:666–669.

Slutsky, B., Staebell, M., Anderson, J., Risen, L., Pfaller, M., and Soll, D. R., 1987, "White-opaque transition": A second high-frequency switching system in *Candida albicans*, *J. Bacteriol.* **169**:189–197.

Smail, E. H., and Jones, J. M., 1984, Demonstration and solubilization of antigens expressed primarily on the surfaces of *Candida albicans* germ tubes, *Infect. Immun.* **45**:74–81.

Snell, R. G., Herman, I. F., Wilkins, R. J., and Conner, B. E., 1987, Chromosomal variations in *Candida albicans*, *Nucleic Acid Res.* **15**:3625.

Sobel, J. D., 1984, Recurrent vulvovaginal candidiasis, what we know and what we don't, *Ann. Intern. Med.* **101**:390–392.

Sobel, J. D., 1985, Epidemiology and pathogenesis of recurrent vulvovaginal candidiasis, *Am. J. Obstet. Gynecol.* **152**:924–935.

Sobel, J. D., Muller, G., and Buckley, H. R., 1984, Critical role of germ tube formation in the pathogenesis of candidal vaginitis, *Infect. Immun.* **44**:576–580.

Soll, D. R., 1984, The cell cycle and commitment to alternate cell fates in *Candida albicans*, in: *The Microbial Cell Cycle* (P. Nurse and E. Streiblova, eds.), CRC Press, Boca Raton, pp. 143–162.

Soll, D. R., 1985, The role of zinc in *Candida* dimorphism, in: *Current Topics in Medical Mycology*, Volume 1 (M. R. McGinnis, ed.), Springer-Verlag, Berlin, pp. 258–285.

Soll, D. R., 1986, The regulation of cellular differentiation in the dimorphic yeast *Candida albicans*, *Bioessays* **5**:5–11.

Soll, D. R., 1988, High frequency switching in *Candida albicans*, in: *Mobile DNA* (M. M. Howe and D. E. Berg, eds.), ASM Press, pp. 791–789.

Soll, D. R., 1990a, Dimorphism and high frequency switching in *Candida albicans*, in: *The Genetics of Candida* (D. R. Kirsch, R. Kelly, and M. B. Kurtz, eds.), CRC Press, Boca Raton, pp. 147–176.

Soll, D. R., 1990b, The regulation of timing in a developing system, in: *Topics in Developmental Biology*, Volume 1 (S. Roth, ed.) (in press).

Soll, D. R., and Herman M., 1983, Growth and the inducibility of mycelium formation in *Candida albicans*: A single cell analysis using a perfusion chamber, *J. Gen. Microbiol.* **129**:2809–2824.

Soll, D. R., and Kraft, B., 1988, A comparison of high frequency switching in *Candida albicans* and the slime mold *Dictyostelium discoideum*, *Dev. Genet.* **9**:615–628.

Soll, D. R., and Mitchell, L. H., 1983, Filament ring formation in the dimorphic yeast *Candida albicans*, *J. Cell Biol.* **96**:486–493.

Soll, D. R., Stasi, M., and Bedell, G., 1978, The regulation of nuclear migration and diversion during pseudo-mycelium outgrowth in the dimorphic yeast *Candida albicans*, *Exp. Cell Res.* **116**:207–215.

Soll, D. R., Bedell, G., Thiel, J., and Brummel, M., 1981, The dependency of nuclear division on volume in the dimorphic yeast *Candida albicans*, *Exp. Cell Res.* **133**:55–62.

Soll, D. R., Herman, M. A., and Staebell, M. A., 1985, The involvement of cell wall expansion in the two modes of mycelium formation of *Candida albicans*, *J. Gen. Microbiol.* **131**:2367–2375.

Soll, D. R., Langtimm, C. J., McDowell, J., Hicks, J., and Galask, R., 1987a, High-frequency switching in *Candida* strains isolated from vaginitis patients, *J. Clin. Microbiol.* **25**:1611–1622.

Soll, D. R., Slutsky, B., Mackenzie, S., Langtimm, C., and Staebell, M., 1987b, Switching systems in *Candida albicans* and their possible roles in oral candidasis, in: *Oral Mucosa Diseases: Biology, Etiology, and Therapy* (J. Mackenzie, C. Squier, and E. Dabelsteen, eds.), Laegeforeningens Folarg, Denmark, pp. 52–59.

Soll, D. R., Staebell, M., Langtimm, C., Pfaller, M., Hicks, J., and Rao, T. V. G., 1988, Multiple *Candida* strains in the course of a single systemic infection, *J. Clin. Microbiol.* **26**:1448–1459.

Soll, D. R., Galask, R., Isley, S., Rao, T. V. G., Stone, D., Hicks, J., Schmid, J., Mac, K., and Hanna, C., 1989, "Switching" of *Candida albicans* during successive episodes of recurrent vaginitis, *J. Clin. Microbiol.* **27**:681–690.

Sonneborn, T. M., 1963, Does preformed cell structure play an essential role in cell heredity? in: *The Nature of Biological Diversity* (J. M. Allen, ed.), McGraw-Hill, New York, pp. 165–221.

Southern, E., 1975, Detection of specific sequences around DNA fragments separated by gel electrophoresis, *J. Mol. Biol.* **98**:503–515.

Staebell, M., and Soll, D. R., 1985, Temporal and spatial differences in cell wall expansion during bud and mycelium formation in *Candida albicans*, *J. Gen. Microbiol.* **131**:1467–1480.

Stewart, E., Gow, N. A. R., and Bowen, D. V., 1988, Cytoplasmic alkalinization during germ tube formation in *Candida albicans*, *J. Gen. Microbiol.* **134**:1079–1087.

Sundstrom, P. M., Nichols, E. J., and Kenny, G. E., 1987, Antigenic differences between mannoproteins of germ tubes and blastophores of *Candida albicans*, *Infect. Immun.* **55**:616–620.

Suzuki, T., Kanbe, T., Kutoiwa, T., and Tanaka, K., 1986, Occurrence of ploidy shift in a strain of the imperfect yeast *Candida albicans*, *J. Gen. Microbiol.* **132**:443–453.

Syverson, R. E., Buckley, H. R., and Campbell, C. C., 1975, Cytoplasmic antigens unique to the mycelial or yeast phase of *Candida albicans*, *Infect. Immun.* **12**:1184–1188.

Timberlake, W. E., 1980, Developmental gene regulation in *Aspergillus nidulans*, *Dev. Biol.* **78**:497–510.

Toussaint, A., and Resibois, A., 1983, Phage Mn: Transposition as a way of life, in: *Mobile Genetic Elements* (J. Shapiro, ed.), Academic Press, New York, pp. 105–158.

Warnock, D. W., Johnson, E. M., Richardson, M. D., and Vickers, C. F. H., 1983, Modified response to ketoconazole of *Candida albicans* from a treatment failure, *Lancet* **1**:642–643.

Wheals, A. E., 1987, Biology of the cell cycle in yeast, in: *The Yeasts*, Volume 1 (A. H. Rose and J. S. Harrison, eds.), Academic Press, New York, pp. 283–377.

Whelan, W. L., 1987, The genetics of medically important fungi, *CRC Crit. Rev. Microbiol.* **12**:99–170.

Whelan, W. L., and Magee, P. T., 1981, Natural heterozygosity in *Candida albicans*, *J. Bacteriol.* **145**:896–903.

Whelan, W. L., and Soll, D. R., 1982, Mitotic recombination in *Candida albicans*: Recessive lethal alleles linked to a gene required for methionine biosynthesis, *Mol. Gen. Genet.* **187**:477–485.

Williamson, M. I., Samaranayake, L. P., and Macfarlane, T. W., 1986, Biotypes of *Candida albicans* using the API 20C system, *FEMS Microbiol. Lett.* **37**:27–29.

Wills, J. W., Lasker, B. A., Sirotkin, K., and Riggsby, W. S., 1984, Repetitive DNA of *Candida albicans*: Nuclear and mitochondrial components, *J. Bacteriol.* **157**:918–924.

Wills, J. W., Troutman, W. B., and Riggsby, W. S., 1985, Circular mitochondrial genome of *Candida albicans* contains a large inverted duplication, *J. Bacteriol.* **164**:7–13.

Taxonomic Index

A page number in italics indicates a figure legend.

Subject Index

Acetylcholinesterase, *227*, 229, 238
N-Acetylglucosaminidase, 278
N-Acetylneuraminic acid, 11
Acquired immunodeficiency syndrome, xii, 403
Acrolein, 34
F-Actin, 28, 504
Activated oxygen, effect on plasma membrane, 373, 374
Adenylate cyclase, 34
Adhesins, *Candida*, 160, 171–174
  composition of, 173, 174
  concentrations of, 161
  floccular, 172
  nonfibrillar, 172
  structure of, 172
Adhesion, fungal, 3–18, 35, 140, 484
  to insect gut, 140, 141, *142*, *143*
  to mammalian gut, 158, 161–167
  significance, 12
  vesicles, 105
Adhesive compounds, production, 13, 14
AIDS: *see* Acquired immunodeficiency syndrome
Air spora, 379, 403
Aleurone layer, 50, 51
Alfalfa mosaic virus, 260, 356
Allergenicity tests, 385–388
  BALISA, 386
  ELISA, 386
  inhalation challenge, 386
  radioallergosorbent test (RAST), 386
  skin prick test, 385
Allergens, 379
  and allergenicity, 385–388
  of ascospores, 397
  of basidiospores, 397
  characterization of, 388–397

Allergens (*cont.*)
  extraction of, 384
  of fungal spores, 380–385, 396, 397
Allergic bronchopulmonary aspergillosis, 386, 462
Allergy, 403
  inhalant, 403
Allochthonous microorganisms, 134
Amino acids, in cutinase, 225
Aniline blue, 57
Animal models, 162
  for candidiasis, 162, 163
Anthracnose fungi, 40
  bean, 250
Antibody screening, cDNA cloning, 488, 489
Anticlinal walls, 31, 41
  appressoria, 35
  plastic replicas, 35
Antifungal agents, 136–140
  insect and diet-derived, 136, 137
  phenolic conjugates, 139, 140
  produced by microbiota, 137–140
Antiphagocytic defense, 475, 477
Apical vesicles, 27
Apple scab, 29, 232
Appressoria, *5*, 13, 25–27, *28*, 34, 37, 39, 74, *80*, 82, *82*, 83, *83*, 205–209, 212, 231–234, 236, 237, 268, 275, 322, 324, 335, 350, 484
  anticlinal walls, 35
  of arthropod mycopathogens, 108, *110*, 115, 141
  on artificial surfaces, 15, 35, 36, *36*
  basidiospore, *76*
  chemostimulation, 34, 38
  controlling genes, 39
  development, 26–29, 39
  form, 26

A page number in italics indicates a figure legend.